MW01405710

Dynamic Asset-Pricing Models

The International Library of Financial Econometrics

Editor: Andrew W. Lo
  Harris & Harris Group Professor and Director
  MIT Laboratory for Financial Engineering
  Massachusetts Institute of Technology, USA

1. Statistical Models of Asset Returns

2. Static Asset-Pricing Models

3. Dynamic Asset-Pricing Models

4. Continuous-Time Methods and Market Microstructure

5. Statistical Methods and Non-Standard Finance

Wherever possible, the articles in these volumes have been reproduced as originally published using facsimile reproduction, inclusive of footnotes and pagination to facilitate ease of reference.

For a list of all Edward Elgar published titles visit our site on the World Wide Web at www.e-elgar.com

Dynamic Asset-Pricing Models

Edited by

Andrew W. Lo

Harris & Harris Group Professor and Director
MIT Laboratory for Financial Engineering
Massachusetts Institute of Technology, USA

THE INTERNATIONAL LIBRARY OF FINANCIAL ECONOMETRICS

An Elgar Reference Collection
Cheltenham, UK • Northampton, MA, USA

© Andrew W. Lo 2007. For copyright of individual articles, please refer to the Acknowledgements.

All rights reserved. No part of this publication may be reproduced, stored in a retrieval system, or transmitted in any form or by any means, electronic, mechanical, photocopying, recording, or otherwise without the prior permission of the publisher.

Published by
Edward Elgar Publishing Limited
Glensanda House
Montpellier Parade
Cheltenham
Glos GL50 1UA
UK

Edward Elgar Publishing, Inc.
William Pratt House
9 Dewey Court
Northampton
Massachusetts 01060
USA

A catalogue record for this book is available from the British Library

ISBN: 978 1 84720 264 2
    978 1 84376 342 0 (5 volume set)

Printed and bound in Great Britain by MPG Books Ltd, Bodmin, Cornwall

Contents

Acknowledgements		vii
Introduction Andrew W. Lo		ix

PART I   VARIANCE-BOUNDS TESTS

1. Stephen F. LeRoy and Richard D. Porter (1981), 'The Present-Value Relation: Tests Based on Implied Variance Bounds', *Econometrica*, **49** (3), May, 555–74   3
2. Robert J. Shiller (1981), 'Do Stock Prices Move Too Much to be Justified by Subsequent Changes in Dividends?', *American Economic Review*, **71** (3), June, 421–36   23
3. Marjorie A. Flavin (1983), 'Excess Volatility in the Financial Markets: A Reassessment of the Empirical Evidence', *Journal of Political Economy*, **91** (6), 929–56   39
4. Terry A. Marsh and Robert C. Merton (1986), 'Dividend Variability and Variance Bounds Tests for the Rationality of Stock Market Prices', *American Economic Review*, **76** (3), 483–98   67
5. Allan W. Kleidon (1986), 'Variance Bounds Tests and Stock Price Valuation Models', *Journal of Political Economy*, **94** (5), 953–1001   83
6. John Y. Campbell and Robert J. Shiller (1987), 'Cointegration and Tests of Present Value Models', *Journal of Political Economy*, **95** (5), 1062–88   132
7. Robert C. Merton (1987), 'On the Current State of the Stock Market Rationality Hypothesis', in Rudiger Dornbusch, Stanley Fischer and John Bossons (eds), *Macroeconomics and Finance: Essays in Honor of Franco Modigliani*, Cambridge, MA: MIT Press, 93–124   159
8. Kenneth D. West (1988), 'Dividend Innovations and Stock Price Volatility', *Econometrica*, **56** (1), January, 37–61   191
9. Christian Gilles and Stephen F. LeRoy (1991), 'Econometric Aspects of the Variance-Bounds Tests: A Survey', *Review of Financial Studies*, **4** (4), 753–91   216

PART II   CONSUMPTION-BASED ASSET-PRICING MODELS

10. Lars Peter Hansen and Kenneth J. Singleton (1983), 'Stochastic Consumption, Risk Aversion, and the Temporal Behavior of Asset Returns', *Journal of Political Economy*, **91** (2), 249–65   257
11. Rajnish Mehra and Edward C. Prescott (1985), 'The Equity Premium: A Puzzle', *Journal of Monetary Economics*, **15**, 145–61   274

	12.	Douglas T. Breeden, Michael R. Gibbons and Robert H. Litzenberger (1989), 'Empirical Tests of the Consumption-Oriented CAPM', *Journal of Finance*, **XLIV** (2), June, 231–62	291
	13.	Lars Peter Hansen and Ravi Jagannathan (1992), 'Implications of Security Market Data for Models of Dynamic Economies', *Journal of Political Economy*, **99** (2), 225–62	323
	14.	John Heaton and Deborah J. Lucas (1996), 'Evaluating the Effects of Incomplete Markets on Risk Sharing and Asset Pricing', *Journal of Political Economy*, **104** (3), 443–87	361
	15.	John Y. Campbell and John H. Cochrane (1999), 'By Force of Habit: A Consumption-Based Explanation of Aggregate Stock Market Behavior', *Journal of Political Economy*, **107** (2), 205–51	406
PART III	**TERM STRUCTURE MODELS AND CREDIT**		
	16.	J. Huston McCulloch (1971), 'Measuring the Term Structure of Interest Rates', *Journal of Business*, **44** (1), January, 19–31	455
	17.	Oldrich A. Vasicek and H. Gifford Fong (1982), 'Term Structure Modeling Using Exponential Splines', *Journal of Finance*, **XXXVII** (2), May, 339–48	468
	18.	Andrew W. Lo (1986), 'Logit Versus Discriminant Analysis: A Specification Test and Application to Corporate Bankruptcies', *Journal of Econometrics*, **31** (2), March, 151–78	478
	19.	Stephen J. Brown and Philip H. Dybvig (1986), 'The Empirical Implications of the Cox, Ingersoll, Ross Theory of the Term Structure of Interest Rates', *Journal of Finance*, **XLI** (3), July, 617–32	506
	20.	Eugene F. Fama and Robert R. Bliss (1987), 'The Information in Long-Maturity Forward Rates', *American Economic Review*, **77** (4), September, 680–92	522
	21.	Darrell Duffie and Kenneth J. Singleton (1993), 'Simulated Moments Estimation of Markov Models of Asset Prices', *Econometrica*, **61** (4), July, 929–52	535
	22.	Michael R. Gibbons and Krishna Ramaswamy (1993), 'A Test of the Cox, Ingersoll, and Ross Model of the Term Structure', *Review of Financial Studies*, **6** (3), 619–58	559
	23.	Darrell Duffie and Kenneth J. Singleton (1997), 'An Econometric Model of the Term Structure of Interest-Rate Swap Yields', *Journal of Finance*, **LII** (4), September, 1287–1321	599

Name Index 635

Acknowledgements

The editor and publishers wish to thank the authors and the following publishers who have kindly given permission for the use of copyright material.

American Economic Association for articles: Robert J. Shiller (1981), 'Do Stock Prices Move too Much to be Justified by Subsequent Changes in Dividends?', *American Economic Review*, **71** (3), June, 421–36; Terry A. Marsh and Robert C. Merton (1986), 'Dividend Variability and Variance Bounds Tests for the Rationality of Stock Market Prices', *American Economic Review*, **76** (3), 483–98; Eugene F. Fama and Robert R. Bliss (1987), 'The Information in Long-Maturity Forward Rates', *American Economic Review*, **77** (4), September, 680–92.

Blackwell Publishing Ltd for articles: Oldrich A. Vasicek and H. Gifford Fong (1982), 'Term Structure Modeling Using Exponential Splines', *Journal of Finance*, **XXXVII** (2), May, 339–48; Stephen J. Brown and Philip H. Dybvig (1986), 'The Empirical Implications of the Cox, Ingersoll, Ross Theory of the Term Structure of Interest Rates', *Journal of Finance*, **XLI** (3), July, 617–32; Douglas T. Breeden, Michael R. Gibbons and Robert H. Litzenberger (1989), 'Empirical Tests of the Consumption-Oriented CAPM', *Journal of Finance*, **XLIV** (2), June, 231–62; Darrell Duffie and Kenneth J. Singleton (1997), 'An Econometric Model of the Term Structure of Interest-Rate Swap Yields', *Journal of Finance*, **LII** (4), September, 1287–1321.

Econometric Society for articles: Stephen F. LeRoy and Richard D. Porter (1981), 'The Present-Value Relation: Tests Based on Implied Variance Bounds', *Econometrica*, **49** (3), May, 555–74; Kenneth D. West (1988), 'Dividend Innovations and Stock Price Volatility', *Econometrica*, **56** (1), January, 37–61; Darrell Duffie and Kenneth J. Singleton (1993), 'Simulated Moments Estimation of Markov Models of Asset Prices', *Econometrica*, **61** (4), July, 929–52.

Elsevier for articles: Rajnish Mehra and Edward C. Prescott (1985), 'The Equity Premium: A Puzzle', *Journal of Monetary Economics*, **15**, 145–61; Andrew W. Lo (1986), 'Logit Versus Discriminant Analysis: A Specification Test and Application to Corporate Bankruptcies', *Journal of Econometrics*, **31** (2), March, 151–78.

MIT Press for excerpt: Robert C. Merton (1987), 'On the Current State of the Stock Market Rationality Hypothesis', in Rudiger Dornbusch, Stanley Fischer and John Bossons (eds), *Macroeconomics and Finance: Essays in Honor of Franco Modigliani*, 93–124.

Oxford University Press for articles: Christian Gilles and Stephen F. LeRoy (1991), 'Econometric Aspects of the Variance-Bounds Tests: A Survey', *Review of Financial Studies*, **4** (4), 753–91; Michael R. Gibbons and Krishna Ramaswamy (1993), 'A Test of the Cox, Ingersoll, and Ross Model of the Term Structure', *Review of Financial Studies*, **6** (3), 619–58.

University of Chicago Press for articles: J. Huston McCulloch (1971), 'Measuring the Term Structure of Interest Rates', *Journal of Business*, **44** (1), January, 19–31; Lars Peter Hansen and Kenneth J. Singleton (1983), 'Stochastic Consumption, Risk Aversion, and the Temporal Behavior of Asset Returns', *Journal of Political Economy,* **91** (2), 249–65; Marjorie A. Flavin (1983), 'Excess Volatility in the Financial Markets: A Reassessment of the Empirical Evidence', *Journal of Political Economy*, **91** (6), 929–56; Allan W. Kleidon (1986), 'Variance Bounds Tests and Stock Price Valuation Models', *Journal of Political Economy*, **94** (5), 953–1001; John Y. Campbell and Robert J. Shiller (1987), 'Cointegration and Tests of Present Value Models', *Journal of Political Economy*, **95** (5), 1062–88; Lars Peter Hansen and Ravi Jagannathan (1992), 'Implications of Security Market Data for Models of Dynamic Economies', *Journal of Political Economy*, **99** (2), 225–62; John Heaton and Deborah J. Lucas (1996), 'Evaluating the Effects of Incomplete Markets on Risk Sharing and Asset Pricing', *Journal of Political Economy*, **104** (3), 443–87; John Y. Campbell and John H. Cochrane (1999), 'By Force of Habit: A Consumption-Based Explanation of Aggregate Stock Market Behavior', *Journal of Political Economy*, **107** (2), 205–51.

Every effort has been made to trace all the copyright holders but if any have been inadvertently overlooked the publishers will be pleased to make the necessary arrangement at the first opportunity.

In addition the publishers wish to thank the Marshall Library of Economics, University of Cambridge, UK, the Library at the University of Warwick, UK, and the Library of Indiana University at Bloomington, USA, for their assistance in obtaining these articles.

Introduction*

Andrew W. Lo

As a discipline, financial econometrics is still in its infancy, and from some economists' perspective, not a separate discipline at all. However, this is changing rapidly, as the publication of these five volumes illustrates. The growing sophistication of financial models requires equally sophisticated methods for their empirical implementation, within academia and in industry, and in recent years financial econometricians have stepped up to the challenge. Indeed, the demand for financial econometricians by investment banks and other financial institutions – not to mention economics departments, business schools, and financial engineering programs throughout the world – has never been greater. Moreover, the toolkit of financial econometrics has also grown in size and sophistication, including techniques such as nonparametric estimation, functional central limit theory, nonlinear time-series models, artificial neural networks, and Markov Chain Monte Carlo methods.

What can explain the remarkable growth and activity of this seemingly small subset of econometrics, which is itself a rather esoteric subset of economics? The answer lies in the confluence of three parallel developments in the last half century.

The first is the fact that the financial system has become more complex over time, not less. This is an obvious consequence of general economic growth and development in which the number of market participants, the variety of financial transactions, and the sums involved have also grown. As the financial system becomes more complex, the benefits of more highly developed financial technology become greater and greater and, ultimately, indispensable.

The second factor is, of course, the set of breakthroughs in the quantitative modeling of financial markets. Pioneered over the past three decades by the giants of financial economics – Fischer Black, John Cox, Eugene Fama, John Lintner, Harry Markowitz, Robert Merton, Franco Modigliani, Merton Miller, Stephen Ross, Paul Samuelson, Myron Scholes, William Sharpe, and others – their contributions laid the remarkably durable foundations on which all of modern quantitative financial analysis is built. Financial econometrics is only one of several intellectual progeny that they have sired.

The third factor is a contemporaneous set of breakthroughs in computer technology, including hardware, software, and data collection and organization. Without these computational innovations, much of the financial technology developed over the past fifty years would be irrelevant academic musings, condemned to the dusty oblivion of unread finance journals in university library basements. The advent of inexpensive and powerful desktop microcomputers and machine-readable real-time and historical data breathed life into financial econometrics, irrevocably changing the way finance is practiced and taught. Concepts like alpha, beta, R^2, correlations, and cumulative average residuals have become concrete objects to be estimated and actively used in making financial decisions. The outcome was nothing short of a new in-

---

* I thank John Cox and John Heaton for helpful discussions.

dustrial revolution in which the 'old-boys network' was replaced by the computer network, where what mattered more was *what* you knew, not *who* you knew, and where graduates of Harvard and Yale suddenly found themselves at a disadvantage to graduates of MIT and Caltech. It was, in short, the revenge of the nerds!

But there is an even deeper reason for the intellectual cornucopia that has characterized financial econometrics in recent years – it is the fact that randomness is central to both finance and econometrics. Unlike other fields of economics, finance is intellectually vapid in the absence of uncertainty; the net-present-value rule and interest-rate compounding formulas are the only major ideas of nonstochastic finance. It is only when return is accompanied by risk that financial analysis becomes interesting, and the same can be said for econometrics. In contrast to many econometric applications where a particular theory is empirically tested by linearizing one of its key equations and then slapping on an error term as an afterthought, the sources and nature of uncertainty are at the core of every financial application. In fact, the error term in financial econometrics is the main attraction, not merely a disturbance to be minimized or averaged away. This approach creates a rich tapestry of models and methods that have genuine practical value because the randomness assumed is more closely related to the randomness observed than in other econometric applications. Indeed, the econometric consequences of uncertainty in financial models usually follow directly from the economics, and are not merely incidental to the empirical analysis.

The papers collected in Volume 1, which consists of the most influential statistical models of financial asset returns, is the starting point for this intimate connection between finance and econometrics. Even without the economic infrastructure of preferences, supply and demand, and general equilibrium, the contributions in Volume 1 shed considerable light on the basic properties of asset returns such as return predictability, fat tails, serial correlation, and time-varying volatilities.

The papers in Volume 2 are able to extract additional information from asset returns and volume by imposing additional structure, for example, specific investor preferences, parametric probability distributions for underlying sources of uncertainty, and general equilibrium. Using a two-period or static framework – the simplest possible context in which price uncertainty exists – the papers of Volume 2 yield remarkably simple yet far-reaching implications for the relation between risk and expected return, the proper economic definition of risk, and methods for evaluating the performance of portfolio managers.

Of course, the static two-period framework is only an approximation to the more complex multi-period case, which is the focus of the articles in Volume 3. By modeling the intertemporal consumption/savings and investment decisions of investors, a wealth of additional testable implications can be derived for the time-series and cross-sectional properties of asset returns and volume.

The dynamics of prices and quantities lead naturally to questions about the fine structure of financial transactions and markets, as well as the notion of continuous-time trading, both of which are examined by the articles of Volume 4. The econometrics of continuous-time stochastic processes is essential for applications of derivatives pricing models to data, and the practical relevance of continuous-time approximations is dictated by the particular market microstructure of the derivative's underlying asset.

Volume 5 is the final volume of the series and contains methodological papers, as well as contributions to finance that are not yet part of the mainstream, but which address important issues nonetheless. In particular, this volume includes papers on quantifying selection and data-snooping biases in tests of financial asset-pricing models, Bayesian methods, event-study analysis, Generalized Method of Moments estimation, technical analysis, neural networks, and some examples from the emerging field of econophysics.

The sheer breadth of topics across the five volumes should give readers a sense of the impact and intellectual vitality of financial econometrics today. Moreover, the articles in these volumes also span a period of four decades – ranging from classic tests of the Random Walk Hypothesis in the 1960s to the application of random matrix theory to portfolio optimization in 2002 – illustrating the remarkable progress that the field has achieved over time. Along with the many innovations produced by global financial markets will be a never-ending supply of wonderful challenges and conundrums for financial econometrics, guaranteeing its importance to economists and investors alike.

1 Statistical Models of Asset Returns

Ever since the publication in 1900 of Louis Bachelier's doctoral thesis in which he modeled stock prices on the Paris Bourse as Brownian motion, finance and statistics have become inextricably linked. In Part I of Volume 1, we begin with four articles that provide some much-needed philosophical background for the role of statistical inference in financial modeling. While statistics now enjoys an independent existence, replete with general and specialized journals, conferences, and professional societies, the financial econometrician has a somewhat different perspective. The uniqueness of financial econometrics lies in the wonderful interplay between financial models and statistical inference, where neither one dominates the other. In particular, Cox (1990, Chapter 3) underscores the importance of models that guide the course of our statistical investigations, but Leamer (1983, Chapter 1), McCloskey and Ziliak (1996, Chapter 4), and Roll (1988, Chapter 2) provide some counterweight to the economist's natural tendency to depend more on models than on facts.

The intimate relationship between financial theory and statistical properties is illustrated perfectly by the Random Walk Hypothesis, which is the subject of the articles in Part II. Unlike the motivation for Brownian motion in physics and biology – the absence of information – the economic justification for randomness in financial asset prices is active information-gathering on the part of all market participants. It is only through the concerted efforts of many investors attempting to forecast asset returns that asset returns become unforecastable. This leads to several testable implications, and much of the early literature in financial econometrics consisted of formal statistical tests of the Random Walk Hypothesis and corresponding empirical results (Working (1960, Chapter 5); Mandelbrot (1963, Chapter 6); Fama (1965, Chapter 7); Lo and MacKinlay (1988, Chapter 8); Poterba and Summers (1988, Chapter 9); Richardson and Stock (1989, Chapter 10)).

However, randomness is not the only interesting characteristic of financial asset returns. Many authors have documented a host of empirical properties unique to financial time series including time-varying moments (Part III: Engle (1982, Chapter 11); Bollerslev (1986, Chapter 12); Nelson (1991, Chapter 14)), fat tails and long-range dependence (Part IV: Mandelbrot and Van Ness (1968, Chapter 15); Greene and Fielitz (1977, Chapter 16); Granger and Joyeux (1980, Chapter 17); Geweke and Porter-Hudak (1983, Chapter 18); Lo (1991, Chapter 19);

Baillie (1996, Chapter 20)), regime shifts (Hamilton (1989, Part III, Chapter 13)), and in some cases, co-integrated price processes (Part V: Engle and Granger (1987, Chapter 21); Phillips (1987, Chapter 22)). Each of these issues is addressed through a series of specific stochastic processes designed to capture these properties, along with empirical evidence that either supports or rejects these models for financial data.

Collectively, the papers in Volume 1 should provide readers with a comprehensive arsenal of statistical descriptions of financial time series, all motivated by particular empirical observations.

2 Static Asset-Pricing Models

The focus of the previous volume was the statistical properties of financial asset returns, without reference to any specific economic model of investors or financial interactions. In Volume 2, we shift our attention to the relative magnitudes of asset returns over a given time period. On average, is the return to one stock or portfolio higher than the return to another stock or portfolio, and if so, to what can we attribute the difference?

These questions are central to financial economics since they bear directly on potential trade-offs between risk and expected return, one of the most basic principles of modern financial theory. This theory suggests that lower-risk investments such as bonds or utility stocks will yield lower returns on average than riskier investments such as airline or technology stocks, which accords well with common business sense: investors require a greater incentive to bear more risk, and this incentive manifests itself in higher expected returns. The issue, then, is whether the profits of successful investment strategies can be attributed to the presence of higher risks – if so, then the profits are compensation for risk-bearing capacity and nothing unusual; if not, then further investigation is warranted. In short, we need a risk/reward benchmark to tell us how much risk is required for a given level of expected return. The first, and perhaps most celebrated financial model that provides an explicit risk/reward trade-off for financial asset returns is the Capital Asset Pricing Model (CAPM) of Sharpe (1964) and Lintner (1965). In the CAPM framework, an asset's 'beta' is the relevant measure of risk – stocks with higher betas should earn higher returns on average. And in many of the recent anomaly studies, the authors argue forcefully that differences in beta cannot fully explain the magnitudes of return differences, hence the term 'anomaly'.

The articles in Part I of this volume provide a comprehensive analysis of the CAPM, and chronicle a fascinating intellectual journey that begins with simple but elegant tests of the CAPM that find support for the theory (Fama and MacBeth (1973, Chapter 1)), leading to more sophisticated statistical tests of the CAPM (Gibbons (1982, Chapter 2); Jobson and Korkie (1982, Chapter 3); MacKinlay (1987, Chapter 4); Gibbons, Ross and Shanken (1989, Chapter 5)), and ends with serious questions about the explanatory power of the CAPM versus other multi-factor models (Fama and French (1992, Chapter 6); Black (1993, Chapter 7); MacKinlay (1995, Chapter 8); Lo and Wang (2000, Chapter 9)). But the historical significance of this literature goes well beyond the CAPM – this line of inquiry was the first to employ rigorous statistical inference, ushering empirical finance into the modern age of financial econometrics.

Part II contains a parallel stream of the Arbitrage Pricing Theory (APT) literature in financial econometrics. Despite the fact that the APT might seem like a close cousin of the CAPM – both are, after all, linear factor models of asset returns – the empirical APT literature was, for a time, stuck in a theoretical quagmire in which the falsifiability of the APT was questioned (Shanken (1982, Chapter 10) and (1985, Chapter 15); Dybvig and Ross (1985, Chapter 14)). While a cynic might argue that the best theory is one that can never be disproved, respectable scientific mores suggest otherwise, and the sometimes bitter debate surrounding this issue yielded many nuggets of theoretical (Chamberlain and Rothschild (1983, Chapter 11)), econometric (Connor and Korajczyk (1993, Chapter 18)), and empirical (Dhrymes, Friend and Gultekin (1984, Chapter 12); Roll and Ross (1984, Chapter 13); Chen, Roll and Ross (1986, Chapter 16); Lehmann and Modest (1988, Chapter 17)) wisdom for the profession.

An important outgrowth of the many econometric innovations surrounding the empirical analysis of the CAPM and APT is the performance attribution literature, the focus of the articles in Part III. It is a truism that one cannot manage what one cannot measure, hence it should come as no surprise that the proper measurement of performance has become an essential part of investment management. In particular, measures of security-selection ability (Treynor and Black (1973, Chapter 19)) and market-timing ability (Merton (1981, Chapter 20); Henriksson and Merton (1981, Chapter 21)), and statistical inference for risk/reward measures such as the Sharpe ratio (Lo (2002, Chapter 22); Getmansky, Lo and Makarov (2004, Chapter 23)) have now become part of the practitioner's lexicon in discussing investment performance. This is another example of how academic research in financial econometrics has made an indelible impact on financial practice.

3 Dynamic Asset-Pricing Models

The static asset-pricing models of Volume 2 are clearly meant to be approximations to a more complex reality in which investors and financial markets interact through time. The challenges of dynamic asset-pricing models are considerable, since they involve many more degrees of freedom for market participants and security prices. It is far easier to model the conditional distribution of tomorrow's stock price than the joint distribution of daily prices over the next five years. However, by imposing sufficient structure on investor preferences and security-price dynamics, it is possible to develop a rich yet testable theory of asset prices over time.

The articles in Part I of Volume 3 illustrate this possibility through the variance-bounds test of market rationality. By assuming that a security's market price is equal to the capitalized value of all future payouts, and by assuming that payouts follow a stationary stochastic process, it is possible to derive an upper bound for the variance of that security's price based on the subsequent stream of payouts. The empirical fact that this variance bound is apparently violated by aggregate historical US equity prices has been interpreted as a violation of market rationality (LeRoy and Porter (1981), Chapter 1); Shiller (1981, Chapter 2)), a conclusion with far-reaching implications for all kinds of financial decisions if it were true. This observation added fuel to the already smoldering debate between proponents of market rationality and its critics, yielding enormously valuable insights into the econometrics of equilibrium asset prices. For example, by replacing the assumption of stationarity for prices and payouts with the Random Walk Hypothesis – which is arguably closer to empirical reality and theoretical consistency (recall Bachelier's model of stock prices on the Paris Bourse) – the upper bound becomes a lower bound, that is, the inequality is reversed (Marsh and Merton (1986, Chapter

4)). Also, because of estimation error, the empirical violation of the variance bound may be attributed to sampling fluctuation (Flavin (1983, Chapter 3); Kleidon (1986, Chapter 5); West (1988, Chapter 8)).

But one of the most interesting outcomes of the variance-bounds literature is its implications for the sociology of scientific inquiry in economics and finance. Like a magnet dropped into a dish of iron filings, the variance-bounds debate polarized the academic community almost immediately, with members of economics departments lining up behind the irrationalists, and members of finance departments in business schools taking the side of market rationality. The debate should have been settled by the weight of econometric analysis and empirical fact, but remarkably, with each new publication that peeled back another layer of this wonderfully controversial challenge, the convictions of the disciples in both camps only grew stronger. To this day, there is no consensus; the response to the title of Shiller's (1981) paper 'Do Stock Prices Move Too Much to be Justified by Subsequent Dividends?' is 'yes' if you teach in an economics department and 'no' if you teach in a business school.

The variance-bounds controversy had another salutary effect on the financial econometrics literature: its focus on aggregate measures sparked additional interest in asset-pricing models based on aggregate measures of consumption. This, in turn, led to a number of significant breakthroughs in asset-pricing theory and econometrics, including the equity premium puzzle (Mehra and Prescott (1985, Chapter 11)), consumption-based asset-pricing models (Hansen and Singleton (1983, Chapter 10); Breeden, Gibbons and Litzenberger (1989, Chapter 12)), stochastic discount factor models (Hansen and Jagannathan (1992, Chapter 13)), asset-pricing models with incomplete markets (Heaton and Lucas (1996, Chapter 14)) and state-dependent preferences (Campbell and Cochrane (1999, Chapter 15)). Although these models have met with limited empirical success, they have generated an enormous literature at the intersection of macroeconomics and finance, enriching our understanding of both in the process.

In Part III, we turn our attention from stock markets to bond markets. Bonds, particularly default-free government bonds, are inherently simpler financial instruments because unlike the dividend streams paid by equity securities, the nominal cash flows of bonds are pre-specified and nonstochastic. There are only three major sources of uncertainty affecting bond prices: interest rates or discount rates over various horizons, realized and expected inflation, and the probability of default. Addressing the first source of uncertainty is the motivation for models of the term structure of interest rates, and one of the earliest models employed curve-fitting techniques to the data (McCulloch (1971, Chapter 16); Vasicek and Fong (1982, Chapter 17)). But the most influential term structure model is the celebrated Cox, Ingersoll and Ross (1985) model, a dynamic general equilibrium model that incorporates investor preferences among other aspects of the macroeconomy. Although empirical implementations of this model have yielded mixed results (Brown and Dybvig (1986, Chapter 19); Gibbons and Ramaswamy (1993, Chapter 22)), it has served as the durable foundation of an extensive literature of more econometrically oriented models of the term structure (Duffie and Singleton (1993, Chapter 21) and (1997, Chapter 23)).

The two remaining sources of uncertainty for bond prices – inflation and default – have rich literatures of their own, much of which is beyond the scope of this series but which has been summarized in other series. Two examples of that literature have been included in Part III for completeness (Fama and Bliss (1987, Chapter 20); Lo (1986, Chapter 18)).

4 Continuous-Time Methods and Market Microstructure

One of the great ironies of modern economics is the fact that most of its theories assume that individuals take prices as given, yet the primary objective is usually to explain how prices are determined. In an economy where everyone takes prices as given, how do prices change? The nineteenth-century mathematical economist Léon Walras hypothesized the existence of an auctioneer who calls out a price, observes the excess demand or supply generated by that price, and then adjusts the price up or down so as to reduce the excess demand or supply. Although a figment of the economist's imagination, this process of 'tâtonnement' was perhaps the first systematic attempt to model the price-discovery process. A more careful examination of how prices are set – from one transaction to the next – has yielded a number of important insights into the fine structure of economic interactions, and this is the purview of the market microstructure literature.

Although market interactions have been the subject of virtually all economic analysis since the publication of Adam Smith's (1776) *An Inquiry into the Nature and Causes of the Wealth of Nations*, market microstructure phenomena are distinct. For example, the impact of price discreteness (Ball (1988, Chapter 7); Hausman, Lo and MacKinlay (1992, Chapter 10)), irregular trading intervals (Scholes and Williams (1977, Chapter 1); Dimson (1979, Chapter 2); Cohen, Hawawini, Maier, Schwartz and Whitcomb (1983), Chapter 3); Lo and MacKinlay (1990, Chapter 9)), and the bid/offer spread (Roll (1984, Chapter 4); Glosten and Harris (1988, Chapter 8)) have only recently been studied thanks to the growing interest in the microstructure of financial markets. Much of this literature owes its genesis to the availability of machine-readable transactions-level data, pioneered by Robert A. Wood and first analyzed in Wood, McInish and Ord (1985, Chapter 5). Since then, other transactions datasets have become available through organized exchanges such as the New York Stock Exchange (the TAQ and TORQ datasets) or through brokerage firms such as Investment Technology Group (limit-order data, analyzed in Lo, MacKinlay and Zhang (2002, Chapter 11)).

The irregular timing of trades at the transaction level highlights the fact that standard discrete-time models cannot fully accommodate the richness of financial markets. This provides part of the motivation for articles in Part II and the set of continuous-time models in the finance literature, in which prices are assumed to evolve continuously through time, typically with sample paths that are everywhere continuous. However, continuous sample paths imply that price movements are smooth, which, in turn, implies that over infinitesimally short time intervals, price changes are completely forecastable. Such an implication is clearly at odds with both reality and basic finance theory – perfectly forecastable prices would mean either unlimited profit opportunities (buy low, sell high), or nonstochastic prices which brings us back to the trivial case of financial markets with no uncertainty.

This conundrum was first recognized and addressed by the French mathematician Louis Bachelier in his 1900 doctoral thesis on warrant pricing (see Volume 1), in which he developed the basic notions of Brownian motion, a continuous-time version of the Random Walk Hypothesis. This mathematical object is strange indeed – it is a continuous-time sequence of random variables where every sample path is continuous but, because even infinitesimal increments are unforecastable, the sample paths are nowhere differentiable. In other words, sample paths are continuous but so jagged that it is impossible to compute their rate of change even over the smallest time intervals. Without the pioneering insights of Bachelier, Albert Einstein (who developed a similar model in 1905 while studying the photoelectric effect), and Nobert Wiener (who was the first to develop a rigorous mathematical formulation of Brownian motion), it is not obvious that a continuous-time random walk should exist.

But it does, and Brownian motion has become the workhorse of modern financial economics thanks to a related mathematical breakthrough achieved by the mathematician Kiyosi Itô (1951): the stochastic calculus and a corresponding theory of stochastic differential equations. The importance of this innovation to modern finance is elegantly described in Merton (1975, Chapter 13). Briefly, the assumption of continuous-time trading, coupled with the ability to derive the exact laws of motion for nonlinear functions of Brownian motion via stochastic calculus, implies that it is possible to replicate the payoffs of complex financial instruments such as options and other derivative securities by dynamically adjusting portfolios of simpler securities such as stocks and bonds. This insight, first hypothesized by Arrow (1964), was given substance in Merton's (1973) option-pricing model which, along with Black and Scholes (1973), revolutionized financial theory and practice by providing not only exact pricing models for derivative securities, but also explicit methods for hedging and synthetically manufacturing such securities through dynamic trading strategies.

These ideas, for which Merton and Scholes shared the 1997 Nobel Prize, have led to numerous breakthroughs in finance theory, but remarkably, they are also partly responsible for the birth of at least three distinct multi-trillion-dollar businesses in the finance industry: organized options exchanges (for example, the Chicago Board Options Exchange, the International Securities Exchange, the Boston Options Exchange), the over-the-counter derivatives business (for example, caps, floors, collars, swaptions, and so on), and today's burgeoning credit-derivatives business (for example, credit default swaps, CDS swaptions, credit-linked notes).

A prerequisite to any application of continuous-time models in financial markets is to estimate the parameters of the stochastic differential equations that describe the prices of the underlying securities. This creates another challenge for financial econometrics: the econometrics of continuous-time stochastic processes. The fact that time is hypothesized to be continuous, yet in practice we observe data only at discrete, and sometimes irregular, time intervals, causes a number of difficulties that are not present in discrete-time models. These issues are addressed in Clark (1973, Chapter 12), Garman and Klass (1980, Chapter 15), Parkinson (1980, Chapter 14), Shiller and Perron (1985, Chapter 17), and Lo (1986, Chapter 18; and 1988, Chapter 20). The root of many of these issues is the subtle relationship between discrete and continuous time – both are, after all, approximations to reality.

The nature of these approximations is derived explicitly by Bertsimas, Kogan and Lo (2000, Chapter 25), where they reverse the standard chain of logic by using option-pricing models to define the notion of 'temporal granularity' and to measure the discrepancies between discrete and continuous time. A number of other econometric applications have arisen from the derivatives pricing literature, including implied binomial trees (Rubinstein (1994, Chapter 21)), nonparametric estimation of state-price densities and risk aversion (Aït-Sahalia and Lo (1998, Chapter 23 and 2000, Chapter 24)), and semi-parametric bounds for option prices (Lo (1987, Chapter 19)).

As useful as Brownian motion is, it cannot capture all aspects of the data, two of which are particularly relevant for financial data: price jumps, and serial correlation. Fortunately, a number of alternatives exist for modeling both in continuous time, including Ball and Torous (1985, Chapter 16) and Lo and Wang (1995, Chapter 22).

Finally, the use of continuous-time stochastic processes in modeling financial markets has led, directly and indirectly, to a number of statistical applications in which functional central limit theory and the notion of *weak convergence* (see, for example, Billingsley, 1968) are used to deduce the asymptotic properties of various estimators, of which Richardson and Stock (1989) in Volume 1 is an excellent example.

5 Statistical Methods and Non-Standard Finance

As discussed in this Introduction, one of the most attractive characteristics of finance from the econometrician's perspective is the central role that uncertainty plays. Therefore, it should come as no surprise that statistical inference and financial models are intimately connected. In Volume 5, we have collected a host of important contributions to financial econometrics that are primarily methodological, though each article is motivated by a particular financial challenge.

Perhaps the most basic challenge to empirical work in finance is the wealth of data available to financial econometrics, and the many false positives that can result from repeated analysis of such data. It is no exaggeration that if one tortures a dataset long enough, it will confess to anything! In Part I, the magnitude of this phenomenon – also known as selection bias, data-snooping bias, and backtest bias – is investigated both analytically and numerically. Brown, Goetzmann, Ibbotson and Ross (1992, Chapter 2) conclude that survivorship bias – the bias induced by including only the surviving corporations or mutual funds in an empirical study – can be quite substantial. Lo and MacKinlay (1990, Chapter 1), and Foster, Smith and Whaley (1997, Chapter 3) come to similar conclusions for other forms of selection biases, and propose new statistical methods to adjust for such biases.

One of the most significant sources of bias is the preconceived notions that empirical researchers cling to as they formulate their experimental designs. Perhaps the only systematic approach to taking these preconceived notions or 'priors' into account is to use a Bayesian statistical framework for conducting inferences, as in the articles of Part II. Bayesian methods are widely used in the statistics literature, hence they can be applied to almost any context in which statistical inference is called for. In the financial context, Bayesian methods have been applied to portfolio optimization (Klein and Bawa (1977, Chapter 4)); Shanken (1987, Chapter 5); Kandel, McCulloch and Stambaugh (1995, Chapter 8)) and tests of the Arbitrage Pricing Theory (McCulloch and Rossi (1991, Chapter 7)) and other asset-pricing models (Harvey and Zhou (1990, Chapter 6)). Although historically quite cumbersome and, as a result, not particularly appealing to the mainstream financial econometrics literature, recent advances in computationally intensive methods for conducting Bayesian inferences, for example, the Gibbs sampler and Markov Chain Monte Carlo methods, have revolutionized the field.

The articles in Part III contain other important techniques that any practicing financial econometrician should be familiar with, including event studies (Fama, Fisher, Jensen and Roll (1969, Chapter 9); Brown and Warner (1985, Chapter 12); Ball and Torous (1988, Chapter 14)), Generalized Method of Moments (Hansen (1982, Chapter 11); Richardson and Smith (1991, Chapter 15)), and robust methods for estimating expected returns and covariance matrices (Merton (1980, Chapter 10); Newey and West (1987, Chapter 13)).

The last part of Volume 5, and the final set of articles for the Financial Econometrics series, is a collection drawn from the underbelly of mainstream finance. Part IV includes articles on nonlinear dynamical systems (Hsieh (1991, Chapter 16)), neural networks (Hutchinson, Lo and Poggio (1994, Chapter 18)), technical analysis (Brock, Lakonishok and LeBaron (1992, Chapter 17); Lo, Mamaysky and Wang (2000, Chapter 20)), and random matrix theory (Pafka and Kondor (2002, Chapter 21); Plerou, Gopikrishnan, Rosenow, Amaral and Stanley (1999, Chapter 19)). Although they are not yet part of the mainstream, they have generated sufficient interest in either academia or industry to deserve inclusion in these volumes. For example, while technical analysis – the practice of forecasting future price movements based on geometric patterns in historical price plots – is generally dismissed by the academic finance community, it is arguably the most widely used set of techniques among commodities and foreign exchange traders. Similarly, random matrix theory is a set of mathematical results originally developed by nuclear physicists for modeling statistical fluctuations of particle interactions, but has recently been applied with some degree of success to estimating covariance matrices for portfolio optimization problems.

These are only a very small and idiosyncratic sample of the tremendously rich and varied literatures that are loosely connected in one way or another to financial econometrics. Although they are at the outskirts of the finance literature today, they may well join the mainstream in the near future once they prove their practical worth.

Concluding Thoughts

Two decades ago, the term 'financial econometrics' did not exist. It is a remarkable testament to the practical value of financial econometrics that we have been able to fill five volumes with pathbreaking articles in this nascent discipline. In contrast to other branches of economics, for example, industrial organization, labor economics, and macroeconomics, the application of econometric analysis to financial markets has given birth to a new and cohesive field of study. Yet the list of unanswered research questions is still much longer than the list of achievements that financial econometrics has produced so far. For example:

- How do we conduct proper statistical inference for financial time series, which are usually non-stationary, non-Gaussian, skewed, leptokurtic, and neither independently nor identically distributed?
- How do we decide which portfolio managers have skill when the standard errors of the usual performance statistics are so large that over 100 years of monthly returns are required to yield any kind of statistical significance?
- Is there a way to adjust simulated portfolio returns to account for backtest bias?
- What is the best way to measure the likelihood of rare events and manage such risks if, by definition, there are so few events in the historical record?
- How should we construct optimal portfolios of securities if estimated means and covariance matrices are subject to so much estimation error?
- How can we estimate the risk preferences of an individual or institutional investor, and are these preferences stable over time and individuals?

- Is the extraordinary investment performance of certain portfolio managers due to their extraordinary risk exposures, or does genuine alpha exist in the investment management business?

These questions are surprisingly simple to state, yet so far no consensus has been reached as to how to answer any of them. They are just a few of the wonderful challenges that lie in store for future generations of financial econometricians.

References

Arrow, K. (1964), 'The Role of Securities in the Optimal Allocation of Risk Bearing', *Review of Economic Studies* **31**, 91–6.
Bachelier, L. (1900), 'Theory of Speculation', in P. Cootner (ed.), *The Random Character of Stock Market Prices*, Cambridge; MIT Press, 1964, reprint.
Billingsley, P. (1968), *Convergence of Probability Measures*, New York: John Wiley and Sons.
Black, F. and M. Scholes (1973), 'The Pricing of Options and Corporate Liabilities', *Journal of Political Economy* **81**, 637–54.
Cox, J., J. Ingersoll and S. Ross (1985), 'A Theory of the Term Structure of Interest Rates', *Econometrica* **53**, 385–408.
Itô, K. (1951), 'On Stochastic Differential Equations', *Memoirs of the American Mathematical Society* **4**, 1–51.
Lintner, J. (1965), 'The Valuation of Risky Assets and the Selection of Risky Investments in Stock Portfolios and Capital Budgets', *Review of Economics and Statistics* **47**, 13–37.
Merton, R. (1973), 'Rational Theory of Option Pricing', *Bell Journal of Economics and Management Science* **4**, 141–83.
Sharpe, W. (1964), 'Capital Asset Prices: A Theory of Market Equilibrium under Conditions of Risk', *Journal of Finance* **19**, 425–42.
Smith, A. (1776), *An Inquiry into the Nature and Causes of the Wealth of Nations*, London: Methuen and Co., Ltd., ed. Edwin Cannan, 1904, fifth edition.

*Dedicated to
Andy Abel, Jerry Hausman, and Whitney Newey, from whom I learned most
of the econometrics I know, and whose high expectations I strive to meet
each day*

— Andrew W. Lo

Part I
Variance-Bounds Tests

ECONOMETRICA

VOLUME 49   MAY, 1981   NUMBER 3

THE PRESENT-VALUE RELATION: TESTS BASED ON IMPLIED VARIANCE BOUNDS[1]

BY STEPHEN F. LeROY AND RICHARD D. PORTER[2]

This paper investigates the implications for asset price dispersion of conventional security valuation models. Successively sharper variance bounds on asset prices are derived. Large-sample tests of the bounds are determined and applied to aggregated and disaggregated price and earnings data of U.S. corporations.

1. INTRODUCTION AND SUMMARY OF CONCLUSIONS

CONSIDER A SCALAR TIME SERIES $\{x_t\}$ which is generated jointly with a vector time series $\{\underline{z}_t\}$ as a stationary multivariate linear stochastic process. We then may define $\{y_t\}$ as another scalar time series related to $\{x_t\}$ and $\{\underline{z}_t\}$ by

(1) $$y_t = \sum_{j=0}^{n} \beta^j x_t^e(j),$$

where $x_t^e(j)$ denotes $E(x_{t+j}|I_t)$, I_t is the realization of $\{x_t\}$ and $\{\underline{z}_t\}$ up to and including time t, and $\beta < 1$. The multivariate time series $\{x_t, \underline{z}_t\}$ may be labeled the independent-variable series, its distribution being taken as exogenous, and $\{y_t\}$ the dependent-variable series. Equation (1) is the present-value relation. It states that the distribution of the dependent-variable process is related to that of the independent-variable process in such a way that the current realization of the dependent-variable process equals the present discounted expected value of one element of the independent-variable process, (x_t), where the expectation is conditional on all information currently available.

The present-value relation is repeatedly encountered in economic theory. The most familiar application is to the theory of stock prices, where $\{x_t\}$ refers to some corporation's earnings, $\{\underline{z}_t\}$ to any variables other than past earnings which are used to predict its future earnings, $\{y_t\}$ to the price of stock, and β to the discount factor.[3] In expectations theories of the term structure of interest rates, the present-value relation also appears (with a finite upper limit in the summation in (1)), although its validity in such applications is based on a linear

---

[1] The analyses and conclusions set forth are those of the authors and do not necessarily indicate concurrence by other members of the research staffs, by the Board of Governors, or by the Federal Reserve Banks.

[2] We wish to thank Evelyn Flynn, Gregory Connor, Juan Perea, and Birch Lee for able assistance and William Barnett, Fischer Black, Christopher Sims, Michael Dooley, Donald Hester, Agustin Maravall, Bennett McCallum, Darrel Parke, David Pierce, William Poole, Jack Rutner, and especially Robert Shiller, and two anonymous referees for helpful criticism. Thanks are also due to Susan Fay Eubank for typing innumerable drafts.

[3] As is well known in the finance literature, the representation of stock prices as the present value of discounted earnings involves double counting if any earnings are retained, since in that case both retained earnings and the revenues generated subsequently by these retentions are counted. However, in our model the maintained hypothesis that earnings are stationary implicitly presumes no retention. In the empirical work examined below an adjustment to the data to correct for earnings retention will be required.

approximation (Shiller [15]). Finally, the permanent income hypothesis of Friedman [2] may also be cast in the framework of equation (1).

In Section 2 of the present paper we state and prove three theorems about the variance of the dependent-variable process as it relates to that of the independent-variable process. The theorems embody successively sharper restrictions on the parameters of the independent and dependent variable processes. Theorem 1 asserts that the coefficient of dispersion (i.e., the ratio of the standard deviation to the mean) of $\{y_t\}$ is less than that of $\{x_t\}$. The second theorem involves two new time series $\{\hat{y}_t\}$ and $\{y_t^*\}$ which are generated by altering the amount of information assumed available about the future innovations of $\{x_t\}$ from that implicit in the specified joint distribution of $\{x_t\}$ and $\{\underline{z}_t\}$. If it is assumed that there is no information about the future innovations in $\{x_t\}$, the derived present value series is defined to be $\{\hat{y}_t\}$. On the other hand, if the future innovations in $\{x_t\}$ are taken as known, the derived present value series is labeled $\{y_t^*\}$. The dependent-variable series of primary interest, $\{y_t\}$, is in a sense an intermediate case since the realizations of $\{\underline{z}_t\}$ may be viewed as in general providing some information about the innovations in the univariate process for $\{x_t\}$, but not complete information. Theorem 2 exploits this fact, asserting that the variances of $\{\hat{y}_t\}$ and $\{y_t^*\}$ constitute lower and upper bounds, respectively, on the variance of $\{y_t\}$. Theorem 3, unlike Theorems 1 and 2 which state that the variance of $\{y_t\}$ lies within an interval, is the basis for an asymptotic point test of the null hypothesis defined by the present-value relation. If $\{\pi_t\}$ is defined as the present value of the forecast errors for $\{x_t\}$, we show that $\text{var}(y_t) + \text{var}(\pi_t) = \text{var}(y_t^*)$. Theorem 3 asserts that the three terms of this variance decomposition can all be estimated from observations on x_t and y_t and, therefore, forms the basis for a large-sample test of the present-value equation.

These theorems furnish a basis for constructing tests of the validity of the present-value relation. Such theorems are necessary because the present-value relation cannot be tested directly without also specifying the variables $\underline{z}_t$ used to predict x_t and then determining the joint distribution of $\{x_t, \underline{z}_t\}$, since in the absence of such a procedure the $x_t^e(j)$ are not measurable. Consequently, direct tests of the present-value relation are always conditional on the specification of the set of variables used to predict x_t, and the difficulty of specifying these variables exhaustively greatly weakens the plausibility of any conclusions based on such direct tests. By contrast, our three theorems are valid for general specifications of the joint distribution of $\{x_t, \underline{z}_t\}$, and do not require identification of the distribution of $\{x_t, \underline{z}_t\}$, or even specification of what the variables $\underline{z}_t$ are. Thus even though we do not measure the expectations $x_t^e(j)$ or the discounted forecast errors π_t, our theorems constitute testable implications of the present-value relation. The fact that the maintained assumptions required for our indirect tests are so much weaker than those required for the direct tests adds to the appeal of our results. Statistical tests of the three theorems are derived in Section 3.

In Section 4 we consider the application of our results to the theory of stock prices. First it is shown that the efficient capital markets hypothesis as conven-

PRESENT-VALUE RELATION

tionally formulated implies (and is implied by) the present-value relation between earnings and stock prices. Consequently, the hypothesis of capital market efficiency implies the validity of the theorems, and the latter may therefore be used to construct tests of market efficiency. That capital market efficiency implies restrictions on the volatility of stock prices is at first surprising because the most commonly-cited implication of market efficiency is that stock prices should move instantaneously rather than gradually in response to news. However, the result that if markets are efficient the coefficient of dispersion of stock prices should be less than that of earnings (Theorem 1) makes sense if it is observed that the present-value equation defines stock prices as a kind of weighted average of earnings, and an average is generally less volatile than its components. Another consideration strengthens this conclusion. Since stock prices are an average of expected rather than actual earnings, and since expected earnings can plausibly be assumed to regress toward a mean (correcting for trend) in the increasingly distant future, it follows that expected earnings should show less dispersion than actual earnings, further reducing the anticipated dispersion of stock prices.

Our first data set is based on Standard & Poor's Composite Index of stock prices and the related earnings and dividends series. The observations are quarterly over the interval 1955 to 1973, and the data are corrected for trend.[4] The estimated coefficients of dispersion of earnings and stock prices are 0.172 and 0.452, indicating that the inequality of Theorem 1 is contradicted (Table III). The point estimates on which the tests of Theorems 2 and 3 are based imply an even more flagrant violation of the model; for example, the estimated variance of y_t is 4.89, whereas the estimated variance of y_t^*, which theoretically should exceed that of y_t, is 0.255. These results, while dramatic, are difficult to interpret in the absence of any indication of the reliability of the test statistics. To provide such an indication, we calculated formal tests based on the asymptotic distribution of the parameter estimates, as described in Section 3. Because the test statistics measuring departures from the null hypothesis are all insignificant, the derived confidence intervals suggest that our statistical tests may have very little power.

The outcome of these tests may reflect the fact that they are based on aggregate data, whereas the theory applies to individual firms; a simple argument (presented in detail in Section 2) demonstrates the possible existence of bias in our tests due to aggregation error across firms. If the earnings of each firm consist of a common factor and an individual uncorrelated term, and if the common factor is forecastable whereas the individual term is not, then our tests will be biased in favor of rejection of market efficiency. In the reverse case, our tests will be biased in favor of acceptance. We do not know which case, if either, is more plausible than the intermediate case under which these biases approximately cancel, but the example provides strong motivation to examine data for individual firms. We collected quarterly earnings and price data for three large

---

[4]For a detailed description of the data, the complete derivation of the statistical tests, and for the data themselves, see LeRoy and Porter [9]; this paper is available upon request from the authors.

corporations—American Telephone & Telegraph, General Electric, and General Motors—then adjusted them for trend in the same way as the aggregate data, and calculated the test statistics for the three theorems.

The empirical results for the firm data show, as might be expected, that sharper hypotheses are more often rejected than blunt hypotheses. The point estimates for the tests of Theorem 1 were somewhat closer to being consistent with the null hypothesis of market efficiency for the individual firms than for the aggregate data. For one firm (GM) the coefficient of dispersion of earnings exceeded that of stock prices, as implied by the theory, while for another (GE) the two were virtually identical. Only AT&T was similar to the aggregate data in that earnings were considerably less volatile than stock prices, although for AT&T as with the aggregate data the test statistic did not allow rejection of the null hypothesis of Theorem 1 at the usual significance levels. Contrary to the implication of Theorem 2, however, the variance of y_t exceeded that of y_t^* by a wide margin for each of the three firms as with the aggregate data. Further, the relevant test statistic was significant for one firm (GE) and of borderline significance for another (GM), although it was insignificant for the third (AT&T). Finally, the test statistics for the more restrictive Theorem 3 all indicated rejection of the null hypothesis at the one per cent level.

Comparison of the results for the three firms with those for the Standard & Poor's series gives no clear reason to suspect that aggregation over firms biases aggregate tests either way, although, of course, this conclusion is not unequivocal since the Standard & Poor's series cannot be viewed as a simple aggregate of the three firms alone. As with the aggregate data, the point estimates for the firms are not consistent with the efficient markets model, although they are somewhat closer to those expected from theory than those for the aggregate data. However, the confidence intervals for the firms are more prone to indicate rejection of the null hypothesis; the smaller confidence regions for the firm data suggest that tests based on firm data may be somewhat more powerful than those for aggregate data.

We see that based on both aggregated and disaggregated data, stock prices appear to be more volatile than is consistent with the efficient capital markets model. This conclusion differs from that of most studies of market efficiency, such as that of Fama [1]. In many studies of market efficiency, it is observed that the martingale assumption requires that measured rates of return be serially uncorrelated; consequently, the efficient capital markets model may be tested by determining whether it is possible to reject the joint hypothesis that all the autocorrelations of rates of return are zero. Typically this null hypothesis cannot be rejected at the usual significance levels; we show this to be the case also for the rates of return earned on the stock of our three firms. Since these tests of the nonautocorrelatedness of rates of return are derivable consequences of the same model used to generate our variance restrictions, we are led to inquire why the nonautocorrelatedness implication is apparently satisfied, whereas the variance implications are not. Although it is possible that the difference in the outcomes of the dispersion and autocorrelation tests is due to differing sensitivity to specification or measurement error, an explanation that appears more attractive to us is that the dispersion tests have greater power than the autocorrelation tests against the hypothesis of market efficiency, given that alternative hypothesis which actually generated the data. This argument suggests that a promising way to investigate the stock market would be to ascertain what kinds of structures in earnings and prices would lead to deviations from market efficiency that would be more readily detected by a dispersion test than by an autocorrelation test. We have not yet pursued this line, however.

It is not clear how to interpret our rejection of the hypothesis we have characterized as "market efficiency." It should be recognized that our theorems are actually tests of a joint hypothesis, some elements of which have only tenuous support. The most important elements in our joint hypothesis are (i) the present-value relation (or, in the stock market application, the equivalent martingale assumption), (ii) the assumption that the real conditional expected rate of return on stock is constant over time, and (iii) the assumption of rational expectations. If our tests are not subject to econometric or measurement difficulties, then our rejection of the theorems implies that one or more of these elements of the joint hypothesis must be rejected. There is no reason to doubt that with further work it will be possible to distinguish which of the components of the rejected joint hypothesis must be revised.

In an important recent paper, Shiller [16] has independently derived and conducted tests of expectations models of the term structure of interest rates based on implied restrictions on the admissible dispersion of long rates relative to short. These restrictions are similar to our Theorems 1 and 2; in addition, Shiller obtained some important frequency-domain implications of the model. Although Shiller's tests, unlike ours, are based on point estimates rather than confidence intervals, implying that there is no way to determine statistical significance, he finds that the expectations model of the term structure appears to be violated, long rates being too volatile relative to short rates. The fact that Shiller's results on interest rates so closely parallel ours on stock prices suggests that neither set of results can be dismissed as a statistical accident. Rather, in our view, the fact that asset prices appear to fluctuate more than is consistent with most financial models in current use should be regarded as a major challenge to those models. As yet, however, it is impossible to determine what changes in financial theory may turn out to be necessary to accommodate our results and those of Shiller.

2. THREE THEOREMS ON THE VARIANCE OF THE DEPENDENT-VARIABLE PROCESS[5]

It is assumed that the $p \times 1$ vector $\{x_t, \underline{z}_t\}$ follows a multivariate linearly regular stationary stochastic process:

$$(2) \quad \begin{bmatrix} x_t \\ \underline{z}_t \end{bmatrix} = \underline{c} + \underline{\varepsilon}_t + D_1 \underline{\varepsilon}_{t-1} + D_2 \underline{\varepsilon}_{t-2} + \cdots = \underline{c} + D(B) \underline{\varepsilon}_t,$$

---

[5] In the remainder of this paper we let the upper limit (n) in the summation in (1) be infinite, as is appropriate for application to the stock market.

where the innovations sequence $\{\underline{\epsilon}_t\}$ is a set of serially uncorrelated vector random variables with zero mean and positive definite covariance matrix Σ, where $\underline{c}$ is a $p \times 1$ vector, the D_i are square matrices of order p, and where B is the lag operator, defined by $B^j\underline{\epsilon}_t = \underline{\epsilon}_{t-j}$.[6] If we delete all but the first element of the vector equation (2), we obtain

(3) $\quad x_t = c + \underline{\delta}'_0 \underline{\epsilon}_t + \underline{\delta}'_1 \underline{\epsilon}_{t-1} + \cdots = c + \underline{\delta}'(B)\underline{\epsilon}_t,$

where c is the first element of $\underline{c}$ and $\underline{\delta}'_i$ is the first row of D_i.[7] We incur no loss of generality by ignoring the distribution of $\underline{z}$ since the information content of current and past values of $\underline{z}$ is contained in the current and past values of $\underline{\epsilon}$, which are known. The conditional expected future values of x are given by

(4) $\quad x_t^e(j) = c + \underline{\delta}'_j \underline{\epsilon}_t + \underline{\delta}'_{j+1} \underline{\epsilon}_{t-1} + \cdots .$

In general, the forecasts $x_t^e(j)$ depend on the $\underline{z}_t$ as well as the x_t, since both are needed to construct the lagged $\underline{\epsilon}_t$. In this case the series $\{\underline{z}_t\}$ is said to be a leading indicator of $\{x_t\}$, in Pierce's [11] usage; see also Granger [3]. In the special case in which all the elements of $\underline{\delta}_j$ are zero except the first, efficient forecasts of $\{x_t\}$ can be constructed from past realizations of $\{x_t\}$ alone since in that case $\{\underline{z}_t\}$ is not a leading indicator of $\{x_t\}$. In this special case $\underline{\epsilon}_t$ and $\underline{\delta}_j$ can be taken as scalars without loss of generality. More generally, when $\underline{z}_t$ is a leading indicator of x_t, we can express the dependent-variable series in terms of the (vector) innovations in the independent-variable series. By substituting (4) into (1), it is easily verified that we have

(5) $\quad y_t = \dfrac{c}{1-\beta} + \sum_{j=0}^{\infty}\left[\sum_{k=j}^{\infty} \beta^{k-j}\underline{\delta}'_k\right]\underline{\epsilon}_{t-j} \equiv \dfrac{c}{1-\beta} + \underline{a}'(B)\underline{\epsilon}_t,$

where $\underline{a}'_j = \sum_{k=j}^{\infty}\beta^{k-j}\underline{\delta}'_k$.

We now state and prove the three theorems restricting the variance of $\{y_t\}$.

THEOREM 1: *The coefficient of dispersion of $\{y_t\}$ is less than that of $\{x_t\}$ for any distribution obeying* (2).

The proof of Theorem 1 is most conveniently presented later. At this point, it is useful to consider a special case in order to render Theorem 1 as intuitive as possible. Suppose that $\{\underline{z}_t\}$ is not a leading indicator of $\{x_t\}$ and that $\{x_t\}$ is distributed by a first-order autoregressive process,

$$x_t - c = \phi(x_{t-1} - c) + \epsilon_t, \quad |\phi| < 1,$$

which has the moving-average representation

$$x_t = c + \epsilon_t + \phi\epsilon_{t-1} + \phi^2\epsilon_{t-2} + \cdots .$$

---

[6] $D(B) = D_0 + D_1 B + D_2 B^2 + \cdots$, with $D_0 = I$.

[7] That is, $\delta(B) = \underline{\delta}_0 + \underline{\delta}_1 B + \underline{\delta}_2 B^2 + \cdots$ where $\underline{\delta}_0 = (1, 0, 0, \ldots, 0)$. See Rozanov [12] or Hannan [4] for a general discussion of the statistical properties of (2).

It is easily verified that the ratio of the coefficients of dispersion takes the simple form

(6) $$\frac{CD(y_t)}{CD(x_t)} = \frac{1-\beta}{1-\beta\phi},$$

which is always less than one since β is bounded by zero and one. Equation (6) shows that the lower ϕ is and the higher β is, the lower will be the ratio of the coefficients of dispersion. The reason is that if ϕ is near zero, expected x_t regresses rapidly toward the trend value c, which means that y_t is approximately equal to current x_t plus the discounted value of a series of constants. The addition of constants to x_t raises the mean of y_t without increasing its standard deviation, thereby lowering its coefficient of dispersion. Similarly, if β is near 1, relatively more weight is given to future expected x_t than to current x_t, compared to the case when β is near zero. Since for any value of ϕ expected future x_t has less dispersion than current x_t, the effect of larger values of β is to lower the ratio of the coefficients of dispersion.

Depending on the actual distribution of $\{x_t\}$, the test implied by Theorem 1 may not be very powerful statistically. Thus, if β is near one in the population and the distribution of $\{x_t\}$ incorporates strong damping, the efficient markets model might imply that the coefficient of dispersion of $\{y_t\}$ is a small fraction of that of $\{x_t\}$, say, one-quarter. In that case, the test implied by Theorem 1 that the ratio of coefficients of dispersion is less than unity would with high probability indicate acceptance of the null hypothesis even when it should be rejected (for example, if in the population the ratio of coefficients of dispersion were 1/2 or 3/4). Again, it is impossible to test this conjecture directly without knowledge of the joint distribution of $\{x_t\}$ and $\{z_t\}$. We seek to derive restrictions on the dispersion of $\{y_t\}$ stronger than those implied by Theorem 1, but still without specifying the distribution of $\{z_t\}$.

To do so we observe that so far we have used only one function of the parameters of the marginal distribution of $\{x_t\}$: its coefficient of dispersion. It might be expected that stronger restrictions on the behavior of $\{y_t\}$ could be derived if all the parameters of the distribution of $\{x_t\}$ were employed. To show that this is in fact possible we consider once again the general leading indicator case under which there exist variables z_t which figure in the forecasts of future x_t, but which do not predict x_t perfectly. Now fix the marginal distribution of $\{x_t\}$ and consider two polar cases: one under which there exist variables z_t which in addition to past x_t allow perfect forecasting of x_t, and the other in which $\{z_t\}$ is not a leading indicator of $\{x_t\}$ (i.e., in which there are no variables other than lagged x_t which assist in the forecasting of future x_t). Define $\{y_t^*\}$ and $\{\hat{y}_t\}$ as the series generated when the present-value relation operates on $\{x_t\}$ in each of these cases, and note that the distributions of these hypothetical price variables, unlike that of $\{y_t\}$, are completely determined by the marginal distribution of $\{x_t\}$.

THEOREM 2: *When z_t is a leading indicator of $\{x_t\}$ the coefficient of dispersion of $\{y_t\}$ is greater than or equal to that of $\{\hat{y}_t\}$, and less than that of $\{y_t^*\}$.*[8]

Theorem 2 gives bounds on the variance of any series $\{y_t\}$ that is generated by some joint distribution of x_t and z_t, and these bounds can be calculated from the marginal distribution of $\{x_t\}$ alone, implying that, as required, the general leading indicator case be tested without actually estimating the joint distribution of $\{x_t\}$ and $\{z_t\}$.

The proof of Theorem 2 is direct. By definition, y_t^* is expressible as

(7) $\quad y_t^* = x_t + \beta x_{t+1} + \beta^2 x_{t+2} + \cdots .$

Now define π_t, the discounted value of forecast errors, as

(8) $\quad \pi_t = \sum_{j=1}^{\infty} \beta^j (x_{t+j} - x_t^e(j)),$

where the $x_t^e(j)$ are the forecasts made under the general leading indicator model, as before. Then we have

$$y_t^* = y_t + \pi_t.$$

Now y_t depends only on the innovations in x_t and z_t up to and including period t, while π_t depends only on the innovations occurring after period t. Accordingly, they are statistically independent, and we have

(9) $\quad \text{var}(y_t^*) = \text{var}(y_t) + \text{var}(\pi_t).$

Equation (9) shows that the higher the variance of the discounted sum of forecast errors, the lower the variance of $\{y_t\}$. Consequently, the variance of $\{y_t^*\}$ provides an upper bound for the variance of $\{y_t\}$. Also, assuming as throughout that the information set always contains at least the past history of $\{x_t\}$, the variance of $\{\hat{y}_t\}$ furnishes a lower bound for the variance of $\{y_t\}$, since the presence of forecasting variables z_t in the information set can never increase the variance of discounted forecasting errors. Stating this conclusion in terms of coefficients of dispersion, we have

$$CD(\hat{y}_t) \leq CD(y_t) < CD(y_t^*).$$

Note that the right-hand side strict inequality follows from the fact that the model is one in which uncertainty cannot be entirely eliminated.

---

[8] As Singleton [18] observed, the proof to follow applies without modification in the case when y_t is given by

(1') $\quad \sum_{k=1}^{K} \sum_{j=0}^{\infty} \beta_k^j x_{kt}^e(j),$

that is, when y_t is the sum of K terms of the form of (1). Since this vector extension is immediate, our proof is restricted to the case $K = 1$. Note also that in economic applications of the present-value relation discussed in this paper, equation (1) is sufficiently general.

We are now in a position to prove Theorem 1. By virtue of Theorem 2, it is sufficient to show that the coefficient of dispersion of $\{y_t^*\}$ is less than that of $\{x_t\}$. But that result may be developed directly from equation (7). We have

$$\operatorname{var}(y_t^*) = E\big[(x_t - c) + \beta(x_{t+1} - c) + \beta^2(x_{t+2} - c) + \cdots \big]^2$$

or

(10) $\quad \operatorname{var}(y_t^*) = \dfrac{1}{1 - \beta^2}\big[\gamma_x(0) + 2\beta\gamma_x(1) + 2\beta^2\gamma_x(2) + \cdots \big],$

where $\gamma_x(i) \equiv \operatorname{covariance}(x_t, x_{t-i})$ for all t. From the Cauchy-Schwartz inequality and stationarity, $\gamma_x(i) < \gamma_x(0)$ if $i > 0$, so

(11) $\quad \operatorname{var}(y_t^*) < \gamma_x(0)\left[\dfrac{1}{1-\beta^2} + \dfrac{2\beta}{(1-\beta)(1-\beta^2)}\right] = \dfrac{\gamma_x(0)}{(1-\beta)^2}.$

From (11) it follows immediately that

$$\dfrac{\sqrt{\gamma_y^*(0)}}{c/(1-\beta)} < \dfrac{\sqrt{\gamma_x(0)}}{c}.$$

THEOREM 3: *When $\{z_t\}$ is a leading indicator of $\{x_t\}$, the variance of $\{y_t^*\}$ is equal to the variance of $\{y_t\}$ plus the variance of $\{\pi_t\}$, the discounted forecast error. Further, all these variances may be estimated directly using only measurements on $\{x_t\}$ and $\{y_t\}$.*[9]

We have already proved the first part of Theorem 3 (see equation (9)). Thus the significant assertion of Theorem 3 is that equation (9) may be used to construct a point test of the efficient markets model which can be applied without specifying the variables z_t and estimating their joint distribution with x_t. This is not obvious since the forecasts $x_t^e(j)$ which are used to calculate the π_t are not directly observable, nor can they be calculated without knowledge of the joint distribution of $\{z_t\}$ and $\{x_t\}$. However, it happens that even though π_t is not directly observable, its variance can be calculated from the distribution of $\{x_t\}$ and $\{y_t\}$ alone, and this is the content of Theorem 3.

To show this, substitute (3) and (4) into (8) to obtain

$$\pi_t = \beta \underline{\delta}_0' \underline{\epsilon}_{t+1} + \beta^2(\underline{\delta}_0' \underline{\epsilon}_{t+2} + \underline{\delta}_1' \underline{\epsilon}_{t+1})$$

$$+ \beta^3(\underline{\delta}_0' \underline{\epsilon}_{t+3} + \underline{\delta}_1' \underline{\epsilon}_{t+2} + \underline{\delta}_2' \underline{\epsilon}_{t+1}) + \cdots$$

---

[9] Singleton [18] also obtained a vector extension of Theorem 3; see footnote 8 supra. His extension, however, assumes that the β_k ($k = 1, 2, \ldots K$) are known, whereas in our model β is estimable.

Collecting terms, squaring, and taking expectations gives

(12) $$\text{var}(\pi_t) = \frac{\beta^2 \underline{a}_0' \Sigma \underline{a}_0}{1 - \beta^2},$$

where $\underline{a}_0$ is as defined in (5). Although $\underline{a}_0$ is not directly estimable, its weighted length is. Equation (5) may be used to derive

(13) $$\underline{a}_0' \underline{\varepsilon}_{t+1} = y_{t+1} + \frac{x_t - y_t}{\beta},$$

from which we calculate an expression for $\underline{a}_0' \Sigma \underline{a}_0$:

(14) $$\underline{a}_0' \Sigma \underline{a}_0 = \text{var}\left[y_{t+1} + 1/\beta(x_t - y_t) \right].$$

Since Σ is positive definite, $\text{var}(\pi_t) > 0$. Combining equations (12) and (14), we have

(15) $$\text{var}(\pi_t) = \frac{\text{var}(\beta y_{t+1} + x_t - y_t)}{1 - \beta^2}$$

which is directly measurable. Since the variances of $\{\hat{y}_t\}$ and $\{y_t^*\}$ are functions of β and a univariate representation for $\{x_t\}$, they are, of course, directly estimable from observations on x_t and y_t.[10]

The theorems just proved apply to individual firms; can they be tested on cross-section averages? A simple example[11] shows that aggregation bias may be a problem, depending on the covariance of x_{it} among firms and on the assumption made about the forecastability of x_{it}. Suppose that x_{it} depends linearly on a common factor z_t, which is perfectly forecastable, and a white noise term w_{it}:

$$x_{it} = \alpha(z_t + w_{it}).$$

Further, suppose that w_{it} is independent across firms, is independent of z_t, has common variance across firms, and is not forecastable. Since the forecastable component of each x_{it} is identical across firms, we have $CD(y_t) = CD(y_{it})$ for all i, as is readily verified. However, upon aggregation, the cancellation of the white noise terms, w_{it}, implies that $CD(y_t^*) < CD(y_{it}^*)$ for all i. If $CD(y_t^*)$ were viewed as an estimate of $CD(y_{it}^*)$, it would be biased toward zero, and a test of the null hypothesis, $CD(y_{it}) < CD(y_{it}^*)$ based on the inequality $CD(y_t) < CD(y_t^*)$ would be biased toward rejection. More generally, the example suggests that our tests will be biased toward rejection if the common component of x_{it} is more forecastable than the independent components. We do not know if this assumption is more reasonable than its opposite, in which case our tests are biased

---

[10] Observe that $E(x_t) = c$ and $E(y_t) = c/(1 - \beta)$ so that β may be readily estimated from the means of the two observed processes.

[11] We are indebted to a referee for this example.

toward acceptance. However, since we do not wish to prejudge the question by presuming that the two components are equally forecastable, as must be implicitly assumed under tests based on aggregated data, we are motivated to conduct our tests on both disaggregated and aggregated data, and thereby to avoid the issue of aggregation error.

3. TEST STATISTICS

The three theorems developed in Section 2 impose nonlinear restrictions on the expected value and autocovariance function of the bivariate process for x_t and y_t. To restate these restrictions in a way that is convenient for testing, we first define

(16) $\quad \gamma_{xy}(k) = E[(x_t - c)(y_{t-k} - c/(1 - \beta))]$

for $k = 0, \pm 1, \pm 2$, and so forth. Theorem 1 states that

(17) $\quad f_1 > 0,$

where

(18) $\quad f_1 = \dfrac{[\gamma_x(0)]^{1/2}}{c} - \dfrac{[\gamma_y(0)]^{1/2}}{c/(1 - \beta)}.$

Theorem 3 imposes the restriction

(19) $\quad f_3 = 0,$

where

(20) $\quad f_3 = \gamma_y^*(0) - \gamma_y(0) - \gamma_\pi(0)$

$= \dfrac{1}{1 - \beta^2}\left[\gamma_x(0) + 2\sum_{j=1}^{\infty} \beta^j \gamma_x(j)\right] - \gamma_y(0)$

$- \dfrac{1}{1 - \beta^2}\left[(1 + \beta^2)\gamma_y(0) + \gamma_x(0) + 2\beta\gamma_{xy}(-1) - 2\beta\gamma_y(1) - 2\gamma_{xy}(0)\right],$

in view of (10) and (15). The upper bound in Theorem 2 may be written as

(21) $\quad f_2^u > 0,$

where

(22) $\quad f_2^u = \dfrac{[\gamma_{y^*}(0)]^{1/2}}{c/(1 - \beta)} - \dfrac{[\gamma_y(0)]^{1/2}}{c/(1 - \beta)}$

$= \dfrac{(1 - \beta)}{c(1 - \beta^2)^{1/2}}\left[\gamma_x(0) + \sum_{k=1}^{\infty} 2\beta^k \gamma_x(k)\right]^{1/2} - \dfrac{[\gamma_y(0)]^{1/2}}{c/(1 - \beta)}.$

Finally, the lower bound restriction in Theorem 2 is

(23) $\quad f_2^l \geq 0,$

where

(24) $\quad f_2^l = \dfrac{[\gamma_v(0)]^{1/2}}{c/(1-\beta)} - \dfrac{[\gamma_{\tilde{y}}(0)]^{1/2}}{c/(1-\beta)}$

and

$$\gamma_{\tilde{y}}(0) = \sum_{j=0}^{\infty}\left[\sum_{k=j}^{\infty} \beta^{k-j} b_k\right]^2 \sigma_\varepsilon^2$$

where σ_ε^2 and $b(B)(= 1 + b_1 B + b_2 B^2 + \cdots)$ may be obtained by factoring the autocovariance generating function of $\{x_t\}$.[12]

Large-sample tests of the nonlinear restrictions in (17), (19), (21), and (23) on the functions in (18), (20), (22), and (24), respectively, may be constructed in a straightforward manner. First, a bivariate stationary and invertible ARMA representation for x and y is specified.[13] To estimate the ARMA model parameters, Wilson's [20] quasi-maximum likelihood algorithm is used except that the means are estimated first and then treated as if they are known.[14] The form of the estimated model is thus

$$\begin{bmatrix} \phi_{11}(B) & \phi_{12}(B) \\ \phi_{21}(B) & \phi_{22}(B) \end{bmatrix} \begin{bmatrix} x_t - \bar{x} \\ y_t - \bar{y} \end{bmatrix} = \begin{bmatrix} \theta_{11}(B) & \theta_{12}(B) \\ \theta_{21}(B) & \theta_{22}(B) \end{bmatrix} \underline{\xi}_t,$$

where $\phi_{ij}(B)$ and $\theta_{ij}(B)$ are polynomials of order p_{ij} and q_{ij}, respectively,

$$\phi_{ij}(B) = k_{ij} - \sum_{s=1}^{p_{ij}} \phi_{ij,s} B^s,$$

$$\theta_{ij}(B) = k_{ij} - \sum_{s=1}^{q_{ij}} \theta_{ij,s} B^s,$$

---
[12] That is, σ_v^2 and $b(B)$ are solutions to

$$\sigma_v^2 b(B) b(B^{-1}) = \sum_{j=-\infty}^{\infty} \gamma_x(j) B^j.$$

[13] See Wilson [19 and 20] for a description of multiple ARMA models. From (3) and (5) it will be seen that the bivariate process for y and x is a linear regular stationary process so that there exists an infinite order moving average representation (Wold decomposition). We assume that this representation can be approximated by a finite parameter bivariate ARMA representation; see Sims [17] for a proof that rational functions provide a mean square approximation to such linear regular processes. We also assume that under the alternative hypothesis, x and y are generated by a linear regular process which can be approximated as under the null hypothesis by a finite parameter ARMA model.

[14] Wilson's procedure maximizes the logarithm of the likelihood function under a normality assumption concerning the error, neglecting effects of initial conditions. The sample means x and y are used to estimate the population means c and $c/(1-\beta)$, respectively.

$k_{ij} = 1$ if $i = j$ and is 0 otherwise, and $\{\zeta_t\}$ is a set of serially uncorrelated bivariate random variables with zero mean and covariance matrix V. Let $\underline{\omega}$ be the vector of ARMA parameters (including intercepts and distinct elements of V) with $\hat{\underline{\omega}}$ denoting the estimate of $\underline{\omega}$. Under general conditions $\hat{\underline{\omega}}$ is asymptotically normally distributed with mean $\underline{\omega}$ and covariance matrix Ω.[15] Next, given $\hat{\underline{\omega}}$ and an estimate of Ω, the associated function $f_i(\underline{\omega})$, i.e., the functions in (18), (20), (22), and (24), and its asymptotic standard error may be evaluated. Since each of the test functions, f_i, is continuous, the ratio of $f_i(\hat{\underline{\omega}})$ to its estimated asymptotic standard error will have a $N(0, 1)$ distribution under the null hypothesis. That is,

$$\sqrt{T}(f_i(\hat{\underline{\omega}}) - f_i(\underline{\omega})) \to N(0, \underline{j}_i' \Omega \underline{j}_i),$$

where

$$\underline{j}_i = \frac{\partial f_i(\underline{\omega})}{\partial \underline{\omega}}$$

and T denotes the sample size.

4. APPLICATION TO THE EFFICIENT MARKETS MODEL OF STOCK PRICES

The efficient markets model may be characterized by the restriction that the (real) rate of return on stock $\{r_t\}$ is a time series obeying the relation

(25) $E(r_t | I_t) = \rho$

for all I_t, where ρ is a positive constant. This relation is the basis for most empirical tests of market efficiency, since it implies that no information contained in I_t is of any assistance in predicting future expected rates of return.[16] The analytical justification for identifying such a restriction with some economic notion of market efficiency, such as Pareto-optimal resource allocation or costless dissemination of information, is not immediate. This point is not pursued here; see, however, LeRoy [6,7,8], Lucas [10], Rubinstein [13], and Woodward [21] for discussion. If all (real) earnings on stock x_t are paid out in dividends and the payout is assumed to occur at the beginning of the period, the rate of return

---

[15] See LeRoy and Porter [9] for a detailed examination of the conditions and our estimate of the Ω based on $\hat{\omega}$. We assume that the fourth cumulants of ζ_t are zero in estimating the covariance matrix of V.

[16] In Fama's review article on the efficient capital markets theory [1], the efficient markets model when $\{z_t\}$ is a leading indicator of x_t is termed the semi-strong-form constant-return model, while the case in which z_t is the empty set is called the weak-form constant return model. In his context, the terminology is appropriate since it appears to be natural to view a model in which the expected return is constant conditional on the broader set of information as involving a stronger restriction on reality. Here, however, these usages would be misleading since in fact neither model is generally a special case of the other. Further, we will derive results that apply over all multivariate stationary earnings distributions, and therefore a fortiori over all distributions in which z_t is not a leading indicator. Thus in Fama's terminology, some of our weak-form results follow as a special case of the strong-form results. We see that Fama's definition, while analytically equivalent to our usage, would be misleading in the present context.

is

(26) $$r_t = \frac{y_{t+1}}{y_t - x_t} - 1,$$

where y_t is the (real) price of stock. Taking expectations conditional on I_t and using (25), this becomes

$$y_t = x_t + \frac{y_t^e(1)}{1+\rho}.$$

Repeating this procedure and assuming convergence, we obtain the present-value relation (1), with $\beta = (1+\rho)^{-1}$.[17]

The fact that stock prices are expressible as the present value of expected earnings means that the theorems derived in Section 2 are consequences of capital market efficiency as defined by (25). These results provide insights into the functions of capital markets that are interesting and not altogether obvious. For example, Theorem 1 says that the coefficient of dispersion of stock prices is necessarily less than that of earnings; this fact was noted and interpreted in the introduction. Additionally, equations (9), (13), and (15) show that the greater the accuracy with which individuals are able to forecast earnings, the higher the variance of stock prices, but the lower the variance of the rate of return on stock. These results are surprisingly powerful considering the generality with which the distribution of earnings has been specified. However, our primary interest is in constructing statistical tests of market efficiency, and not in providing extended interpretation of the properties of efficient markets, so we turn now to the empirical implementation.

Earnings and price data for Standard & Poor's Composite Index, AT&T, GE, and GM were assembled, and an attempt was made to correct for trends induced by inflation and earnings retention.[18] The question remains whether the resulting

---

[17] This argument, of course, does no more than motivate the connection between equation (25) defining an efficient capital market and the present-value relation. A formal derivation is found in Samuelson [14]. Note that even though under certainty the present-value relation is an immediate consequence of the definition of the rate of return, under uncertainty the strong restriction (25) on the distribution of rates of return is required in order to derive the present-value relation from the definition of the rate of return. Under general conditions of uncertainty (i.e., without assuming (25)), the present-value relation does not obtain.

[18] To correct for inflation, we divided all variables by the GNP deflator. The correction for retained earnings was somewhat more involved. First, we calculated a new variable, k_t, which may be viewed as a quantity index of the physical capital to which corporate equity is title. This index was assumed to equal unity at the initial time period and was augmented in proportion to the amount of retained corporate earnings in each quarter:

$$k_t = \begin{cases} 1, & t=1, \\ k_{t-1} + \dfrac{E_t - D_t}{P_0}, & t = 2, 3, \ldots, \end{cases}$$

where E_t is real earnings, D_t is real dividends, and P_t is real stock value. Finally, the adjusted earnings and equity value series, x_t and y_t, were calculated by dividing the actual earnings and equity value series by k_t:

$$x_t = E_t/k_t, \quad y_t = P_t/k_t.$$

See LeRoy and Porter [9] for the original data and adjusted series.

series can be assumed to obey the stationarity requirement. There appears to be some evidence of downward trends, although they are not clearly significant. We have decided to neglect such evidence and simply assume that the series are stationary since otherwise it is necessary to address such difficult questions as ascertaining to what degree stockholders can be assumed to have foreseen the assumed trend in earnings. It seems preferable to assume instead that there exist long cycles in the earnings series, implying that a sample of only a few decades may well appear nonstationary. On this interpretation, no correction for nonstationarity is indicated, but we must expect that, as with any statistical test based on a small sample, high Type II error will occur. We do not argue that this treatment is entirely adequate, nor do we in any way minimize the problem of nonstationarity; the dependence of our results on the assumption of stationarity is probably their single most severe limitation.

Table I presents the bivariate ARMA estimates for the four different data sets as well as the large-sample standard errors.[19] Table II shows the chi-square statistics $C(i, j)$ for the overall adequacy of the bivariate model.[20] The results in Table II suggest that the overall specification is adequate. The lefthand panel of Table III displays estimates of the four statistics f_1, f_2', f_2'', and f_3, and of the asymptotic standardized normal ratios (z ratios) for $f_1, f_2'', $ and f_3, namely:

$$z_1 = f_1 / \left(\hat{f}_1' \left(\frac{1}{T} \right) \hat{\Omega} \hat{f}_1 \right)^{1/2}, \quad z_2^u = f_2^u / \left(\hat{f}_2^{u\prime} \left(\frac{1}{T} \right) \hat{\Omega} \hat{f}_2^u \right)^{1/2},$$

$$z_3 = f_3 / \left(\hat{f}_3' \left(\frac{1}{T} \right) \hat{\Omega} \hat{f}_3 \right)^{1/2}.[21]$$

The middle and right panels of Table III present estimates of the variance and coefficients of dispersion, respectively, of y_t, $\hat{y}_t$, y_t^*, and π_t. For GM the coefficient of dispersion of earnings exceeds that of prices, as required by Theorem 1. However, for GE the two statistics are virtually identical, while for AT&T and the Standard & Poor's Index the coefficients of dispersion of prices are several times higher that those of earnings. Despite these apparently pronounced inequalities, none of the three z-statistics for the associated test $H_0: f_1 = 0$ are even nearly significant, so we can conclude that at the 5 percent level the data are consistent with Theorem 1. These results indicate that, as reported in the introduction, our tests have very wide confidence intervals. As expected, the hypothesis $H_0: f' > 0$ that stock price variance exceeds its theoretical lower bound is accepted; since the point estimate indicates acceptance, it is unnecessary to calculate the z statistics associated with f'.

---

[19] As indicated earlier, a circumflex over a parameter denotes an estimate. Only the nonzero lags are reported in Table I. Selection of the nonzero lags followed the identification procedures suggested by Haugh [5].

[20] See Wilson [19, 20].

[21] To conserve space we have listed the estimates of j_i and Ω in LeRoy and Porter [9]. The sample periods for the four data sets were 1955:1 to 1973:4 (Standard & Poor), 1955:1 to 1977:4 (AT&T); 1955:1 to 1978:2 (GE); and 1955:4 to 1977:4 (GM). To let starting transients damp out, the first ten observations in each sample were used to provide initial conditions; see Wilson [19]. The sample means $\bar{x}$ and $\bar{y}$ were also based on this truncated sample.

TABLE I
PARAMETER ESTIMATES OF THE BIVARIATE ARMA PROCESS

Firm or Aggregate	$\bar{x}$	ρ	$\hat{\beta}$	$\hat{v}_{11}$	$\hat{v}_{12}$	$\hat{v}_{22}$	Lag	$\hat{\phi}_{11}$ Coeff.	Lag	$\hat{\theta}_{11}$ Coeff.	Lag	$\hat{\theta}_{12}$ Coeff.	Lag	$\hat{\theta}_{21}$ Coeff.	Lag	$\hat{\phi}_{22}$ Coeff.	Lag	$\hat{\theta}_{22}$ Coeff.
Standard and Poor																		
Estimate	.285	4.89	.942	$.542 \times 10^{-3}$	$.990 \times 10^{-2}$	.280	1	.814	4	−.182					1	.761	3	.158
Standard Error	.0343	3.79	.0456					.072		.082						.068		.062
Estimate							4	.099	5	.338					4	.240	4	.082
Standard Error								.068		.082						.072		.080
Estimate									12	−.237							5	.447
Standard Error										.072								.075
Estimate									17	.268							12	−.306
Standard Error										.076								.079
Estimate																	17	.246
Standard Error																		.086
American Telephone and Telegraph																		
Estimate	.783	46.8	.983	3.21×10^{-3}	-3.28×10^{-2}	1.09×10^{-2}	1	.988	7	.231			3	-4.85×10^{-2}	1	.966	1	−.298
Standard Error	.095	13.7	5.29×10^{-3}					.022		.086				1.64×10^{-2}		.032		.102
Estimate									13	−.179			14	-3.06×10^{-2}			3	−.265
Standard Error										.087				1.58×10^{-2}				.103
Estimate													16	3.67×10^{-2}				
Standard Error														1.61×10^{-2}				
General Electric																		
Estimate	.497	44.7	.989	.0139	.0416	18.1	1	.273	9	.288	8	−.008	3	12.16	1	.944	12	−.290
Standard Error	.00173	6.91	.00169					.090		.095		.002		3.63		.048		.100
Estimate									11	.194			5	5.37			13	.359
Standard Error										.100				3.48				.102
Estimate									12	−.451			10	5.04				
Standard Error										.094				3.36				
Estimate													11	9.75				
Standard Error														3.41				
General Motors																		
Estimate	1.37	69.5	.980	.217	.590	41.5	4	.632	1	−.144	2	−.0209			1	.965	6	−.147
Standard Error	.104	18.2	4.86×10^{-3}					.082		.114		.0082				.029		.105
Estimate									9	.262	3	−.0181					9	.066
Standard Error										.118		.0083						.102
Estimate									10	.178	16	−.0158						
Standard Error										.121		.0082						

PRESENT-VALUE RELATION

TABLE II
"CHI-SQUARE" STATISTICS FOR OVERALL ADEQUACY OF BIVARIATE SPECIFICATION

Firm or Aggregate	$C(1,1)$	$C(1,2)$	$C(2,1)$	$C(2,2)$
Standard & Poor	49.3 (38)	32.2 (38)	28.7 (38)	20.7 (38)
AT&T	30.2 (46)	27.2 (46)	28.4 (46)	22.2 (46)
GE	35.2 (47)	35.8 (47)	27.1 (47)	21.8 (47)
GM	22.0 (44)	41.2 (44)	21.0 (44)	32.8 (44)

NOTE:
$$C(i,j) = T\left(\sum_{k=1}^{df} r_{ij}^2(k)\right)$$
where
$$r_{ij}(k) = \frac{1}{T}\left(\sum_{t=1}^{T-k} \hat{\varepsilon}_{it}\hat{\varepsilon}_{jt+k}\right)$$
and df, the degrees of freedom, is reported in parentheses beneath each "chi-square" statistic.

On the basis of point estimates, the Theorem 2 upper bound test, $\gamma_y(0) < \gamma_y^*(0)$, or, equivalently, $f_2^u > 0$, is flagrantly violated for all four data sets. However, as before, the asymptotic variances of the test statistics are very high—only for GE is rejection of the null hypothesis clearly called for. GM is a borderline case at the 5 per cent level, while for AT&T and the Standard & Poor's Index acceptance is indicated. Finally, for the more restrictive Theorem 3 test that $\gamma_y^*(0) = \gamma_y(0) + \gamma_\pi(0)$, the z statistic for the hypothesis $H_0: f_3 = 0$ indicates clear rejection of market efficiency for the three firms; for the aggregate index the test statistic was not significantly different from zero.

Our results may be summarized as follows: the point estimates corresponding to our three theorems all indicate that the bounds on price dispersion implied by the efficient markets model are dramatically violated empirically, although the confidence intervals on our tests are so wide that the departures are not always statistically significant. This conclusion differs from that of most tests of restriction (25), which generally indicate acceptance of the null hypothesis (Fama [1]).[22]

In order to interpret this discrepancy, we computed the standard autocorrelation test of the sort that has led to the acceptance of market efficiency (Table IV). The statistic appropriate for testing the joint hypothesis that the population autocorrelation of the rate of return up to lag k equal zero is

(27) $$\chi^2(k) = T\sum_{i=1}^{k}[\hat{\gamma}_r(i)/\hat{\gamma}_r(0)]^2,$$

---

[22] It should be noted that we have tested the model in real terms in contrast to most work in which nominal magnitudes are examined.

TABLE III
TEST STATISTICS, VARIANCES, AND COEFFICIENTS OF DISPERSION

Firm or Aggregate Index	Test Statistics				Variances			Coefficients of Dispersion			
	f_1	f_2^1	f_2^a	f_3	$\gamma_y(0)$	$\gamma_{\hat{y}}(0)$	$\gamma_y^*(0)$	$\gamma_w(0)$	$CD(x)$	$CD(\hat{y})$	$CD(y)$	$CD(y^*)$
Standard & Poor	$-.280$	.396	$-.348$	-8.63	4.89	1.64×10^{-1}	.255	3.99	.172	8.28×10^{-2}	.452	.052
z Statistic	$-.193$		$-.242$	$-.254$								
AT&T	$-.281$	.420	$-.314$	-828.7	385.7	9.77×10^{-6}	24.6	467.6	.139	6.68×10^{-5}	.420	.106
z Statistic	-1.096		-1.223	-2.006								
GE	-6.84×10^{-4}	.288	$-.264$	-1478.4	165.9	3.81×10^{-4}	1.12	1313.6	.287	4.36×10^{-4}	.288	.024
z Statistic	$-.0056$		-2.57	-4.41								
GM	.103	.375	$-.314$	-1773.9	690.5	3.37×10^{-2}	19.90	1103.3	.481	2.64×10^{-3}	.378	.064
z Statistic	.596		-1.84	-2.76								

TABLE IV
TESTS OF OVERALL MARKET EFFICIENCY

Firm or Aggregate	Chi-Square Statistics for Rates of Return $\chi^2(12)$	$\chi^2(24)$
Standard & Poor	88.0	155.4
American Telephone and Telegraph	9.4	15.2
General Electric	10.2	17.0
General Motors	10.8	14.4

where the term in brackets is the sample autocorrelation of rates of return, equation (26), at lag i. Under the null hypothesis of market efficiency, (27) is distributed as a chi-square statistic with k degrees of freedom. We calculated $\chi^2(k)$ for $k = 12$ and $k = 24$; for $k = 12$ ($k = 24$) the critical value of the chi-square statistic at the twenty-five per cent level is 14.8 (28.2), while at the one per cent level the critical value is 26.2 (43.0). Comparison of the sample statistics with the critical values indicates that the hypotheses that all lagged autocorrelations in rates of return are zero is accepted at the 25 per cent level for either $k = 12$ or $k = 24$ for the firm data, although it is rejected at the one per cent level for the Standard & Poor's Index for either $k = 12$ or $k = 24$.

As indicated earlier in Section 1, we are not able to resolve this difference between our results in which market efficiency is rejected with the standard results in which the opposite conclusion is reached. As suggested in the introduction, one possibility is that our test has greater power than the standard test for the particular dispersion restriction embodied in Theorem 3.[23]

University of California, Santa Barbara and Federal Reserve Board

Manuscript received March, 1978; revision received January, 1980.

---

[23] Both our test and the standard test may be derived from (13). Theorem 3 tests only one of the restrictions contained in (13), while the standard test is a simultaneous test of all the restrictions. The situation is analogous to a multivariate test that all the coefficients in a linear model are simultaneously zero versus a t test on an individual coefficient. If one particular coefficient is nonzero, the t test for that coefficient would have greater power than the multivariate test.

REFERENCES

[1] FAMA, EUGENE F.: "Efficient Capital Markets: A Review of Theory and Empirical Work," *Journal of Finance*, 25 (1970), 383–416.
[2] FRIEDMAN, MILTON: *A Theory of the Consumption Function*. Princeton: Princeton University Press, 1957.
[3] GRANGER, C. W. J.: "Investigating Causal Relations by Econometric Models and Cross-Spectral Methods," *Econometrica*, 37 (1969), 424–438.

[4] HANNAN, E. J.: *Multiple Time Series.* New York: John Wiley and Sons, 1970.
[5] HAUGH, LARRY D.: *The Identification of Time Series Interrelationships with Special Reference to Dynamic Regression Models,* Ph.D. Dissertation, Department of Statistics, University of Wisconsin, 1972.
[6] LeRoy, STEPHEN F.: "Risk Aversion and the Martingale Property of Stock Prices," *International Economic Review,* 14 (1973), 436–446.
[7] ———: "Efficient Capital Markets: Comment," *Journal of Finance,* 31 (1976), 139–141.
[8] ———: "Securities Prices Under Risk-Neutrality and Near Risk-Neutrality," reproduced, University of Chicago, 1979.
[9] LeRoy, STEPHEN F., AND RICHARD D. PORTER: "The Present-Value Relation: Tests Based on Implied Variance Bounds," Federal Reserve Board Special Studies Paper, 1980.
[10] LUCAS, ROBERT E., JR.: "Asset Prices in an Exchange Economy," *Econometrica,* 46 (1978), 1426–1446.
[11] PIERCE, DAVID A.: "Forecasting Dynamic Models with Stochastic Regressors," *Journal of Econometrics,* 3 (1975), 349–374.
[12] ROZANOV, YU. A.: *Stationary Random Processes,* Tr. by A. Feinstein. San Francisco: Holden-Day, 1963.
[13] RUBINSTEIN, MARK: "Securities Market Efficiency in an Arrow-Debreu Economy," *American Economic Review,* 65 (1975), 812–824.
[14] SAMUELSON, PAUL A.: "Proof that Properly Anticipated Prices Fluctuate Randomly," *Industrial Management Review,* 6 (1965), 41–49.
[15] SHILLER, ROBERT J.: *Rational Expectations and the Structure of Interest Rates,* unpublished Ph.D. dissertation, Department of Economics, M.I.T., 1972.
[16] ———: "The Volatility of Long-Term Interest Rates and Expectations of the Term Structure," *Journal of Political Economy,* 87 (1979), 1190–1219.
[17] SIMS, CHRISTOPHER A.: "Approximate Price Restrictions in Distributed Lag Estimation," *Journal of the American Statistical Association,* 67 (1972), 169–175.
[18] SINGLETON, KENNETH J.: "Expectations Models of the Term Structure and Implied Variance Bounds," *Journal of Political Economy,* 88 (1980), 1159–1176.
[19] WILSON, G. TUNNICLIFFE: Unpublished Ph.D. dissertation, Lancaster University, 1970.
[20] ———: "The Estimation of Parameters in Multivariate Time Series Models," *Journal of the Royal Statistical Society,* Series B, 35 (1973), 76–85.
[21] WOODWARD, S. E.: "Properly Anticipated Prices Do Not, In General, Fluctuate Randomly," reproduced, University of California, Santa Barbara, 1979.

[2]

Do Stock Prices Move Too Much to be Justified by Subsequent Changes in Dividends?

By ROBERT J. SHILLER*

A simple model that is commonly used to interpret movements in corporate common stock price indexes asserts that real stock prices equal the present value of rationally expected or optimally forecasted future real dividends discounted by a constant real discount rate. This valuation model (or variations on it in which the real discount rate is not constant but fairly stable) is often used by economists and market analysts alike as a plausible model to describe the behavior of aggregate market indexes and is viewed as providing a reasonable story to tell when people ask what accounts for a sudden movement in stock price indexes. Such movements are then attributed to "new information" about future dividends. I will refer to this model as the "efficient markets model" although it should be recognized that this name has also been applied to other models.

It has often been claimed in popular discussions that stock price indexes seem too "volatile," that is, that the movements in stock price indexes could not realistically be attributed to any objective new information, since movements in the price indexes seem to be "too big" relative to actual subsequent events. Recently, the notion that financial asset prices are too volatile to accord with efficient markets has received some econometric support in papers by Stephen LeRoy and Richard Porter on the stock market, and by myself on the bond market.

To illustrate graphically why it seems that stock prices are too volatile, I have plotted in Figure 1 a stock price index p_t with its *ex post* rational counterpart p_t^* (data set 1).[1] The stock price index p_t is the real Standard and Poor's Composite Stock Price Index (detrended by dividing by a factor proportional to the long-run exponential growth path) and p_t^* is the present discounted value of the actual subsequent real dividends (also as a proportion of the same long-run growth factor).[2] The analogous series for a modified Dow Jones Industrial Average appear in Figure 2 (data set 2). One is struck by the smoothness and stability of the *ex post* rational price series p_t^* when compared with the actual price series. This behavior of p^* is due to the fact that the present value relation relates p^* to a long-weighted moving average of dividends (with weights corresponding to discount factors) and moving averages tend to smooth the series averaged. Moreover, while real dividends did vary over this sample period, they did not vary long enough or far enough to cause major movements in p^*. For example, while one normally thinks of the Great Depression as a time when business was bad, real dividends were substantially below their long-run exponential growth path (i.e., 10–25 percent below the

---

*Associate professor, University of Pennsylvania, and research associate, National Bureau of Economic Research. I am grateful to Christine Amsler for research assistance, and to her as well as Benjamin Friedman, Irwin Friend, Sanford Grossman, Stephen LeRoy, Stephen Ross, and Jeremy Siegel for helpful comments. This research was supported by the National Bureau of Economic Research as part of the Research Project on the Changing Roles of Debt and Equity in Financing U.S. Capital Formation sponsored by the American Council of Life Insurance and by the National Science Foundation under grant SOC-7907561. The views expressed here are solely my own and do not necessarily represent the views of the supporting agencies.

[1] The stock price index may look unfamiliar because it is deflated by a price index, expressed as a proportion of the long-run growth path and only January figures are shown. One might note, for example, that the stock market decline of 1929–32 looks smaller than the recent decline. In real terms, it was. The January figures also miss both the 1929 peak and 1932 trough.

[2] The price and dividend series as a proportion of the long-run growth path are defined below at the beginning of Section I. Assumptions about public knowledge or lack of knowledge of the long-run growth path are important, as shall be discussed below. The series p^* is computed subject to an assumption about dividends after 1978. See text and Figure 3 below.

FIGURE 1

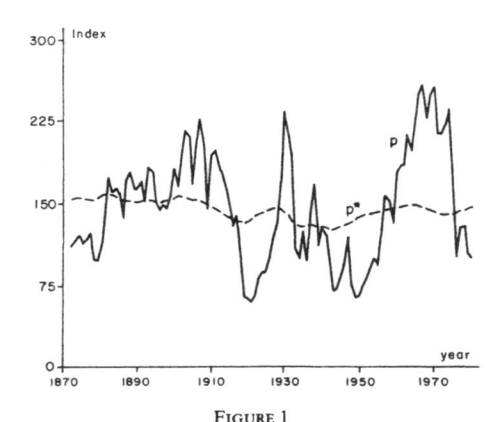

Note: Real Standard and Poor's Composite Stock Price Index (solid line p) and *ex post* rational price (dotted line p^*), 1871–1979, both detrended by dividing a long-run exponential growth factor. The variable p^* is the present value of actual subsequent real detrended dividends, subject to an assumption about the present value in 1979 of dividends thereafter. Data are from Data Set 1, Appendix.

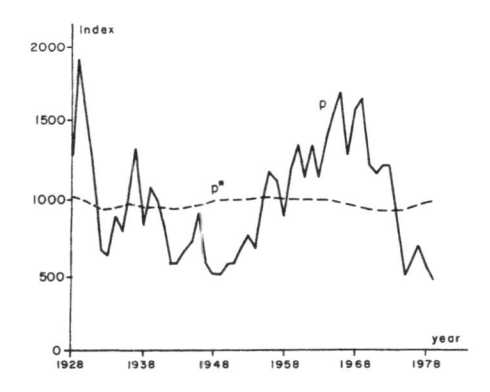

FIGURE 2

Note: Real modified Dow Jones Industrial Average (solid line p) and *ex post* rational price (dotted line p^*), 1928–1979, both detrended by dividing by a long-run exponential growth factor. The variable p^* is the present value of actual subsequent real detrended dividends, subject to an assumption about the present value in 1979 of dividends thereafter. Data are from Data Set 2, Appendix.

growth path for the Standard and Poor's series, 16–38 percent below the growth path for the Dow Series) only for a few depression years: 1933, 1934, 1935, and 1938. The moving average which determines p^* will smooth out such short-run fluctuations. Clearly the stock market decline beginning in 1929 and ending in 1932 could not be rationalized in terms of subsequent dividends! Nor could it be rationalized in terms of subsequent earnings, since earnings are relevant in this model only as indicators of later dividends. Of course, the efficient markets model does not say $p=p^*$. Might one still suppose that this kind of stock market crash was a rational mistake, a forecast error that rational people might make? This paper will explore here the notion that the very volatility of p (i.e., the tendency of big movements in p to occur again and again) implies that the answer is no.

To give an idea of the kind of volatility comparisons that will be made here, let us consider at this point the simplest inequality which puts limits on one measure of volatility: the standard deviation of p. The efficient markets model can be described as asserting that $p_t = E_t(p_t^*)$, i.e., p_t is the mathematical expectation conditional on all information available at time t of p_t^*. In other words, p_t is the optimal forecast of p_t^*. One can define the forecast error as $u_t = p_t^* - p_t$. A fundamental principle of optimal forecasts is that the forecast error u_t must be uncorrelated with the forecast; that is, the covariance between p_t and u_t must be zero. If a forecast error showed a consistent correlation with the forecast itself, then that would in itself imply that the forecast could be improved. Mathematically, it can be shown from the theory of conditional expectations that u_t must be uncorrelated with p_t.

If one uses the principle from elementary statistics that the variance of the sum of two uncorrelated variables is the sum of their variances, one then has $var(p^*) = var(u) + var(p)$. Since variances cannot be negative, this means $var(p) \leq var(p^*)$ or, converting to more easily interpreted standard deviations,

(1) $\sigma(p) \leq \sigma(p^*)$

This inequality (employed before in the

papers by LeRoy and Porter and myself) is violated dramatically by the data in Figures 1 and 2 as is immediately obvious in looking at the figures.[3]

This paper will develop the efficient markets model in Section I to clarify some theoretical questions that may arise in connection with the inequality (1) and some similar inequalities will be derived that put limits on the standard deviation of the innovation in price and the standard deviation of the change in price. The model is restated in innovation form which allows better understanding of the limits on stock price volatility imposed by the model. In particular, this will enable us to see (Section II) that the standard deviation of Δp is highest when information about dividends is revealed smoothly and that if information is revealed in big lumps occasionally the price series may have higher kurtosis (fatter tails) but will have *lower* variance. The notion expressed by some that earnings rather than dividend data should be used is discussed in Section III, and a way of assessing the importance of time variation in real discount rates is shown in Section IV. The inequalities are compared with the data in Section V.

This paper takes as its starting point the approach I used earlier (1979) which showed evidence suggesting that long-term bond yields are too volatile to accord with simple expectations models of the term structure of interest rates.[4] In that paper, it was shown how restrictions implied by efficient markets on the cross-covariance function of short-term and long-term interest rates imply inequality restrictions on the spectra of the long-term interest rate series which characterize the smoothness that the long rate should display. In this paper, analogous implications are derived for the volatility of stock prices, although here a simpler and more intuitively appealing discussion of the model in terms of its innovation representation is used. This paper also has benefited from the earlier discussion by LeRoy and Porter which independently derived some restrictions on security price volatility implied by the efficient markets model and concluded that common stock prices are too volatile to accord with the model. They applied a methodology in some ways similar to that used here to study a stock price index and individual stocks in a sample period starting after World War II.

It is somewhat inaccurate to say that this paper attempts to contradict the extensive literature of efficient markets (as, for example, Paul Cootner's volume on the random character of stock prices, or Eugene Fama's survey).[5] Most of this literature really examines different properties of security prices. Very little of the efficient markets literature bears directly on the characteristic feature of the model considered here: that expected *real* returns for the aggregate stock market are constant through time (or approximately so). Much of the literature on efficient markets concerns the investigation of nominal "profit opportunities" (variously defined) and whether transactions costs prohibit their exploitation. Of course, if real stock prices are "too volatile" as it is defined here, then there may well be a sort of real profit opportunity. Time variation in expected real interest rates does not itself imply that any

---

[3] Some people will object to this derivation of (1) and say that one might as well have said that $E_t(p_t) = p_t^*$, i.e., that forecasts are correct "on average," which would lead to a reversal of the inequality (1). This objection stems, however, from a misinterpretation of conditional expectations. The subscript t on the expectations operator E means "taking as given (i.e., nonrandom) all variables known at time t." Clearly, p_t is known at time t and p_t^* is not. In practical terms, if a forecaster gives as his forecast anything other than $E_t(p_t^*)$, then high forecast is not optimal in the sense of expected squared forecast error. If he gives a forecast which equals $\hat{E}_t(p_t^*)$ only on average, then he is adding random noise to the optimal forecast. The amount of noise apparent in Figures 1 or 2 is extraordinary. Imagine what we would think of our local weather forecaster if, say, actual local temperatures followed the dotted line and his forecasts followed the solid line!

[4] This analysis was extended to yields on preferred stocks by Christine Amsler.

[5] It should not be inferred that the literature on efficient markets uniformly supports the notion of efficiency put forth there, for example, that no assets are dominated or that no trading rule dominates a buy and hold strategy, (for recent papers see S. Basu; Franco Modigliani and Richard Cohn; William Brainard, John Shoven and Lawrence Weiss; and the papers in the symposium on market efficiency edited by Michael Jensen).

trading rule dominates a buy and hold strategy, but really large variations in expected returns might seem to suggest that such a trading rule exists. This paper does not investigate this, or whether transactions costs prohibit its exploitation. This paper is concerned, however, instead with a more interesting (from an economic standpoint) question: what accounts for movements in real stock prices and can they be explained by new information about subsequent real dividends? If the model fails due to excessive volatility, then we will have seen a new characterization of how the simple model fails. The characterization is not equivalent to other characterizations of its failure, such as that one-period holding returns are forecastable, or that stocks have not been good inflation hedges recently.

The volatility comparisons that will be made here have the advantage that they are insensitive to misalignment of price and dividend series, as may happen with earlier data when collection procedures were not ideal. The tests are also not affected by the practice, in the construction of stock price and dividend indexes, of dropping certain stocks from the sample occasionally and replacing them with other stocks, so long as the volatility of the series is not misstated. These comparisons are thus well suited to existing long-term data in stock price averages. The robustness that the volatility comparisons have, coupled with their simplicity, may account for their popularity in casual discourse.

I. The Simple Efficient Markets Model

According to the simple efficient markets model, the real price P_t of a share at the beginning of the time period t is given by

$$(2) \quad P_t = \sum_{k=0}^{\infty} \gamma^{k+1} E_t D_{t+k} \qquad 0 < \gamma < 1$$

where D_t is the real dividend paid at (let us say, the end of) time t, E_t denotes mathematical expectation conditional on information available at time t, and γ is the constant real discount factor. I define the constant real interest rate r so that $\gamma = 1/(1+r)$. Information at time t includes P_t and D_t and their lagged values, and will generally include other variables as well.

The one-period holding return $H_t \equiv (\Delta P_{t+1} + D_t)/P_t$ is the return from buying the stock at time t and selling it at time $t+1$. The first term in the numerator is the capital gain, the second term is the dividend received at the end of time t. They are divided by P_t to provide a rate of return. The model (2) has the property that $E_t(H_t) = r$.

The model (2) can be restated in terms of series as a proportion of the long-run growth factor: $p_t = P_t/\lambda^{t-T}$, $d_t = D_t/\lambda^{t+1-T}$ where the growth factor is $\lambda^{t-T} = (1+g)^{t-T}$, g is the rate of growth, and T is the base year. Dividing (2) by λ^{t-T} and substituting one finds[6]

$$(3) \quad p_t = \sum_{k=0}^{\infty} (\lambda\gamma)^{k+1} E_t d_{t+k}$$

$$= \sum_{k=0}^{\infty} \bar{\gamma}^{k+1} E_t d_{t+k}$$

The growth rate g must be less than the discount rate r if (2) is to give a finite price, and hence $\bar{\gamma} \equiv \lambda\gamma < 1$, and defining $\bar{r}$ by $\bar{\gamma} \equiv 1/(1+\bar{r})$, the discount rate appropriate for the p_t and d_t series is $\bar{r} > 0$. This discount rate $\bar{r}$ is, it turns out, just the mean dividend divided by the mean price, i.e, $\bar{r} = E(d)/E(p)$.[7]

---

[6]No assumptions are introduced in going from (2) to (3), since (3) is just an algebraic transformation of (2). I shall, however, introduce the assumption that d_t is jointly stationary with information, which means that the (unconditional) covariance between d_t and z_{t-k}, where z_t is any information variable (which might be d_t itself or p_t), depends only on k, not t. It follows that we can write expressions like $var(p)$ without a time subscript. In contrast, a realization of the random variable the conditional expectation $E_t(d_{t+k})$ is a function of time since it depends on information at time t. Some stationarity assumption is necessary if we are to proceed with any statistical analysis.

[7]Taking unconditional expectations of both sides of (3) we find

$$E(p) = \frac{\bar{\gamma}}{1-\bar{\gamma}} E(d)$$

using $\bar{\gamma} = 1/1+\bar{r}$ and solving we find $\bar{r} = E(d)/E(p)$.

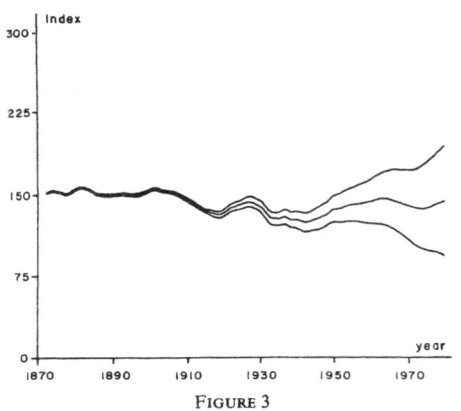

[Figure 3: Index plot from 1870 to 1970, three curves]

FIGURE 3

Note: Alternative measures of the *ex post* rational price p^*, obtained by alternative assumptions about the present value in 1979 of dividends thereafter. The middle curve is the p^* series plotted in Figure 1. The series are computed recursively from terminal conditions using dividend series d of Data Set 1.

TABLE 1 — DEFINITIONS OF PRINCIPAL SYMBOLS

γ = real discount factor for series before detrending; $\gamma = 1/(1+r)$
$\bar{\gamma}$ = real discount factor for detrended series; $\bar{\gamma} \equiv \lambda \gamma$
D_t = real dividend accruing to stock index (before detrending)
d_t = real detrended dividend; $d_t \equiv D_t / \lambda^{t+1-T}$
Δ = first difference operator $\Delta x_t \equiv x_t - x_{t-1}$
δ_t = innovation operator; $\delta_t x_{t+k} \equiv E_t x_{t+k} - E_{t-1} x_{t+k}$; $\delta x \equiv \delta_t x_t$
E = unconditional mathematical expectations operator. $E(x)$ is the true (population) mean of x.
E_t = mathematical expectations operator conditional on information at time t; $E_t x_t \equiv E(x_t | I_t)$ where I_t is the vector of information variables known at time t.
λ = trend factor for price and dividend series; $\lambda \equiv 1 + g$ where g is the long-run growth rate of price and dividends.
P_t = real stock price index (before detrending)
p_t = real detrended stock price index; $p_t = P_t / \lambda^{t-T}$
p_t^* = *ex post* rational stock price index (expression 4)
r = one-period real discount rate for series before detrending
$\bar{r}$ = real discount rate for detrended series; $\bar{r} = (1-\bar{\gamma})/\bar{\gamma}$
$\bar{r}_2$ = two-period real discount rate for detrended series; $\bar{r}_2 = (1+\bar{r})^2 - 1$
t = time (year)
T = base year for detrending and for wholesale price index; $p_T = P_T$ = nominal stock price index at time T

We may also write the model as noted above in terms of the *ex post* rational price series p_t^* (analogous to the *ex post* rational interest rate series that Jeremy Siegel and I used to study the Fisher effect, or that I used to study the expectations theory of the term structure). That is, p_t^* is the present value of actual subsequent dividends:

(4) $$p_t = E_t(p_t^*)$$

where $$p_t^* = \sum_{k=0}^{\infty} \bar{\gamma}^{k+1} d_{t+k}$$

Since the summation extends to infinity, we never observe p_t^* without some error. However, with a long enough dividend series we may observe an approximate p_t^*. If we choose an arbitrary value for the terminal value of p_t^* (in Figures 1 and 2, p^* for 1979 was set at the average detrended real price over the sample) then we may determine p_t^* recursively by $p_t^* = \bar{\gamma}(p_{t+1}^* + d_t)$ working backward from the terminal date. As we move back from the terminal date, the importance of the terminal value chosen declines. In data set (1) as shown in Figure 1, $\bar{\gamma}$ is .954 and $\bar{\gamma}^{108} = .0063$ so that at the beginning of the sample the terminal value chosen has a negligible weight in the determination of p_t^*. If we had chosen a different terminal condition, the result would be to add or subtract an exponential trend from the p^* shown in Figure 1. This is shown graphically in Figure 3, in which p^* is shown computed from alternative terminal values. Since the only thing we need know to compute p^* about dividends after 1978 is p^* for 1979, it does not matter whether dividends are "smooth" or not after 1978. Thus, Figure 3 represents our uncertainty about p^*.

There is yet another way to write the model, which will be useful in the analysis which follows. For this purpose, it is convenient to adopt notation for the innovation in a variable. Let us define the innovation operator $\delta_t \equiv E_t - E_{t-1}$ where E_t is the conditional expectations operator. Then for any variable X_t the term $\delta_t X_{t+k}$ equals $E_t X_{t+k} - E_{t-1} X_{t+k}$ which is the change in the conditional expectation of X_{t+k} that is made in response to new information arriving between $t-1$ and t. The time subscript t may be dropped so that δX_k denotes $\delta_t X_{t+k}$ and

δX denotes δX_0 or $\delta_t X_t$. Since conditional expectations operators satisfy $E_j E_k = E_{\min(j,k)}$ it follows that $E_{t-m}\delta_t X_{t+k} = E_{t-m}(E_t X_{t+k} - E_{t-1} X_{t+k}) = E_{t-m} X_{t+k} - E_{t-m} X_{t+k} = 0$, $m \geqslant 0$. This means that $\delta_t X_{t+k}$ must be uncorrelated for all k with all information known at time $t-1$ and must, since lagged innovations are information at time t, be uncorrelated with $\delta_{t'} X_{t+j}$, $t' < t$, all j, i.e., innovations in variables are serially uncorrelated.

The model implies that the innovation in price $\delta_t p_t$ is observable. Since (3) can be written $p_t = \bar{\gamma}(d_t + E_t p_{t+1})$, we know, solving, that $E_t p_{t+1} = p_t/\bar{\gamma} - d_t$. Hence $\delta_t p_t \equiv E_t p_t - E_{t-1} p_t = p_t + d_{t-1} - p_{t-1}/\bar{\gamma} = \Delta p_t + d_{t-1} - \bar{r} p_{t-1}$. The variable which we call $\delta_t p_t$ (or just δp) is the variable which Clive Granger and Paul Samuelson emphasized should, in contrast to $\Delta p_t \equiv p_t - p_{t-1}$, by efficient markets, be unforecastable. In practice, with our data, $\delta_t p_t$ so measured will approximately equal Δp_t.

The model also implies that the innovation in price is related to the innovations in dividends by

$$(5) \qquad \delta_t p_t = \sum_{k=0}^{\infty} \bar{\gamma}^{k+1} \delta_t d_{t+k}$$

This expression is identical to (3) except that δ_t replaces E_t. Unfortunately, while $\delta_t p_t$ is observable in this model, the $\delta_t d_{t+k}$ terms are not directly observable, that is, we do not know when the public gets information about a particular dividend. Thus, in deriving inequalities below, one is obliged to assume the "worst possible" pattern of information accrual.

Expressions (2)–(5) constitute four different representations of the same efficient markets model. Expressions (4) and (5) are particularly useful for deriving our inequalities on measures of volatility. We have already used (4) to derive the limit (1) on the standard deviation of p given the standard deviation of p^*, and we will use (5) to derive a limit on the standard deviation of δp given the standard deviation of d.

One issue that relates to the derivation of (1) can now be clarified. The inequality (1) was derived using the assumption that the forecast error $u_t = p_t^* - p_t$ is uncorrelated with p_t. However, the forecast error u_t is not serially uncorrelated. It is uncorrelated with all information known at time t, but the lagged forecast error u_{t-1} is not known at time t since p_{t-1}^* is not discovered at time t. In fact, $u_t = \sum_{k=1}^{\infty} \bar{\gamma}^k \delta_{t+k} p_{t+k}$, as can be seen by substituting the expressions for p_t and p_t^* from (3) and (4) into $u_t = p_t^* - p_t$, and rearranging. Since the series $\delta_t p_t$ is serially uncorrelated, u_t has first-order autoregressive serial correlation.[8] For this reason, it is inappropriate to test the model by regressing $p_t^* - p_t$ on variables known at time t and using the ordinary t-statistics of the coefficients of these variables. However, a generalized least squares transformation of the variables would yield an appropriate regression test. We might thus regress the transformed variable $u_t - \bar{\gamma} u_{t+1}$ on variables known at time t. Since $u_t - \bar{\gamma} u_{t+1} = \bar{\gamma} \delta_{t+1} p_{t+1}$, this amounts to testing whether the innovation in price can be forecasted. I will perform and discuss such regression tests in Section V below.

To find a limit on the standard deviation of δp for a given standard deviation of d_t, first note that d_t equals its unconditional expectation plus the sum of its innovations:

$$(6) \qquad d_t = E(d) + \sum_{k=0}^{\infty} \delta_{t-k} d_t$$

If we regard $E(d)$ as $E_{-\infty}(d_t)$, then this expression is just a tautology. It tells us, though, that d_t, $t = 0, 1, 2, \ldots$ are just different linear combinations of the same innovations in dividends that enter into the linear combination in (5) which determine $\delta_t p_t$, $t = 0, 1, 2, \ldots$. We can thus ask how large $var(\delta p)$ might be for given $var(d)$. Since innovations are serially uncorrelated, we know from (6) that the variance of the sum is

---

[8]It follows that $var(u) = var(\delta p)/(1 - \bar{\gamma}^2)$ as LeRoy and Porter noted. They base their volatility tests on our inequality (1) (which they call theorem 2) and an equality restriction $\sigma^2(p) + \sigma^2(\delta p)/(1 - \bar{\gamma}^2) = \sigma^2(p^*)$ (their theorem 3). They found that, with postwar Standard and Poor earnings data, both relations were violated by sample statistics.

the sum of the variances:

(7) $$var(d) = \sum_{k=0}^{\infty} var(\delta d_k) = \sum_{k=0}^{\infty} \sigma_k^2$$

Our assumption of stationarity for d_t implies that $var(\delta_{t-k}d_t) \equiv var(\delta d_k) \equiv \sigma_k^2$ is independent of t.

In expression (5) we have no information that the variance of the sum is the sum of the variances since all the innovations are time t innovations, which may be correlated. In fact, for given $\sigma_0^2, \sigma_1^2, \ldots$, the maximum variance of the sum in (5) occurs when the elements in the sum are perfectly positively correlated. This means then that so long as $var(\delta d) \neq 0$, $\delta_t d_{t+k} = a_k \delta_t d_t$, where $a_k = \sigma_k/\sigma_0$. Substituting this into (6) implies

(8) $$\hat{d}_t = \sum_{k=0}^{\infty} a_k \varepsilon_{t-k}$$

where a hat denotes a variable minus its mean: $\hat{d}_t \equiv d_t - E(d)$ and $\varepsilon_t \equiv \delta_t d_t$. Thus, if $var(\delta p)$ is to be maximized for given $\sigma_0^2, \sigma_1^2, \ldots$, the dividend process must be a moving average process in terms of its own innovations.[9] I have thus shown, rather than assumed, that if the variance of δp is to be maximized, the forecast of d_{t+k} will have the usual ARIMA form as in the forecast popularized by Box and Jenkins.

We can now find the maximum possible variance for δp for given variance of d. Since the innovations in (5) are perfectly positively correlated, $var(\delta p) = (\sum_{k=0}^{\infty} \bar{\gamma}^{k+1} \sigma_k)^2$. To maximize this subject to the constraint $var(d) = \sum_{k=0}^{\infty} \sigma_k^2$ with respect to $\sigma_0, \sigma_1, \ldots$, one may set up the Lagrangean:

(9) $$L = \left(\sum_{k=0}^{\infty} \bar{\gamma}^{k+1} \sigma_k \right)^2 + \nu \left(var(d) - \sum_{k=0}^{\infty} \sigma_k^2 \right)$$

[9] Of course, all indeterministic stationary processes can be given linear moving average representations, as Hermann Wold showed. However, it does not follow that the process can be given a moving average representation in terms of its own innovations. The true process may be generated nonlinearly or other information besides its own lagged values may be used in forecasting. These will generally result in a less than perfect correlation of the terms in (5).

where ν is the Lagrangean multiplier. The first-order conditions for $\sigma_j, j=0,\ldots\infty$ are

(10) $$\frac{\partial L}{\partial \sigma_j} = 2 \left(\sum_{k=0}^{\infty} \bar{\gamma}^{k+1} \sigma_k \right) \bar{\gamma}^{j+1} - 2\nu \sigma_j = 0$$

which in turn means that σ_j is proportional to $\bar{\gamma}^j$. The second-order conditions for a maximum are satisfied, and the maximum can be viewed as a tangency of an isoquant for $var(\delta p)$, which is a hyperplane in $\sigma_0, \sigma_1, \sigma_2, \ldots$ space, with the hypersphere represented by the constraint. At the maximum $\sigma_k^2 = (1-\bar{\gamma}^2) var(d) \bar{\gamma}^{2k}$ and $var(\delta p) = \bar{\gamma}^2 var(d)/(1-\bar{\gamma}^2)$ and so, converting to standard deviations for ease of interpretation, we have

(11) $$\sigma(\delta p) \leq \sigma(d)/\sqrt{\bar{r}_2}$$

where $$\bar{r}_2 = (1+\bar{r})^2 - 1$$

Here, $\bar{r}_2$ is the two-period interest rate, which is roughly twice the one-period rate. The maximum occurs, then, when d_t is a first-order autoregressive process, $\hat{d}_t = \bar{\gamma}\hat{d}_{t-1} + \varepsilon_t$, and $E_t \hat{d}_{t+k} = \bar{\gamma}^k \hat{d}_t$, where $\hat{d} \equiv d - E(d)$ as before.

The variance of the innovation in price is thus maximized when information about dividends is revealed in a smooth fashion so that the standard deviation of the new information at time t about a future dividend d_{t+k} is proportional to its weight in the present value formula in the model (5). In contrast, suppose all dividends somehow became known years before they were paid. Then the innovations in dividends would be so heavily discounted in (5) that they would contribute little to the standard deviation of the innovation in price. Alternatively, suppose nothing were known about dividends until the year they are paid. Here, although the innovation would not be heavily discounted in (5), the impact of the innovation would be confined to only one term in (5), and the standard deviation in the innovation in price would be limited to the standard deviation in the single dividend.

Other inequalities analogous to (11) can also be derived in the same way. For example, we can put an upper bound to the standard deviation of the change in price (rather than the innovation in price) for given standard deviation in dividend. The only difference induced in the above procedure is that Δp_t is a different linear combination of innovations in dividends. Using the fact that $\Delta p_t = \delta_t p_t + \bar{r} p_{t-1} - d_{t-1}$ we find

$$(12) \quad \Delta p_t = \sum_{k=0}^{\infty} \bar{\gamma}^{k+1} \delta_t d_{t+k}$$

$$+ \bar{r} \sum_{j=1}^{\infty} \delta_{t-j} \sum_{k=0}^{\infty} \bar{\gamma}^{k+1} d_{t+k-1} - \sum_{j=1}^{\infty} \delta_{t-j} d_{t-1}$$

As above, the maximization of the variance of δp for given variance of d requires that the time t innovations in d be perfectly correlated (innovations at different times are necessarily uncorrelated) so that again the dividend process must be forecasted as an ARIMA process. However, the parameters of the ARIMA process for d which maximize the variance of Δp will be different. One finds, after maximizing the Lagrangean expression (analogous to (9)) an inequality slightly different from (11),

$$(13) \quad \sigma(\Delta p) \leq \sigma(d)/\sqrt{2\bar{r}}$$

The upper bound is attained if the optimal dividend forecast is first-order autoregressive, but with an autoregressive coefficient slightly different from that which induced the upper bound to (11). The upper bound to (13) is attained if $\hat{d}_t = (1-\bar{r})\hat{d}_{t-1} + \varepsilon_t$ and $E_t d_{t+k} = (1-\bar{r})^k \hat{d}_t$, where, as before, $\hat{d}_t \equiv d_t - E(d)$.

II. High Kurtosis and Infrequent Important Breaks in Information

It has been repeatedly noted that stock price change distributions show high kurtosis or "fat tails." This means that, if one looks at a time-series of observations on δp or Δp, one sees long stretches of time when their (absolute) values are all rather small and then an occasional extremely large (absolute) value. This phenomenon is commonly attributed to a tendency for new information to come in big lumps infrequently. There seems to be a common presumption that this information lumping might cause stock price changes to have high or infinite variance, which would seem to contradict the conclusion in the preceding section that the variance of price is limited and is maximized if forecasts have a simple autoregressive structure.

High sample kurtosis does not indicate infinite variance if we do not assume, as did Fama (1965) and others, that price changes are drawn from the stable Paretian class of distributions.[10] The model does not suggest that price changes have a distribution in this class. The model instead suggests that the existence of moments for the price series is implied by the existence of moments for the dividends series.

As long as d is jointly stationary with information and has a finite variance, then p, p^*, δp, and Δp will be stationary and have a finite variance.[11] If d is normally distributed, however, it does not follow that the price variables will be normally distributed. In fact, they may yet show high kurtosis.

To see this possibility, suppose the dividends are serially independent and identically normally distributed. The kurtosis of the price series is defined by $K = E(\hat{p})^4/(E(\hat{p})^2)^2$, where $p \equiv \hat{p} - E(p)$. Suppose, as an example, that with a probability of $1/n$

---

[10] The empirical fact about the unconditional distribution of stock price changes in not that they have infinite variance (which can never be demonstrated with any finite sample), but that they have high kurtosis in the sample.

[11] With any stationary process X_t, the existence of a finite $var(X_t)$ implies, by Schwartz's inequality, a finite value of $cov(X_t, X_{t+k})$ for any k, and hence the entire autocovariance function of X_t, and the spectrum, exists. Moreover, the variance of $E_t(X_t)$ must also be finite, since the variance of X equals the variance of $E_t(X_t)$ plus the variance of the forecast error. While we may regard real dividends as having finite variance, innovations in dividends may show high kurtosis. The residuals in a second-order autoregression for d_t have a studentized range of 6.29 for the Standard and Poor series and 5.37 for the Dow series. According to the David-Hartley-Pearson test, normality can be rejected at the 5 percent level (but not at the 1 percent level) with a one-tailed test for both data sets.

the public is told d_t at the beginning of time t, but with probability $(n-1)/n$ has no information about current or future dividends.[12] In time periods when they are told d_t, $\hat{p}_t$ equals $\bar{\gamma}\hat{d}_t$, otherwise $\hat{p}_t = 0$. Then $E(\hat{p}_t^4) = E((\bar{\gamma}\hat{d}_t)^4)/n$ and $E(\hat{p}_t^2) = E((\bar{\gamma}\hat{d}_t)^2)/n$ so that kurtosis equals $nE(\bar{\gamma}\hat{d}_t)^4)/E((\bar{\gamma}\hat{d}_t)^2)$ which equals n times the kurtosis of the normal distribution. Hence, by choosing n high enough one can achieve an arbitrarily high kurtosis, and yet the variance of price will always exist. Moreover, the distribution of $\hat{p}_t$ conditional on the information that the dividend has been revealed is also normal, in spite of high kurtosis of the unconditional distribution.

If information is revealed in big lumps occasionally (so as to induce high kurtosis as suggested in the above example) $var(\delta p)$ or $var(\Delta p)$ are not especially large. The variance loses more from the long interval of time when information is not revealed than it gains from the infrequent events when it is. The highest possible variance for given variance of d indeed comes when information is revealed smoothly as noted in the previous section. In the above example, where information about dividends is revealed one time in n, $\sigma(\delta p) = \bar{\gamma}n^{1/2}\sigma(d)$ and $\sigma(\Delta p) = \bar{\gamma}(2/n)^{1/2}\sigma(d)$. The values of $\sigma(\delta p)$ and $\sigma(\Delta p)$ implied by this example are for all n strictly below the upper bounds of the inequalities (11) and (13).[13]

III. Dividends or Earnings?

It has been argued that the model (2) does not capture what is generally meant by efficient markets, and that the model should be replaced by a model which makes price the present value of expected earnings rather than dividends. In the model (2) earnings may be relevant to the pricing of shares but only insofar as earnings are indicators of future dividends. Earnings are thus no different from any other economic variable which may indicate future dividends. The model (2) is consistent with the usual notion in finance that individuals are concerned with returns, that is, capital gains plus dividends. The model implies that expected total returns are constant and that the capital gains component of returns is just a reflection of information about future dividends. Earnings, in contrast, are statistics conceived by accountants which are supposed to provide an indicator of how well a company is doing, and there is a great deal of latitude for the definition of earnings, as the recent literature on inflation accounting will attest.

There is no reason why price per share ought to be the present value of expected earnings per share if some earnings are retained. In fact, as Merton Miller and Franco Modigliani argued, such a present value formula would entail a fundamental sort of double counting. It is incorrect to include in the present value formula both earnings at time t and the later earnings that accrue when time t earnings are reinvested.[14] Miller and Modigliani showed a formula by which price might be regarded as the present value of earnings corrected for investments, but that formula can be shown, using an accounting identity to be identical to (2).

Some people seem to feel that one cannot claim price as present value of expected dividends since firms routinely pay out only a fraction of earnings and also attempt somewhat to stabilize dividends. They are right in the case where firms paid out *no* dividends, for then the price p_t would have to grow at the discount rate $\bar{r}$, and the model (2) would not be the solution to the difference equation implied by the condition $E_t(H_t) = r$. On the other hand, if firms pay out a fraction of dividends or smooth short-run fluctuations in dividends, then the price of the firm will grow at a rate less than the

---

[12] For simplicity, in this example, the assumption elsewhere in this article that d_t is always known at time t has been dropped. It follows that in this example $\delta_t p_t \neq \Delta p_t + d_{t-1} - rp_{t-1}$ but instead $\delta_t p_t = p_t$.

[13] For another illustrative example, consider $\hat{d}_t = \bar{\gamma}\hat{d}_{t-1} + \varepsilon_t$, as with the upper bound for the inequality (11) but where the dividends are announced for the next n years every $1/n$ years. Here, even though $\hat{d}_t$ has the autoregressive structure, ε_t is not the innovation in d_t. As n goes to infinity, $\sigma(\delta p)$ approaches zero.

[14] LeRoy and Porter do assume price as present value of earnings but employ a correction to the price and earnings series which is, under additional theoretical assumptions not employed by Miller and Modigliani, a correction for the double counting.

discount rate and (2) is the solution to the difference equation.[15] With our Standard and Poor data, the growth rate of real price is only about 1.5 percent, while the discount rate is about 4.8%+1.5%=6.3%. At these rates, the value of the firm a few decades hence is so heavily discounted relative to its size that it contributes very little to the value of the stock today; by far the most of the value comes from the intervening dividends. Hence (2) and the implied p^* ought to be useful characterizations of the value of the firm.

The crucial thing to recognize in this context is that once we know the terminal price and intervening dividends, we have specified all that investors care about. It would not make sense to define an *ex post* rational price from a terminal condition on price, using the same formula with earnings in place of dividends.

IV. Time-Varying Real Discount Rates

If we modify the model (2) to allow real discount rates to vary without restriction through time, then the model becomes untestable. We do not observe real discount rates directly. Regardless of the behavior of P_t and D_t, there will always be a discount rate series which makes (2) hold identically. We might ask, though, whether the movements in the real discount rate that would be required aren't larger than we might have expected. Or is it possible that small movements in the current one-period discount rate coupled with new information about such movements in future discount rates could account for high stock price volatility?[16]

[15]To understand this point, it helps to consider a traditional continuous time growth model, so instead of (2) we have $P_0 = \int_0^\infty D_t e^{-rt} dt$. In such a model, a firm has a constant earnings stream I. If it pays out all earnings, then $D=I$ and $P_0 = \int_0^\infty I e^{-rt} dt = I/r$. If it pays out only s of its earnings, then the firm grows at rate $(1-s)r$, $D_t = sIe^{(1-s)rt}$ which is less than I at $t=0$, but higher than I later on. Then $P_0 = \int_0^\infty sIe^{(1-s)rt}e^{-rt}dt = \int_0^\infty sIe^{-srt}dt = sI/(rs)$. If $s \neq 0$ (so that we're not dividing by zero) $P_0 = I/r$.

[16]James Pesando has discussed the analogous question: how large must the variance in liquidity premia be in order to justify the volatility of long-term interest rates?

The natural extension of (2) to the case of time varying real discount rates is

$$(14) \quad P_t = E_t \left(\sum_{k=0}^{\infty} D_{t+k} \prod_{j=0}^{k} \frac{1}{1+r_{t+j}} \right)$$

which has the property that $E_t((1+H_t)/(1+r_t)) = 1$. If we set $1+r_t = (\partial U/\partial C_t)/(\partial U/\partial C_{t+1})$, i.e., to the marginal rate of substitution between present and future consumption where U is the additively separable utility of consumption, then this property is the first-order condition for a maximum of expected utility subject to a stock market budget constraint, and equation (14) is consistent with such expected utility maximization at all times. Note that while r_t is a sort of *ex post* real interest rate not necessarily known until time $t+1$, only the conditional distribution at time t or earlier influences price in the formula (14).

As before, we can rewrite the model in terms of detrended series:

$$(15) \quad p_t = E_t(p_t^*)$$

where

$$p_t^* \equiv \sum_{k=0}^{\infty} d_{t+k} \prod_{j=0}^{k} \frac{1}{1+\bar{r}_{t+j}}$$

$$1+\bar{r}_{t+j} \equiv (1+r_t)/\lambda$$

This model then implies that $\sigma(p_t) \leq \sigma(p_t^*)$ as before. Since the model is nonlinear, however, it does not allow us to derive inequalities like (11) or (13). On the other hand, if movements in real interest rates are not too large, then we can use the linearization of p_t^* (i.e., Taylor expansion truncated after the linear term) around $d = E(d)$ and $\bar{r} = E(\bar{r})$; i.e.,

$$(16) \quad \hat{p}_t^* \cong \sum_{k=0}^{\infty} \bar{\gamma}^{k+1} \hat{d}_{t+k} - \frac{E(d)}{E(\bar{r})} \sum_{k=0}^{\infty} \bar{\gamma}^{k+1} \hat{\bar{r}}_{t+k}$$

where $\bar{\gamma} = 1/(1+E(\bar{r}))$, and a hat over a variable denotes the variable minus its mean. The first term in the above expression is just the expression for p_t^* in (4) (demeaned). The second term represents the effect on p_t^* of movements in real discount rates. This second term is identical to the expression for p^* in (4) except that d_{t+k} is replaced by $\bar{\bar{r}}_{t+k}$ and the expression is premultiplied by $-E(d)/E(\bar{r})$.

It is possible to offer a simple intuitive interpretation for this linearization. First note that the derivative of $1/(1+\bar{r}_{t+k})$, with respect to $\bar{r}$ evaluated at $E(\bar{r})$ is $-\bar{\gamma}^2$. Thus, a one percentage point increase in $\bar{r}_{t+k}$ causes $1/(1+\bar{r}_{t+k})$ to drop by $\bar{\gamma}^2$ times 1 percent, or slightly less than 1 percent. Note that all terms in (15) dated $t+k$ or higher are premultiplied by $1/(1+\bar{r}_{t+k})$. Thus, if $\bar{r}_{t+k}$ is increased by one percentage point, all else constant, then all of these terms will be reduced by about $\bar{\gamma}^2$ times 1 percent. We can approximate the sum of all these terms as $\bar{\gamma}^{k-1}E(d)/E(\bar{r})$, where $E(d)/E(\bar{r})$ is the value at the beginning of time $t+k$ of a constant dividend stream $E(d)$ discounted by $E(\bar{r})$, and $\bar{\gamma}^{k-1}$ discounts it to the present. So, we see that a one percentage point increase in $\bar{r}_{t+k}$, all else constant, decreases p_t^* by about $\bar{\gamma}^{k+1}E(d)/E(\bar{r})$, which corresponds to the kth term in expression (16). There are two sources of inaccuracy with this linearization. First, the present value of all future dividends starting with time $t+k$ is not exactly $\bar{\gamma}^{k-1}E(d)/E(\bar{r})$. Second, increasing $\bar{r}_{t+k}$ by one percentage point does not cause $1/(1+\bar{r}_{t+k})$ to fall by exactly $\bar{\gamma}^2$ times 1 percent. To some extent, however, these errors in the effects on p_t^* of $\bar{r}_t, \bar{r}_{t+1}, \bar{r}_{t+2}, \cdots$ should average out, and one can use (16) to get an idea of the effects of changes in discount rates.

To give an impression as to the accuracy of the linearization (16), I computed p_t^* for data set 2 in two ways: first using (15) and then using (16), with the same terminal condition p_{1979}^*. In place of the unobserved $\bar{r}_t$ series, I used the actual four–six-month prime commercial paper rate plus a constant to give it the mean $\bar{r}$ of Table 2. The commercial paper rate is a *nominal* interest rate, and thus one would expect its fluctuations represent changes in inflationary expectations as well as real interest rate movements. I chose it nonetheless, rather arbitrarily, as a series which shows much more fluctuation than one would normally expect to see in an expected *real* rate. The commercial paper rate ranges, in this sample, from 0.53 to 9.87 percent. It stayed below 1 percent for over a decade (1935–46) and, at the end of the sample, stayed generally well above 5 percent for over a decade. In spite of this erratic behavior, the correlation coefficient between p^* computed from (15) and p^* computed from (16) was .996, and $\sigma(p_t^*)$ was 250.5 and 268.0 by (15) and (16), respectively. Thus the linearization (16) can be quite accurate. Note also that while these large movements in $\bar{r}_t$ cause p_t^* to move much more than was observed in Figure 2, $\sigma(p^*)$ is still less than half of $\sigma(p)$. This suggests that the variability $\bar{r}_t$ that is needed to save the efficient

TABLE 2—SAMPLE STATISTICS FOR PRICE AND DIVIDEND SERIES

		Data Set 1: Standard and Poor's	Data Set 2: Modified Dow Industrial
Sample Period:		1871–1979	1928–1979
1)	$E(p)$	145.5	982.6
	$E(d)$	6.989	44.76
2)	$\bar{r}$	.0480	0.456
	$\bar{r}_2$	.0984	.0932
3)	$b = \ln\lambda$	.0148	.0188
	$\hat{\sigma}(b)$	(.0011)	(1.0035)
4)	$cor(p, p^*)$	.3918	.1626
	$\sigma(d)$	1.481	9.828
Elements of Inequalities:			
Inequality (1)			
5)	$\sigma(p)$	50.12	355.9
6)	$\sigma(p^*)$	8.968	26.80
Inequality (11)			
7)	$\sigma(\Delta p + d_{-1} - \bar{r}p_{-1})$	25.57	242.1
	$min(\sigma)$	23.01	209.0
8)	$\sigma(d)/\sqrt{\bar{r}_2}$	4.721	32.20
Inequality (13)			
9)	$\sigma(\Delta p)$	25.24	239.5
	$min(\sigma)$	22.71	206.4
10)	$\sigma(d)/\sqrt{2\bar{r}}$	4.777	32.56

Note: In this table, E denotes sample mean, σ denotes standard deviation and $\hat{\sigma}$ denotes standard error. $Min(\sigma)$ is the lower bound on σ computed as a one-sided χ^2 95 percent confidence interval. The symbols $p, d, \bar{r}, \bar{r}_2, b$, and p^* are defined in the text. Data sets are described in the Appendix. Inequality (1) in the text asserts that the standard deviation in row 5 should be less than or equal to that in row 6, inequality (11) that σ in row 7 should be less than or equal to that in row 8, and inequality (13) that σ in row 9 should be less than that in row 10.

markets model is much larger yet, as we shall see.

To put a formal lower bound on $\sigma(\bar{r})$ given the variability of Δp, note that (16) makes $\hat{p}_t^*$ the present value of $z_t, z_{t+1}, \ldots$ where $z_t \equiv \hat{d}_t - \hat{r}_t E(d)/E(\bar{r})$. We thus know from (13) that $2E(\bar{r}) var(\Delta p) \leq var(z)$. Moreover, from the definition of z we know that $var(z) \leq var(d) + 2\sigma(d)\sigma(\bar{r})E(d)/E(\bar{r}) + var(\bar{r})E(d)^2/E(\bar{r})^2$ where the equality holds if d_t and $\bar{r}_t$ are perfectly negatively correlated. Combining these two inequalities and solving for $\sigma(\bar{r})$ one finds

(17)
$$\sigma(\bar{r}) \geq \left(\sqrt{2E(\bar{r})}\, \sigma(\Delta p) - \sigma(d)\right) E(\bar{r})/E(d)$$

This inequality puts a lower bound on $\sigma(\bar{r})$ proportional to the discrepancy between the left-hand side and right-hand side of the inequality (13).[17] It will be used to examine the data in the next section.

V. Empirical Evidence

The elements of the inequalities (1), (11), and (13) are displayed for the two data sets (described in the Appendix) in Table 2. In both data sets, the long-run exponential growth path was estimated by regressing $ln(P_t)$ on a constant and time. Then λ in (3) was set equal to e^b where b is the coefficient of time (Table 2). The discount rate $\bar{r}$ used to compute p^* from (4) is estimated as the average d divided by the average p.[18] The terminal value of p^* is taken as average p.

With data set 1, the nominal price and dividend series are the real Standard and Poor's Composite Stock Price Index and the associated dividend series. The earlier observations for this series are due to Alfred Cowles who said that the index is

> intended to represent, ignoring the elements of brokerage charges and taxes, what would have happened to an investor's funds if he had bought, at the beginning of 1871, all stocks quoted on the New York Stock Exchange, allocating his purchases among the individual stocks in proportion to their total monetary value and each month up to 1937 had by the same criterion redistributed his holdings among all quoted stocks. [p. 2]

In updating his series, Standard and Poor later restricted the sample to 500 stocks, but the series continues to be value weighted. The advantage to this series is its comprehensiveness. The disadvantage is that the dividends accruing to the portfolio at one point of time may not correspond to the dividends forecasted by holders of the Standard and Poor's portfolio at an earlier time, due to the change in weighting of the stocks. There is no way to correct this disadvantage without losing comprehensiveness. The original portfolio of 1871 is bound to become a relatively smaller and smaller sample of U.S. common stocks as time goes on.

With data set 2, the nominal series are a modified Dow Jones Industrial Average and associated dividend series. With this data set, the advantages and disadvantages of data set 1 are reversed. My modifications in the Dow Jones Industrial Average assure that this series reflects the performance of a single unchanging portfolio. The disadvantage is that the performance of only 30 stocks is recorded.

Table 2 reveals that all inequalities are dramatically violated by the sample statistics for both data sets. The left-hand side of the inequality is always at least five times as great as the right-hand side, and as much as thirteen times as great.

The violation of the inequalities implies that "innovations" in price as we measure them can be forecasted. In fact, if we regress $\delta_{t+1} p_{t+1}$ onto (a constant and) p_t, we get significant results: a coefficient of p_t of $-.1521$ ($t=-3.218$, $R^2=.0890$) for data set 1 and a coefficient of $-.2421$ ($t=-2.631$, $R^2=.1238$) for data set 2. These results are

---

[17] In deriving the inequality (13) it was assumed that d_t was known at time t, so by analogy this inequality would be based on the assumption that r_t is known at time t. However, without this assumption the same inequality could be derived anyway. The maximum contribution of $\bar{r}_t$ to the variance of ΔP occurs when $\bar{r}_t$ is known at time t.

[18] This is not equivalent to the average dividend price ratio, which was slightly higher (.0514 for data set 1, .0484 for data set 2).

not due to the representation of the data as a proportion of the long-run growth path. In fact, if the holding period return H_t is regressed on a constant and the dividend price ratio D_t/P_t, we get results that are only slightly less significant: a coefficient of 3.533 ($t=2.672$, $R^2=.0631$) for data set 1 and a coefficient of 4.491 ($t=1.795$, $R^2=.0617$) for data set 2.

These regression tests, while technically valid, may not be as generally useful for appraising the validity of the model as are the simple volatility comparisons. First, as noted above, the regression tests are not insensitive to data misalignment. Such low R^2 might be the result of dividend or commodity price index data errors. Second, although the model is rejected in these very long samples, the tests may not be powerful if we confined ourselves to shorter samples, for which the data are more accurate, as do most researchers in finance, while volatility comparisons may be much more revealing. To see this, consider a stylized world in which (for the sake of argument) the dividend series d_t is absolutely constant while the price series behaves as in our data set. Since the actual dividend series is fairly smooth, our stylized world is not too remote from our own. If dividends d_t are absolutely constant, however, it should be obvious to the most casual and unsophisticated observer by volatility arguments like those made here that the efficient markets model must be wrong. Price movements cannot reflect new information about dividends if dividends never change. Yet regressions like those run above will have limited power to reject the model. If the alternative hypothesis is, say, that $\hat{p}_t = \rho \hat{p}_{t-1} + \varepsilon_t$, where ρ is close to but less than one, then the power of the test in short samples will be very low. In this stylized world we are testing for the stationarity of the p_t series, for which, as we know, power is low in short samples.[19] For example, if post-war data from, say, 1950–65 were chosen (a period often used in recent financial markets studies) when the stock market was drifting up, then clearly the regression tests will not reject. Even in periods showing a reversal of upward drift the rejection may not be significant.

Using inequality (17), we can compute how big the standard deviation of real discount rates would have to be to possibly account for the discrepancy $\sigma(\Delta p) - \sigma(\bar{d})/(2\bar{r})^{1/2}$ between Table 2 results (rows 9 and 10) and the inequality (13). Assuming Table 2 $\bar{r}$ (row 2) equals $E(\bar{r})$ and that sample variances equal population variances, we find that the standard deviation of $\bar{r}_t$ would have to be at least 4.36 percentage points for data set 1 and 7.36 percentage points for data set 2. These are very large numbers. If we take, as a normal range for $\bar{r}_t$ implied by these figures, a ± 2 standard deviation range around the real interest rate $\bar{r}$ given in Table 2, then the real interest rate $\bar{r}_t$ would have to range from -3.91 to 13.52 percent for data set 1 and -8.16 to 17.27 percent for data set 2! And these ranges reflect lowest possible standard deviations which are consistent with the model only if the real rate has the first-order autoregressive structure and perfect negative correlation with dividends!

These estimated standard deviations of *ex ante* real interest rates are roughly consistent with the results of the simple regressions noted above. In a regression of H_t on D_t/P_t and a constant, the standard deviation of the fitted value of H_t is 4.42 and 5.71 percent for data sets 1 and 2, respectively. These large standard deviations are consistent with the low R^2 because the standard deviation of H_t is so much higher (17.60 and 23.00 percent, respectively). The regressions of $\delta_t p_t$ on p_t suggest higher standard deviations of expected real interest rates. The standard deviation of the fitted value divided by the average detrended price is 5.24 and 8.67 percent for data sets 1 and 2, respectively.

VI. Summary and Conclusions

We have seen that measures of stock price volatility over the past century appear to be far too high—five to thirteen times too

---

[19] If dividends are constant (let us say $d_t = 0$) then a test of the model by a regression of $\delta_{t+1} p_{t+1}$ on p_t amounts to a regression of p_{t+1} on p_t with the null hypothesis that the coefficient of p_t is $(1+\bar{r})$. This appears to be an explosive model for which t-statistics are not valid yet our true model, which in effect assumes $\sigma(d) \neq 0$, is nonexplosive.

high—to be attributed to new information about future real dividends if uncertainty about future dividends is measured by the sample standard deviations of real dividends around their long-run exponential growth path. The lower bound of a 95 percent one-sided χ^2 confidence interval for the standard deviation of annual changes in real stock prices is over five times higher than the upper bound allowed by our measure of the observed variability of real dividends. The failure of the efficient markets model is thus so dramatic that it would seem impossible to attribute the failure to such things as data errors, price index problems, or changes in tax laws.

One way of saving the general notion of efficient markets would be to attribute the movements in stock prices to changes in expected real interest rates. Since expected real interest rates are not directly observed, such a theory can not be evaluated statistically unless some other indicator of real rates is found. I have shown, however, that the movements in expected real interest rates that would justify the variability in stock prices are very large—much larger than the movements in nominal interest rates over the sample period.

Another way of saving the general notion of efficient markets is to say that our measure of the uncertainty regarding future dividends—the sample standard deviation of the movements of real dividends around their long-run exponential growth path—understates the true uncertainty about future dividends. Perhaps the market was rightfully fearful of much larger movements than actually materialized. One is led to doubt this, if after a century of observations nothing happened which could remotely justify the stock price movements. The movements in real dividends the market feared must have been many times larger than those observed in the Great Depression of the 1930's, as was noted above. Since the market did not know in advance with certainty the growth path and distribution of dividends that was ultimately observed, however, one cannot be sure that they were wrong to consider possible major events which did not occur. Such an explanation of the volatility of stock prices, however, is "academic," in that it relies fundamentally on unobservables and cannot be evaluated statistically.

APPENDIX

A. *Data Set 1: Standard and Poor Series*

Annual 1871–1979. The price series P_t is Standard and Poor's Monthly Composite Stock Price index for January divided by the Bureau of Labor Statistics wholesale price index (January *WPI* starting in 1900, annual average *WPI* before 1900 scaled to 1.00 in the base year 1979). Standard and Poor's Monthly Composite Stock Price index is a continuation of the Cowles Commission Common Stock index developed by Alfred Cowles and Associates and currently is based on 500 stocks.

The Dividend Series D_t is total dividends for the calendar year accruing to the portfolio represented by the stocks in the index divided by the average wholesale price index for the year (annual average *WPI* scaled to 1.00 in the base year 1979). Starting in 1926 these total dividends are the series "Dividends per share...12 months moving total adjusted to index" from Standard and Poor's statistical service. For 1871 to 1925, total dividends are Cowles series Da-1 multiplied by .1264 to correct for change in base year.

B. *Data Set 2: Modified Dow Jones Industrial Average*

Annual 1928–1979. Here P_t and D_t refer to real price and dividends of the portfolio of 30 stocks comprising the sample for the Dow Jones Industrial Average when it was created in 1928. Dow Jones averages before 1928 exist, but the 30 industrials series was begun in that year. The published Dow Jones Industrial Average, however, is not ideal in that stocks are dropped and replaced and in that the weighting given an individual stock is affected by splits. Of the original 30 stocks, only 17 were still included in the Dow Jones Industrial Average at the end of our sample. The published Dow Jones Industrial Average is the simple sum of the price per share of the 30 companies divided by a divisor which changes through time. Thus, if a stock splits two for one, then Dow Jones continues to include only one share but changes the divisor to prevent a sudden drop in the Dow Jones average.

To produce the series used in this paper, the *Capital Changes Reporter* was used to trace changes in the companies from 1928 to 1979. Of the original 30 companies of the Dow Jones Industrial Average, at the end of our sample (1979), 9 had the identical names, 12 had changed only their names, and 9 had been acquired, merged or consolidated. For these latter 9, the price and dividend series are continued as the price and dividend of the shares exchanged by the acquiring corporation. In only one case was a cash payment, along with shares of the acquiring corporation, exchanged for the shares of the acquired corporation. In this case, the price and dividend series were continued as the price and dividend of the shares exchanged by the acquiring corporation. In four cases, preferred shares of the acquiring corporation were among shares exchanged. Common shares of equal value were substituted for these in our series. The number of shares of each firm included in the total is determined by the splits, and effective splits effected by stock dividends and merger. The price series is the value of all these shares on the last trading day of the preceding year, as shown on the Wharton School's Rodney White Center Common Stock tape. The dividend series is the total for the year of dividends and the cash value of other distributions for all these shares. The price and dividend series were deflated using the same wholesale price indexes as in data set 1.

REFERENCES

C. Amsler, "An American Consol: A Reexamination of the Expectations Theory of the Term Structure of Interest Rates," unpublished manuscript, Michigan State Univ. 1980.

S. Basu, "The Investment Performance of Common Stocks in Relation to their Price-Earnings Ratios: A Test of the Efficient Markets Hypothesis," *J. Finance*, June 1977, *32*, 663–82.

G. E. P. Box and G. M. Jenkins, *Time Series Analysis for Forecasting and Control*, San Francisco: Holden-Day 1970.

W. C. Brainard, J. B. Shoven, and L. Weiss, "The Financial Valuation of the Return to Capital," *Brookings Papers,* Washington 1980, *2*, 453–502.

Paul H. Cootner, *The Random Character of Stock Market Prices*, Cambridge: MIT Press 1964.

Alfred Cowles and Associates, *Common Stock Indexes, 1871–1937,* Cowles Commission for Research in Economics, Monograph No. 3, Bloomington: Principia Press 1938.

E. F. Fama, "Efficient Capital Markets: A Review of Theory and Empirical Work," *J. Finance*, May 1970, *25*, 383–420.

_____, "The Behavior of Stock Market Prices," *J. Bus., Univ. Chicago*, Jan. 1965, *38*, 34–105.

C. W. J. Granger, "Some Consequences of the Valuation Model when Expectations are Taken to be Optimum Forecasts," *J. Finance*, Mar. 1975, *30*, 135–45.

M. C. Jensen et al., "Symposium on Some Anomalous Evidence Regarding Market Efficiency," *J. Financ. Econ.*, June/Sept. 1978, *6*, 93–330.

S. LeRoy and R. Porter, "The Present Value Relation: Tests Based on Implied Variance Bounds," *Econometrica*, forthcoming.

M. H. Miller and F. Modigliani, "Dividend Policy, Growth and the Valuation of Shares," *J. Bus., Univ. Chicago*, Oct. 1961, *34*, 411–33.

F. Modigliani and R. Cohn, "Inflation, Rational Valuation and the Market," *Financ. Anal. J.*, Mar./Apr. 1979, *35*, 24–44.

J. Pesando, "Time Varying Term Premiums and the Volatility of Long-Term Interest Rates," unpublished paper, Univ. Toronto, July 1979.

P. A. Samuelson, "Proof that Properly Discounted Present Values of Assets Vibrate Randomly," in Hiroaki Nagatani and Kate Crowley, eds., *Collected Scientific Papers of Paul A. Samuelson*, Vol. IV, Cambridge: MIT Press 1977.

R. J. Shiller, "The Volatility of Long-Term Interest Rates and Expectations Models of the Term Structure," *J. Polit. Econ.*, Dec. 1979, *87*, 1190–219.

_____ and J. J. Siegel, "The Gibson Paradox and Historical Movements in Real Interest Rates," *J. Polit. Econ.*, Oct. 1979, *85*, 891–907.

H. Wold, "On Prediction in Stationary Time Series," *Annals Math. Statist.* 1948, *19*, 558–67.

Commerce Clearing House, *Capital Changes Reporter*, New Jersey 1977.

Dow Jones & Co., *The Dow Jones Averages 1855–1970*, New York: Dow Jones Books 1972.

Standard and Poor's *Security Price Index Record*, New York 1978.

[3]

Excess Volatility in the Financial Markets: A Reassessment of the Empirical Evidence

Marjorie A. Flavin
University of Virginia

Numerous authors, including Shiller, LeRoy and Porter, and Singleton, have reported empirical evidence that stock prices and long interest rates are more volatile than can be justified by standard asset-pricing models. This paper shows that in small samples the "volatility" or "variance-bounds" tests tend to be biased, often severely, toward rejection of the null hypothesis of market efficiency. Thus the apparent violation of market efficiency may be reflecting the sampling properties of the volatility measures, rather than a failure of the market efficiency hypothesis itself. The paper also reports some unbiased estimates of the bounds on holding period yields and long interest rates. Much of the evidence of excess volatility disappears when the tests are corrected for small sample bias.

In recent papers, Shiller (1979) and LeRoy and Porter (1981) have reported empirical evidence that stock prices and long interest rates are more volatile than can be justified within the standard asset-pricing models. Further empirical evidence on excess volatility in the financial markets has been reported in numerous studies, including Pesando (1979), Amsler (1980), Singleton (1980), Grossman and Shil-

I am grateful for comments and suggestions received during workshops at the Board of Governors, Brown, Chicago, CUNY, Duke, Michigan, University of Rochester, and Virginia Polytechnic Institute. For helpful discussions and comments on earlier drafts I also thank Richard Ashley, Robert Barro, Edwin Burmeister, Matthew Cushing, Wake Epps, Robert Hall, James Hamilton, Lars Hansen, Stephen LeRoy, Ronald Michener, Richard Porter, Robert Shiller, Kenneth Singleton, Jonathan Skinner, James Stock, William Poole, and Charles Whiteman. Earlier versions of the paper were circulated under the title "Small Sample Bias in Tests of Excess Volatility in the Financial Markets."

ler (1981), Shiller (1981a, 1981b, 1981c), and Blanchard and Watson (1982). According to the empirical evidence reported in these papers, the variance of stock prices, holding yields on long-term bonds, and long interest rates exceed the upper bounds implied by the variance of dividends and short interest rates. Further, the variances of stock prices and long interest rates exceed their estimated upper bounds by very large margins in many cases.

This paper argues that in small samples the "volatility" or "variance-bounds" tests tend to be strongly biased toward rejection of the null hypothesis of no excess volatility. Thus the apparent violation of the market efficiency hypothesis may be reflecting the sampling properties of the volatility measures in small samples rather than a failure of the market efficiency hypothesis.

The innovative tests developed by Shiller and LeRoy and Porter are formulated according to the following line of reasoning. If stock prices are modeled as the present discounted value of rationally forecasted future dividends, the volatility, or variance, of the stock price is limited by the volatility, or variance, of the dividend series. Similarly, under the expectations theory of the term structure of interest rates, which asserts that the long-term interest rate is equal to an average of rationally expected future short-term interest rates, the variance of the long rate is limited by the variance of the short rate. The upper bound on the volatility of long rates, or stock prices, has been tested either (1) by comparing a point estimate of the upper bound with a point estimate of the variance being bounded or (2) by calculating both point estimates and the asymptotic covariance matrix of the estimates and testing whether the estimated variance of long rates or stock prices exceeds the estimated upper bound by an amount that is statistically significantly greater than zero. In either procedure, the test statistics may be misleading if, for samples of the size typically used in the variance-bounds tests, the point estimates are biased or, more generally, if the asymptotic distributions are not close approximations to the finite sample distributions. This paper argues that the estimate of the upper bound in these tests is biased downward in small samples and that the magnitude of the bias is large enough to provide a potential explanation of the apparent violation of the bounds.

To see intuitively why the variance-bounds tests tend to be biased against the null hypothesis, consider the basic bound on the volatility of long interest rates: var (R_t) < var (R_t^*), where R_t is the actual long rate and R_t^* is the perfect-foresight long rate, defined as the value the long rate would take if agents had perfect foresight concerning the path of the short rate. Under the market efficiency hypothesis, R_t is equal to the expectation of R_t^* conditional on currently available in-

EXCESS VOLATILITY

formation and therefore must have a variance smaller than the variance of R_t^*. If the population means of R_t and R_t^* were known a priori, unbiased estimates of var (R_t) and var (R_t^*) could be obtained by taking squared deviations of R_t and R_t^* from their population means. The empirical applications of the variance-bounds tests have relied on sample variances of R_t and R_t^* that were computed by taking deviations from sample means. Taking deviations from the sample mean induces downward bias in the sample variance, however, since the sample mean has the following property: the sample variance of a data series, expressed in deviations from some constant, is minimized when that constant is set equal to the sample mean. The greater the variance of the sample mean, the greater is the extent to which the sample mean will "fit" some of the stochastic components of the data series and the greater is the bias in the sample variance. Because R_t^* is a long moving average of a variable (the short rate), which is itself highly serially correlated, the variance of $\bar{R}_t^*$ tends to exceed the variance of $\bar{R}_t$, and as a result the sample variance of R_t^* tends to be more strongly downward biased than the sample variance of R_t. Since var (R_t^*) is the upper bound on var (R_t), the net effect is that the difference vâr (R_t^*) − vâr (R_t) is biased toward rejection of the null hypothesis of no excess volatility. This bias toward rejection of the null hypothesis also arises in tests of the upper bound on the variance of stock prices and the variance of holding period yields on long-term bonds.

Section I considers an economy in which the short rate is generated by an AR1 process with the autoregressive parameter equal to 0.95 for quarterly observations. Investors are risk neutral and form expectations of future short rates rationally, with the result that yields on 20-year discount bonds are generated exactly as hypothesized by the pure (i.e., no liquidity premium) expectations hypothesis. The exact finite sample distributions of the sample statistics, vâr (R_t), vâr (R_t^*), and vâr (R_t^*) − vâr (R_t), are calculated for a sample of size 100 in such an economy.

In Section II, some of the test procedures implemented in Shiller (1979, 1981b) and Singleton (1980) are reviewed in light of the findings concerning the small sample distributions of the variance-bounds statistics. Depending on the bound being tested and the estimation method used, the bias toward rejection of no excess volatility ranges from modest to strong to severe. Section II also reports some unbiased estimates of the bounds on the variances of holding period yields and long interest rates. Much of the evidence of excess volatility in the bond market disappears when the tests are corrected for small sample bias.

I. Comparison of Small Sample and Asymptotic Distributions in an Efficient-Market Economy

In order to keep the problem as simple as possible, the model economy studied in this section is one in which the short rate follows a first-order autoregressive (AR1) process:

$$r_t = \rho r_{t-1} + \epsilon_t, \tag{1}$$

where r_t is the short-term interest rate, expressed in deviations from the mean, and ϵ_t is an independently and identically distributed disturbance; $\epsilon \sim N(0, \sigma_\epsilon^2)$.

According to the expectations hypothesis of the term structure, the linearized long rate on a pure discount bond is simply the average of current and future expected short rates:

$$R_t = \frac{1}{n} \sum_{j=0}^{n-1} {}_t r_{t+j}, \tag{2}$$

where R_t is the n-period long rate and ${}_t r_{t+j}$ is the expectation, in period t, of r_{t+j}. Note that the long rate does not include a liquidity premium.

Using the assumption that the short rate follows an AR1 process, we see that all expected future short rates are proportional to the current short rate:

$$_t r_{t+j} = \rho^j r_t = \rho^j \sum_{i=0}^{\infty} \rho^i \epsilon_{t-i}. \tag{3}$$

Thus the long rate is also proportional to the short rate:

$$R_t = \frac{1 - \rho^n}{n(1 - \rho)} r_t = \frac{1 - \rho^n}{n(1 - \rho)} \sum_{i=0}^{\infty} \rho^i \epsilon_{t-i}. \tag{4}$$

Define the "perfect-foresight" long rate, R_t^*, as the value the long rate would take if agents had perfect foresight concerning the path of the short rate:[1]

$$R_t^* = \frac{1}{n} \sum_{j=0}^{n-1} r_{t+j}. \tag{5}$$

By straightforward but tedious manipulation, the perfect-foresight long rate can be expressed as a linear combination of past, current, and future disturbances:

---

[1] Shiller uses the terminology "ex post rational" long rate in referring to this variable.

EXCESS VOLATILITY

$$R_t^* = \frac{1}{n(1-\rho)} \sum_{j=1}^{n-1} (1 - \rho^{n-j})\epsilon_{t+j} + \frac{1-\rho^n}{n(1-\rho)} \sum_{i=0}^{\infty} \rho^i \epsilon_{t-i}. \quad (6)$$

Equation (6) reflects the basic hypothesis of efficiency in the bond market: the actual long rate, R_t, is the expectation of the perfect-foresight long rate, R_t^*, conditional on all available information.

Substituting equation (4) into equation (6) yields

$$R_t^* = R_t + \theta_t, \quad (7)$$

where

$$\theta_t = \frac{1}{n(1-\rho)} \sum_{j=1}^{n-1} (1 - \rho^{n-j})\epsilon_{t+j}.$$

R_t, which depends only on current and past disturbances, and θ_t, which depends only on future disturbances, are distributed independently, with the implication that

$$\text{var}(R_t^*) = \text{var}(R_t) + \text{var}(\theta_t). \quad (8)$$

Since the variance of the forecast error must be nonnegative, the variance of R_t^* constitutes an upper bound on the variance of R_t:

$$\text{var}(R_t^*) \geq \text{var}(R_t). \quad (9)$$

The upper bound on the variance of the long rate, equation (9), is, of course, a restriction on the population moments of R_t^* and R_t. Assuming that r_t, and therefore R_t^* and R_t, are stationary and ergodic time-series processes, the population variances of R_t^* and R_t can be consistently estimated from a single realization of the process over time.[2]

The upper bound on the volatility of long rates, or stock prices, has been tested either by comparing point estimates of var (R_t^*) and var (R_t) or by calculating both point estimates of var (R_t^*) and var (R_t) and the asymptotic covariance matrix of the estimates and testing whether the difference $\hat{\text{var}}(R_t^*) - \hat{\text{var}}(R_t)$ is significantly less than zero. In either procedure, the test statistics may be misleading if, for samples of the size typically used in the variance-bounds tests, the point estimates are biased or, more generally, if the asymptotic distributions are not close approximations to the finite sample distributions.

---

[2] If the short rate is nonstationary (i.e., if $\rho \geq 1$), the variances of r_t, R_t, and R_t^* are undefined and the theoretical variance bounds must be reformulated. Almost all of the empirical volatility literature, including LeRoy and Porter (1981), Shiller (1979, 1981b), and Singleton (1980), has been based on the assumption that the short rate (or dividends) is a stationary process.

This section of the paper studies the properties of the variance-bounds statistics in samples of 100 quarterly observations on the yields of 20-year discount bonds and 3-month bills. These observations are assumed to be drawn from an efficient-market economy in which the short rate is generated by an AR1 process with $\rho = 0.95$. The exact small sample distributions that the sample statistics, vâr (R_t^*), vâr (R_t), and vâr (R_t^*) − vâr (R_t), would have in such an economy are then calculated and compared to the asymptotic distributions.

Calculation of the Small Sample Distributions[3]

In order to avoid having to refer to "the variance of the variance of R_t," the following notation will be used: $V = $ vâr (R_t), $V^* = $ vâr (R_t^*), and $D = $ vâr (R_t^*) − vâr (R_t).

In equations (1)–(6), the first observation on the short rate, r_1, was expressed as a function of disturbances from the infinite past. For the purpose of calculating the small sample distributions, it is more convenient to model the first observation on the short rate as a random draw from the stationary distribution of the short rate:

$$r_1 = \frac{\epsilon_1}{\sqrt{1 - \rho^2}}, \tag{10}$$

where $\epsilon_1 \sim N(0, \sigma_\epsilon^2)$. By modeling r_1 as a drawing from the stationary distribution of the short rate, a sample of T observations on r can be expressed as a function of T disturbances rather than an infinite number. For the purpose of characterizing the distributions of V, V^*, and D, the stochastic specification of r_1 given by equation (10) is completely equivalent to the specification of r_1 as a function of disturbances from the infinite past.

With this modification, each of the random variables V, V^*, and D can be expressed as a quadratic form in the disturbance of the short-rate process, ϵ. To construct the quadratic form representing the variance of the long rate, V, recall that in this example the long rate is proportional to the short rate,

$$R_t = \alpha r_t, \tag{11}$$

where $\alpha = (1 - \rho^n)/n(1 - \rho)$. The vector of T observations on the short rate can be expressed as a linear transformation of the disturbances:

---

[3] I am grateful to Robert Hall for suggesting this approach for computing the exact finite sample distributions of the variances.

EXCESS VOLATILITY

$$\begin{bmatrix} r_1 \\ r_2 \\ r_3 \\ \cdot \\ \cdot \\ \cdot \\ r_T \end{bmatrix} = \begin{bmatrix} \frac{1}{\sqrt{1-\rho^2}} & 0 & 0 & 0 & \cdots & 0 \\ \frac{\rho}{\sqrt{1-\rho^2}} & 1 & 0 & 0 & \cdots & 0 \\ \frac{\rho^2}{\sqrt{1-\rho^2}} & \rho & 1 & 0 & \cdots & 0 \\ \cdot & & & & & \\ \cdot & & & & & \\ \cdot & & & & & \\ \frac{\rho^{T-1}}{\sqrt{1-\rho^2}} & \rho^{T-2} & \rho^{T-3} & \rho^{T-4} & \cdots & 1 \end{bmatrix} \begin{bmatrix} \epsilon_1 \\ \epsilon_2 \\ \epsilon_3 \\ \cdot \\ \cdot \\ \cdot \\ \epsilon_T \end{bmatrix}. \quad (12)$$

Using the notation S (for short rate) for the $T \times T$ matrix in equation (12) and ϵ for the T-element column vector of disturbances, the sample variance of the long rate, V, can be expressed as a quadratic form in ϵ,

$$V = \epsilon'A\epsilon, \quad (13)$$

where A is the $T \times T$ symmetric matrix, $A = \alpha^2 T^{-1} S'S$. At this point, the mean of the short rate, which is also the mean of the long rate, is assumed to be known a priori; the quadratic form A represents the variance of R_t around the population mean. The variance of R_t around its sample mean will be studied later.

The quadratic form representing the sample variance of the perfect-foresight long rate, V^*, will be of order $T + n - 1$, where n is the number of periods in the long rate, because the last observation on R_t^* depends on r_T and $n - 1$ subsequent observations on the short rate. Let L denote the $T \times T + n - 1$ matrix that transforms the $T + n - 1$ observations on the short rate into T observations on the perfect-foresight long rate:

$$\begin{bmatrix} R_1^* \\ R_2^* \\ R_3^* \\ \cdot \\ \cdot \\ \cdot \\ R_T^* \end{bmatrix} = \frac{1}{n} \begin{bmatrix} 1 & 1 & 1 & \cdots & 1 & 0 & 0 & 0 & \cdots & 0 \\ 0 & 1 & 1 & 1 & \cdots & 1 & 0 & 0 & \cdots & 0 \\ 0 & 0 & 1 & 1 & 1 & \cdots & 1 & 0 & \cdots & 0 \\ \cdot & & & & & & & & & \\ \cdot & & & & & & & & & \\ \cdot & & & & & & & & & \\ 0 & 0 & 0 & \cdots & 0 & 1 & 1 & 1 & \cdots & 1 \end{bmatrix} \begin{bmatrix} r_1 \\ r_2 \\ r_3 \\ \cdot \\ \cdot \\ \cdot \\ r_{T+n-1} \end{bmatrix}, \quad (14)$$

where the width of the band of ones is n.

The sample variance of R_t^* around its population mean can be expressed as the quadratic form of order $T + n - 1$,

$$V^* = \epsilon'B\epsilon, \tag{15}$$

where $B = T^{-1}S'L'LS$; S is the lower-triangular matrix defined as in equation (12), except with order $T + n - 1$ instead of T, and ϵ is the $T + n - 1$ element column vector of disturbances.

The difference between the variances, $D = V^* - V$, is given by the difference of the quadratic forms for V^* and V:

$$D = \epsilon'[B - A]\epsilon. \tag{16}$$

(Of course the $T \times T$ matrix A as defined above must be augmented by adding $n - 1$ rows and $n - 1$ columns of zeros so that it conforms with the matrix B.)

The problem now becomes one of calculating the distribution of a quadratic form in normal deviates. Let Λ denote the diagonal matrix with the eigenvalues of the quadratic form A on the main diagonal, and P the matrix of eigenvectors; $P'AP = \Lambda$. Since $PP' = I$, $\epsilon'A\epsilon = \epsilon'PP'APP'\epsilon = \epsilon'P\Lambda P'\epsilon$. Define a new disturbance term η such that $\eta = P'\epsilon$. Since P is an orthonormal matrix, the new disturbances are independently distributed, $\eta \sim N(0, \sigma_\epsilon^2)$, with the same variance as the original disturbance, ϵ. Thus the sample variance of the long rate, V, is a weighted sum of squared normal deviates:

$$V = \epsilon'A\epsilon = \eta'\Lambda\eta = \sum_{j=1}^{T} \lambda_j \eta_j^2, \tag{17}$$

where λ_j are the eigenvalues of the quadratic form A. The characteristic function of $\sum_{j=1}^{T} \lambda_j \eta_j^2$ is

$$\phi(t) = \prod_{1}^{T}(1 - 2i\lambda_j\sigma_\epsilon^2 t)^{-1/2}. \tag{18}$$

By inverting the characteristic function, one can obtain the cumulative distribution function of the random variable, V. The value of the distribution function, evaluated at x, is given by

$$F(x) = \frac{1}{2} - \frac{1}{\pi}\int_0^\infty t^{-1}I[e^{-itx}\phi(t)]dt, \tag{19}$$

where $I[\cdot]$ denotes the imaginary part of the expression in the brackets.

All of the small sample distributions reported in the next section were computed by the following procedure: (1) The symmetric matrix defining the quadratic form was generated, after assigning numerical values to the parameters T, n, and ρ; (2) the eigenvalues of the matrix were obtained using numerical methods; and (3) the inversion

EXCESS VOLATILITY

formula (eq. [19]), which is a function of the eigenvalues, was integrated numerically.[4]

Asymptotic Distributions

Unlike the small sample distributions, the asymptotic distributions can be derived analytically. Let V, V^*, and D denote the sample statistics calculated by taking deviations from the population mean, and V_s, V_s^*, and D_s denote the corresponding statistics computed by taking deviations from the sample mean. The bias induced in V_s, V_s^*, and D_s by taking deviations from the sample mean is of order $1/T$ (Anderson 1971, p. 463). Similarly, $T \cdot \text{var}(V_s)$, $T \cdot \text{var}(V_s^*)$, and $T \cdot \text{var}(D_s)$ differ from $T \cdot \text{var}(V)$, $T \cdot \text{var}(V^*)$, and $T \cdot \text{var}(D)$, respectively, by terms of order $1/T$ (Anderson 1971, p. 471). Thus, in deriving the means and variances of the asymptotic distributions of V_s, V_s^*, and D_s, we can analyze the simpler case in which the statistics are calculated by taking deviations from the population mean.

Using equations (4) and (6) for R_t and R_t^*, respectively, we see that straightforward calculation of the means of the asymptotic distributions of V, V^*, and D yields

$$\mu_V = E(R_t^2) = \frac{\alpha^2 \sigma_\epsilon^2}{1 - \rho^2},$$

$$\mu_{V^*} = E(R_t^{*2}) = \left(\frac{\alpha^2}{1 - \rho^2} + \sum_{i=1}^{n-1} \alpha_i^2\right)\sigma_\epsilon^2, \qquad (20)$$

$$\mu_D = E[(R_t^* - R_t)^2] = \sigma_\epsilon^2 \sum_{i=1}^{n-1} \alpha_i^2,$$

where $\alpha = (1 - \rho^n)/n(1 - \rho)$ and $\alpha_i = [1 - \rho^{(n-i)}]/n(1 - \rho)$.

Singleton (1980) showed that the sample statistics V, V^*, and D are consistent estimators of μ_V, μ_{V^*}, and μ_D. Further, V, V^*, and D are asymptotically normally distributed, with variances given by

$$\lim_{T \to \infty} T \text{ var}(V) = 2 \sum_{s=-\infty}^{\infty} [E(R_t R_{t+s})]^2,$$

$$\lim_{T \to \infty} T \text{ var}(V^*) = 2 \sum_{s=-\infty}^{\infty} [E(R_t^* R_{t+s}^*)]^2, \qquad (21)$$

$$\lim_{T \to \infty} T \text{ var}(D) = 2 \sum_{s=-\infty}^{\infty} [E(\theta_t \theta_{t+s})]^2.$$

(Recall that $\theta_t = R_t^* - R_t$.)

---

[4] The eigenvalues were computed using the International Math and Science Library (IMSL) routine EIGRS; the numerical integration was performed using IMSL routine DCADRE.

Evaluating these variances for the model posited in this paper gives

$$\text{var}(V) = \frac{2\alpha^4(1+\rho^2)}{T(1-\rho^2)^3}\sigma_\epsilon^4,$$

$$\text{var}(V^*) = \frac{2}{T}\left\{\left(\frac{\alpha^2}{1-\rho^2} + \sum_{i=1}^{n-1}\alpha_i^2\right)^2\right.$$

$$+ 2\sum_{j=1}^{n-2}\left(\sum_{i=1}^{n-1-j}\alpha_i\alpha_{i+j} + \frac{\rho^j\alpha^2}{1-\rho^2}\right)$$

$$+ \left(\sum_{i=1}^{J}\alpha\alpha_i\rho^{j-1}\right)^2 \qquad (22)$$

$$+ \frac{2}{1-\rho^2}\left[\frac{\rho^{(n-1)}\alpha^2}{1-\rho^2} + \sum_{i=1}^{n-1}\alpha\alpha_i\rho^{(n-1-i)}\right]^2\right\}\sigma_\epsilon^4,$$

$$\text{var}(D) = \frac{2}{T}\left[\left(\sum_{i=1}^{n-1}\alpha_i^2\right)^2 + 2\sum_{j=1}^{n-2}\left(\sum_{i=1}^{n-1-j}\alpha_i\alpha_{i+j}\right)^2\right]\sigma_\epsilon^4.$$

Comparison of Small Sample and Asymptotic Distributions

Recall that the numerical example was constructed to mimic quarterly data in which the short rate was a 3-month rate and the term of the long rate was 20 years. The autoregressive parameter of the short-rate process, ρ, was set at 0.95.[5] For the small sample distributions, the sample size, T, was set at 100 (quarterly) observations. Table 1 reports the means of the asymptotic distributions of V, V^*, and D, for $\rho = 0.95$, $T = 100$, and $n = 80$. The variance of the short-rate innovation is normalized at one ($\sigma_\epsilon^2 = 1$).

The asymptotic standard deviations reported in table 1 were obtained by evaluating the expressions (22) for the asymptotic variances of V, V^*, and D for a sample size of 100. Before turning to the calculations of the actual small sample distributions, it should be noted that table 1 itself contains some evidence that the asymptotic distributions are not close approximations to the small sample distributions for samples of 25 years of quarterly data. Because V and V^* are both sample variances, neither random variable can take on negative values. Using asymptotic distribution theory to approximate the distribution of V^*, however, one would conclude that V^* is normally distributed with mean 3.802 and standard deviation 4.619, implying that V^* is less than zero for over 20 percent of its distribution.

[5] In a first-order autoregression of quarterly observations on 3-month Treasury bill yields (sample period 1950:I–1982:I), the estimated autoregressive parameter was 0.953, with a standard error of .03. (These data were obtained from Salomon Brothers, Inc. 1982.)

EXCESS VOLATILITY

TABLE 1
Means and Standard Deviations of Asymptotic Distributions

Variable	Mean	Asymptotic Standard Deviation
V^*	3.802	4.619
V	.620	.150
D	3.182	3.139

The actual small sample distributions of V, V^*, and D are plotted in figure 1. In each of the three panels of figure 1, the distributions labeled a represent the small sample distributions of V^*, V, and D, assuming that the mean of the short-rate process is known a priori. The distributions labeled b represent the small sample distributions of V_s^*, V_s, and D_s when R_t and R_t^* are each expressed in deviations from their respective sample means instead of the population mean. The distributions labeled c in panels 1 and 3 represent the small sample distributions of V_s^* and D_s, respectively, when R_t^* is calculated using a terminal condition R_T^* and both R_t^* and R_t are expressed in deviations from the sample mean. The distributions labeled d represent the small sample distributions of spectral estimates. The distributions b, c, and d are explained more fully below.

Consider the distributions labeled a. If the mean of the underlying process is known, the sample variance is an unbiased estimate of the population variance (Anderson 1971, p. 448).[6] However, even when the mean is known, the sample variances are not closely approximated by the normal distribution. All three random variables have strongly skewed small sample distributions; the probability that V will take on a value less than its mean is 60 percent, and V^* and D each have a 65 percent probability of taking on values less than their respective means.

It is important to keep in mind that two unrealistically strong assumptions concerning the information available to the econometrician have been maintained in computing the small sample distributions represented by the a curves. First, the mean of the short-rate process has been assumed to be known a priori. Second, the perfect-foresight variance V^* has been calculated assuming that all of the $n - 1$ postsample observations on the short rate, r_{T+1} to r_{T+n-1}, are available, enabling the econometrician to construct the perfect-foresight

---

[6] The means of the small sample distributions of V, V^*, and D were calculated making use of the fact that the mean of a quadratic form in normal deviates is equal to the sum of the eigenvalues of the quadratic form. For each of the three quadratic forms, the sum of the eigenvalues matched the analytically derived population mean (reported in table 1) to four decimal places.

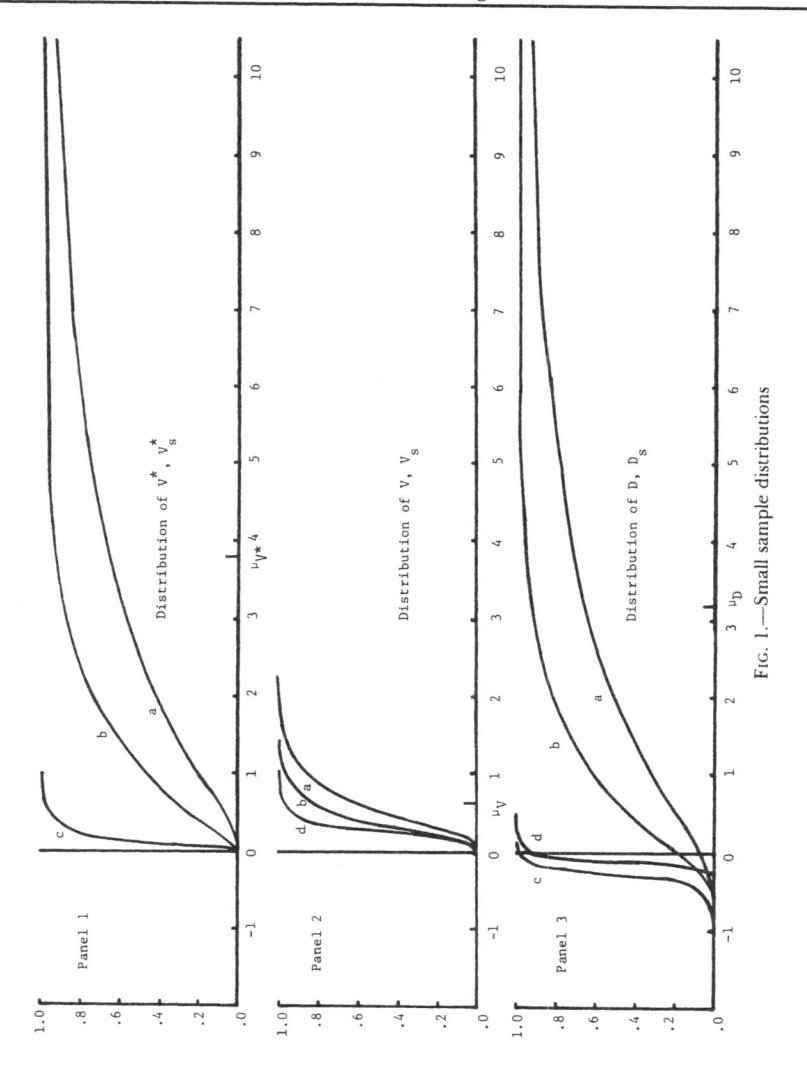

FIG. 1.—Small sample distributions

EXCESS VOLATILITY 941

long-rate series, R_1^* to R_T^*, without having to resort to any form of extrapolation of the short-rate data. Even under these unrealistically favorable assumptions concerning the availability of prior information and data, there is a 6.6 percent chance that the sample variance, V, will exceed its upper bound, V^*, if the null hypothesis of market efficiency is true.

In practice, the variance-bounds tests have been implemented using data on r_t and R_t in deviations from their sample means rather than deviations from population means. Let A_s denote the quadratic form representing the variance of R_t expressed in deviations from its sample mean.[7] The distribution of $V_s = \epsilon' A_s \epsilon$ is given by curve b in panel 2 of figure 1. When the population mean of R_t is not known a priori and the sample variance is expressed in deviations from the sample mean, the sample variance is a downward biased estimator of the population mean of var (R_t); the mean of V_s is 0.4251, as compared to the population mean of var (R_t) of 0.6200.

Similarly, the curve labeled b in panel 1 is the small sample distribution of V_s^* in which the vector of observations on the short rate is expressed in deviations from the sample mean before constructing the series on R_t^*. Again, expressing the data in deviations from the sample mean creates a downward bias: the mean of the distribution b is 1.537, less than half the value of the population moment, var (R_t^*), of 3.802. Because V_s^* is more strongly downward biased than V_s, expressing the data in deviations from sample means results in a net downward bias to $D_s = V_s^* - V_s$. When the data are expressed in deviations from sample means, D_s has a mean of 1.112, as compared to a population value of var (R_t^*) − var (R_t) of 3.182. Further, there is a 16.8 percent chance that the sample variance, V_s, will exceed its upper bound, V_s^*, if the null hypothesis of market efficiency is true.

II. Review of Previous Tests of Excess Volatility

In this section some of the test procedures implemented by Shiller and Singleton are reviewed in light of the findings concerning the small sample distributions of the variance-bounds test statistics.

---

[7] To construct the quadratic form A_s, take the matrix S and calculate the sum of the elements in each column. Denote the sum of the elements in the jth column as $m(j)$. Subtract $m(j)/100$ from each element of the jth column of the original matrix S, for $j = 1, 2, \ldots, 100$. Premultiplying this matrix by its transpose and multiplying by the scalar $\beta^2 T^{-1}$ gives the quadratic form A_s. In modifying A to form A_s, the degrees of freedom correction is automatically incorporated into A_s, since the rank of the quadratic form is reduced by one by the modification. Thus A has T nonzero eigenvalues, while A_s has $T - 1$ nonzero eigenvalues.

Shiller's Approach

In his empirical work on the volatility of stock prices and long interest rates (Shiller 1979, 1981a, 1981b, 1981c; Grossman and Shiller 1981), Shiller not only examines the basic upper bound on the variance of the long interest rate or stock prices, var (R_t) < var (R_t^*), but also formulates and presents empirical evidence on an upper bound on the variance of holding period yields.

In illustrating the apparent excess volatility of long interest rates in his 1979 paper, Shiller graphs actual AAA utility bond yields against a perfect-foresight long rate constructed from data on the 4–6-month prime commercial paper rate. In these graphs, the perfect-foresight long rate moves smoothly and remains within the band between 6.25 percent and 6.75 percent, while the actual long rate moves sharply and varies between 4.5 percent and 11.5 percent over the same period (1966:I–1977:III). Since the variance of the perfect-foresight long rate places an upper bound on the variance of the actual long rate, these graphs do appear to "stand in glaring contradiction" (p. 1213) to the efficient markets model.

In the absence of actual data on the postsample values of the short rate, Shiller computed the perfect-foresight long-rate series, R_t^*, recursively from an assumed terminal value, R_T^*,

$$R_t^* = \gamma R_{t+1}^* + (1 - \gamma) r_t, \tag{23}$$

where γ is a constant close to but less than one.[8] In the case of a pure discount bond, $\gamma = (n - 1)/n$. The terminal value, R_T^*, was assumed to be equal to the average short rate over the sample period. It is a simple matter to grind out the distribution of Shiller's approximation to the perfect-foresight long rate using the methods of the previous section. Let $\tilde{R}_t^*$ denote Shiller's approximation to R_t^* and $\tilde{V}^*$ the sample variance of $\tilde{R}_t^*$ expressed in deviations from the sample mean; $\tilde{V}^* = \epsilon' \tilde{B}_s \epsilon$, where $\tilde{B}_s$ is a symmetric matrix of order 100.[9]

---

[8] The parameter γ arises in Shiller's linearization of the basic term structure equation relating the long rate to future expected short rates. If a coupon bond is selling near par, the mean of the long rate ($\bar{R}$) will be approximately equal to the coupon rate (C). Shiller takes Taylor series expansions around $R_t = \bar{R} = C$ and $r_{t+j} = \bar{R} = C, j = 0, 1, \ldots, n - 1$, to obtain $R_t = [(1 - \gamma)/(1 - \gamma^n)] \Sigma_{j=0}^{n-1} \gamma^j r_{t+j}$, where $\gamma = 1/(1 + R_0)$ and R_0 is the point around which the equation is linearized. In practice, Shiller sets $R_0 = \bar{R}$ and thus linearizes around the mean of the long rate. For a pure discount bond, $C = 0$ and $\gamma = 1$ so that the long rate is a simple (unweighted) average of future short rates in the linearized model: $R_t = (1/n) \Sigma_{j=0}^{n-1} r_{t+j}$ (see Shiller 1979, pp. 1194–99).

[9] The quadratic form representing V^* was generated in the following way. The 100 × 100 matrix, $\tilde{L}$, which transforms the 100 observations on the short rate into 100 observations on R_t^*, was constructed by setting its first 21 rows equal to the first 21 rows of the previously defined matrix L. (This reflects the fact that the first 21 observations on R_t^* do not depend on the assumed terminal value $\tilde{R}_T^*$.) The assumption that R_T^* is equal to the average short rate over the sample period is imposed by setting each

EXCESS VOLATILITY

The distribution of the quadratic form $\epsilon'\bar{B}_s\epsilon$, which represents the sample variance of the perfect-foresight long rate constructed using Shiller's assumption concerning the terminal value of R_t^*, is given by the curve c in panel 1 of figure 1. The small sample variance $\epsilon'\bar{B}_s\epsilon$ has a mean of 0.1521 and a zero probability of taking on values greater than 1.1. As illustrated in panel 1, Shiller's method of obtaining an approximate series for R_t^*, when applied to this numerical example, results in an estimated variance of R_t^* that is severely biased downward. Not only is the expectation of $\epsilon'\bar{B}_s\epsilon$ far below the population mean of var (R_t^*) of 3.802, there is a zero probability that $\epsilon'\bar{B}_s\epsilon$ will take on a value even one-third the value of the population mean of var (R_t^*).

Curve c in panel 3 shows the distribution of $\epsilon'[\bar{B}_s - A_s]\epsilon$, which represents the difference between Shiller's approximation to the variance of R_t^* and the variance of R_t, both expressed in deviations from sample means. The measure of the perfect-foresight variance is more strongly biased than the sample variance of R_t, and the difference between the sample variances is negative throughout 99.9 percent of its distribution. The mean of $\epsilon'[\bar{B}_s - A_s]\epsilon$ is -0.2729, as compared to the population mean of the difference var (R_t^*) − var (R_t) of 3.182. Thus, even though markets are efficient in this example and the *population* variance of R_t^* is several times the population variance of R_t, *estimating* the variance of R_t^* by imposing the terminal condition that R_T^* equal the sample mean of the short rate induces so much downward bias that the sample variance of R_t exceeds its estimated upper bound with probability .999.

It is important to point out that Shiller's 1979 paper uses the constructed variable $\bar{R}_t^*$ only for the purpose of illustrating the notion of excess volatility of long interest rates; none of his formal statistical tests of market efficiency in the bond market use the constructed variable. In his subsequent paper addressing the volatility of stock prices (1981b), however, Shiller does use a perfect-foresight stock price variable that parallels $\bar{R}_t^*$ in its construction. That is, the perfect-foresight stock price variable is constructed by assuming a terminal value equal to the sample mean of the (detrended) actual stock price

---

element of the last (100th) row equal to 0.01. Using $\hat{L}(i, j)$ to denote the element in the ith row and the jth column of $\hat{L}$, rows 99–22 were generated recursively by setting $\hat{L}(i, j) = (79/80)\hat{L}(i + 1, j)$ for $i \neq j$ and $\hat{L}(i, j) = (79/80)\hat{L}(i + 1, j) + (1/80)$ for $i = j$. The 100 observations on $\bar{R}_t^*$ are given by $\bar{R}_t^* = \hat{L}S\epsilon$, where S is the previously defined matrix that transforms the vector of disturbances into the vector of realizations on the short rate. In order to reflect the fact that $\bar{R}_t^*$ is expressed in deviations from the sample mean, take the matrix product $\hat{L}S$ and calculate, for each column, the sum of the elements in the column. Denote the sum of the elements in the jth column as $m(j)$. Subtract $m(j)/100$ from each element of the jth column of the original matrix product $\hat{L}S$, for $j = 1, 2, \ldots, 100$. Premultiplying this matrix by its transpose and dividing by T gives the matrix $\bar{B}_s$.

series and solving backward recursively. While the numerical values in this paper were chosen with the bond market rather than the stock market in mind, the two problems are similar enough that the evidence of downward bias in Shiller's estimate of the perfect-foresight long rate may indicate that his estimate of the standard deviation of the perfect-foresight stock price index could be seriously downward biased as well.

The formal statistical evidence of excess volatility of long interest rates presented in Shiller's 1979 paper was based primarily on a comparison of the variance of the holding period return of a long-term bond with the variance of the short interest rate. As derived by Shiller, the linearized holding yield ($\tilde{H}_t$) is given by

$$\tilde{H}_t = \frac{R_t - \gamma_n R_{t+1}}{1 - \gamma_n}, \tag{24}$$

where $\gamma_n = \gamma(1 - \gamma^{n-1})/(1 - \gamma^n)$, γ is as previously defined, and n is the number of periods in the long-term bond. Directly from equation (24), var $(\tilde{H}_t)$ can be expressed as a function of var (R_t) and cov (R_t, R_{t+1}). The cov (R_t, R_{t+1}) term is then substituted out to obtain an expression for var $(\tilde{H}_t)$ in terms of var (R_t), var (r_t), γ_n, and ρ_{rR} (the correlation coefficient between R_t and r_t). The upper bound on var $(\tilde{H}_t)$ is obtained by maximizing this expression with respect to var (R_t); thus the bound itself is not a function of var (R_t):

$$\max_{\text{var}(R_t)} \text{var}(\tilde{H}_t) = \frac{\text{var}(r_t)\rho_{rR}^2}{1 - \gamma_n^2}. \tag{25}$$

Since ρ_{rR}^2 must be less than one, the variance of the short rate places an upper bound on the variance of $\tilde{H}_t$. Shiller's basic inequality restriction is

$$\sigma(\tilde{H}_t) \leq a\sigma(r_t), \tag{26}$$

where $a = 1/\sqrt{1 - \gamma_n^2}$.

Using the fact that $\tilde{H}_t$ is approximately serially uncorrelated (and assuming that $\tilde{H}_t$ is normally distributed), Shiller uses the χ^2 distribution to compute a one-sided 95 percent confidence interval for the sample statistic $\hat{\sigma}(\tilde{H}_t)$. A lower bound on $\hat{\sigma}(\tilde{H}_t)$, denoted $\sigma_m(\tilde{H}_t)$, is then calculated from the confidence interval. Since the small sample distribution of the estimated standard deviation of the short rate, $\hat{\sigma}(r_t)$, is not known, Shiller does not conduct a formal statistical test of the hypothesis that the standard deviation of the holding period return satisfies the upper bound in equation (26). However, comparison of the point estimates of the standard deviation of the short rate with the lower bound on the holding period yield seems discouraging from

EXCESS VOLATILITY

the point of view of proponents of market efficiency. In four of the six data sets studied by Shiller, the lower bound on the variability of holding period yields, $\sigma_m(\tilde{H}_t)$, was twice as large as the point estimate of its upper bound. For the other two data sets, $\sigma_m(\tilde{H}_t)$ was narrowly within the estimated upper bound.[10]

To see that inequality (26) will tend to be biased toward rejection of the null hypothesis of market efficiency, note that under Shiller's assumptions an unbiased estimate of the variance of the holding period yield can be obtained simply by taking the sum of squares of the deviations of $\tilde{H}_t$ from its sample mean and dividing by degrees of freedom $(T - 1)$ rather than sample size (T). However, the short rate, r_t, is highly serially correlated, so that the same correction for degrees of freedom will not eliminate the downward bias in the sample variance of r_t. Recall that in the numerical example studied in this paper the actual long rate is proportional to the short rate. Thus the small sample distribution of the variable $V = \epsilon' A_s \epsilon$ characterizes the sample variances of either R_t or r_t, under different normalizations of the error variance, σ_ϵ^2. For the numerical values examined above, the downward bias was substantial even for a sample of 25 years worth of data; the sample standard deviation of r_t is $\sqrt{.4251/.6200} = 82.8$ percent of the population standard deviation of r_t.

Using the notation var (r_t) for the population variance of r_t and vâr (r_t) for the sample variance of r_t (computed by taking deviations from the sample mean and dividing by T), we denote the relative bias of vâr (r_t) by[11]

$$\frac{E[\text{vâr}(r_t)]}{\text{var}(r_t)} = 1 - \frac{\text{var}(\bar{r}_t)}{\text{var}(r_t)}, \tag{27}$$

where var $(\bar{r}_t)$ is the variance of the sample mean of r_t. If the short rate is generated by an AR1 process, equation (27) can be evaluated by straightforward algebra:

$$\frac{E[\text{vâr}(r_t)]}{\text{var}(r_t)} = 1 - \frac{1 + \rho}{(1 - \rho)T} + \frac{2\rho(1 - \rho^T)}{(1 - \rho)^2 T^2}. \tag{28}$$

[10] In the paper discussed above, Shiller's long-term interest rate data consisted of data on bonds with very long terms to maturity; in some data sets, the bonds were 25-year bonds, in other data sets, the bonds were consols. In a subsequent paper (1981c), Shiller reports the sample standard deviations of the 6-month Treasury bill rate and the holding period yield on medium term bonds (1-year–4.5-year Treasury notes). For the sample period 1955:II–1972:II, the sample standard deviation of the holding period yield did not exceed the point estimate of its upper bound for Treasury notes with 1 year or 1.5 years to maturity. For Treasury notes with 2–4.5 years to maturity, the sample standard deviation of the holding period yield did exceed the point estimate of the upper bound, but the violation was smaller in magnitude than the violations reported in Shiller (1979), based on the very long-term bonds.

[11] I am grateful to James H. Stock for suggesting the closed-form expressions (eqq. [27] and [28]) for the bias of the sample variance of the short rate.

TABLE 2

Data Set	T (1)	n (2)	γ_n (3)	ρ (4)	$\hat{\sigma}(r)$ (5)	$a\hat{\sigma}(r)$ (6)	k (7)	$\frac{a\hat{\sigma}(r)}{k}$ (8)	$\hat{\sigma}(\bar{H})$ (9)	$\sigma_m(\bar{H})$ (10)
1. U.S., quarterly 1966:I–1977:II	46	100	.978	.95	1.78	8.55	.70	12.21	19.5	16.5
2. U.S., monthly 1969:1–1974:1	61	288	.992	.983	1.77	14.03	.53	26.47	27.4	23.6
3. U.S., annual 1960–76	17	25	.925	.815	1.39	3.66	.79	4.63	9.82	7.65
4. U.S., annual 1919–58	40	25	.940	.815	1.86	5.44	.90	6.04	5.48	4.58
5. U.K., quarterly 1956:I–1977:II	86	∞	.980	.95	2.84	14.3	.81	17.65	34.4	30.4
6. U.K., annual 1824–1929	106	∞	.968	.815	1.17	4.66	.96	4.85	4.95	4.43

NOTE.—Explanation of symbols: T = sample size; n = number of periods in the long-term bond; γ_n = constant involved in the linearization of the model; ρ = autoregressive parameter in the short-rate process; $\hat{\sigma}(r)$ = estimated standard deviation of the short rate, calculated by taking deviations from the sample mean and dividing by $T - 1$; $a\hat{\sigma}(r)$ = estimated upper bound, where $a = 1/(1 - \gamma_n^2)^{1/2}$. k = relative bias of $\hat{\sigma}(r)$; $\hat{\sigma}(\bar{H})$ = estimated standard deviation of the linearized holding period yield; $\sigma_m(\bar{H})$ = critical value for the lower 5 percent tail of $\hat{\sigma}(\bar{H})$, assuming that $\hat{\sigma}^2(\bar{H})$ is distributed χ^2

Using equation (28) we can calculate and correct for the bias to the upper bound of the holding period yield. The accuracy of the bias calculation depends, of course, on the validity of the assumption concerning the time-series process generating the short rate. Table 2 calculates the bias to the upper bound on the holding period yield for each of the six data sets studied in Shiller's 1979 paper. In calculating the bias, I retain the assumption that quarterly data on short-term interest rates (3–6-month maturities) are well approximated by an AR1 process with an autoregressive parameter of 0.95. I also assume that the AR1 parameter for monthly, as opposed to quarterly, observations on the short rate is $0.95^{1/3} = 0.983$ and that the AR1 parameter for annual observations is $(0.95)^4 = 0.815$.

Columns 2, 3, 5, 6, 9, and 10 of table 2 reproduce certain columns of table 1 in Shiller (1979).[12] Column 7 reports the value of k, the relative bias of $\hat{\sigma}(r)$; $E[\hat{\sigma}(r)] = k\sigma(r)$, which was calculated as

$$k = \left\{\left(\frac{T}{T-1}\right)\left[1 - \frac{1+\rho}{(1-\rho)T} + \frac{2\rho(1-\rho^T)}{(1-\rho)^2 T^2}\right]\right\}^{1/2}. \qquad (29)$$

As indicated by column 7, the small sample bias in $\hat{\sigma}(r)$ ranges from trivial for the data set with a sample period of 100 years ($k = 0.96$) to substantial for the monthly data set with a sample period of 5 years

---

[12] Shiller also reports a "tighter" upper bound on $\sigma(\bar{H})$ as well as bounds that are applicable when r_t is nonstationary in his table 1.

($k = 0.53$). Column 8 reports the estimated upper bound, corrected for small sample bias: $a\hat{\sigma}(r)/k$.

Without the bias correction, the upper bound on the volatility of holding period yields was violated on the basis of the sample statistics, that is, $\sigma_m(H) > a\hat{\sigma}(r)$, for four of the six data sets. Further, in the four data sets that violate the bound, $\sigma_m(H)$ is roughly twice its estimated upper bound.

The bias correction changes the result of only one data set (data set 2) from violation to nonviolation of the bound. However, the three data sets that still violate the bound are not independent observations against the null hypothesis since they cover substantially the same historical period: each of the three contains the period 1966–76.

After the upper bound has been corrected for bias, the evidence of excessive volatility of holding period yields is considerably less dramatic: for three virtually nonoverlapping sample periods (U.S., 1969–74; U.S., 1919–58; and U.K., 1824–1929), the standard deviation of holding period yields is narrowly within the upper bound. The standard deviation of holding period yields exceeds the upper bound by a margin of 35–75 percent for the three data sets that contain the period 1966–76.[13]

Singleton's Approach

While Singleton (1980) reports some results based on holding period returns, his paper focuses primarily on the upper bound on the long interest rate:

$$\text{var}(R_t) \leq \text{var}(R_t^*). \tag{30}$$

Using spectral analysis, Singleton computes consistent estimates not only of var (R_t) and var (R_t^*) but also of the covariance matrix of the estimates of var (R_t) and var (R_t^*). Like Shiller, Singleton finds that his point estimates of var (R_t) exceed the point estimates of the upper bound, var (R_t^*). Further, Singleton conducts asymptotic tests of whether the variance of the long rate satisfies the upper bound. For

---

[13] When the bound on the volatility of holding period yields is applied to the stock market, Shiller finds that the estimated standard deviation of the holding period yield is more than five times the upper bound, even with a sample period of over one hundred years (Shiller 1981b). Considering the sample size, correcting for the bias induced by eliminating the sample mean will not substantially change the magnitude of the violation. However, the exponential trend in the stock price series had been removed from the stock price data as well as the dividend data in order to achieve stationarity. One would have to assess the biases potentially induced by the detrending in order to reliably interpret the strength of the evidence against the market efficiency hypothesis in the context of the stock market. Shiller emphasized that the results could be sensitive to detrending, stating that "assumptions about public knowledge or lack of knowledge of the long run growth path are important" (1981b, p. 421, n. 2).

each of the three data sets analyzed, the violation of the upper bound is statistically significant at the 5 percent level.

Singleton does not use postsample data on the short rate to literally construct a data series on the perfect-foresight long rate R_t^*. Instead, his estimates of var (R_t^*) are computed on the basis of observations on the short rate and on the theoretical relationship between the short rate and the perfect-foresight long rate. The perfect-foresight long rate in Shiller's linearized model is given by

$$R_t^* = \delta \sum_{s=0}^{n-1} \gamma^s r_{t+s}, \tag{31}$$

where γ is a constant (see n. 8), and $\delta = (1 - \gamma)/(1 - \gamma^n)$. Based on the known linear relationship between R_t^* and r_t,[14] the spectral density of R_t^* can be expressed as a function of the spectral density of r_t:

$$S_{R^*}(\lambda) = g^2(\lambda) S_r(\lambda), \quad |\lambda| \leq \pi, \tag{32}$$

where $g^2(\lambda) = \delta^2[1 - 2\gamma^n \cos(n\lambda) + \gamma^{2n}]/[1 + \gamma^2 - 2\gamma \cos(\lambda)]$ and $S_{R^*}(\lambda)$ and $S_r(\lambda)$ are the spectral density functions of R_t^* and r_t, respectively. The variance of R_t^* is equal to the integral of the spectral density of R_t^*:

$$\text{var } R_t^*) = \int_{-\pi}^{\pi} S_{R^*}(\lambda) d\lambda. \tag{33}$$

Singleton estimated the variance of the perfect-foresight long rate by estimating $S_r(\lambda)$ from the short-rate data, calculating the function $S_{R^*}(\lambda)$ implied by equation (32), and integrating the estimate of $S_{R^*}(\lambda)$ over the interval $-\pi$ to π. Before computing the spectral densities of r_t and R_t, Singleton removed the sample mean and sample (linear) trend from the data, which consisted of monthly observations over the sample period 1959:1–1971:6.

In his appendix, "Model Restrictions on the Spectral Densities of Interest Rates," Shiller discusses the theoretical relationship (eq. [32]) between the spectral densities of R_t^* and r_t in the limiting case in which the long bond is a consol. He notes that $g^2(\lambda)$ is equal to one at $\lambda = 0$ and declines monotonically as λ increases. Further, $g^2(\lambda)$ drops rapidly as λ increases for γ close to one.

The fact that $g^2(\lambda)$ declines monotonically implies that the low fre-

---

[14] For purposes of illustration, I assume that the parameters as well as the form of the relationship between R_t^* and r_t are known. In the context of the term structure of interest rates, treating γ as known is, in my view, a sensible interpretation of the model, since γ is defined as $\gamma = 1/(1 + R_0)$, where R_0 is the point around which the linearization is taken. In practice, both Shiller and Singleton set $R_0 = \bar{R}_t$ and linearize around the sample mean of the long rate. Empirical results reported below indicate that the results are perceptibly but not dramatically affected by varying the value of γ.

EXCESS VOLATILITY

quency components of $S_r(\lambda)$ account for a greater proportion of the variance of R_t^* than of the variance of r_t. A sample mean and sample trend will tend to "fit" much of the low frequency movement in a small sample of time-series data. Thus, taking deviations from the sample mean and trend will bias a sample variance downward by underestimating the low frequency movements of the series. This downward bias in the sample variance of the short rate will be "amplified" by the filter function $g^2(\lambda)$ to create a (proportionally) greater downward bias in the estimate of var (R_t^*).[15] Elimination of the sample mean and sample trend will also create a downward bias in the sample variance of the actual long rate, R_t. However, because the spectrum of the actual long rate is much less concentrated at the low frequencies than the perfect-foresight long rate, the bias in the sample variance of R_t will tend to be smaller than the bias to the estimated variance of R_t^*.

The Appendix describes a procedure for constructing quadratic forms that represent the sample statistics calculated by applying the spectral estimators to 100 quarterly observations generated by the model economy specified in Section I.[16] The finite sample distributions of these quadratic forms are plotted in figure 1. The distribution of the spectral estimate of var (R_t^*) was so close to the distribution of Shiller's estimator that it could not be plotted distinctly from curve c in panel 1. The mean of the spectral estimate of var (R_t^*) was 0.183, as compared to the population value of var (R_t^*) of 3.802. The curve labeled d in panel 2 gives the small sample distribution of the spectral estimate of var (R_t), which has mean 0.288, as compared to the population value of var (R_t) of 0.62. The small sample distribution of the spectral estimate of the difference is given by curve d in panel 3. In the numerical example, $\mu_D = \text{var}(R_t^*) - \text{var}(R_t) = 3.2$. However, because of the severe bias to the spectral estimate of var (R_t^*), the spectral estimate of D has mean -0.104 and has a 92.12 percent probability of taking on values less than zero. That is, the estimated variance of R_t exceeds the estimated upper bound 92 percent of the time, even though the null hypothesis of market efficiency holds, by construction, in the numerical example.

---

[15] The point that Singleton's estimate of the upper bound was biased downward by the detrending is stated in Shiller (1981c): "Perhaps [Singleton's] more dramatic results stem from his decision to subtract linear trends from the data, and in effect assume the trends were known by the market in advance. Any such assumption has the effect of reducing the uncertainty about future interest rates and thus reducing the permissible volatility of long rates according to the expectations model. Ultimately the inequality tests must hinge on our priors as to the reasonableness of such assumptions" (p. 76).

[16] Because the long rate in the numerical example is assumed to be the yield on a pure discount bond, the parameters in the linearized term structure relation (eq. [31]) can be specified a priori; $\gamma = 1$ and $\delta = 1/n$.

TABLE 3

ESTIMATED NONCENTRAL SECOND MOMENT OF R_t^* AND R_t

TERM OF LONG BOND	SAMPLE PERIOD	ESTIMATED NONCENTRAL SECOND MOMENT OF R_t^*	R_t
10 years	1950:I–1973:I	22.03	19.00
20 years	1950:I–1963:I	17.79	10.90

NOTE.—Noncentral second moments were calculated as $(1/T) \sum_{t=1}^{T} x_t^2$, where T denotes the number of observations.

Some Unbiased Estimates

The paper closes by reporting some unbiased estimates of the variance-bounds statistic var (R_t^*) − var (R_t). Following a suggestion of Richard Porter, the noncentral second moments of R_t^* and R_t were calculated. Assuming that the long rate does not contain a liquidity premium, R_t^* and R_t have the same mean, μ. Since

$$E(R_t^{*2}) = \text{var}(R_t^*) + \mu^2$$

and (34)

$$E(R_t^2) = \text{var}(R_t) + \mu^2,$$

the difference between the noncentral moments is an unbiased estimate of var (R_t^*) − var (R_t).

Data series on R_t^* for 10- and 20-year Treasury bonds were constructed from data on the 3-month Treasury bill rate using Shiller's linearized term structure relation:

$$R_t^* = \frac{1-\gamma}{1-\gamma^n} \sum_{s=0}^{n-1} \gamma^s r_{t+s}, \quad \gamma = \frac{1}{1+R_0}, \tag{35}$$

where $n = 40$ for the 10-year bond and $n = 80$ for the 20-year bond. The term structure equation was linearized around the point $R_0 = 0.01$. Data were available on the short rate for 1950:I–1982:IV.[17] In order to avoid using a terminal condition, the R_t^* series for the 10-year bond was calculated for $t = 1950:I–1973:I$ and the R_t^* series for the 20-year bond was calculated for $t = 1950:I–1963:I$. The estimated noncentral second moments of R_t^* and R_t are reported in table 3.[18]

[17] The data were from Salomon Brothers, Inc. (1982).
[18] If the term structure relation is linearized around $R_0 = 0.02$ instead of 0.01, the sample noncentral second moment of R_t^* is 21.19 for the 10-year bond and 15.22 for the 20-year bond. With quarterly data, R_0 is a quarterly (nonannualized) interest rate.

EXCESS VOLATILITY

According to table 3, the difference between the estimated noncentral second moments of R_t^* and R_t, which is an unbiased estimate (assuming no liquidity premium) of var (R_t^*) − var (R_t), is 3.03 for the 10-year bonds and 6.89 for the 20-year bonds.[19] The empirical finding that var (R_t) is within the upper bound imposed by var (R_t^*) will not be reversed if the model is generalized to include a (constant) positive liquidity premium in the long rate. If R_t does contain a liquidity premium, the population mean of R_t would exceed the population mean of R_t^*. Thus the assumed absence of a liquidity premium biases the estimate of var (R_t^*) − var (R_t) downward.

The average of the squared observations on the short rate for the sample, 1950:I–1982:IV, was 34.53. Since r_t, R_t, and R_t^* all have the same population mean in the absence of a liquidity premium, these estimates of the noncentral second moments imply that the variances are in the order predicted by the efficient markets model:

$$\text{var }(r_t) > \text{var }(R_t^*) > \text{var }(R_t). \tag{36}$$

Since var (R_t^*) − var (R_t) = var (θ_t), table 3 provides estimates of the market's standard error in predicting R_t^*. For the 10-year bonds, the market's standard error in forecasting R_t^* was 174 basis points; for 20-year bonds, the standard error was 262 basis points.

The postwar quarterly data on the 3-month Treasury bill rate, 10-year Treasury bond yield, and perfect-foresight Treasury bond yield are plotted in figure 2. The perfect-foresight 10-year bond yield, denoted by the dotted line, starts to rise steeply in the early 1970s, reflecting the unusually high short rates in 1979–82. The rise in the long rate may have appeared, in 1969, to indicate overreaction to the contemporaneous rise in short rates, or excess volatility. However, history has clearly exonerated the sharp rise in the long rate in the late 1960s and early 1970s. In studying figure 2, one is struck, not by the volatility of the long rate, but by the accuracy of the long rate in predicting the explosion of short rates in the early 1980s.

---

[19] In calculating the noncentral second moment of R_t, the sample period was limited to the exact sample available for the corresponding perfect-foresight long rate; i.e., the 1973:II–1982:IV data on the 10-year long rate and the 1963:II–1982:IV data for the 20-year long rate were not used. If the noncentral second moment of R_t itself were of primary interest, using all of the available data would be efficient. However, in the variance-bounds problem, one is primarily interested in obtaining a precise estimate of the *difference* of the two moments. The sampling variability of the difference of the two sample moments is an increasing function of the variance of the sample second moment of R_t and a decreasing function of the covariance of the sample second moments of R_t and R_t^*. Including the additional data on R_t reduces the variance of the sample moment of R_t but also reduces the covariance of the two sample moments. Based on the conjecture that the effect of the covariance term dominates, the additional observations available for R_t were excluded.

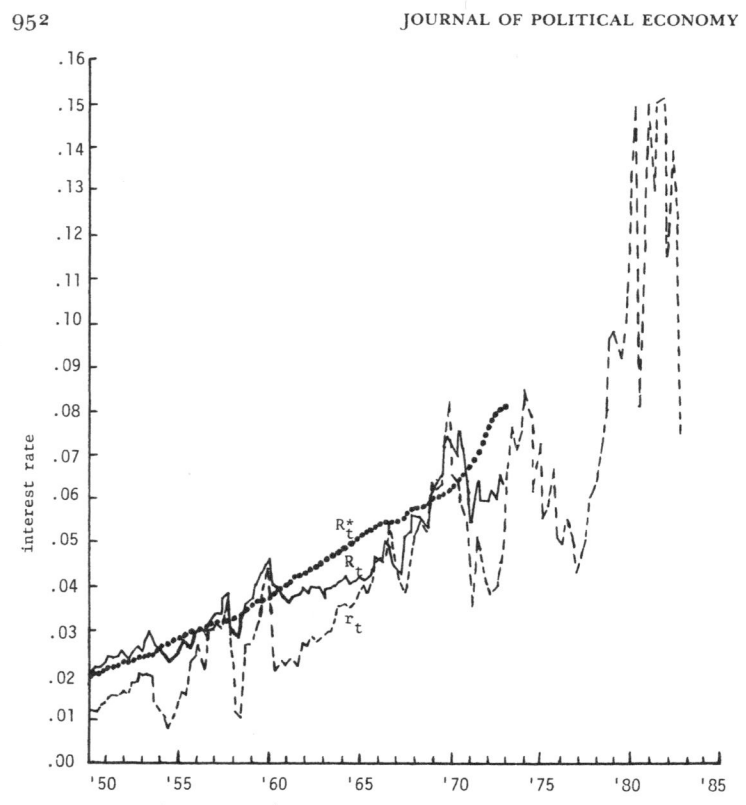

FIG. 2.—Plot of r_t, R_t, and R_t^*, U.S. Treasury securities: r_t = 3-month Treasury bill rate, denoted by dashed line; R_t = 10-year Treasury bond yield, denoted by solid line; R_t^* = perfect-foresight 10-year rate, denoted by dotted line. The sample period for the short-rate data is 1950:I–1982:IV. The perfect-foresight long rate was computed using Shiller's linearized term structure relation, $R_t^* = [(1 - \gamma)/(1 - \gamma^n)] \sum_{s=0}^{n-1} \gamma^s r_{t+s}$, for $\gamma = 1/(1 + R_0)$, where R_0 is the point around which the term structure relation is linearized. For the R_t^* series plotted in figure 2, $R_0 = 0.01$ and $n = 40$. The series on R_t^* was computed only up to 1973:I, the last observation for which the necessary postsample data were available. The data, from Salomon Brothers (1982), consist of observations taken during January, April, July, and October of each year.

III. Conclusions

The basic problem addressed by this paper—that the upper bound on the volatility of long interest rates or stock prices is difficult to measure in small samples—was certainly recognized by the authors who formulated the variance-bounds tests. In fact, Shiller refrains from conducting formal statistical tests of the hypothesis that the holding period yield on long interest rates is within its upper bound

EXCESS VOLATILITY

on the grounds that the small sample distribution of the upper bound is unknown. In the conclusion to his 1979 paper, Shiller acknowledges that he cannot rule out the possibility that the population variance of the short rate exceeds the sample variance by a sufficiently large margin to exonerate the market efficiency hypothesis, "since we have no real information in small samples about possible trends or long cycles in interest rates. Indeed, some would claim that short-term interest rates may be unstationary and hence have infinite variance. The fact that the lower bound on the left-hand side exceeds the sample value of the right-hand side may be interpreted as safely telling us, then, that we must rely on such unobserved variance or expected explosive behavior of short rates if we wish to retain expectations models" (pp. 1213–14).

Shiller's subsequent papers on long interest rate volatility (1981a, 1981c) reach the same general conclusion: the observed volatility of long interest rates can be justified as the rational response to new information about future short-term interest rates only if the population variance of short interest rates is much larger than the sample variance. In addition to random sampling error, Shiller cites several situations in which the population variance of the short rate would tend to exceed the sample variance: the short-rate process is nonstationary; the short-rate process is stationary but inappropriately detrended;[20] or the short-rate data suffer from what Krasker (1980) has termed the "peso problem"—market participants rationally perceived the possible occurrence of a major disturbance that was not realized within the sample period.[21]

---

[20] LeRoy and Porter (1981) also recognized the importance of the treatment of trends. They write, "The question remains whether the resulting series [earnings and price data for Standard and Poor's Composite Index, ATT, GE, and GM, corrected for inflation and earnings retention] can be assumed to obey the stationarity requirement. There appears to be some evidence of downward trends, although they are not clearly significant. We have decided to neglect such evidence and simply assume that the series are stationary since otherwise it is necessary to address such difficult questions as ascertaining to what degree stockholders can be assumed to have foreseen the assumed trend in earnings. It seems preferable to assume instead that there exist long cycles in the earnings series, implying that a sample of only a few decades may well appear nonstationary.... We do not argue that this treatment is entirely adequate, nor do we in any way minimize the problem of nonstationarity; the dependence of our results on the assumption of stationarity is probably their single most severe limitation" (pp. 568–69).

[21] Krasker (1980) examines an "apparent" failure of market efficiency in which the forward rate for Mexican pesos persistently underpredicted the future spot rate. Krasker argued that market participants rationally perceived a significant probability that the peso would be devalued. Since the devaluation did not occur within the sample period, the rational discounting of the peso in forward contracts gave rise to strong serial correlation in the spot-rate forecast errors.

This paper is focused primarily on the small sample properties of the variance-bounds test when the variances are expressed in deviations from the sample mean rather than the population mean. Even in the absence of other problems, such as nonstationarity, inappropriate detrending, and the peso problem, the results of the paper indicate that the variance-bounds test statistics tend to be seriously biased toward rejection of the null hypothesis of market efficiency when the variances are computed in deviations from sample means. For samples of the size typically used in the variance-bounds tests, the magnitude of the bias is substantial. The strategy of focusing on the consequences of taking deviations from the sample mean was not motivated by a judgment that other potential problems with the data, such as the peso problem, are empirically unimportant. To the contrary, my guess is that, for some data sets, the peso problem is probably very important. However, the effects of the peso problem are extremely difficult to assess empirically since by definition it involves the effects of unrealized possible outcomes.

By taking into account the small sample properties of the variance-bounds statistics, the evidence of excess volatility of holding period returns and long rates is attenuated along several dimensions. First, the upper bound on the variance of 10- and 20-year long rates is not violated in the postwar U.S. quarterly data. Second, the violation of the upper bound on holding period yields is not robust with respect to sample period. Third, in data sets for which the variance of holding period yields still exceeds the upper bound, the magnitude of the violation is smaller, and no evidence has been presented that the violation of the upper bound is statistically significant.

Appendix

Procedure for Obtaining the Small Sample Distributions of the Spectral Estimates

A vector of 101 observations on the short rate is given by $S\epsilon$, where S is a square matrix of order 101 as given in equation (12) and ϵ is a vector of 101 observations on the disturbance. Before computing the spectra, Singleton transformed the data by removing the sample mean and sample trend and by prewhitening by the filter $1 - .85L$. Defining X as the 101×2 matrix consisting of a column of ones and the column vector $[1, 2, 3, \ldots, 101]'$, and I the 101×101 identity matrix, construct $M = [I - X(X'X)^{-1}X']$. Construction of a 100×101 matrix, denoted H, which quasi differences the data with the filter $1 - .85L$ is straightforward. A vector of 100 observations on the transformed short-rate data is represented as $HMS\epsilon$. Denote the matrix product HMS as the 100×101 matrix D: $D = HMS$.

EXCESS VOLATILITY

Construct the complex-valued matrix P,

$$P = \frac{1}{\sqrt{2\pi T}} \begin{bmatrix} e^{i1\lambda_0} & e^{i2\lambda_0} & \cdots & e^{iT\lambda_0} \\ e^{i1\lambda_1} & e^{i2\lambda_1} & \cdots & e^{iT\lambda_1} \\ \vdots & \vdots & & \vdots \\ e^{i1\lambda_{T-1}} & e^{i2\lambda_{T-1}} & \cdots & e^{iT\lambda_{T-1}} \end{bmatrix}, \quad (A1)$$

where T is the sample size (in this case 100) and $\lambda_j = 2\pi j/T, j = 0, \ldots, T-1$. The Fourier transform of the data is given by $PD\epsilon$.

The unweighted sum of the periodogram ordinates could be obtained by computing $\epsilon'D'\bar{P}'PD\epsilon$, where $\bar{P}$ is the conjugate of P. Singleton's estimates were weighted averages of the periodogram ordinates, however; he first smoothed the ordinates with an inverted V window (of width nine ordinates) and multiplied by a filter function, denoted $f(\lambda)$. The filter function is used to "recolor" the data and, in the case of the perfect-foresight long rate, also incorporates the theoretical filter $g^2(\lambda)$ given in equation (32). Denote the column vector of T periodogram ordinates as $\mathbf{z}$. Applying the window can be represented by premultiplying $\mathbf{z}$ by a square matrix V of order T. Define a diagonal matrix F, also of order T, in which the kth diagonal element is $f(\lambda_j)$, $\lambda_j = 2\pi(k-1)/T$. The sum of the weighted periodogram ordinates is given by $\mathbf{u}FV\mathbf{z}$, where $\mathbf{u}$ is a T-element row vector of ones. Denote the T-element row vector $\mathbf{u}FV$ as $\mathbf{w}$ and construct the $T \times T$ diagonal matrix W, in which the ith diagonal element of W is the ith element of the vector $\mathbf{w}$. Thus the weighted sum of periodogram ordinates is given by

$$\epsilon'D'\bar{P}'WPD\epsilon. \quad (A2)$$

Multiply the matrix product $D'\bar{P}'WPD$ by the scalar $2\pi/T$ and denote the resulting matrix as C:

$$C = \frac{2\pi}{T} D'\bar{P}'WPD. \quad (A3)$$

The matrix C is a real, symmetric matrix of order 101.

Thus the sum of the weighted spectral density function can be expressed as a quadratic form in normal deviates. The small sample distributions of the spectral estimates of the variances can be obtained by applying the procedure described in Section I to the quadratic form $\epsilon'C\epsilon$.

All that remains is to specify the filters used to weight the smoothed periodogram ordinates. In the case of the spectral estimate of var (R_t^*), the filter was

$$f_{v^*}(0) = \frac{1}{(1 - .85)^2},$$

$$f_{v^*}(\lambda_j) = \frac{1 - \cos(n\lambda_j)}{n^2(1 - \cos\lambda_j)[1 - 2(.85)\cos\lambda_j + (.85)^2]}, \quad (A4)$$

where $\lambda_j = 2\pi j/T, j = 1, 2, \ldots, T-1$, and $n = 80$.

To obtain the spectral estimate of var (R_t), the filter was

$$f_v(\lambda_j) = \frac{\alpha^2}{1 - 2(.85)\cos\lambda_j + (.85)^2}, \quad (A5)$$

where $\lambda_j = 2\pi j/T$, $j = 0, 1, \ldots, T - 1$, and α is the factor of proportionality between the short rate and the long rate in the numerical example; $\alpha = (1 - \rho^n)/n(1 - \rho)$.

References

Amsler, Christine. "The Term-Structure of Interest Rates in an Expanded Market Model." Ph.D. dissertation, Univ. Pennsylvania, 1980.

Anderson, Theodore W. *The Statistical Analysis of Time Series*. New York: Wiley, 1971.

Blanchard, Olivier J., and Watson, Mark W. "Bubbles, Rational Expectations, and Financial Markets." Working Paper no. 945. Cambridge, Mass.: Nat. Bur. Econ. Res., July 1982.

Flavin, Marjorie A. "Time Series Evidence on the Expectations Hypothesis of the Term Structure." *Carnegie-Rochester Conference Series on Public Policy*, vol. 20. Amsterdam: North-Holland, in press.

Grossman, Sanford J., and Shiller, Robert J. "The Determinants of the Variability of Stock Market Prices." *A.E.R. Papers and Proc.* 71 (May 1981): 222–27.

Krasker, William. "The 'Peso Problem' in Testing the Efficiency of Forward Exchange Markets." *J. Monetary Econ.* 6 (April 1980): 269–76.

LeRoy, Stephen F., and Porter, Richard D. "The Present-Value Relation: Tests Based on Implied Variance Bounds." *Econometrica* 49 (May 1981): 555–74.

Michener, Ronald W. "Variance Bounds in a Simple Model of Asset Pricing." *J.P.E.* 90 (February 1982): 166–75.

Nerlove, Marc; Grether, David M.; and Carvalho, José L. *Analysis of Economic Time Series: A Synthesis*. New York: Academic Press, 1979.

Pesando, James. "Time Varying Term Premiums and the Volatility of Long-Term Interest Rates." Unpublished manuscript. Toronto: Univ. Toronto, Dept. Econ., July 1979.

Salomon Brothers, Inc. *Analytical Record of Yields and Yield Spreads*. New York: Salomon Brothers, April 1982.

Shiller, Robert J. "The Volatility of Long-Term Interest Rates and Expectations Models of the Term Structure." *J.P.E.* 87 (December 1979): 1190–1219.

―――. "The Use of Volatility Measures in Assessing Market Efficiency." *J. Finance* 36 (May 1981): 291–304. (*a*)

―――. "Do Stock Prices Move Too Much to Be Justified by Subsequent Changes in Dividends?" *A.E.R.* 71 (June 1981): 421–36. (*b*)

―――. "Alternative Tests of Rational Expectations Models: The Case of the Term Structure." *J. Econometrics* 16 (May 1981): 71–87. (*c*)

Singleton, Kenneth J. "Expectations Models of the Term Structure and Implied Variance Bounds." *J.P.E.* 88 (December 1980): 1159–76.

[4]

Dividend Variability and Variance Bounds Tests for the Rationality of Stock Market Prices

By Terry A. Marsh and Robert C. Merton*

Perhaps for as long as there has been a stock market, economists have debated whether or not stock prices rationally reflect the "intrinsic" or fundamental values of the underlying companies. At one extreme on this issue is the view expressed in well-known and colorful passages by Keynes that speculative markets are no more than casinos for transferring wealth between the lucky and unlucky. At the other is the Samuelson-Fama Efficient Market Hypothesis that stock prices fully reflect available information and are, therefore, the best estimates of intrinsic values. Robert Shiller has recently entered the debate with a series of empirical studies which claim to show that the volatility of the stock market is too large to be consistent with rationally determined stock prices. In this paper, we analyze the variance-bound methodology used by Shiller and conclude that this approach cannot be used to test the hypothesis of stock market rationality.

Resolution of the debate over stock market rationality is essentially an empirical matter. Theory may suggest the correct null hypothesis—in this case, that stock market prices are rational—but it cannot tell us whether or not real-world speculative prices as seen on Wall Street or LaSalle Street are indeed rational. As Paul Samuelson wrote in his seminal paper on efficient markets: "You never get something for nothing. From a nonempirical base of axioms, you never get empirical results. Deductive analysis cannot determine whether the empirical properties of the stochastic model I posit come close to resembling the empirical determinants of today's real-world markets" (1965, p. 42).

On this count, the majority of empirical studies report results that are consistent with stock market rationality.[1] There is, for example, considerable evidence that, on average, individual stock prices respond rationally to surprise announcements concerning firm fundamentals, such as dividend and earnings changes, and that prices do not respond to "noneconomic" events such as cosmetic changes in accounting techniques. Stock prices are, however, also known to be considerably more volatile than either dividends or accounting earnings. This fact, perhaps more than any other, has led many, both academic economists and practitioners, to the belief that prices must be moved by waves of "speculative" optimism and pessimism beyond what is reasonably justified by the fundamentals.[2]

---

*Sloan School of Management, MIT, Cambridge, MA 02139 and Hoover Institution, Stanford University; and Sloan School of Management, MIT, respectively. This paper is a substantial revision of a part of our 1983 paper which was presented in seminars at Yale and Harvard. We thank the participants for helpful comments. We also thank G. Gennotte, S. Myers, R. Ruback, and R. Shiller; and for his advice on the econometric issues, J. Hausman. We are especially grateful to F. Black, both for his initial suggestion to explore this topic, and for sharing with us his deep insights into the problem. We are pleased to acknowledge financial support from the First Atlanta Corporation for computer services. We dedicate the paper to the scientific contributions and the memory of John V. Lintner, Jr.

[1] To be sure, of the hundreds of tests of efficient markets, there have been a few which appear to reject market efficiency (see "Symposium on Some Anomalous Evidence on Capital Market Efficiency," *Journal of Financial Economics*, June-September 1978). For the most part, however, these studies are joint tests of both market efficiency and a particular equilibrium model of differential expected returns across stocks such as the Capital Asset Pricing Model and, therefore, rejection of the joint hypothesis may not imply a rejection of market efficiency. Even in their strongest interpretation, such studies have at most rejected market efficiency for select segments of the market. For further discussions, see Merton (1986).

[2] For example, in discussing the problems of Tobin's Q theory in explaining investment, Barry Bosworth

Until recently, the belief that stock prices exhibit irrationally high volatility had not been formally tested. In a series of papers (1981a, b, and 1982), Shiller uses seemingly powerful variance bounds tests to show that variations in aggregate stock market prices are much too large to be justified by the variation in subsequent dividend payments.[3] Under the assumption that the expected real return on the market remains essentially constant over time, he concludes that the excess variation in stock prices identified in his tests provides strong evidence against the Efficient Market Hypothesis. Even if the expected real return on the market does change over time, Shiller further concludes that the amount of variation in that rate necessary to "save" the Efficient Market Hypothesis is so large that the measured excess variation in stock prices cannot reasonably be attributed to this source.

We need hardly mention the significance of such a conclusion. If Shiller's rejection of market efficiency is sustained, then serious doubt is cast on the validity of this cornerstone of modern financial economic theory. Although often discussed in the context of profit opportunities for the agile and informed investor, the issue of stock market rationality has implications far beyond the narrow one of whether or not some investors can beat the market. As Keynes noted long ago (1936, p. 151), and as is evident from the modern Q theory of investment, changes in stock prices—whether rationally determined or not—can have a significant impact on real investment by firms.[4] To reject the Efficient Market Hypothesis for the whole stock market and at the level suggested by Shiller's analysis implies broadly that production decisions based on stock prices will lead to inefficient capital allocations. More generally, if the application of rational expectations theory to the virtually "ideal" conditions provided by the stock market fails, then what confidence can economists have in its application to other areas of economics where there is not a large central market with continuously quoted prices, where entry to its use is not free, and where shortsales are not feasible transactions?

The strength of Shiller's conclusions is derived from three elements: (i) the apparent robustness of the variance bound methodology; (ii) the length of the data sets used in the tests—one set has over 100 years of dividend and stock price data; and (iii) the large magnitude of the empirical violation of his upper bound for the volatility of rational stock prices. Shiller in essence relies upon elements (ii) and (iii) to argue that his rejection of the efficient market model cannot be explained away by "mere" sampling error alone.[5] Nevertheless, Marjorie Flavin (1983) and Allan Kleidon (1983a, b) have shown that such sampling error can have a nontrivial effect on the variance bound test statistics.

In this paper, we focus exclusively on element (i) and conclude that Shiller's variance bound methodology is wholly unreliable for the purpose of testing stock market rationality. Thus, even if his estimates contained no sampling error at all, his findings do not constitute a rejection of the efficient market model. To support our claim, we present an alternative variance bound test which has the feature that observed prices will, *of necessity*, be judged rational if they fail the

---

writes: "Nor does it seem reasonable to believe that the present value of expected corporate income actually fell in 1973–1974 by the magnitude implied by the stock-market decline of that period, when q declined by 50 percent. ...As long as management is concerned about long-run market value and believes that this value reflects 'fundamentals.' it would not scrap investment plans in response to the highly volatile short-run changes in stock prices" (1975, p. 286).

[3] Using the variance bound methodology, Stephen LeRoy and Richard Porter (1981) claim to show that stock prices are "too volatile" relative to accounting earnings. For a similar discussion of their analysis, see our 1984 paper.

[4] For a recent discussion of the "causal" effect of stock price changes on investment, see Stanley Fischer and Merton (1984).

[5] Shiller notes on this general point: "The lower bound of a 95 percent one-sided χ^2 confidence interval for the standard deviation of annual changes in real stock prices is over five times higher than the upper bound allowed by our measure of the observed variability of real dividends. The failure of the efficient markets model is thus so dramatic that it would seem impossible to attribute the failure to such things as data errors, price index problems, or changes in tax laws" (1981a, p. 434).

Shiller test. That is, if observed stock prices were to satisfy Shiller's variance bound test, then they would be deemed irrational by our test. It would seem, therefore, that for any set of stock market price data, the hypothesis of market rationality can be rejected by some variance bound test.

This seeming paradox arises from differences in assumptions about the underlying stochastic processes used to describe the evolution of dividends and rational stock prices. Affirmative empirical evidence in support of the class of aggregate dividend processes postulated in our variance bound test is presented in our forthcoming article. The specific model derived and tested in that paper significantly outperforms the univariate autoregressive model associated with the Shiller analysis.

The Shiller variance bound test and our alternative test share in common the null hypothesis that stock prices are rational, but differ as to the assumed stochastic process for dividends. Since Shiller's data sets strongly reject the joint hypothesis of his test and sustain our's, we conclude that his variance bound test results might better be interpreted as an impressive rejection of his model of the dividend process than as a rejection of stock market rationality.

I. On the Reliability of the Dividend Variance Bound Test of Stock Market Rationality

In his 1981a article, Shiller concludes that:

> measures of stock price volatility over the past century appear to be far too high—five to thirteen times too high—to be attributed to new information about future real dividends if uncertainty about future dividends is measured by the sample standard deviation of real dividends around their long-run exponential path.   [p. 434]

In reaching this conclusion, he relies upon a variance bound test—hereafter called the "p^* test"—which establishes an upper bound on the variance of the level of detrended real stock prices in terms of the variance of a constructed "*ex post* rational" detrended real price series.[6] In this section, we begin with a brief review of the development of his test and then present an alternative variance bound test which actually *reverses* the direction of the inequality established in the p^* test. That is, the *upper* bound on the variance of rationally determined stock prices in the Shiller test is shown to be the *lower* bound on that same variance in the alternative test.

The key assumptions underlying the p^* test can be summarized as follows:

(S1) Stock prices reflect investor beliefs which are rational expectations of future dividends.
(S2) The "real" (or inflation-adjusted) expected rate of return on the stock market, r, is constant over time.
(S3) Aggregate real dividends on the stock market, $\{D(t)\}$, can be described by a finite-variance stationary stochastic process with a deterministic exponential trend (or growth rate) which is denoted by g.

To develop the p^* test from these assumptions, Shiller defines an *ex post* rational detrended price per share in the market portfolio at time t:

$$(1) \quad p^*(t) \equiv \sum_{k=0}^{\infty} \eta^{k+1} d(t+k),$$

where $d(s) \equiv D(s)/(1+g)^{s+1}$ is the detrended dividend per share paid at the end of period s and $\eta \equiv (1+g)/(1+r)$. $p^*(t)$ is called an *ex post* (detrended) rational price because it is the present value of *actual* subsequent (to time t) detrended dividends. If as posited in (S1), actual stock prices, $\{P(t)\}$, are *ex ante* rational prices, then it follows from (1) that

$$(2) \quad p(t) = \varepsilon_t[p^*(t)],$$

---

[6]Shiller also develops a second variance bound test that establishes an upper bound on the variance of unanticipated changes in detrended real stock prices in terms of the variance of detrended real dividends. An analysis of this "innovations test" is presented later in this section.

for each t where $p(t) \equiv P(t)/(1+g)^t$ is the detrended real stock price per share of the market portfolio at the beginning of period t and ε_t is the expectation operator conditional on all information available to the market as of time t.

If, as Shiller (1981a, p. 422) points out, $p(t)$ is an *ex ante* rational price, then it is also an optimal forecast of $p^*(t)$. If $p(t)$ is such an optimal forecast, then the forecast error, $u(t) \equiv p^*(t) - p(t)$, should be uncorrelated with $p(t)$. It follows therefore that under this hypothesis, $\text{Var}[p^*(t)] = \text{Var}[p(t)] + \text{Var}[u(t)] > \text{Var}[p(t)]$. That is, in a set of repeated experiments where a forecast $p(t)$ and a sequence of subsequent dividends, $d(t+k)$, $k = 0,1,\ldots$, are "drawn," it should turn out that the sample variance of $p^*(t)$ exceeds the sample variance of the forecast $p(t)$.

If (detrended) dividends follow a regular stationary process, then rationally determined (detrended) stock prices must also. Hence, from assumption (S3), it follows by the Ergodic Theorem that time-series ensembles of $\{p(t)\}$ and $\{p^*(t)\}$ can be used to test the "cross-sectional" proposition that $\text{Var}[p^*(t)] > \text{Var}[p(t)]$.[7]

To compute an estimate of $p^*(t)$ with a finite sample time period, it is, of course, necessary to truncate the summation in (1). If, as Shiller (1981a, p. 425) notes, the time-series sample is "long enough," then a reasonable estimate of the variance of $p^*(t)$ can be obtained from that truncated summation. At the point of truncation, Shiller assigns a "terminal" value, $p^*(T)$, which is the average of the detrended stock prices over the sample period. That is,

$$(3) \quad p^*(T) = \left[\sum_{t=0}^{T-1} p(t)\right]/T,$$

where T is the number of years in the sample period.

---

[7]That is, the time-series estimator $\sum_{0}^{T-1}[p(t)-\bar{p}]^2/T$ can be used to estimate $\text{Var}[p(t)]$ and similarly for $p^*(t)$.

Under the posited conditions (S1)–(S3), the null hypothesis of the p^* test for rational stock prices can be written as

$$(4) \quad \text{Var}[p^*] \geq \text{Var}[p],$$

where from (1) and (3), the constructed $p^*(t)$ series used to test the hypothesis is given by

$$(5) \quad p^*(t) = \sum_{k=0}^{T-t-1} \eta^{k+1} d(t+k)$$

$$+ \eta^{T-t} p^*(T), \quad t = 0,\ldots,T-1.$$

As summarized by Shiller in the paragraph cited at the outset of this section, the results reported in his Table 2 (1981a, p. 431) show that the variance bound in (4) is grossly violated by both his Standard and Poor's 1871–1979 data set and his modified Dow Industrial 1928–79 data set.

Although widely interpreted as a rejection of stock market rationality (S1),[8] these findings are more precisely a rejection of the joint hypothesis of (S1), (S2), and (S3). As noted in our introduction, Shiller (1981, pp. 430–33) argues that a relaxation of (S2) to permit a time-varying real discount rate would not produce sufficient additional variation in prices to "explain" the large magnitude of the violation of the derived variance bound. However, even if (S2) were known to be true, this violation of the bound is not a valid rejection of stock market rationality unless (S3) is also known to be true. Nevertheless, to some, (S3) may appear to encompass such a broad class of stochastic processes that any plausible real-world time-series of dividends can be well-approximated by some process within its domain.[9] If this were so, then, of course, the p^* test, viewed as a test of stock market rationality, would be robust. In fact, however, this test is very sensitive to

---

[8]As a recent example, see James Tobin (1984).

[9]Perhaps this belief explains why Shiller devotes 20 percent of his paper (1981a) to justifying the robustness of his findings with respect to assumption (S2) and virtually no space to justifying (S3).

the posited dividend process. We show this by deriving a variance bound test of rational stock prices that reverses the key inequality (4). While maintaining assumptions (S1) and (S2) of the p^* test, this alternative test replaces (S3) with the assumption of a different, but equally broad, class of dividend processes. As background for the selection of this alternative class, we turn now to discuss some of the issues surrounding dividend policy and the sense in which rational stock prices are a reflection of expected future dividends, this to be followed by the derivation of our test.

If the required expected real rate of return on the firm is constant, then its intrinsic value per share at time t, $V(t)$, is defined to be the present value of the expected future real cash flows of the firm that will be available for distribution to each of the shares currently outstanding. From the well-known accounting identity,[10] it follows that the firm's dividend policy must satisfy the constraint:

(6) $$V(t) = \varepsilon_t \left[\sum_{k=0}^{\infty} D(t+k)/(1+r)^{k+1} \right].$$

Although management can influence the intrinsic value of its firm by its investment decisions, management has little, if any, control over the stochastic or unanticipated changes in $V(t)$. In sharp contrast, management has sole responsibility for, and control over, the dividends paid by the firm. There are, moreover, no important legal or accounting constraints on dividend policy. Hence, subject only to the constraint given in (6), managers have almost complete discretion and control over the choice of dividend policy.

This constraint on dividend choice is very much like the intertemporal budget constraint on rational consumption choice in the basic lifetime consumption decision problem for an individual. In this analogy, the intrinsic value of the firm, $V(t)$, corresponds to the capitalized permanent income or wealth of the individual, and the dividend policy of the firm corresponds to the consumption policy of the individual. Just as there are an uncountable number of rational consumption plans which satisfy the consumer's budget constraint for a given amount of wealth, so there are an uncountable number of distinct dividend policies that satisfy (6) for a given intrinsic value of the firm. Hence, like rational consumers in selecting their plans, rational managers have a great deal of latitude in their choice of dividend policy.[11]

If stock prices are rationally determined, then

(7) $$P(t) = V(t) \quad \text{for all } t.$$

Hence, the only reason for a change in rational stock price is a change in intrinsic value. Since a manager can choose any number of different dividend policies that are consistent with a particular intrinsic value of the firm, the statement that "rational stock prices reflect expected future dividends" needs careful interpretation. It follows from (6) and (7) that rational stock prices will satisfy

(8) $$P(t) = \varepsilon_t \left[\sum_{k=0}^{\infty} D(t+k)/(1+r)^{k+1} \right].$$

Thus, rational stock prices reflects expected future dividends through (8) in the same sense that an individual's current wealth reflects his expected future consumption through the budget constraint. Pursuing the analogy further: if because of an exogenous event (for example, a change in preferences), a consumer changes his planned pattern of

---

[10] The cash flow accounting identity applies only to dividends paid *net* of any issues or purchases of its outstanding securities. "Gross" dividends are, of course, subject to no constraint. Hence, all references to "dividends" throughout the paper are to "net" dividends paid.

[11] The fact that individual firms pursue dividend policies which are vastly different from one another is empirical evidence consistent with this view.

consumption, then it surely *does not* follow from the budget constraint that this change in the expected future time path of his consumption will cause his current wealth to change. Just so, it does not follow from (8) that a change in dividend policy by managers will cause a change in the current rationally determined prices of their shares.[12] For a fixed discount rate, r, it does however follow from (8) that an unanticipated change in a rationally determined stock price must necessarily cause a change in expected future dividends, and this is so for the same feasibility reason that with a constant discount rate, an unanticipated change in a consumer's wealth must necessarily cause a change in his planned future consumption. *In short, (8) is a constraint on future dividends and not on current rational stock price.*

Since management's choice of dividend policy clearly affects the time-series variation in observed dividends, the development of the relation between the volatility of dividends and rational stock prices requires analysis of the linkage between the largely controllable dividend process and the largely uncontrollable process for intrinsic value.

Unlike the theory of consumer choice, there is no generally accepted theory of optimal dividend policy.[13] Empirical researchers have, therefore, relied on positive theories of dividend policy to specify their models. The prototype for these models is John Lintner's model (1956) based on stylized facts first established by him in a classic set of interviews of managers about their dividend policies. Briefly, these facts are: (L1) Managers believe that their firms should have some long-term target payout ratio; (L2) In setting dividends, they focus on the change in existing payouts and not on the level; (L3) A major unanticipated and nontransitory change in earnings would be an important reason to change dividends; (L4) Most managers try to avoid making changes in dividends which stand a good chance of having to be reversed within the near future. In summary, managers set the dividends that their firms pay to have a target payout ratio as a long-run objective, and they choose policies which smooth the time path of the changes in dividends required to meet that objective.

As most textbook discussions seem to agree, these target payout ratios are measured in terms of long-run sustainable ("permanent") earnings rather than current earnings per share. In the special case where the firm's cost of capital r is constant in real terms, real permanent earnings at time t, $E(t)$, are related to the firm's intrinsic value per share by $E(t) = rV(t)$.

With this as background, we now develop a model of the dividend process as an alternative to the p^* test's (S3) process. A class of dividend policies which captures the behavior described in the Lintner interviews is given by the rule:

$$(9) \quad \Delta D(t) = gD(t) + \sum_{k=0}^{N} \gamma_k [\Delta E(t-k) - gE(t-k)],$$

where Δ is the forward difference operator, $\Delta X(t) \equiv X(t+1) - X(t)$, and it is assumed that $\gamma_k \geq 0$ for all $k = 0, 1, \ldots, N$. In words, managers set dividends to grow at rate g, but deviate from this long-run growth path in response to changes in permanent earnings that deviate from their long-run growth path. Describing the policies in terms of the

---

[12] By the accounting identity, net dividend policy (as described in fn. 10) cannot be changed without changing the firm's investment policy. However, changes in investment policy need not change the current intrinsic value of the firm. Managers can implement virtually any change in net dividends per share (without affecting the firm's intrinsic value) by the purchase or sale of financial assets held by the firm or by marginal changes in the amount of investment in any other "zero net present value" asset held by the firm (for example, inventories). Such transactions will change the composition of the firm's assets and the time pattern of its future cash flows, but not the present value of the future cash flows. Since these "trivial" changes in investment policy will not affect the intrinsic value of the firm, they will not affect the current level of rationally determined stock price. See our forthcoming article (Section 6.3) for further discussion of the difficulties of measuring net dividends.

[13] Indeed, the classic Miller-Modigliani (1961) theory of dividends holds that dividend policy is irrelevant, and hence, in this case, there is no optimal policy.

change in dividends rather than the levels, and having these changes depend on changes in permanent earnings, is motivated by Lintner's stylized facts (L2) and (L3). His behavioral fact (L4) is met in (9) by specifying the change in dividends as a moving average of current and past changes in permanent earnings over the previous N periods.

Equation (9) can be rewritten in terms of detrended real dividends and permanent earnings as

$$(10) \quad \Delta d(t) = \sum_{k=0}^{N} \lambda_k \Delta e(t-k),$$

where $e(s) \equiv E(s)/(1+g)^s$ and $\lambda_k \equiv \gamma_k/(1+g)^{k+1}$. By integrating (10),[14] we can express the level of detrended dividends at time t in terms of current and past detrended permanent earnings as

$$(11) \quad d(t) = \sum_{k=0}^{N} \lambda_k e(t-k).$$

By inspection of (11), the dividend policies in (9) satisfy Lintner's (L1) condition of a long-run target payout ratio where this ratio is given by $\delta \equiv \sum_{0}^{N} \lambda_k$.

Consider an economy in which the p^* test assumptions (S1) and (S2) *are known* to hold, but instead of (S3), assume that (9) describes the stochastic process for aggregate real dividends on the market portfolio. From the assumption of a constant discount rate (S2) and the definition of permanent earnings, we have from (11) that detrended real dividends at time t can be written as

$$(12) \quad d(t) = r \sum_{k=0}^{N} \lambda_k v(t-k)$$

$$= r\delta \sum_{k=0}^{N} \theta_k v(t-k)$$

where $v(s) \equiv V(s)/(1+g)^s$ is the detrended real intrinsic value per share of the firm at time s and $\theta_k \equiv \lambda_k/\delta \geq 0$ with $\sum_{0}^{N} \theta_k = 1$.

From (S1), stock prices are known to be rationally determined, and therefore, it follows from (7) that $p(t) = v(t)$ for all t. Hence, from (12), current detrended dividends can be expressed as a function of current and past detrended stock prices: namely,

$$(13) \quad d(t) = \rho \sum_{k=0}^{N} \theta_k p(t-k),$$

where $\rho = r\delta$ is the long-run or steady-state dividend-to-price ratio on the market portfolio.[15] Thus, from (S1), (S2), and (9), detrended aggregate real dividends are a moving average of current and past detrended real stock prices. Moreover, under these posited conditions, the *ex post* rational price series constructed for the sample period $[0, T]$ can be expressed as a convex combination of the observed detrended stock prices, $p(t)$, $t = -N, \ldots, 0, 1, \ldots, T-1$. That is, from (3) and (13), (5) can be rewritten as

$$(14) \quad p^*(t) = \sum_{k=-N}^{T-1} w_{tk} p(k),$$

$$t = 0, 1, \ldots, T-1,$$

where, as can be easily shown, the derived weights satisfy

$$\sum_{k=-N}^{T-1} w_{tk} = 1$$

and $w_{tk} \geq 0$ with $w_{tk} = 0$ for $k < t - N$.

THEOREM 1: *If, for each t, $p^*(t) = \sum_{k=0}^{T-1} \pi_{tk} p(k)$ where $\sum_{k=0}^{T-1} \pi_{tk} = 1$; $\sum_{t=0}^{T-1} \pi_{tk} \leq 1$ and $\pi_{tk} \geq 0$, then for each and every sample path of stock price realizations, $\text{Var}(p^*) \leq \text{Var}(p)$, with equality holding if and only if*

---

[14] The constant of integration must be zero since $e(t) = 0$ implies that $V(t) = 0$, which implies that $e(t+s) = 0$ and $d(t+s) = 0$ for all $s \geq 0$.

[15] The target payout ratio δ and the long-run growth rate g are related by $g = (1-\delta)r/[1+r\delta]$.

all realized prices are identical in the sample $t = 0, \ldots, T-1$.

The formal proof is in the Appendix. However, a brief intuitive explanation of the theorem is as follows: for each $t, t = 0, \ldots, T-1$, $p^*(t)$ is formally similar to a conditional expectation of a random variable p with possible outcomes $p(0), \ldots, p(T-1)$ where the $\{\pi_{tk}\}$ are interpreted as conditional probabilities. $\text{Var}(p^*)$ is, therefore, similar to the variance of the conditional expectations of p which is always strictly less than the variance of p itself (unless, of course, $\text{Var}(p) = 0$).

The variance inequality in Theorem 1 is the exact opposite of inequality (4) which holds that $\text{Var}(p^*) \geq \text{Var}(p)$. That is, if the ex post rational price series satisfies the hypothesized conditions of Theorem 1, then the p^* test inequality will be violated whether or not actual stock prices are ex ante rational. Because Theorem 1 applies to each and every time path of prices, its derived inequality $\text{Var}(p^*) \leq \text{Var}(p)$ holds in-sample. A fortiori, it will obtain for any distribution of prices. Thus, even for a "bad draw," $\text{Var}(p^*)$ will not exceed $\text{Var}(p)$.

Although the inequality in Theorem 1 is an analytic result, it does not strictly hold for all possible sample paths of the $p^*(t)$ series generated by the dividend process (9) and rational stock prices. By inspection of (3) and (14), for each $t, N \leq t \leq T$, $p^*(t)$ is a convex combination of the sample stock prices $\{p(0), \ldots, p(T-1)\}$ that satisfies the hypothesized conditions of Theorem 1. However, for $0 \leq t \leq N-1$, $p^*(t)$ will depend upon both the sample period's stock prices and one or more "out-of-sample" stock prices $\{p(-N), \ldots, p(-1)\}$. Hence, with the exception of one member of the class of processes given by (9),[16] $\text{Var}(p^*) \leq \text{Var}(p)$ need not obtain for each and every sample path of prices.[17] The problem created here by out-of-sample prices is similar to the general "start-up" problem in using a finite sample to estimate a moving average or distributed lag process. Because only the first N of the T sample elements in the p^* series depend on out-of-sample prices, the influence of these prices on the sample variance of p^* becomes progressively smaller as the length of the sample period is increased. Indeed, as proved in the Appendix, we have that

THEOREM 2: *If* (S1) *and* (S2) *hold and if the process for aggregate real dividends is given by* (9), *then in the limit as* $T/N \to \infty$, $\text{Var}(p^*)/\text{Var}(p) \leq 1$ *will hold almost certainly.*

As noted in the introduction, the Shiller variance bound theorem has been widely interpreted as a test of stock market rationality. However, as with Theorem 1, Theorem 2 concludes that $\text{Var}(p^*)$ is a *lower* bound on $\text{Var}(p)$ whereas, the corresponding Shiller theorem concludes that $\text{Var}(p^*)$ is an *upper* bound on $\text{Var}(p)$. Both Theorem 2 and the Shiller theorem are mathematically correct and both share in common the hypothesis (S1) that stock prices are rationally determined. Therefore, if these variance bound theorems are interpreted as tests of stock market rationality, then we have the empirical paradox that this hypothesis can always be rejected. That is, if observed stock prices were to satisfy the p^* test of stock market rationality, then this same sample of prices must fail our test, and conversely. This finding alone casts considerable doubt on the reliability of such variance bound theorems as tests of stock market rationality.

The apparent empirical paradox is, of course, resolved by recognizing that each of the variance bound theorems provides a test of a different joint hypothesis. In addition to

---

[16] The exception is the polar case of (9) where $N = 0$ and mangers choose a dividend policy so as to maintain a target payout ratio in both the short and long run. In this case, with $d(t) = \rho p(t)$ for all t, we have the stronger analytic proposition that the Shiller variance bound inequality (4) must be violated in all samples if stock prices are rational.

[17] For example, if all in-sample prices happened to be the same (i.e., $p(t) = \bar{p}, t = 0, \ldots, T-1$), but the out-of-sample prices were not, then for that particular sample path, $\text{Var}(p^*) > \text{Var}(p) = 0$.

(S1), both theorems also assume that the real discount rate is constant. Hence, neither (S1) nor (S2) of the respective joint hypotheses is the source of each theorem's contradictory conclusion to the other.[18] It therefore follows necessarily that the class of aggregate dividend processes (9) postulated in Theorem 2 is incompatible with the Shiller theorem assumption (S3) of a regular stationary process for detrended aggregate dividends.[19] That is, given that (S1) and (S2) hold, nonstationarity of the dividend process is a necessary condition for the validity of Theorem 2[20] whereas stationarity of the dividend process is a sufficient condition for the validity of the p^* test inequality (4). Thus, the diametrically opposite conclusions of these variance bound theorems follow directly from the differences in their posited dividend processes.

In this light, it seems to us that if the p^* test is to be interpreted as a test of any single element of its joint hypothesis, (S1), (S2), and (S3), then it is more appropriately viewed as a test of (S3) than of (S1). Viewed in this way, the previously cited empirical findings of a large violation of inequality (4) would appear to provide a rather impressive rejection of the hypothesis that aggregate real dividends follow a stationary stochastic process with a trend. As noted, Shiller has argued extensively that his results are empirically robust with respect to assumption (S2). In a parallel fashion, we would argue that they are also robust with respect to (S1). That is, even if stock prices were irrationally volatile, the amount of irrationality required to "save" the stationarity hypothesis (S3) is so large that the measured five-to-thirteen times excess variation in stock prices cannot reasonably be attributed to this source.

Perhaps the p^* test might still be saved as a test of stock market rationality if there were compelling a priori economic reasons or empirical evidence to support a strong prior belief that aggregate dividends follow a stationary process with a trend. We are, however, unaware of any strong theoretical or empirical foundation for this belief. Indeed, the standard models in the theoretical and empirical literature of both financial economics and accounting assume that stock prices, earnings, and dividends are described by nonstationary processes.[21] In his analyses of the Shiller and other variance bounds tests, Kleidon (1983a, b) uses regression and other time-series methods to show that the hypothesis of stationarity for the aggregate Standard and Poor's 500 stock price, earnings, and dividend series can be rejected.

We (in our forthcoming article) develop and test an aggregate dividend model based on the same Lintner stylized facts used to motivate (9) here. In this model, the dividend-to-price ratio follows a stationary process, but both the dividend and stock price

---

[18] Since the two theorems share the assumption (S2) and for any sample of prices, one must fail, they cannot reliably be used to test this hypothesis either. However, as Eugene Fama (1977) and Stewart Myers and Stuart Turnbull (1977) have shown, we note that a constant discount rate is inconsistent with a stationary process for dividends when investors are risk averse. Hence, the assumptions (S2) and (S3) are a priori mutually inconsistent.

[19] If $V(t)$ follows a stationary process and the dividend process is given by (9), then the innovations or unanticipated changes in intrinsic value, $\Delta V(t) + D(t) - rV(t)$, will not form a martingale as is required by (6). If, as is necessary for the validity of (9), the intrinsic value follows a nonstationary process, then from (6) and (7), both dividends and rational stock prices must also be nonstationary.

[20] If $p(t)$ and $d(t)$ follow nonstationary processes, the variances of the price and dividend are, of course, not well-defined in the time-series sense that they were used in Shiller's variance bound test. However, $\text{Var}(p^*)$ and $\text{Var}(p)$ can be simply treated as sample statistics constructed from the random variables $\{p(t)\}$ and $\{d(t)\}$, and for any finite T, the conditional moments of their distributions will exist. If, moreover, the processes are such that the dividend-to-price ratio coverages to a finite-variance steady-state distribution, then the conditional expectation of the variance bound inequality as expressed in Theorem 2, $\varepsilon_0[\text{Var}(p^*)/\text{Var}(p)]$, will exist even in the limit as $T \to \infty$.

[21] In financial economics, the prototypical assumption is that the per period rates of return on stocks are independently and identically distributed over time. Together with limited liability on stock ownership, this implies a geometric Brownian motion model for stock prices which is not, of course, a stationary process. There is a long-standing and almost uniform agreement in the accounting literature that accounting earnings (either real or nominal) can best be described by a nonstationary process (see George Foster, 1978, ch. 4).

processes are themselves nonstationary. This model is shown to significantly outperform empirically the univariate autoregressive model (with a trend) normally associated with a stationary process. These results not only cast further doubt on the stationarity assumption, but also provide affirmative evidence in support of the class of dividend processes hypothesized in Theorems 1 and 2.

Our model can also be used to reinterpret other related empirical findings which purport to show that stock prices are too volatile. For example, to provide a more-visual (if less-quantitatively precise) representation of the "excess volatility" of stock prices, Shiller (1981a, p. 422) plots the time-series of the levels of actual detrended stock prices and the constructed *ex post* rational prices, $p^*(t)$. By inspection of these plots, it is readily apparent that $p(t)$ is more volatile than $p^*(t)$. Instead of implying "too much" stock price volatility, these plots can be interpreted as implying that the p^* series has "too little" volatility to be consistent with a dividend process which is not smoothed. They are, however, entirely consistent with rational and nonstationary stock prices and dividend policies like (9) which smooth the dividend process.

It also appears in these plots that the levels of actual prices "revert" toward the p^* trend line. In the context of (14), this apparent correspondence in trend should not be surprising since $p^*(t)$ is in effect a weighted sum of *future* actual prices that were, of course, not known to investors at time t. The *ex post* "mistakes" in forecasts of these future prices by the market at time t are, thus, "corrected" when the subsequent "right" prices (which were already contained in $p^*(t)$) are revealed.[22]

In his latest published remarks on the plots of these time-series, Shiller concludes:

> The near-total lack of correspondence, except for trend, between the aggregate stock price and its *ex post* rational counterpart (as shown in Figure 1 of my 1981a paper) means that essentially no observed movements in aggregate dividends were ever correctly forecast by movements in aggregate stock prices! [1983, p. 237]

This conclusion does not, however, appear to conform to the empirical facts. As shown in our forthcoming article, the single variable that provides, by far, the most significant and robust forecasting power of the subsequent year's change in aggregate dividends is the previous year's unanticipated change in aggregate stock price.[23]

Shiller (1981a, pp. 425–27) presents a second variance bound test of rational stock prices that uses the time-series of "price innovations" that he denotes by $\delta p(t) \equiv p(t) - p(t-1) + d(t-1) - \rho p(t-1)$. Under the assumption that detrended dividends have a stationary distribution, he derives as a condition for rational stock prices that

$$(15) \quad \text{Var}(d) \geq \text{Var}(\delta p)\left[(1+\rho)^2 - 1\right],$$

where Var(d) and Var(δp) denote the sample variances of the level of detrended dividends and the innovations of price changes, respectively. As reported in Shiller's cited Table 2, the null hypothesis of rational stock prices seems, once again, to be grossly violated by both his data sets.

If, however, dividends are generated by a process like (9) and rational stock prices follow a nonstationary process, then the inequality (15) is no longer valid. Moreover, in this case, it is likely that inequality (15) will be violated. Suppose, for example, that the innovations in (detrended) stock prices follow a geometric Brownian motion

---

[22] The strength of this apparent reversion to trend is further accentuated by using the *ex post* or in-sample trend of stock prices to detrend both the actual stock price and the $p^*(t)$ time-series.

[23] As shown in Fischer and Merton, in addition to predicting dividend changes, aggregate real stock price changes are among the better forecasters of future changes in business cycle variables including *GNP*, corporate earnings, and business fixed investment. These empirical findings might also be counted in the support of the hypothesis of stock market rationality.

given by

(16) $$\delta p(t) = \sigma p(t-1) Z(t),$$

where $\{Z(t)\}$ are independently and identically distributed random variables with $\varepsilon_{t-1}[Z(t)] = 0$; $\varepsilon_{t-1}[Z^2(t)] = 1$; σ, a positive constant and where ε_t is the expectation operator, conditional on information available at time t. It follows from (16) and the properties of $\{Z(t)\}$ that

(17) $$\varepsilon_0[\text{Var}(\delta p)] = \varepsilon_0 \left[\sum_{t=0}^{T-1} \sigma^2 p^2(t) \right] / T,$$

$$= \sigma^2 \varepsilon_0 \left[\text{Var}(p) + (\bar{p})^2 \right],$$

where $\bar{p} = p^*(T)$ given in (3).

From the posited dividend process given in (13), $d(t)/\rho$ is a distributed lag of past stock prices where the distribution weights are nonnegative and sum to unity. Thus, the ensemble $\{d(t)/\rho\}$ satisfies the hypothesized conditions of Theorems 1 and 2. It follows, therefore, that $\text{Var}(d/\rho) \leq \text{Var}(p)$. Factoring out the constant ρ and rearranging terms, the inequality can be rewritten as

(18) $$\text{Var}(d) \leq \rho^2 \text{Var}(p),$$

with equality holding only in the special limiting case of (13) with $d(t) = \rho p(t)$. Combining (17) and (18) and rearranging terms, we have that

(19)
$$\varepsilon_0[\text{Var}(d)] \leq \rho^2 \varepsilon_0 [\text{Var}(\delta p) - \sigma^2 \bar{p}^2] / \sigma^2.$$

In sharp contrast to the stationary dividend case in (15) where the variance of dividends provides an upper bound on the volatility of rational stock price innovations, inspection of (19) shows that this variance provides only a *lower* bound on that volatility for our dividend process.[24] Inequalities (15) and (19) are not, of course, mutually exclusive for all parameter values. However, using the estimated values of $\text{Var}(\delta p)$, $\bar{p}$, $\text{Var}(p)$, and ρ reported by Shiller for his 1871–1979 Standard and Poor's data set, and our equation (17), we have that $\hat{\rho} = 0.0480$ and $\hat{\sigma}^2 = \text{Var}(\delta p)/[\text{Var}(p) + \bar{p}^2] = 0.0276$. Substitution of these values in (15) implies that $\varepsilon_0[\text{Var}(d)] \geq .0983 \varepsilon_0[\text{Var}(\delta p)]$ whereas the same values substituted in (19) implies that $\varepsilon_0[\text{Var}(d)] \leq .0835 \varepsilon_0[\text{Var}(\delta p) - .0276(\bar{p})^2]$. Thus, given these parameter values it would appear that any recorded values for $\text{Var}(\delta p)$ and $\text{Var}(d)$ would violate one or the other variance bound inequalities for rational stock price innovations. Hence, the empirical finding that $\text{Var}(d) \ll .0983 \text{Var}(\delta p)$—although inconsistent with the stationarity assumption (S3)— is entirely consistent with rational stock prices and the aggregate dividend process (9).

We are not alone in questioning the specification of the dividend process in the Shiller model. In addition to the cited Kleidon analyses, Basil Copeland (1983) has commented on the assumption of a deterministic trend. In his reply to Copeland, Shiller had this to say on the specification issue:[25] "Of course, we do not literally believe with certainty all the assumptions in the model which are the basis of testing. I did not intend to assert in the paper that I know dividends were indeed stationary around the historical trend" (1983, p. 236). We have shown, however, that variance bound inequality (4) is critically sensitive to the assumption of a stationary process for aggregate dividends. If aggregate dividend policy is described by a smoothing or averaging of intrinsic values that follow a nonstationary process, then the misspecification of stationarity in the dividend process does not

---

erally, the larger is N and the more evenly distributed the weights $\{\theta_k\}$ in (13), the smaller will be $\text{Var}(d)$ with no corresponding reduction in $\text{Var}(\delta p)$.

[25] We surely echo this view with respect to our own dividend model (9). We do not however, assert that the variance bound condition of Theorems 1 and 2 provides a reliable method for testing stock market rationality.

[24] By inspection of (18) and (19), it is evident that the strength of the inequality (19) will depend on the degree of "dividend smoothing" undertaken by managers. Gen-

just weaken the power of this bound as a test of stock market rationality—it destroys it—because in that case the fundamental inequality is exactly reversed.

In summary, the story that dividends follow a stationary process with a trend leads to the empirical conclusion that aggregate stock prices are *grossly* irrational. It has, therefore, the deep and wide-ranging implications for economic theory and policy that follow from this conclusion. The majority of empirical tests of the efficient market theory do not, however, concur with this finding. Hence, to accept this dividend story, we must further conclude that the methodologies of these tests were sufficiently flawed that they failed to reject this hypothesis in spite of the implied substantial irrationality in stock prices. Similar flaws must also be ascribed to the extensive studies in finance and accounting that claim to show earnings, dividends, and stock prices follow nonstationary processes. If, however, this dividend story is rejected, then the empirical violation of inequalities (4) and (15) implies nothing at all about stock market rationality. In the spirit of Edward Leamer's (1983) discussion of hypothesis testing, we therefore conclude that the Shiller variance bound theorem is a wholly unreliable test of stock market rationality because, as Leamer said, "...there are assumptions within the set under consideration that lead to radically different conclusions" (p. 38).

II. Overview and Conclusion

In the previously cited reply to Copeland, Shiller proclaims:

> The challenge for advocates of the efficient markets model is to tell a convincing story which is consistent both with observed trendiness of dividends for a century and with the high volatility of stock prices. They can certainly tell a story which is within the realm of possibility, but it is hard to see how they could come up with the inspiring evidence for the model. [1983, p. 237]

We believe that the theoretical and empirical analysis presented here provides such "inspiring evidence."

In general, the statistical properties of estimators drawn from a nonstationary population are an important matter in evaluating the significance of variance bound inequality violations.[26] As it happens, our reconciliation of the gross empirical violations of Shiller's variance bound inequalities is not based on sampling arguments. That is, we do not merely show that it is *possible* to have a chance run of history where the measured volatility of dividends is greatly exceeded by the measured volatility of rational stock prices. Instead, our "reversals" of variance bound inequalities (4) and (15) are based on expected values over the population. Thus, over repeated runs of history, we expect that, on average, these inequalities would be violated.

If our results seem counterintuitive to some, then perhaps this indicates that such intuitions about volatility relations between optimal forecasts and realizations are implicitly based on the assumption of stationary and linear processes. If so, we hope that this analysis serves to illustrate the potential for cognitive misperceptions from applying such intuitions to nonstationary systems.

Economists have long known that fluctuations in stock prices are considerably larger than the fluctuations in aggregate consumption, national income, the money supply, and many other similar variables whose expected future values presumably play a part in the rational determination of stock prices. Indeed, as noted in the introduction, we suspect that the sympathetic view held by some economists toward the proposition of excess stock market volatility can largely be traced to this long-established observation. Those who make this inference implicitly assume that the level of variability observed in these economic variables provides the appropriate

---

[26] Kenneth West (1984) and N. Gregory Mankiw et al. (1985) derive variance bound inequalities which appear to apply to nonstationary processes for prices and dividends. Using the Shiller data, Mankiw et al. find their inequalities are grossly violated. By studying the statistical properties of their estimates, Merton (1986) shows that the Mankiw et al. tests cannot be reliably applied when stock prices follow diffusion processes such as the geometric Brownian motion.

frame of reference from which to judge the rationality of observed stock price volatility. Although quantitatively more precise, the Shiller analysis adopts this same perspective when it asks: "If stock prices are rational, then why are they so volatile (relative to dividends)?" The apparent answer is that stock prices are not rational.

Our analysis turns this perspective "on its head" by asking: "If stock prices are rational, then why do dividends exhibit so little volatility (relative to stock prices)?" Our answer is simply that managers choose dividend policies so as to smooth the effect of changes in intrinsic values (and hence, rational stock prices) on the change in dividends. The a priori economic arguments and empirical support presented for this conclusion surely need no repeating. We would note, however, that this explanation is likely to also apply to the time-series of other economic flow variables. There are, for example, good economic reasons for believing that aggregate accounting earnings, investment, and consumption have in common with dividends that their changes are smoothed either by the behavior of the economic agents that control them or by the statistical methods which are used to measure them. An initial examination of the data appears to support this belief. If a thorough empirical evaluation confirms this finding, then our analysis casts doubt in general over the use of volatility comparisons between stock prices and economic variables which are not also speculative prices, as a methodology to test stock market rationality.

In summary of our view of the current state of the debate over the efficient market theory, Samuelson said it well when he addressed the practicing investment managers of the financial community over a decade ago:

> Indeed, to reveal my bias, the ball is in the court of the practical men: it is the turn of the Mountain to take a first step toward the theoretical Mohammed....
>
> ...If you oversimplify the debate, it can be put in the form of the question, Resolved, that the best of money managers cannot be demonstrated to be able to deliver the goods of superior portfolio-selection performance. Any jury that reviews the evidence, and there is a great deal of relevant evidence, must at least come out with the Scottish verdict:
>
> Superior Investment performance is unproved.   [1974, pp. 18–19]

Just so, our evidence does not prove that the market is efficient, but it does at least warrant the Scottish verdict:

Excess stock price volatility is unproved. The ball is once again in the court of those who doubt the Efficient Market Hypothesis.

APPENDIX

PROOF of Theorem 1:

Define Π as the $T \times K$ matrix of elements π_{tk} in Theorem 1, so that $p^* = \Pi p$. We show that the following three conditions

(A1) $\pi_{tk} > 0$

(A2) $\sum_{k=1}^{T} \pi_{tk} = 1$ for all $t = 1, \ldots, T$

(A3) $\sum_{t=1}^{T} \pi_{tk} \leq 1$ for all $k = 1, \ldots, T$

are sufficient for

(A4) $\mathrm{Var}(\Pi p) \leq \mathrm{Var}(p)$,

where $\mathrm{Var}(x)$ is defined as the *sample* variance operator applied to the elements of x.

LEMMA: *Any given p can be decomposed as $\bar{p}\iota + \tilde{p}$ where $\bar{p}$ is the sample mean, ι is a vector of ones, and $\tilde{p}$ is the vector of deviations of the elements of p about $\bar{p}$. Then,*

(A5) $\tilde{p}\Pi'\left(I - \frac{1}{T}\iota\iota'\right)\Pi\tilde{p} \leq \tilde{p}'\tilde{p}$

implies

(A6) $\quad \text{Var}(\Pi p) = \dfrac{\left[p'\Pi'\left(I - \dfrac{1}{T}\iota\iota'\right)\Pi p \right]}{T}$

$\leq \dfrac{\left[p'\left(I - \dfrac{1}{T}\iota\iota'\right)p \right]}{T} = \text{Var}(p)$

under condition (A2) where $\text{Var}(\Pi p)$ is computed with respect to the T elements of Πp, and $\text{Var}(p)$ is defined with respect to the T elements of p. If the matrix Π is rectangular of dimension $T \times X$, then the additional constraint $\sum_{t=1}^{T} \pi_{tk} \leq T/K$ is sufficient for equation (A6).

PROOF:
Substitute the decomposition $p = \bar{p}\iota + \tilde{p}$ into (A5) and realize that $(I - 1/T\iota\iota')\bar{p}\iota = 0$ and $(I - 1/T\iota\iota')\Pi\bar{p}\iota = 0$.

To prove Theorem 1, define the norm:

(A7) $\quad \|\Pi\tilde{p}\|^2 = \sum_t \left(\sum_k \pi_{tk}\tilde{p}_k \right)^2.$

Since the function $f(u) = u^2$ is convex, it follows that if $p_{tk} > 0$ and $\sum_k \pi_{tk} = 1$, then,

(A8) $\quad \left(\sum_k \pi_{tk}\tilde{p}_k \right)^2 \leq \sum_k \pi_{tk}\tilde{p}_k^2,$

so

(A9) $\quad \|\Pi\tilde{p}\|^2 \leq \sum_t \sum_k \pi_{tk}\tilde{p}_k^2$

$= \sum_k \tilde{p}_k^2 \sum_t \pi_{tk} \leq \sum_k \tilde{p}_k^2.$

(The last inequality in (A10) is strict if $\sum_t \pi_{tk} < 1$). Equation (A9) can be rewritten as

(A10) $\quad \|\Pi\tilde{p}\| \leq \|\tilde{p}\|.$

Also,

(A11) $\quad \tilde{p}\Pi'\left(I - \dfrac{1}{T}\iota\iota'\right)\Pi\tilde{p} \leq \|\Pi\tilde{p}\|.$

(A10) and (A11) together imply (A5), which,

by the Lemma, implies (A6), that is,

(A12) $\quad \text{Var}(\Pi p) \leq \text{Var}(p).$

PROOF of Theorem 2:
Using the definition of the (detrended) ex post rational price, $p^*(t)$, given in (5), and allowing (detrended) dividends to be a general distributed lag of (detrended) prices as in (13), ex post rational prices can be expressed in terms of the observed and presample (detrended) prices as

(A13) $\begin{bmatrix} p^*(T-1) p^*(T-2) \\ p(T-3) \\ \vdots \\ p^*(1) \end{bmatrix}$

$= \left\{ \dfrac{1}{T} \begin{bmatrix} \eta \ldots \eta & 0 \ldots 0 \\ \eta^2 \ldots \eta^2 & 0 \ldots 0 \\ \eta^3 \ldots \eta^3 & 0 \ldots 0 \\ \vdots & \vdots \\ \eta^T \ldots \eta^T & 0 \ldots 0 \end{bmatrix} \right.$

$+ \rho \begin{bmatrix} \eta & 0 & 0 \ldots 0 \\ \eta^2 & \eta & 0 \ldots 0 \\ \eta^3 & \eta^2 & \eta \ldots 0 \\ \vdots & \vdots & \vdots \\ \eta^T & \eta^{T-1} & \eta^{T-2} \ldots 0 \end{bmatrix}$

$\times \begin{bmatrix} \theta_0 & \theta_1 \ldots \theta_N & 0 \ldots 0 & 0 \ldots 0 \\ 0 & \theta_0 \ldots \theta_{N-1} & \theta_N & 0 & 0 \ldots 0 \\ \vdots & \vdots & \theta_N & 0 & 0 \\ \vdots & \vdots & & & \vdots \\ 0 & 0 & & \theta_0\theta_1 & 0 \ldots 0 \\ 0 & 0 \ldots 0 & 0 \ldots \theta_0 & \theta_1 \ldots \theta_N \end{bmatrix} \right\}$

$\times \begin{bmatrix} p(T-1) \\ p(T-2) \\ \vdots \\ p(1) \\ p(-1) \\ \vdots \\ p(-N) \end{bmatrix}$

where $\eta \equiv (1+g)/(1+r) \equiv 1/(1+r\delta) \equiv 1/(1-\rho)$ (the first identity follows from the definition of η in (1), the second from fn. 15, and the third from the definition of ρ in (15)) and the level of dividends is a distributed lag of the level of past prices, as in (13), that is, $d(t) = \rho \sum_{k=0}^{N} \theta_k p(t-k)$.

Equation (A13) may be conveniently rewritten as

(A14) $\qquad p^* = [A_1 + A_2\Theta] p$

where A_1 is the first matrix on the right-hand side of (A13), that is, the matrix that involves multiplication by the scalar $1/T$, A_2 is the next matrix, that involves multiplication by the scalar ρ, and Θ is the matrix that contains the elements $\theta_1, \theta_2, \ldots, \theta_N$. The weights in the matrix A_1 reflect the contribution $1/T$ of each of the observed prices $[p(T-1), \ldots p(0)]$ to $p^*(T)$ in accordance with (3), together with the weight $[1/(1+\rho)]^{T-t}$ attached to $p^*(T)$ in the determination of $p^*(t)$ in (5). The matrix A_2 contains the discount weights that (5) places on dividends as components of each $p^*(t)$, while Θ contains the distributed lag weights of dividends on past prices, as given in (13). Using these definitions of A_1, A_2, and Θ, (A14) is equivalent to

(A15) $\qquad p^* = Wp,$

where $W = [w_{tk}]$, the w_{tk} being those defined in (14).

It may be verified that the component A_1 of the transformation matrix W is irrelevant to the application of Theorem 1 (because the proof proceeds in terms of $\tilde{p}$, the deviations of the elements of p about $\bar{p}$). The elements of $A_2\Theta$ are positive and sum to unity or less across the rows, and if $\Theta = I$, or Θ is such that column elements of $A_2\Theta$ sum to less that $T/(T+N)$ if $\Theta \neq I$, then the conditions of Theorem 1 are satisfied, and we have

(A16) $\qquad \mathrm{Var}(p^*) \leq \mathrm{Var}(p).$

In the market rationality tests, the variance of the *ex post* rational prices is compared not to the variance of the $(T+N)$ vector of T observed *and* N presample prices, but to the variance of only the T observed prices. Partitioning of the $(T+N)$ prices into in-sample and out-of-sample prices, it is straightforward to show that

(A17) $\quad \mathrm{Var}(p^*)/\mathrm{Var}(p_T) \leq 1$

$$+ \frac{T \cdot N}{(N+T)^2} \frac{(\bar{p}_T - \bar{p}_N)^2}{\mathrm{Var}(p_T)}$$

$$- \frac{N}{(N+T)} \left[\frac{\mathrm{Var}(p_T) - \mathrm{Var}(p_N)}{\mathrm{Var}(p_T)} \right],$$

where $\bar{p}_T = \sum_{t=0}^{T-1} p(t)/T;$

$$\bar{p}_N = \sum_{t=-N}^{-1} p(t)/N;$$

$$\mathrm{Var}(p_T) = \sum_{t=0}^{T-1} [p(t) - \bar{p}_T]^2/T;$$

$$\mathrm{Var}(p_N) = \sum_{t=-N}^{-1} [p(t) - \bar{p}_N]^2/N.$$

The sum of the last two terms on the right-hand side of (A17) can be positive for some sample paths. However, if N is finite, and the nonstationary process for prices is not degenerate, then it is clear that the startup adjustment terms in (A17) converge in mean square to zero as $T \to \infty$.

REFERENCES

Bosworth, Barry, "The Stock Market and the Economy," *Brookings Papers on Economic Activity*, 2:1975, 257–300.

Copeland, Basil L., Jr., "Do Stock Prices Move Too Much to be Justified by Subsequent Changes in Dividends?: Comment," *American Economic Review*, March 1983, 73, 234–35.

Fama, Eugene F., "Risk-Adjusted Discount Rates and Capital Budgeting under Uncertainty," *Journal of Financial Economics*, August 1977, 5, 3–24.

Fischer, Stanley and Merton, Robert C., "Macroeconomics and Finance: The Role of the Stock Market," *Carnegie-Rochester Conference Series on Public Policy*, Vol. 21, *Essays on Macroeconomic Implications of Financial and Labor Markets and Political Processes*, Autumn 1984, 57–108.

Flavin, Marjorie A., "Excess Volatility in the Financial Markets: A Reassessment of the Empirical Evidence," *Journal of Political Economy*, December 1983, *91*, 929–56.

Foster, George, *Financial Statement Analysis*, Englewood Cliffs: Prentice-Hall, 1978.

Keynes, John Maynard, *The General Theory of Employment Interest and Money*, New York: Harcourt, Brace, 1936.

Kleidon, Allan W., (1983a) "Variance Bounds Tests and Stock Price Valuation Models," unpublished working paper, Graduate School of Business, Stanford University, January 1983.

_____, (1983b) "Bias in Small Sample Tests of Stock Price Rationality," unpublished paper, University of Chicago, 1983.

Leamer, Edward E., "Let's Take the Con Out of Econometrics," *American Economic Review*, March 1983, *73*, 31–43.

LeRoy, Stephen F. and Porter, Richard D., "The Present-Value Relation: Tests Based on Implied Variance Bounds," *Econometrica*, May 1981, *49*, 555–74.

Lintner, John, "Distribution of Incomes of Corporations Among Dividends, Retained Earnings, and Taxes," *American Economic Review Proceedings*, May 1956, *66*, 97–113.

Mankiw, N. Gregory, Romer, David and Shapiro, Matthew D., "An Unbiased Reexamination of Stock Market Volatility, *Journal of Finance*, July 1985, *40*, 677–78.

Marsh, Terry A. and Merton, Robert C., "Aggregate Dividend Behavior and Its implications for Tests of Stock Market Rationality," Working Paper 1475-83, Sloan School of Management, September 1983.

_____ and _____, "Earnings Variability and Variance Bounds Tests for the Rationality of Stock Market Prices," Working Paper 1559-84, Sloan School of Management, March 1984.

_____ and _____, "Dividend Behavior for the Aggregate Stock Market," *Journal of Business*, forthcoming.

Merton, Robert C., "On the Current State of the Stock Market Rationality Hypothesis," in Stanley Fischer et al., eds., *Macroeconomics and Finance: Essays in Honor of Franco Modigliani*, Cambridge: MIT Press, 1986.

Miller, Merton H. and Modigliani, Franco, "Dividend Policy, Growth and the Valuation of Shares," *Journal of Business*, October 1961, *34*, 411–33.

Myers, Stewart C. and Turnbull, Stuart M., "Capital Budgeting and the Capital Asset Pricing Model: Good News and Bad News," *Journal of Finance*, May 1977, *32*, 321–33.

Samuelson, Paul A., "Proof That Properly Anticipated Prices Fluctuate Randomly," *Industrial Management Review*, Spring 1965, *6*, 41–49.

_____, "Challenge to Judgment," *Journal of Portfolio Management*, Fall 1974, *1*, 17–19.

_____, "Optimality of Sluggish Predictors Under Ergodic Probabilities, *International Economic Review*, February 1976, *17*, 1–7.

Shiller, Robert J., (1981a) "Do Stock Prices Move Too Much to be Justified by Subsequent Changes in Dividends?," *American Economic Review*, June 1981, *71*, 421–36.

_____, (1981b) "The Use of Volatility Measures in Assessing Market Efficiency," *Journal of Finance*, May 1981, *36*, 291–311.

_____, "Consumption, Asset Markets, and Macroeconomic Fluctuations," *Carnegie-Rochester Conference Series on Public Policy*, Vol. 17, *Essays on Economic Policy in a World of Change*, 1982, 203–38.

_____, "Do Stock Prices Move Too Much to be Justified by Subsequent Changes in dividends?: Reply," *American Economic Review*, March 1983, *73*, 236–37.

Tobin, James, "On the Efficiency of the Financial System," Hirsch Memorial Lecture, New York, May 15, 1984.

West, Kenneth D., "Speculative Bubbles and Stock Price Volatility," Financial Research Center, Memo. No. 54, Princeton University, December 1984.

[5]

Variance Bounds Tests and Stock Price Valuation Models

Allan W. Kleidon

Stanford University

Previous use of plots of stock prices and "perfect-foresight" prices p_t^* as evidence of either "excess volatility" or nonconstant discount rates is invalid since by construction p_t^* will differ from and be much smoother than rational prices if discount rates are constant. Further, prices appear nonstationary, which can account for the previously reported gross violations of variance bounds. Conditional variance bounds that are valid under nonstationarity are not violated for Standard and Poor's data. The results are consistent with changes in expectations of future cash flows causing changes in stock prices.

I. Introduction

The question what determines changes in stock prices has long intrigued economists. The suggested answers cover the range from the "animal spirits" of Keynes (1936, p. 161) to models of market efficiency and rational expectations, for example, in Fama (1970b). A fundamental problem in testing rational expectation models is the well-known identification issue: If the implications of a particular model are not supported empirically, is it the fault of the assumptions of market efficiency and rational expectations, the fault of the particular model being tested, or both?

I am grateful for the assistance and encouragement of my dissertation committee, Merton Miller (chairman), Craig Ansley, George Constantinides, Eugene Fama, John Gould, Jon Ingersoll, and Richard Leftwich, and the others who have given helpful comments on this paper, especially Fischer Black, Michael Gibbons, Robert Korajczyk, David Modest, Paul Pfleiderer, Myron Scholes, and the referees (particularly Stephen LeRoy). Partial financial support was provided by the Program in Finance, Stanford University.

Another possibility is that the model has not been adequately tested either because additional assumptions required to conduct the tests are violated empirically or because the data used simply do not correspond to the theory. It is argued here that these problems are found in much of a recent literature that has led to a resurgence in stated opposition to the belief that stock prices represent a rational valuation of future cash flows. Tobin (1984, p. 26), for example, cites Shiller (1981b) as showing that "asset markets [do not] in fact generate fundamental valuations. The speculative content of market prices is all too apparent in their excessive volatility." He continues: "Keynes's classic description of equity markets as casinos where assessments of long-term investment prospects are overwhelmed by frantic short-term guesses about what average opinion will think average opinion will think . . . rings as true today as when he wrote it" (see also Ackley 1983, p. 13; Arrow 1983, p. 12). These are strong statements; however, this paper will argue that they are not justified by the work offered in their support.

The variance bounds literature referred to by Tobin uses a deceptively simple idea to test stock price valuation models based on Miller and Modigliani (1961), with an assumption of constant discount rates. As shown in Miller and Modigliani, there are several equivalent representations in terms of dividends, earnings, and investments. Shiller uses the following dividend model:[1]

$$p_t = \sum_{\tau=1}^{\infty} \frac{E\{d_{t+\tau}|\Phi_t\}}{(1+r)^\tau}, \qquad (1)$$

where r is an assumed constant discount rate, d_t is dividends in time t, and $\{X|\Phi\}$ denotes the conditional distribution of the random variable X given the information Φ. The "perfect-foresight price" p_t^{*}[2] is defined as

$$p_t^{*} \equiv \sum_{\tau=1}^{\infty} \frac{d_{t+\tau}}{(1+r)^\tau}. \qquad (2)$$

A comparison of (1) and (2) shows that

$$p_t = E\{p_t^{*}|\Phi_t\}, \qquad (3)$$

which forms the basis for the variance bound

$$\text{var}(p_t) \leq \text{var}(p_t^{*}). \qquad (4)$$

---

[1] Grossman and Shiller (1981) and Shiller (1981a, 1981b, 1981c) work with this model, while LeRoy and Porter (1981) use the earnings stream approach.

[2] See Shiller (1981c, p. 292). The term "perfect-foresight price" is unfortunate since p_t^{*} as defined in (2) will not necessarily be the price that would prevail under certainty. See Sec. IV below for futher comment.

VARIANCE BOUNDS TESTS

The logic behind the bound is the simple and general notion that the variance of the conditional mean of a distribution is less than that of the distribution itself. Since the price p_t is a forecast of p_t^*, the variance of the forecast p_t should be less than that of the variable being forecast.

Figure 1 plots Standard and Poor's (1980) annual composite stock price index 1926–79 augmented with the Cowles et al. (1938) common stock index 1871–1925 (the solid line) and p_t^* calculated from the following recursion implied by definition (2):

$$p_t^* = \frac{p_{t+1}^* + d_{t+1}}{1 + r}, \tag{5}$$

subject to a condition that equates the terminal p_T^* to the terminal price p_T. It seems obvious from figure 1 that the bound in (4) is grossly violated, with the consequent implication that prices cannot be set by the model (1). Since (1) implies that changes in price are driven by changes in expectations of future cash flows, it seems reasonable to infer that something else must be causing the large variation in prices. Tobin relies on speculation unrelated to fundamental values.

The data shown in figure 1 were used in Shiller (1981b), but similar characteristics are apparent in other data as well. Consider figure 2, which also plots prices p_t (the solid line) and corresponding p_t^* series. The relevant characteristics are very similar to those in figure 1.

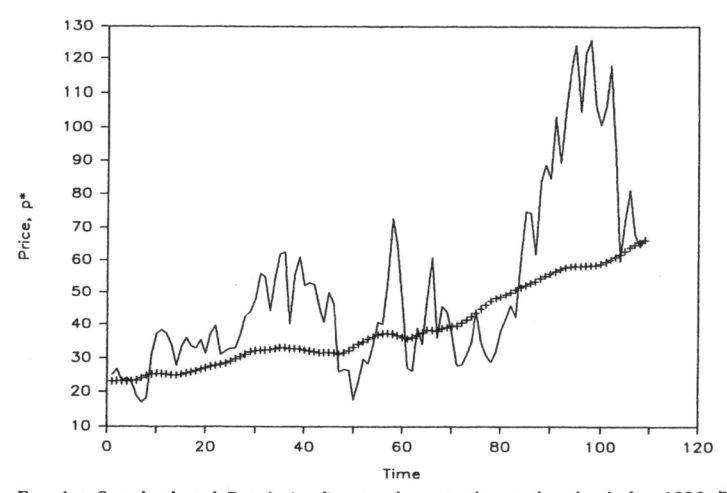

Fig. 1.—Standard and Poor's (real) annual composite stock price index 1926–79 augmented with Cowles Commission common stock index 1871–1925 (solid line) and corresponding perfect-foresight series, including terminal condition $p_T^* = p_T$.

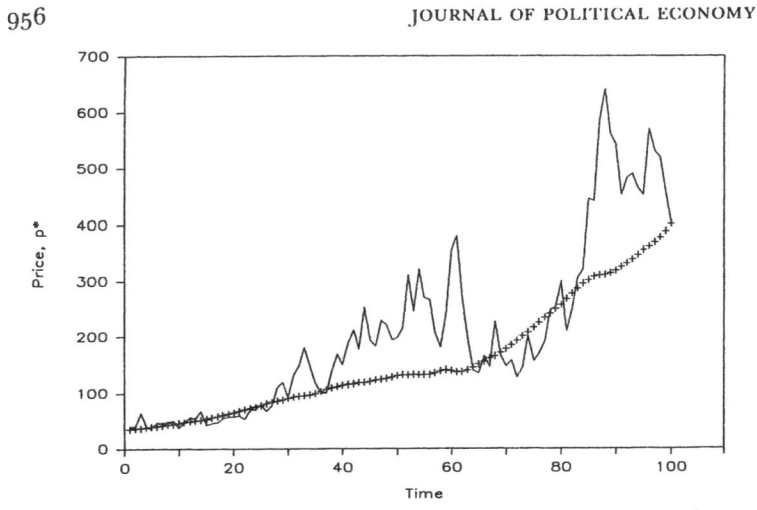

FIG. 2.—Nonstationary (geometric random walk) price series (solid line) and corresponding perfect-foresight series, including terminal condition $p_T^* = p_T$.

Again, it seems obvious that the bound (4) is violated and that consequently the valuation model (1) is empirically untenable.

However, such conclusions based on figure 2 are absolutely unfounded. This figure is based not on real data but on simulated data that by construction are generated by the rational valuation model (1). The variance bound (4) is *not* violated, and absolutely nothing can be inferred from the plots about the validity of the model (1).

This seems startling at first glance. Much of the impact of the variance bounds literature has come from the apparent clear violation of the inequality (4) by plots such as figure 1. Indeed, it has been claimed that an inspection of these plots provides such obvious evidence against the inequality (4) and the valuation model (1) that formal empirical tests of (4) need not be relied on (see Shiller 1981a, pp. 4, 7; 1984). Tirole (1985, p. 1085) also claims: "Simply by looking at Figures 1 and 2 in Shiller [1981b], this inequality [i.e., (4)] is not satisfied." This interpretation is clearly false if plots virtually identical to figure 1 can be readily created when (1) holds by construction.

More important, the price process used in figure 2 is not an unusual or artificial construct, but rather is the (geometric) random walk traditionally regarded in finance as an excellent empirical description of the price process in actual data.[3] This paper examines Standard and

[3] For construction details, see Sec. IIA below, particularly n. 7. Note also that the primary characteristics of time-series plots such as figs. 1 and 2 do not depend on the nonstationarity assumption and are present even in stationary AR(1) processes for prices, as demonstrated in Sec. II below. See Kleidon (1986) for more detail on the stationary case.

VARIANCE BOUNDS TESTS

Poor's series in some detail and demonstrates empirically that the traditional process used to construct figure 2 is consistent with Standard and Poor's price series in figure 1.

The economic intuition behind the compatibility of plots such as figures 1 and 2 with the variance bound (4) is simple, once one sees it. The fundamental flaw in the current interpretation is that the inequality (4) is essentially a *cross-sectional* relation across different economies, but figures 1 and 2 give time-series plots for a single economy. The bound (4) is derived with respect to values of p^* that differ from each other at date t because different realizations of future dividends have different present values at date t. These different realizations occur across the different economies or worlds that may possibly occur in the future, looking forward from date t. If future realizations of dividends are unexpectedly good, the realized value of p_t^* will be greater than what is expected at t, which by (3) is simply the current price p_t. If the future is unexpectedly bad, p_t^* is less than p_t.

Consider the possible values of p^* and price that may occur at some particular date t. If the price p_t predicts p_t^*, the theory given by (4) states that there should be greater variation across all possible realizations of p_t^* than in p_t. The problem with using real data is that ex post we can observe only one of the ex ante possible economies, and so we cannot look across different values of p_t^*, each corresponding to a different economy, to see if the theory is correct. We can do this by simulation, however, and it is shown below that precisely the predicted relation across different possible economies holds for the process used to construct figure 2, which is a time-series plot of only one of the ex ante possible outcomes.

Given that we observe only one world in practice, it is important to examine what should be expected in plots of time series of price and p^* for a single economy. First, note that we would not expect the series to look like each other if there is uncertainty at t about future dividends since the price p_t will be the expected value of p_t^* across possible economies and the ex post value of p_t^* once the future is revealed will in general differ from its expected value at t. How much difference will exist between plots of p_t^* and p_t depends on the amount of information available when prices are set, and it is shown below that figures 1 and 2 are consistent with a reasonable assumption about information available when Standard and Poor's prices are set.

The second insight, which is crucial to an interpretation of plots such as figures 1 and 2, is that the dividend stream being forecast at dates $t-1, t-2, \ldots$ and $t+1, t+2, \ldots$ is essentially identical to the stream forecast at t, and hence the present value of the ex post realizations will be highly correlated. Consequently the time series of p_t^* will be highly correlated, which translates into the "smooth" time-series path given in figures 1 and 2.

Of course, since p_t^* depends on information about future dividends not known at t, it is not part of the information used to set p_t or, indeed, any other price. At each date the best available information is used to set prices, and as information changes, the price will change. If, for example, the information Φ_t comprises current and past dividends, any change in dividends at t will in general imply changes in all future dividends, and the price will change by the present value of the change in expected dividends. Empirically, changes in dividends tend to persist for a very long time, and so the implied revisions in price can be very large relative to the change in current dividends.

But since by construction p_t^* is always calculated using all realized future dividends, there are no unexpected changes in dividends with implications for changes in p_t^* as there are for prices. In fact, the ex post return from both dividends and capital gains will always exactly equal the discount rate r for the p_t^* series, by the definition (2). Therefore, the possible change in consecutive values of p_t^* is limited to the capital gain required to give the ex post return r, which is another way of stating why the time series p_t^* can be much smoother than that of price. Consequently, one should expect time-series plots of p_t^* and p_t for a single economy to look like figures 1 and 2, even if *across possible economies* the variability of p_t^* exceeds that of p_t.

These arguments are established more rigorously in Section II, which demonstrates that plots such as figures 1 and 2 cannot be used to replace more formal tests of the inequality (4). Further, it is clear that, since (4) is derived by considering alternative possible economies, extra assumptions must be made to test (4) using time-series data for only one economy. Section III shows empirically that the traditional assumption in finance of nonstationary (random walk) prices is not rejected for Standard and Poor's series and that the gross violations of (4) currently reported in the literature are consistent with incorrect assumptions of stationarity in the time-series tests conducted. Section III also derives and tests inequalities similar to (4) that are implied by the (geometric random walk) time-series process for prices. It is shown that Standard and Poor's price and dividend data do not violate these bounds. Section IV contains a summary and concluding remarks.

II. Interpretation of Plots of Price and p_t^*

The current interpretation of plots such as figures 1 and 2 is that they demonstrate that prices are not set by the valuation model (1). Although the literature is not always clear on the reasoning, there appear to be two related arguments based on these plots. The first, examined in Section II*A*, relies on the undisputed smoothness of a time-series plot of p_t^* relative to prices as evidence against (1). Section IIB discusses the second, which attempts to infer the plausibility of the model from the degree of correspondence between the series p_t and p_t^*. The argument based on smoothness is clearly a less stringent test than that based on correspondence since two series may be drawn from similar stochastic processes and hence show similar time-series properties, yet not show correspondence between the observations. The conclusion reached here is that neither argument is valid.

A. Variance Bounds and "Short-Term Variation"

The characteristic of the time-series plots of price p_t and p_t^* that seems most at odds with the claim that $\text{var}(p_t^*) \geq \text{var}(p_t)$ is the striking "smoothness" of p_t^* compared with the price series. The current interpretation in the literature is that this is evidence against the inequality. However, this interpretation is incorrect, and in fact the bound does not address the issue of how smooth one time series is compared with the other. The literature has incorrectly identified the variances used in the inequality (4) with smoothness or "short-term variation" in time-series plots of price and p_t^*.

Examples of this argument occur frequently in the variance bounds literature. For example, Shiller (1981b, p. 421) states that "one is struck by the smoothness and stability of the *ex post* rational price series p_t^* when compared with the actual price series." Grossman and Shiller (1981), in one of the most influential papers using the argument, assume a constant relative risk aversion utility of consumption function,

$$U(c) = \frac{1}{1-A} c^{1-A}, \quad 0 < A < \infty, \tag{6}$$

and calculate (p. 223) the "perfect-foresight stock price" p_t^{*}[4] with constant and nonconstant discount rates. Under the assumption that investors know the whole future path of consumption (p. 223), they calculate implied discount rates from (6) for different values of the risk aversion parameter and attempt to infer the parameter value that makes the observed stock price series consistent with market efficiency (p. 224). The risk neutrality case ($A = 0$) gives constant discount rates (assuming constant time preference), and p_t^* appears much closer to the actual price series for $A = 4$ (nonconstant rates), at least for the period up to about 1950. Their results are reproduced here as figure 3.

---

[4] Some papers use the notation P_t^* and P_t, as in Grossman and Shiller (1981), while others use the lower-case notation p_t^* and p_t, which is used throughout this paper.

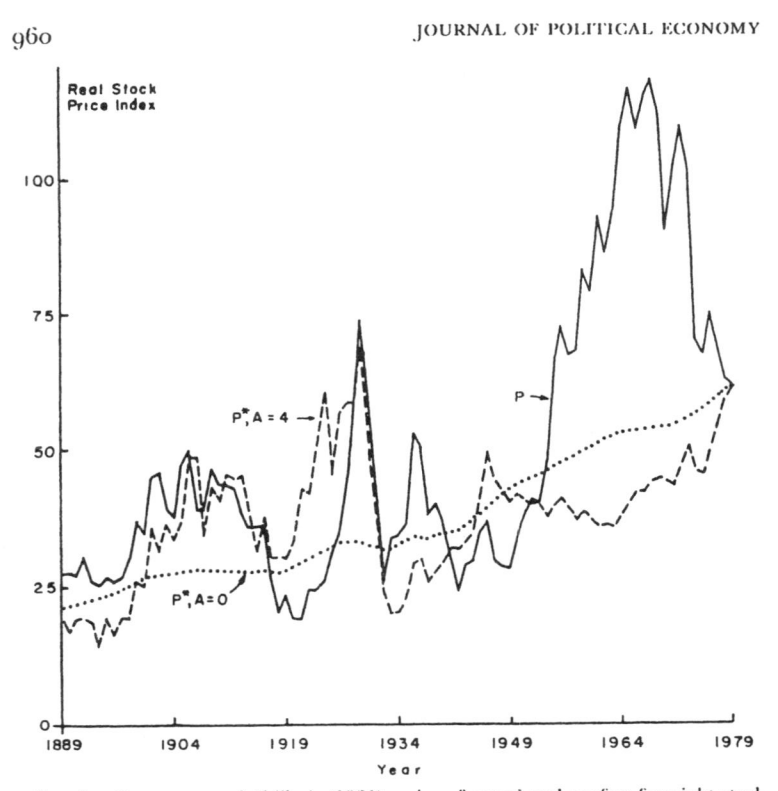

FIG. 3.—Grossman and Shiller's (1981) series of actual and perfect-foresight stock prices, 1889–1979 (reproduced from p. 225). The solid line P_t is the real Standard and Poor's composite stock price average. The other lines are: P_t^* (as defined by their expressions 6 and 7), the present value of actual subsequent real dividends using the actual stock price in 1979 as a terminal value. With $A = 0$ (dotted line) the discount rates are constant, while with $A = 4$ (dashed line) they vary with consumption.

Grossman and Shiller select the risk aversion parameter $A = 4$ in figure 3 (1981, p. 224) because of the smoothness of p_t^* when discount rates are assumed constant: "Notice that with a constant discount factor, P_t^* just grows with the trend in dividends; it shows virtually none of the *short-term variation* of actual stock prices. The larger A is, the bigger the variations of P_t^* and $A = 4$ was shown here because for this A, P, and P^* have movements of very similar magnitude" (emphasis added).

It has been shown in figure 2 that p_t^* is much smoother than price even if the constant discount rate model (1) holds by construction. The primary cause of the confusion shown in the current literature is related to the construction of p_t^* using ex post information not available when prices are set. The variance bound (4) is essentially a cross-sectional restriction on the prices that would prevail across different economies at date t. Tests of the bound using time-series data for a single economy, which are found throughout the literature, require additional strong assumptions beyond those needed to derive (4), and care must be exercised to ensure that the "variances" discussed with respect to time-series data correspond to those in the variance inequality. This section first highlights the cross-sectional nature of the inequality, then shows exactly how the argument in the literature fails.

1. Cross-sectional Variance Bounds

The equations used to derive the bound are (1)–(3) above. Equation (3) implies

$$p_t^* = p_t + \xi_t, \tag{7}$$

where $E\{\xi_t|p_t\} = 0$ by rational expectations. Clearly $\mathrm{var}(p_t^*) \geq \mathrm{var}(p_t)$, which gives the variance bound (4) in terms of the unconditional variances of p_t^* and p_t. This illustrates the essentially cross-sectional nature of the bound. At any date t the realized information Φ_t restricts the possible economies that may occur, and the possible values of the present value of dividends in those economies are given by the conditional distribution $\{p_t^*|\Phi_t\}$, with expectation p_t by (3). Each possible realization for Φ_t implies a (possibly different) conditional distribution for p_t^*, including the conditional expectation p_t. Integration over all possible economies results in the distribution of prices with variance $\mathrm{var}(p_t)$ used in the bound (4) and the unconditional distribution of p_t^*.

This argument also applies to distributions other than the unconditional distributions that result when all possible realizations of Φ_t are considered. For example, knowledge of Φ_{t-1} may restrict the possible economies at t relative to the total set. More generally, (7) implies that

$$\begin{aligned}\mathrm{var}\{p_t^*|\Phi_{t-k}\} &= \mathrm{var}\{p_t|\Phi_{t-k}\} + \mathrm{var}\{\xi_t|\Phi_{t-k}\} \\ &\geq \mathrm{var}\{p_t|\Phi_{t-k}\}, \quad k = 0, \ldots, \infty,\end{aligned} \tag{8}$$

where information at $t - k$ is included in information at t (traders do not forget), and rational expectations require that $\mathrm{cov}\{\xi_t, p_t|\Phi_{t-k}\} = 0$. The inequalities in (8) are clearly useful if conditional variances ($k < \infty$) are defined but unconditional variances ($k = \infty$) are not—for example, for the case of a random walk in prices, which is shown below to be empirically relevant. Further, it is shown below that confusion in interpretation of time-series plots of price and p_t^* stems from comparing the conditional variance of price, var$\{p_t|p_{t-k}\}$, with an inappropriate conditional variance of p_t^*, var$\{p_t^*|p_{t-k}^*\}$, which does not limit the conditioning information to information available to traders at time $t - k$.

To illustrate the distinctions, consider the following dividend process (which ignores irrelevant means for current purposes):

$$d_t = \rho d_{t-1} + \eta_t, \tag{9}$$

where η_t is independently and identically distributed (i.i.d.) $(0, \sigma_\eta^2)$. Then we have the following proposition.

PROPOSITION 1. If prices are set by (1) and information comprises current and past dividends given by (9), then

$$\begin{aligned} p_t &= a d_t \\ &= \rho p_{t-1} + a\eta_t, \end{aligned} \tag{10}$$

where $a \equiv \rho/(1 + r - \rho)$.

Proof. Follows directly from substitution in (1) for expected future dividends given (9), with simplification of the resulting infinite series. Q.E.D.

This process includes both stationary dividends ($|\rho| < 1.0$) and nonstationary random walk dividends ($\rho = 1.0$). We proceed by giving the variances of the conditional distributions $\{p_t|\Phi_{t-k}\}$ and $\{p_t^*|\Phi_{t-k}\}$, where Φ_{t-k} is limited to current and past dividends or, equivalently from (10), to p_{t-k}. The limit as $k \to \infty$ gives the unconditional distributions. The variances of the appropriate conditional distributions verify (8), but for the random walk case when $\rho = 1.0$, the conditional variances are well defined but the unconditional variances are not.

PROPOSITION 2. Assume prices are set by (1) with current and past dividends given by (9) as information. Then

$$\begin{aligned} \text{var}\{p_t|\Phi_{t-k}\} &= \text{var}\{p_t|p_{t-k}\} \\ &= \sigma_\eta^2 a^2 \left(\frac{1 - \rho^{2k}}{1 - \rho^2}\right). \end{aligned} \tag{11}$$

Proof. Given the dividend process (9), the result follows directly from (10) conditioned on p_{t-k} with simplification of the resulting infinite series. Q.E.D.

PROPOSITION 3. Assume that prices are set by (1) with current and past dividends given by (9) as information and that $|1/(1 + r)| < 1.0$.

VARIANCE BOUNDS TESTS

Then

$$\operatorname{var}\{p_t^* | \Phi_{t-k}\} = \operatorname{var}\{p_t^* | p_{t-k}\}$$

$$= \sigma_\eta^2 a^2 \left[\left(\frac{1-\rho^{2k}}{1-\rho^2}\right) + \frac{(1+r)^2}{\rho^2(2r+r^2)}\right] \quad (12)$$

$$= \operatorname{var}\{p_t | p_{t-k}\} + \frac{\sigma_\eta^2 a^2 (1+r)^2}{\rho^2(2r+r^2)}.$$

Proof. Follows from the definition (2) and the dividend process (9), conditioning on d_{t-k}, and simplifying the resulting infinite series. Q.E.D.

Note that in (12) the difference between the conditional variances $\operatorname{var}\{p_t^* | \Phi_{t-k}\}$ and $\operatorname{var}\{p_t | \Phi_{t-k}\}$, which by (8) equals $\operatorname{var}\{\xi_t | \Phi_{t-k}\}$, is for this case a constant equal to $\operatorname{var}\{p_t^* | \Phi_t\}$. Note also that the restriction on r in proposition 3 prohibits $-2 \leq r \leq 0$, which ensures that the denominator in the expression for $\operatorname{var}\{p_t^* | \Phi_t\}$ in (12) is positive.[5]

It can be verified that the limits (as $k \to \infty$) of the conditional variances in (11) and (12) equal the corresponding unconditional variances:

$$\operatorname{var}(p_t) = \frac{\sigma_\eta^2 a^2}{1-\rho^2}, \quad (13)$$

$$\operatorname{var}(p_t^*) = \frac{\sigma_\eta^2 (1+r+\rho)}{(1+r-\rho)(1-\rho^2)(2r+r^2)}. \quad (14)$$

Further, for the random walk case ($\rho = 1.0$), we have

$$\lim_{\rho \to 1} \operatorname{var}\{p_t | p_{t-k}\} = \frac{\sigma_\eta^2 k}{r^2} \quad (15)$$

and

$$\lim_{\rho \to 1} \operatorname{var}\{p_t^* | p_{t-k}\} = \frac{\sigma_\eta^2}{r^2}\left[k + \frac{(1+r)^2}{2r+r^2}\right]. \quad (16)$$

This shows that the unconditional variances of p_t and p_t^* are not defined for the random walk, so that strictly speaking the bound (4) involves undefined terms. However, the corresponding variances satisfy inequality (8).

Throughout this section, the interpretation of the variances has been in the cross-sectional sense of (unobserved) variances at t across different possible economies. To illustrate this notion, we now show

---

[5] See Kleidon (1986) for an interpretation of this condition in terms of the time-series process for p_t^*.

the values for p_1 and p_1^* for 20 replications of the simulated economy used to generate figure 2. The model used is the (geometric) random walk for prices traditionally used in finance, and it is shown in Section III below that this model is consistent with Standard and Poor's prices used in figures 1 and 3. The dividend process is[6]

$$\ln d_t = \mu + \ln d_{t-1} + \epsilon_t, \qquad (17)$$

where ϵ_t is i.i.d. $N(0, \sigma^2)$. We then have the following proposition.

PROPOSITION 4. Assume that prices are generated by (1) with current and past dividends given by (17) as information. Then the implied price is

$$p_t = \left(\frac{1+g}{r-g}\right) d_t, \qquad (18)$$

where $1 + g \equiv \exp[\mu + (\sigma^2/2)]$.

Proof. From (17), the lognormality of $\exp(\mu + \epsilon_t)$ and the standard result for its expectation, and the independence of $\epsilon_t, \epsilon_\tau, \tau \neq t$, we have

$$E\{d_{t+\tau}|d_t\} = d_t(1+g)^\tau,$$

where g is defined in (18). Substitution into (1) gives (18) directly. Q.E.D.

Figure 4 shows price and p^* *at the same date* $t = 1$ across 20 economies that were identical at $t = 0$ but are different at $t = 1$. In each economy the starting price is set as $p_0 = 40.0$, and the same dividend process given by (17) is used in each replication—all that change are the random innovations ϵ_t.[7] The first seed chosen arbitrarily for the random number generator produces the observations for "economy 1" used for figure 2, and subsequent seeds are produced internally by the IMSL generator.

From (8), we know that the variance of p_1 given p_0 should be less than the variance of p_1^* given p_0, and figure 4 shows precisely this result. Values of p_1 vary across the 20 economies from a low of 30.48 for economy 10 to a high of 61.35 for economy 17. Much greater variability across economies is seen in p_1^*, as the theory predicts, and values range from 8.99 (economy 4) to 477.83 (economy 6).

To complete the picture, figure 5 shows *time-series* plots of 100

---

[6] No dividend smoothing is assumed, which is conservative since Standard and Poor's dividend series since about 1950 appears much smoother than either prices or (accounting) earnings. Section III discusses the implications of dividend smoothing in more detail.

[7] The values for the drift μ and the innovation variance σ^2 are estimated from first differences of logs of Standard and Poor's (real) price series for 1926–79, and Standard and Poor's (real) price index in 1926 is approximately 40.0. The series ϵ_t are generated using the IMSL subroutine GGNPM. For more details, see Kleidon (1983).

VARIANCE BOUNDS TESTS

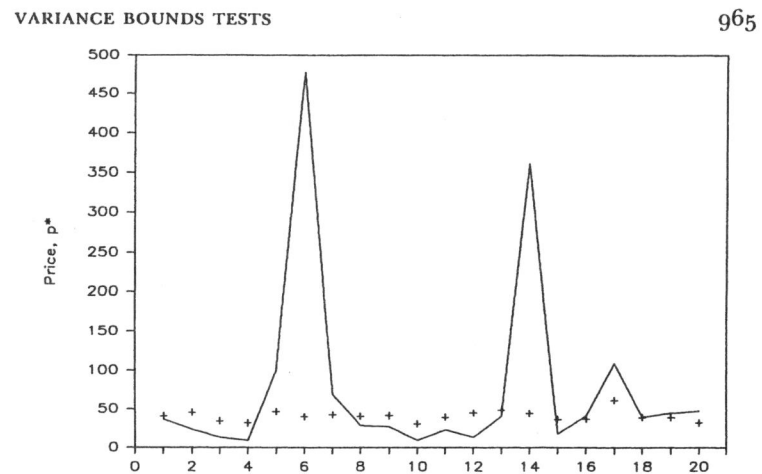

Fig. 4.—Distribution of p^* (solid line) and prices at time 1 across 20 economies that are identical at time 0. Note that this is not a time-series plot for one economy but the values at time $t = 1$ across 20 different economies.

observations of p_t and p_t^* for three of the 20 economies shown in figure 4.[8] The three economies are 2, 4, and 6; the latter two are chosen because they give the lowest and highest values of p_t^*, respectively. It is obvious from figure 5 that the wide variation in p_t^* is simply the result of different ex post draws of dividends over time for the different economies. Each is possible at time 0 since the same stochastic process and same initial price p_0 prevail in each economy. Ex post, quite different worlds could be encountered, and each implies its own value of p_t^*. The variance bounds hold across these different economies.

Although figures 4 and 5 show clearly the notion underlying variance bounds tests, the luxury of observing different worlds that may unfold through time is limited to theory or simulation. In reality we observe only one world. I now consider the properties one should expect to find in time-series plots for one economy.

2. Resolution of the Apparent Paradox

The current consensus has interpreted "smoothness" or lack of "short-term variation" in p_t^* relative to price as evidence against the

---

[8] The first economy is shown in fig. 2, and plots for the first 10 economies are given in Kleidon (1983, app. A).

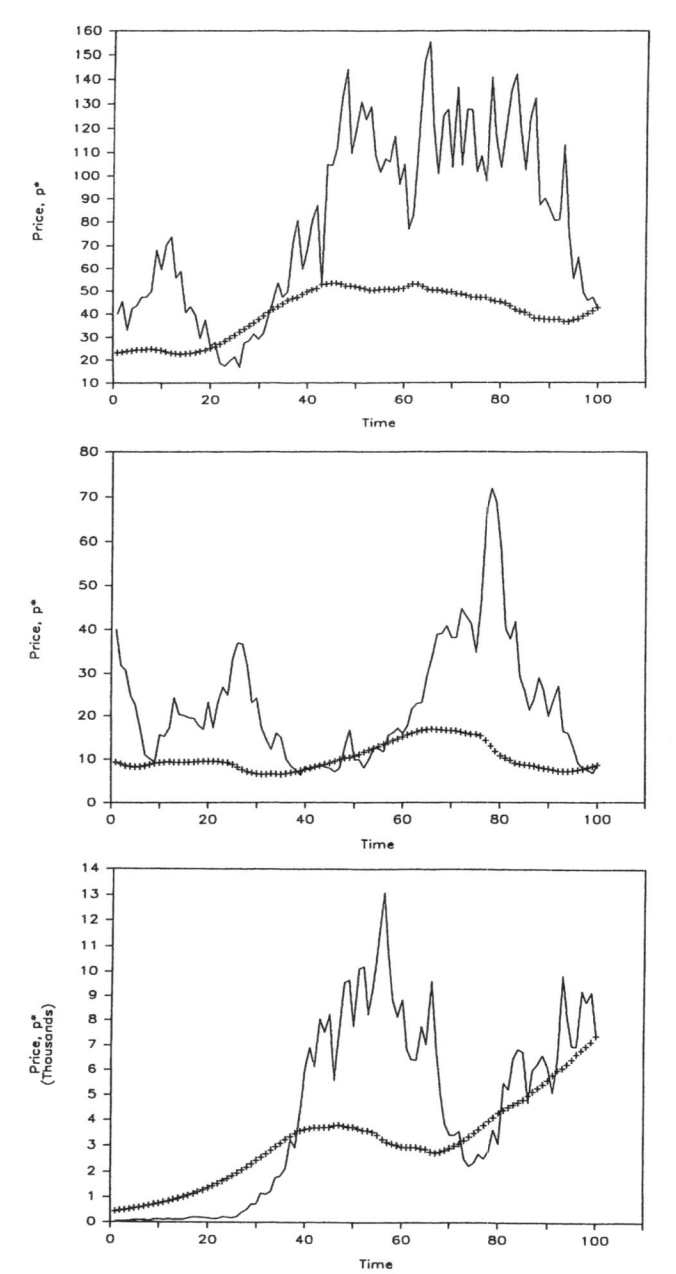

FIG. 5.—Plots of time series of nonstationary price series (solid line) and corresponding p_t^* series, for economies 2, 4, and 6 from 20 replications shown in fig. 4.

VARIANCE BOUNDS TESTS

inequality (4). Although the terms are not explicitly defined in the literature, it seems reasonable to interpret the comments about smoothness or short-term variation as relating to the conditional variance of the series, given past values of that series. Thus I interpret the smoothness of price and p_t^* to be determined by $\text{var}\{p_t|p_{t-k}\}$ and $\text{var}\{p_t^*|p_{t-k}^*\}$, respectively. Lack of short-term variation in p_t^* versus p_t, which led Grossman and Shiller (1981) to reject the valuation model (1), is consequently defined here to mean that, for small k,

$$\text{var}\{p_t^*|p_{t-k}^*\} < \text{var}\{p_t|p_{t-k}\}. \tag{19}$$

Since the issue concerns conditional variances, it is natural to examine the general bound (8), which is written in terms of conditional variances.[9] It is immediately apparent that the conditional distribution $\{p_t^*|p_{t-k}^*\}$ does not appear in (8)! Given the cross-sectional nature of these bounds, it could not since the variable p_t^* by (2) uses future dividend realizations that are not known at t and hence cannot be used as part of conditioning information at t to derive a valid bound. Consequently, despite the numerous references in the literature to the relative smoothness of price and p_t^*, this is a red herring with respect to variance inequalities.

It is clear that as $k \to \infty$ the conditional distribution $\{p_t^*|p_{t-k}^*\}$ approaches the unconditional distribution of p_t^*, so that the bound (4) will indeed hold for sufficiently large k (assuming the variances of the unconditional distributions exist). What is not obvious is the behavior of $\{p_t^*|p_{t-k}^*\}$ for k small. We now show exactly what happens to the three conditional variances that appear in (8) and (19) as k changes for the dividend model (9). We have already seen that, consistent with (8), $\text{var}\{p_t|p_{t-k}\} < \text{var}\{p_t^*|p_{t-k}\}$. It remains to show the relation between $\text{var}\{p_t|p_{t-k}\}$, which determines the smoothness of prices, and $\text{var}\{p_t^*|p_{t-k}^*\}$, which determines the smoothness of p_t^*.

PROPOSITION 5. Assume that prices are set by (1) with current and past dividends given by (9) as information, that η_t is normally distributed, and that $|1/(1+r)| < 1.0$. Then

$$\text{var}\{p_t^*|p_{t-k}^*\} = \text{var}(p_t^*)(1 - \rho_k^2), \tag{20}$$

---

[9] An earlier version of this paper distinguished between conditional variances similar to (19) and the unconditional variances in (4), and this argument is adopted in LeRoy (1984) using the conditional variances in (19). The current comparison of the conditional variances in (8) with those in (19) has the advantage of showing that the problem is not primarily with the use of conditional vs. unconditional variances, but with the use of incorrect conditional variances in (19).

where

$$\rho_k \equiv \frac{\text{cov}(p_t^*, p_{t-k}^*)}{\text{var}(p_t^*)}$$

$$= \frac{\rho^{k+1}(2r + r^2) - (1 - \rho^2)(1 + r)^{1-k}}{(1 + r + \rho)(\rho + r\rho - 1)}.$$

Proof. Equation (20) follows directly from the normality of p_t^*, the definition of var(p_t^*) is given in (14) above, and cov(p_t^*, p_{t-k}^*) is straightforward to calculate given the definition (2) and the dividend process (9). Q.E.D.

It can be verified that the limit (as $k \to \infty$) of var$\{p_t^*|p_{t-k}^*\}$ in (20) is var(p_t^*) and that for the random walk case ($\rho = 1.0$)

$$\lim_{\rho \to 1} \text{var}\{p_t^*|p_{t-k}^*\} =$$

$$\frac{\sigma_\eta^2[(k + 1)(2r + r^2) - (1 + r)(3 + r) + 2(1 + r)^{1-k} + 1]}{r^2(2r + r^2)}. \quad (21)$$

Again in this case, the conditional variances var$\{p_t^*|p_{t-k}^*\}$ are well defined for $k < \infty$.

Figure 6 shows the relevant conditional variances var$\{p_t|p_{t-k}\}$, var$\{p_t^*|p_{t-k}\}$, and var$\{p_t^*|p_{t-k}^*\}$, assuming $r = 0.065$ and $\sigma_\eta^2 = 1$. Parts *a*, *b*, and *c* each show the three conditional variances for k from 0 to 100, for values of $\rho = 0.80, 0.99$, and 1.0, respectively. As k increases, both var$\{p_t|p_{t-k}\}$ and var$\{p_t^*|p_{t-k}\}$ increase, and by (12) the difference is the constant var$\{p_t^*|p_t\}$. The inequalities in (8) are never violated, although for the random walk in part *c* of figure 6 both variances increase without bound.

Particularly interesting is the behavior of var$\{p_t^*|p_{t-k}^*\}$ relative to var$\{p_t|p_{t-k}\}$, which determines the relative smoothness of the series. Both equal zero at $k = 0$, and for some value k (which increases in ρ) it must be the case that var$\{p_t^*|p_{t-k}^*\} > $ var$\{p_t|p_{t-k}\}$ since we know that eventually the unconditional variances of p_t^* and p_t satisfy this inequality (assuming they exist). The key result, however, is that short-term variances show the opposite result, just as noted by Grossman and Shiller (1981). For k small, we see that var$\{p_t^*|p_{t-k}^*\} < $ var$\{p_t|p_{t-k}\}$, and this can hold for quite large k depending on the parameter ρ in the dividend process.

This implies that plots such as figures 1 and 3 above *should* show greater smoothness in p_t^* than in the price series if prices are given by (1). Such smoothness provides no evidence against either the bound (4) or the valuation model (1) but, on the contrary, is to be expected. Consequently the evidence used by Grossman and Shiller (1981) to

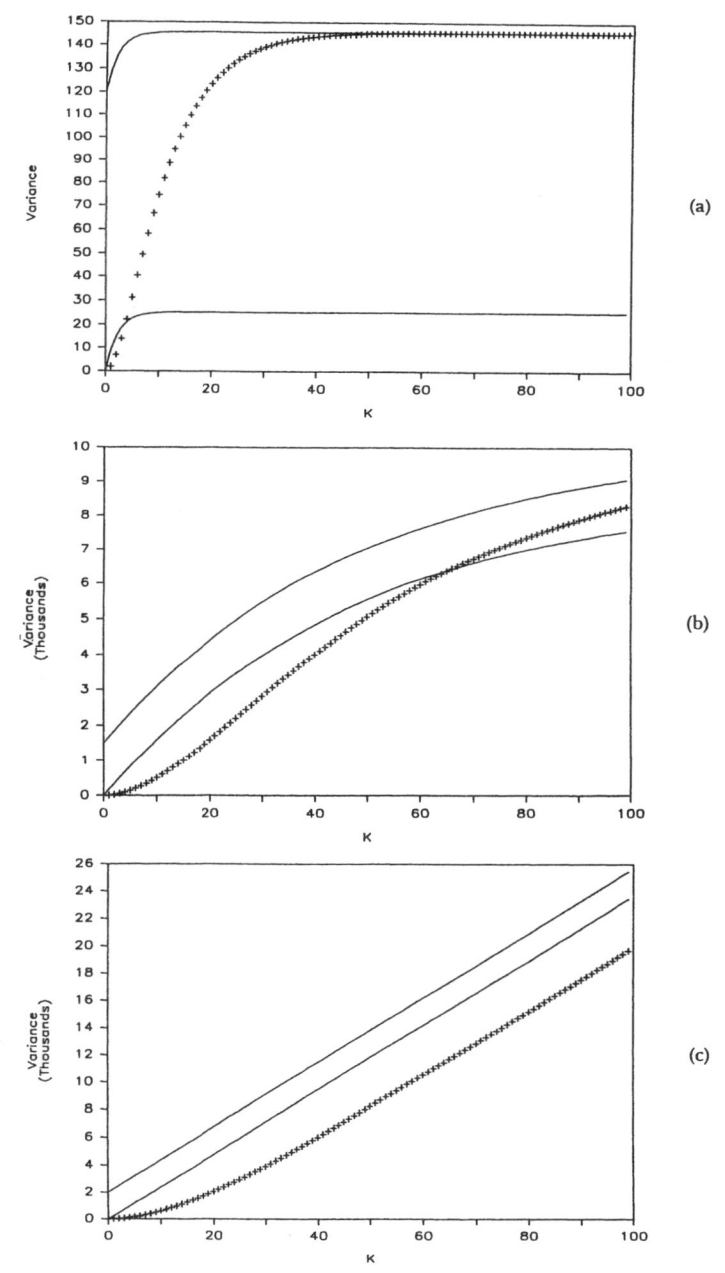

FIG. 6.—Conditional variances var$\{p_t^*|p_{t-k}\}$ (upper solid line), var$\{p_t|p_{t-k}\}$ (lower solid line), and var$\{p_t^*|p_{t-k}^*\}$ (broken line), $k = 0, \ldots, 100$, for AR(1) prices and dividends with (a) $\rho = 0.80$, (b) $\rho = 0.99$, and (c) $\rho = 1.0$.

conclude that prices cannot be given by (1) does not support their conclusion.

The intuition behind this result is straightforward. The series p_t^* is constructed so that ex post the sum of dividend yield and capital gain always gives exactly the rate r by (2). Consequently, changes in p_t^* will by construction give just the capital gain, which, together with the dividend d_t, ensures the total return r. Prices, however, can and frequently will show short-term changes of an order of magnitude larger than this since changes in current dividends in general imply changes in expected dividends for the infinite future. The price will change by the present value of these revisions in expected future dividends. Since by assumption the series p_t^* is already calculated using the ex post infinite dividend series, changes in current dividends imply no new information and no unexpected changes in p_t^*.

Given an understanding of what should be expected in time-series plots of price and p_t^*, we turn now to the issue of correspondence between the series.

B. Correspondence between p_t^* and p_t

1. The Argument

Grossman and Shiller (1981) rely on the relative degree of correspondence between two p_t^* series (with constant and nonconstant rates) and the price series p_t to determine which model is preferable, and they argue (p. 224) that "the rough correspondence between $[p_t^*, A = 4]$ and $[p_t]$ (except for the recent data) shows that if we accept a coefficient of relative risk aversion of 4, we can to some extent reconcile the behavior of $[p_t]$ with economic theory *even under the assumption that future price movements are known with certainty*" (emphasis added).

The statement concerning a certainty assumption is crucial, and we return to it shortly. A more recent claim that the price series should correspond to the p_t^* series is one of the strongest. Shiller (1984) relies exclusively on plots such as figure 3 as a "particularly striking way of presenting the evidence" that stock price changes cannot be explained in terms of "some new information about future earnings" (p. 30). He uses virtually the same plot as figure 3 (extended to 1981) and claims (p. 31):

> [Figure 3] shows that actual dividend movements of the magnitude "forecast" by price movements never appeared in nearly a century of data. We *might* have observed big movements in $[p_t^*, A = 0]$ that correspond to big movements in $[p_t]$ and that would mean that movements in $[p_t]$ really did appropriately forecast movements in future dividends. On

the other hand, this just did not happen. Look, for example, at the stock market decline of the Great Depression, from 1929 to 1932. $[p_t^*, A = 0]$ did go down then, but only very slightly, far less than the decline in $[p_t]$. The reason is that real dividends declined substantially only for the few worst years of the Depression. These few lean years have little impact on $[p_t^*, A = 0]$, which depends in effect on the longer-run outlook for stocks.

2. Analysis

Section IIA demonstrates that, even if cross-sectional variance bounds are satisfied, time-series plots of price and p_t^* will frequently show the series p_t^* as being much smoother than the price series if there is uncertainty about future dividends when prices are set. Consequently, it is not surprising that the series do not correspond to each other. What is crucial is how much information is available, which determines the degree of correspondence that should be expected. It is clear from the simulations in figures 2 and 5 that the amount of uncertainty about future cash flows implicit in the traditional geometric random walk is sufficient to imply the degree of divergence between p_t^* and p_t shown in Standard and Poor's series in figure 1.

Shiller's (1984) argument that the stock price should not have declined as much as it did between 1929 and 1932 because dividends declined substantially only in the few worst years of the depression assumes that stockholders knew that the lower dividends they were seeing would not last far into the future. Grossman and Shiller (1981) are more explicit and add an assumption of certainty about future prices. This assumption is not part of the model ostensibly being tested. The original model, given as (1) above, writes price in terms of expected future dividends, in contrast to p_t^*, which uses the ex post outcomes. In a world of certainty we would expect p_t^* to correspond to the actual price series—if discount rates were estimated correctly and the price series were rational, they should be identical.

But of course the actual stock prices shown in figure 3 were not set in a world of omniscience. If Grossman and Shiller's p_t^* series with nonconstant discount rates exactly corresponded to the actual price series, it would be misleading to claim that the series were consistent with economic theory "even under the assumption that future prices are known with certainty." Rather, there would be consistency with economic theory *only* under certainty since the price series will follow the ex post series exactly only if shareholders have perfect information about the future dividend series. If they do not—which is surely the state of things—then one should expect deviation between ex ante and ex post prices.

The question then is not whether the p_t and p_t^* series deviate, but rather how much they deviate. It is initially tempting to regard the p_t^*, $A = 4$ series as preferable to the $A = 0$ series because it more closely resembles actual prices p_t. But until we specify how much the p_t^* series *should* deviate from the price series—that is, until we specify the amount of uncertainty in the market about future cash flows—we cannot decide which plot deviates by the correct amount. The issue is addressed by Shiller (1984, p. 35), but he does not present sufficient evidence to allow inference about the degree of divergence to be expected: "Of course, people do not have perfect foresight, and so actual stock prices [p_t] need not equal [p_t^*]. We [i.e., Grossman and Shiller (1981)] argue that even under imperfect information we might expect [p_t] to resemble [p_t^*], though if information is very bad the resemblance could be very weak." This illustrates precisely the difficulty in examining plots such as figures 1 and 3. Until we know how imperfect the information is, we cannot interpret how weak the resemblance should be. A fundamental misinterpretation of such figures has been to make inferences about the validity of the valuation model (1) without specifying the yardstick necessary to allow such inferences.

To see whether the degree of correspondence between p_t^* and price in figure 1 is consistent with the valuation equation (1), we need a model that specifies the information available to the market about future cash flows. One possibility—favored by Grossman and Shiller—is to assume that shareholders have a large amount of information about future dividends. Then the only way prices could be rational is if discount rates vary greatly because of changes in aggregate consumption, which is their solution. Unfortunately, as discussed in more detail below, this solution fails when applied to other data.

An alternative explanation is much more consistent with the data. Using the (geometric) random walk for prices traditionally used in finance and assuming that the only information available at time t is the past history of dividends, we see in figures 2 and 5 that there is sufficient uncertainty about future cash flows to imply the large divergence between prices and p_t^* seen in Standard and Poor's data in figures 1 and 3. The procedures used to construct figures 2 and 5 are conservative since discount rates are strictly constant by construction and no dividend smoothing is assumed.[10]

---

[10] Hence Marsh and Merton (1984a, p. 19) are incorrect in claiming that "[fig. 3] can be interpreted as implying that the p_t^* series has 'too little' volatility to be consistent with a dividend process which is not smoothed."

C. Conclusion

This section has demonstrated that plots such as figures 1 and 3 cannot be regarded as inconsistent with the valuation model (1), although at first they appear to be convincing evidence against its validity. It is tempting to look at the p_t^* series as the "true" price, which does not vary much through time, and the actual price as (correlated) deviations from the true price. Such an interpretation is incorrect because the price at t can only be assessed relative to the information Φ_t. Thus in figure 2 the actual price series is by construction the conditional expectation of p_t^* given Φ_t, and by construction the prediction error ξ_t in (7) (i.e., the difference between this conditional expectation and the ex post outcome for p_t^*) is uncorrelated with p_t or with past prices, which are also in Φ_t.

What is potentially misleading from figure 2 is that successive prediction errors are highly correlated with each other, which appears to contradict the previous statement. Again, however, the problem lies in the information that is implicitly being used for conditioning. Previous forecast errors ξ_{t-k} are not in the information set at t since previous p_{t-k}^* that depend on the ex post outcomes for future dividends are unknown at t. Clearly the errors will be correlated since almost the same future set of dividends are being forecast at, say, t and $t+1$.[11] As seen in figures 4 and 5, the errors across economies at time t are indeed unrelated to prices at t.

Despite the potential for confusion in plots such as figure 3, they have been heavily relied on in the literature and have even been treated as stronger evidence against (1) than formal tests of the bound (4). Shiller (1981a, pp. 4, 7; 1984) claims that figure 3 alone is sufficient to show that stock prices are inconsistent with the valuation model (1), as does Tirole (1985). This is simply incorrect. However (as Shiller [1981a] points out), the more formal tests of (4) based on time-series data for a single economy are also problematic, and I now turn to them.

III. Time-Series Tests of Variance Bounds

The assumption typically made to test the bound (4) using time-series data is that the relevant variables (namely, dividends and prices for the dividend discount model being discussed here) follow stationary and ergodic processes. If this is true, then the sample moments are consistent estimators for the moments of the unobservable distribu-

---

[11] The issue of overlapping forecast errors also arises in other contexts, e.g., spot and forward foreign exchange rates (see Hansen and Hodrick 1980).

tions used in the inequality, assuming a sufficiently long time series of realizations from those distributions.[12]

Shiller (1981b, 1981c) tests the bound (4) with Standard and Poor's and Dow Jones Industrial Average indexes of annual stock prices and dividends (1981b, pp. 434–35), using sample variances of price and p_t^* as estimators of unconditional population variances. He reports that the bound appears grossly violated but does not conduct formal significance tests. LeRoy and Porter (1981) also test (4) but derive it from Miller and Modigliani's (1961) model based on future earnings X_t and investments I_t:

$$p_t = \sum_{\tau=1}^{\infty} \frac{E\{(X_{t+\tau} - I_{t+\tau})/n_t|\Phi_t\}}{(1+r)^\tau}, \qquad (22)$$

where r is an assumed constant discount rate and n_t is the number of shares outstanding at t.[13] LeRoy and Porter conduct formal tests of the bound under the assumption of stationarity of their series. The point estimates imply violation for Standard and Poor's data, but sampling error is sufficiently high that the bound is not rejected at conventional significance levels (p. 557). Tests on individual stocks indicate rejection.

However, there are at least two important reasons to question whether the extra assumptions underlying these tests are valid empirically. First, as documented in Section IIIC below, the data used in many of the variance bounds tests are consistent with the assumption that prices follow a nonstationary random walk. If so, the unconditional variances in (4) do not exist, and the use of sample variances of p_t and p_t^* as estimators of population unconditional variances is invalid. Section IIIA shows that the apparent gross violations of the variance bound (4) reported in the current literature using sample variances of p_t and p_t^* are consistent with an incorrect assumption of stationarity of prices and dividends. However, it is valid to estimate conditional variances even if prices are nonstationary, and Section

---

[12] See Fuller (1976, p. 230). Just how long is "sufficient" in this context, even assuming stationarity and ergodicity, is investigated in detail in Kleidon (1983, chap. 5; 1986). See also Flavin (1983).

[13] They do not use exactly this model but use $n_{t+\tau}$ as the divisor, which implicitly assumes that the net benefits of future investments do not accrue to current shareholders. In private correspondence, Stephen LeRoy indicates that this adjustment makes little difference. Two other issues are of greater potential significance. First, LeRoy and Porter (1979, pp. 2, 3) adjust prices and earnings to account for earnings retention. Although this is feasible under certain conditions, their procedure uses an incorrect timing assumption that violates the dividend irrelevance proposition. Second, their results are based on incorrect data since in effect they create an artificial Standard and Poor's price series with a spurious seasonal at lag 4, as shown in their table 4 (1981, p. 570). For details, see Kleidon (1983, chap. 3).

VARIANCE BOUNDS TESTS

IIIB shows that Standard and Poor's data do not violate the conditional variance inequalities in (8).

LeRoy and Porter discuss the assumption of stationarity of earnings and prices in some detail and make adjustments for earnings retention. Shiller (1981c, p. 293) claims that "the resulting series appear stationary," but LeRoy and Porter report that, after their adjustments, there remains evidence of nonstationarity.[14] They continue (1981, p. 569): "We have decided to neglect such evidence and simply assume that the series are stationary. . . . We do not argue that this treatment is entirely adequate, nor do we in any way minimize the problem of nonstationarity; the dependence of our results on the assumption of stationarity is probably their single most severe limitation."

The second problem, that of dividend smoothing, has important implications for all research that attempts to infer the properties of an infinite stream of future dividends from some finite ex post set of dividends that are under some control of management. Empirical evidence suggests that management takes care to create a smooth short-run dividend series that may not reflect one for one the fortunes of the firm as determined primarily by its earnings and investment opportunities.[15] Ceteris paribus, the less variable the dividend stream, the more variable will be the price series that comprises the present value of future dividends. For example, a firm seeking to finance expansion internally may withhold all dividends over some finite period, with an implicit promise of some future (perhaps liquidating) dividend.[16]

If dividends are smoothed, the time series may be covariance nonstationary and violate the assumption of ergodicity necessary to allow estimation of valid cross-sectional variance bounds with time-series data. To illustrate, suppose that at t there exists a firm that has future cash flows per share D composed of earnings (paid out fully as dividends) at only one period, say T, and suppose for convenience that the discount rate $r = 0$.[17] This implies that the conditional distribution $\{p_t^* | \Phi_t\}$ is just the conditional distribution $\{D | \Phi_t\}$. Clearly the bound (4) holds at t assuming that the relevant variances are defined.

---

[14] Since LeRoy and Porter attempt to correct for nonstationarity, the results in this section based on original data apply to their work only to the extent that nonstationarity remains.

[15] See, e.g., Lintner (1956) and Fama and Babiak (1968). This does not deny that dividends may contain some information, as in the signaling hypotheses of Ross (1977) and Bhattacharya (1980).

[16] Note that Marsh and Merton's (1984a) definition of dividend smoothing does not deal with this case since it does not allow firms to pay zero dividends in any period when the price is positive (see their eq. [7], p. 13).

[17] Paul Pfleiderer suggested this example, say for the case of a firm drilling for oil.

However, the ex post time series p_t^*, calculated from the recursion (5) and based on the terminal payment, will be a constant with zero sample variance. The price series will show positive variance if information about the terminal payment becomes available through time so that the bound (4) will appear violated if estimated by sample variances.

The problem is more severe for inequalities that, unlike (4) or (8), are invalid if an assumption of ergodicity of dividends is violated. The variance inequality that has received most attention in the literature is (4), but others exist. Some, such as LeRoy and Porter's bound (1981, p. 560) on the coefficient of dispersion (i.e., the ratio of the standard deviation to the mean), are similar to (4) in that they rely on stationarity and ergodicity assumptions for testing, but not for the intrinsic validity of the bound. Others such as in Shiller (1981c) are based on the time series of prices and dividends, and so rely on some form of stationarity for their validity, even aside from issues of testing. Shiller's alternate bounds are given as (1981c, p. 296, eqq. I-2, I-3)

$$\sigma(\Delta p) \leq \frac{\sigma(d)}{\sqrt{2r}}, \qquad (23)$$

$$\sigma(\Delta p) \leq \frac{\sigma(\Delta d)}{\sqrt{2r^3/(1 + 2r)}}, \qquad (24)$$

where $\sigma(\cdot)$ is standard deviation, Δp and Δd are first differences of price and dividends, respectively, and r is the (assumed constant) one-period discount rate. The derivation of (23) assumes joint covariance stationarity of the time series p_t and d_t, while that of (24) assumes joint stationarity of Δp_t and Δd_t, with information variables contained in the information set (Shiller 1981c, pp. 295–97). Only (24) is consistent with nonstationary (random walk) prices and dividends, and only (24) is not violated by point estimates.[18] However, the assumptions underlying both (23) and (24) may be violated if dividends are smoothed.

The issue of dividend smoothing can have striking implications for some more recent tests that attempt to overcome criticisms of early variance bounds tests. For example, West (1984) derives and tests the inequality that the variance of changes through time in the present value of expected dividends will be greater when the information set comprises current and past dividends than when it comprises a larger set. Although he regards the necessary assumption that dividends

---

[18] See Shiller (1981c, p. 297). He continues: "Of course, we do not expect the data to violate all inequalities even if the model is wrong" and notes that, although this inequality is not violated for first differences of the data, the relevant bound is violated when the data are differenced using an interval of 10 years (i.e., $x_t - x_{t-10}$). Section IIIB discusses this claim in the context of comparable results based on conditional variances.

follow an autoregressive integrated moving average (ARIMA) process as "relatively mild" (p. 3), this can be violated if dividend smoothing implies changes in a future residual dividend that do not show up in the currently observed dividend series. In the extreme, if the finite and observed dividend series were constant, the use of only this stream to predict future dividends would imply constant future dividends, and so the present value would be constant through time and the innovation zero. In West's terminology, this would appear to be evidence that 100 percent of price changes could be attributed to speculative bubbles—in fact, the violation of the assumption of an ARIMA process for dividends simply means that the theoretical inequality is invalid.[19]

Although the issue of dividend smoothing is potentially very important in interpreting the results from any particular test, the remainder of this section assumes nonsmoothed dividends as in figures 2 and 5 to highlight the implications of nonstationarity, which is most crucial in the current context. First, Section IIIA discusses the nonstationary price model used in figure 2, derives consistent dividend and earnings models, and shows that the current gross violations of the bound (4) are not surprising if prices follow this process with parameters corresponding to Standard and Poor's price data. Section IIIB derives conditional variance inequalities that are valid for the nonstationary price process and demonstrates that Standard and Poor's series do not violate these bounds. Section IIIC completes the argument by showing empirically that the assumed process is consistent with Standard and Poor's data.

A. Nonstationary Prices and Tests of Unconditional Variances

Stationarity of stock prices is vital to the validity of much of the variance bounds literature. The cited tradition in finance for treating stock prices as nonstationary random walks goes back to at least 1934 when it was recognized "that stock prices resemble cumulations of purely random changes" (Working [1934]; cited in Roberts [1959, p. 2]). Annual accounting earnings also appear to be well described as

---

[19] Similar issues arise in recent attempts to account for nonconstant discount rates in (1). For example, Scott (1984) uses Hansen's (1982) generalized method of moments estimator and assumes in one specification that dividends are not mean-reverting to avoid criticisms concerning assumed stationarity of dividends. However, in this case he assumes (p. 8) that "the percentage change in dividends and stock prices as well as the price-dividend ratio ($\Delta D/D$, $\Delta P/P$, P/D) are stationary." Such an assumption may be violated if dividends are smoothed.

random walks.[20] However, most variance bounds tests assume stationarity of stock prices, dividends, or earnings, usually after deflation by some price index to account for inflation, and "detrending" to remove a perceived deterministic time trend. Nelson and Plosser (1982) compare these two approaches and cannot reject that stock prices (as well as several other macroeconomic variables) are "nonstationary stochastic processes with no tendency to return to a trend line" (p. 139).[21]

The simplest random walk model, (10) above with $\rho = 1.0$, implies a zero expected capital gain component in stock returns, which historically is not true given less than full payout of earnings as dividends. An alternate model,

$$p_t = \mu + p_{t-1} + \epsilon_t, \tag{25}$$

where ϵ_t is i.i.d. with mean zero and variance σ^2, implies an expected capital gain rate that varies inversely with the price level. The most plausible economic model in this context is a geometric random walk,

$$\ln p_t = \mu + \ln p_{t-1} + \epsilon_t, \tag{26}$$

or

$$p_t = p_{t-1} \exp(\mu + \epsilon_t). \tag{27}$$

If the capital gain rate is defined as $(p_t - p_{t-1})/p_{t-1}$, then the (conditional) expected capital gain rate (g) is constant and is given by

$$\begin{aligned} g &\equiv E\left\{\frac{p_t - p_{t-1}}{p_{t-1}}\bigg|p_{t-1}\right\} \\ &= \exp\left(\mu + \frac{\sigma^2}{2}\right) - 1.0, \end{aligned} \tag{28}$$

assuming lognormality of $\exp(\mu + \epsilon_t)$ in (27). Expected capital gain rates are calculated below using (28).

1. Consistent Price, Dividend, and Earnings Processes

The valuation models based on dividends (1) and earnings or net cash flows (22) preclude any necessary one-to-one relation between the time-series process for price and the time-series process for earnings or dividends. As discussed with respect to dividend smoothing, any

---

[20] For empirical studies on annual earnings as a random walk, see, e.g., Little (1962), Ball and Watts (1972), Albrecht, Lookabill, and McKeown (1977), and Watts and Leftwich (1977). For quarterly earnings see Foster (1978).
[21] They note the implications of nonstationarity for variance bounds tests (pp. 142, 143), as do Black (1980) and Copeland (1983). See also Kling (1982).

particular set of observations may be unrepresentative of the total expected dividend stream. In principle, the same phenomenon could occur in the earnings stream (net cash flow) approach since the pattern of net cash flows does not necessarily correspond to changes in the present value of expected future net cash flows through time. Further, we do not observe the requisite "economic earnings" but accounting earnings. One cannot infer that a rational price series must be generated by a particular stochastic process just because dividends or earnings follow the process in a finite set of observations, or vice versa. However, one can infer a (nonunique) process for dividends or earnings that is compatible with the observed price series and see if the process is confirmed in dividend/earnings data.

This section assumes that prices follow the geometric random walk (26), defines consistent dividend and earnings processes, and discusses the underlying economic models. We have seen from proposition 4 above that one dividend process consistent with the price process (26) is

$$\ln d_t = \mu + \ln d_{t-1} + \epsilon_t, \qquad (17)$$

where μ and ϵ_t are identical to those in (26), since

$$p_t = \left(\frac{1+g}{r-g}\right) d_t. \qquad (18)$$

Not surprisingly, one consistent earnings process also has constant expected growth and is analogous to the price (26) and dividend per share (17) processes. However, the earnings stream approach involves investment as well as earnings. To specify the expected growth rate in earnings and its relation to that in prices and dividends per share, two issues are important: Is investment financed internally (via retained earnings) or externally (via new capital issues), and how profitable are the investments?

PROPOSITION 6. If investment is a constant proportion δ of earnings each period, is financed internally, and earns the rate of return r, then an earnings per share process consistent with the price process (26) and the dividend per share process (17) is

$$\ln e_t = \mu + \ln e_{t-1} + \epsilon_t, \qquad (29)$$

and the relation between p_t and e_t is given by

$$p_t = \frac{1}{r} E\{e_{t+1} | e_t\}$$
$$= \frac{1+g}{r} e_t, \qquad (30)$$

where $e_t \equiv X_t/n_t$ and $1 + g \equiv \exp[\mu + (\sigma^2/2)]$.

Proof. Equation (29) follows from (26) if (30) holds, and the expected growth rate in earnings is g. But given the assumed investment process, $g = \delta r$ (cf. Copeland and Weston 1983, p. 485). Further, since all financing is internal, $d_t = (1 - \delta)e_t$, and substitution into (18) gives (30).[22] Q.E.D.

Note that, although consistent processes for price and dividends were derived in terms of an unspecified empirical growth rate g, the earnings and investment model defines this rate in terms of fundamental variables, $g = \delta r$.

2. Tests of Unconditional Variances: Simulation Results

I now demonstrate that gross violations of bounds based on unconditional variances can result if the procedures of, say, Shiller (1981b) are applied to a series that by construction is rationally set by Miller-Modigliani valuation models with constant discount rates but that is nonstationary. This section reports the results of Monte Carlo simulations of the nonstationary price and dividend processes given above. The parameter values $\mu = .0095$, $\sigma = .218$, and $p_0 = 40.0$ are set to correspond to estimates for Standard and Poor's (deflated) annual price series, 1926–79. A series of disturbances ϵ_t are generated by the IMSL subroutine GGNPM, and the dividend and corresponding rational price series are generated by (17) and (18). The p_t^* series is generated recursively by (5) with the terminal condition $p_T^* = p_T$.[23] The sample variances of the price series p_t and the p_t^* series are then calculated, and the variance bound (4) is deemed violated if for sample variances $\text{var}(p_t) > \text{var}(p_t^*)$. The procedure is carried out for two different but related price series. The first is the series constructed by (18), and the second "detrends" prices and dividends before calculating the corresponding p_t^* series following Shiller (1981b, p. 432).

Table 1 gives the percentage (across 100 replications) of violations of the variance bound (4) for the simulated price series (18) and its

---

[22] If external financing (from new securities) is raised for the investment, the growth rate in earnings will exceed g, the growth rate for prices and dividends per share (see Miller and Modigliani 1961, pp. 421–26). The assumption of normal returns on investment is less likely to be violated for the economy as a whole, as reflected in Standard and Poor's index, than for some particular "growth company." If investment earns abnormal returns, compare Miller and Modigliani (1961, p. 423, eq. 25).

[23] Marsh and Merton (1984a) show that, if the terminal value p_T^* is set equal to the average sample price, the bound (4) is always violated if prices follow (26). This result does not hold for the terminal condition imposed by Grossman and Shiller (1981) and examined here (although Marsh and Merton [1984b, p. 12, n. 4] state the contrary). Their analysis does not show whether the "gross violations" of the bound that are reported in the literature can be explained by nonstationarity of prices, which is examined in table 2 below.

TABLE 1

PERCENTAGE OF VIOLATIONS OF VARIANCE BOUNDS CALCULATED FROM: (i) SIMULATED RATIONAL (GEOMETRIC) RANDOM WALK SERIES, (ii) THE SAME SERIES WITH EXPONENTIAL "DETRENDING," AND (iii) THE MATCHING CONSTRUCTED "PERFECT-FORESIGHT" PRICE SERIES

	\multicolumn{6}{c	}{VIOLATIONS (%) OF VARIANCE BOUND: $\text{var}(p^*) \geq \text{var}(p)$}	\multicolumn{2}{c}{DURBIN-WATSON STATISTIC FOR RESIDUALS FROM DETRENDING IN (ii)}
	\multicolumn{3}{c	}{(i) Random Walk Series}	\multicolumn{3}{c	}{(ii) "Detrended" Random Walk Series}		
SAMPLE SIZE (T)	$r = .05$	$r = .065$	$r = .075$	$r = .05$	$r = .065$	$r = .075$	Mean	Standard Deviation
5	100	100	100	95	95	95	2.37	.54
10	100	100	100	97	99	99	1.44	.52
50	83	83	88	89	90	90	.38	.19
100	86	87	87	90	92	92	.20	.10
200	86	87	88	92	92	93	.10	.05
1,000	95	95	95	100	100	100	.02	.01
2,000	91	91	91	96	96	96	.01	.006
3,000	72	71	71	99	99	99	.007	.005

NOTE.—Based on 100 replications. The relevant processes are: (i) $\ln p_t = \mu_p + \ln p_{t-1} + \epsilon_t$, ϵ_t i.i.d. $N(0, \sigma^2)$, $\ln d_t = \mu_p + \ln d_{t-1} + \epsilon_t$, $p_t = \alpha d_t$, where $\alpha = (1+g)/(r-g)$, $(1+g) \equiv \exp[\mu_p + (\sigma^2/2)]$, and $r \equiv$ a constant discount rate. Parameter values are $\mu_p = 0.0095$, $\sigma = 0.218$, and $p(0) = 40.0$. (ii) The series are "detrended" (following Shiller [1981b, p. 432]): $\hat{d}_t = d_t/e^{bt-T}$, $\hat{p}_t = p_t/e^{bt-T}$, where T is the base year, and $\hat{b}$ is estimated as the coefficient on time in a regression of the log of price on a constant and time. (iii) The "perfect-foresight" series corresponding to cases i and ii are defined as $p_t^* = (p_{t+1}^* + D_{t+1})/(1+r)$, where $D_{t+1} \equiv$ the dividend from i or ii, and $p_T^* = p_T$, the terminal price in i or ii.

detrended counterpart. Results are shown for three different (real) discount rates r, namely 0.05, 0.065, and 0.075. Over 1926–81, Ibbotson and Sinquefield (1982, p. 15, exhibit 3) report an arithmetic mean nominal return per annum on the Center for Research in Security Prices (CRSP) file of common stocks of 0.114, with mean inflation of 0.031 per annum. Over the same period, the mean return for small stocks was 0.181, and Standard and Poor's index is composed of larger stocks. Shiller (1981b, p. 431, table 2) uses a discount rate of 0.048 per annum (in real terms) for detrended data or 0.063 per annum (p. 430) for nondetrended data. The rate 0.075 is given for comparison.

The most striking result of table 1 is the very high number of violations of the variance bound (4). The detrending procedure appears to exacerbate the tendency to reject the inequality, but the discount rates examined here do not appear to have much effect on the frequency of violation.[24] Table 1 also gives the mean and standard deviation (across replications) of the Durbin-Watson statistic from ordinary least squares (OLS) regression of prices on time, which is part of the detrending procedure. As noted below (n. 33), the average value for sample size 50 is almost identical to that obtained for the actual Standard and Poor's price series.

Although table 1 establishes that the variance bounds test procedures overwhelmingly result in violations of the bound (4) when applied to a nonstationary series generated by (1), it does not show whether gross violations are likely to occur. For 1,000 Monte Carlo replications for the sample size 100, which corresponds to Shiller (1981b, 1981c) and Grossman and Shiller (1981), table 2 gives both the number of violations of (4) (i.e., when the ratio of the sample standard deviation of price to the sample standard deviation of p_t^* exceeds 1.0) and the number of gross violations (when the ratio exceeds 5.0). Shiller (1981b, p. 341, table 5) reports a gross violation ratio of 5.59 for Standard and Poor's data. For rational, nonstationary series that are detrended, 397 replications out of 1,000 (or about 40 percent) give violation ratios greater than 5.0 using a discount rate r of 0.05, and 148 replications (almost 15 percent) for $r = 0.065$. Even for $r = 0.075$, almost 5 percent of the replications result in gross violations.

In short, the results of table 2 show that the gross violations of the bound (4) are not surprising if test procedures that assumed the existence of population unconditional variances were incorrectly applied

---

[24] As Joerding (1984) points out, the discount rate can in principle affect the frequency of violation. Further, table 2 shows that the discount rates examined here do affect the average amount by which the bound is violated.

TABLE 2

Summary Statistics of the Distribution of the Ratio of Sample Standard Deviation of Price to Sample Standard Deviation of "Perfect-Foresight" Price, across 1,000 Replications for Sample Size 100, for (i) Simulation Rational (Geometric) Random Walk Series and (ii) the Same Series with Exponential "Detrending"

	Number of Violations (Ratio > 1.0)	Number of "Gross Violations" (Ratio > 5.0)	Ratio Mean*	Ratio Standard Deviation	Minimum Ratio	Maximum Ratio	Percentile 50th (Median)	90th	95th
Case i, random walk:									
$r = .05$	855	307	3.56	2.70	.51	15.59	2.62	7.22	8.40
$r = .065$	865	62	2.65	1.50	.53	8.46	2.55	4.55	5.25
$r = .075$	869	21	2.33	1.15	.57	7.48	2.39	3.75	4.31
Case ii, "detrended" random walk:									
$r = .05$	894	397	4.29	3.04	.49	14.65	3.31	8.53	9.62
$r = .065$	915	148	3.30	1.66	.51	8.86	3.42	5.39	6.09
$r = .075$	925	46	2.90	1.24	.52	6.89	2.98	4.44	4.94

Note.—The relevant processes are defined in table 1.
* All summary statistics are for the distribution across 1,000 replications.

to nonstationary price data. Note that the parameter values used in these simulations are chosen as those estimated for Standard and Poor's price series, 1926–79, and that the nonstationary process (26) used here is consistent with the data. Note also that the simulations assume a dividend process with the same innovation variance as the price series, in (17) and (26). At least since 1950, Standard and Poor's dividend series is much smoother than the corresponding price or earnings series. Consequently, the simulation results are biased against finding gross violations relative to a dividend series with a lower sample innovation variance, and one would expect even greater rejection in actual empirical tests.

B. Tests of Inequalities Based on Conditional Variances

This subsection tests the conditional variance inequalities given by (8), which are valid for nonstationary price series. The results show that the conditional bounds are not violated by Standard and Poor's data and provide both confirmation and interpretation of tests of inequality (24) above based on differences of prices and dividends.

We test variances of p_t and p_t^* conditional on past prices p_{t-k}, for $k = 1, 2, 5,$ and 10. The assumed price process is the geometric random walk (26), which implies that

$$\text{var}\{p_{t+k}|p_t\} = \text{var}\{p_t \exp(\mu + \epsilon_{t+1})\exp(\mu + \epsilon_{t+2}) \ldots \exp(\mu + \epsilon_{t+k})|p_t\}$$

$$= p_t^2 \text{var}\left[\exp\left(k\mu + \sum_{n=1}^{k} \epsilon_{t+n}\right)\right]$$

$$\equiv p_t^2 c_k,$$

where c_k is constant through time given ϵ_t i.i.d. $N(0, \sigma^2)$.[25] Hence the conditional variances are constant through time except for scaling by p_t^2, and to avoid the resulting heteroscedasticity the inequality tested here is

$$\text{var}\left\{\frac{p_{t+k}}{p_t}\bigg|p_t\right\} \leq \text{var}\left\{\frac{p_{t+k}^*}{p_t}\bigg|p_t\right\}. \tag{31}$$

---

[25] As noted by Gary Chamberlain, more general distributional assumptions that allow conditional heteroscedasticity (i.e., nonconstant c_k) are consistent with the tests conducted here, although in that case the bounds are in terms of the expected value of the conditional variances.

VARIANCE BOUNDS TESTS

The population variances in (31) are constant through time for the price and dividend processes (26) and (17).[26] Note the equality of the conditional means,

$$E\left\{\frac{p_{t+k}}{p_t}\bigg|p_t\right\} = E\left\{\frac{p^*_{t+k}}{p_t}\bigg|p_t\right\}$$
$$= (1 + g)^k, \quad (32)$$

where $1 + g$ is the growth rate in prices as above. The conditional variances in (31) are estimated by the corresponding sample mean square deviations from the conditional means, using the sample estimated growth for the conditional means of p_{t+k} and p^*_{t+k} by (32).[27]

Table 3 reports the ratio of the conditional standard deviation of price to the conditional standard deviation of p^*_t for Standard and Poor's series, 1926–79, together with a sampling distribution for a sample size of 54, for discount rates of 0.05, 0.065, and 0.075. This distribution was constructed as in Section IIIA by simulation (here over 2,000 replications) of the price and dividend processes given by equations (26), (17), and (18), for parameter values corresponding to Standard and Poor's series.

There are two main results of interest in table 3. First, comparison of Standard and Poor's statistics with the corresponding sampling distribution shows that none of the inequalities given by (31) is violated at even a 10 percent significance level. Second, note that the point estimates do not violate (31) for $k = 1, 2$, or 5 but do violate for $k = 10$. It is significant that Shiller (1981c, p. 297) reports that the bound (24), which is consistent with nonstationary prices and dividends, is not violated when the data are differenced with a lag (k) of 1 but are violated when $k = 10$. Although he treats this as an important rejection, he presents no formal significance tests. The simulation results here show that violation of the bound (31) by point estimates for $k = 10$ is consistent with the valuation model (1).

Again, note that this sampling distribution is generated under the assumption of no dividend smoothing and, consequently, is conservative if dividends are smoothed. Even if smoothing is ignored, however, these tests show that Standard and Poor's price and dividend

---

[26] The use of a sample p^*_t that is constructed subject to a terminal condition such as $p_T = p^*_T$ implies that the conditional variances are equal at T, but the estimation in this section does not explicitly account for this time dependence. However, the sampling distribution constructed by Monte Carlo simulation implicitly accounts for this since the same procedures are carried out there as for Standard and Poor's data.

[27] The sensitivity of results to the use of sample growth rates was checked in the simulations by repeating the analysis using the true (known) growth rate, and the results were essentially unchanged.

TABLE 3

RATIO OF CONDITIONAL STANDARD DEVIATION OF PRICE TO CONDITIONAL STANDARD DEVIATION OF p^* BY DIFFERENT CONDITIONING LAGS k, FOR STANDARD AND POOR'S SERIES 1926–79, AND SUMMARY STATISTICS OF THE DISTRIBUTION OF THIS RATIO IN 2,000 REPLICATIONS OF SIMULATED GEOMETRIC RANDOM WALK WITH SAMPLE SIZE 54

			SIMULATION							
									Percentile	
LAG	STANDARD AND POOR'S RATIO	Number of Simulation Violations (Ratio > 1.0)	Ratio Mean*	Ratio Standard Deviation	Minimum Ratio	Maximum Ratio	50th (Median)	90th	95th
Rate $r = .05$:										
$k = 1$	.46	0	.38	.15	.01	.92	.38	.57	.64
$k = 2$	.64	31	.53	.20	.01	1.23	.53	.79	.87
$k = 5$	.83	435	.78	.30	.02	2.09	.76	1.16	1.29
$k = 10$	1.17	816	.98	.41	.04	4.24	.91	1.48	1.76
Rate $r = .065$:										
$k = 1$	.59	0	.41	.15	.02	.93	.41	.60	.66
$k = 2$	.81	32	.56	.20	.02	1.26	.57	.82	.90
$k = 5$	.96	495	.82	.29	.03	2.11	.80	1.19	1.30
$k = 10$	1.21	888	1.00	.38	.06	3.12	.95	1.49	1.73
Rate $r = .075$:										
$k = 1$	.58	0	.43	.15	.02	.94	.43	.62	.68
$k = 2$	.79	34	.58	.20	.02	1.26	.59	.84	.90
$k = 5$	.92	528	.83	.28	.04	2.04	.83	1.19	1.30
$k = 10$	1.13	924	1.02	.37	.07	2.76	.97	1.48	1.69

NOTE.—The conditional variances at lag k are: $\text{var}(p_{t+k}/p_t|p_t) = E[(p_{t+k} - E(p_{t+k}|p_t))/p_t]^2$, $\text{var}(p^*_{t+k}/p_t|p_t) = E[(p^*_{t+k} - E(p^*_{t+k}|p_t))/p_t]^2$, where $E(p_{t+k}|p_t) = E(p^*_{t+k}|p_t) = p_t(1 + g)^k$. The growth rate g is estimated as $g = \exp[\hat{\mu} + (\hat{\sigma}^2/2)]$, where μ and σ are mean and standard deviation of the series in $p_t - \ln p_{t-1}$. The relevant processes are defined in table 1. No detrending is done here.

* All summary statistics are for the distribution across 2,000 replications.

VARIANCE BOUNDS TESTS

series do not violate the well-defined conditional variance inequalities implied by the valuation model (1). I now complete the argument by showing the empirical validity of the nonstationary price process used to derive these tests.

C. Evidence on Nonstationarity of Prices, Earnings, and Dividends

This subsection applies the tests in Nelson and Plosser (1982) for nonstationarity of various macroeconomic variables to prices, earnings, and dividends used in variance bounds tests. First, autocorrelation functions for levels and first differences of random walks, and for residuals ("deviations from trend") from a regression of a random walk on time, are compared with the sample autocorrelations. Second, two different specifications of the autoregressive representation are tested directly for unit roots, as discussed in Fuller (1976) and Dickey and Fuller (1979).[28] The first specification is the simple autoregression (Dickey and Fuller 1979, p. 428, eq. 2.1)

$$Y_t = \mu + \rho Y_{t-1} + e_t, \tag{33}$$

where e_t is assumed i.i.d. $N(0, \sigma^2)$ and $\rho = 1.0$ under the null hypothesis. Fuller (1976, pp. 371, 373) tabulates empirical distributions for two test statistics for this model under the null hypotheses $\rho = 1.0$ and $\mu = 0.0$. The first statistic is $n(\hat{\rho}_\mu - 1)$, where $\hat{\rho}_\mu$ is the least-squares estimate of ρ in (33) and n is the sample size. The second test statistic, $\hat{\tau}_\mu$, is the "t-statistic" under the null hypothesis $\rho = 1$.[29] The second specification (Dickey and Fuller 1979, p. 428, eq. 2.2) includes time as a regressor:

$$Y_t = \mu + \beta t + \rho Y_{t-1} + e_t. \tag{34}$$

The statistics are similar to those for (33) and are denoted $n(\hat{\rho}_\tau - 1)$ and $\hat{\tau}_\tau$ for the model (34) (Fuller 1976, pp. 371, 373).

1. Stock Prices

The primary price data used here are Standard and Poor's annual composite stock price index for 1926–79 and quarterly composite

---

[28] Other procedures for testing for the existence of more than one unit root are discussed in Hasza and Fuller (1979) and applied in Meese and Singleton (1982). The hypothesis of multiple unit roots is rejected for the series examined here.

[29] For a given significance level the critical value of the statistic $\hat{\tau}_\mu$ is larger (in absolute value) than for the usual t-distribution since the sampling distribution of $\hat{\rho}_\mu$ is centered at values less than 1.0 in finite samples. Dickey and Fuller (1979, pp. 429–30) indicate that, if $\mu \neq 0.0$ in (33), the statistic $\hat{\tau}_\mu$ will imply acceptance of the hypothesis $\rho = 1$ with probability greater than the nominal level, although they do not indicate the amount of discrepancy.

stock price index for 1947:I to 1978:IV (1980, pp. 134–37). Diagnostic plots of both levels and first differences show that the raw (nominal) data reflect price level changes in later periods. Consequently the series are deflated by the gross national product (GNP) implicit price deflator, and diagnostic checks indicate that this procedure seems adequate. Unless otherwise stated, p_t here refers to deflated prices.[30]

Tables 4 and 5 give results for autocorrelation and Dickey and Fuller (1979) tests, respectively. Section A in table 4 gives results for sample autocorrelations for the levels of seven series. The first three series are taken from Nelson and Plosser (1982, table 2). Series 1 and 2 are constructed as a random walk and a time-aggregated random walk,[31] and the autocorrelations are those expected in a sample of size T (here, 100). The third series is the log of nominal stock price. Series 4 and 5 (6 and 7) are deflated price and log of deflated price for Standard and Poor's annual (quarterly) series cited above. Section B in table 4 gives corresponding autocorrelations for first differences of the series, while section C gives autocorrelations for the residuals from a regression of the series on time, following Nelson and Kang (1981).

The major result from table 4 is that the autocorrelation functions for the stock price data show marked similarity to those for the constructed random walks. Several other results are also apparent. First, there seems little difference in the autocorrelation functions of the deflated price series, P_t/GNP_t, and its logarithm, $\ln(P_t/GNP_t)$. Second, the sample size affects the degree to which the first-order sample autocorrelation coefficient r_1 is less than 1.0 in levels of the series. For the constructed random walk, r_1 is 1.0 with infinite observations but 0.95 for sample size 100. For Standard and Poor's annual data (series 4, 5) r_1 is approximately 0.90 ($T = 54$), while for quarterly data (series 6, 7) r_1 is 0.97 ($T = 128$). Third, although first differences of Nelson and Plosser's nominal series (table 4, series 10) indicate large first-order autocorrelation ($r_1 = 0.22$, standard error .10), which is consistent with time aggregation, the deflated annual and quarterly Standard and Poor's data do not show such high first-order autocorrelation.[32]

---

[30] The stock price series in Nelson and Plosser (1982) is not deflated, and their results are reported in table 4 for comparison.

[31] That is, the series is constructed by averaging sets of observations from a random walk with a smaller observation interval than the resulting series. Working (1960) demonstrates that, as the number of shorter interval observations averaged to produce the resultant time-aggregated series becomes large, the first-order serial correlation in the latter series approaches .25.

[32] For the annual data, r_1 is virtually zero, while the quarterly series shows $r_1 = 0.14$ with a standard error of .09. The autocorrelation in the nominal series may reflect price level changes rather than temporal aggregation.

VARIANCE BOUNDS TESTS

Section C in table 4 shows that not only do the stock price series match the constructed random walk data, but OLS regression of stock prices on time is very poorly specified. For series 19 (ln P_t/GNP_t, annual data), the Durbin-Watson statistic is only 0.38,[33] reflecting the very high autocorrelation in the residuals. Nelson and Kang (1981) show that this is to be expected if a random walk is inappropriately regressed on time, and the results are consistent with those of Nelson and Plosser (1982).

Table 5 gives the results of the Fuller (1976) and Dickey and Fuller (1979) tests for prices. In no case is the null hypothesis $\rho = 1$ rejected at the 10 percent level, and especially for the quarterly data the test statistics are well above the 10 percent critical value (rejection is indicated by small values of the statistics). Further, when time is included as a regressor (eq. [34] above), the null hypothesis $\beta = 0$ is not rejected at conventional significance levels.[34] Although the intercept $\hat{\mu}$ from (33) to (34) is not statistically far from zero in table 5, the implied economic magnitudes are very large, which is consistent with sample values of the slope coefficient $\hat{\rho}$ less than 1.0 if the true coefficient equals 1.0. For example, the estimated intercept for series 2 (annual data, ln P_t/GNP_t) is 0.44, which implies an expected capital gain rate of over 0.44 per annum (in real terms). If one imposes the null hypothesis $\rho = 1$ from the geometric random walk model (26), $\hat{\mu}$ is given by the sample mean of the first differences in log of price ($\nabla \ln p_t \equiv \ln p_t - \ln p_{t-1}$). For Standard and Poor's annual index, 1926–79, $\hat{\mu}$ is 0.0095, and the point estimate of σ^2 (the sample variance of $\nabla \ln p_t$) is 0.048. This implies using (28) a (real) expected capital gain rate of 0.033 per annum, which is reasonable.

In short, tables 4 and 5 show that the random walk models (25) and (26) cannot be rejected for Standard and Poor's price series.

2. Earnings and Dividends

We examine whether the nonstationarity of dividends and earnings implied by (17) and (29) is supported empirically. The earnings per share and dividend per share series are Standard and Poor's annual series corresponding to the composite stock price index, 1926–79. Note that the accounting earnings series is only a proxy for the eco-

---

[33] Table 1 above gives the average Durbin-Watson statistic across 100 replications, for different sample sizes, of the regression of a (geometric) random walk on time. For samples of size 50 (table 4, series 19, has 54 observations), the average Durbin-Watson statistic is 0.375.

[34] For table 5, series 6 (ln P_t/GNP_t, annual data), although the "t-statistic" is 2.26, this statistic does not have a true t-distribution for the sample size 54. For the corresponding quarterly series ($T = 128$), the "t-statistic" is -1.05.

TABLE 4

SAMPLE AUTOCORRELATIONS FOR LEVELS, FIRST DIFFERENCES, AND DEVIATIONS FROM TIME TREND FOR RANDOM WALKS AND STOCK PRICES
(Annual and Quarterly Data)

SERIES*	PERIOD	T	r_1†	r_2	r_3	r_4	r_5	r_6	S.E.‡	ADJUSTED R^2	DURBIN-WATSON
A. Sample autocorrelations: levels:											
1. Random walk		100	.95	.90	.85	.81	.76	.70	...	...	...
2. Time-aggregated random walk		100	.96	.91	.86	.82	.77	.73	...	...	...
3. Log of price (ln P_t): annual	1871–1970	100	.96	.90	.85	.79	.75	.71	.10	...	...
4. Deflated price (P_t/GNP$_t$): annual	1926–79	54	.91	.82	.81	.77	.71	.63	.13	...	...
5. Log of deflated price (ln P_t/GNP$_t$): annual	1926–79	54	.89	.79	.75	.71	.68	.64	.13	...	...
6. Deflated price (P_t/GNP$_t$): quarterly	1947:I–1978:IV	128	.97	.93	.90	.86	.83	.79	.09	...	...
7. Log of deflated price (ln P_t/GNP$_t$): quarterly	1947:I–1978:IV	128	.97	.94	.91	.87	.84	.81	.09	...	...

B. Sample autocorrelations: first differences:

		Large	.00	.00	.00	.00	.00	...	...	...
8. Random walk		Large								
9. Time-aggregated random walk		Large								
10. ln P_t: annual	1871–1970	100	.25	.00	.00	.00	.02	...	...	...
11. P_t/GNP_t: annual	1926–79	54	.22	−.13	−.08	−.18	−.23	.10	...	...
12. ln(P_t/GNP_t): annual	1926–79	54	−.03	−.31	.06	.15	.00	.13	...	...
13. P_t/GNP_t: quarterly	1947:I–1978:IV	128	.02	−.25	−.02	−.13	.01	.13	...	...
14. ln(P_t/GNP_t): quarterly			.14	−.09	.01	−.02	−.01	.09	...	...
quarterly	1947:I–1978:IV	128	.13	−.07	.02	.02	−.01	.09	...	...

C. Autocorrelations of residuals ("deviations from trend")—Model:
$Y_t = \beta_0 + \beta_1 t + \epsilon_t$:

15. Random walk		61	.85	.71	.58	.47	.36	.27	...	...
16. Random walk		101	.91	.82	.74	.66	.58	.51	...	...
17. ln P_t: annual	1871–1970	100	.90	.76	.64	.53	.46	.43	.10	...
18. P_t/GNP_t: annual	1926–79	54	.81	.63	.55	.46	.33	.19	.13	.45	.33
19. ln(P_t/GNP_t): annual	1926–79	54	.80	.57	.44	.33	.28	.22	.13	.46	.38
20. P_t/GNP_t: quarterly	1947:I–1978:IV	128	.95	.89	.83	.78	.72	.67	.09	.38	.07
21. ln(P_t/GNP_t): quarterly	1947:I–1978:IV	128	.95	.90	.85	.80	.75	.70	.09	.46	.05

* The source for series (1)–(3), (8)–(10), and (15)–(17) is Nelson and Plosser (1982) and references therein. The other series are Standard and Poor's annual (fourth quarter) and quarterly composite stock price indexes. GNP_t is the gross national product implicit price deflator.
† r_n is the nth-order sample autocorrelation coefficient.
‡ S.E. gives the approximate standard error of r for the sample size T under the null hypothesis of zero autocorrelation.

TABLE 5
TEST STATISTICS FOR SMALL SAMPLE TESTS FOR RANDOM WALKS IN STANDARD AND POOR'S ANNUAL PRICE SERIES (1926–79) AND QUARTERLY PRICE SERIES (1947:I–1978:IV)

SERIES	$\hat\mu$ (t-Statistic)*	$\hat\beta$ (t-Statistic)*	$\hat\rho$	$n(\hat\rho - 1)$†	t-Statistic‡ ($H_0: \hat\rho = 1$)	ADJUSTED R^2	Durbin-Watson	RESIDUALS r_1§	r_2	r_3	r_4
				Model A: $Y_t = \mu + \rho Y_{t-1} + e_t$							
Annual (4th quarter):											
1. Deflated price (P_t/GNP_t)‖	6.63 (1.62)	...	.90	−5.04	−1.65	.83	1.96	.01	−.25	.10	.18
2. Log of deflated price (ln P_t/GNP_t)	.44 (1.75)	...	.89	−5.57	−1.72	.80	1.84	.07	−.19	.03	−.08
Quarterly:											
3. P_t/GNP_t	2.54 (1.70)	...	.97	−3.77	−.16	.96	1.69	.15	−.08	.01	−.02
4. ln(P_t/GNP_t)	.12 (1.92)	...	.97	−3.56	−.19	.97	1.72	.13	−.07	.02	.02
				Model B: $Y_t = \mu + \beta t + \rho Y_{t-1} + e_t$							
Annual (4th quarter):											
5. P_t/GNP_t	3.42 (.81)	.28 (1.73)	.82	−9.22	−2.29	.84	1.93	.03	−.18	.20	.25
6. ln(P_t/GNP_t)	.74 (2.60)	.006 (2.26)	.77	−11.59	−2.76	.83	1.86	.07	−.13	.18	.04
Quarterly:											
7. P_t/GNP_t	2.75 (1.77)	−.02 (−.81)	.98	−2.45	−.82	.96	1.72	.14	−.10	−.00	−.04
8. ln(P_t/GNP_t)	.09 (1.13)	−.0003 (−1.05)	.98	−2.03	−.76	.97	1.77	.11	−.10	−.00	−.01

* Large sample t-statistics (in parentheses) under the null hypotheses that μ and β equal zero.
† The 10 percent critical values for this statistic under the null hypothesis $\hat\rho = 1$ are −10.7 (−11.0) for $n = 50$ (100) under model A (i.e., with no time regressor) and −16.8 (−17.5) under model B (which includes time as a regressor) (Fuller 1976, p. 371).
‡ The 10 percent critical values under the null hypothesis $\hat\rho = 1$ are −2.60 (−2.58) for $n = 50$ (100) under model A and −3.18 (−3.15) under model B (Fuller 1976, p. 373).
§ r_n is the nth-order autocorrelation coefficient. The approximate standard errors are .13 and .09 for annual and quarterly data, respectively, under the null hypothesis of zero correlation.
‖ GNP$_t$ is the gross national product implicit price deflator.

nomic earnings series X_t in the Miller-Modigliani valuation models. The deflation procedure and nonstationarity tests used for prices are applied to the earnings and dividend series.

Table 6 gives the results from the Fuller (1976) and Dickey and Fuller (1979) tests, and the autocorrelation tests give similar results. Section A tests directly for unit roots in the simple autoregression (33), and the null hypothesis $\hat{\rho} = 1$ is not rejected at even the 10 percent level for either earnings or dividends. Section B gives results for the autoregression (34), which includes time as an additional regressor. This model adds virtually no extra explanatory power over the simple autoregression (in terms of R^2), and for the dividend series (series 7 and 8, table 6) the null hypothesis $\hat{\rho} = 1$ is not rejected at the 10 percent level. However, the null hypothesis $\hat{\rho} = 1$ is rejected at the 5 percent level for both earnings series (5 and 6) when time is included.

The earnings results produce an interesting question in interpretation and are similar to results for a dividend series that Shiller (1981c) relies on to conclude that dividends are stationary. When looking just at the simple earnings autoregression without time, the random walk model fits well. When time is included, although there is virtually no increase in R^2, the coefficient on time appears significantly different from zero and the coefficient on lagged earnings seems significantly less than one. On balance, the simple autoregression seems preferable. First, it is consistent with the results of other studies of earnings per share, including those based on individual securities.[35] Second, it is consistent with the price process established above and economically seems more reasonable than (34).

The evidence relied on in Shiller (1981c, 1983) for stationarity of dividends (and consequently prices) is more tenuous. He considers a combination of the Standard and Poor's data used in table 6 with earlier Cowles Commission data, which together extend from 1871 to 1978. For this series, he reports (1981c, p. 299, n. 7) that the autoregression of $\log d_t$, including time as a regressor, gives a coefficient on $\log d_{t-1}$ of 0.807 and a standard error of .058, which has a probability value of .05 using Fuller's (1976) tabulations. On the basis of this result, he concludes (1983, p. 237) that "we can reject a random walk at the 5 percent level in favor of stationary fluctuations around a trend."

There are several problems with this interpretation. First, table 6

---

[35] See the references in n. 20 above. Note also the result in Watts and Leftwich (1977) that, although for particular samples some ARIMA models apparently fit better than the simple random walk (e.g., to accommodate the residual autocorrelation in table 6), there is little evidence that such models are better at prediction.

TABLE 6
TEST STATISTICS FOR SMALL SAMPLE TESTS FOR RANDOM WALKS IN STANDARD AND POOR'S ANNUAL EARNINGS AND DIVIDENDS SERIES (1926–79)

SERIES	$\hat{\mu}$ (t-Statistic)*	$\hat{\beta}$ (t-Statistic)*	$\hat{\rho}$	$n(\hat{\rho}-1)$†	t-STATISTIC‡ ($H_0: \hat{\rho}=1$)	ADJUSTED R^2	Durbin-Watson	r_1§	r_2	r_3	r_4
				Model A: $Y_t = \mu + \rho Y_{t-1} + e_t$							
1. Deflated earnings (E_t/GNP$_t$)‖	.24 (.84)	...	.97	−1.54	−.51	.85	1.85	.06	−.14	−.14	−.11
2. Log of deflated earnings (ln E_t/GNP$_t$)	.12 (1.30)	...	.93	−3.78	−1.16	.81	1.64	.17	−.09	−.24	−.09
3. Deflated dividends (D_t/GNP$_t$)	.27 (1.61)	...	.91	−4.85	−1.49	.81	1.60	.19	−.19	−.05	−.02
4. Log of deflated dividends (ln D_t/GNP$_t$)	.12 (1.80)	...	.89	−6.09	−1.71	.77	1.63	.18	−.20	−.06	.01
				Model B: $Y_t = \mu + \beta t + \rho Y_{t-1} + e_t$							
5. E_t/GNP$_t$	.45 (1.72)	.05 (3.86)	.61	−19.86	−3.64	.88	1.72	.06	−.11	−.19	−.16
6. ln(E_t/GNP$_t$)	.26 (2.81)	.01 (3.76)	.61	−20.04	−3.84	.86	1.58	.15	−.09	−.30	−.13
7. D_t/GNP$_t$	.35 (2.06)	.0009 (2.28)	.78	−11.25	−2.59	.82	1.59	.17	−.24	−.09	.01
8. ln(D_t/GNP$_t$)	.13 (2.03)	.0004 (2.33)	.75	−12.71	−2.78	.79	1.62	.16	−.25	−.11	.02

* Large sample t-statistics (in parentheses) under the null hypotheses that μ and β equal zero.
† The 10 percent (5 percent) critical values for this statistic under the null hypothesis $\hat{\rho}=1$ are −10.7 (−13.3) for $n=50$ under model A (i.e., with no time regressor) and −16.8 (−19.8) under model B (which includes time as a regressor) (Fuller 1976, p. 371).
‡ The 10 percent (5 percent) critical values under the null hypothesis $\hat{\rho}=1$ are −2.60 (−2.93) for $n=50$ under model A and −3.18 (−3.50) under model B (Fuller 1976, p. 373).
§ r_n is the nth-order autocorrelation coefficient. The approximate standard error is .13 under the null hypothesis of zero correlation.
‖ GNP$_t$ is the gross national product implicit price deflator.

VARIANCE BOUNDS TESTS 995

shows that, even for the autoregression including time, the dividend series since 1926 does not reject the random walk. Although it is true that a longer data set gives greater power in such tests, it is likely that the very early data are less reliable than Standard and Poor's series. Second, the results for the longer dividend series are not as clear-cut as Shiller implies. Replication of his results (over 1871–1979) gives a value for $n(\hat{\rho}_\tau - 1)$ of -21.61, which barely rejects at the .05 level, and a value for $\hat{\tau}_\tau$ of -3.41, which in Fuller's tabulations is not significant at the .05 level. Further, the longer data do not reject the random walk in either prices or dividends using the simple autoregression (33) (without time as an additional regressor) at even the .10 level for either test statistic, and the price data do not reject the hypothesis $\hat{\rho} = 1$ at the .10 level even when time is included.

3. Conclusion

In summary, the price data never reject nonstationarity, even for long time series, and although there are some cases in which nonstationarity in earnings or dividends appears rejected when time is included as a regressor, there is no rejection of nonstationarity of these series for the simple autoregression (33). Of course, even if the series were stationary, this does not indicate that the price series should be stationary because of the possible dividend (and accounting earnings) smoothing discussed above. In fact, time-series plots show that the dividend series since 1950 is much smoother than either the price or earnings series. This is consistent with the argument that earnings (and investments) are the fundamental variables and that a finite set of derived dividends may not be representative of the information used to set stock prices.[36] Even when smoothing is ignored, however, Sections IIIA and IIIB demonstrate that, once nonstationarity of prices is accounted for, valid variance bounds tests are not rejected in Standard and Poor's price and dividend data.

IV. Conclusions

This paper demonstrates that reliance on plots of price and p_t^* to determine whether changes in expectations of future cash flows cause

[36] This argument casts some doubt on the procedures of Granger (1975), who combines a dividend smoothing model from Fama and Babiak (1968) with a random walk in earnings, to generate predictions of future dividends for use with the dividend valuation model (1). What is not verified in his example is that the short-run properties of his smoothed dividend series are sufficient to derive the price process implied by his earnings model. If the smooth dividend stream is not representative of all future dividends, then the use in (1) of optimal forecasts based on the smooth process will not necessarily give the true rational price.

price changes is very misleading since by construction p_t^* will not correspond to p_t and will be much smoother than p_t if prices are set by (1) and the future is not known with certainty. Further, it is shown empirically that one cannot reject the hypothesis that prices are nonstationary and that the "gross violations" of the bound (4) that have been reported in the literature are consistent with incorrect application of estimation techniques that assume stationarity to nonstationary series. The conditional variance bounds (8) derived and tested here are valid if prices are nonstationary and are not violated for Standard and Poor's price and dividend series.

The implications of these results can best be seen with reference to the conclusions drawn in the literature from plots of price and p_t^* and the apparent violations of the inequality (4). Early conclusions were that stock prices cannot be reconciled with rational valuation models, as in Shiller (1981b, p. 422). Although Shiller (1981b, 1981c) recognizes that discount rates need not be constant, he argues that there is so little variation in the cash flow variables in such valuation models that discount rate movements must be very large if prices are rational. Moreover, he regards this possibility as at least counter to generally held views and states (1981a, p. 1) that "most people feel that stock price changes are due primarily to changing expectations about future dividends rather than changing rates of discount."

Attempts have been made to explain stock price movements in terms of nonconstant discount rates. The most influential work is that of Grossman and Shiller (1981),[37] whose primary claim, as noted by Shiller (1981a, p. 2), is that "most of the variability of stock prices might be attributed to information about consumption," which causes changes in discount rates. However, subsequent work has not been successful in extending their results. Hansen and Singleton (1983), for example, are able to explain only a small portion of the variability of stock prices in terms of nonconstant discount rates. Shiller (1981a) notes that, if price changes are driven by changes in expectations about aggregate consumption, then changes across assets should show a degree of contemporaneous correlation that is absent for the assets he examines. In general, even within the same industry and with very clean stock price data, there are wide cross-sectional differences in returns for any given period that seem difficult to reconcile purely in terms of changes in expectations of aggregate consumption.

Given the discouraging evidence on the ability of changes in expectations about consumption to explain changes in asset prices, Shiller

---

[37] See also LeRoy and LaCivita (1981), Shiller (1981a), Michener (1982), Hansen and Singleton (1983), Joerding (1983), Litzenberger and Ronn (1985), and Mehra and Prescott (1985).

VARIANCE BOUNDS TESTS

(1981a, p. 40) suggests that it might be possible to develop a "psychological model of asset prices" that preserves large discount rate movements, although he argues that it seems equally plausible that there are "temporary fads or speculative bubbles." He concludes: "If ... the reader goes back to a rational expectations model in which information about potential dividend movements, rather than discount rate movements, causes stock prices to move, then since actual aggregate dividend movements of such magnitude have never been observed, what is the source of information about such potential movements? Can we be satisfied with a model which attributes stock price movements and their business cycle correlation to public rational expectations about movements in a variable which has, in effect, never yet been observed to move?"

This paper demonstrates a plausible solution to the apparent puzzle: The assertion that price changes cannot be attributed to changes in expectations of future cash flows, based on plots such as figure 3 and the results of tests of the bound (4), has simply not been established. Recall that figure 2 displays similar characteristics to Standard and Poor's data in figure 3, yet by construction prices in figure 2 are set by the valuation model (1). Further, Kleidon (1983, chap. 6) shows that a large part of observed price changes can be associated with changes in expectations of future cash flows, using simple models and a few information variables.

Nevertheless, the question whether or not discount rates are constant as in (1) is a different issue. The variance bounds methodology may not be very powerful in detecting departures from constancy, as shown in Stock (1982). Further, even if the constant rate model performs relatively well empirically, there are still important theoretical questions about the conditions under which (1) will hold exactly. Although (1) does not require risk neutrality, the derivation of temporally constant expected rates of return for discounting expected cash flows—"risk-adjusted" discount rates—requires restrictive conditions in models of expected return that allow for risk aversion, such as the capital asset pricing model (CAPM) of Sharpe (1964) and Lintner (1965) or more general models.[38]

One implication is that the construct p_t^* will not in general be the

---

[38] Although LeRoy (1973) demonstrates that discount rates are not necessarily constant with risk aversion, he does not show the converse. See Fama (1970a) and Constantinides (1980) for sufficient conditions for a constant discount rate (across time for a given security) with risk aversion, in the context of the CAPM. Note also that financial economists typically do not reserve the term "expected present value" model for (1) with constant discount rates but include the use of nonconstant risk-adjusted rates. More general models of asset pricing include Merton (1973), Rubinstein (1976), Lucas (1978), Breeden (1979), Brock (1982), and Grossman and Shiller (1982).

price that would prevail if investors had perfect foresight, and so the term "perfect-foresight" price is unfortunate. If investors were risk neutral, the rate r used to discount the uncertain flows in (1) would be the same as that used to discount the certain flows in (2), but in general the appropriate expected rates of return will be different. However, the analysis in this paper does not depend on whether p_t^* is truly the price that would prevail under perfect foresight or whether the definition (2) just gives the present value of the ex post dividends discounted at the (possibly risk-adjusted) rate r from (1).

The major impact of the variance bounds literature has been to suggest that virtually no stock price changes are related to changes in expectations of future cash flows and further that prices may be irrational. This impact has been widespread; for example, Arrow (1983) discusses the volatility of securities markets as compatible with "irrational judgements about uncertainty" (p. 13) and states (p. 12) that "[a] very rigorous analysis for the bond and stock markets (Shiller, 1979, 1981[b]) has shown the incompatibility of observed behavior with rational expectations models, at least in a simple form." At least one published paper explicitly presumes excess volatility in stock prices. Pakes (1985, p. 395, n. 3) states: "Note that the presence of the error term, $\eta_{1,t}$, implies that there may be more variance in stock market evaluations than can be justified by the variance in earnings (which accords with the results of LeRoy and Porter [1981] and Shiller [1981b])."

The results of this paper suggest that such modifications to our theories are, at best, premature.

References

Ackley, Gardner. "Commodities and Capital: Prices and Quantities." *A.E.R.* 73 (March 1983): 1–16.

Albrecht, W. Steve; Lookabill, Larry L.; and McKeown, James C. "The Time-Series Properties of Annual Earnings." *J. Accounting Res.* 15 (Autumn 1977): 226–44.

Arrow, Kenneth J. "Behavior under Uncertainty and Its Implications for Policy." Technical Report no. 399. Stanford, Calif.: Stanford Univ., Center Res. Organizational Efficiency, February 1983.

Ball, Ray, and Watts, Ross. "Some Time Series Properties of Accounting Income." *J. Finance* 27 (June 1972): 663–81.

Bhattacharya, Sudipto. "Nondissipative Signaling Structures and Dividend Policy." *Q.J.E.* 95 (August 1980): 1–24.

Black, Fischer. "Notes on 'The Determinants of the Variability of Stock Market Prices.'" Manuscript. Cambridge: Massachusetts Inst. Tech., Sloan School Management, October 1980.

Breeden, Douglas T. "An Intertemporal Asset Pricing Model with Stochastic Consumption and Investment Opportunities." *J. Financial Econ.* 7 (September 1979): 265–96.

VARIANCE BOUNDS TESTS 999

Brock, William A. "Asset Prices in a Production Economy." In *The Economics of Information and Uncertainty*, edited by John J. McCall. Chicago: Univ. Chicago Press (for N.B.E.R.), 1982.

Constantinides, George M. "Admissible Uncertainty in the Intertemporal Asset Pricing Model." *J. Financial Econ.* 8 (March 1980): 71–86.

Copeland, Basil L., Jr. "Do Stock Prices Move Too Much to Be Justified by Subsequent Changes in Dividends? Comment." *A.E.R.* 73 (March 1983): 234–35.

Copeland, Thomas E., and Weston, J. Fred. *Financial Theory and Corporate Policy*. 2d ed. Reading, Mass.: Addison-Wesley, 1983.

Cowles, Alfred, et al. *Common Stock Indexes, 1871–1937*. Cowles Commission Monograph no. 3. Bloomington, Ind.: Principia, 1938.

Dickey, David A., and Fuller, Wayne A. "Distribution of the Estimators for Autoregressive Time Series with a Unit Root." *J. American Statis. Assoc.* 74, pt. 1 (June 1979): 427–31.

Fama, Eugene F. "Efficient Capital Markets: A Review of Theory and Empirical Work." *J. Finance* 25 (May 1970): 383–417. (*a*)

———. "Multiperiod Consumption-Investment Decisions." *A.E.R.* 60 (March 1970): 163–74. (*b*)

Fama, Eugene F., and Babiak, Harvey. "Dividend Policy: An Empirical Analysis." *J. American Statis. Assoc.* 63 (December 1968): 1132–61.

Flavin, Marjorie A. "Excess Volatility in the Financial Markets: A Reassessment of the Empirical Evidence." *J.P.E.* 91 (December 1983): 929–56.

Foster, George. *Financial Statement Analysis*. Englewood Cliffs, N.J.: Prentice-Hall, 1978.

Fuller, Wayne A. *Introduction to Statistical Time Series*. New York: Wiley, 1976.

Granger, Clive W. J. "Some Consequences of the Valuation Model When Expectations Are Taken to Be Optimum Forecasts." *J. Finance* 30 (March 1975): 135–45.

Grossman, Sanford J., and Shiller, Robert J. "The Determinants of the Variability of Stock Market Prices." *A.E.R. Papers and Proc.* 71 (May 1981): 222–27.

———. "Consumption Correlatedness and Risk Measurement in Economies with Non-traded Assets and Heterogeneous Information." *J. Financial Econ.* 10 (July 1982): 195–210.

Hansen, Lars Peter. "Large Sample Properties of Generalized Method of Moments Estimators." *Econometrica* 50 (July 1982): 1029–54.

Hansen, Lars Peter, and Hodrick, Robert J. "Forward Exchange Rates as Optimal Predictors of Future Spot Rates: An Econometric Analysis." *J.P.E.* 88 (October 1980): 829–53.

Hansen, Lars Peter, and Singleton, Kenneth J. "Stochastic Consumption, Risk Aversion, and the Temporal Behavior of Asset Returns." *J.P.E.* 91 (April 1983): 249–65.

Hasza, David P., and Fuller, Wayne A. "Estimation for Autoregressive Processes with Unit Roots." *Ann. Statis.* 7 (September 1979): 1106–20.

Ibbotson, Roger G., and Sinquefield, Rex A. *Stocks, Bonds, Bills, and Inflation: The Past and the Future*. Charlottesville, Va.: Financial Analysts Res. Found., 1982.

Joerding, Wayne. "Variable Risk Factors and Excess Volatility in the Stock Market." Manuscript. Pullman: Washington State Univ., Dept. Econ., October 1983.

———. "Stock Market Volatility and a Finite Time Horizon." Manuscript. Pullman: Washington State Univ., Dept. Econ., February 1984.

Keynes, John Maynard. *The General Theory of Employment, Interest and Money.* London: Macmillan, 1936.

Kleidon, Allan W. "Stock Prices as Rational Forecasters of Future Cash Flows." Ph.D. dissertation, Univ. Chicago, 1983.

———. "Bias in Small Sample Tests of Stock Price Rationality." *J. Bus.* 59 (April 1986): 237–61.

Kling, Arnold. "What Do Variance Bounds Tests Show?" Manuscript. Washington: Bd. Governors, Fed. Reserve Sys., May 1982.

LeRoy, Stephen F. "Risk Aversion and the Martingale Property of Stock Prices." *Internat. Econ. Rev.* 14 (June 1973): 436–46.

———. "Efficiency and the Variability of Asset Prices." *A.E.R. Papers and Proc.* 74 (May 1984): 183–87.

LeRoy, Stephen F., and LaCivita, C. J. "Risk Aversion and the Dispersion of Asset Prices." *J. Bus.* 54 (October 1981): 535–47.

LeRoy, Stephen F., and Porter, Richard D. "The Present-Value Relation: Technical Supplement." Manuscript. Santa Barbara: Univ. California, 1979.

———. "The Present-Value Relation: Tests Based on Implied Variance Bounds." *Econometrica* 49 (May 1981): 555–74.

Lintner, John. "Distribution of Incomes of Corporations among Dividends, Retained Earnings, and Taxes." *A.E.R. Papers and Proc.* 46 (May 1956): 97–113.

———. "The Valuation of Risk Assets and the Selection of Risky Investments in Stock Portfolios and Capital Budgets." *Rev. Econ. and Statis.* 47 (February 1965): 13–37.

Little, Ian M. D. "Higgledy Piggledy Growth." *Bull. Oxford Univ. Inst. Econ. and Statis.* 24 (November 1962): 387–412.

Litzenberger, Robert H., and Ronn, Ehud I. "A Utility-based Model of Common Stock Price Movements." Working Paper no. 791. Stanford, Calif.: Stanford Univ., Grad. School Bus., May 1985.

Lucas, Robert E., Jr. "Asset Prices in an Exchange Economy." *Econometrica* 46 (November 1978): 1429–45.

Marsh, Terry A., and Merton, Robert C. "Dividend Variability and Variance Bounds Tests for the Rationality of Stock Market Prices." Working Paper no. 1584-84. Cambridge: Massachusetts Inst. Tech., Sloan School Management, August 1984. (*a*)

———. "Earnings Variability and Variance Bounds Tests for the Rationality of Stock Market Prices." Working Paper no. 1559-84. Cambridge: Massachusetts Inst. Tech., Sloan School Management, March 1984. (*b*)

Meese, Richard A., and Singleton, Kenneth J. "On Unit Roots and the Empirical Modeling of Exchange Rates." *J. Finance* 37 (September 1982): 1029–35.

Mehra, Rajnish, and Prescott, Edward C. "The Equity Premium: A Puzzle." *J. Monetary Econ.* 15 (March 1985): 145–61.

Merton, Robert C. "An Intertemporal Capital Asset Pricing Model." *Econometrica* 41 (September 1973): 867–87.

Michener, Ronald W. "Variance Bounds in a Simple Model of Asset Pricing." *J.P.E.* 90 (February 1982): 166–75.

Miller, Merton H., and Modigliani, Franco. "Dividend Policy, Growth, and the Valuation of Shares." *J. Bus.* 34 (October 1961): 411–33.

Nelson, Charles R., and Kang, Heejoon. "Spurious Periodicity in Inappropriately Detrended Time Series." *Econometrica* 49 (May 1981): 741–51.

Nelson, Charles R., and Plosser, Charles I. "Trends and Random Walks in

Macroeconomic Time Series: Some Evidence and Implications." *J. Monetary Econ.* 10 (September 1982): 139–62.

Pakes, Ariel. "On Patents, R & D, and the Stock Market Rate of Return." *J.P.E.* 93 (April 1985): 390–409.

Roberts, Harry V. "Stock-Market 'Patterns' and Financial Analysis: Methodological Suggestions." *J. Finance* 14 (March 1959): 1–10.

Ross, Stephen A. "The Determination of Financial Structure: The Incentive-Signalling Approach." *Bell J. Econ.* 8 (Spring 1977): 23–40.

Rubinstein, Mark. "The Valuation of Uncertain Income Streams and the Pricing of Options." *Bell J. Econ.* 7 (Autumn 1976): 407–25.

Scott, Louis O. "The Present Value Model of Stock Prices: Empirical Tests Based on Instrumental Variables Estimators." Manuscript. Urbana: Univ. Illinois, October 1984.

Sharpe, William F. "Capital Asset Prices: A Theory of Market Equilibrium under Conditions of Risk." *J. Finance* 19 (September 1964): 425–42.

Shiller, Robert J. "Consumption, Asset Markets and Macroeconomic Fluctuations." Paper presented at the Money and Banking Workshop. Chicago: Univ. Chicago, September 1981. (*a*)

———. "Do Stock Prices Move Too Much to Be Justified by Subsequent Changes in Dividends?" *A.E.R.* 71 (June 1981): 421–36. (*b*)

———. "The Use of Volatility Measures in Assessing Market Efficiency." *J. Finance* 36 (May 1981): 291–304. (*c*)

———. "Do Stock Prices Move Too Much to Be Justified by Subsequent Changes in Dividends? Reply." *A.E.R.* 73 (March 1983): 236–37.

———. "Theories of Aggregate Stock Price Movements." *J. Portfolio Management* 10 (Winter 1984): 28–37.

Standard and Poor's. *Security Price Index Record*. New York: Standard and Poor's, 1980.

Stock, James H. "Tests of Market Efficiency When Consumers Are Risk Averse." Manuscript. Berkeley: Univ. California, Dept. Econ., November 1982.

Tirole, Jean. "Asset Bubbles and Overlapping Generations." *Econometrica* 53 (September 1985): 1071–1100.

Tobin, James. "A Mean-Variance Approach to Fundamental Valuation." *J. Portfolio Management* 11 (Fall 1984): 26–32.

Watts, Ross L., and Leftwich, Richard W. "The Time Series of Annual Accounting Earnings." *J. Accounting Res.* 15 (Autumn 1977): 253–71.

West, Kenneth D. "Speculative Bubbles and Stock Price Volatility." Manuscript. Princeton, N.J.: Princeton Univ., Dept. Econ., 1984.

Working, Holbrook. "A Random-Difference Series for Use in the Analysis of Time Series." *J. American Statis. Assoc.* 29 (March 1934): 11–24.

———. "Note on the Correlation of First Differences of Averages in a Random Chain." *Econometrica* 28 (October 1960): 916–18.

[6]

Cointegration and Tests of Present Value Models

John Y. Campbell
Princeton University

Robert J. Shiller
Yale University

Application of some advances in econometrics (in the theory of cointegrated vector autoregressive models) enables us to deal effectively with two problems in rational expectations present value models: nonstationarity of time series and incomplete data on information of market participants. With U.S. data, we find some relatively encouraging new results for the rational expectations theory of the term structure and some puzzling results for the present value model of stock prices.

Present value models are among the simplest dynamic stochastic models of economics. A present value model for two variables, y_t and Y_t, states that Y_t is a linear function of the present discounted value of expected future y_t:

$$Y_t = \theta(1 - \delta) \sum_{i=0}^{\infty} \delta^i E_t y_{t+i} + c, \qquad (1)$$

We are grateful to Don Andrews, Gregory Chow, Rob Engle, Dick Meese, Peter Phillips, Ken West, and an anonymous referee and to participants in seminars at the University of California, Berkeley, Columbia University, the Federal Reserve Bank of Philadelphia, the National Bureau of Economic Research, Princeton University, Rice University, and the University of Virginia for helpful comments on an earlier version of this paper. We are responsible for any remaining errors. We acknowledge support from the National Science Foundation.

COINTEGRATION

where c, the constant, θ, the coefficient of proportionality, and δ, the discount factor, are parameters that may be known a priori or may need to be estimated. Here and in what follows, E_t denotes mathematical expectation, conditional on the full public information set $\mathbf{I}_t$, which includes y_t and Y_t themselves and in general exceeds the information set $\mathbf{H}_t$ available to the econometrician. Models of this form include the expectations theory for interest rates (Y_t is the long-term yield and y_t the one-period rate), the present value model of stock prices (Y_t is the stock price and y_t the dividend), and, with some modification, the permanent income theory of consumption.[1]

Despite the simplicity of their structure, there is a surprising degree of controversy about the validity of present value models for bonds, stocks, and other economic variables.[2] The controversy seems to be stimulated by three problems that arise in testing equation (1). First, there are several test procedures in the literature: these include single-equation regression tests, tests of cross-equation restrictions on a vector autoregression (VAR), and variance bounds tests. It is not clear how these alternative approaches are related to one another.

Second, a statistical rejection of the model (1) may not have much economic significance. It is entirely possible that the model explains most of the variation in Y_t even if it is rejected at the 5 percent level. Most work on present value models concentrates on statistical testing rather than informal evaluation of the "fit" of the models.

Finally, the variables y_t and Y_t usually require some transformation before the theory of stationary stochastic processes can be applied. One approach is to remove a deterministic linear trend, but this can bias test procedures against the model (1) if in fact y_t and Y_t are nonstationary in levels.[3]

In this paper we develop a test of the present value relation that is valid when the variables are stationary in first differences.[4] Hansen and Sargent (1981a), Mankiw et al. (1985), and West (1986, 1987)

---

[1] The discounted sum in eq. (1) extends to an infinite horizon. Most of the methods in this paper can be applied to the finite horizon case, at the cost of some additional complexity. Throughout this paper we will treat conditional expectations as equivalent to linear projections on information.

[2] For bonds, see Sargent (1979), Shiller (1979, 1981a, 1987), Hansen and Sargent (1981a), Shiller, Campbell, and Schoenholtz (1983), and Campbell and Shiller (1984). For stocks, see LeRoy and Porter (1981), Shiller (1981b, 1984), Mankiw, Romer, and Shapiro (1985), Scott (1985), Marsh and Merton (1986), and West (1986, 1987).

[3] This point is made for stocks by Kleidon (1986) and Marsh and Merton (1986). Mankiw and Shapiro (1985) present a similar argument for the permanent income theory of consumption.

[4] It might be attractive to model the variables y and Y as stationary in log first differences. However, since the model (1) is linear in levels, a log specification is intractable unless one is willing to focus on a special case (Kleidon 1986) or to approximate the model (Campbell and Shiller 1986).

have also studied this case. We follow Hansen and Sargent and differ from Mankiw et al. and West by using a relatively large information set $\mathbf{H}_t$. We include in $\mathbf{H}_t$ current and lagged values not just of y_t but also of Y_t.

Our choice of information set has several advantages. By including Y_t in the vector stochastic process for analysis, we in effect include *all* relevant information of market participants, even if we econometricians do not observe all their information variables. We can test *all* the implications of the model for the bivariate (y_t, Y_t) process, giving a natural extension of Fama's (1970) notion of a "weak-form" test. We can exploit the recently developed theory of cointegrated processes (Phillips and Durlauf 1986; Phillips and Ouliaris 1986; Engle and Granger 1987; Stock 1987). Our test procedure can be interpreted as a single-equation regression or as a test of restrictions on a VAR. We propose a way to assess the economic significance of deviations from (1), comparing the forecast of the present value of future y_t embodied in Y_t with an unrestricted VAR forecast. Because the information set $\mathbf{H}_t$ includes Y_t, the two forecasts should be equal if the model is true.

We examine the present value models for bonds and stocks, while a companion piece by one of us (Campbell 1987) studies the permanent income theory of consumption. The paper is organized as follows. Section I discusses alternative tests of the present value relation when y_t and Y_t are stationary in first differences rather than levels. Section II is an introduction to the literature on cointegration, summarizing the results we use in testing the present value model. Section III applies the method to data on bonds and stocks. Section IV presents conclusions.

I. Alternative Tests of the Present Value Relation

One straightforward way to test the model (1) is to use it to restrict the behavior of the variable $\xi_t \equiv Y_t - (1/\delta)[Y_{t-1} - \theta(1 - \delta)y_{t-1}]$. Substitution from (1) shows that

$$\xi_t = Y_t - E_{t-1}Y_t + c\left(1 - \frac{1}{\delta}\right). \tag{2}$$

Apart from a constant, ξ_t is the true innovation at time t in Y_t (i.e., the innovation with respect to the full market information set $\mathbf{I}_t$). The model has the striking implication that this innovation is observable when only Y_t, Y_{t-1}, y_{t-1}, and the parameters c, θ, and δ are known.[5] In

---

[5] The variable ξ_t can also be written as a constant plus the true innovation in the expected present value of all future y_t. We note, however, that in general the model does *not* identify the true innovation in y_t itself.

COINTEGRATION 1065

the applications of this paper, ξ_t has the economic interpretation of an asset return. In the term structure it is the excess return on long bonds over short bills, while in the stock market it is the excess return on stocks over a constant mean, multiplied by the stock price.

Since the right-hand side of (2), adjusted for a constant, is orthogonal to all elements of the information set $\mathbf{I}_{t-1}$, one can test the present value relation by regressing ξ_t on variables in this set and testing that the coefficients are jointly zero. This approach is standard in the literature and seems attractively simple. However, there are some econometric pitfalls and issues of interpretation that need careful handling.

First, the regressors used to predict ξ_t must be stationary if conventional asymptotic distribution theory is to apply. Of course, there are many stationary elements of $\mathbf{I}_{t-1}$, but one may want to choose variables that summarize the joint history of y_t and Y_t. It is not clear how the stationarity requirement can be reconciled with this objective if y_t and Y_t are themselves nonstationary.

Second, while (1) implies (2), the reverse is not true. Equation (2) is consistent with a more general form of (1) that includes a "rational bubble," a random variable b_t satisfying $b_t = \delta E_t b_{t+1}$. Recently there has been considerable interest in testing (1) against the alternative that Y_t is influenced by a rational bubble (Blanchard and Watson 1982; Hamilton and Whiteman 1985; Quah 1986; West 1987).

Third, it is not clear what are the implications for Y_t of nonzero coefficients in a regression of ξ_t on information. Predictability of returns has consequences for asset price behavior, and one may want to calculate these explicitly.

Further insight into these issues can be gained by defining a new variable $S_t \equiv Y_t - \theta y_t$. We will refer to S_t as the "spread." In the case of the term structure, it is just the spread between long- and short-term interest rates; for stocks, it is the difference between the stock price and a multiple of dividends. The spread can also be written as a linear combination of the variables ΔY_t, Δy_t, and ξ_t: $S_t = [1/(1-\delta)]\Delta Y_t - \theta \Delta y_t - [\delta/(1-\delta)]\xi_t$.

The present value model (1) implies two alternative interpretations of the spread. Subtracting θy_t from both sides of equation (1) and rearranging, one obtains

$$S_t = E_t S_t^* + c, \tag{3}$$

where

$$S_t^* = \theta \sum_{i=1}^{\infty} \delta^i \Delta y_{t+i},$$

and

$$S_t = \left(\frac{\delta}{1-\delta}\right) E_t \Delta Y_{t+1} + c. \tag{4}$$

Equation (3) says that the spread is a constant plus the optimal forecast of S_t^*, a weighted average of future changes in y; equation (4) says that the spread is linear in the optimal forecast of the change in Y.

Equation (4) can be used in an alternative test of the present value model, in which one regresses ΔY_t on a constant, S_{t-1}, and other variables. The coefficient on S_{t-1} should be $(1 - \delta)/\delta$, and the coefficients on the other variables should be zero. This regression is just a linear transformation of the regression that has ξ_t as the dependent variable, and it yields the same test statistic.

Equations (3) and (4) help to resolve the issues raised above. If Δy_t is stationary, it follows from (3) that S_t is stationary; (4) then implies that ΔY_t is stationary. Thus one can use S_t and Δy_t, or S_t and ΔY_t, as stationary variables that summarize the bivariate history of y_t and Y_t in a regression test of the model. (The pair Δy_t and ΔY_t is also stationary, but by using these one would lose information on the relative levels of y_t and Y_t.) Our strategy is to work with S_t and Δy_t.

The effect of a "rational bubble" alternative is easily seen using (3) and (4). If a term b_t is added to the right-hand side of equation (1), satisfying $b_t = \delta E_t b_{t+1}$, it appears on the right-hand side of (3) but does not affect equations (2) and (4). The term b_t is explosive by construction, so it causes explosive behavior of S_t by (3), and this is passed through to ΔY_t by (4).[6]

One way to test for the importance of rational bubbles is therefore to test the stationarity of S_t and ΔY_t. This approach has been proposed by Diba and Grossman (1984), among others. As we noted above, S_t can be written as a linear combination of ΔY_t, Δy_t, and ξ_t. Therefore, independent of any model, if three of the variables S_t, ΔY_t, Δy_t, and ξ_t are stationary, the fourth must be also. This linear dependence needs to be taken into account in testing for stationarity.

Finally, (3) and (4) suggest a way to compute the implications for Y_t of predictable ξ_t. Consider estimating a VAR representation for Δy_t and S_t (with their means removed):

$$\begin{bmatrix} \Delta y_t \\ S_t \end{bmatrix} = \begin{bmatrix} a(L) & b(L) \\ c(L) & d(L) \end{bmatrix} \begin{bmatrix} \Delta y_{t-1} \\ S_{t-1} \end{bmatrix} + \begin{bmatrix} u_{1t} \\ u_{2t} \end{bmatrix}, \tag{5}$$

---

[6] Quah (1986) gives an example in which b_t satisfies $b_t = \delta E_t b_{t+1}$ but is stationary. However, this example violates the equivalence of conditional expectations and linear projections, which we assume here.

where the polynomials in the lag operator $a(L)$, $b(L)$, $c(L)$, and $d(L)$ are all of order p. This VAR can be used for multiperiod forecasting of Δy_t, and it includes the variable S_t, which, according to (3), is the optimal forecast of the present value of future Δy_t.

To simplify notation, (5) can be stacked into a first-order system

$$\begin{bmatrix} \Delta y_t \\ \cdot \\ \cdot \\ \cdot \\ \Delta y_{t-p+1} \\ S_t \\ \cdot \\ \cdot \\ \cdot \\ S_{t-p+1} \end{bmatrix} = \begin{bmatrix} a_1 \ldots a_p & b_1 \ldots b_p \\ 1 & \\ & \cdot \\ & \cdot \\ 1 & \\ c_1 \ldots c_p & d_1 \ldots d_p \\ & 1 \\ & \cdot \\ & \cdot \\ & 1 \end{bmatrix} \begin{bmatrix} \Delta y_{t-1} \\ \cdot \\ \cdot \\ \cdot \\ \Delta y_{t-p} \\ S_{t-1} \\ \cdot \\ \cdot \\ \cdot \\ S_{t-p} \end{bmatrix} + \begin{bmatrix} u_{1t} \\ 0 \\ \cdot \\ \cdot \\ 0 \\ u_{2t} \\ 0 \\ \cdot \\ \cdot \\ 0 \end{bmatrix}, \quad (6)$$

where blank elements are zero. This can be written more succinctly as $\mathbf{z}_t = \mathbf{A}\mathbf{z}_{t-1} + \mathbf{v}_t$. The matrix $\mathbf{A}$ is called the companion matrix of the VAR. For all i, $E(\mathbf{z}_{t+i}|\mathbf{H}_t) = \mathbf{A}^i \mathbf{z}_t$, where $\mathbf{H}_t$ is the limited information set containing current and lagged values of y_t and Y_t or, equivalently, of $\mathbf{z}_t$. As elsewhere in the paper, we are taking conditional expectations to be linear projections on information.

We can now discuss the implications of the present value relation for the VAR system. A rather weak implication is that S_t must linearly Granger-cause Δy_t unless S_t is itself an exact linear function of current and lagged Δy_t (which is a stochastic singularity we do not observe in the data; it would require, e.g., that the variance-covariance matrix of u_{1t} and u_{2t}, Ω, be singular).

The intuitive explanation for this result is that S_t is an optimal forecast of a weighted sum of future values of Δy_t, conditional on agents' full information set. Therefore, S_t will have incremental explanatory power for future Δy_t if agents have information useful for forecasting Δy_t beyond the history of that variable. If agents do not have such information, they form S_t as an exact linear function of current and lagged Δy_t.[7]

The full set of restrictions of the present value model is more demanding. We obtain these restrictions by projecting equation (3) onto the information set $\mathbf{H}_t$, noting that the left-hand side is unchanged because S_t is in $\mathbf{H}_t$ and rewriting as

$$\mathbf{g}'\mathbf{z}_t = \theta \sum_{i=1}^{\infty} \delta^i \mathbf{h}' \mathbf{A}^i \mathbf{z}_t,$$

---

[7] A formal proof is as follows. Suppose that S_t does not Granger-cause Δy_t. Then $E(\Delta y_{t+i}|\mathbf{H}_t) = E(\Delta y_{t+i}|\Delta y_t, \Delta y_{t-1}, \ldots)$ for all i, and from (3), $E(S_t|\mathbf{H}_t) = E(S_t|\Delta y_t, \Delta y_{t-1}, \ldots)$, an exact linear function of current and lagged Δy_t. But because S_t is itself in the information set $\mathbf{H}_t$, $S_t = E(S_t|\mathbf{H}_t)$.

where $\mathbf{g}'$ and $\mathbf{h}'$ are row vectors with $2p$ elements, all of which are zero except for the $p + 1$st element of $\mathbf{g}'$ and the first element of $\mathbf{h}'$, which are unity. If this expression is to hold for general $\mathbf{z}_t$ (i.e., for nonsingular $\mathbf{\Omega}$), it must be the case that

$$\mathbf{g}' = \theta \sum_{i=1}^{\infty} \delta^i \mathbf{h}' \mathbf{A}^i = \theta \mathbf{h}' \delta \mathbf{A} (\mathbf{I} - \delta \mathbf{A})^{-1}. \quad (7)$$

Here the second equality follows by evaluating the infinite sum, noting that it must converge because the variables Δy_t and S_t are stationary under the null.[8]

The restrictions of equation (7) appear to be highly nonlinear cross-equation restrictions of the type described by Hansen and Sargent (1981b) as the "hallmark" of rational expectations models. However, it turns out that (7) can be simplified so that (taking θ and δ as given) its restrictions are linear and easily interpreted. Postmultiplying both sides of (7) by $(\mathbf{I} - \delta \mathbf{A})$, one obtains

$$\mathbf{g}'(\mathbf{I} - \delta \mathbf{A}) = \theta \mathbf{h}' \delta \mathbf{A}. \quad (8)$$

From the structure of the matrix $\mathbf{A}$, the constraints imposed by (8) on individual coefficients are $c_i = -\theta a_i$, $i = 1, \ldots, p$; $d_1 = (1/\delta) - \theta b_1$; and $d_i = -\theta b_i$, $i = 2, \ldots, p$. By adding $\theta \Delta y_t$ to S_t, one can interpret these restrictions. They state that $\xi_t = S_t - (1/\delta) S_{t-1} + \theta \Delta y_t$ is unpredictable given lagged Δy_t and S_t, which is what equation (2) implies for the information set $\mathbf{H}_t$. In our empirical application, we obtain a Wald test statistic for equation (8) that is numerically identical to the Wald test statistic for a regression of ξ_t on lagged Δy_t and S_t.[9]

The major advantage of the VAR framework is that it can be used to generate alternative measures of the economic importance, not merely the statistical significance, of deviations from the present value relation. To see this more clearly, suppose that the present value model is false so that $E_t \xi_{t+i} \neq 0$ for $i \geq 1$. Then equations (3) and (4) no longer hold. We define the "theoretical spread," S'_t, as the optimal forecast, given the information set $\mathbf{H}_t$, of the present value of all future changes in y:

$$S'_t \equiv E(S^*_t | \mathbf{H}_t) = \theta \mathbf{h}' \delta \mathbf{A} (\mathbf{I} - \delta \mathbf{A})^{-1} \mathbf{z}_t. \quad (9)$$

---

[8] Under an explosive bubble alternative this infinite sum will not converge, and the matrix $(\mathbf{I} - \delta \mathbf{A})$ will be singular.

[9] However, this statistic is not numerically identical to the Wald statistic for a test of eq. (7), even though (7) and (8) are algebraically equivalent restrictions. Nonlinear transformations of restrictions can change the numerical values of Wald statistics and, as Gregory and Veall (1985) point out, can dramatically alter their power. We report Wald statistics for (8) in the tables that summarize our empirical results and Wald statistics for (7) in notes.

COINTEGRATION 1069

We then have, ignoring constant terms,

$$S_t - S'_t = \sum_{i=1}^{\infty} \delta^i E(\xi_{t+i}|\mathbf{H}_t) \qquad (10)$$

and

$$S_t - \left(\frac{\delta}{1-\delta}\right) E(\Delta Y_{t+1}|\mathbf{H}_t) = \left(\frac{1}{1-\delta}\right) E(\xi_{t+1}|\mathbf{H}_t). \qquad (11)$$

Equations (10) and (11) measure deviations from the model in two different ways. The metric of equation (11) is the difference between S_t and the optimal forecast, given the information set $\mathbf{H}_t$, of the one-period change in Y. Equation (11) shows that this difference is large if excess returns are predictable one period in advance.

The metric of equation (10) is the difference between S_t and the theoretical spread, which is large if the present value of all future excess returns is predictable. By this measure, a large deviation from the model requires not only that movements in ξ be predictable one period in advance but that they be predictable many periods in advance. Loosely speaking, predictable excess returns must be persistent as well as variable.[10]

We use the VAR framework not only to conduct statistical tests of the present value relation but also to evaluate its failures using the metric of equation (10). We display time-series plots of the spread S_t and the theoretical spread S'_t, the unrestricted VAR forecast of the present value of future changes in y. If the present value model is true, these variables should differ only because of sampling error. Large observed differences in the time-series movements of the two variables imply (subject to sampling error) economically important deviations from the model.

The VAR framework can also be used to test the present value model against more specific alternatives. Volatility tests, for example, are designed to test against the alternative that Y_t or some transformation of it "moves too much."

We present two different volatility tests. The first is just a test that the ratio $\text{var}(S_t)/\text{var}(S'_t)$ is unity. This ratio, together with its standard error, can be computed from the VAR system. Under the present value model, the ratio should be one but would be larger than one if the spread is too volatile relative to information about future y. A statistic that complements this is the correlation between S_t and S'_t

---

[10] The terminology of our earlier paper (Campbell and Shiller 1984) may be helpful in understanding (10) and (11). The right-hand side of (11) is proportional to what we called the one-period "holding premium," and the right-hand side of (10) is what we called the "rolling premium."

since if the variance ratio and correlation both equal one, then S_t must equal S'_t and the model is satisfied.[11]

We obtain a second volatility test, following West (1987), as follows. Let us define ξ'_t as θ times the innovation from $t - 1$ to t in the expected present value of Δy, conditional on the VAR information set:

$$\xi'_t \equiv \theta \sum_{i=0}^{\infty} \delta^i [E(\Delta y_{t+i}|\mathbf{H}_t) - E(\Delta y_{t+i}|\mathbf{H}_{t-1})]$$

$$= S'_t - \left(\frac{1}{\delta}\right) S'_{t-1} + \theta \Delta y_t. \qquad (12)$$

Under the present value model, $\xi'_t = \xi_t$ since $S'_t = S_t$. We construct the ratio $\mathrm{var}(\xi_t)/\mathrm{var}(\xi'_t)$, again with standard error.[12] The model implies that this ratio should be one, while the notion that stock prices are too volatile suggests that it will be greater than one. We call the first of our variance ratios the "levels variance ratio" and the second the "innovations variance ratio."

The fact that a linear combination S_t of y_t and Y_t is stationary in its level, even though y_t and Y_t are individually stationary only in first differences, turns out to be important for understanding present value models. In the language of time-series analysis, the vector $\mathbf{x}_t = (y_t \ Y_t)'$ is cointegrated. Cointegrated vectors have a number of important properties, which we now discuss.

II. Properties of Cointegrated Vectors

In this section we summarize the theory of cointegrated processes and show how it applies to present value models.

DEFINITION (Engle and Granger 1987). A vector $\mathbf{x}_t$ is said to be cointegrated of order (d, b), denoted $\mathbf{x}_t$ CI(d, b), if (i) all components of $\mathbf{x}_t$ are integrated of order d (stationary in dth differences) and (ii) there exists at least one vector $\boldsymbol{\alpha}$ ($\neq 0$) such that $\boldsymbol{\alpha}'\mathbf{x}_t$ is integrated of order $d - b$, $b > 0$.

When y_t is stationary in first differences, the vector $\mathbf{x}_t = (y_t \ Y_t)'$ is CI(1, 1) if the present value model holds. The CI(1, 1) case is the one

---

[11] We compute the levels variance ratio and correlation from the sample moments of S_t and S'_t. We report numerical standard errors that are conditional on the sample moments of $\mathbf{z}_t$ and take account of sampling error only in the coefficients of the estimated VAR.

[12] We use the estimated variance-covariance matrix of the VAR to compute the innovations variance ratio. The standard error takes account of sampling error in this matrix as well as in the VAR coefficients.

that has been studied almost exclusively in the theoretical literature, and the results that follow apply to it.

Cointegrated systems of order (1, 1) have two unusual properties. These concern the existence of well-behaved vector time-series representations for the cointegrated variables and the estimation of unknown elements of the vector α. Both properties turn out to be relevant for testing present value models.

The first important property of a cointegrated vector is that the vector moving average (VMA) representation of the first difference $\Delta \mathbf{x}_t$ is noninvertible. Equivalently, the spectral density matrix of $\Delta \mathbf{x}_t$ is singular at zero frequency. This singularity is what "holds together" the elements of $\mathbf{x}_t$ so that a linear combination is stationary.

More formally, write $\Delta \mathbf{x}_t = \mathbf{K}(L)\boldsymbol{\epsilon}_t = \mathbf{I}\boldsymbol{\epsilon}_t + \mathbf{K}_1\boldsymbol{\epsilon}_{t-1} + \ldots$. The matrix $\mathbf{M} = \mathbf{K}(1)\mathbf{K}(1)'$, where $\mathbf{K}(1) = \mathbf{I} + \mathbf{K}_1 + \mathbf{K}_2 + \ldots$, is the spectral density matrix of $\Delta \mathbf{x}_t$ at zero frequency. Now if the variance of $\boldsymbol{\alpha}'\mathbf{x}_t$ exists, it will be given by

$$\operatorname{var}(\boldsymbol{\alpha}'\mathbf{x}_t) = \sum_{i=0}^{\infty} \boldsymbol{\alpha}'\mathbf{C}_i \mathbf{V} \mathbf{C}_i' \boldsymbol{\alpha},$$

where $\mathbf{V}$ is the variance-covariance matrix of $\boldsymbol{\epsilon}_t$ and $\mathbf{C}_i = \mathbf{I} + \mathbf{K}_1 + \ldots + \mathbf{K}_i$. Ignoring the degenerate case in which $\mathbf{V}$ is singular, the summation above converges only if $\boldsymbol{\alpha}'\mathbf{C}_i$ converges to zero. But the limit of $\mathbf{C}_i$ as $i \to \infty$ is $\mathbf{K}(1)$, so for convergence we must have $\boldsymbol{\alpha}'\mathbf{K}(1) = 0$, which requires $\mathbf{K}(1)$, and hence $\mathbf{M}$, to be singular.

It follows from this that if an economic theory imposes cointegration on a set of nonstationary variables, simple first differencing of all the variables can lead to econometric problems. Noninvertibility of the VMA destroys the usual argument for using a finite VAR representation, that a finite VAR can approximate the true VMA arbitrarily well. Intuitively, the problem arises because a cointegrated system has fewer unit roots than variables, so first differencing all the variables amounts to overdifferencing the system.[13]

Fortunately, there is a simple solution to the difficulty, which is to include $\boldsymbol{\alpha}'\mathbf{x}_t$ in a VAR along with a subset of the elements of $\Delta \mathbf{x}_t$. An equation that relates the change in an element of $\mathbf{x}_t$ to its own lags and lags of $\boldsymbol{\alpha}'\mathbf{x}_t$ is called an error-correction model for that element of $\mathbf{x}_t$. The VAR proposed in the previous section to test present value models is an error-correction model for y_t, along with an equation describing the evolution of $\boldsymbol{\alpha}'\mathbf{x}_t$.

---

[13] Shiller (1981b) and Melino (1983) criticized Sargent (1979) on this ground (and on the ground that he failed to test the implications of the model for the relative levels of y_t and Y_t). Baillie, Lippens, and McMahon (1983) also overdifferenced their system. Hansen and Sargent (1981a) corrected the problems with Sargent's procedure.

The second major result from the theory of cointegration concerns the "cointegrating vector" α. In a present value model, α is unique up to a scalar normalization and is proportional to $(-\theta\ 1)'$. Stock (1987) and Phillips and Ouliaris (1986) prove that a variety of methods provide estimates that converge to the true parameter at a rate proportional to the sample size T (rather than $\sqrt{T}$ as in ordinary cases). The reason for this is that, asymptotically, all linear combinations of the elements of $\mathbf{x}_t$ other than $\alpha'\mathbf{x}_t$ have infinite variance.

The practical implication is that an unknown element of α may be estimated in a first-stage regression and then treated as known in second-stage procedures, whose asymptotic standard errors will still be correct. This is extremely useful in carrying out the VAR tests of the previous section. In the case of stock prices, for example, the present value model constrains $\theta = \delta/(1 - \delta)$, so one can estimate the discount factor from a preliminary regression and then treat it as known in testing the model.

Two types of preliminary regression have been proposed for estimating the unknown parameter θ. The first, called the cointegrating regression by Engle and Granger (1987), is just a regression of Y_t on y_t. The second is an "error-correction" regression of Δy_t or ΔY_t on lagged changes in and levels of y_t and Y_t. In the first case, one estimates θ as the coefficient on y_t, while in the second case one takes the ratio of the coefficient on lagged y_t to that on lagged Y_t.

One might argue that use of the error-correction regression is preferable because it accounts more fully for the short-run dynamics of Y_t and y_t. However, it has an important disadvantage. For any cointegrated vector with two elements, there are two possible error-correction regressions, one for Δy_t and one for ΔY_t. Cointegration alone does not rule out that, in one of these regressions, lagged Y_t and y_t have zero coefficients in the population, so that the coefficient ratio fails to identify the desired parameter.[14] Of course, under the present value model the error-correction equation for Δy_t has nonzero coefficients (because $\alpha'\mathbf{x}_t$ Granger-causes Δy_t), but this is not implied by all plausible alternatives. Accordingly, we rely primarily on the cointegrating regression to identify θ.

One may want to conduct a formal statistical test of the null hypothesis that $\mathbf{x}_t$ is not cointegrated. This turns out to pose some difficult statistical problems. If a candidate for the cointegrating vector α is available, the null hypothesis is that $\alpha'\mathbf{x}_t$ is nonstationary, and one can use a modified Dickey-Fuller (1981) test, regressing the change in $\alpha'\mathbf{x}_t$ on a constant and a single lagged level. The t-statistics

---

[14] Cointegration does rule out that the coefficients are zero in both error-correction regressions.

COINTEGRATION

and F-statistic are corrected for serial correlation in the equation residual as proposed by Phillips and Perron (1986) and Phillips (1987) and then compared with significance levels computed numerically by Dickey and Fuller. If the statistics are sufficiently high, the null hypothesis is rejected.

If the cointegrating vector is not known but must be estimated from a cointegrating regression, the Dickey-Fuller significance levels are no longer appropriate. Engle and Granger (1987) analyze a variety of tests that use the residual from the cointegrating regression, an estimate of $\alpha'x_t$. We report two of their test statistics, one based on the Dickey-Fuller regression and one that augments that regression with four lagged dependent variables. Engle and Granger provide significance levels for these tests, based on a Monte Carlo study.[15]

Phillips and Ouliaris (1986) propose an alternative test procedure for the null hypothesis of no cointegration. Their method involves computing the matrix $\mathbf{M}$, the spectral density matrix at zero frequency, nonparametrically. As discussed above, this matrix will be nonsingular under the null and singular under the alternative of cointegration. Unlike the Engle-Granger procedures, their test statistics have a distribution that is asymptotically free of nuisance parameters. They applied their methods to our data, and we note their results below.

III. Testing the Model in Bond and Stock Markets

In this section we apply the methods developed above to test present value models for bonds and stocks. The model for bonds, usually referred to as the "expectations theory of the term structure," is a special case of equation (1) in which the parameters θ and δ are known a priori (θ equals one, and δ is a parameter of linearization), while the constant c is a liquidity premium unrestricted by the model.[16]

We test the present value model for bonds on a monthly U.S. Treasury 20-year yield series, available from 1959 to 1983 from Salomon Brothers' *Analytical Record of Yields and Yield Spreads*. The short rate used is a 1-month Treasury bill rate, obtained from the *Treasury Bulletin* for dates prior to 1982 and from the *Wall Street Journal* thereafter.[17] These data were previously studied in Campbell and Shiller

---

[15] The Monte Carlo results are based on 10,000 replications of 100 observations of independent random walks, with four lagged residual changes included in the test.

[16] The linearization required to write the expectations theory in this form is explained in Shiller (1979) and Shiller et al. (1983).

[17] The *Treasury Bulletin* and *Wall Street Journal* data are consistent with one another at dates when they are both available.

(1984); Shiller et al. (1983) worked with very similar data. We present empirical results both for the full sample 1959:1–1983:10 and for a short sample ending in 1978:8, which is more likely to correspond to a single interest rate regime.[18]

The present value model for stocks is a special case of equation (1) in which θ is known to equal $\delta/(1 - \delta)$. The model restricts the constant c to be zero. The discount factor δ is not known a priori but can be inferred by estimating the cointegrating vector for stock prices and dividends; a consistent estimate is also provided by the sample mean return on stocks.[19]

One difficulty with this formulation for stocks is that Y_t and y_t are not measured contemporaneously. The term Y_t is a beginning-of-period stock price, and y_t is paid sometime within period t. Literal application of the methods outlined in Section I would require us to assume that y_t is known to the market at the start of period t; but, as pointed out by West (1987) and others, this might lead us to a spurious rejection of the model if in fact y_t is known only at the start of period $t + 1$. Intuitively, it is not hard to "predict" excess returns using ex post information. In order to avoid this problem, we modify the procedures of Section I by constructing a variable $SL_t \equiv Y_t - \theta y_{t-1}$. We use this variable in our tests and alter the cross-equation restrictions appropriately. The dependent variables in the VAR are now SL_t and Δy_{t-1}, both of which are in the information set at the start of time t but not at the start of time $t - 1$ under our conservative assumption about the market's information.[20] Since $SL_t = S_t + \theta \Delta y_t$, it is of course stationary if S_t and Δy_t are.

We tested the model for stocks using time-series data for real annual prices and dividends on a broad stock index from 1871 to 1986. The term Y_t is the Standard and Poor's composite stock price index for January, divided by the January producer price index scaled so that the 1967 producer price index equals 100. (Before 1900 an annual average producer price index was used.) The nominal dividend series is, starting in 1926, dividends per share adjusted to index, four-quarter total, for the Standard and Poor's composite index. The nominal dividend before 1926 was taken from Cowles (1939), who ex-

---

[18] For both samples, the parameter of linearization δ is set equal to $1/(1 + R)$, with R at 0.0587/12 (the mean 20-year bond rate in the short sample, expressed at a monthly rate). Our subsequent empirical results are conditional on a fixed value of δ.

[19] The sample mean return converges to the population mean only at rate $\sqrt{T}$ and therefore should not strictly be taken as known in second-stage procedures. However, we ignore this problem in our empirical work.

[20] Engle and Watson (1985) did some regressions similar to ours, using a similar data set on stock prices and dividends. They used the variable S_t rather than SL_t. Their results differ from ours in that they found no evidence of Granger causality from S_t to Δy_t, but they did not reject the present value model more strongly than we do.

COINTEGRATION

TABLE 1
Unit Root Tests (Test Statistic $Z t\alpha$)

Variable	With Trend	Without Trend
	A. In the Term Structure	
Sample 1959–78:		
y_t	−2.78	−1.72
Y_t	−2.76	−.46
Δy_t	−17.40 (1%)	−17.44 (1%)
ΔY_t	−15.30 (1%)	−15.32 (1%)
S_t	−3.15 (10%)	−3.08 (5%)
ξ_t	−15.22 (1%)	−15.25 (1%)
Sample 1959–83:		
y_t	−3.83 (2.5%)	−2.32
Y_t	−2.51	−.50
Δy_t	−17.05 (1%)	−17.08 (1%)
ΔY_t	−15.27 (1%)	−15.29 (1%)
S_t	−4.77 (1%)	−4.67 (1%)
ξ_t	−15.18 (1%)	−15.19 (1%)
	B. In the Stock Market	
y_t	−2.88	−1.28
Y_t	−2.19	−1.53
Δy_t	−8.40 (1%)	−8.44 (1%)
ΔY_t	−9.91 (1%)	−9.96 (1%)
$\theta = 31.092$:		
SL_t	−4.35 (1%)	−4.31 (1%)
ξ_t	−9.93 (1%)	−9.99 (1%)
$\theta = 12.195$:		
SL_t	−2.68	−2.15
ξ_t	−9.76 (1%)	−9.69 (1%)

Note.—Test statistics for a variable X_t are based on the t-statistics on α in the regression $\Delta X_t = \mu + \beta t + \alpha X_{t-1}$ (with trend) or the regression $\Delta X_t = \mu + \alpha X_{t-1}$ (without trend). The t-statistic is corrected for serial correlation in the equation residual in the manner proposed by Phillips and Perron (1986) and Phillips (1987). Significance levels are: with trend: 10%, −3.12; 5%, −3.41; 2.5%, −3.66; 1%, −3.96; without trend: 10%, −2.57; 5%, −2.86; 2.5%, −3.12; 1%, −3.43.

tended the Standard and Poor's series back in time.[21] Finally, y_t is the nominal dividend series, divided by the annual average producer price index scaled so that the 1967 producer price index equals 100.

As shown in table 1, parts A and B, we ran unit root tests on our raw data and the various linear combinations discussed in Section I. This is an important preliminary because our approach is appropriate only if y_t is integrated of order one. We present test statistics that are based on the t-statistic on the lagged level in a Dickey-Fuller regres-

---

[21] The dividend data differ slightly from those used in Shiller (1981b), Mankiw et al. (1985), West (1987), and others. It has recently come to our attention that the second (1939) edition of Cowles's book contains some corrections to the dividend series presented in the original 1938 edition, and these corrections have been incorporated here.

sion, corrected for fourth-order serial correlation as proposed by Phillips and Perron (1986) and Phillips (1987).[22] We ran the Dickey-Fuller regression with and without a time trend; the former is appropriate when the alternative hypothesis is that the series is stationary around a trend, the latter when the alternative is that the series is stationary around a fixed mean.

The results in part A of table 1 are generally supportive of the view that short- and long-term interest rates are cointegrated, with the cointegrating vector equal to (-1 1) as implied by the expectations theory. Over the short sample 1959–78, one cannot reject the hypothesis that short and long rates have a unit root at even the 10 percent level; however, there is strong evidence that *changes* in interest rates are stationary. The hypothesis that the long-short spread has a unit root is rejected at the 10 percent level when a trend is estimated and at the 5 percent level when the trend is excluded from the regression. Finally, the excess return ξ_t also appears stationary; this, together with the results for Δy_t and ΔY_t, is indirect evidence for stationarity of the spread because of the linear dependence discussed in Section I.

Results are fairly similar over the full sample 1959–83. There is even stronger evidence that the spread is stationary, and the unit root hypothesis for short rates can be rejected unless a trend in interest rates is ruled out on a priori grounds.[23]

In part B of the table, we repeated these tests for the stock market data. Once again y_t and Y_t appear to be integrated of order one. In the stock market, the parameter θ is not determined by the present value model as it is in the term structure. Therefore, we must compute SL_t and ξ_t using estimates of θ obtained from the data. Strictly speaking, this invalidates the Phillips-Perron tests for SL_t and ξ_t, but we report the statistics as data description.

Table 2 gives details of alternative estimation procedures for θ. The cointegrating regression estimates θ at 31.092; the corresponding real discount rate (the reciprocal of θ) is 3.2 percent, which is lower than the average dividend-price ratio and considerably lower than the sample mean return of 8.2 percent.[24] The error-correction regression

---

[22] The results are qualitatively unchanged by looking at other statistics from the Dickey-Fuller regression or by varying the order of the serial correlation correction between one and 10.

[23] The results in table 1, pt. A, are more favorable to the hypothesis of cointegration between long and short rates than are the results reported by Phillips and Ouliaris (1986). They reject the null hypothesis of no cointegration at only the 15 percent level (their table 6). However, their procedure does not impose the cointegrating vector a priori, and this may involve a loss of power.

[24] The estimate of θ that corresponds to the sample mean return is 12.195. The higher estimate in the cointegrating regression is associated with a negative constant

COINTEGRATION

TABLE 2
ESTIMATION OF THE COINTEGRATING VECTOR AND TEST FOR COINTEGRATION IN THE STOCK MARKET

	R^2	Estimate of θ	Implied Discount Rate
1. $Y_t = -12.979 + 31.092y_t$	.842	31.092	3.2%
2. $\Delta y_t = .101 + .165\Delta y_{t-1} + .010\Delta Y_t$ $\quad -.157y_{t-1} + .004Y_t$	.373	37.021	2.7%
3. Sample mean return = 8.2%	..	12.195	8.2%

Tests of no cointegration: Engle and Granger (1987) ξ_2 statistic for eq. (1) residual, 3.58; significance levels: 10%, 3.03; 5%, 3.37; 1%, 4.07. Engle and Granger (1987) ξ_3 statistic for eq. (1) residual, 2.64., significance levels: 10%, 2.84; 5%, 3.17; 1%, 3.77.

delivers a fairly similar estimate of θ, 37.021 with an implied real discount rate of 2.7 percent. We proceed to construct SL_t using discount rates of 8.2 percent and 3.2 percent as a check on the robustness of our methods.

Engle and Granger's tests for no cointegration, based on the residual from the cointegrating regression, give mixed results: the ξ_2 statistic rejects at the 5 percent level, while the ξ_3 statistic narrowly fails to reject at the 10 percent level. The Phillips-Perron tests in part B of table 1 are also mixed. Both SL_t and ξ_t appear to be stationary when the 3.2 percent discount rate is used, but at an 8.2 percent discount rate the tests fail to reject the unit root null for SL_t even though they reject for Δy_t, ΔY_t, and ξ_t. There seems to be some evidence for cointegration between stock prices and dividends, but it is weaker than the evidence for cointegration in the term structure.[25]

The results in table 1 do not suggest that a "rational bubble" is present in the term structure or the stock market since a bubble would cause both ΔY_t and S_t to be nonstationary. Accordingly, we interpret the test statistics below in terms of predictable excess returns.

In table 3, part A, we report summary statistics for a VAR test of the expectations theory of the term structure. The VAR includes Δy_t

---

term; under the present value model, the constant should be proportional to the unconditional mean change in dividends, so it should be positive rather than negative. An estimated discount rate lower than the mean dividend-price ratio is consistent with the model only if dividends are expected to decline through time, the historical rise being due to sampling error.

[25] Phillips and Ouliaris (1986) did not reject the null hypothesis of no cointegration between stock prices and dividends at even the 25 percent level (their table 6). Campbell and Shiller (1986) report unit root tests for log dividends, log prices, and the log dividend-price ratio. There is some evidence for trend stationarity of log dividends, no evidence against the unit root null for log prices, and strong evidence for stationarity of the dividend-price ratio.

TABLE 3
Tests of Present Value Model

A. In the Term Structure

Sample 1959–78:
 Akaike criterion selects 11-lag VAR
 Δy equation $R^2 = .216$; S Granger-causes Δy at 0.01% level
 S equation $R^2 = .877$; Δy Granger-causes S at 0.3% level
 Test of present value model: $\chi^2(22) = 83.02$; P-value $< 0.005\%$
 Summary statistics:

$E(\Delta y)$	=	.016	$\sigma(S)$	=	1.060
$E(S)$	=	1.144	var(S)/var(S')	=	.987 (.360)
$E(S')$	=	.016	corr(S, S')	=	.978 (.011)
$\sigma(\Delta y)$	=	.442	var(ξ)/var(ξ')	=	1.160 (1.146)

Sample 1959–83:
 Akaike criterion selects six-lag VAR
 Δy equation $R^2 = .171$; S Granger-causes Δy at 0.3% level
 S equation $R^2 = .772$; Δy Granger-causes S at 1.3% level
 Test of present value model: $\chi^2(12) = 35.63$; P-value $= 0.03\%$
 Summary statistics:

$E(\Delta y)$	=	.021	$\sigma(S)$	=	1.320
$E(S)$	=	1.138	var(S)/var(S')	=	3.394 (3.948)
$E(S')$	=	.021	corr(S, S')	=	.956 (.098)
$\sigma(\Delta y)$	=	.793	var(ξ)/var(ξ')	=	.502 (.506)

B. In the Stock Market

Sample 1871–1986:
 $\theta = 12.195$ (8.2% discount rate): Akaike criterion selects four-lag VAR
 Δy equation $R^2 = .400$; SL Granger-causes Δy at $< 0.001\%$ level
 SL equation $R^2 = .837$; Δy Granger-causes SL at 63.3% level
 Test of present value model with mean restriction: $\chi^2(9) = 15.74$; P-value $= 7.2\%$
 Test of present value model without mean restriction: $\chi^2(8) = 15.72$; P-value $= 4.7\%$
 Summary statistics:

$E(\Delta y)$	=	.017	$\sigma(SL)$	=	15.51
$E(SL)$	=	16.07	var(SL)/var(SL')	=	67.22 (86.04)
$E(SL')$	=	2.563	corr(SL, SL')	=	$-.459$ (.801)
$\sigma(\Delta y)$	=	.168	var(ξ)/var(ξ')	=	11.27 (4.49)

 $\theta = 31.092$ (3.2% discount rate): Akaike criterion selects two-lag VAR
 Δy equation $R^2 = .378$; SL Granger-causes Δy at $< 0.001\%$ level
 SL equation $R^2 = .516$; Δy Granger-causes SL at 1.8% level
 Test of present value model with mean restriction: $\chi^2(5) = 14.90$; P-value $= 1.1\%$
 Test of present value model without mean restriction: $\chi^2(4) = 5.75$; P-value $= 21.8\%$
 Summary statistics:

$E(\Delta y)$	=	.017	$\sigma(SL)$	=	9.937
$E(SL)$	=	-12.52	var(SL)/var(SL')	=	4.786 (5.380)
$E(SL')$	=	16.66	corr(SL, SL')	=	.911 (.207)
$\sigma(\Delta y)$	=	.167	var(ξ)/var(ξ')	=	1.414 (.441)

COINTEGRATION 1079

and S_t as variables, and the number of lags is chosen by the Akaike information criterion (AIC).[26] White's (1984) heteroscedasticity-consistent covariance matrix estimator is used in constructing standard errors and test statistics. The VARs are estimated for the short sample 1959–78 and the full sample 1959–83; they have 11 and six lags, respectively.

In both sample periods the lagged variables have a fair degree of explanatory power for the change in short rates. The R^2 for the Δy_t equation is 21.6 percent in the short sample and 17.1 percent in the full sample. This argues against the view of Mankiw and Miron (1986) that short-rate changes are essentially unpredictable in the postwar period in the United States. Furthermore, there is strong evidence that spreads Granger-cause short-rate changes, as they should do if the expectations theory is true. The hypothesis of no Granger causality can be rejected at the 0.01 percent level for the short sample and the 0.3 percent level for the full sample.

A formal test of the expectations theory restrictions in equation (8) rejects very strongly. The null that excess returns on long bonds are unpredictable can be rejected at less than the 0.005 percent level in the short sample and at the 0.03 percent level in the full sample. The R^2 values for excess returns are 26.3 percent and 16.7 percent, respectively.[27] In the corresponding regression (4), which has the change in the long rate as its dependent variable, the coefficient on the spread has the wrong sign (-0.020 in the short sample and -0.039 in the full sample).[28]

Despite these negative results, the summary statistics in table 3, part A, suggest that there is an important element of truth to the expectations theory of the term structure. The spread does seem to move very closely with the theoretical spread, the unrestricted forecast of the present value of future short-rate changes. In both sample periods the variance of the spread is insignificantly different from the variance of the theoretical spread (i.e., our "levels variance ratio" does not reject), and the two variables have similar innovation variances and an extremely high correlation. In the 1959–78 period the correlation between the actual and theoretical spreads is 0.978 with a standard error of 0.011, while in the 1959–83 period it is 0.956 with a

---

[26] That is, we pick the number of lags to minimize ($-\ln$ likelihood + number of parameters) in the VAR. Sawa (1978) has argued that the AIC tends to choose models of higher order than the true model but states that the bias is negligible when $p < T/10$, as it is here. The test statistics in table 3 are not highly sensitive to small changes in the lag length of the VAR system.

[27] Nonlinear Wald tests of eq. (7) reject at significance levels of less than 0.005 percent in the short sample and 8.4 percent in the full sample.

[28] This is consistent with the results of Shiller et al. (1983).

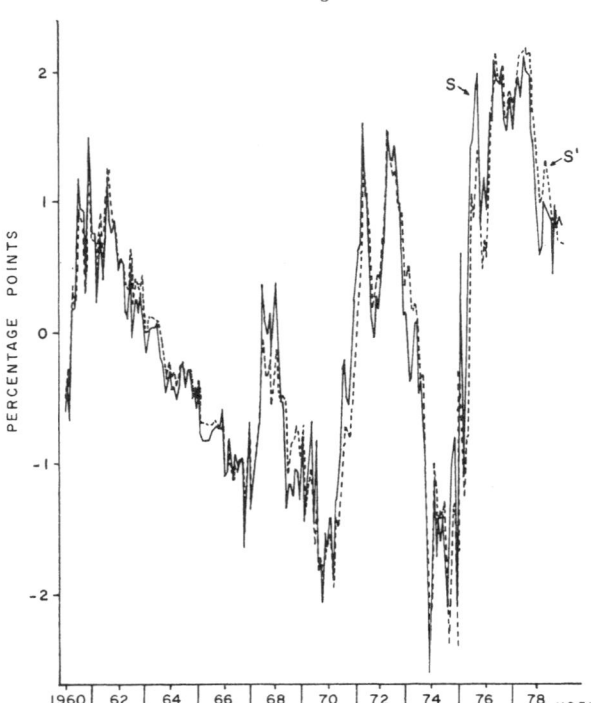

Fig. 1.—Term structure: deviations from means of long-short spread S_t and theoretical spread S'_t.

standard error of 0.098. Figure 1 illustrates the comovement of S_t and S'_t in the short sample.[29]

What this suggests is that tests of predictability of returns are highly sensitive to deviations from the expectations theory—so sensitive, in fact, that they may obscure some of the merits of the theory. An example illustrates the point. Suppose long and short rates differ from the expectations theory in the following manner: $S_t = S'_t + w_t$,

---

[29] The high correlation of these variables in postwar U.S. data might also have been inferred from results in Modigliani and Shiller (1973) (see particularly their fig. 6). Despite the evidence reported in Modigliani and Shiller and in the present paper, one of us (Shiller 1979) presented evidence suggesting that long-term interest rates are too volatile to accord with the expectations theory. By contrast with Modigliani and Shiller and the present paper, Shiller (1979) assumed that *levels* of short rates are stationary, an assumption more clearly appropriate for prewar data sets.

where w_t is serially uncorrelated noise. As Campbell and Shiller (1984) point out, excess bond returns will be predicted by S_t, and a regression of ΔY_{t+1} on S_t may find that the coefficient has the opposite sign from that predicted by (4), even if the variance of w_t is quite small. However, a regression of S_t^* on S_t will find that the coefficient has the same sign as predicted by (3), and downward bias caused by w_t will be small if the variance of w_t is small. Moreover, the variance ratios $\text{var}(S_t)/\text{var}(S_t')$ and $\text{var}(\xi_t)/\text{var}(\xi_t')$ may not be much greater than one. In this example the spread predicts short-rate movements almost correctly, even though it badly misforecasts long-rate movements. Deviations from the present value model are transitory rather than persistent, so the metric of equation (10) reveals the strengths of the expectations theory that are obscured by the metric of equation (11).[30]

In part B of table 3, we repeated the exercises above for stock prices and dividends. We worked with one sample period but two discount rates. The Akaike criterion selected a four-lag representation for the data when the sample mean discount rate 8.2 percent was used and a two-lag representation when the cointegrating regression discount rate 3.2 percent was used.

The VAR estimates suggest that dividend changes are rather highly predictable; the R^2 values for the equations that explain them are around 40 percent. There is very strong evidence that price-dividend spreads Granger-cause dividend changes, which is what one would expect if there is any truth to the present value model for stock prices.

We conducted two formal tests of the model. The first restricted the mean of the price-dividend difference, while the second left the mean unconstrained and restricted only the dynamics of the variable. (In the case of the term structure, the mean spread is always unconstrained because we allowed a constant risk premium.)

The results of these tests include some statistical rejections at conventional significance levels, but they are not nearly as strong as the rejections in the term structure. The pattern of results is sensitive to the choice of discount rate. When the sample mean return is used, the mean restriction on SL_t is satisfied almost exactly. Therefore, the test of only the dynamic restrictions in equation (8) rejects more strongly, at the 4.7 percent level as compared with the 7.2 percent level for the full set of restrictions. When the discount rate from the cointegrating regression is used, the complete set of restrictions is rejected at the 1.1 percent level while the significance level for the dynamic restrictions is

---

[30] We do not claim that this example is literally correct for our data. The model $S = S' + w$ can be tested, for any MA(q) process for w, by regressing ξ on information known $q + 2$ periods earlier. We found that this test rejected the model for q up to 8 using the bond data for 1959–78.

only 21.8 percent.[31] For both discount rates, a regression of ΔY_{t+1} on SL_t gives a coefficient estimate with a negative sign rather than the positive sign implied by the present value model.[32]

These tests are "portmanteau" tests of the present value model against an unspecified alternative. We also present variance ratios in order to test against the specific alternative that stock prices "move too much" in levels or innovations. The point estimate of the levels variance ratio $\mathrm{var}(SL_t)/\mathrm{var}(SL'_t)$ is dramatically different from unity, at 67.22, when the sample mean discount rate is used. Not surprisingly, the variance ratio is smaller when future dividend changes are discounted at the lower rate estimated by the cointegrating regression, but it is still considerable at 4.79. However, the asymptotic standard errors on these ratios are huge, and one cannot reject the hypothesis that both of them equal unity.

The innovations variance ratios $\mathrm{var}(\xi_t)/\mathrm{var}(\xi'_t)$ are also estimated larger than unity, and here the standard errors are less extreme. In the sample mean discount rate case, one can reject at the 5 percent level the hypothesis that the innovation variance ratio is unity; it is estimated to be 11.27, with a standard error of 4.49. With the lower discount rate, the ratio is estimated at 1.41, with a standard error of 0.44.

Plots of the price-dividend difference and the unrestricted VAR forecast of dividend changes give a visual image of these variance results. At an 8.2 percent discount rate (fig. 2), SL_t and SL'_t are negatively correlated (but there is a very large standard error on the correlation) and the excess volatility of the spread is very dramatic. At a 3.2 percent discount rate (fig. 3), SL_t and SL'_t have a correlation of 0.911 (with standard error 0.207) and the excess volatility is much less dramatic.[33]

To compare our results on volatility with results using earlier methods, we also computed sample values of S_t^* using the terminal condition $S_T^* = S_T$, where T is the last observation in our sample. We computed SL_t^* analogously. Equation (3) implies $\sigma(S_t^*) > \sigma(S_t)$ and $\sigma(SL_t^*) > \sigma(SL_t)$. For the bond data in the period 1959–78, $\sigma(S_t^*) = 1.217$, while $\sigma(S_t) = 1.060$, so the inequality is satisfied. For the stock

---

[31] Nonlinear Wald tests of the dynamic restrictions in the form (7), rather than (8), reject at less than the 0.005 percent level for the 8.2 percent discount rate and at the 7.3 percent level for the 3.2 percent discount rate.

[32] The coefficient is -0.064 for the 8.2 percent discount rate and -0.079 for the 3.2 percent discount rate.

[33] It should be emphasized that excess volatility of the spread SL_t is not quite the same as the excess volatility discussed in Shiller (1981b). That analysis suggested that stock prices should very nearly follow a trend. If that were in fact what was observed, the spread SL_t would be quite volatile because of dividend movements.

COINTEGRATION

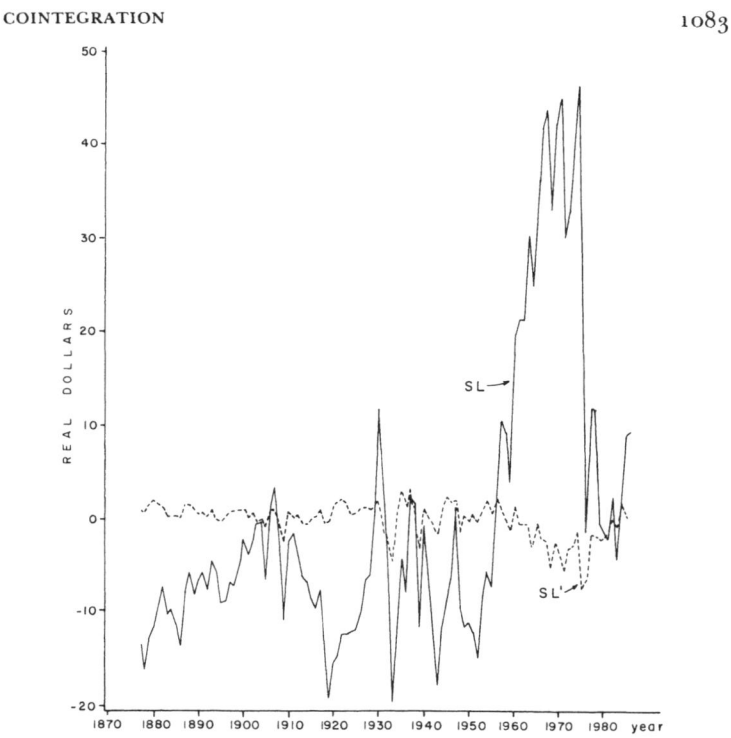

[Figure 2: plot with y-axis "REAL DOLLARS" ranging from -20 to 50, x-axis years from 1870 to 1980, showing two curves labeled SL and SL']

Fig. 2.—Stock market: deviations from means of actual spread (SL_t = Price$_t$ − θ · Dividend$_{t-1}$) and theoretical spread SL'_t, θ = 12.195.

data at an 8.2 percent discount rate, $\sigma(SL_t^*) = 7.928$, while $\sigma(SL_t) = 15.506$, so the inequality is sharply violated. The inequality is again satisfied by the stock data at a 3.2 percent discount rate, where $\sigma(SL_t^*) = 12.888$ and $\sigma(SL_t) = 9.937$.

Following Scott (1985), we also regressed S_t^* on S_t and a constant. If the present value model is true, the coefficient on S_t should be one. The same holds for the corresponding regression with SL_t^* and SL_t. For bonds in 1959–78, we estimated the coefficient at 0.81; for stocks at an 8.2 percent discount rate we estimated it at 0.16, while for stocks at a 3.2 percent discount rate we estimated it at 0.02. Thus the results using S_t^* and SL_t^* generally support the conclusion that the present value model for bonds fits the data comparatively well, whereas the model for stocks has a poor fit even though it cannot be rejected statistically at high levels of confidence.

We close with a caveat about the plots and summary statistics gener-

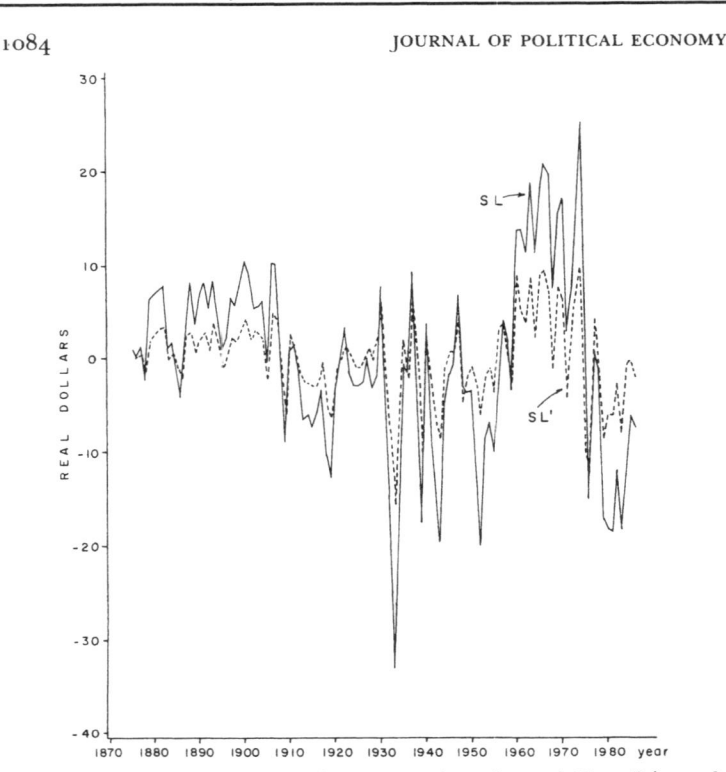

FIG. 3.—Stock market: deviations from means of actual spread (SL_t = Price$_t$ − θ · Dividend$_{t-1}$) and theoretical spread SL'_t, θ = 31.092.

ated by the VAR system. The VAR simulation method may be misleading if the wrong value of θ is chosen so that the spread variable is nonstationary. For example, if θ is chosen too large, the movements of S_t are dominated by the movements of $-\theta y_t$. The VAR results are then approximately those one would get if one regressed Δy_t and $-\theta y_t$ on lagged values of these variables. It is well known that in finite samples estimates of autoregressive parameters for nonstationary variables are biased downward, and this problem will afflict the VAR if θ is too large.

In a simple case in which y_t follows an AR(1) process with a unit root and the VAR includes one lag only, one can show that the estimated VAR companion matrix will have first column zero and second column $((1 - \rho)/\theta \quad \rho)'$, where ρ is a downward-biased estimate of the unit root. This companion matrix satisfies the restrictions of equation

COINTEGRATION

(9) almost exactly, whatever the behavior of the variable Y_t. A symptom of this misspecification would be that mean returns would not obey the model, even though the dynamics of returns would appear to satisfy the restrictions.

It is possible that a problem of this sort affects our results for the stock market when we use a low 3.2 percent discount rate corresponding to a high θ of 31.092. The cointegrating regression that generates this θ estimate—a regression of the level of Y on the level of y—is dominated by the enormous postwar hump in stock prices. Since this hump coincided with a much milder hump in real dividends, the regression estimates a coefficient for y that is much larger than the historical average price-dividend ratio. The negative intercept prevents the fitted value from overpredicting Y over the sample period as a whole. As a result, over the bulk of the sample period, the spread SL_t is distinctly negatively correlated with the lagged dividend.[34] The VAR estimates place considerable weight on this earlier part of the sample period because the dividend equation is specified in terms of dividend changes that are more variable before 1946. Thus the high correlation of SL_t and SL'_t may be to some extent spurious.

This view is supported by the results from regressing SL_t^* on SL_t. This is a levels regression that is dominated by the postwar hump in stock prices, and here we find the coefficient to be essentially zero rather than one as required by the model. Further support comes from the fact that we strongly reject the implications of the model for the mean of the data when we impose a 3.2 percent discount rate.

IV. Conclusion

In this paper we have shown how a present value model may be tested when the variables of the model, y_t and Y_t, follow linear stochastic processes that are stationary in first differences rather than in levels. If the present value model is true, a linear combination of the variables—which we call the spread—is stationary. Thus y_t and Y_t are cointegrated. The model implies that the spread is linear in the optimal forecast of the one-period change in Y_t and also in the optimal forecast of the present value of all future changes in y_t. We have shown how to conduct formal Wald tests of these implications.

We have also proposed an informal method for evaluating the "fit" of a present value model. A VAR is used to construct an optimal unrestricted forecast of the present value of future y_t changes, and this is compared with the spread. If the model is true, the unrestricted

---

[34] Over the period 1871–1946, the spread has a correlation of −0.7 with the lagged dividend when θ is set equal to 31.092.

forecast or "theoretical spread" should equal the actual spread. We computed the variances and correlation of the two variables and plotted their historical movements.

We applied our methods to the controversial present value models for stocks and bonds. We found that both models can be rejected statistically at conventional significance levels, with much stronger evidence for bonds. However, in our data set, the spread between long- and short-term interest rates seems to move quite closely with the unrestricted forecast of the present value of future short-rate changes. This can be interpreted as evidence that deviations from the present value model for bonds are transitory. In contrast, our evaluation of the present value model for stocks indicates that the spread between stock prices and dividends moves too much and that deviations from the present value model are quite persistent, although the strength of the evidence for this depends sensitively on the discount rate assumed in the test.

References

Baillie, Richard T.; Lippens, Robert E.; and McMahon, Patrick C. "Testing Rational Expectations and Efficiency in the Foreign Exchange Market." *Econometrica* 51 (May 1983): 553–63.

Blanchard, Olivier J., and Watson, Mark W. "Bubbles, Rational Expectations and Financial Markets." In *Crises in the Economic and Financial Structure: Bubbles, Bursts and Shocks*, edited by Paul Wachtel. Lexington, Mass.: Lexington, 1982.

Campbell, John Y. "Does Saving Anticipate Declining Labor Income? An Alternative Test of the Permanent Income Hypothesis." *Econometrica* (1987), in press.

Campbell, John Y., and Shiller, Robert J. "A Simple Account of the Behavior of Long-Term Interest Rates." *A.E.R. Papers and Proc.* 74 (May 1984): 44–48.

———. "The Dividend-Price Ratio and Expectations of Future Dividends and Discount Factors." Working Paper no. 2100. Cambridge, Mass.: NBER, 1986.

Cowles, Alfred. *Common-Stock Indexes*. 2d ed. Bloomington, Ind.: Principia, 1939.

Diba, Behzad T., and Grossman, Herschel I. "Rational Bubbles in the Price of Gold." Working Paper no. 1300. Cambridge, Mass.: NBER, 1984.

Dickey, David A., and Fuller, Wayne A. "Likelihood Ratio Statistics for Autoregressive Time Series with a Unit Root." *Econometrica* 49 (July 1981): 1057–72.

Engle, Robert F., and Granger, Clive W. J. "Cointegration and Error-Correction: Representation, Estimation and Testing." *Econometrica* 55 (March 1987): 251–76.

Engle, Robert F., and Watson, Mark W. "Applications of Kalman Filtering in Econometrics." Paper presented at the World Congress of the Econometric Society, Cambridge, Mass., August 1985.

Fama, Eugene F. "Efficient Capital Markets: A Review of Theory and Empirical Work." *J. Finance* 25 (May 1970): 383–417.
Gregory, Allan W., and Veall, Michael R. "Formulating Wald Tests of Nonlinear Restrictions." *Econometrica* 53 (November 1985): 1465–68.
Hamilton, James D., and Whiteman, Charles H. "The Observable Implications of Self-fulfilling Expectations." *J. Monetary Econ.* 16 (November 1985): 353–73.
Hansen, Lars Peter, and Sargent, Thomas J. "Exact Linear Rational Expectations Models: Specification and Estimation." Staff Report no. 71. Minneapolis: Fed. Reserve Bank, November 1981. (a)
———. "Linear Rational Expectations Models for Dynamically Interrelated Variables." In *Rational Expectations and Econometric Practice*, edited by Robert E. Lucas, Jr., and Thomas J. Sargent. Minneapolis: Univ. Minnesota Press, 1981. (b)
Kleidon, Allan W. "Variance Bounds Tests and Stock Price Valuation Models." *J.P.E.* 94 (October 1986): 953–1001.
LeRoy, Stephen F., and Porter, Richard D. "The Present-Value Relation: Tests Based on Implied Variance Bounds." *Econometrica* 49 (May 1981): 555–74.
Mankiw, N. Gregory, and Miron, Jeffrey A. "The Changing Behavior of the Term Structure of Interest Rates." *Q.J.E.* 101 (May 1986): 211–28.
Mankiw, N. Gregory; Romer, David; and Shapiro, Matthew D. "An Unbiased Reexamination of Stock Market Volatility." *J. Finance* 40 (July 1985): 677–87.
Mankiw, N. Gregory, and Shapiro, Matthew D. "Trends, Random Walks, and Tests of the Permanent Income Hypothesis." *J. Monetary Econ.* 16 (September 1985): 165–74.
Marsh, Terry A., and Merton, Robert C. "Dividend Variability and Variance Bounds Tests for the Rationality of Stock Market Prices." *A.E.R.* 76 (June 1986): 483–98.
Melino, Angelo. "Essays on Estimation and Inference in Linear Rational Expectations Models." Ph.D. dissertation, Harvard Univ., 1983.
Modigliani, Franco, and Shiller, Robert J. "Inflation, Rational Expectations and the Term Structure of Interest Rates." *Economica* 40 (February 1973): 12–43.
Phillips, P. C. B. "Time Series Regression with Unit Roots." *Econometrica* 55 (March 1987): 277–302.
Phillips, P. C. B., and Durlauf, S. N. "Multiple Time Series Regression with Integrated Processes." *Rev. Econ. Studies* 53 (August 1986): 473–95.
Phillips, P. C. B., and Ouliaris, S. "Testing for Cointegration." Discussion Paper no. 809. New Haven, Conn.: Yale Univ., Cowles Found., October 1986.
Phillips, P. C. B., and Perron, P. "Testing for Unit Roots in Time Series Regression." Discussion Paper. New Haven, Conn.: Yale Univ., Cowles Found., June 1986.
Quah, Danny. "Stationary Rational Bubbles in Asset Prices." Manuscript. Cambridge: Massachusetts Inst. Tech., 1986.
Sargent, Thomas J. "A Note on Maximum Likelihood Estimation of the Rational Expectations Model of the Term Structure." *J. Monetary Econ.* 5 (January 1979): 133–43.
Sawa, Takamitsu. "Information Criteria for Discriminating among Alternative Regression Models." *Econometrica* 46 (November 1978): 1273–91.

Scott, Louis O. "The Present Value Model of Stock Prices: Regression Tests and Monte Carlo Results." *Rev. Econ. and Statis.* 67 (November 1985): 599–605.

Shiller, Robert J. "The Volatility of Long-Term Interest Rates and Expectations Models of the Term Structure." *J.P.E.* 87 (December 1979): 1190–1219.

———. "Alternative Tests of Rational Expectations Models: The Case of the Term Structure." *J. Econometrics* 16 (May 1981): 71–87. (a)

———. "Do Stock Prices Move Too Much to Be Justified by Subsequent Changes in Dividends?" *A.E.R.* 71 (June 1981): 421–36. (b)

———. "Stock Prices and Social Dynamics." *Brookings Papers Econ. Activity*, no. 2 (1984), pp. 457–98.

———. "Conventional Valuation and the Term Structure of Interest Rates." In *Macroecomonics and Finance: Essays in Honor of Franco Modigliani*, edited by Rudiger Dornbusch, Stanley Fischer, and John Bossons. Cambridge, Mass.: MIT Press, 1987.

Shiller, Robert J.; Campbell, John Y.; and Schoenholtz, Kermit L. "Forward Rates and Future Policy: Interpreting the Term Structure of Interest Rates." *Brookings Papers Econ. Activity*, no. 1 (1983), pp. 173–217.

Stock, James H. "Asymptotic Properties of Least Squares Estimates of Cointegrating Vectors." *Econometrica* (1987), in press.

West, Kenneth D. "A Specification Test for Speculative Bubbles." Manuscript. Princeton, N.J.: Princeton Univ., 1986.

———. "Dividend Innovations and Stock Price Volatility." *Econometrica* (1987), in press.

White, Halbert. *Asymptotic Theory for Econometricians.* Orlando, Fla.: Academic Press, 1984.

[7]

On the Current State of the Stock Market Rationality Hypothesis

Robert C. Merton

1 Introduction

The foundation for valuation in modern financial economics is the rational market hypothesis. It implies that the market price of a security is equal to the expectation of the present value of the future cash flows available for distribution to that security where the quality of the information embedded in that expectation is high relative to the information available to the individual participants in the market. As has been discussed at length elsewhere,[1] the question whether this hypothesis is a good approximation to the behavior of real-world financial markets has major substantive implications for both financial and general economic theory and practice.

The rational market hypothesis provides a flexible framework for valuation. It can, for example, accommodate models where discount rates are stochastic over time and statistically dependent on future cash flows. It can also accommodate nonhomogeneity in information and transactions costs among individual market participants. The theory is not, however, a tautology. It is not consistent with models or empirical facts that imply that either stock prices depend in an important way on factors other than the fundamentals underlying future cash flows and discount rates, or that the quality of information reflected in stock prices is sufficiently poor that investors can systematically identify significant differences between stock price and fundamental value.

Although the subject of much controversy at its inception more than two decades ago, the rational market hypothesis now permeates virtually every part of finance theory. It has even become widely accepted as the "rule" (to which one must prove the exception) for finance practice on Wall Street, LaSalle Street, and in courtrooms and corporate headquarters. However, recent developments in economic theory and empirical work have again cast doubts on the validity of the hypothesis. Representing one view, Summers (1985) sees much of the renewed controversy as little more than a case of financial economists and general economists engaging in a partisan diversion of intellectual effort over methodological questions instead of focusing on sound research on major substantive questions.[2] He sees this development as only hastening an apparent secular trend toward inefficient disjunction between the fields of finance and economics on subjects of conjoint research interest. Perhaps that is so. But I must confess to having quite the opposite view on these same research efforts with regard to both their substance and their presumed dysfunctional effects on the fields of finance and economics. However, to pursue this issue further would only be an exercise in self-refutation. Thus, it suffices to say that whether market rationality is viewed as a "hot topic" or as merely a "topic with too much heat," an analysis of the current state of research on this issue would appear timely—especially so on this occasion honoring Franco Modigliani, past president of both the American Economic Association and the American Finance Association and prime counterexample to the Summers doctrine.

This paper focuses on the central economic question underlying the issue of stock market rationality: Do real-world capital markets and financial intermediaries, as a practical matter, provide a good approximation to those ideal-world counterparts that are necessary for efficient investor risk bearing and efficient allocation of physical investment? Although satisfaction of the rational market hypothesis is surely not sufficient to ensure efficient allocations, its broad-based rejection is almost certainly sufficient to rule out efficient allocations.[3]

From this perspective on the issue, it matters little whether or not real-world dealers and deal makers can "scalp" investors and issuers as long as their profits are a small fraction of aggregate transactions in important and well-established markets. Similarly, it matters little for this issue whether, as suggested by Van Horne (1985), promoters often make large-percentage profits during the transient period of time between the inception of a new financial product (or market) and the widespread acceptance (or rejection) of the product by investors and issuers.

In evaluating market rationality as it bears on economic efficiency, it matters very much whether stock prices generally can be shown to depend in an important way on factors other than fundamentals. It also matters very much whether it can be shown that either academic economists or practitioners systematically provide better forecasts of fundamental values than stock prices do. Thus, this analysis focuses on empirical work on aggregate stock price behavior, and especially the new volatility test methodologies, which appear to provide evidence of this very sort.

Although these empirical findings have had the most immediate effect in reviving the controversy over stock market rationality, some of the emerging developments in theory may prove, in the longer run, to be more important in resolving the controversy. Before proceeding with the analysis of empirical work, therefore, I pause briefly to comment on two of the more promising candidates to supersede the rational market theory.

Grounded in the sociological behavioral theory of the self-fulfilling prophecy, the theory of rational expectations speculative bubbles[4] in effect provides a theoretical foundation for answering the "If you are so smart, why aren't you rich?" question underlying the rational market argument that fullyrecognized, sizable, and persistent deviations between market price and fundamental value must necessarily provide "excess profit" opportunities for either investors or issuers. As we know, however, from the work of Tirole (1982), the interesting conditions under which such rational bubble equilibria can exist are still to be determined. In particular, if the theory is to be applied to the aggregate stock market in realistic fashion, then it must accommodate both "positive" and "negative" bubbles in a rational expectations framework. Such application would seem to require a satisfactory process to explain both the limits on share repurchase by firms when prices are persistently below marginal production cost and the limits on the creation of new firms with "instant profits" for the promoters in periods when general stock market prices significantly exceed that marginal cost.

Although few economists would posit irrational behavior as the foundation of their models, many, of course, do not subscribe to the sort of "super-rational" behavior implied by the rational expectations theory (with or without bubbles). Based on the pioneering work of Kahneman and Tversky (1979, 1982), the theory of cognitive misperceptions (by which I mean the observed set of systematic "errors" in individual decision making under uncertainty) may become a base from which economic theory formally incorporates nonrational (or as some economists have described it, "quasi-rational") behavior.

As discussed in Arrow (1982), the empirical findings of such systematic misperceptions in repeated laboratory experiments appear sound, and there would also appear to be many test cases within economics. In terms of the current state of empirical evidence in both cognitive psychology and financial economics, it would seem somewhat premature, however, to conclude that cognitive misperceptions are an important determinant of aggregate stock market behavior. Specifically, the same sharp empirical findings of cognitive misperceptions have not (at least to my knowledge) been shown to apply to individual decision making *when the individual is permitted to interact with others (as a group) in analyzing an important decision and when the group is repeatedly called upon to make similar types of important decisions*. But, this is, of course, exactly the environment in which professional investors make their stock market decisions.

If professional investors are not materially affected by these cognitive misperceptions, then it would seem that either competition among professional investors would lead to stock prices that do not reflect the cognitive errors of other types of investors, or professional investors should earn substantial excess returns by exploiting the deviations in price from fundamental value. Unlike the theory of rational expectations bubbles with its self-fulfilling prophecy, there is no a priori reason in this theory to believe that investment strategies designed to exploit significant deviations of price from fundamental value will not be successful. However, as shown in the following section, rather robust evidence indicates that professional investors do not earn substantial excess returns.

These two theories, along with Shiller's (1984) theory of fads, explicitly incorporate in an important way positive theories of behavior derived from other social sciences. In doing so, they depart significantly from the "traditional" approach of mainstream *modern* economic theory: namely, to derive the positive theories of "how we do behave" almost exclusively from normative economic theories of "how we should behave." Whether these theories throw light on the specific issue of aggregate stock market rationality, it will surely be interesting to follow the impact on economic theory generally from these attempts to bring economics "back into line" with the rest of the social sciences.

2 Empirical Studies of Stock Market Rationality

In his seminal 1965 paper proving the martingale property of rationally determined speculative price changes, Paul Samuelson was careful to warn readers against interpreting conclusions drawn from his model about markets as empirical statements: "You never get something for nothing. From a nonempirical base of axioms, you never get empirical results. Deductive analysis cannot determine whether the empirical properties of the stochastic model I posit come close to resembling the empirical determinants of today's real-world markets" (p. 42). One can hardly disagree that the question whether stock market rationality remains a part of economic theory should be decided empirically. There is, however, a complication: we have no statute of limitations for rejecting a theory. To the extent that one assumes the advancement of knowledge, it is the fate of all theory to be encompassed, superseded, or outright rejected in the long run. Nevertheless, at any moment, one must choose: either to continue to use the theory or to discard it. It is with this choice in mind that I examine the empirical evidence to date on stock market rationality.

As economists have cause to know well, the "long run" in economic behavior can indeed be long. Having already sustained itself for at least twenty years,[5] the rational market theory exemplifies this same fact—here in the history of economic science instead of in the history of economic behavior. The longevity of the theory can surely not be attributed to neglect on the part of economists bent on putting it to empirical test. I have not made any formal comparisons, but I suspect that over these twenty years, few, if any, maintained hypotheses in economic theory have received as much empirical attention as the rational market hypothesis. Indeed, there have probably been *too many* such tests. Although it is likely that this claim could be supported on the grounds of optimal resource allocation alone, the case is made here solely on statistical grounds. In preparation for this and other matters that bear on the testing of market rationality, I briefly review the history of these tests.

2.1 Early Tests of Stock Market Returns

About the time that Samuelson's fundamental paper appeared in print, what has since become the Chicago Center for Research in Security Pricing completed the construction of a file of prices and related data on all New York Stock Exchange-listed stocks from 1926 to 1965. This file has been periodically updated and expanded to include other exchanges so that there are now available almost sixty years of monthly data and more than twenty years of daily data on thousands of stocks. In addition, Robert Shiller of Yale has created a return file for the aggregate stock market with data going back to 1872.

There had been some earlier empirical studies of the randomness of speculative price changes, but the availability of a large-scale, easily accessible data base caused a flurry of such studies beginning in the mid-1960s. From simple runs and serial correlation tests to sophisticated filtering and spectral analysis, the results were virtually uniform in finding no significant serial dependencies in stock returns. The few cases of significant serial correlation were small in magnitude and short-lived (disappearing over a matter of a few days), and they could largely be explained by specialist activities for individual stocks or "non-contemporaneous trading effects" for portfolios of stocks. These findings were, of course, consistent with the Samuelson martingale property as a necessary condition for rationally determined prices.

Financial researchers at this time were aware of the possibility that a significant part of this randomness could be from random "animal spirits," which would cause prices to deviate from fundamental values. There was, however, a widespread belief that the empirical evidence did not support this alternative to market rationality. The foundation for this belief was the assumption that even with animal spirits, in the long run, stock prices will converge in the statistical equilibrium sense to their fundamental values. From this assumption, it follows that deviations from fundamental values will, by necessity, induce serial dependences in stock returns.[6] If such deviations were significant, then these dependences should be detectable as, for example, systematic patterns in the long-wave frequencies of the spectral analysis of stock returns. Moreover, there had been empirical studies of "relative strength" portfolio strategies that should do well if the market "underreacts" to information and of "relative weakness" (contrary opinion) portfolio strategies that should do well if the market tends to "overreact" to information. Neither of these produced significant results.[7] Working along similar lines were the studies of stocks that appear on the most active trading list or that had moved up or down by unusually large amounts, designed to look for evidence of under- or overreaction. Once again, no significant findings. Thus, it appeared at the time that the empirical evidence not only gave support to Samuelson's necessary condition for rationally determined prices, but also failed to lend support to the alternative hypothesis of random animal spirits.

As we know today from the work of Summers (1986) and others, many of these studies provided rather weak tests for detecting the types of generalized serial correlations that random animal spirits might generate, especially when the speed of reversion to fundamental values is slow. However, the concern in the 1960s was over another issue surrounding the power of these tests: the selective bias inherent in "secret models."

2.2 Tests of Professional Investor Performance

As the cynical version of the story goes, one could not lose by testing market rationality. If, indeed, significant empirical violations were found, one could earn gold, if not glory, by keeping this discovery private and developing portfolio strategies to be sold to professional money managers who would take advantage of these violations. If, instead, one found no significant violations, then this (financial) "failure" could be turned into academic success by publishing the results in the scientific journals. Thus, while each study performed might represent an unbiased test, the collection of such studies *published* were likely to be biased in favor of not rejecting market rationality. Unlike the more generally applicable claim for "quality" bias that studies that are consistent with the accepted theory are subject to less scrutiny by reviewers than ones that purport to reject it, the potential for material effects from "profit-induced" biases is probably specialized within economic analyses to studies of speculative prices.

One need not, however, accept this cynical characterization of academic financial researchers to arrive at much the same conclusion. The portfolio strategies tested by academics were usually simple and always mechanical; therefore, the fact that they yielded no evidence of significant profit opportunities is perhaps no great surprise. However, real-world professional investors with significant resources might well have important information sources and sophisticated models (be they of fundamentals or market psychology) that are used to beat the market systematically. As this version of the story goes, *if only* the academics could gain access to these proprietary models, they would quickly be able to reject the rational market hypothesis. Unfortunately, one assumes that few successful professional investors are likely to reveal their hypothetically profitable models, and thereby risk losing their source of income, simply to refute publicly the rationally determined price hypothesis of economists (which by hypothesis they have, of course, already determined privately to be false). Thus, it would seem that the possibility of proprietary models would, at least, significantly weaken, and in all likelihood, bias, the academic tests of market rationality.

Concern over the "secret model" problem led to the next wave of empirical tests for which the pioneering study of the mutual fund industry by Jensen (1968) serves as a prototype. The basic assumptions underlying these tests hold that if such models exist, then professional investors have them, and if they have them, then the results should show themselves in superior performance (at least, before expenses charged to investors) of the funds they managed. Tracking the performance of 115 investment companies over the period 1945–1964, Jensen found no significant evidence of superior performance for the fund industry as a whole. Later work by Jensen and others also found no evidence that individual investment companies within the industry had superior performance. That is, it was found that for any fund which had outperformed the naive market strategy of investing in the past, the odds of the same fund doing so in the future were essentially fifty-fifty. Similar studies subsequently made of the performance of other professional investor groups (e.g., insurance company equity funds, bank trust departments) came to much the same results. Moreover, as I have indicated in my preliminary remarks, these findings have remained robust to date.[8]

To be sure, the variances of the returns on these managed portfolios are sufficiently large that although the point estimates of the excess returns in these studies support the null hypothesis of no superior performance, they cannot reject the alternative null hypothesis that the managers do provide sufficient performance to earn the 25–100 basis points they charge. This fact may be important to the economics of the money management industry, but is inconsequential for the broader question of market rationality as a good approximation to the real-world stock market. That is, the undiscovered existence of proprietary models is not likely to provide an important explanation for the rational market hypothesis having remained unrejected for so long a time.

2.3 Anomalous Evidence on Stock Market Rationality

During the period of the 1960s and early 1970s, the overwhelming majority of empirical findings continued to support the market rationality theory (cf. Fama, 1970). Indeed, editors of both finance and broader economic journals, quite understandably, became increasingly reluctant to allot scarce journal space to yet another test that did not reject market rationality. Despite the mountain of accumulated evidence in support of the hypothesis, there were a relatively few of the empirical studies conducted during this period that did not seem to fit the rational market model. For example, low-price-to-earnings-ratio stocks seem systematically to earn higher average returns (even after correcting for risk differences) than high-price-to-earnings-ratio stocks. This "PE effect," later renamed the "small stock effect" after it was shown to be more closely associated with firm size than PE ratios, still remains a puzzle. Some other anomalies were the finding of various seasonal regularities such as the "January effect" and "the-day-of-the-week effect," and still another is the behavior of stock returns after a stock split. As the number of such puzzles gradually accumulated, the apparently closed gate on the empirical issue of market rationality began to reopen. Indeed, by 1978, even the *Journal of Financial Economics* (with its well-known editorial view in support of market rationality) devoted the entire June–September issue to a symposium on anomalous evidence bearing on market efficiency.

During this period, there were a number of empirical findings in the general economics literature that also cast doubt on the hypothesis of market rationality. Time series calculations of Tobin's Q appeared to suggest that stock market prices were too high at times while much too low at others, to be explained by economic fundamentals alone. Modigliani and Cohn (1979) presented a theory and empirical evidence that stock prices were irrationally low during the 1970s because investors failed to take correct account of the radically increased levels of the inflation rate in assessing expected future corporate profits and the rate at which they should be discounted.

Collectively these findings raised questions about the validity of stock market rationality, but they were hardly definitive. Some were found to be significant in one time period, but not in another. Others, such as Long's (1978) study on the market valuation of cash dividends, focused on a small sample of obscure securities. Virtually all shared the common element of testing a joint hypothesis with other important and unproven assumptions in addition to stock market rationality. There is, for example, the common joint hypothesis of stock market rationality *and* prices that are formed according to (one or another tax version of) the Capital Asset Pricing Model. Thus, at most, these tests rejected a hypothesis including stock market rationality but also other assumptions that, on a priori grounds, could reasonably be argued as less likely to obtain than market rationality.

During the past five years, a series of tests based upon the volatility of stock prices has produced seemingly new evidence of market non-rationality that some consider relatively immune to these criticisms of the earlier apparent rejections. One group of these tests pioneered by LeRoy and Porter (1981) and Shiller (1981) has focused on the volatility of aggregate stock market price relative to either aggregate earnings or dividends over long time periods (in the case of the former for the postwar period, and in the latter since before the turn of the century). Their findings have been interpreted as confirming the long-felt-but-unproved belief among some economists that stock prices are far more volatile than could ever be justified on fundamental evaluations alone.

A second group of tests examines the short-run volatility of stock price changes from one trading day to the next. It was known in the 1960s that the measured variance rate on stock returns is significantly lower over short time periods including weekends and holidays when the market is closed than over the same-length time periods when the market is open every day. The "rational" explanation given for this "seasonal" observation on volatility held that with businesses and many government activities closed, less new information is produced on these nontrading days than on trading days when they are open. However, using a period in the 1960s when the stock market was closed on every Wednesday, French and Roll (1984) show that the previously identified lower stock return volatility over short time periods that include a nontrading day applied to the Wednesday closings as well. Because nonspeculative market activities were generally open on these Wednesdays, the earlier presumed explanation was thus plainly inadequate. It would appear that market trading itself seems to cause increased volatility in market prices, and some interpret this finding as evidence against market prices being based on fundamentals alone.[9]

2.4 *Ex Ante* and *Ex Post* Predictions of the Theory

Explaining why rationally determined speculative price changes would exhibit the martingale property even though the underlying economic variables upon which these prices are formed may have considerable serial dependences, Samuelson (1965, p. 44) writes, "We would expect people in the market place, in pursuit of avid and intelligent self-interest, to take account of those elements of future events that in a probability sense may be discerned to be casting their shadows before them." The empirical evidence to date has been remarkably robust in finding no important cases of either lagged variables explaining stock price returns or of real-world investors (who make their decisions without benefit of even a peek into the future) being able to beat the market. This impressive success in confirming the *ex ante* component of the theory's prophecy has not, however, been matched in confirming its *ex post* component: namely, one should be able to find current or future economic events related to the fundamentals that, on average, explain current and past changes in stock prices.

As has been discussed elsewhere (cf. Fama, 1981; Fischer and Merton, 1984; Marsh and Merton, 1983, 1985), the change in aggregate stock prices is an important leading indicator of macro economic activity. Indeed, it is the best single predictor of future changes in business fixed investment, earnings, and dividends. Moreover, the forecast errors in the realization of future earnings changes are significantly correlated with the then-contemporaneous changes in stock prices. Nevertheless, although the writers for the popular financial press try hard, they often cannot identify the specific economic events that are important enough to cause the aggregate value of the stock market to change by as much as 2% in a single day.

At the micro level, the accounting and finance literatures are populated with studies of the behavior of individual stock prices, on, before, and after, the date of some potentially important event such as an earnings or tender offer announcement. These "event" studies lend some support to the *ex post* component of market rationality by showing that stock price changes predict many such events, respond quickly and in an unbiased fashion to surprises, and do not respond to seemingly important events that, in fact, should be affect the fundamentals (e.g., "cosmetic" changes in accounting earnings that have no impact on current or future cash flows). However, some of these studies (cf. Ohlson and Penman, 1985, who find that stock price return volatility appears to increase significantly after a stock split) provide conflicting evidence that indicates that stock prices may be affected by factors other than fundamentals.

Just as the strong empirical support for the *ex ante* component of market rationality has moved the focus of theoretical research from models of differential information to models of rational expectations bubbles, animal spirits, and fads, so the relative lack of closure on the *ex post* component seems to be the driving force behind the methodological focus of current empirical tests of the hypothesis. Finance specialists seem to favor short-term volatility or event studies, while general economists favor long-term studies, but both appear to agree that the statistical properties of volatility tests make them the most promising approach for rejecting the hypothesis of aggregate stock market rationality. The bulk of the formal analysis in this paper is focused on the long-term volatility tests, leaving for another occasion the examination of the event-study approach. Before undertaking that task, I digress to comment on a few, perhaps prosaic, but nevertheless important issues that frame the testing of this hypothesis.

2.5 Some Methodological Problems in Testing the Theory

As we all know, what the stock market actually did from 1872 to 1985 is an enumerable fact. As such, those numbers do not change even as the number of tests of the rational market hypothesis on these same data continues to grow. As we also know, the standard test statistics used in these studies do not reflect that fact. While, of course, the same comment could be made about virtually every area of economic model testing (cf. Leamer, 1983), it perhaps warrants more than usual attention in this case because of the unusually large number of studies, the large number of observations in the data set, and the magnitudes of unexplained volatility in stock prices.

As a case study of the problem, let us consider the regression study of the hypothesis that the expected real rate of return on the market is a constant, which is discussed in the Summers (1985) article. He writes, "Simple regression of real ex post stock returns on lagged dividend yields find that the null hypothesis that the real ex ante rate is constant can be rejected at almost any confidence level" (p. 635). Although hardly a proponent of this null hypothesis in either theory or practice (cf. Merton, 1973, 1980), I would nevertheless argue that in making his statement for apparently clear rejection, Summers does not take account of the number of regressions, *collectively*, researchers have run of stock returns on various contemporaneous and lagged variables. That some adjustment for this fact could have material implications for the strength of his conclusion is readily apparent from the negligible R^2 or explanatory power of these lagged yields. While one could perhaps argue on a priori grounds that dividend yield is a reasonable surrogate variable for expected return, I can report that much the same statistical significance results obtain (on the same data set, of course) if one regresses returns on the reciprocal of current stock price alone, omitting the dividend series altogether.[10]

If knowledge is to advance, we must seek out the exceptions, the puzzles, the unexplained residuals and attempt to explain them. But, before problem solution must come problem identification. Thus, economists place a premium on the discovery of puzzles, which in the context at hand amounts to finding apparent rejections of a widely accepted theory of stock market behavior. All of this fits well with what the cognitive psychologists tell us is our natural individual predilection to focus, often disproportionately so, on the unusual. As I have hinted earlier, this emphasis on the unusual has been institutionalized by responsible and knowledgeable journal editors who understandably look more favorably upon empirical studies that find anomalous evidence with respect to a widely accepted theory than upon studies that merely serve to confirm that theory yet again. This focus, both individually and institutionally, together with little control over the number of tests performed, creates a fertile environment for both unintended selection bias and for attaching greater significance to otherwise unbiased estimates than is justified.

To clarify the point, consider this parable on the testing of coin-flipping abilities. Some three thousand students have taken my finance courses over the years, and suppose that each had been asked to keep flipping a coin until tails comes up. At the end of the experiment, the winner, call her A, is the person with the longest string of heads. Assuming no talent, the probability is greater than a half that A will have flipped 12 or more straight heads. As the story goes, there is a widely believed theory that no one has coin-flipping ability, and, hence, a researcher is collecting data to investigate this hypothesis. Because one would not expect everyone to have coin-flipping ability, he is not surprised to find that a number of tests failed to reject the null hypothesis. Upon hearing of A's feat (but not of the entire environment in which she achieved it), the researcher comes to MIT where I certify that she did, indeed, flip 12 straight heads. Upon computing that the probability of such an event occurring by chance alone is 2^{-12}, or .00025, the researcher concludes that the widely believed theory of no coin-flipping ability can be rejected at almost any confidence level.

Transformed to the context of tests of stock market rationality, what empirical conclusion about the theory can be reached if we are told of a certified discovery of a particular money manager who outperformed the market in each and every year for twelve years? Even if the individual researcher can further certify that the discovery of this apparently gifted manager was by a random drawing, the significance of the finding cannot be easily assessed. We know that the population size of (past and present) money managers is quite large. We also know that the number of researchers (past and present) studying professional money management performance is not small. However, as indicated, for quite legitimate individual and institutional reasons, results that simply confirm the "norm" (of no significant performance capability) tend not to be reported. Thus, the number of such random drawings undertaken *collectively* by researchers is unknown, and this makes the assessment of significance rather difficult.

As we surely could do in the case of A's purported coin-flipping talent, we might try to resolve this problem by testing the money manager's talent "out of sample." Because of survivorship bias, this cannot be done easily with data from years prior to the money manager's run. If the run is still current, then we must wait many years to accumulate the new data needed to test the hypothesis properly.

The problem of assessing significance becomes, therefore, especially acute for testing theories of stock market behavior where very long observation periods (e.g., fifty to one hundred years) are required. One such class of examples is theories where price and fundamental value deviate substantially and where it is further posited that the speed of convergence of price to value is slow.

If, as is not unusual (cf. Shiller, 1984), a theory is formulated as a possible solution for an empirical puzzle previously found in the data, then the construction of a proper significance test of the theory on these same data becomes quite subtle.

Consider, for example, the following sequence of empirical studies and theories, which followed the finding in the early 1970s, that low-price-to-earnings-ratio stocks seem significantly to outperform high-price-to-earnings-ratio stocks when performance is adjusted for risk according to the Capital Asset Pricing Model. Because there was already theory and evidence to suggest that the CAPM was inadequate to explain all the cross-sectional differentials in average security returns and because price-to-earnings ratios are not statistically independent of other firm characteristics (e.g., industry, dividend yield, financial and business risks), early explanations of the puzzle centered on additional dimensions of risk as in the arbitrage pricing and intertemporal capital asset pricing theories and on the tax effects from the mix of the pretax returns between dividends and capital gains. Further empirical analysis of the same data suggested that the aberration was more closely related to the size of the firm than to price-to-earnings ratios, although this claim is still subject to some dispute. Although firm size is also not statistically independent of other firm characteristics, this finding added the prospect of market segmentation or "tiering" to the original list of possible explanations for the puzzle.

Still further empirical analysis of the same data found a "seasonal" effect in stock returns that appeared to produce systematically larger returns on the market in the month of January. Closer inspection of these data pinpointed the source in place and time to be smaller firms in the early part of January. Moreover, by combining these two studies, it seems that the original PE/small-firm puzzle is almost entirely the result of stock price behavior in January. This result shifted the emphasis of theoretical explanation from risk factors and segmentation to "temporary" depressions in prices caused by year-end tax-loss sales of stocks that have already declined in price.

In the growing list of theoretical explanations of this puzzle (followed by tests on the same data set), perhaps the most recent entry is the "overreaction behavioral theory" of DeBondt and Thaler (1985) which implies that a "contrary opinion" portfolio strategy will outperform the market. It is particularly noteworthy because it also represents an early attempt at a formal test of cognitive misperceptions theories as applied to the general stock market.[11] To test their theory, they construct two portfolios (each containing 35 stocks): one contains extreme winners based on past returns and the other extreme losers. They find that in a series of nonoverlapping three-year holding periods, the "winners," on average, underperformed the market by 1.7% per year and the "losers" overperformed the market by 6.5% per year. The difference between the two, 8.2% per year, was judged to be significant with a *t*-statistic of 2.20.

Do the empirical findings of DeBondt and Thaler, using over a half-century of data, really provide significant evidence for their theory? Is it reasonable to use the standard *t*-statistic as a valid measure of significance when the test is conducted on the same data used by many earlier studies whose results influenced the choice of theory to be tested? As it happens in this particular case, the former substantive question can be answered without addressing the latter methodological one. That is, Franco Modigliani is fond of the saying, "If, for a large number of observations, you have to consult the tables to determine whether or not your *t*-statistic is significant, then it is not significant." This expressed concern over the delicate issue of balancing type I and type II errors would seem to apply here. Moreover, consider the additional findings of the study as described by the authors (p. 799): "First, the overreaction effect is asymmetric; it is much larger for losers than for winners. Secondly, consistent with previous work on the turn-of-the-year effect and seasonality, most of the excess returns are realized in January." As the authors later put it. (p. 804), "Several aspects of the results remain without adequate explanation." It is at this moment difficult to see a clear theoretical explanation for overreaction being asymmetric and, even more so, for the excesses tending to be corrected at the same time each year.

Suppose, however, that the authors had found no such unexplained anomalies with respect to their theory and a larger *t*-statistic. Would their test, considered in methodological terms, have fulfilled their expressed goal? Namely, "... our goal is to test whether the overreaction hypothesis is *predictive* [their emphasis]. In other words, whether it does more for us than merely to explain, ex post, the P/E effect or Shiller's results on asset price dispersion" (p. 795). When a theory is formulated as an explanation of a known empirical puzzle and then tested on the same data from which the puzzle arose, it would appear that the distinction between "prediction" and "ex post explanation" can be quite subtle.

These same concerns, of course, apply equally to the many empirical studies that do not reject market rationality. The early tests of serial dependences in stock returns that used the newly created data bases in the 1960s may have been sufficiently independent to satisfy the assumptions underlying the standard test statistics. It is, however, difficult to believe in the same level of independence for the practically countless subsequent runs used to test closely related hypotheses on the same data.

Although there is no obvious solution to these methodological problems in testing the rational market hypothesis, it does not follow that the controversies associated with the hypothesis cannot be empirically resolved. It does follow, however, that the reported statistical significance of the evidence, both for and against the hypothesis, is likely to overstate—perhaps, considerably so—the proper degree of precision to be attached to these findings. As noted at the outset, although common to all areas of economic hypothesis testing, these methodological problems appear to be especially acute in the testing of market rationality. Thus, it would seem that in evaluating the evidence on this matter, "more-than-usual" care should be exercised in examining the substantive economic assumptions and statistical methodologies used to present the evidence. In this spirit, I try my hand at examining the recent volatility tests of aggregate stock market rationality.

2.6 Volatility Tests of Stock Market Rationality

Having already expressed my views on the LeRoy and Porter (1981) and Shiller (1981) variance bound studies as tests of stock market rationality,[12] I provide only a brief summary of those views as background for the discussion of more recent volatility tests that have evolved from their work.

In formulating his variance bound tests, Shiller (1981) makes three basic economic assumptions: (S.1) stock prices reflect investor beliefs, which are rational expectations of future dividends; (S.2) the real expected rate of return on the stock market is constant over time; (S.3) aggregate real dividends on the stock market can be described by a finite-variance stationary stochastic process with a deterministic exponential growth rate. From these assumptions, Shiller derives two variance bound relations: the first is that the variance of real and detrended stock prices is bounded above by the variance of real and detrended "perfect-foresight" stock prices constructed by discounting *ex post* the realized stream of dividends at the estimated average expected rate of return on the stock market. The second is that the variance of the innovations (or unanticipated changes) in stock prices is bounded from above by the product of the variance of dividends and a constant that parametrically depends on the long-run or statistical equilibrium expected dividend-to-price ratio. Using 109 years of data, Shiller found that the sample statistics violated by a very large margin both of his variance bounds on stock price behavior. Although he did not derive the sampling properties of his estimates, Shiller argued that the magnitude of the violations together with the long observation period make sampling error an unlikely candidate to explain these violations. Nevertheless, subsequent simulations by Flavin (1983) and Kleidon (1983a,b) have shown that sampling error, and, in addition, sample bias, could be important factors.

Some economists interpret the Shiller findings as strong evidence against the theory that stock prices are based upon fundamentals alone. Others, most recently Summers (1985) and Marsh and Merton (1986), are more careful in noting that even if the results are "true" rejections, then they reject the joint hypothesis (S.1), (S.2), and (S.3), which need not, of course, imply rejection of (S.1).[13] As noted earlier in this section, there are a priori economic reasons as well as empirical evidence leading us to reject the hypothesis (S.2) that the expected real rate is constant. While these are perhaps sufficient to reconcile the test findings with market rationality, some economists (including Shiller, 1981, 1982) have presented analyses suggesting that fluctuations in the expected real rate might have to be "unreasonably large" to make this accommodation.

If (S.2) were modified to permit the expected real rate to follow a stochastic but stationary process, then, together with (S.3), detrended rational stock prices must follow a stationary process. The prototype processes for stock prices and dividends used by both finance academics and practitioners are not stationary, and this raises a priori questions about assumption (S.3). Kleidon (1983a,b) reports time series evidence against stationarity for both stock prices and dividends, and, using simulations, shows that Shiller's findings can occur for nonstationary dividend processes and rationally determined stock prices.

Marsh and Merton (1986) show that if the stationarity assumption is replaced by a Lintner-like dividend model where the dividend is a positive distributed lag of past stock prices, then the inequality in Shiller's first variance bounds test is exactly reversed. Thus, for any given time series of stock prices, this variance bound will always be violated by one or the other assumption about the dividend process. Hence, they conclude that the bound is wholly unreliable as a test of stock market rationality. They further show that for this class of dividend processes, there is no easily identified bound between the variance of dividends and the variance of stock price innovations.

Judging from these studies, the amount of light that these variance bounds tests can shed on the issue of market rationality seems to depend critically on the way in which we model the uncertainty surrounding future economic fundamentals. That is, if the underlying economic fundamentals are such that the levels of rationally determined, real (and detrended) stock prices can be described by a stationary process, then they have power. If, instead, it is the percentage change in stock prices that is better described by a stationary process, then they have no power. This observation was surely one of the important driving forces in the development of the "second-generation" volatility tests beginning with West (1983, 1984) and represented most recently by Mankiw, Romer, and Shapiro (1985). Although closely related to the original Shiller-LeRoy-Porter formulations, these tests appear to be far more robust because they do not require the stationarity assumption. Since the Mankiw, Romer, and Shapiro (MRS) study is the most recent version of these tests, the analysis here focuses on it.

As with the original Shiller variance bound test, which derived an inequality between the variance of rational stock prices $\{P(t)\}$ and the variance of ex post, perfect-foresight stock prices $\{P^*(t)\}$, MRS also use these series, together with a time series of "naive forecast" stock prices $\{P^0(t)\}$, to test the following derived bounds [p. 679, (11') and (12')]:

$$E_0[P^*(t) - P^0(t)]^2 \geq E_0[P^*(t) - P(t)]^2 \tag{1}$$

and

$$E_0[P^*(t) - P^0(t)]^2 \geq E_0[P(t) - P^0(t)]^2, \tag{2}$$

where E_0 denotes the expectation operator, conditional on initial conditions at $t = 0$. Although MRS do retain what has been called here Shiller's assumptions (S.1) and (S.2), they do not make the stationarity assumption (S.3). Hence, this conditioning of the expectations is necessary to make sense of (1) and (2) when the series are not stationary processes.

To test the bounds (1) and (2), they form the test statistics [p. 683, (16), (17)],

$$S_1 = \frac{1}{T} \sum_{t=1}^{T} [P^*(t) - P^0(t)]^2 - \frac{1}{T} \sum_{t=1}^{T} [P^*(t) - P(t)]^2 \tag{3}$$

and

$$S_2 = \frac{1}{T} \sum_{t=1}^{T} [P^*(t) - P^0(t)]^2 - \frac{1}{T} \sum_{t=1}^{T} [P(t) - P^0(t)]^2, \tag{4}$$

and show that $E[S_1] \geq 0$ and $E[S_2] \geq 0$. With the same data set used by Shiller (1981) but now extended to run from 1872 to 1983, and a "naive forecast" $\{P^0(t)\}$ based on the current dividend, MRS find that these second moment inequalities are substantially violated by the point estimates of both (3) and (4).

The MRS analysis appears to address all the cited criticisms of the first-generation volatility tests with two exceptions, both of which they point out (p. 686): the assumption of a constant discount rate and the statistical significance of their estimates. Since the former has already been discussed in the literature on the first-generation tests, I examine only the latter here.

As with the original Shiller analysis, it is understandable that MRS did not examine the significance issue formally. After all, it is no easy task to derive the necessary mathematical relations for general processes. In the Shiller case, the assumption of stationarity for the underlying processes make somewhat credible the heuristic argument that with a 109-year observation period, the sample statistic is not likely to differ from its expected value by the large magnitudes necessary to void his apparent rejection. Such creditability does not, however, extend to nonstationary processes. Because the extension to include nonstationary processes is the most important contribution of the MRS and other second-generation volatility tests, it is appropriate to examine the sampling properties of their statistics in such an environment.

As noted, deriving these properties in general is no easy task. Thus, I focus here on a simple example that fits their conditions and is easy to solve for the sampling properties.

Suppose there is a rationally priced stock that we know as of today ($t = 0$) will not pay a dividend until at least time T in the future. Suppose (as is often assumed in representative finance models) the dynamics for stock price in real terms, $P(t)$, follows a geometric Brownian motion, which we can describe by the Itô stochastic differential equation:

$$\frac{dP}{P} = r\,dt + \sigma\,dZ, \qquad (5)$$

where r is the required expected real return on the stock; σ^2 is the instantaneous variance rate; and dZ is a Weiner process. r and σ^2 are positive constants.

Suppose further that we decide to perform an MRS type experiment using price data from today until year T in the future. Since none of us knows today what stock prices will be in the future, it is clear that the test statistic is conditional only on the current price, $P(0) = P_0$, and the date at which we end the test, T.

By the MRS definition, the ex post perfect-foresight stock price series,

$\{P^*(t)\}$, will be constructed according to the rule

$$dP^*(t) = rP^*(t)\,dt \tag{6}$$

with the further terminal or boundary condition that

$$P^*(T) = P(T). \tag{7}$$

From (6) and (7), it follows immediately that

$$P^*(t) = e^{-r(T-t)}P(T). \tag{8}$$

From the posited dynamics (5), we can represent the random variable for the stock price at time t in the future, conditional on $P(0) = P_0$, by

$$P(t) = P_0 \exp[\mu t + \sigma Z(t)], \tag{9}$$

where $\mu \equiv (r - \sigma^2/2)$ and $Z(t) = \int_0^t dZ(s)$ is a normally distributed random variable with the properties that

$$E_0[Z(t)] = 0,$$

$$E_0[Z(t)Z(s)] = \min(s, t). \tag{9a}$$

It follows from (9) and (9a) that $(0 \le t \le T)$

$$E_0[P(t)] = P_0 e^{rt} \tag{10a}$$

and

$$E_0[P^2(t)] = P_0^2 \exp[(2r + \sigma^2)t]. \tag{10b}$$

It follows from (8), (10a), and (10b) that

$$E_0[P^*(t)] = P_0 e^{rt} \tag{11a}$$

and

$$E_0\{[P^*(t)]^2\} = P_0^2 \exp[2rt + \sigma^2 T]. \tag{11b}$$

By comparison of (10) with (11), we see that the conditional expectation of the "forecast," $P(t)$, is equal to the conditional expectation of the "realization," $P^*(t)$, and the conditional noncentral second moment of the forecast is always less than the corresponding second moment of the realization. This verifies in this model the fundamental principle underlying both the first- and second-generation volatility tests, the principle that rational forecasts should exhibit less volatility than the realizations.

For analytic convenience, suppose that in performing this test, we choose our "naive forecast," $P^0(t)$, equal to zero for all t (which is acceptable within the MRS methodology). In this case, the MRS volatility bound statistic (3) can be rewritten as

$$E_0(X_1) \geq E_0(X_3) \tag{12}$$

and the MRS volatility bound statistic (4) can be rewritten as

$$E_0(X_1) \geq E_0(X_2), \tag{13}$$

where

$$X_1 \equiv \frac{1}{T}\int_0^T [P^*(t)]^2 \, dt,$$

$$X_2 \equiv \frac{1}{T}\int_0^T [P(t)]^2 \, dt, \tag{14}$$

$$X_3 \equiv X_1 + X_2 - \frac{2}{T}\int_0^T P^*(t)P(t)\, dt.$$

with the MRS $S_1 = X_1 - X_3$ and the MRS $S_2 = X_1 - X_2$.

Substituting from (8) and (9) and computing the conditional expectations, we have that

$$E_0[S_1] = E_0[X_1 - X_3]$$
$$= (P_0)^2[e^{(2r+\sigma^2)T} - 1]/[2r + \sigma^2]T \tag{15}$$

and

$$E_0[S_2] = E_0[X_1 - X_2]$$
$$= \frac{(P_0)^2 e^{\sigma^2 T}}{2r(2r+\sigma^2)T}[\sigma^2(e^{2rT} - 1) - 2r(1 - e^{-\sigma^2 T})]. \tag{16}$$

By inspection of (15) and (16), we confirm the MRS inequalities $E_0[S_1] \geq 0$ and $E_0[S_2] \geq 0$, and moreover, we see that for $\sigma^2 > 0$, they are strict inequalities whose magnitudes grow without bound as the observation period T becomes large. Unfortunately, the standard deviations of both statistics also grow without bound as the observation period becomes large, and moreover, the rates of growth are at a larger exponential rate than the expected values. Hence, for large T, virtually any realized sample values for S_1 and S_2 are consistent with the ex ante inequalities (12) and (13).

In noting the upward trend in their series and the prospect for heteroskedasticity, MRS (pp. 685–686) attempt to correct for this possible inefficiency by weighting each observation by the inverse of the market price of the stock. However, such a scaling of the data does not rectify the sampling problem. For example, using their scheme, the new statistic S_2', replacing S_2 in (16), can be written as

$$S_2' = \frac{1}{T}\int_0^T \{[P^*(t)]/P(t)\}^2\, dt - \frac{1}{T}\int_0^T [P(t)/P(t)]^2\, dt. \tag{17}$$

Again computing expectations, we have

$$E_0[S_2'] = [e^{\sigma^2 T} - 1]/\sigma^2 T - 1, \tag{18}$$

which is positive and growing in magnitude without bound. Again, the standard deviation of S_2' also grows at a larger exponential rate than $E_0[S_2']$.

Because $E_0(S_1) \geq 0$ and $E_0(S_2) \geq 0$, it follows that $E_0(X_3)/E_0(X_1) \leq 1$ and $E_0(X_2)/E_0(X_1) \leq 1$. A perhaps tempting alternative method for testing the inequalities (12) and (13) would be to use the ratios X_3/X_1 and X_2/X_1 instead of the differences S_1 and S_2. However, as we now show, unless the real discount rate is considerably larger than the volatility parameter σ^2, the ex ante expected values of both these ratios produce exactly the reverse of the inequalities for the ratios of their individual expectations.

Define the statistics $Q_1 \equiv X_3/X_1$ and $Q_2 \equiv X_2/X_1$. By substituting from (8), (9), and (14), we can write the expressions for Q_1 and Q_2 as

$$Q_1 = 1 + Q_2 - 4r\left\{\int_0^T \exp[-(r+\mu)(T-t)\right.$$
$$\left. - \sigma[Z(T) - Z(t)]]\, dt\right\}\bigg/[1 - e^{-2rT}] \tag{19}$$

and

$$Q_2 = 2r\left\{\int_0^T \exp[-2\mu(T-t) - 2\sigma[Z(T) - Z(t)]]\, dt\right\}\bigg/[1 - e^{-2rT}]. \tag{20}$$

Taking expectations and integrating (20), we have

$$E_0[Q_2] = 2r[1 - e^{-(2r-3\sigma^2)T}]/\{(2r - 3\sigma^2)[1 - e^{-2rT}]\}. \tag{21}$$

By inspection of (21), if $2r > 3\sigma^2$, then $E_0[Q_2] \to 2r/[2r - 3\sigma^2] > 1$ as T gets large. If $0 < r \leq 3\sigma^2$, then $E_0[Q_2] \to \infty$ as T gets large. Thus, for large T, the expectation of the ratio X_2/X_1 satisfies exactly the reverse of the inequality satisfied by the ratio of their expectations $E_0[X_2]/E_0[X_1]$, and this is the case for all positive parameter values r and σ^2.

Taking expectations in (19) and substituting from (21), we have

$$E_0[Q_1] = 1 + 2r\left\{\frac{[1-e^{-(2r-3\sigma^2)T}]}{(2r-3\sigma^2)} - \frac{2[1-e^{-(2r-\sigma^2)T}]}{(2r-\sigma^2)}\right\}\bigg/[1-e^{-2rT}]. \quad (22)$$

From (22), if $0 < 2r \leq 3\sigma^2$, then $E_0[Q_1] \to \infty$ as T gets large. For $2r > 3\sigma^2$ and large T, we have that $E_0[Q_1] \to 1 + 2r(5\sigma^2 - 2r)/(2r - 3\sigma^2)(2r - \sigma^2)$ which only becomes less than one if $2r > 5\sigma^2$. As described in Merton (1980, p. 353, table 4.8), the average monthly variance rate on the market between 1926 and 1978 was estimated to be 0.003467, which amounts to a $\sigma^2 = 0.0416$ in annual units. Hence, an expected annual real rate of return on the market of the order of 10% would be required to make $E_0[Q_1]$ satisfy the inequality $E_0[Q_1] < 1$. Thus, in addition to being indicative of the sampling problems, the expectation of these ratios are largely consistent with the empirical evidence reported by MRS.

The choice of $P^0(t) \equiv 0$ as the "naive forecast" in this example does not explain these findings. If, for example, we chose $P^0(t) = P_0 e^{rt}$, the "true" conditional expected value for both $P(t)$ and $P^*(t)$, the large T results will remain essentially unchanged because the ratios of second central and noncentral moments tend to one for both $P(t)$ and $P^*(t)$. Indeed, in this case, the MRS inequality just reduces to the original Shiller variance bound defined here in terms of conditional variances and using the "true" ex ante expected values for $P^*(t)$ and $P(t)$. For much the same reason, the selection of almost any naive forecast whose volatility is considerably less than that of stock price is unlikely to change these results. As shown by example in the appendix, the asymptotic distributions for S_1 and S_2 need not converge even if the naive forecast is unbiased and follows a nonstationary process quite similar to the one posited for stock prices.

The example presented here assumes that the underlying stock pays no interim dividends, and therefore one might wonder whether perhaps this polar case is also pathological with respect to the MRS analysis. Although unable to solve fully the dividend-paying case analytically, I offer the following analysis to suggest that the fundamental sampling problems identified by this example will not be significantly changed.

The MRS analysis appears to be impeccable with respect to bias (i.e., the expected value conditions on their inequalities). The problem is that the standard deviation of their estimate for the noncentral second moments grows at an exponential rate greater than the growth of the expected value of the estimate. Thus, the important characteristic to examine is the relation between the second moment and the square root of the fourth moment of future stock prices. Suppose that the dividend paid is a constant proportion ρ of the current stock price. The noncentral second moment of $P(T)$, given $P(0)$, can be written as $[P(0)]^2 \exp[2(r - \rho + \sigma^2/2)T]$. The square root of the noncentral fourth moment of $P(T)$ can be written as $[P(0)]^2 \cdot \exp[(2(r - \rho) + 3\sigma^2)T]$. Thus, as long as $2r + \sigma^2 > 2\rho$, the expected second-moment estimate grows exponentially. However, the ratio of the expected value of the estimate to its standard deviation will, for large T, always decline according to $\exp[-2\sigma^2 T]$, independently of the payout ratio, ρ. Because the MRS estimates involve simple averages of sums (or integrals) of squared stock prices, it thus seems unlikely that the sampling properties of the estimators for large T will be significantly affected by appending dividends to the model. To the extent that dividend changes are more sticky than proportional to stock price changes (which as an empirical matter, they seem to be),[14] the model presented here becomes an even better approximation.

In this light and given that Shiller (1981) had already found enormous empirical violations of the central second-moment bounds between actual stock prices and ex post perfect-foresight prices, it is not altogether surprising to find that the measured noncentral second moments of these same two series also exhibit large violations when estimated on the same data set. In that sense, the Mankiw-Romer-Shapiro study provides no important new empirical findings about the magnitudes of stock market volatility. Nevertheless, their study (together with the West, 1984, analysis) is central to the controversy over the rational market hypothesis because of its claim to rule out the interpretation of Shiller's empirical findings as simply a rejection of the assumption of a stationary process for dividends and stock prices. As shown here, this claim remains to be proved.

3 Conclusion

In summary, I believe that when the heat of the controversy dissipates, there will be general agreement that the rejection or acceptance of the rational market hypothesis as a good approximation to real-world stock market behavior will turn on how we model uncertainty. If, in fact, the levels of expected real corporate economic earnings, dividends, and discount rates in the future are, *ex ante*, well-approximated by a long average of the past levels (plus perhaps a largely deterministic trend), then it is difficult to believe that observed volatilities of stock prices, in both the long and not-so-long runs, are based primarily on economic fundamentals. This assertion can be confirmed by simulations using economic models of the nonfinancial sector with stationary processes for the levels of outputs generating the uncertainty.

Thus, if the well-informed view among economists and investors in the 1930–1934 period was that corporate profits and dividends for *existing*[15] stockholders would return in the reasonably near future to their historical average levels (plus say a 6% trend), then market prices in that period were not based upon fundamentals. If this were the view, then it is surely difficult to explain on a rational basis why the average standard deviation of stock returns during this period was almost three times the corresponding average for the forty-eight other years between 1926 and 1978 (cf. Merton, 1980, pp. 353–354). If once again in the 1962–1966 period, the informed view was that required expected returns and the levels and growth rates of real profits in the future would be the same as in the long past, then stock prices were (ex ante) too high.[16]

If, as is the standard assumption in finance, the facts are that the future levels of expected real corporate economic earnings, dividends, and discount rates are better approximated by nonstationary stochastic processes, then even the seemingly extreme observations from these periods do not violate the rational market hypothesis.

In light of the empirical evidence on the nonstationarity issue, a pronouncement at this moment that the rational market theory should be discarded from the economic paradigm can, at best, be described as "premature." However, no matter which way the issue is ultimately resolved, the resolution itself promises to identify fruitful new research paths for both the finance specialists and the general economist. Just as the break-throughs of more than two decades ago by Lintner, Markowitz, Miller, Modigliani, Samuelson, Sharpe, and Tobin dramatically changed every aspect of both finance theory and practice, so the rejection of market rationality together with the development of the new theory to supersede it would, once again, cause a complete revision of the field. If, however, the rationality hypothesis is sustained, then instead of asking the question "Why are stock prices so much more volatile than (measured) consumption, dividends, and replacement costs?" perhaps general economists will begin to ask questions like "Why do (measured) consumption, dividends, and replacement costs exhibit so little volatility when compared with rational stock prices?" With this reversed perspective may come the development of refined theories of consumer behavior (based upon intertemporally dependent preferences, adjustment costs for consumption, the nontradability of human capital, and cognitive misperceptions) that will explain the sluggish changes in aggregate consumption relative to permanent income. They may also see new ways of examining the question of sticky prices that has long been an important issue in the analysis of the business cycle.

Because rational speculative prices cannot be sticky, comparisons of the volatilities of such prices with nonspeculative prices may provide a useful yardstick for measuring the stickiness of nonspeculative prices and their impact on aggregate economic activity.

Appendix

In the text, it was shown that if rational stock prices follow a geometric Brownian motion and if the naive forecast $P^0(t) = 0$, then the MRS sample statistics, S_1 and S_2, will have asymptotic distributions whose dispersions are growing at an exponential rate greater than their expected values. As noted, the choice of a naive forecast that follows a stationary process with an exponential trend does not change this conclusion about the asymptotic distributions. Using the model of the text, we now show that selection of a naive forecast variable that is both unbiased and follows a nonstationary process very much like the rational stock price need not alter this conclusion. Thus, it would appear that conditions under which the MRS statistics will exhibit proper distributional properties for long observation periods are quite sensitive to the choice of the naive forecast variable and, therefore, are not robust.

Suppose that the naive forecast is given by $P^0(t) = \lambda(t)P(t)$, where $\{\lambda(t)\}$ are independently and identically distributed positive random variables with

$$E[\lambda(t)] = 1,$$
$$\text{var}[\lambda(t)] = \delta^2,$$
$$E[\lambda^3(t)] = m_3,$$
$$E[\lambda^4(t)] = m_4.$$
(A.1)

$\lambda(t)$ describes the "noise" component of the naive forecast relative to the optimal forecast, which by assumption is the stock price, $P(t)$. It is further assumed that the noise is independent of all stock prices [i.e., $\lambda(t)$ and $P(s)$ are independent for all t and s]. Therefore, $E[P^0(t)|P(t)] = P(t)$, and hence, $P^0(t)$ is an unbiased forecast. Because, moreover, the $\{\lambda(t)\}$ follow a stationary process, the nonstationary part of the process describing the naive forecast is perfectly correlated with the optimal forecast, $P(t)$.

Substituting for $P^0(T)$ in (3) and rearranging terms, we can write the continuous-time form for the MRS statistics S_1 as

$$S_1 = \frac{1}{T} \int_0^T P(t)[2[1 - \lambda(t)]P^*(t) - [1 - \lambda^2(t)]P(t)]\,dt.$$
(A.2)

From (A.1) and (A.2), we can write the expectation of S_1 conditional on the sample path $\{P(t)\}$, $\overline{S}_1$, as

$$\overline{S}_1 = \frac{\delta^2}{T} \int_0^T [P(t)]^2\,dt,$$
(A.3)

because $\lambda(t)$ is independent of both $\{P(t)\}$ and $\{P^*(t)\}$. Note: $\overline{S}_1$ does not depend on the sample path of $P^*(t)$. From (A.3) and (10.b), we have

$$E_0[S_1] = \delta^2 P_0^2 [e^{(2r+\sigma^2)T} - 1]/[(2r + \sigma^2)T], \tag{A.4}$$

which satisfies the MRS strict inequality $E_0[S_1] > 0$ provided the naive forecast is not optimal (i.e., $\delta^2 > 0$).

Define the random variable $Y_1 \equiv [S_1 - \bar{S}_1]^2$. From (A.2) and (A.3), we write Y_1 as

$$Y_1 = \frac{1}{T^2} \int_0^T \int_0^T P(t)P(s)[2[1 - \lambda(t)]P^*(t) - [1 + \delta^2 - \lambda^2(t)]P(t)] \tag{A.5}$$
$$\cdot [2[1 - \lambda(s)]P^*(s) - [1 + \delta^2 - \lambda^2(s)]P(s)] \, ds \, dt.$$

Because $\lambda(t)$ is independent of $\lambda(s)$ for $t \neq s$, we have from (A.1) and (A.5) that the expectation of Y_1, conditional on the sample path $\{P(t)\}$, $\bar{Y}_1$, can be written as

$$\bar{Y}_1 = \frac{1}{T^2} \int_0^T P^2(t)[4\delta^2[P^*(t)]^2 + 4[1 + \delta^2 - m_3]P(t)P^*(t) \tag{A.6}$$
$$+ [m_4 - (1 + \delta^2)^2]P^2(t)] \, dt.$$

Note that the integrand of (A.6) is always positive. From (8) and (9), we have, for $k = 2, 3, 4$,

$$E_0\{[P(t)]^k[P^*(t)]^{4-k}\} = [P_0 e^{rt}]^4 \exp[6\sigma^2 T + \frac{\sigma^2}{2}k(k-7)(T-t)]. \tag{A.7}$$

Taking expectations in (A.6) and substituting from (A.7), we have that $E_0[Y_1] = E_0[\bar{Y}_1]$ grows exponentially as

$$E_0[Y_1] \sim \exp[(4r + 6\sigma^2)T]/T^2. \tag{A.8}$$

Therefore, the standard deviation of the MRS sample statistics S_1 given by $\sqrt{E_0[Y_1]}$ grows exponentially according to $\exp[(2r + 3\sigma^2)T]/T$. By inspection of (A.4), we have that the ratio of $E_0[S_1]$ to $\sqrt{E_0[Y_1]}$ declines exponentially at the rate $(-2\sigma^2 T)$. Thus, for large T, virtually any sample result for S_1 is consistent with the population condition $E_0[S_1] > 0$. By a similar analysis, the reader can verify that the same result obtains for the MRS statistic S_2.

In contrasting their tests with the earlier Shiller (1981) analysis, MRS (1985, p. 683) point out that their statistics do not require detrending "... because the 'naive forecast' P_t^0 can grow as dividends grow ...." On p. 684, they further their case for robustness by noting "... that the naive forecast need not be efficient in any sense." The naive forecast analyzed here does not seem to be pathological with respect to the conditions they set forth. Thus, it would appear that the naive forecasts necessary to provide proper asymptotic distributional properties for their statistics are anything but naive.

Notes

1. See Fischer and Merton (1984), Marsh and Merton (1983, 1986), and Merton (1983).

2. As may come as a great surprise to those financial economists who regularly publish papers on capital budgeting problems, earnings estimation, financing decisions, and dividend policy, Summers (1985, p. 634) finds it rather "... unfortunate that financial economists remain so reluctant to accept any research relating asset prices and fundamental values." In making this remark, perhaps Summers has in mind those financial economists who might select the closing price on the New York Stock Exchange of a ketchup company's common stock as a better estimate of that firm's fundamental value than an estimate provided by a general economist who computes a present value based on a linear regression model 1 of the supply and demand for ketchup; autoregressive forecasts of future costs of tomatoes, wages, prices of ketchup substitutes, and consumer incomes; and a "reasonable" discount rate.

3. As is well known, even with well-functioning (although not complete) markets and rational, well-informed consumer-investors, the competitive market solution may not be a pareto optimum, and thus, market rationality is not a sufficient condition for efficiency. Using the neoclassical model with overlapping generations, Tirole (1985) has shown that financial security prices that deviate from fundamentals can lead to better allocations than "rational" prices. However, I would argue that those cases in which stock prices both deviate substantially from fundamental values *and* lead to a pareto optimum allocation of investment are, at best, rare.

4. On the self-fulfilling prophecy, see R. K. Merton, (1948). On the rational expectations speculative bubble theory, see Blanchard (1979), Blanchard and Watson (1982), Tirole (1982), and Van Horne (1985).

5. This assumes as a "base date" the publication of Samuelson's 1965 paper, which first set forth the theory in rigorous form. There was, of course, the oral publication of his ideas for at least fifteen years before 1965, as well as many studies of speculative prices and their random properties, extending back as far as the early 1900s.

6. See, for example, the model analyzed in Merton (1971, pp. 403–406), which examines price behavior and optimal portfolio selection when instantaneous stock price changes are random, but the level of stock price regresses toward a "normal price level" with a trend.

7. As will be discussed, the recent study by DeBondt and Thaler (1985) presents evidence that seemingly contradicts these earlier findings.

8. Jensen (1968) found that the average "excess return" per year (including management expenses) across all funds in his sample and all the years from 1945 to 1964, was -1.1%, and 66% of the funds had negative average excess returns. When expenses were excluded, the corresponding statistics were -0.4% per year and 48%. As reported in a recent Business Week article (February 4, 1985, pp. 58–59), based on the industry standard data from SEI Funds Evaluation Services, 74% of managed equity portfolios underperformed the Standard & Poor's 500 Index in 1984; 68% underperformed for the period 1982–1984; 55% underperformed for 1980–1984; and 56% underperformed for 1975–1984.

9. To the extent that stock market prices themselves are an important source of information for investors in calibrating and evaluating other data used to make their assessments of the fundamentals, the original argument that systematically less information is produced on days when the market is closed can be extended to include the Wednesday closings.

10. See Marsh and Merton (1985). Miller and Scholes (1982) find the same result for individual stock returns.

11. As perhaps some indication of the tentative nature of the evidence drawn to support behavioral theories of the stock market, we have, on the one hand, DeBondt and Thaler concluding that investors make cognitive mistakes that result in the underpricing of stocks that have declined (losers) and overpricing of stocks that have risen (winners) and, on the other, Shefrin and Statman (1985) concluding that the evidence supports (different) cognitive mistakes that cause investors to sell their winners "too early" and hold on to their losers "too long." It would seem, therefore, that even a "rational" investor, fully cognizant of his natural tendency to make these mistakes, would, nevertheless, find himself "convicted" by his actions of one or the other cognitive failures.

12. As junior author of Marsh and Merton (1983, 1986).

13. More precisely, Summers (1985, p. 635) refers to the joint hypothesis involving what has been called here "(S.1) and (S.2)." I do not know whether his failure to note the stationarity condition (S.3) as well was intended or not.

14. See Marsh and Merton (1985).

15. Some investors in 1930–1934 may have believed that there was a significantly changed probability of broad-based nationalization of industry than in the past. Given the substantially increased levels of business and financial leverage, there were perhaps others who saw a different prospect for widespread bankruptices than was the case in the past.

16. There were, however, some economists and professional investors who apparently believed that the government had finally found both the will and the means to avoid major macroeconomic disruptions from high unemployment, erratic growth rates, and unstable inflation. Their best guesses for the future may have been formulated with less weight on the distant past.

References

Arrow, K. J., 1982, "Risk Perception in Psychology and Economics," *Economic Inquiry* 20 (January), 1–9.

Blanchard, O., 1979, "Speculative Bubbles, Crashes, and Rational Expectations," *Economic Letters* 3, 387–389.

Blanchard, O., and M. W. Watson, 1982, "Bubbles, Rational Expectations, and Financial Markets," in *Crises in the Economic and Financial Structure*, P. Wachtel (ed.), Lexington Books, pp. 295–315.

DeBondt, W. F. M., and R. Thaler, 1985, "Does the Stock Market Overreact?" *Journal of Finance* 40(3) (July), 793–805.

Fama, E., 1970, "Efficient Capital Markets: A Review of Theory and Empirical Work," *Journal of Finance* 25 (May), 383–417.

Fama, E., 1981, "Stock Returns, Real Activity, Inflation and Money," *American Economic Review*, 71, 545–565.

Fischer, S., and R. C. Merton, 1984, "Macroeconomics and Finance: The Role of the Stock Market," in *Essays on Macroeconomic Implications of Financial and Labor Markets and Political Processes*, K. Brunner and A. H. Meltzer (eds.), Carnegie-Rochester Conference Series on Public Policy, Vol. 21 (Autumn), pp. 57–108.

Flavin, M. A., 1983, "Excess Volatility in the Financial Markets: A Reassessment of the Empirical Evidence," *Journal of Political Economy* 91 (December), 929–956.

French, K., and R. Roll, 1984, "Is Trading Self-Generating?" unpublished paper, Graduate School of Business, University of Chicago (February).

Jensen, M. C., 1968, "The Performance of the Mutual Funds in the Period 1945–1964," *Journal of Finance* 23 (May), 384–416.

Kahneman, D. and A. Tversky, 1979, "Prospect Theory: An Analysis of Decision under Risk," *Econometrica* 47 (March), 263–291.

Kahneman, D. and A. Tversky, 1982, "Intuitive Prediction: Biases and Corrective Procedures," in *Judgement under Uncertainty: Heuristics and Biases*, D. Kahneman, P. Slovic, and A. Tversky (eds.), Cambridge University Press.

Kleidon, A. W., 1983a, "Variance Bounds Tests and Stock Price Valuation Models," working paper, Graduate School of Business, Stanford University (January).

Kleidon, A. W., 1983b, "Bias in Small Sample Tests of Stock Price Rationality," unpublished, University of Chicago.

Leamer, E. E., 1983, "Let's Take the Con Out of Econometrics," *American Economic Review* 73(1), 31–43.

LeRoy, S. F., and R. D. Porter, 1981, "The Present-Value Relation: Tests Based on Implied Variance Bounds," *Econometrica* 49(3), 555–574.

Long, Jr., J. B., 1978, "The Market Valuation of Cash Dividends: A Case to Consider," *Journal of Financial Economics* 6(2/3) (June/September), 235–264.

Mankiw, N. G., D. Romer, and M. D. Shapiro, 1985, "An Unbiased Reexamination of Stock Market Volatility," *Journal of Finance* XL(3) (July), 677–687.

Marsh, T. A., and R. C. Merton, 1983, "Aggregate Dividend Behavior and Its Implications for Tests of Stock Market Rationality," working paper No. 1475–83, Sloan School of Management, MIT (September).

Marsh, T. A., and R. C. Merton, 1985, "Dividend Behavior for the Aggregate

Stock Market," working paper No. 1670–85, Sloan School of Management, MIT (May).

Marsh, T. A., and R. C. Merton, 1986, "Dividend Variability and Variance Bounds Tests for the Rationality of Stock Market Prices," *American Economic Review.* 76(3) (June), 483–498.

Merton, R. C., 1971, "Optimum Consumption and Portfolio Rules in a Continuous Time Model," *Journal of Economic Theory* 3 (December), 373–413.

Merton, R. C., 1973, "An Intertemporal Capital Asset Pricing Model," *Econometrica* 41 (September), 867–887.

Merton, R. C., 1980, "On Estimating the Expected Return on the Market: An Exploratory Investigation," *Journal of Financial Economics* 8, 323–361.

Merton, R. C., 1983, "Financial Economics," in *Paul Samuelson and Modern Economic Theory*, E. C. Brown and R. M. Solow (eds.), McGraw-Hill, pp. 105–138.

Merton, R. K., 1948, "The Self-Fulfilling Prophecy," *Antioch Review* (Summer), 193–210.

Miller, M. H., and M. S. Scholes, 1982, "Dividends and Taxes: Some Empirical Evidence," *Journal of Political Economy* (90), 1118–1142.

Modigliani, F., and R. Cohn, 1979, "Inflation, Rational Valuation and the Market," *Financial Analysts Journal* (March–April), 3–23.

Ohlson, J. A., and S. H. Penman, 1985, "Volatility Increases Subsequent to Stock Splits: An Empirical Abberation," *Journal of Financial Economics* 14(2) (June), 251–266.

Samuelson, P. A., 1965, "Proof That Properly Anticipated Prices Fluctuate Randomly," *Industrial Management Review* 6 (Spring), 41–49.

Shefrin, H., and M. Statman, 1985, "The Disposition to Sell Winners Too Early and Ride Losers Too Long: Theory and Evidence," *Journal of Finance* 40(3) (July), 777–790.

Shiller, R. J., 1981, "Do Stock Prices Move Too Much to be Justified by Subsequent Changes in Dividends?" *American Economic Review* 71 (June), 421–436.

Shiller, R. J., 1982, "Consumption, Asset Markets, and Macroeconomic Fluctuations," *Carnegie-Rochester Conference on Public Policy* 17, 203–250.

Shiller, R. J., 1984, "Stock Prices and Social Dynamics," *Brookings Papers on Economic Activity* 2, 457–498.

Summers, L. H., 1982, "Do We Really Know That Financial Markets are Efficient?" National Bureau of Economic Research, working paper No. 994 (September).

Summers, L. H., 1985, "On Economics and Finance," *Journal of Finance* XL(3) (July), 633–635.

Summers, L. H., 1986, "Does the Stock Market Rationally Reflect Fundamental Values?" *Journal of Finance* 41(3) (July), 591–600.

Tirole, J., 1982, "On the Possibility of Speculation Under Rational Expectations," *Econometrica* 59 (September), 1163–1181.

Tirole, J., 1985, "Asset Bubbles and Overlapping Generations," *Econometrica* 53(5) (September), 1071–1100.

Van Horne, J. C., 1985, "On Financial Innovations and Excesses," *Journal of Finance* XL(3) (July), 621–631.

West, K. D., 1983, "A Variance Bounds Test of the Linear-Quadratic Inventory Model," in *Inventory Models and Backlog Costs: An Empirical Investigation*, unpublished Ph.D. dissertation, Massachusetts Institute of Technology (May).

West, K. D., 1984, "Speculative Bubbles and Stock Price Volatility," Financial Research Center, Memorandum No. 54, Princeton University (December).

[8]

DIVIDEND INNOVATIONS AND STOCK PRICE VOLATILITY

By Kenneth D. West[1]

A standard efficient markets model states that a stock price equals the expected present discounted value of its dividends, with a constant discount rate. This is shown to imply that the variance of the innovation in the stock price is smaller than that of a stock price forecast made from a subset of the market's information set. The implication follows even if prices and dividends require differencing to induce stationarity. The relation between the variances appears not to hold for some annual U.S. stock market data. The rejection of the model is both quantitatively and statistically significant.

KEYWORDS: Volatility test, efficient markets, stock price, nonstationary, random walk.

1. INTRODUCTION

THE SOURCES OF FLUCTUATIONS in stock prices have long been argued. Some observers have suggested that a major part of the fluctuations results from self fulfilling rumors about potential price fluctuations. In a famous passage, Keynes, for example, described the stock market as a certain type of beauty contest in which judges try to guess the winner of the contest: speculators devote their "intelligence to anticipating what average opinion expects average opinion to be" (1964, p. 136). An examination of practically any modern finance text (e.g., Brealey and Myers (1981)) indicates that the economics profession tends to hold the opposite view. Stock price fluctuations are argued to result solely from changes in the expected present discounted value of dividends.

The subject has received increased attention in recent years because of the volatility tests of Leroy and Porter (1981) and, especially, Shiller (1981a). These tests seem to indicate that stock price fluctuations are too large to result solely from changes in the expected present discounted value (PDV) of dividends. There is, however, some question as to the validity of this conclusion. Marsh and Merton (1986) have objected to the tests' assumption that dividends are stationary around a time trend; Flavin (1983) and Kleidon (1985, 1986) have argued that in small samples the tests are biased toward finding excess volatility.

This paper develops and applies a stock market volatility test that is not subject to these criticisms. The test is based on an inequality on the variance of the innovation in the expected PDV of a given stock's dividend stream, and was first suggested by Blanchard and Watson (1982).[2] The inequality states that if discount rates are constant this variance is smaller when expectations are conditional on

---

[1] I thank A. Blinder, J. Campbell, G. Chow, S. Fischer, R. Flood, L. P. Hansen, W. Newey, J. Rotemberg, R. Trevor, J. Taylor, the referees, and an editor of this journal for helpful comments, my own. This paper was revised while I was a National Fellow at the Hoover Institution. An earlier National Science Foundation for financial support. Responsibility for remaining errors is version of this paper was circulated under the title "Speculative Bubbles and Stock Price Volatility."

[2] While Blanchard and Watson (1982) do suggest examining the inequality that is the focus of this paper, they do not formally establish the validity of the inequality, consider possible nonstationarity of dividends or prices, or test the inequality rigorously. Subsequent to the initial circulation of this paper, however, M. Watson sent me a proof of this inequality that is valid when prices and dividends are stationary.

the market's information set than when expectations are conditional on a smaller information set. It may be shown that this implies that the variance of the innovation in a stock price is bounded above by a certain function of the variance of the innovation in the corresponding dividend.

The paper checks whether the bound is satisfied by some long term annual data on the Standard and Poor 500 and the Dow Jones indices. It is not. The estimated variance of the stock price innovation is about four to twenty times its theoretical upper bound. The violation of the inequality is in all cases highly statistically significant.

It is to be emphasized that the inequality is valid even when prices and dividends are an integrated ARIMA process with infinite variances, and that the empirical work allows for such nonstationarity. In addition, the test procedure does not require calculation of a perfect foresight price; this price appears to be central to the small sample biases that are argued by Flavin (1983), Kleidon (1985, 1986), and Marsh and Merton (1986) to plague the Shiller (1981a) volatility test. The paper nonetheless performs some small Monte Carlo experiments to check whether under certain simple circumstances small sample bias in this paper's test procedure is likely to explain the results of the test. The answer is no.

While one of the purposes of this paper is to apply a volatility test with a relatively weak set of maintained statistical assumptions, that is not its only aim. It also considers the consistency of some of the test's maintained economic assumptions with the data, to help determine which among these should be relaxed, so that the excess price volatility might be explained. To that end, the paper uses a battery of formal diagnostic tests on the regressions that must be estimated to calculate the inequality. The test results are in general quite consistent with the test's maintained hypotheses of rational expectations and, perhaps surprisingly, of a constant rate for discounting future dividends. Some additional, less formal analysis, which considers further the constant discount rate hypothesis, does not suggest that the excessive price variability results solely from variation in discount rates.

The evidence, then, does not suggest that the excess volatility is caused by a simple failure of the rational expectations or constant discount rate assumptions. This suggests the possibility that the volatility is due either to rational bubbles (e.g., Blanchard and Watson (1982)) or nearly rational "fads" (e.g., Summers (1986)), whose profit opportunities (if any) are difficult to detect. The paper does not, however, attempt to make a case for bubbles, fads or, for that matter, any other factor, as the explanation of the excess volatility. Instead what is emphasized are two empirical regularities that seem to characterize the data studied here. The first is that prices appear to be too variable to be set as the expected PDV of dividends, with a constant discount rate; this holds even if prices and dividends are nonstationary. The second is that it is difficult to attribute the excess variability to variations in discount rates. Reconciliation of these two points is a task left for future research.

Before turning to the details of the subject at hand, two final introductory remarks seem worth making. The first is that the inequality established here may be of general interest in that it could be used to test other infinite horizon present value models. Possible examples include testing whether consumption is too variable to be consistent with the permanent income hypothesis (see Deaton (1985), West (1988)) or whether exchange rates are too variable to be consistent with a standard monetary model (West (1986a)). That the inequality is valid even in a nonstationary environment makes it particularly appealing in these and perhaps other contexts. The second remark concerns the estimation technique. This is in part an application of West's (1986b) result that it is not always necessary to difference regressions on nonstationary variables, to obtain asymptotically normal parameter estimates. The key requirement is that the nonstationary variables have a drift. Since this is plausible for not only stock prices but for many other macroeconomic variables as well, the estimation technique applied in this paper may be of general interest.

The plan of the paper is as follows. Section 2 establishes the basic inequality. Section 3 explains how the inequality may be used to test a rational expectations, constant discount rate stock price model. Section 4 presents formal econometric results. Section 5 considers informally whether small sample bias or discount rate variation are likely to explain the Section 4 results. Section 6 has conclusions. The Appendix has some econometric and algebraic details.

2. THE BASIC INEQUALITY

The following proposition is the basis of this paper.

PROPOSITION 1. *Let I_t be the linear space spanned by the current and past values of a finite number of random variables, with I_t a subset of I_{t+1} for all t. It is assumed that after s differences, all random variables in I_t jointly follow a covariance stationary ARMA (q, r) process, for some finite s, q, $r \geq 0$. This s'th difference is assumed without loss of generality to have zero mean. All variables are assumed to be identically zero for $t \leq q$.*

Let d_t be one of these variables. Let H_t be a subset of I_t consisting of the space spanned by current and past values of some subset of the variables in I_t, including at a minimum current and past values of d_t. Let b be a positive constant, $0 \leq b < 1$. Let $P(\cdot|\cdot)$ denote linear projections, calculated for $s > 0$ as in Hansen and Sargent (1980). Let

$$x_{tI} = \lim_{k \to \infty} P\left(\sum_0^k b^j d_{t+j} \bigg| I_t\right), \quad x_{tH} = \lim_{k \to \infty} P\left(\sum_0^k b^j d_{t+j} \bigg| H_t\right).$$

(All summations in this section run over j.) Let E denote mathematical expectations. Then

(1) $\quad E[x_{tH} - P(x_{tH}|H_{t-1})]^2 \geq E[x_{tI} - P(x_{tI}|I_{t-1})]^2.$

PROOF:[3] Since d_t is in I_t,

(2) $$x_{tI} = d_t + bP\left(\sum_0^\infty b^j d_{t+j+1} \mid I_t\right)$$
$$= d_t + bx_{t+1,I} - be_{t+1},$$

$$e_{t+1} = x_{t+1,I} - P\left(\sum_0^\infty b^j d_{t+j+1} \mid I_t\right) = x_{t+1,I} - P(x_{t+1,I} \mid I_t).$$

Equation (2) may be rewritten as
$$x_{tI} - d_t = bx_{t+1,I} - be_{t+1}.$$

Recursive substitution for $x_{t+1,I}$, then for $x_{t+2,I}$, etc., yields

(3) $$x_{tI} - \sum_0^{k-1} b^j d_{t+j} = b^k x_{t+k,I} - \sum_1^k b^j e_{t+j}.$$

The assumptions of the proposition insure that as $k \to \infty$, $b^k x_{t+k,I} \to 0$ in mean square. Consider first the ARIMA $(q, s, 0)$ case. The formulas in Hansen and Sargent (1980) state that x_{tI} is a finite distributed lag on the variables in I_t. Since these variables started up at a finite date in the past, and some arithmetic difference of each variable has a finite variance, the rate of growth of the variance of each variable, and therefore of the variance of x_{tI} as well, is some power of t. Since $\lim_{k \to \infty} b^{2k}(t+k)^n = 0$ for any fixed $n \geq 0$, $\lim_{k \to \infty} \text{var}(b^k x_{t+k,I}) = 0$. The argument for the ARIMA (q, s, r) case is implied by Hansen and Sargent (1980) since for $j > r$, $P(d_{t+j} \mid I_t)$ is determined by a difference equation that depends only on s and the autoregressive parameters.

The assumptions of the proposition also guarantee that e_t has constant, finite variance and is serially uncorrelated. For the ARIMA $(q, s, 0)$ case, this follows directly from inspection of the formula for x_{tI} in Hansen and Sargent (1980). Once again, this argument immediately extends to the ARIMA (q, s, r) case. Therefore, $\sum_1^\infty b^j e_{t+j}$ exists, in the sense that $\lim_{k \to \infty} E(\sum_1^k b^j e_{t+j} - \sum_1^\infty b^j e_{t+j})^2 = 0$, $\text{var}(\sum_1^\infty b^j e_{t+j}) = b^2(1-b^2)^{-1} Ee_t^2$ (Fuller (1976, p. 36)). Equation (3) therefore implies

(4) $$x_{tI} - \sum_0^\infty b^j d_{t+j} = -\sum_1^\infty b^j e_{t+j}.$$

By a similar argument, involving projections onto H_t,

(5) $$x_{tH} - \sum_0^\infty b^j d_{t+j} = -\sum_1^\infty b^j f_{t+j},$$

$$f_{t+j} = x_{t+j,H} - P(x_{t+j,H} \mid H_{t+j-1}), \quad \text{var}\left(-\sum_1^\infty b^j f_{t+j}\right) = b^2(1-b^2)^{-1} Ef_t^2.$$

---

[3] J. Campbell suggested the basic idea of this proof. L. P Hansen and M. Watson have provided alternative proofs. S. Leroy has pointed out to me that a similar proposition is implied in the stationary case in Leroy and Porter (1981, p. 568). My own, rather tedious, proof may be found in an earlier version of this paper (West (1984)).

Now,

(6) $\quad \text{var}\left(-\sum_{1}^{\infty} b^j f_{t+j}\right) = \text{var}\left(x_{tH} - \sum_{0}^{\infty} b^j d_{t+j}\right)$

$\quad = \text{var}\left(x_{tH} - x_{tl} + x_{tl} - \sum_{0}^{\infty} b^j d_{t+j}\right)$

$\quad = \text{var}\left(x_{tH} - x_{tl} - \sum_{1}^{\infty} b^j e_{t+j}\right)$

$\quad = \text{var}(x_{tH} - x_{tl}) + \text{var}\left(-\sum_{1}^{\infty} b^j e_{t+j}\right)$

$\quad \geq \text{var}\left(-\sum_{1}^{\infty} b^j e_{t+j}\right).$

The last equality follows since for $j \geq 1$, e_{t+j} is uncorrelated with anything in I_t, including, in particular, $x_{tH} - x_{tl}$. It follows from (6) that $b^2(1-b^2)^{-1} Ef_t^2 \geq b^2(1-b^2)^{-1} Ee_t^2 \Rightarrow Ee_t^2 \geq Ef_t^2$, i.e., $E[x_{tH} - P(x_{tH}|H_{t-1})]^2 \geq E[x_{tl} - P(x_{tl}|I_{t-1})]^2$.

Q.E.D.

A verbal restatement of Proposition 1 is as follows. Suppose we are forecasting the present discounted value of d_t, by calculating x_{tl} and x_{tH}. Each period as new data become available we revise our forecast. $E(x_{tl} - P(x_{tl}|I_{t-1}))^2$ and $E(x_{tH} - P(x_{tH}|H_{t-1}))^2$ are measures of the average size of this period to period revision. Proposition 1 says that with less information the size of the revision tends to be larger. That is, when less information is used, the variance of the innovation in the expected present discounted value of d_t is larger.

It is worth making four comments on the conditions under which (1) is valid. First, if the random variables in I_t are stationary without differencing, Proposition (1) does not require that the variables follow a finite parameter ARMA (q, r) process. The ARIMA assumption is maintained because to my knowledge infinite horizon prediction for nonstationary variables has been developed only for ARIMA processes. Second, (1) may not extend immediately if logarithms or logarithmic differences are required to induce stationarity in d_t, even if linear projections are replaced with mathematical expectations. If, for example, $\log(d_t) = \log(d_{t-1}) + \varepsilon_t$, $\varepsilon_t \sim N(0, \sigma^2)$, and H_t is the information set generated by past d_t, it may be shown that $[x_{tH} - E(x_{tH}|H_{t-1})]^2$ is proportional to d_{t-1}^2. Third, the inequality need not hold for a finite horizon. That is, it need not hold if we consider the variance of the innovation in the expected PDV of $\sum_0^n b^j d_{t+j}$ instead of $\sum_0^\infty b^j d_{t+j}$. An example is given in footnote 4 of West (1986c). The reason is that terms of the form $b^{n+1} x_{t+n+1,l}$ and $b^{n+1} x_{t+n+i,H}$ are present. See equation (3). Fourth, (1) does not hold for arbitrary subsets of I_t. If, for example, H_t were the empty set, x_{tH} would also be the empty set, and the left-hand side of (1) would be identically zero.

Before developing the implications of (1) for stock price volatility, it may be helpful to work through a simple example. Suppose I_t consists of lags of d_t and of one other variable, z_t. Let H_t consist simply of lags of d_t. Let the bivariate (d_t, z_t) representation be

(7) $$\begin{bmatrix} d_t \\ z_t \end{bmatrix} = \begin{bmatrix} \phi & 1 \\ 0 & 0 \end{bmatrix} \begin{bmatrix} d_{t-1} \\ z_{t-1} \end{bmatrix} + \begin{bmatrix} \varepsilon_{1t} \\ \varepsilon_{2t} \end{bmatrix}$$

with $|\phi| \leq 1$, ε_{1t} and ε_{2t} i.i.d., $E\varepsilon_{1t}\varepsilon_{2s} = 0$ for all t, s. Let $E\varepsilon_{1t}^2 = \sigma_1^2$, $E\varepsilon_{2t}^2 = \sigma_2^2$. The univariate representation of d_t clearly is $d_t = \phi d_{t-1} + v_t$, $v_t = \varepsilon_{1t} + z_{t-1} = \varepsilon_{1t} + \varepsilon_{2t-1}$, $Ev_t^2 = \sigma_1^2 + \sigma_2^2$. Let us calculate both sides of (1).

(8) $\quad P(d_{t+j}|H_t) = \phi^j d_t$

$$\Rightarrow x_{tH} = P\left(\sum_0^\infty b^j d_{t+j} \middle| H_t\right) = (1 - b\phi)^{-1} d_t$$

$$\Rightarrow E[x_{tH} - P(x_{tH}|H_{t-1})]^2 = E[(1 - b\phi)^{-1} v_t]^2 = (1 - b\phi)^{-2}(\sigma_1^2 + \sigma_2^2).$$

$P(d_t | I_t) = d_t$.

$P(d_{t+j}|I_t) = \phi^j d_t + \phi^{j-1} z_t, \quad j > 0,$

$$\Rightarrow x_{tI} = P\left(\sum_0^\infty b^j d_{t+j} \middle| I_t\right) = (1 - b\phi)^{-1}(d_t + bz_t)$$

$$\Rightarrow E[x_{tI} - P(x_{tI}|I_{t-1})]^2 = E[(1 - b\phi)^{-1}(\varepsilon_{1t} + bz_t)]^2$$
$$= (1 - b\phi)^{-2}(\sigma_1^2 + b^2\sigma_2^2).$$

Since $b^2 < 1$, $\sigma_1^2 + \sigma_2^2 > \sigma_1^2 + b^2\sigma_2^2$, so (1) holds. Observe that (1) holds even when $\phi = 1$ so that d_t is nonstationary.

3. THE MODEL

According to a standard efficient markets model, a stock price is determined by the relationship (9) (Brealey and Myers (1981, pp. 42-45)):

(9) $\quad p_t = bE[(p_{t+1} + d_{t+1})|I_t],$

where p_t is the real stock price at the end of period t, b the constant ex-ante real discount rate, $0 < b = 1/(1+r) < 1$, r the constant expected return, E denotes mathematical expectations, d_{t+1} the real dividend paid to the owner of the stock in period $t+1$, and I_t the information set common to traders in period t. I_t is assumed to contain, at a minimum, current and past dividends, and, in general, other variables that are useful in forecasting dividends.

Equation (9) may be solved recursively forward to get

(10) $\quad p_t = \sum_1^n b^j E(d_{t+j}|I_t) + b^n E(p_{t+n}|I_t).$

DIVIDEND INNOVATIONS

If the transversality condition

(11) $\lim_{n\to\infty} b^n E(p_{t+n}|I_t) = 0$

holds, then

(12) $p_t = \sum_1^\infty b^j E(d_{t+j}|I_t).$

It will be assumed that in forecasts of $\sum_1^\infty b^j d_{t+j}$ from I_{t-k}, for any $k \geq 0$, mathematical expectations conditional on the market's information set are the same as linear projections. So x_{tl}, defined in Proposition 1 as the linear projection of $\sum_0^\infty b^j d_{t+j}$ onto a period t set of random variables equals $E(\sum_0^\infty b^j d_{t+j}|I_t)$. Similarly, the linear projection of x_{tl} onto the market's period $t-1$ set of random variables equals $E(x_{tl}|I_{t-1})$.

Proposition 1 is used to test the model (12) as follows. Since $x_{tl} = E(\sum_0^\infty b^j d_{t+j}|I_t)$, (12) implies that $x_{tl} = p_t + d_t$. So $E[x_{tl} - E(x_{tl}|I_{t-1})]^2 = E[p_t + d_t - E(p_t + d_t|I_{t-1})]^2$, and, therefore,

(13) $E[x_{tH} - P(x_{tH}|H_{t-1})]^2 \geq E[p_t + d_t - E(p_t + d_t|I_{t-1})]^2.$

The intuitive reason that the model (12) implies (13) is as follows. $E(x_{tH} - P(x_{tH}|H_{t-1}))^2$ is by definition a measure of the average size of the innovation in the expected present discounted value (PDV) of dividends, when expectations are conditional on H_t. According to (12), price adjusts unexpectedly only in response to news about dividends. $E[p_t + d_t - E(p_t + d_t|I_{t-1})]^2$ is a measure of the average size of the innovation in the expected PDV of dividends, with expectations conditional on the market's information set I_t. Since the market is presumed to use the variables in I_t to forecast optimally, the market's forecasts tend to be more precise, i.e., (13) holds.[4]

To make (13) operational, both sides of it must be calculated. Consider first $E[p_t + d_t - E(p_t + d_t|I_{t-1})]^2$. A consistent estimate of this is easily obtained by estimating (9) with the instrumental variables method of McCallum (1976) and Hansen and Singleton (1982). Rewrite (9) as

(14) $p_t = b(p_{t+1} + d_{t+1}) - b[p_{t+1} + d_{t+1} - E(p_{t+1} + d_{t+1}|I_t)]$

$= b(p_{t+1} + d_{t+1}) + u_{t+1},$

$\sigma_u^2 = b^2 E[p_t + d_t - E(p_t + d_t|I_{t-1})]^2.$

Equation (14) can be estimated by instrumental variables, using as instruments variables known at time t. An estimate of $E[p_t + d_t - E(p_t + d_t|I_{t-1})]^2$ is then obtainable as $\hat{b}^{-2}\hat{\sigma}_u^2$.

---

[4] Note that inequality (13) holds even for the class of dividend and price processes studied by Marsh and Merton (1986), as long as arithmetic differences suffice to induce stationarity. This is because the March and Merton (1984) model implies that $H_t = I_t$. When $H_t = I_t$, inequality (13) holds trivially, as a strict equality. See the discussion in West (1984).

Estimation of $E[x_{tH} - P(x_{tH}|H_{t-1})]^2$ is slightly more involved. It requires first of all specification of H_t. The simplest possible one is $H_t = \{1, d_{t-j}|j \geq 0\}$, and H_t defined this way is what is used in this paper's empirical work.[5] Choices of H_t that include lags of additional variables might produce sharper results, but would also entail more complex calculations. With $H_t = \{1, d_{t-j}|j \geq 0\}$, $E[x_{tH} - P(x_{tH}|H_{t-1})]^2$ can be calculated as a function of d_t's univariate ARIMA parameters. Suppose $d_t \sim$ ARIMA $(q, s, 0)$,

(15) $\quad \Delta^s d_{t+1} = \mu + \phi_1 \Delta^s d_t + \cdots + \phi_q \Delta^s d_{t-q+1} + v_{t+1},$

where $\Delta^s = (1-L)^s$, L the lag operator. (A moving average component to d_t is assumed absent for notational and computational simplicity.) Then $x_{tH} = P(\sum b^j d_{t+j}|H_t) = m + \sum_1^{q+s} \delta_i d_{t-i+1}$. The δ_i are complicated functions of b and the ϕ_i. Hansen and Sargent (1980) provide explicit formulas for the δ_i. In particular, given b and the ARIMA parameters of d_t, one can use the Hansen and Sargent (1980) formula for δ_1 to calculate $\delta_1^2 \sigma_v^2 = E[x_{tH} - E(x_{tH}|H_{t-1})]^2$. To test the null hypothesis that prices are determined according to (12), then we calculate

(16) $\quad \delta_1^2 \sigma_v^2 - b^{-2} \sigma_u^2$

and test $H_0: \delta_1^2 \sigma_v^2 - b^{-2} \sigma_u^2 \geq 0$. If the estimate of (16) is negative (that is, the implications of (12) for the innovation variances are not borne out by the data), a convenient way to quantify the extent of the failure of the model (12) is to calculate

(17) $\quad -100(\delta_1^2 \sigma_v^2 - b^{-2} \sigma_u^2)/(b^{-2} \sigma_u^2).$

When (16) is negative, (17) yields a number between 0 and 100. I will refer to this somewhat loosely as the percentage of the variance of the innovation in p_t that is excessive. This is of course somewhat imprecise in that $b^{-2} \sigma_u^2$ is the variance of the innovation in the *sum* of dividends and prices. But given that price innovations are much larger than dividend innovations (see the empirical results below), this terminology does not seem misleading.[6]

What alternatives might explain a rejection of the null hypothesis that (16) is positive? Three have figured prominently in discussions of related work: gross expectational irrationality, of the sort that systematically leads to profit opportunities (e.g., Ackley (1983)); variation in discount rates (e.g., Leroy (1984)); and rational or nearly rational bubbles or fads (e.g., Blanchard and Watson (1982), Summers (1986)), whose profit opportunities (if any) are very difficult to detect. In light of some empirical evidence yet to be presented, it is of interest to note that diagnostic tests on the estimates of (14) and (15) can help to distinguish between bubbles and fads on the one hand, and gross expectational irrationality

---

[5] Proposition 1 assumed that variables had zero mean. If not, H_t and I_t must be expanded to include suitable deterministic terms. In the annual data used here, a constant is the only relevant such term.

[6] In fact, in some empirical work the variable that is here called d_{t+1} is assumed known at time t and thus has an innovation of zero when forecast at time t (Shiller (1981a), Leroy and Porter (1981)).

and time varying discount rates on the other, as possible explanations of any excess price volatility. Technically, when rational bubbles are absent (i.e., the transversality condition (11) holds), equation (14) and the dividend equation (15) together imply that (16) is positive. But when rational bubbles are present, (14) and (15) need not imply that (16) is positive (West (1986c)). So bubbles provide a logical explanation of any excess price volatility if (14) and (15) appear to be well specified. More generally, since it may be difficult to detect small departures from the rational bubble alternative in a given finite sample, evidence that (14) and (15) appear legitimate, despite excess price volatility, is consistent as well with nearly rational bubbles or fads of the sort considered in Summers (1986). So an essential part of the strategy used here to distinguish between rational or nearly rational bubbles or fads versus other alternatives as explanations of excess price variability is to perform diagnostic tests on equations (14) and (15). The greater the extent to which these two equations appear to be well specified, the more persuasive is the inference that bubbles or fads explain the excess volatility.[7]

4. EMPIRICAL EVIDENCE

A. *Data and Estimation Technique*

The data used were those used by Shiller (1981a) in his study of stock price volatility, and were supplied by him. There were two data sets, both containing annual aggregate price and dividend data. One had the Standard and Poor 500 for 1871–1980 (p_t is price in January divided by producer price index (1979 = 100), d_{t+1} is the sum of dividends from that same January to the following December, deflated by the average of that year's producer price index). The other data set was a modified Dow Jones index, 1928–1978 (p_t, d_{t+1} as above). See Shiller (1981a) for a discussion of the data.

The following aspects of estimation will be discussed in turn: (i) selection of the lag length q of the dividend process, (ii) estimation of (14), (15), and (16), (iii) calculation of the variance-covariance matrix of the parameters estimated, and (iv) diagnostic tests performed.

(i) It was assumed that the univariate d_t process required at most one difference to induce stationarity. That is, in equation (15), $s = 0$ (the original series was used) or $s = 1$ (first difference of original series used). No other values of s were tried.

For both the differenced and undifferenced versions of each data set's dividend process, two values of the lag length q were used. One was arbitrarily selected

---

[7] Unless, of course, one has a theoretical presumption that bubbles are not present: a consensus view on how general are the equilibria that admit bubbles is far from established. For a general equilibrium model that allows bubbles, see Tirole (1985). For an argument that bubbles are inconsistent with rationality, see Diba and Grossman (1985). For discussions on the use of volatility tests versus other techniques in studying present value models, see Hamilton and Whiteman (1984), Hansen and Sargent (1981), and Shiller (1981b). See West (1987) on the interpretation of the Summers (1986) alternative as a nearly rational bubble.

as $q=4$. The other was the q selected by the information criterion of Hannan and Quinn (1979). This criterion chooses the value of q that minimizes a certain function of the estimated parameters. Conditional on q being no greater than some fixed upper bound, which I set to 4, the correct q will be chosen asymptotically if the process truly has a finite order autoregressive representation.[8]

Thus, for each data set up to four sets of parameter estimates were calculated: $q=4$, where $q=$ lag length selected by the information criterion, for differenced and undifferenced data. In one case (Dow Jones, differenced), the Hannan and Quinn (1979) criterion chose $q=4$. So only three sets of parameters were calculated for the Dow Jones.

(ii) Calculation of (16) required estimation of the bivariate system consisting of equations (14) and (15). Equation (14) was estimated by Hansen's (1982) and Hansen and Singleton's (1982) two-step, two-stage least squares. The first step obtained the optimal instrumental variables estimator. The $q+1$ instruments used were the variables on the right hand side of (15), i.e., a constant term and q lags of $\Delta^s d_t$ ($s=0$ or $s=1$). Equation (15) was estimated by OLS, with the covariance matrix of the parameter estimates adjusted for conditional heteroskedasticity as described in (iii).

With $\Delta^s d_t \sim AR(q)$, the δ_1 parameter in the formula (16) is $[(1-b)^s \Phi(b)]^{-1}$, $\Phi(b) = 1 - \sum_1^q b^j \phi_j$ (Hansen and Sargent (1980)). Thus, formula (16) was calculated as $[(1-\hat{b})^s (1-\sum_1^q \hat{b}^j \hat{\phi}_j)]^{-1} \hat{\sigma}_v^2 - \hat{b}^{-2} \hat{\sigma}_u^2$.

The innovation variances $\hat{\sigma}_v^2$ and $\hat{\sigma}_u^2$ were calculated from the moments of the residuals of the regressions, with a degree of freedom correction used for $\hat{\sigma}_v^2$:

(18) $\quad \hat{\sigma}_u^2 = (T-s)^{-1} \sum_{t=1}^{T-s} \hat{u}_{t+1}^2,$

$\hat{\sigma}_v^2 = (T-s-q-1)^{-1} \sum_{t=1}^{T-s} \hat{v}_{t+1}^2.$

T is the number of observations; $T=110$ for the Standard and Poor's index, $T=51$ for the Dow Jones index.

The parameter vector estimated was thus $\hat{\theta} = (\hat{b}, \hat{\mu}, \hat{\phi}_1, \ldots, \hat{\phi}_q, \hat{\sigma}_v^2, \hat{\sigma}_u^2)$. $\hat{\theta}$ is asymptotically normal with an asymptotic covariance matrix V (see the Appendix and (iii) below).[9] Let $f(\theta)$ be formula (16) above. The standard error on the

---

[8] The Hannan-Quinn procedure selects the q that minimizes

$\ln \hat{\sigma}_v^2 + T^{-1} 2qk \ln \ln T,$

for $q<Q$ for some fixed Q, with $k>1$. I set $Q=4$, $k=1.001$. The choice of $k=1.001$ was made because Hannan and Quinn (1979, p. 194) seem to suggest a value very close to one is appropriate for sample sizes such as those used in this paper.

[9] A referee has suggested that I emphasize that West (1986b), the references for the asymptotic distribution of parameter estimates for differenced specifications, requires $E(\Delta d_t + \Delta p_t) \neq 0$. While this certainly seems reasonable a priori given that the data are from a stock market in a growing economy, the upward drift in the data is not particularly well reflected in formal statistical tests. See, for example, the insignificant constant terms in all the differenced specifications in Table IB. It is reassuring, then, that the Monte Carlo simulations in West (1986b), which assume data as noisy as those used here, suggest that the asymptotic normal approximation can be useful even with the sample sizes used here.

estimate of (16) was calculated as $[(\partial f/\partial \theta) V(\partial f/\partial \theta)']$. The derivatives of f were calculated analytically.

(iii) The estimate of V, the variance covariance matrix of $\hat{\theta}$, was calculated by the methods of Hansen (1982), Newey and West (1987), and West (1986c), so that the estimate would be consistent for an *arbitrary* ARMA process for u_t and v_t. This is necessary because, for example, the correlation between u_t and v_{t+j} may in principle be nonzero for all $j \geq 0$. The Newey and West (1987) procedure was used to insure that V was positive semidefinite. Details may be found in the Appendix. It suffices to note here that the procedure for calculating the standard error on (16) properly accounts for the uncertainty in the estimates of both the regression parameters and the variances of the residuals.

(iv) The final item discussed before results are presented is diagnostic tests on equations (14) and (15). Four diagnostic checks were performed.

The first checked for serial correlation in the residuals to the equations, using a pair of tests. As noted above, u_{t+1}, the disturbance to equation (14), is an expectational error. If expectations are rational, then u_{t+1} will be serially uncorrelated. Equation (15)'s disturbance v_{t+1} should also be serially uncorrelated, since v_{t+1} is the innovation in the dividend process.

The first of the pair of serial correlation tests checked for first order serial correlation in u_{t+1} and v_{t+1}. The calculation of the standard errors for this is described in the Appendix. The second of the pair of serial correlation tests, performed only for (15), calculated the Box-Pierce Q statistic for the residuals. This statistic of course simultaneously tests for first and higher order serial correlation. See Granger and Newbold (1977, p. 93).

The second of the four diagnostic checks was performed only on equation (14). This was a test of instrument-residual orthogonality, basically checking whether the residual to (14) is uncorrelated with lagged dividends (Hansen and Singleton (1982)). Let Z_t be the $(q+1) \times 1$ vector of instruments and $\hat{b}$ the estimate of b. The orthogonality test is computed as:

$$(19) \quad \left\{ \sum_{t=1}^{T-s} Z_t'[p_t - \hat{b}(p_{t+1} + d_{t+1})] \right\} (T\hat{S}_z) \left\{ \sum_{t=1}^{T-s} Z_t[p_t - \hat{b}(p_{t+1} + d_{t+1})] \right\}.$$

$\hat{S}_z$ is an estimate of $E(Z_t u_{t+1})(Z_t u_{t+1})'$ and was calculated as $T^{-1}(\sum Z_t Z_t' \bar{u}_{t+1}^2)$, where $\bar{u}_t$ is the 2SLS residual to (14). The statistic (19) is asymptotically distributed as a chi-squared random variable with q degrees of freedom. This test has the power to detect irrational expectational errors and variations in discount rates that are correlated with dividends.

The third of the four diagnostic checks tested for the stability of the regression coefficients in (14) and (15). Each sample was split in half, a pair of regression estimates was obtained, and equality of the pair was tested. The resulting statistic is asymptotically chi-squared, with one degree of freedom for (14) and $(q+1)$ degrees of freedom for (15). This test clearly has the power to detect shifts in the discount rate, as well as in the dividend process.

The fourth and final diagnostic check performed is implicit in the estimation procedure described above. A variety of specifications for the dividend process were used—differenced and undifferenced, with a variety of lag lengths. Since the results did not prove sensitive to the specification of the dividend process, the likelihood is relatively small that changes in the specification of the dividend process will affect the results.

B. *Empirical Results*

Regression results for (14) and (15) are reported in Tables I-A and I-B. The results in Table I-A suggest that the basic arbitrage equation (1) is a sensible one. The entries in column (4) do not reject the null hypothesis of no serial correlation in u_{t+1}, the disturbance to equation (14). The test statistic in all cases is far from significant at the .05 level. The equation (19) test for instrument-residual orthogonality also does not reject the null hypothesis of no correlation between the instruments and the residuals at the .05 level, for any specification. See column (5).[10]

TABLE 1-A: EQUATION 14
REGRESSION RESULTS

Data Set	(1) Differenced	(2)[c] q	(3)[c] b	(4)[c] ρ	(5)[c] H/sig	(6)[c] Stability/sig
S and P						
1873–1980	no	2[a]	.9311 (.0186)[b]	.0695 (.0766)	5.50/.064	4.55/.033
1874–1980	yes	2[a]	.9413 (.0170)	.0670 (.0974)	2.87/.238	.33/.566
1875–1980	no	4	.9315 (.0158)	.0661 (.0754)	6.96/.138	3.69/.055
1876–1980	yes	4	.9449 (.0136)	.0671 (.0984)	3.15/.533	.28/.594
Modified Dow Jones						
1931–1978	no	3[a]	.9402 (.0301)	−.1040 (.0806)	5.42/.144	1.56/.211
1933–1978	yes	4[a]	.9379 (.0188)	−.1182 (.0752)	5.20/.267	2.02/.154
1932–1978	no	4	.9271 (.0253)	−.1112 (.1493)	6.08/.108	.49/.483

See notes to Table I-B.

[10] Some results of Flood, Hodrick, and Kaplan (1986) should, however, be noted. They apply the test of instrument-residual orthogonality to these data using three lags of d_t/p_t as instruments, and estimating (14) in the form $1 = b(p_{t+1} + d_{t+1})/p_t +$ error. They report $\chi^2(3)$ test statistics with significance levels of .03 for the S and P and .08 for the Dow Jones. This suggests some mild evidence against the model. They also report stronger rejections using some indirect tests of the constant expected return model. See Section 5B for further discussion.

TABLE I-B: Equation 15 Regression Results

DIVIDEND INNOVATIONS

Data Set	(1) Differenced	(2)[c] q	(3)[c] μ	(4)[c] ϕ_1	(5)[c] ϕ_2	(6)[c] ϕ_3	(7)[c] ϕ_4	(8)[c] ρ	(9)[c] Q/sig	(10)[c] Stability/sig
S and P										
1873–1980	no	2[a]	.168 (.084)[b]	1.196 (.114)	−.238 (.103)			.045 (.025)	36.87/.181	12.93/.005
1874–1980	yes	2[a]	.034 (.029)	.262 (.118)	−.214 (.071)			.002 (.023)	22.79/.824	2.71/.438
1875–1980	no	4	.150 (.080)	1.247 (.116)	−.480 (.093)	.227 (.113)	−.029 (.066)	.001 (.010)	21.39/.875	33.49/.000
1876–1980	yes	4	.036 (.031)	.264 (.115)	−.230 (.094)	.026 (.080)	−.006 (.153)	.001 (.011)	23.98/.773	4.34/.501
Modified Dow Jones										
1931–1978	no	3[a]	1.945 (1.037)	1.265 (.112)	−.664 (.108)	.333 (.098)		.002 (.054)	4.05/1.000	7.53/.111
1933–1978	yes	4[a]	.275 (.405)	.302 (.119)	−.351 (.133)	.051 (.093)	.050 (.176)	−.024 (.067)	9.77/.939	8.06/.153
1932–1978	no	4	1.925 (1.900)	1.263 (.111)	−.662 (.208)	.330 (.209)	.004 (.134)	.005 (.022)	4.06/1.000	10.22/0.69

[a] Lag length q chosen by Hannan and Quinn (1979) procedure
[b] Asymptotic standard errors in parentheses
[c] Symbols: q is lag length of dividend autoregression (15); parameters b, μ, ϕ_i are defined in equations (9) and (15); ρ is the first order serial correlation coefficient of disturbance; H is the statistic in equation (19), $H \sim \chi^2(q)$; "stability" is test for stability of coefficients, as described in text, distributed $\chi^2(q+1)$ in Table I-A and $\chi^2(q+1)$ in Table I-B; Q is Box-Pierce Q statistic, $Q \sim \chi^2(30)$ for S and P, $Q \sim \chi^2(18)$ for Dow Jones. For the "H", "stability", and "Q" columns, "sig" refers to the probability of seeing the statistic under the null hypothesis

Most important, the discount rate b is estimated plausibly and extremely precisely in all regressions. See column (3). The implied annual real interest rates are about six to seven per cent. These rates are quite near the arithmetic means for ex post returns: 8.1 per cent for the S and P index (1872–1981) and 7.4 per cent for the Dow Jones index (1929–1979). The estimates of the discount rate therefore are reasonable. Moreover, there is little evidence that the rate was different in the two halves of either sample. As indicated in column (6), the null hypothesis of equality cannot be rejected at the five per cent level for any specification except the S and P, undifferenced, $q = 2$. In addition, no evidence against the constancy of the discount rate may be found in a comparison of the two halves' mean ex post returns. For the S and P index, these were (in per cent) 8.09 (1872–1926) versus 8.12 (1927–1981); for the Dow Jones the figures are 7.87 (1929–1954) versus 6.92 (1955–1979).

In general, then, the specification of the arbitrage equation (14) seems acceptable, with the possible exception of the S and P data set with dividends undifferenced. Let us now turn to the estimates for the dividend process, reported in Table I-B. Once again, the entries in columns (8) and (9) allow comfortable acceptance of the null hypothesis of no serial correlation in the disturbance to equation (15). With one exception, both test statistics in all regressions are far from significant. The only possible exception was the estimate of the first order serial correlation coefficient $\hat{\rho}$ for the S and P index, undifferenced, lag length = 2. Note, however, that this regression's Q statistic in column (9) comfortably accepts the null hypothesis of no serial correlation. Overall, then, no serial correlation to the residual to (15) is apparent. Also, the estimates of most regression coefficients are fairly precise, at least when the lag length q was chosen by the Hannan and Quinn (1979) procedure. Finally, the null hypothesis that the parameters of the dividend process are the same in the two halves of each sample cannot be rejected for any specification except the Standard and Poor's, undifferenced. See column (10). Overall, then, the specification of the dividend process seems quite acceptable, again with the possible exception of the S and P data set, undifferenced.

The null hypothesis that price is the expected present discounted value of dividends, with a constant discount rate, does not, however, appear acceptable, for any specification. As may be seen from column (7) in Table II, formula (16) is always negative, and significantly so. The asymptotic z-stat (ratio of parameter to asymptotic standard error) was always larger than 2.5. This means that the column (7) entries are always significant at the one-half per cent level, using a one-tailed test. The null hypothesis may therefore be rejected at traditional significance levels. Furthermore, the fraction of the variance of the price innovation that is excessive is substantial, about 80 to 95 per cent (column (8) of Table II).

The residual price fluctuation might reflect grossly irrational reaction to news about dividends, variation in discount rates, or some combination of these and other factors. For the S and P undifferenced specifications, the econometric evidence perhaps is not particularly helpful in discriminating among the various possibilities. It is worth noting, however, that for the other specifications, the

DIVIDEND INNOVATIONS

TABLE II
TEST STATISTICS

Data Set	(1) Differenced	(2) q	(3)[c] b	(4)[c] δ_1	(5)[d] σ_v^2	(6)[d] σ_u^2	(7)[d] Eqn (16)	(8) Eqn (17)
S and P	No	2[a]	.9311 (.0186)[b]	10.82 (3.47)	215.2 (79.0)	.1501 (.0543)	−230.66 (87.10)	92.92
	Yes	2[a]	.9413 (.0170)	18.06 (6.22)	214.1 (80.2)	.1485 (.0523)	−193.22 (71.07)	79.95
	No	4	.9315 (.0158)	10.76 (3.10)	219.4 (73.4)	.1502 (.0510)	−235.51 (90.12)	93.12
	Yes	4	.9449 (.0136)	18.45 (5.64)	218.2 (81.1)	.1538 (.0511)	−192.05 (73.63)	78.58
Modified Dow-Jones	No	3[a]	.9402 (.0301)	8.28 (2.85)	19,653 (5,836)	9.980 (3.383)	−21,545 (5,978)	96.92
	Yes	4[a]	.9379 (.0188)	15.78 (8.13)	19,342 (5,871)	9.014 (2.655)	−19,740 (5,852)	89.79
	No	4	.9271 (.0253)	7.55 (3.12)	19,228 (3,912)	10.453 (2.427)	−21,777 (4,309)	97.34

[a] Lag length q chosen by Hannan-Quinn (1979) criterion
[b] Asymptotic standard errors in parentheses
[c] Symbols: q = lag length in dividend regression; b defined in equation (9); δ_1 defined above equation (18); σ_v^2 and σ_u^2 defined in equation (18)
[d] Units for columns (5)–(7) are 1979 dollars squared For the S and P, $P_{1979} = 99.71$, $d_{1979} = 5.65$; for the Dow-Jones, $P_{1979} = 468.94$, $d_{1978} = 30.91$

results of the diagnostic tests were more consistent with the residual volatility being due to bubbles or fads whose profit opportunities are difficult to detect, than to a misspecification of the arbitrage or dividend equations.[11]

5. SOME ADDITIONAL ANALYSIS

This section considers the possibilities that the previous section's results are due to (A) small sample bias, or (B) variation in discount rates. It is to be emphasized that the analysis is informal, and the conclusions are far from definitive. The goal here is simply to gather some evidence on whether either possibility explains the results; a complete, rigorous econometric examination of either possibility would require a separate paper.

A. Small Sample Bias

This section uses two small Monte Carlo experiments to get a feel for the importance of two types of bias. Part (a) below considers whether under certain simple circumstances small sample bias is likely to account for the finding of excess variability. Part (b) studies whether under equally simple circumstances low small sample power of the equation (19) test of instrument residual volatility is likely to explain the generally favorable results of the diagnostic tests.

a. Bias in Estimate of Excess Volatility

It is important to consider whether small sample bias explains the finding of excess variability, in light of the evidence in Kleidon (1985, 1986) and Marsh and Merton (1986) suggesting that if prices and dividends are nonstationary, the Shiller (1981a) variance bounds test is strongly biased towards finding excess variability. To see whether there is a similar bias in the present paper's test, an environment similar to that in Kleidon (1985, 1986) and Marsh and Merton (1986) was assumed. Two Monte Carlo experiments were performed. The first assumed that dividends follow a random walk, $\Delta d_t = \mu + v_t$, the second that dividends follow a lognormal random walk, $\Delta(\log d_t) = f + w_t$. In both experiments, it was assumed that only lagged dividends were used to forecast future dividends, so that $H_t = I_t$.

---

[11] This seems an appropriate place to give the results of another test of this model, suggested to me by a referee. Equation (6) states that $\text{var}(x_{tH} - \sum_0^\infty b^j d_{t+j}) - \text{var}(x_{tI} - \sum_0^\infty b^j d_{t+j}) - \text{var}(x_{tH} - x_{tI}) = 0$. So, under the null hypothesis that $x_{tI} = p_t + d_t$,

$$\delta_1^2 \sigma_v^2 - b^{-2}\sigma_u^2 - b^{-2}(1-b^2)\text{var}[p_t + d_t - (m + \sum_1^{q+s} \delta_i d_{t-i+1})] = 0.$$

The formulas for $m, \delta_1, \ldots, \delta_{q+s}$, which are needed to calculate x_{tH} under the null, may be found in West (1987).

I tested this equality constraint for all seven specifications, with the number of lags used in the calculation of the matrix $\hat{S}$ (defined in the Appendix) set to 11. The z-statistics for the seven specifications, presented in the same order as in Table II, were: 1.88, 2.07, 1.71, 2.23, 1.85, 2.17, 1.71. Thus this suggests some mild evidence against the null hypothesis.

The basic reason for the relatively low statistics was a very noisy estimate of $\text{var}[p_t + d_t - (m + \sum_1^{q+s} \delta_i d_{t-i+1})]$. This was insignificantly different from zero at the five per cent level, for all seven specifications.

DIVIDEND INNOVATIONS 53

In the first experiment, μ and σ_v^2 were matched to the S and P sample values of the mean and variance of Δd_t, $\mu = .0373$, $\sigma_v^2 = .1574$. b was set to .9413, the value estimated in line 2 of Table I-A. For each of 1000 samples, the following was done: A vector of 100 independent normal shocks was drawn, $(v_1, \ldots, v_{100})$, using the IMSL routine GGNPM. Dividends and prices were calculated as $\Delta d_t = .0373 + v_t$; $d_t = d_0 + \sum \Delta d_s$ ($d_0 = 1.3$); $p_t = \sum (.9413)^j Ed_{t+j} | I_t = m + \delta_1 d_t$, $m = (.0373)*(.9413)/(1 - .9413)$, $\delta_1 = .9413/(1 - .9413)$. $\hat{\mu}$ and $\hat{\sigma}_v^2$ were then estimated by an OLS regression of Δd_t on a constant, $\hat{b}$ and $\hat{\sigma}_u^2$ by an instrumental variables regression of equation (14), with a constant as the only instrument. Finally, formula (17), the percentage of price variability that is excessive, was calculated from the estimated parameters. Since $H_t = I_t$, the population value of (17) is zero.

Table III-A presents the empirical distribution of the estimates of formula (17). Ideally, the median value of this distribution would be zero, with half the samples yielding a positive value to (17). Instead, there appears to be a very slight bias towards finding excess variability, with 53 per cent of the estimates being positive. The bias is not, however, particularly marked, and fewer than 5 per cent of the simulated regressions produced the extreme values of the sort found in all of the Table II specifications.

The second experiment assumed that log differences are required to induce stationarity, as in the Monte Carlo evidence in Kleidon (1986). It was noted earlier that the proof of Proposition 1 assumes that arithmetic differences suffice to induce stationarity. Since this is not true in the present Monte Carlo experiment, it is not clear what value (if any) formula (17) will converge to as the sample size grows.[12] The aim of the experiment, then, is not to evaluate the small sample divergence of estimates of (17) from a population value, but to see if this form of nonstationarity is likely to account for the large values found in column (8) of Table II.

The experiment assumed that $\Delta (\log d_t) = f + w_t$, with f and σ_w^2 matched to the S and P sample values for the mean and variance of $\Delta(\log d_t)$, $f = .013$, $\sigma_w^2 = .016$. b was again set to .9413. The $\log d_t$ data were generated by the obvious analogue to the procedure described above for the first experiment, with $d_t = \exp[\log(d_t)]$ and $p_t = \delta d_t$, $\delta = \exp(f + \sigma_w^2/2)/[b^{-1} - \exp(f + \sigma_w^2/2)] = 24.82$. This experiment used a different seed than did the first to initiate the generation of the random w_t. The parameters needed to calculate (17) were estimated exactly as in the first

TABLE III-A

MONTE CARLO DISTRIBUTION OF FORMULA (17) FOR ARITHMETIC RANDOM WALK

Percentile	1	5	50	53
Formula (17)	100.0	31.1	1.7	0.0

[12] In this model one can, however, place a theoretical lower bound on the variance of the innovation in the first difference of log dividends. I tested this in West (1987) and, once again, found that this variance is so small that it is unlikely that a lognormal random walk model generates the data.

experiment. So, for example, arithmetic first differences of d_t were regressed on a constant, even though logarithmic first differences were in fact required to induce stationarity.

Table III-B presents the results of the experiment. This time, over four fifths of the estimates of (17) were positive, suggesting a tendency to find excess volatility. Once again, however, fewer than 5 per cent of the simulated regressions produced the extreme values found in all the Table II specifications. This indicates that it is unlikely that the basic results of the empirical work are attributable to the small sample effects of the misspecification considered in this experiment. More generally, in conjunction with the other Monte Carlo experiment and the empirical results in the previous section, this indicates that the apparent inconsistency of the simple efficient markets model with the S and P and Dow Jones data is unlikely to result from the nonstationarity that is central to Kleidon's (1986) critique of Shiller (1981a).

That the estimates for the artificial data rarely display the extreme Table II excess variability suggests more strongly than might be immediately apparent that small sample bias does not explain the Table II results. This is because Table III-A, and possibly Table III-B as well, contain worst case figures, since they are based on simulations in which $H_t = I_t$. Proposition 1 implies that for any given b and univariate Δd_t process, σ_u^2 will be smaller when I_t contains additional variables useful in forecasting d_t than when $I_t = H_t$. This suggests that when I_t contains these variables estimates of σ_u^2 and of formula (17) will be smaller as well. But a simulation with such variables in I_t does not seem worth undertaking, because even under worst case circumstances assumed here, there is little to suggest that small sample bias explains the extreme excess variability reported in Table II.

b. *Bias in Test of Instrument-Residual Orthogonality*

It is possible that the diagnostic tests reported basically favorable results because the tests have low power against some interesting alternatives; see Summers (1986), for example, on tests for serial correlation. It is particularly difficult to consider this comprehensively, even if only one of the diagnostic tests is analyzed. This is because Monte Carlo experiments here are potentially quite burdensome computationally. This will be true if p_t or d_t are generated nonlinearly under the alternative, as will be the case, for example, in most formulations of the Lucas (1978) asset pricing model.

TABLE III-B
MONTE CARLO DISTRIBUTION OF FORMULA (17) FOR LOGNORMAL RANDOM WALK

Percentile	1	5	50	83
Formula (17)	81.0	73.9	41.1	0.0

So this section has a relatively modest aim, of using a single diagnostic test and a single, simple form of misspecification, to suggest whether the data and sample size are such that the diagnostic tests are unlikely to detect plausible misspecifications. The test that is used is the equation (19) test of instrument residual orthogonality. The misspecification that is assumed is that expectations are static rather than rational, $Ed_{t+j}|I_t = d_t$. In such a case, the disturbance to the arbitrage equation (14) is $-b(\Delta p_{t+1} + \Delta d_{t+1})$. So the test must pick up a correlation between $\Delta p_{t+1} + \Delta d_{t+1}$ on the one hand and lagged Δd_t (the instruments, assuming a differenced specification) on the other. Note that there are variations in (mathematically) expected returns.

Under this alternative, $p_t = [b/(1-b)]d_t$; $b = .9413$ was again assumed. Dividends were assumed to be generated by an ARIMA (2, 1, 0) process, with the parameters given by line (2) of Table I-B. The following was done 1000 times. A vector of 100 independent normal disturbances was generated, with the variance of the disturbances equal to that reported in line (2), column (6) of Table II, and with a different random number seed than those used in the other experiments. One hundred Δd_t's, and then one hundred d_t's and p_t's, were computed, with initial conditions matching the initial values of the S and P ($\Delta d_{-1} = .16$, $\Delta d_0 = .11$, $d_0 = 1.61$). $\hat{b}$ was then estimated by two-step, 2SLS, with a constant, Δd_t, and Δd_{t-1} as instruments. Finally, the equation (19) statistic was calculated.

The distribution of this statistic, which is a $\chi^2(2)$ random variable under the null, is reported in Table III-C. In over three fourths of the samples, the statistic was above 5.99, the ninety-five per cent level for a $\chi^2(2)$ random variable. In over nine tenths of the samples, the statistic was over 2.87, the value reported in line (2), column (5), in Table I-A.

Against this alternative, then, the test of instrument residual orthogonality appears to have reasonable power. Whether this applies to other alternatives or to the other diagnostic tests performed is uncertain. But the limited amount of evidence presented here at any rate does not suggest that the favorable results of the diagnostic tests result solely from low power of the tests.

B. Variation in Discount Rates

One possible explanation for the excess variability found in Section 4 is that discount rates are time varying, so that the error in equation (14) reflects not only news about dividends but also about discount rates (or, equivalently, expected returns). Special consideration of the plausibility of this variation as an explanation seems warranted, given theoretical work such as Lucas (1978) and

TABLE III-C
MONTE CARLO DISTRIBUTION OF EQUATION (19)

Percentile	5	10	50	77	95
Equation (19)	∞6.57	14.83	8.86	5.99	2.89

empirical evidence such as in Shiller (1984) and Flood, Hodrick, and Kaplan (1986).

This will be done in two separate exercises. The first (part (a) below) assumes as in, e.g., Hansen and Singleton (1982) that a consumption based asset pricing model determines expected returns, with the representative consumer's utility function displaying constant relative risk aversion. For small values of the coefficient of relative risk aversion, this permits exact calculation of formula (17), the percentage excess variability. The second (part (b) below) does not model expected returns parametrically but instead uses Shiller's (1981a) linearized version of a completely general model. This permits calculation of a lower bound to how large a standard deviation in expected returns is required to explain the excess variability reported in Table II.

a. *Consumption Based Asset Pricing Model*

Consider the class of models (e.g., Hansen and Singleton (1982)) in which the first order condition for the return on a stock is $E\{\{\beta C_{t+1}/C_t\}^{-\alpha}[(p_{t+1}+d_{t+1})/p_t]\}|I_t\}=1$, where β, $0<\beta<1$, is the representative consumer's subjective discount rate, C_t is his real consumption, α his coefficient of relative risk aversion, with E, d_t, p_t, and I_t defined as above. This may be rearranged as

(20) $\quad \tilde{p}_t = \beta E[(\tilde{p}_{t+1} + \tilde{d}_{t+1})|I_t],$

$\tilde{p}_t = p_t C_t^{-\alpha}, \quad \tilde{d}_t = d_t C_t^{-\alpha}.$

Equation (20) is of the same form as equation (9). R. Flood has pointed out to me that if $\tilde{d}_t$ is stationary, perhaps after one or more differences are taken, the statistics computed in the constant discount rate case can be computed in this model as well. Repetition of the entire procedure is beyond the scope of this paper (and, in light of the results about to be presented, seems pointless). Instead, I will focus on obtaining a point estimate of formula (17), the percentage excess variability, for various imposed values of β and α.

The C_t variable used in these calculations was the Grossman and Shiller (1981) annual figure on real, per Capita consumption of nondurables and services, 1889–1978. $\tilde{d}_t$ and $\tilde{p}_t$ were calculated using the S and P data for various values of α. A simple plot of $\tilde{d}_t$ suggested that $\tilde{d}_t$ in neither levels nor first nor higher differences is stationary for α much bigger than one. The problem is that for big α, $\tilde{d}_t$ displays a marked secular decline, because annual C_t growth was slightly higher than annual d_t growth.

I nonetheless calculated (17), the percentage excess variability, for a wide range of α, just in case $\tilde{d}_t$ really is stationary for large α. This was done for $\beta = .95$ and $\beta = .98$, with very similar results. In all cases the lag length of the $\tilde{d}_t$ autoregression was set to four. Table IV-A contains the figures that resulted for some of the α, with $\beta = .98$. Since (17) was not only positive but large, the price and dividend data are as inconsistent with the model implied by (20) as they are with the constant expected return model assumed in Sections 3 and 4. There is

TABLE IV-A
PERCENTAGE EXCESS PRICE VARIABILITY

α	.5	1	2	3	10	25	50
Formula (17)	96.5	97.5	80.9	88.4	99.6	100.0	100.0

therefore no evidence supporting the hypothesis that the excess variability displayed in Table II is explained solely by the sort of variation in expected returns predicted by this asset pricing model.[13]

Since $\tilde{d}_t$ does not appear stationary for α much bigger than unity, it is equally true that Table IV-A contains no evidence against the hypothesis that the Table II excess variability is explained by variation in expected returns associated with a coefficient of risk aversion greater than, say, one. Table IV-A does, however, suggest if the model of expected returns assumed here is correct, that the Table II excess variability is unlikely to be due to variation in expected returns associated with a coefficient of relative risk aversion of less than, say, one.

b. Linearized Model

Let us now consider a general model that does not parameterize expected returns, linearized as in Shiller (1981a) to make the analysis tractable. Let r_{t+j} be the one period return expected by the market at period $t+j$, assumed covariance stationary. Suppose $p_t = E\{\{\sum_{j=1}^{\infty} [\prod_{i=1}^{j} (1+r_{t+i-1})^{-1}] d_{t+j}\} | I_t\}$. Let us linearize the quantity in braces around $\bar{r}$ and $\bar{d}$. $\bar{r}$ is the mean of r_t; selection of $\bar{d}$ is discussed below. Define $b = (1+\bar{r})^{-1}$, $a = -\bar{d}/\bar{r}$. Then (Shiller (1981a)) $p_t \approx E\{\{\sum_{j=1}^{\infty} b^j [a(r_{t+j-1} - \bar{r}) + d_{t+j}]\} | I_t\}$. Let $u_{t+1} = p_t - b(p_{t+1} + d_{t+1})$. Proposition 1 may be used to show that in this linearized model

(21) $\delta_1^2 \sigma_v^2 - b^{-2} \sigma_u^2 \geq -[a^2 + (1-b^2)^{-1} a^2] \sigma_r^2 - [2(1-b^2)^{-1/2} a \delta_1 \sigma_v] \sigma_r,$

where σ_r is the standard deviation of r_t, and δ_1 and σ_v are as defined in formula (16). The algebra to derive (21) is in the Appendix.

The left-hand side of (21) is precisely the quantity studied in Sections 3 and 4. If this is positive, as it will be in the model (12), $\sigma_r = 0$ would of course satisfy the inequality. The empirical estimates of (16), in Table II, column (7), however, were negative; the minimum return variability needed to explain the Table II results is given by the positive σ_r that makes (21) hold with equality.

This lower bound σ_r was calculated for all seven of the specifications. σ_u^2, σ_v^2, δ_1, and b were set equal to the estimated values reported in Table II. When dividends were assumed stationary, $\bar{d}$ was set equal to mean dividends, $\bar{d} = T^{-1} \sum d_t$. When dividends were assumed nonstationary, $\bar{d}$ was set equal to average

---

[13] Note that the entries in the table are not a monotonic function of α. To make sure that the entries were representative, I calculated the percentage excess variability for α in steps of 0.1 from 0 to 3.0, in steps of 1.0 from 3.0 to 10.0, and in steps of 5.0 from 10.0 to 50.0. The results were quite similar to those reported in the table. The lowest percentage happened to occur at $\alpha = 2.0$.

TABLE IV-B
MINIMUM σ_r NEEDED TO EXPLAIN EXCESS VARIABILITY

Data Set	S&P	S&P	S&P	S&P	DJ	DJ	DJ
Differenced	no	yes	no	yes	no	yes	no
Lags	2	2	4	4	3	4	4
σ_r	.146	.222	.146	.201	.127	.176	.169

expected discounted dividends, $\bar{d} = (1-b)\sum_{t=1}^{\infty} b^{t-1} E_0 d_t$, where: $E_0 d_t = E_0 d_0 + t E \Delta d_t$, $E_0 d_0 = d_0$, d_0 the level of dividends at the beginning of the sample, and $E \Delta d_t$ calculated as $T^{-1} \sum \Delta d_t$. The parameter a was in all cases set to $-\bar{d}/\bar{r}$, with $\bar{r}$ defined implicitly by $(1+\bar{r})^{-1} = b$.

The resulting lower bound values may be found in Table IV-B. They are rather large. None of the estimates are less than .12. With $\sigma_r = .12$ and $\bar{r} = .07$, a two standard deviation confidence interval for the (real) expected return is about −17 per cent to +31 per cent. This would seem to be an implausibly large range.

In the linearized model considered here, then, Table IV-B suggests that variations in ex ante discount rates do not plausibly explain the excess variability of stock prices. How well this conclusion applied to any given nonlinear model of course depends on how well the linear model approximates the nonlinear one. An example in Shiller (1981a) suggests that if dividends are stationary the approximation can be quite good, even when changes in expected returns are larger than are typically considered reasonable. It is of course debatable that the approximation makes any sense, let alone is very accurate, if dividends are nonstationary. But the results here can in any case be said not to lend support to the hypothesis that the excess price variability reported in Table II is solely due to variation in expected returns.

6. CONCLUSIONS

This paper has derived and applied a stock price volatility test. The test required neither of two strong assumptions required by the Shiller (1981a) volatility test: that prices and dividends have finite variance, and that a satisfactory approximation to a perfect foresight price can be calculated from a finite data series.

The test indicated that stock prices are too volatile to be the expected present discounted value of dividends, with a constant discount rate. Among the explanations for the test results are that discount rates vary and that there are rational or nearly rational bubbles or fads. The possibility that the excess volatility is caused by discount rate fluctuations has been considered in detail, with largely negative results. The possibility that the excess volatility is due to bubbles has received little direct attention. But since this alternative is consistent with the econometric diagnostics, it seems worthy of further investigation.

A detailed case for bubbles, or, for that matter, any other factor as the explanation of the excess volatility is, however, beyond the scope of this paper.

A challenging task for future research is to make such a case, explaining the apparently excessive price volatility.

Woodrow Wilson School, Princeton University, Princeton, NJ 08544, U.S.A.

Manuscript received December, 1984; final revision received March, 1987.

APPENDIX

A. CALCULATION OF THE VARIANCE-COVARIANCE MATRIX

This describes the calculation of the variance-covariance matrix of the parameter vector $\theta = (b, \phi, \sigma_u^2, \sigma_v^2) = (b, \mu, \phi_1, \ldots, \phi_q, \sigma_u^2, \sigma_v^2)$.

Let $Z_t = (1, \Delta^s d_t, \ldots, \Delta^s d_{t-q+1})'$ be the $(q+1) \times 1$ vector of instruments, $s = 0$ or $s = 1$, $n_{t+1} = (d_{t+1} + p_{t+1})$ be the right-hand side variable in (14). One way of describing the estimation technique is to note that $\hat{\theta}$ was chosen to satisfy the orthogonality condition

$$0 = T^{-1} \sum h_t(\hat{\theta}) = \begin{bmatrix} T^{-1}(T^{-2s} \sum n_{t+1} Z_t')(\hat{S}_z)^{-1} \sum Z_t(p_t - n_{t+1}\hat{b}) \\ T^{-1} \sum Z_t(\Delta^s d_{t+1} - Z_t'\hat{\phi}) \\ \hat{\sigma}_u^2 - T^{-1} \sum (p_t - n_{t+1}\hat{b})^2 \\ \hat{\sigma}_v^2 - T^{-1} \sum (\Delta^s d_{t+1} - Z_t'\hat{\phi})^2 \end{bmatrix}.$$

(The degrees of freedom corrections in $\hat{\sigma}_u^2$ and $\hat{\sigma}_v^2$ are suppressed for notational simplicity.) The summations in the orthogonality condition run over t, from 1 to $T-s$. $\hat{S}_z$ is an estimate of $EZ_t Z_t' u_{t+1}^2$, calculated as described below equation (19). Thus $\hat{b}$ is estimated by two-step, 2SLS, $\hat{\phi}$ by OLS, $\hat{\sigma}_u^2$ and $\hat{\sigma}_v^2$ from moments of the residuals.

Since $Eh_t(\theta) = 0$, where θ is the true but unknown parameter vector, it may be shown that under fairly general conditions, $C_T(\hat{\theta} - \theta)$ is asymptotically normal with a covariance matrix $V = (\text{plim } F_T^{-1} \sum h_{t\theta} F_T^{-1})^{-1} S(\text{plim } F_T^{-1} \sum h_t' F_T^{-1})^{-1}$ (Hansen (1982), West (1986c)). C_T and F_T are $(q+4) \times (q+4)$ diagonal normalizing matrices, $C_T = F_T = \text{diag}(T^{1/2}, \ldots, T^{1/2})$ for undifferenced specifications, $C_T = \text{diag}(T^{3/2}, T^{1/2}, \ldots, T^{1/2})$ and $F_T = \text{diag}(T, T^{1/2}, \ldots, T^{1/2})$ for differenced specifications. $h_{t\theta}$ is the $(q+4) \times (q+4)$ matrix of derivatives of h_t with respect to θ and $S = Eh_t h_t' + \sum_{j=1}^{\infty} [Eh_t h_{t-j}' + (Eh_t h_{t-j}')']$. $h_{t\theta}$ is straightforward to calculate. Calculation of S is slightly more involved. Newey and West (1986) and West (1986c) show that in general S and thus V are consistently estimated if $\hat{S} = \hat{\Omega}_0 + \sum_{i=1}^{m} w(i, m)(\hat{\Omega}_i + \hat{\Omega}_i')$, where $m \to \infty$ as $T \to \infty$ and m is $o(T^{1/2})$; $w(i, m) = 1/(m+1)$; $\hat{\Omega}_i = T^{-1} \sum_{t=i+1}^{T} \hat{h}_t \hat{h}_{t-i}'$, $\hat{h}_t = h_t(\tilde{\theta})$, $\tilde{\theta}$ an initial consistent estimate (2SLS and OLS). The weights $w(i, m)$ insure that $\hat{S}$ is positive semidefinite. In the absence of any theoretical or Monte Carlo evidence on the small sample properties of various choices of m, I tried various values: $m = 3, 7,$ or 11. The value of m that led to the *largest* standard error in column (7) of Table II is what is reported in Table II. For all specifications, this turned out to be $m = 11$.

It is easy to show that the $T^{3/2}$ rate for $\hat{b}$ when $s = 1$ implies that uncertainty about b can be ignored when calculating the standard error for $\hat{\rho}$ in column (5) of Table I-A. I therefore did so, and used the OLS standard error of the regression of the 2SLS residual on a lagged residual. The standard errors for the undifferenced specifications in Table I-A were calculated according to equation (50) in Pagan and Hall (1983). All standard errors in column (8) of Table I-B were calculated according to Theorem 3 in Pagan and Hall (1983).

B. DERIVATION OF EQUATION (21)

In the linearized model the analogue to equation (9) is $p_t = bE[a(r_t - \bar{r}) + d_{t+1} + p_{t+1}|I_t]$. Let $y_{t+i} = a(r_{t+i-1} - \bar{r}) + d_{t+i}$ and redefine $x_{it} = E(\sum b^j y_{t+j}|I_t)$. (Of course, if expected returns are constant, $r_t = \bar{r}$ for all t, x_{it} as defined here reduces to its Proposition 1 counterparts.) To simplify the argument, it will be assumed throughout this section that linear projections and mathematical expectations are equivalent. The efficient markets model considered in Section 3 implies $x_{it} = d_t + p_t$;

the one currently under consideration implies $x_{tt} = y_t + p_t = a(r_{t-1} - \bar{r}) + d_t + p_t$. So with r_{t-1} an element of I_{t-1}, $x_{tt} - E(x_{tt}|I_{t-1}) = d_t + p_t - E(d_t + p_t|I_{t-1})$. Now,

(A.1) $\quad u_{t+1} = p_t - b(d_{t+1} + p_{t+1}) = [ba(r_t - \bar{r}) + bE(p_{t+1} + d_{t+1}|I_t) - b(d_{t+1} + p_{t+1})]$

$= b\{a(r_t - \bar{r}) - [x_{t+1,I} - E(x_{t+1,I}|I_t)]\}$

$\Rightarrow b^{-2}\sigma_u^2 = a^2\sigma_r^2 + E[x_{t+1,I} - E(x_{t+1,I}|I_t)]^2$

$\Rightarrow E[x_{t+1,I} - E(x_{t+1,I}|I_t)]^2 = b^{-2}\sigma_u^2 - a^2\sigma_r^2.$

Now define J_t as the information set determined by a constant and all current and lagged dividends and expected returns, $x_{tJ} = E(\sum b^j y_{t+j}|J_t)$. Let $x_{tJ} - E(x_{tJ}|J_{t-1}) = aw_{1t} + w_{2t}$, where w_{1t} and w_{2t} are the innovations in the expected present discounted values of r_t and d_t. Shiller (1981a) shows that $\sigma_{w_1}^2 \le \sigma_r^2/(1-b^2)$. Assume that d_t or Δd_t follows the autoregression (15). Then since H_t is a subset of J_t, Proposition 1 tells us that $\sigma_{w_2}^2 \le \delta_1^2 \sigma_v^2$, where, as previously, σ_v^2 is the variance of the univariate dividend innovation and δ_1 is defined above formula (16). So

(A.2) $\quad E[x_{tJ} - E(x_{tJ}|J_{t-1})]^2 = a^2\sigma_{w_1}^2 + 2a\sigma_{w_1 w_2} + \sigma_{w_2}^2$

$\le a^2\sigma_{w_1}^2 + 2a\sigma_{w_1}\sigma_{w_2} + \sigma_{w_2}^2$

$\le (1-b^2)^{-1}a^2\sigma_r^2 + 2a(1-b^2)^{-1/2}\delta_1\sigma_v\sigma_r + \delta_1^2\sigma_v^2.$

Since J_t is a subset of I_t, Proposition 1 tells us that $E[x_{tt} - E(x_{tt}|I_{t-1})]^2 \le E[x_{tJ} - E(x_{tJ}|J_{t-1})]^2$. With a little rearrangement, (A.1) and (A.2) together imply equation (21) in the text.

REFERENCES

ACKLEY, G. (1983): "Commodities and Capital: Prices and Quantities," *American Economic Review*, 73, 1-16.
BLANCHARD, O., AND M. WATSON (1982): "Bubbles, Rational Expectations and Financial Markets," NBER Working Paper No. 945.
BREALEY, R., AND S. MYERS (1981): *Principles of Corporate Finance*. New York, NY: McGraw Hill.
DIBA, B. T., AND H. I. GROSSMAN (1985): "The Impossibility of Rational Bubbles," NBER Working Paper No. 1615.
DEATON, A. (1985): "Life Cycle Models of Consumption: Is the Evidence Consistent with the Theory?" manuscript, Princeton University.
FLAVIN, M. (1983): "Excess Volatility in the Financial Markets: A Reassessment of the Empirical Evidence," *Journal of Political Economy*, 91, 929-956.
FLOOD, R., R. HODRICK, AND P. KAPLAN (1986): "An Evaluation of Recent Evidence on Stock Market Bubbles," manuscript, Northwestern University.
FULLER, W. A. (1976): *Introduction to Statistical Time Series*. New York: John Wiley and Sons.
GRANGER, C. W. J., AND P. NEWBOLD (1977): *Forecasting Economic Time Series*. New York, NY: Academic Press.
GROSSMAN, S., AND R. SHILLER (1981). "The Determinants of the Variability of Stock Prices," *American Economic Review*, 71, 222-227.
HAMILTON, J., AND C. WHITEMAN (1984): "The Observable Implications of Self Fulfilling Prophecies," manuscript, University of Virginia.
HANNAN, E. J., AND B. G. QUINN (1979): "The Determination of the Order of an Autoregression," *Journal of the Royal Statistical Society Series B*, 41, 190-195.
HANSEN, L. P. (1982): "Large Sample Properties of Generalized Method of Moments Estimators," *Econometrica*, 50, 1029-1054.
HANSEN, L. P., AND T. J. SARGENT (1980): "Formulating and Estimating Dynamic Linear Rational Expectations Models," *Journal of Economic Dynamics and Control*, 2, 7-46.
——— (1981): "Exact Linear Rational Expectations Models: Specification and Estimation," Federal Reserve Bank of Minneapolis Staff Report 71.
HANSEN, L. P., AND K. J. SINGLETON (1982): "Generalized Instrumental Variables Estimation of Nonlinar Rational Expectations Models," *Econometrica*, 50, 1269-1286.
KEYNES, J. M. (1964): *The General Theory of Employment, Interest and Money*. New York, NY: Harcourt, Brace, and World.
KLEIDON, A. W. (1985): "Bias in Small Sample Tests of Stock Price Rationality," Stanford University Graduate School of Business Research Paper 819R.

——— (1986): "Variance Bounds Tests and Stock Price Valuation Models," *Journal of Political Economy*, 94, 953-1001.
LEROY, S. (1984): "Efficiency and Variability of Asset Prices," *American Economic Review*, 74, 183-187.
LEROY, S. AND R. PORTER (1981): "The Present Value Relation: Tests Based on Implied Variance Bounds," *Econometrica*, 64, 555-574.
LUCAS, R. E., JR. (1978): "Asset Prices in an Exchange Economy," *Econometrica*, 46, 1429-1445.
MARSH, T. A., AND R. C. MERTON (1986): "Dividend Variability and Variance Bounds Tests for the Rationality of Stock Market Prices," *American Economic Review*, 76, 483-498.
MCCALLUM, B. (1976): "Rational Expectations and the Natural Rate Hypothesis: Some Consistent Estimates," *Econometrica*, 44, 43-52.
NEWEY, W. K., AND K. D. WEST (1987): "A Simple, Positive, Semidefinite, Heteroskedasticity and Autocorrelation Consistent Covariance Matrix," *Econometrica*, 55, 703-708.
PAGAN, A. R., AND A. D. HALL (1983): "Diagnostic Tests as Residual Analysis," *Econometric Reviews*, 2, 159-218.
SHILLER, R. J. (1981a): "Do Stock Prices Move Too Much to be Justified by Subsequent Changes in Dividends?" *American Economic Review*, 71, 421-436.
——— (1981b): The Use of Volatility Measures in Assessing Market Efficiency," *Journal of Finance*, 35, 291-304.
——— (1984): "Stock Prices and Social Dynamics," *Brookings Papers on Economic Activity*, 457-498.
SUMMERS, L. H. (1986): "Does the Stock Market Rationally Reflect Fundamental Values?" *Journal of Finance*, 41, 591-600.
TIROLE, J. (1985): "Asset Bubbles and Overlapping Generations," *Econometrica*, 53, 1071-1100.
WEST, K. D. (1984): "Speculative Bubbles and Stock Price Volatility," Princeton University Financial Research Memorandum No. 54.
——— (1987): "A Specification Test for Speculative Bubbles," *Quarterly Journal of Economics*, 102, 553-580.
——— (1986a): "A Standard Monetary Model and the Variability of the Deutschemark-Dollar Exchange Rate," forthcoming, *Journal of International Economics*.
——— (1986b): "Asymptotic Normality, When Regressors Have a Unit Root," Princeton University Woodrow Wilson School Discussion Paper No. 110.
——— (1986c): "Dividend Innovations and Stock Price Volatility," NBER Working Paper No. 1833.
——— (1988): "The Insensitivity of Consumption to News About Income," forthcoming, *Journal of Monetary Economics*.

[9]

Econometric Aspects of the Variance-Bounds Tests: A Survey

Christian Gilles
Board of Governors of the Federal Reserve System

Stephen F. LeRoy
University of Minnesota

We survey the variance-bounds tests of asset-price volatility, stressing the econometric aspects of these tests. The first variance-bounds tests of the present-value relation reported apparently striking evidence of excess volatility of asset prices. The statistical significance of the results, however, was either marginal or, in the case of model-free tests, impossible to assess. Moreover, the tests were soon criticized for a number of biases. Various other tests of the present-value relation were later developed, avoiding in different degrees the econometric problems attending the first-generation tests. The majority of these second-generation tests also found excess volatility, though sometimes of borderline statistical significance. This finding of excess volatility is robust and is difficult to explain within the representative-consumer, frictionless-market model.

The variance-bounds debate—whether asset prices are more volatile than traditional models imply—has

---

We have received helpful comments from Marjorie Flavin, Wayne Joerding, Hashem Pesaran, Richard Porter, Gary Shea, Robert Shiller, Jon Sonstelie, Douglas Steigerwald, and Kenneth West. This article should not be interpreted as reflecting the views of the Board of Governors of the Federal Reserve System or its staff.

engaged a number of economists over the past decade. At first glance, the variance-bounds controversy seems rooted in ideological considerations: as in debates in other areas of economics, it raises questions about the rationality of markets. In all such debates, the issue is the success of current neoclassical theory in explaining economic phenomena. Some argue for fundamental departures from the neoclassical program.

Fortunately, the variance-bounds debate has largely avoided the meaningless verbal exchanges that are typical of ideological disagreements. Genuine scientific differences, which are resolvable by further theoretical and empirical work, characterize this debate. As a result, progress has been rapid and a convergence of opinion appears to be occurring. While it is an exaggeration to characterize the variance-bounds debate as a model of how scientific work should proceed, the case could be made.

Econometric, not ideological or theoretical, issues have dominated the variance-bounds debate. As a consequence, evaluating the empirical evidence of excess volatility is difficult. Further, the debate is based on simulations in settings where straightforward analytical demonstrations are available. In many cases, these simulations obscure rather than clarify the main lines of development. As things stand, the time required to gain a foothold in this literature is excessive.

In this article, we introduce the variance-bounds debate and stress econometric issues—unlike LeRoy (1984, 1989) and West (1988b) who focus on theoretical issues. While the exposition is intended to be self-contained, readers should consult the original papers.

The theory and first-generation tests due to Shiller (1979, 1981) and LeRoy and Porter (1981) are reviewed in Section 1. Discussion of theory is brief to avoid duplication of LeRoy (1984, 1989) and West (1988b), and because those readers who wish to master the econometric aspects of the debate will already be familiar with the underlying theory. In Section 2, we present Flavin's (1983) and Kleidon's (1986b) econometric critiques of the Shiller–LeRoy–Porter findings of excess volatility. Here theoretical analysis of the econometric properties of two-observation samples drawn from the autoregressive dividend model replaces the simulation methods used by Flavin and Kleidon. By relying on analytical methods, readers should see the intuition underlying Flavin and Kleidon's claim that the original tests are biased toward rejection. In Section 3, we turn to the "second-generation" variance-bounds tests, which also found excess volatility of asset prices using alternative econometric methods. We believe that these second-generation tests reinforce the conclusion based on the first-generation tests; asset-price volatility exceeds that predicted by the simplest present-value model. Granted this provisional con-

Variance-Bounds Tests

clusion, the question becomes why asset-price volatility is high. Here the literature has not arrived at even a provisional conclusion. Several possibilities are discussed in Section 4. Conclusions are offered in Section 5.

1. First-Generation Tests

1.1 The variance-bounds theorems

The variance-bounds inequalities are implications of the present-value relation. In our usage, this means that an asset's price is equal to the expected present value of the asset's payments, discounted at a constant rate.[1] Two questions arise: (i) Why focus on testing the present-value relation rather than consumption-based asset pricing more generally? (ii) Why test only the volatility implications of the present-value relation, rather than all its implications for the covariances between asset yields and prices? With regard to (i), whether yields of financial assets are sufficient to explain their prices is an important substantive question. If so, the principle of parsimony can be invoked in future modeling to justify ignoring the aggregate effects of risk aversion.

With regard to (ii), the same question—Why focus on one particular implication of the theory to the exclusion of all others?—could be directed equally well to tests of market efficiency that focus on return autocorrelations. The answer then would be that if, contrary to the present-value relation, future price changes can be inferred from past prices, analysts can recommend trading rules that outperform buy and hold in expected return. With regard to the variance-bounds tests, the alternative to the present-value relation might be that stock prices are driven by protracted waves of optimism or pessimism. The impli-

---

[1] The constancy assumption has caused misinterpretation: in the finance literature, it is not customary to include constancy of the discount rate in the definition of the present-value relation [e.g., Merton (1987)]. This element of generality in the finance literature's definition of the present-value relation is an outgrowth of the practice in elementary finance instruction of portraying net present value as the single unambiguously correct way to think about all problems in finance (in preference to such alternatives as internal rate of return as a criterion for capital budgeting, for example). In practice, the important problem in applying the present-value relation is to determine the magnitude of the adjustment for risk that is appropriate in the setting under consideration.

The problem of determining asset prices generally and that of determining what discount rate to use in the present-value relation are related tautologically. In general equilibrium, it makes no difference whether the analyst determines the price of financial assets directly, or indirectly from the present-value relation after determining the equilibrium discount factor. We follow here the practice, common among nonfinancial economists, of analyzing asset pricing without explicit reference to the present-value relation, and then formulating the present-value relation as a falsifiable special case by assuming constancy of the discount rate. Empirical testing then determines whether the present-value relation appears in fact to be false.

The difference between these two usages is entirely semantic—our purpose is not to urge one framework or the other, but only to alert readers to a possible source of misunderstanding.

755

cation of excessive volatility is tested directly by variance-bounds tests. Such informal discussion suggests that variance-bounds tests have different power than regression-based tests under different alternative hypotheses and therefore merit independent study.

The main variance-bounds theorem is now reviewed. Define p_t^*, the ex post rational price of stock at date t (the term is Shiller's), as the present discounted value of future dividends:[2]

$$p_t^* \triangleq \sum_{i=1}^{\infty} \beta^i d_{t+i}. \qquad (1)$$

The present-value relation says that actual stock price is the conditional expectation of ex post rational price

$$p_t = E(p_t^* | I_t), \qquad (2)$$

where I_t is investors' information.

Under assumptions to be specified, the variance of ex post rational price and that of actual price are related by

$$V(p_t^*) = V(p_t) + \frac{\beta^2 V(r_t)}{1 - \beta^2}, \qquad (3)$$

where r_t represents excess return:

$$r_t \triangleq d_t + p_t - E(d_t + p_t | I_{t-1}). \qquad (4)$$

(Note our terminology: we use "return" and "rate of return" where others sometimes use "payoff" and "return," respectively.) To derive (3), note that (1) and (2) imply

$$p_t = \beta E(d_{t+1} + p_{t+1} | I_t)$$

or

$$p_t = \beta(d_{t+1} + p_{t+1} - r_{t+1}), \qquad (5)$$

using (4). Now update (5) by one period and use the result to substitute out p_{t+1} in (5). Iterating and using (1), there results

$$p_t = p_t^* - \sum_{j=1}^{\infty} \beta^j r_{t+j}.$$

Assume now that $V(r_t)$ is constant over time, as occurs, for example, under a linear covariance-stationary dividends process when information revelation is regular (as in the example below). Shifting the

---

[2] We let the symbol $\triangleq$ denote the defining equality.

Variance-Bounds Tests

rightmost term to the left side and taking variances (noting that the r_{t+j}'s are mutually uncorrelated), there results

$$V(p_t) + 2\operatorname{Cov}\left(p_t, \sum_{j=1}^{\infty} \beta^j r_{t+j}\right) + \frac{\beta^2 V(r_t)}{1 - \beta^2} = V(p_t^*). \qquad (6)$$

Because rational forecasting implies in addition that p_t and r_{t+j} have zero covariance, (6) implies (3). Dropping subscripts, the latter equation becomes

$$V(p^*) = V(p) + \frac{\beta^2 V(r)}{1 - \beta^2}. \qquad (7)$$

Equation (7) underlies all the variance-bounds tests discussed in this paper. It says that the variance of ex post rational price equals the sum of the variance of actual price and that of returns (where the latter is multiplied by a constant which depends on the discount rate). If investors have little information about future dividends, then the variance of price is low. Returns, however, buffeted by the largely unexpected realizations of dividends, are highly volatile. In the opposite case in which agents can accurately predict dividends into the distant future, the variance of price is almost as high as that of ex post rational price. Returns, on the other hand, are very smooth.

LeRoy and Porter outlined two types of volatility tests: bounds tests and orthogonality tests [the terms are due to Durlauf and Hall (1989)]. The simplest bounds test is

$$V(p) \leq V(p^*),$$

a direct consequence of (7). A less straightforward bounds test is the LeRoy–Porter lower bound on $V(p)$. Define H_t to be the information set consisting of current and past dividends alone and take $\hat{p}_t$ to be the price that would prevail under H_t:

$$\hat{p}_t \triangleq E(p_t^* | H_t).$$

Then, assuming that I_t is at least as informative as H_t, the rule of iterated expectations implies that

$$\hat{p}_t = E(p_t | H_t),$$

but the conditional expectation of any random variable x is less volatile than x itself, implying

$$V(\hat{p}_t) \leq V(p_t)$$

or, assuming stationarity,

$$V(\hat{p}) \leq V(p). \qquad (8)$$

This lower bound on $V(p)$ is of interest primarily because it is equivalent to West's inequality on return volatility, discussed below.

The simplest orthogonality test of asset-price volatility determines whether $V(p^*)$, $V(p)$, and $V(r)$ are consistent with (7). If instead $V(p)$ and $V(r)$ are too large to be consistent with (7), the orthogonality of price and returns is rejected in favor of the hypothesis that these variables are negatively correlated [in view of (6), if $V(p_t) + \beta^2 V(r_t)/(1 - \beta^2)$ exceeds $V(p_t^*)$, the covariance term must be negative].

1.2 An example

An extended example will facilitate understanding of the variance-bounds theorems and will aid in the exposition of the sampling problems that attend their empirical implementation. Suppose that dividends follow a first-order Markov process

$$d_{t+1} = \lambda d_t + \epsilon_{t+1}, \tag{9}$$

where $E(\epsilon_t) = 0$, $V(\epsilon_t) = 1$, $|\lambda| < 1$, and the ϵ_t are serially independent. Here the mean of the dividend process is set to zero since a nonzero mean would drop out of all variance expressions. Further, agents are assumed to have sufficient information to forecast dividends perfectly up to m periods in advance. However, agents have no information about dividends beyond $t + m$ other than the information conveyed by dividends up to $t + m$. This all-or-nothing specification is restrictive if taken literally; its advantage is to provide a simple parameterization of the amount of information available to investors. As a consequence, comparison of actual and ex post rational stock prices is very easy, since the ex post rational stock price process is a limiting case ($m = \infty$) of the class of actual stock price processes.

The present-value relation yields

$$p_t^m = \beta d_{t+1} + \cdots + \beta^m d_{t+m} + \beta^{m+1} E_{t+m}(d_{t+m+1}) + \beta^{m+2} E_{t+m}(d_{t+m+2}) + \cdots, \tag{10}$$

where p_t^m is the actual price of stock. Using (9) to form expectations and simplifying, (10) becomes

$$p_t^m = \frac{\beta \lambda}{1 - \beta \lambda} d_t + \sum_{i=1}^{m} \frac{\beta^i \epsilon_{t+i}}{1 - \beta \lambda}. \tag{11}$$

We now can compute the variance of p_t^m:

$$V(p_t^m) = \frac{\beta^2 \lambda^2}{(1 - \lambda^2)(1 - \beta \lambda)^2} + \frac{\beta^2 - \beta^{2(m+1)}}{(1 - \beta^2)(1 - \beta \lambda)^2}. \tag{12}$$

$V(p_t^m)$ rises with m toward an asymptote,

Variance-Bounds Tests

$$V(p_t^\infty) = \frac{\beta^2(1 + \beta\lambda)}{(1 - \lambda^2)(1 - \beta\lambda)(1 - \beta^2)}, \quad (13)$$

which is just the variance of the ex post rational stock price.

Now we evaluate the variance of the one-period returns. Equation (5) implies that

$$r_{t+1}^m = d_{t+1} + p_{t+1}^m - \beta^{-1}p_t^m.$$

Substituting (11) for p_t^m and p_{t+1}^m and taking the variance, we get

$$V(r_t^m) = \frac{\beta^{2m}}{(1 - \beta\lambda)^2}. \quad (14)$$

Notice that $V(p_t^\infty)$, $V(p_t^m)$, and $V(r_t^m)$ as given by (13), (12), and (14), respectively, satisfy

$$V(p_t^\infty) = V(p_t^m) + \frac{\beta^2 V(r_t^m)}{1 - \beta^2},$$

agreeing with (7).

1.3 Implementation

Shiller's (1981) implementation of an operational test of the inequality $V(p_t) \leq V(p_t^*)$ was simple and direct—more so than that of LeRoy and Porter, as will be seen below. To correct for trend, Shiller divided through by constant growth rate trends. To understand Shiller's resolution of the problem that p_t^* is not observable, notice that the ex post rational price (1) is the solution to the recursion

$$p_t^* = \beta(p_{t+1}^* + d_{t+1}) \quad (15)$$

that satisfies the condition

$$\lim_{t \to \infty} \beta^t p_t^* = 0. \quad (16)$$

Shiller simply replaced p_t^* by the solution to (15) that satisfies instead the terminal condition

$$p_T^* = \frac{1}{T}\sum_{t=1}^{T} p_t. \quad (17)$$

Denote by $p_{t|\bar{p}}^*$ the observable version of p_t^* generated by (15) and (17).

Shiller constructed sample estimates of the variances of p_t and $p_{t|\bar{p}}^*$ in the usual way as the average squared deviation of each variable from its own mean. Using a century-long data set, Shiller found that the standard deviation of actual stock prices exceeded that of the ex post rational stock prices by a factor of 5.59. Although no significance

759

tests were reported, he interpreted this result as constituting rejection of the variance-bounds inequalities.

As noted in Section 1.1, LeRoy and Porter defined their null hypothesis from the equality version (7) of the variance-bounds relation. They assumed that dividends and stock prices, adjusted for trend as described below, were generated by a covariance-stationary bivariate linear process, with parameters restricted by (7). A simplified and intuitive version of the LeRoy–Porter implementation goes as follows: (i) estimate a linear autoregressive model for dividends and estimate β as the reciprocal of 1 plus the average rate of return on stock; (ii) estimate $V(p_t^*)$ by applying the present-value relation (1) directly to the model for dividends (thus avoiding the problem that p_t^* is unobservable); (iii) estimate $V(r_t)$ from the observable series of one-period returns; and (iv) estimate $V(p_t)$ from a linear model for p_t. It was found that the point estimate of each of the two terms on the right-hand side of (7) by itself exceeds the term on the left-hand side, indicating rejection of the variance bounds. However, the associated confidence intervals based on asymptotic distributions were so wide that the rejection of (7) was of borderline statistical significance.

LeRoy and Porter corrected for trend, not by implementing a mechanical trend adjustment, but by reversing the effect of the variables which cause the trend: inflation and corporate retained earnings.[3] Assuming that the returns per unit of capital are covariance stationary, the LeRoy–Porter algorithm should produce stationary series for price and dividends. Instead, LeRoy and Porter noted (1981, p. 569) that the adjusted series showed some evidence of a downward trend, but the LeRoy–Porter implementation of the trend correction was flawed, as Kleidon (1986b, note 4) pointed out.[4]

---

[3] Specifically, inflation was removed from the data by dividing a commodity price index into dividends, earnings, and prices. To reverse the effect of earnings retention, it was assumed that the value of stock equals the product of a quantity index representing the amount of physical capital held by firms and a price index representing the dollar value per unit of this capital. If it is assumed that the physical capital acquired with retained earnings is purchased at the same price as currently available capital is valued, and is equally productive, then data on retained earnings can be used to compute recursively the price and quantity variables.

[4] In the 1988 version of this paper, we provided a corrected exposition of the algorithm and LeRoy and Parke implemented it in the 1988 version of their 1990 paper, using the 112-year sample of Shiller instead of the LeRoy–Porter postwar sample. The downward trend in the supposedly trend-adjusted series that LeRoy and Parke reported was even more pronounced in the 112-year sample than that which LeRoy and Porter had reported. We do not know the reason for this downward trend, but it clearly invalidates the LeRoy–Porter trend correction and renders it of secondary interest (except, perhaps, to researchers interested in investigating the cause of the downward trend, which is worthy of study in its own right). Because the trend correction failed at its assigned task, detailed description of it was deleted from this paper and that of LeRoy and Parke (the earlier versions of these papers are available from the authors).

Variance-Bounds Tests

2. The Critics

2.1 Flavin[5]

Flavin's (1983) paper made two criticisms of Shiller's econometric tests. The first was that both the variance of p_t and that of p_t^* are estimated with downward bias in small samples. Further, the effect is more severe for p_t^* than for p_t, implying possible reversal of the empirical counterpart of the variance-bounds inequalities even if the present-value relation is true. The second is that Shiller's procedure for calculating an observable version of p_t^* also induces bias toward rejection.

To understand Flavin's first criticism, we review the elementary statistics involved in estimating the variance of a population. If a sample $\{x_1, x_2, \ldots, x_n\}$ is drawn from a common distribution with known mean μ, the average squared deviation around the population mean yields an unbiased estimator of the variance:

$$\frac{E\left[\sum_{i=1}^{n}(x_i - \mu)^2\right]}{n} = \sigma^2. \tag{18}$$

Note that it is not necessary to assume that the x_i are mutually uncorrelated.

If the mean must be estimated, however, the variance estimator constructed by substituting the sample mean for the population mean in (18) is biased toward zero. The standard remedy is to reduce the degrees of freedom by 1:

$$\hat{\sigma}^2 = \sum_{i=1}^{n} \frac{(x_i - \bar{x})^2}{n - 1}.$$

The resulting estimator is unbiased only under the additional assumption that the x_i are uncorrelated. Flavin pointed out that both p_t and p_t^* are autocorrelated positively. Therefore reducing degrees of freedom by 1 will not provide an unbiased estimator of the variance of either p_t or p_t^*. Further, p_t^* is more strongly autocorrelated than p_t, implying that the variance of p_t^* is estimated with a greater downward bias than that of p_t. Flavin's argument suggests that if the present-value model is rejected when the volatility statistic $\hat{V}(p^*) - \hat{V}(p)$ is negative, type I error (rejecting the null when it is true) will occur with high probability. Therefore, the rejection region should be chosen smaller. Ideally, one would prespecify the probability of type I error and choose the rejection region accordingly. Under model-free

---

[5] Kleidon's (1986a) paper, written contemporaneously with Flavin's (1983) paper, made independently many of the same points as Flavin.

tests, however, this is impossible because the distribution of the volatility statistic is unspecified.

To illustrate these assertions, we consider the sampling distribution of the sample variance of p_t^* and p_t, based on samples of size 2, from the example in Section 1.2. The assumption of two-element samples is less restrictive than it might seem: a sample of size 2, but with observations separated by n periods, is qualitatively similar to a consecutive sample of size n. This is so because for large n, p_{t+n} is virtually independent of p_t, and similarly for p_{t+n}^* and p_t^*. We follow Flavin in assuming that the discount rate β is known. One would like to avoid this restrictive assumption, but the calculations to be presented would be rendered intractable if sampling variability of the estimate of β were considered. It is also assumed that p_t^* is measured correctly, pending discussion below.

Assume, as in Section 1.2, that individuals can forecast dividends up to m periods in the future. The sample variance of p_t^m is given by

$$\hat{V}_n(p_t^m) = (p_{t+n}^m - \bar{p})^2 + (p_t^m - \bar{p})^2, \tag{19}$$

where $\bar{p} = (p_{t+n}^m + p_t^m)/2$. Expression (19), which reflects the usual correction for degrees of freedom (the sum of squares is divided by the number of observations, 2, less 1) simplifies to

$$\hat{V}_n(p_t^m) = \frac{(p_{t+n}^m - p_t^m)^2}{2}.$$

We wish to take the expectation of the sample variance under the assumption that dividends follow (9). To do so, use (11) to write p_{t+n}^m as

$$p_{t+n}^m = \frac{\beta \lambda d_{t+n} + \sum_{i=1}^{m} \beta^i \epsilon_{t+n+i}}{1 - \beta \lambda}$$

$$= \frac{\beta \lambda (\epsilon_{t+n} + \cdots + \lambda^{n-1} \epsilon_{t+1} + \lambda^n d_t) + \sum_{i=1}^{m} \beta^i \epsilon_{t+n+i}}{1 - \beta \lambda}.$$

Using (11) again to obtain an expression for p_t^m, subtracting this from p_{t+n}^m, squaring, taking the expectation, and dividing by 2, we obtain

$$E[\hat{V}_n(p_t^m)] = \left[\frac{\beta^2 \lambda^2 (\lambda^n - 1)^2}{1 - \lambda^2} + \sum_{i=1}^{n} (\beta \lambda^{n+1-i} - \beta^i)^2 \right.$$

$$\left. + \sum_{i=n+1}^{m} (\beta^{i-n} - \beta^i)^2 + \sum_{i=m+1}^{m+n} \beta^{2(i-n)} \right]$$

$$\times [2(1 - \beta \lambda)^2]^{-1}, \tag{20}$$

Variance-Bounds Tests

for $n \leq m$, and

$$E[\hat{V}_n(p_t^m)] = \left[\frac{\beta^2 \lambda^2 (\lambda^n - 1)^2}{1 - \lambda^2} + \sum_{i=1}^{m} (\beta \lambda^{n+1-i} - \beta^i)^2 \right.$$

$$\left. + \sum_{i=m+1}^{n} (\beta \lambda^{n+1-i})^2 + \sum_{i=n+1}^{m+n} \beta^{2(i-n)} \right]$$

$$\times \left[2(1 - \beta \lambda)^2 \right]^{-1}, \qquad (21)$$

for $n > m$, and

We now use these expressions for the expectation of the sample variance to verify some properties of the bias of the variance estimators. First, in large samples, the variance of p_t^m is estimated without bias:

$$\lim_{n \to \infty} E[\hat{V}_n(p_t^m)] = V(p_t^m),$$

for all m and λ [let n go to infinity in (21) and observe that the result coincides with (12)]. It makes sense for the sample variance of p_t^m (corrected for degrees of freedom) to be an asymptotically unbiased estimator of $V(p_t^m)$ since p_{t+n}^m is approximately independent of p_t^m when n is large, even when λ is high (but less than unity).

Second, we have as a special case ($m = \infty$) of (20)

$$\lim_{n \to \infty} E[\hat{V}_n(p_t^*)] = V(p_t^*),$$

so the sample variance of p_t^* gives an asymptotically unbiased estimate of its population variance as well. With both actual and ex post rational stock price estimated without bias, there is no bias problem in large-sample tests.

For small samples and unrestricted m and λ, the sample variance of p_t^m underestimates its population variance for two reasons. First, with $\lambda > 0$, the positive autocorrelation of dividends induces a corresponding positive correlation between nearby values of p_t^m. The simple correction for degrees of freedom takes no account of this dependence. Second, since agents can foresee dividends up to m periods ahead, the dividend realizations upon which p_{t+n}^m is based overlap those on which p_t^m is based if $m > n$. Hence again bias results. If both of these conditions are absent (i.e., if $\lambda = 0$ and $m < n$), the sample variance is unbiased for any sample size:

$$E[\hat{V}_n(p_t^m)] = V(p_t^m)$$

---

[6] The (unavoidably tedious) verification of these equations involves isolating the coefficient of d_t and each of the ϵ_{t+i}, squaring term-by-term, and using the facts (i) $E[(\epsilon_t)^2] = 1$; (ii) $E[(d_t)^2] = 1/(1 - \lambda^2)$; and (iii) d_t is independent of the ϵ_{t+i}.

[verify the equality of the appropriately restricted versions of (21) and (12)].

Even if these conditions are met, the test statistic of the variance-bounds inequality is biased toward rejection—$V(p_t^m)$ is estimated without bias, but with n finite, $V(p_t^*)$ is underestimated as a result of the second effect noted above.[7] Thus, rejection of the variance-bounds inequality is likely in finite samples even if the null hypothesis is true. We see that whether dividends are autocorrelated or not, and however many periods ahead agents can forecast dividends, the sample variances provide an unbiased estimate of the corresponding population variances only in large samples, and that in finite samples this problem is more severe for the ex post rational price than for the actual price.

We turn now to Flavin's second criticism of Shiller's empirical implementation of the variance-bounds tests: bias is induced in the test statistic when an observable proxy is substituted for the unobservable p_t^* series [see also Shea (1989)]. As noted in Section 1, Shiller generated an observable version $p_{t|\bar{p}}^*$ of the p_t^* series from the recursion (15), supplying the remaining degree of freedom by imposing terminal condition (17). For each t, $p_{t|\bar{p}}^*$ in general gives a biased estimate of p_t^*, conditional on the realization of the sample. To show this, compare $p_{t|\bar{p}}^*$ to the series $p_{t|T}^*$ constructed as the solution to the recursion (15) satisfying the terminal condition $p_{T|T}^* = p_T$. By (2), $p_{T|T}^* = E(p_T^*|I_T)$. And if at any date $t \le T$, $p_{t|T}^* = E(p_t^*|I_T)$, then

$$p_{t-1|T}^* = \beta(d_t + p_{t|T}^*) = \beta[d_t + E(p_t^*|I_T)] = E(p_{t-1}^*|I_T),$$

where the last equality follows from (15). This proves that at each date t in the sample, $p_{t|T}^* = E(p_t^*|I_T)$, but if $p_{t|T}^*$ is an unbiased estimate of p_t^*, then $p_{t|\bar{p}}^*$ must be biased to the extent that the two series differ. The fact that $p_{t|T}^*$ is a better proxy for p_t^* than $p_{t|\bar{p}}^*$ is now well understood and more recent variance-bounds papers, such as Mankiw, Romer, and Shapiro (1985) and Durlauf and Hall (1989), use the unbiased series.

We just established that $p_{t|T}^*$ estimates p_t^* without bias, but this does not mean that the sample variance of $p_{t|T}^*$ estimates the expectation of the sample variance of p_t^* without bias.[8] To see that the sample variance of $p_{t|T}^*$ will not be closely related to the expected sample variance of p_t^*, it suffices to observe that (except in the trivial

---

[7] For example, (13) implies that $V(p_t^*) = \beta^2/(1 - \beta^2)$, while (20) implies that $E[\hat{V}_n(p_t^*)] = \beta^2(1 - \beta^n)/(1 - \beta^2)$, which is smaller than $V(p_t^*)$.

[8] The latter, rather than the population variance of p_t^*, is the relevant parameter for comparison if we wish to separate the bias induced by incorrect measurement of p_t^*, under discussion here, from the general problem of small-sample bias discussed above.

Variance-Bounds Tests

certainty case) the population variance of $p^*_{t|T}$ is not constant over time even if the population variance of p^*_t is constant. Because $p^*_{T|T} = p_T$, we have $V(p^*_{T|T}) = V(p_T) < V(p^*_T)$, by the variance-bounds theorem, but $\lim_{t \to -\infty} V(p^*_{T-t|T}) = V(p^*_T)$ since the $p^*_{T-t|T}$ series converges toward the true p^*_t going backward in time.[9]

It is clear intuitively why substituting the observable $p^*_{t|T}$ for the unobservable p^*_t will reduce the estimated variance of ex post rational price. The two variables are equal, except that in $p^*_{t|T}$ the innovations in dividends occurring after the end of the sample are set equal to zero, which can only reduce the variance of ex post rational price. Of course, the effect is strongest near the end of the sample. It follows that in long samples the downward bias induced by substituting $p^*_{t|T}$ for p^*_t will be negligible since at most dates $p^*_{t|T}$ will be very close to p^*_t. For smaller samples, however, the effect may be far from negligible.

Flavin restricted her discussion to Shiller (1979, 1981) and Singleton (1980); therefore it remains to determine whether her criticisms apply to LeRoy and Porter (1981). The criticism just discussed does not apply since LeRoy and Porter made no use of an observable proxy for p^*_t. However, Flavin's first criticism of Shiller, that autocorrelation between successive elements of p^*_t induces downward bias in the variance estimate, does have an analog that applies to LeRoy and Porter. Suppose that dividends follow the first-order autoregression (9). In small samples, the estimated value of the autoregression parameter will be biased toward zero. The first-order term of the bias is $2/n$. This bias induces a corresponding downward small-sample bias in the estimated variance of p^*_t.

However, the bias problem is more complicated than this. As (13) indicates, the variance of p^*_t is a convex function of λ, but then, by Jensen's inequality, the expectation of the sample variance will be greater than the value indicated by (13), with the expected sample value substituted for the population value of each parameter. In sum then, the bias will depend on two effects: λ is estimated with downward bias, inducing downward bias in $V(p^*_t)$; sample variability in the estimated λ will induce upward bias in $V(p^*_t)$ because of the convexity of the variance expression. The outcome will depend on which effect is stronger.

Thus, about the LeRoy–Porter estimation procedure, Flavin's analysis tells us only what is in any case evident from first principles: any nonlinear estimation procedure will in general be subject to bias, with the sign of the bias impossible to determine *a priori*.

---

[9] Shea (1989) pointed out that the population variance of $p^*_{t|T}$ is nonconstant.

2.2 Kleidon

Kleidon's (1986b) criticism focused on Grossman and Shiller's (1981) contention that the smoothness of a time-series plot of p_t^* relative to p_t contradicts the variance-bounds theorems. To Kleidon such a conclusion is completely unwarranted. The variance-bounds inequalities refer to cross sections, not time series. That is, if we could run a large number of replications of the economy, the variance-bounds theorems imply that we would find that

$$\hat{V}(p_t) = \frac{\Sigma_i (p_{it} - \bar{p}_t)^2}{n-1} \leq \hat{V}(p_t^*) = \frac{\Sigma_i (p_{it}^* - \bar{p}_t^*)}{n-1}, \quad (22)$$

for each t, where i runs over replications of the economy, not time. Unfortunately, history occurs only once, so we have only a single value for i. The time-series inequality

$$\hat{V}(p) = \frac{\Sigma_t (p_t - \bar{p})^2}{n-1} \leq \hat{V}(p^*) = \frac{\Sigma_t (p_t^* - \bar{p}^*)^2}{n-1}, \quad (23)$$

for a single economy is what the data appear to contradict, but (22) and (23) are very different. Kleidon observed that nothing about the variance-bounds theorems guarantees that (23) will be satisfied, even in large samples. Hence reversal of (23) does not contradict the present-value relation.

To see Kleidon's point, we present two examples. The first is adapted from Kleidon (1986b, pp. 975–976). Suppose that a stock pays a dividend d_T only at some terminal date T. Take d_T to equal $\Sigma_{i=1}^T \epsilon_i$, where ϵ_t is an identically and independently distributed random variable with mean zero and variance unity (as above, the mean is suppressed). Assuming no discounting, we have

$$p_t^* = d_T = \sum_{i=1}^T \epsilon_i,$$

for all t, so

$$V(p_t^*) = T, \quad \text{for } t = 1,2,\ldots,T.$$

Actual price is given by

$$p_t = E_t(p_t^*) = \sum_{i=1}^t \epsilon_i,$$

implying that

$$V(p_t) = t, \quad \text{for } t = 1,2,\ldots,T.$$

Since $V(p_t) = t \leq V(p_t^*) = T$, if we look across economies, the variance-bounds inequality is satisfied at each date.

Variance-Bounds Tests

But imagine that only one replication of this experiment is available. The distributions of p_t and p_t^* corresponding to time-series plots of these variables are those conditional on the one realization of d_T. Since $p_t^* = d_T$, for all t, the conditional variance of p_t^* is zero, while p_t evolves randomly from $p_0 = 0$ to $p_T = d_T$. Plainly, p_t^* appears smoother than p_t.[10]

The second example is less transparent, but is nonetheless useful because it aids in the interpretation of Kleidon's simulations. Suppose that dividends follow a geometric random walk,

$$\ln(d_{t+1}) = \ln(d_t) + \epsilon_{t+1},$$

or, equivalently,

$$d_{t+1} = d_t \times \exp(\epsilon_{t+1}), \tag{24}$$

where ϵ_{t+1} is distributed independently as normal with mean μ and variance σ^2 and $d_0 = 1$. Further, if $\mu + \sigma^2/2 = 0$, then $E(d_{t+1}|d_t) = d_t$, for all t, so that d_t is a martingale without trend.[11] Now take as a measure of ex post rational price the variable $p_t^*(1)$, defined by

$$p_t^*(1) \triangleq \beta(p_{t+1} + d_{t+1}), \tag{25}$$

so that ex post rational price equals the present value of the position next year. Set p_t in the usual way:

$$p_t = E(p_t^*(1)|d_t, d_{t-1}, \ldots). \tag{26}$$

Using the rule of iterated expectations, (25) and (26) imply that

$$p_t = \frac{\beta}{1-\beta} d_t, \tag{27}$$

where we have assumed away bubbles and have used the fact that dividends have an expected growth rate of zero. Therefore p_t is a martingale as well. Because of (26), the variance-bounds theorem implies that p_t has lower volatility than $p_t^*(1)$ for each t. If we take the second moment around zero, we therefore have

$$E[p_t^2] < E[(p_t^*(1))^2],$$

---

[10] Analytically, we have $V(p_t^*|d_T) = 0$, for all t. As to the actual price, its conditional variance is given by

$$V(p_t|d_T) = V(p_t) - \frac{[\text{Cov}(p_t, d_T)]^2}{V(d_t)} = t - \frac{t^2}{T}.$$

This parabola has zeros at $t = 0$ and $t = T$, and is positive between the two.

[11] This is so because (24) implies that $E(d_{t+1}|d_t) = d_t E[\exp(\epsilon_{t+1})]$. The lognormal distribution has the property that $E[\exp(\epsilon_{t+1})] = \exp(\mu + \sigma^2/2) = 1$ if $\mu + \sigma^2/2 = 0$.

for each t, implying that if $\hat{\theta}$ is defined by

$$\hat{\theta} \triangleq \frac{\sum_{t=1}^{T}(p_t^*(1))^2}{T} - \frac{\sum_{t=1}^{T} p_t^2}{T},$$

then

$$E(\hat{\theta}) > 0. \tag{28}$$

The variance-bounds test associated with this inequality—check whether $\hat{\theta}$ is in fact positive—appears straightforward. Neither of Flavin's criticisms applies: by taking second moments around zero instead of the sample mean her first criticism is avoided,[12] while defining ex post rational price so that it is observable avoids her second criticism as well. Further, dividends have a constant mean by assumption, so it would seem that trend adjustment is not a problem. Yet if the variance-bounds test is conducted on simulated data, $\hat{\theta}$ turns out to be negative in almost every sample which has at least 50 or 100 draws.

Joerding (1986), who originated this example, interpreted it as reflecting a problem with random number generators [see also Joerding (1988), Kleidon and Koski (1992), and Joerding (1992)]. We take a different line. Note that, from (25) and (27), we have $p_t^*(1) = p_{t+1}$. Therefore $\hat{\theta}$ takes the simple form

$$\hat{\theta} = \frac{p_{T+1}^2 - p_1^2}{T}.$$

Because p_t is a martingale, the martingale convergence theorem [Billingsley (1979, p. 416)] implies that p_t will approach a constant on almost every sample path. In view of

$$p_{t+1} = p_t \exp(\epsilon_{t+1})$$

[an implication of (24) and (27)], the only way p_{t+1} can equal p_t is for the constant to which they converge to be zero. Thus the martingale convergence theorem implies that, in a sufficiently large sample, $T\hat{\theta}$ will be arbitrarily close to $-p_1^2$ with arbitrarily high probability. Even though the variance-bounds inequality is satisfied in the population, its sample counterpart is reversed with arbitrarily high probability in large samples.

The preceding examples give ample indication that a single time series, no matter how long, may fail to provide reliable estimates of the relevant population variances, but both examples were constructed under the assumption that dividends are nonstationary, a

---

[12] This device was used by Mankiw, Romer, and Shapiro (1985); see Section 3.

Variance-Bounds Tests

specification explicitly excluded by Shiller, and LeRoy and Porter. We have already seen that if dividends are generated by (9) with $|\lambda| < 1$, then the sample variances $\hat{V}(p)$ and $\hat{V}(p^*)$ based on a single time series are asymptotically unbiased estimates of $V(p)$ and $V(p^*)$. However, suppose that $\lambda = 1$, so that dividends are a random walk. Analysis of the geometric version of this case occupies the bulk of Kleidon's article.

Two preliminary points: first, a difference equation such as (9) does not completely characterize dividends, whatever the value of λ. Also needed is an initial condition. In taking $V(p_t)$ and $V(p_t^*)$ as constants, we adopted tacitly the usual convention that the drawing d_0 of dividends at some initial date is from the limiting distribution to which d_t converges. Under this convention, d_t is stationary, implying that its variance is constant. In the random-walk case, however, there is no limiting distribution. It follows that no matter what distribution d_0 is drawn from, d_t is not stationary. Accordingly, $V(p_t)$ and $V(p_t^*)$, although finite, are functions of time. Thus by taking $\lambda = 1$, Kleidon undertook to analyze the robustness of Shiller's conclusions when the underlying assumption of stationarity is a misspecification. This approach is in sharp contrast to Flavin, who accepted Shiller's stationarity assumption and analyzed small-sample problems.

Second, nonstationarity does not invalidate the variance-bounds theorems. The equation $p_t = E(p_t^*|I_t)$ implies that $V(p_t) \leq V(p_t^*)$, for each t, whether or not dividends are stationary. Rather, it is the assumption that these variances are constant over time, which is adopted in econometric implementation of the variance-bounds theorems, that is violated if stationarity fails.

Now we return to Kleidon's discussion of the Grossman–Shiller contention that the smoothness of p^* relative to p indicates violation of the variance-bounds inequalities. We have seen that if dividends are a random walk, the sample variances of p and p^* over a single time series do not correspond to the (nonconstant) cross-section variances that enter the variance-bounds theorems. Hence, the apparent smoothness of p^* relative to p is uninformative about whether the variance bounds are satisfied.

Sample variances from a single time series, as we have seen, are not good estimators of the corresponding (unconditional) population variances, but that does not mean that they are uninterpretable. We now prove that in small samples

$$E[\hat{V}_n(p_t^m)] > E[\hat{V}_n(p_t^*)] \qquad (29)$$

is true if and only if

$$V(p_{t+n}^m|p_t^m) > V(p_{t+n}^*|p_t^*) \qquad (30)$$

is true.[13] Thus, loosely, sample unconditional variances behave like population conditional variances.[14]

To prove the above assertion, recall that in two-element samples we have

$$\hat{V}_n(p_t^m) = \frac{(p_{t+n}^m - p_t^m)^2}{2}.$$

Taking expectations, using the fact that $V(p_{t+n}^m) = V(p_t^m)$, and simplifying, this becomes

$$E[\hat{V}_n(p_t^m)] = \gamma_0^m - \gamma_n^m, \qquad (31)$$

where $\gamma_n^m \triangleq \text{Cov}(p_t^m, p_{t+n}^m)$. Now, by the linearity of the dividend process and the present-value relation, the variance of p_{t+n}^m conditional on p_t^m is given by

$$V(p_{t+n}^m | p_t^m) = \gamma_0^m - \frac{(\gamma_n^m)^2}{\gamma_0^m}, \qquad (32)$$

the formula for residual variance from linear regression theory. Dividing (31) for finite m by (31) for $m = \infty$ and applying the same procedure to (32), we get

$$\frac{E[\hat{V}_n(p_t^m)]}{E[\hat{V}_n(p_t^*)]} = \frac{V(p_{t+n}^m | p_t^m)}{V(p_{t+n}^* | p_t^*)} \times A,$$

where $A \triangleq \gamma_0^m(\gamma_0^\infty + \gamma_n^\infty)/\gamma_0^\infty(\gamma_0^m + \gamma_n^m)$.

For small samples, γ_n^m/γ_0^m (the regression coefficient of p_{t+n}^m on p_t^m) is approximately equal to unity, leading to $A \cong 1$. Note here that "small samples" means that n is vanishingly small, not just finite. We conclude that

$$\frac{E[\hat{V}_n(p_t^m)]}{E[\hat{V}_n(p_t^*)]} \cong \frac{V(p_{t+n}^m | p_t^m)}{V(p_{t+n}^* | p_t^*)},$$

for all m and λ, and small n, implying that (30) is true if and only if (29) is true, as asserted.

Returning to Kleidon, consider the population inequality

$$V(p_{t+n} | p_t) > V(p_{t+n}^* | p_t^*), \qquad (33)$$

corresponding to the observed choppiness of p relative to p^*. In view of the result just reported, this inequality is consistent with the vari-

---

[13] This is a new result, the only one in this paper.

[14] In early drafts of his paper, Kleidon provided informal discussion along these lines, but deleted this material from the published version; see Kleidon (1986b, p. 967, note 9).

Variance-Bounds Tests

ance-bounds theorems, which state that the variance of p is less than that of p^* only if the respective variances are conditioned on the same information set. In (33), the conditioning sets are different, so the variance-bounds theorems do not contradict (33). In particular, the variance of p^*_{t+n} conditional on p^*_t has no economic meaning since agents cannot observe p^*_t at t (or, for that matter, at any later date).

Since the variance-bounds theorems do not apply to conditional variances, the inequality (33) must be studied from scratch. This analysis is done easily using the linear autoregression model for dividends. Explicit evaluation of the conditional variances shows that for small n, the inequality (33) is exactly what one should expect. Essentially because the recursion $p^*_t = \beta(d_{t+1} + p^*_{t+1})$ has no error, nearby values of p^* are much more highly correlated than nearby values of p [see LeRoy (1984, pp. 185–186)], leading to conditional variances lower for p^* than for p. If $\lambda = 0.99$, this effect persists for n as high as 50, while if $\lambda = 1$, any finite-sized sample will have the qualitative properties of a small sample.

The foregoing considerations suggest that if dividends follow a random walk, the sample variance of p is likely to exceed the sample variance of p^* for samples of any size, but is the random-walk model consistent with a sample standard deviation of p which exceeds that of p^* by a factor as high as 5.59, the value Shiller reported? To gauge the meaning of the coefficient reported by Shiller, Kleidon ran Monte Carlo studies in which the present-value relation holds by construction and where dividends are generated by a geometric random walk with parameters chosen to give the best fit to Shiller's data. The variance inequality was deemed violated by the sample if the sample variance of p_t exceeded that of p^*_t, and to be "grossly violated" if the ratio of sample variances exceeded 5. Kleidon reported a frequency of violations of about 90 percent (when following Shiller's detrending procedure); the frequency of gross violations varied from 4.6 percent, for a rate of interest of 0.075, to almost 40 percent, for a rate of interest of 0.05.

Shiller (1988) responded that when the rate of interest varies, so does the dividend–price ratio if the dividend growth rate μ stays constant. The combination of the rate of growth μ chosen by Kleidon and a rate of interest of 5 percent implies an implausible value of the dividend–price ratio. Shiller ran experiments similar to those of Kleidon, but specified μ to vary with the rate of interest in order to leave the implied dividend–price ratio at its average historical level. He obtained much lower frequencies of gross violations.

Note that in the debate between Kleidon and Shiller about the appropriate parameter values to assume in Monte Carlo experiments, we have completely lost contact with the variance-bounds theorems.

If dividends follow a random walk (whether arithmetic or geometric), the sample variance bears no relation to any simple unconditional population parameter. Consequently, the robustness of the original variance-bounds theorems is sacrificed. For example, Kleidon and Shiller assumed that agents have no information beyond current dividends. How would their rejection frequencies be affected if this unrealistic assumption were relaxed? In the absence of any underlying theory, it is impossible to say.

Kleidon's paper emphasized that the first-generation empirical tests of the variance-bounds relation depended on the stationarity of the underlying series. The Shiller and LeRoy–Porter papers were clear about this dependence, so the question becomes whether the trend adjustment used in each case in fact produced a stationary series. Shiller's trend removal—take residuals from a constant growth rate path—will produce a stationary series only if the original series is trend stationary. The empirical evidence for the existence of unit roots in dividends—i.e., against the stationarity of dividends—is weaker than for most macroeconomic time series; thus, based on the evidence on dividends alone, the stationarity question remains open. However, in most macroeconomic models, dividends are cointegrated with variables such as GNP, investment, consumption, and so forth. Therefore the relevant evidence is that concerning unit roots for macroeconomic variables generally, not just dividends. The evidence now indicates that most macroeconomic variables do have unit roots. Of course, even if so, it remains open whether the bias induced by nonstationarity is sufficient to explain the variance-bounds violations.

2.3 Summary

The reason Shiller's rejections of the variance-bounds inequality were so striking is that the results did not depend on a particular specification of the model for dividends (as noted above, the LeRoy–Porter version did not share this attractive "model-free" property). Flavin and Kleidon pointed out that, while it is true that the variance-bounds inequality itself is model free, the properties of any econometric test of that inequality can only be investigated conditional on a particular dividends model. They argued that under reasonable specifications of the dividends model, variance-bounds tests will reject with high probability even if the present-value model is true. There can be no doubt that, at a minimum, the critics established that econometric problems with variance-bounds tests are potentially severe. Whether these problems are severe enough to account for the extent of the apparent excess volatility, however, remained controversial.

Variance-Bounds Tests

3. Second-Generation Tests

3.1 West
West (1988a) derived a variance-bounds test that (i) is valid even if dividends are nonstationary, and (ii) does not require a proxy for p_t^*. To understand West's test, recall the definition of H_t as the information set consisting of current and past dividends. West defined x_{tH} as the expectation of the cum dividend ex post rational price conditional on H_t, and x_{tI} as the corresponding variable for investors' actual information set I_t:

$$x_{tH} \triangleq \sum_{i=0}^{\infty} \beta^i E(d_{t+i}|H_t), \qquad x_{tI} \triangleq \sum_{i=0}^{\infty} \beta^i E(d_{t+i}|I_t).$$

West's result was that under the weak assumption that I_t contains H_t, the innovation variance of x_{tH} exceeds that of x_{tI}:

$$E[(x_{t+1,H} - E[x_{t+1,H}|H_t])^2] \geq E[(x_{t+1,I} - E[x_{t+1,I}|I_t])^2].$$

The meaning of West's inequality becomes apparent when we recast it in more familiar notation. Recalling the definitions of the ex dividends price series $\hat{p}_t$ and p_t,

$$\hat{p}_t \triangleq \sum_{i=1}^{\infty} \beta^i E(d_{t+i}|H_t), \qquad p_t \triangleq \sum_{i=1}^{\infty} \beta^i E(d_{t+i}|I_t),$$

we have

$$x_{t+1,H} = \hat{p}_{t+1} + d_{t+1} \quad \text{and} \quad E(x_{t+1,H}|H_t) = \beta^{-1}\hat{p}_t.$$

Similarly,

$$x_{t+1,I} = p_{t+1} + d_{t+1} \quad \text{and} \quad E(x_{t+1,I}|I_t) = \beta^{-1}p_t.$$

Thus the innovations in x_{tH} and x_{tI} are recognized as just the returns

$$\hat{r}_{t+1} = \hat{p}_{t+1} + d_{t+1} - \beta^{-1}\hat{p}_t \quad \text{and} \quad r_{t+1} = p_{t+1} + d_{t+1} - \beta^{-1}p_t,$$

and West's inequality reduces to

$$V(\hat{r}_t) \geq V(r_t). \tag{34}$$

Thus, investors' actual returns must have lower variance than the returns that would obtain if investors' information consisted of H_t.

West's upper bound on return variance is a direct implication of the LeRoy–Porter lower bound on price variance. To see this, observe that the LeRoy–Porter equality (3) holds for any information set. Therefore it holds for H_t, implying

773

$$V(p_t^*) = V(\hat{p}_t) + \frac{\beta^2 V(\hat{r}_t)}{1 - \beta^2}. \tag{35}$$

The LeRoy–Porter lower bound (8) on price variance, combined with (3) and (35), yields West's inequality (34).

West argued in favor of his test that, unlike the first-generation variance-bound tests, it is valid whether or not dividends are stationary. Even if dividends are generated by a linear process with a unit root, so that dividends and prices are cointegrated rather than stationary, the population return variances will be constant. Therefore their sample counterparts provide consistent estimates of population values. This argument is correct, but it must be remembered that it applies only for nonstationary dividend processes that are linear. We believe that a more natural treatment of trend is to specify a log-linear, rather than linear, dividend process, so that dividend growth rates are stationary. Although West's variance-bounds test does not apply directly, it can be adapted to the log-linear case; see LeRoy and Parke (1990), discussed below.

3.2 Mankiw, Romer, and Shapiro

Mankiw, Romer, and Shapiro (1985, 1991) (hereafter MRS) claimed to have provided an unbiased volatility test. The derivation is as follows: Let p_t^0 be any variable that can be constructed from information available to agents at t, and let $p_{t|T}^*$, as in Section 2.1, be the observable version of p_t^*. Consider the identity

$$p_{t|T}^* - p_t^0 \equiv (p_{t|T}^* - p_t) + (p_t - p_t^0). \tag{36}$$

Under unbiased forecasting, $p_{t|T}^* - p_t$ must be uncorrelated with any variable observable at t; in particular, with $p_t - p_t^0$. Hence by squaring both sides of (36), taking expectations, and making use of $E[(p_{t|T}^* - p_t)(p_t - p_t^0)] = 0$, we get

$$S \triangleq E[(p_{t|T}^* - p_t^0)^2] - E[(p_{t|T}^* - p_t)^2] - E[(p_t - p_t^0)^2] = 0. \tag{37}$$

The last equality is valid for any definition of p_t^0, but if p_t^0 is interpreted as a "naive forecast" of $p_{t|T}^*$, then the actual price must outperform p_t^0 as a forecast of $p_{t|T}^*$ which implies (37).

The sample counterpart of (37) is defined in the obvious way:

$$\hat{S} \triangleq \frac{1}{T}\left\{\sum_{t=1}^{T}(p_{t|T}^* - p_t^0)^2 - \sum_{t=1}^{T}(p_{t|T}^* - p_t)^2 - \sum_{t+1}^{T}(p_t - p_t^0)^2\right\}.$$

MRS set p_t^0 as a constant multiple of dividends in the preceding year. They found that $\hat{S}$ was negative, indicating excess volatility.

Variance-Bounds Tests

MRS claimed that because this volatility test uses noncentral rather than central variances, it is unbiased.[15] To evaluate this claim, we first point out a redefinition of terms that crept into the variance-bounds literature with Flavin's paper. By "test," econometricians ordinarily mean something very precise. Besides specifying a test statistic, it is necessary to identify a rejection region such that, if the null hypothesis is true, it will be rejected with preassigned probability α. A test is biased toward rejection if the probability of rejection exceeds α.

In the variance-bounds literature, however, the terms "test" and "unbiased test" have been used more loosely. Because of the difficulty of evaluating small-sample distributions, many have dispensed with any formal attempt to specify a rejection region as defined above. The variance-bounds inequalities are then rejected if the sample variance of p_t is much higher than that of p_t^*, with no attempt to define "much." In the variance-bounds literature, a test is said to be biased toward rejection if the expected value of the test statistic is in the direction of excess volatility, relative to the corresponding population parameter. Because this property concerns only the first moment of the distribution of the test statistic, it is usually easy to investigate. For example, the equality $E(\hat{S}) = S = 0$ is the basis for MRS's claim to have provided an unbiased test. However, the result that $\hat{S}$ gives an unbiased estimate of S does not establish that the test is unbiased in the usual sense.

Given the difficulty in evaluating type I error probabilities, the redefinitions of the terms "test" and "unbiased test" that have taken place are understandable. The two definitions are related: in showing small-sample bias (as redefined), Flavin created a strong presumption that the apparently dramatic rejections of the variance-bounds inequalities that Shiller reported are suspect, but when MRS went on to report the results of what they called an unbiased test, it is important to note that by this they meant only that the expectation of their test statistic equals the corresponding population parameter.

Shea (1989) pointed out two major problems with MRS's tests, both attributable to the use of $p_{t|T}^*$ in place of p_t^*. First, the outcome of the tests is very sensitive to the choice of terminal date. Second, because of the nonstationarity induced by the dependence of both population parameters and statistics on $p_{t|T}^*$ (which is nonstationary even if p_t^* is stationary), there is no prospect of using asymptotic theory to derive confidence intervals for the tests. To avoid both problems, Shea suggested using a rolling terminal date. Under Shea's

---

[15] Flavin had already noted that noncentral variances have this property and had reported the results of such tests. She credited Richard Porter with the insight that use of noncentral variances provides a way around her criticism of Shiller's tests (1983, p. 950).

recommended procedure, for each time period the same value of τ is used to calculate the ex post rational price

$$p^*_{t|\tau} \triangleq \sum_{i=1}^{\tau} \beta^i d_{t+i} + \beta^\tau p_{t+\tau},$$

instead of defining τ as $T - t$. However, the estimates of $p^*_{t|\tau}$ calculated in this way are not equal to the expectation of p^*_t conditional on the entire sample, lessening the appeal of Shea's procedure. Also, Shea's procedure is wasteful of data: with sample size T, a rolling terminal date τ allows construction of the price series only until $t = T - \tau$ since $p^*_{t|\tau}$ requires an observation of $p_{t+\tau}$. A larger value of τ reduces the observation error in the series $p^*_{t|\tau}$, but also reduces its length. In any case, Shea's results were much more favorable to the variance-bounds theorems than MRS's.

As noted above, in their 1985 paper, MRS made no attempt to determine whether their rejection of the present-value model was statistically significant—they concluded only that their point estimates suggested excess volatility. However, in their 1991 paper, MRS calculated asymptotic standard errors, using the rolling horizon to assure stationarity. They concluded that the excess volatility was of moderate but not overwhelming statistical significance. This was exactly the conclusion LeRoy and Porter had reached, also based on asymptotic standard errors.

Marsh and Merton (1986) and Merton (1987) observed that dividend smoothing by management could bias variance-bounds tests in general, and MRS's test in particular, toward rejection. If dividends are slow to reflect changes in underlying profitability, measured dividend volatility could give an impression that fundamentals had remained stable even when the opposite was the case. To illustrate the problem, Merton (1987) analyzed an example in which dividends were assumed to be zero within the sample and showed that econometric problems with the MRS test would result. Merton's example is a hybrid of models analyzed by Flavin and Kleidon. The nonstationarity of prices implied by the zero-dividend assumption results in econometric problems, as Kleidon showed; the substitution of the observable $p^*_{t|T}$ for the unobservable p^*_t induces small-sample bias, as Flavin showed. In fact, the latter problem is magnified in the context of Merton's example as a consequence of his assumption that within-sample dividends are zero. This observation formed the basis for MRS's (1991) response to Merton's criticism. They noted that, empirically, end-of-sample price is a small contributor to within-sample price relative to within-sample dividends at most dates of a hundred-year sample. Therefore, they argued, the econometric problem Merton pointed out is of little practical importance.

Variance-Bounds Tests

Merton's criticism and MRS's reply failed to distinguish between two separate phenomena: low dividends and smoothed dividends. It is true that firms retain a large fraction of their earnings, implying that capital gains typically exceed dividends, but that does not necessarily cause problems for model-based variance-bounds tests (since these estimate the variance of ex post rational price directly from dividends, rather than relying on some observable version of p_t^*). For example, if dividends follow a geometric random walk, variance-bounds tests can be constructed by estimating the mean and variance of the dividend growth rate and comparing the implied volatility of ex post rational price with that of actual price [see LeRoy and Parke (1990), discussed below]. Nothing about Merton's example suggests that these parameters are more difficult to estimate when the mean dividend growth rate is high (as will occur when corporations retain most of their earnings) than when it is low. Thus low dividend payout rates, even though they imply that stock prices depend mostly on out-of-sample dividends, do not necessarily cause problems for variance-bounds tests. Smoothed dividends, however, are another story. If dividend growth rates have important low-frequency components, small-sample problems are exacerbated. MRS's reply did not address this aspect of Merton's criticism.

3.3 Scott, and Durlauf and Hall

Scott (1985) observed that nothing about the hypothesized equality between p_t and $E(p_t^* | I_t)$ requires that this equality be tested by comparing variances. Under the null hypothesis, the error e_t in

$$p_t^* = a + bp_t + e_t$$

is uncorrelated with the explanatory variable, so the present value model can be tested equally well by determining whether a equals zero and b equals unity in an ordinary least squares regression. Scott argued that such regression tests of the present-value model are essentially free of the econometric problems associated with volatility tests. He found empirically that b was near zero rather than unity, implying that p_t^* and p_t are virtually uncorrelated, conflicting with the present-value model.

Recent work by Durlauf and Hall (1989) clarified the relation between Scott's regression test and the volatility tests of the present-value relation. Continuing along the line initiated by Scott, Durlauf and Hall pointed out that it is possible to translate the restriction on parameters implied by volatility tests into a corresponding coefficient restriction in regression tests. To achieve this translation, Durlauf and Hall modeled stock prices p_t as the sum of q_t, the expected value of discounted dividends, and a noise variable s_t, which is unrestricted.

As usual, p_t^* is written as q_t plus an orthogonal forecast error f_t. We have

$$p_t = q_t + s_t, \qquad p_t^* = q_t + f_t, \qquad (38)$$

where the left-hand side variables are observed by the econometrician and the right-hand side variables are unobserved. The present-value model corresponds to $s_t = 0$. Now consider the regression

$$p_t = \theta(p_t - p_t^*) + u_t.$$

We wish to obtain the restriction on θ implied by $V(p) \leq V(p^*)$. To do so, write θ as $\text{Cov}(p_t, p_t - p_t^*)/V(p_t - p_t^*)$ and use (38) to eliminate p_t and p_t^*:

$$\theta = \frac{\text{Cov}(q_t + s_t, s_t - f_t)}{V(s_t - f_t)}$$

$$= \frac{\text{Cov}(q_t, s_t) + V(s_t) - \text{Cov}(s_t, f_t)}{V(s_t) - 2\text{Cov}(s_t, f_t) + V(f_t)} \qquad (39)$$

[using the orthogonality condition $\text{Cov}(q_t, f_t) = 0$]. Similarly, express the variance bounds inequality $V(p_t) \leq V(p_t^*)$ in terms of the moments of s_t, q_t, and f_t:

$$V(q_t) + 2\text{Cov}(q_t, s_t) + V(s_t) \leq V(q_t) + V(f_t). \qquad (40)$$

Combining (39) and (40), Durlauf and Hall obtained $\theta \leq \frac{1}{2}$.

Thus, rejection of $V(p) \leq V(p^*)$ is equivalent to rejecting $\theta \leq \frac{1}{2}$, but the present-value model should be rejected for any θ that differs significantly from zero, not just for values of θ that significantly exceed $\frac{1}{2}$. Durlauf and Hall argued that regression-based tests such as that of Scott, which reject for any nonzero value of θ, are likely to be superior to volatility tests, which do not detect small amounts of model noise.[16]

This argument is incorrect. The reason is that so far we have been vague about the choice of rejection region. If in a model-based variance-bounds test the rejection region is computed via simulation, then the rejection region associated with the chosen probability of type I error depends on the variance of the forecast error. In the Durlauf–Hall regression context, the same statistic, θ, is involved in both the (bounds) test of the inequality $\theta \leq \frac{1}{2}$ and the (orthogonality) test of the equality $\theta = 0$. Therefore type I error can be set equal in the two cases only if both tests have the same rejection region, but then the probability of type II error will also be equal under any

---

[16] LeRoy and Porter (1981, p. 561), Frankel and Stock (1987), and Froot (1989), reasoning similarly, also concluded that bounds tests have lower power than orthogonality tests.

Variance-Bounds Tests

alternative hypothesis so that, contrary to the Durlauf–Hall argument, in the regression context orthogonality tests do not have greater power than bounds tests.

The main purpose of the Durlauf–Hall paper was not to compare bounds and orthogonality tests of stock price volatility, but to measure the variance of model noise as a contributor to the total variance of stock prices. The fact that noise cannot be assumed to be orthogonal to any of the variables causes ambiguity in the measurement of noise variance. However, Durlauf and Hall showed that there exist lower bounds on the amount of model noise consistent with the data. Durlauf and Hall showed that at least 84 percent of the variance of prices can be attributed to noise under the estimator associated with the bounds test $V(p) \leq V(p^*)$.[17] Under the noise variance estimator associated with the more stringent orthogonality test, noise variance is essentially equal to that of stock prices. Durlauf and Hall concluded from these results that "movements in expected discounted dividends, with a constant discount rate, have essentially nothing to do with the actual movements of the stock market" (p. 17).

3.4 Campbell and Shiller

In three recent papers, Campbell and Shiller (1987, 1988a, 1988b) introduced material not found in Shiller's early papers on asset-price volatility. Campbell and Shiller noted that if the present-value model is true, (i) an optimal prediction of the present value of future expected dividends can be formed using current price alone and (ii) this optimal prediction coincides with current price. It follows that the present-value model implies testable restrictions of the coefficients of a bivariate vector autoregression of stock prices and dividends. This insight was not new to the variance-bounds literature (the LeRoy–Porter volatility test consisted exactly of tests of restrictions on the coefficients of a bivariate autoregression).[18] However, in implementing their tests, Campbell and Shiller (1988a) introduced a method for trend correction that, although not perfect, is superior to anything that went before. Campbell and Shiller assumed that dividends and whatever other variables predict dividends form a multivariate log-

---

[17] Apparently, Durlauf and Hall did not correct for trend in any way. Therefore, given that the forecast error is a martingale (facing toward the past), if the price process has a unit root, then spurious correlation will result, implying upward bias in the Durlauf–Hall estimate of noise variance. We are indebted to Douglas Steigerwald for pointing this out.

[18] In 1987, Campbell and Shiller observed that "there are several [procedures for testing the present-value relation] in the literature: these include single-equation regression tests, tests of cross-equation restrictions on a vector autoregression (VAR), and variance bounds tests" (p. 1063). In taking the latter two categories to be distinct, Campbell and Shiller appeared to be unaware that the idea of testing the volatility implications of the present-value relation by investigating the validity of cross-equation restrictions in a bivariate autoregression had already been introduced into the variance-bounds literature.

linear model. To reconcile the log-linearity of the dividend model with the linearity of the present-value relation, Campbell and Shiller log-linearized the expression defining the rate of return (the opposite procedure, linearizing the dividend model, would introduce heteroskedasticity). Comparison of the actual rate of return with its log-linearized counterpart—the two are correlated almost perfectly—allowed Campbell and Shiller to argue that the error introduced by the log-linearization is negligible. The present-value model that results from iterating the log-linearized version of the definition of the rate of return expresses the log price–dividend ratio as the present value of the discounted expected dividend growth rates. All these variables are essentially free of trend, implying that the econometric problems attending the first-generation volatility tests by reason of the nonstationarity of the underlying series are avoided.

Campbell and Shiller reported the results of a variety of tests of the equality of the log price–dividend ratio and the present value of future dividend growth rates, and for the most part found evidence of significant violation. Most relevant for the variance-bounds question, for example, Campbell and Shiller compared the standard deviation of the log actual price–dividend ratio with the standard deviation of the present value of future expected dividend growth rates. If the present value model is correct, it must be the case that, since these variables are the same, their standard deviations are equal. In fact, Campbell-Shiller found that the standard deviation of the actual price–dividend ratio was twice that of its theoretical counterpart, indicating significant excess volatility.

In 1988, Campbell and Shiller (1988b) added corporate earnings to the price–dividend autoregression. They found that earnings is a strong predictor of dividend growth even conditional on the current log price–dividend ratio. This finding contradicts the present-value model, according to which current price is a sufficient statistic for future dividend growth. This article also contains a valuable discussion of the relation between the variance-bounds tests and the return autocorrelation tests conducted by Fama and French (1988), Poterba and Summers (1988), and others. Recall that LeRoy and Porter had based their variance-bounds test on the fact that the forecast error for ex post rational price equals a weighted average of future total returns, so rejection of the variance-bounds inequality implies that returns are autocorrelated. Consequently, in the language of Campbell and Shiller, "excess volatility and predictability of multiperiod returns are not two phenomena, but one" (p. 663).

Unfortunately, direct comparison between the first-generation variance-bounds tests and the return autocorrelation results is impossible because the former tests imply autocorrelatedness of total return,

Variance-Bounds Tests

whereas the latter tests imply autocorrelatedness of the rate of return (i.e., total return per dollar of stock value). However, under the Campbell–Shiller linearization, the analog of total return is exactly the rate of return. Thus, since the same return measure is involved in both variance-bounds and return autocorrelation tests, one can go beyond the mere observation that rejection of the variance-bounds inequality implies returns are autocorrelated. In addition, one can translate any pattern of autocorrelation of rates of return into the implied correlation between the log actual price–dividend ratio and the forecast error for log ex post rational price–dividend ratio. Further, rejection of the orthogonality of the log ex post rational price–dividend ratio and the log actual price–dividend ratio implies exactly that a weighted average of future rates of return, with weights that decline geometrically according to a factor related to the average price–dividend ratio, is forecastable. Given the numerical value of this weighting factor, it turns out that the variance-bounds violations correspond to the finding that long-term (five- and ten-year) rates of return are highly autocorrelated. This was exactly the finding of Fama and French and Poterba and Summers.

3.5 LeRoy and Parke, and Cochrane

LeRoy and Parke (1990), in work initially conducted independently of the Campbell–Shiller research discussed above, proposed correcting for the trend in stock prices by dividing prices by dividends. They assumed that dividends follow a geometric random walk, their purpose being to derive variance-bounds tests that would be valid in the setting used by Kleidon to criticize Shiller. LeRoy and Parke began by reporting a model-free bounds test of the volatility of the price–dividend ratio along the lines of Shiller (1981), and also a model-based bounds test as developed by LeRoy and Porter. Taking the rejection region to be defined by a negative value of the volatility statistic $\hat{V}(p^*) - \hat{V}(p)$, they found that the model-free bounds test rejected the model, whereas the model-based test accepted it. This discrepancy reflected primarily the radically different estimates of ex post rational price volatility generated under the two procedures: the volatility estimate produced by the model-free procedure [using $p^*_{t|T}$, the observable version of ex post rational price obtained from the recursion (15) and the terminal condition $p^*_{T|T} = p_T$] was less than one-quarter that produced by the model-based procedure (estimating the discount factor and the mean and variance of the dividend growth rate and entering these as arguments in the expression for the variance of ex post rational price).

The obvious explanation for the disparity between model-free and model-based estimates of ex post rational price volatility is estimation bias. Flavin's arguments established the strong presumption that the model-free estimates are biased downward. The model-based estimates are also subject to bias: even though the mean and variance of the dividend growth rate and discount factor can be estimated without bias, sampling variance in the estimates of these parameters will generally induce bias in the estimated variance of ex post rational price because of the nonlinearity of the variance expression.

This differential effect of estimation bias suggests that type I error is much higher under model-free tests than under model-based tests. This reflects the arbitrary specification that the rejection region consists of negative values of $\hat{V}(p^*) - \hat{V}(p)$ regardless of sampling error. A better procedure would be to use the Monte Carlo simulations to specify a rejection region that fixes the estimated probability of type I error at some constant like 10 percent. LeRoy and Parke found that the 10 percent rejection region (based on the Monte Carlo experiments) depended critically on how much information agents are assumed to have, which is not restricted under the null hypothesis. The first conclusion of LeRoy and Parke, therefore, was that the inequality test of the variance-bounds relation, being subject to a "nuisance parameter" problem, is uninformative about the null hypothesis. This is true of both the model-free and model-based versions.

LeRoy and Parke went on to consider orthogonality tests. The orthogonality test (7), devised by LeRoy and Porter, takes the form of an equality, not an inequality. Because under the null hypothesis this equality is valid for any specification of agents' information, there is no nuisance parameter problem (at least with regard to population parameters; there remains the possibility that sampling distributions depend on agents' information). The problem with (7) is that its derivation depends on the assumption that dividends are generated by a linear process, which is inconsistent with the log-linear geometric random walk. To get around this problem, LeRoy and Parke derived an analog to (7) that is valid under log-linear dividend processes. The associated null hypothesis was significantly rejected for any specification of agents' information.

Finally, LeRoy and Parke considered West's test. The Campbell–Shiller log-linearization implies that West's inequality takes a particularly simple form under the geometric random walk: it says that the variance of the rate of return on stock is bounded above by the variance of the dividend growth rate. As an inequality test, West's test is in principle subject to the same nuisance parameter problem as the price variance inequality test. In practice, however, it turned out that

Variance-Bounds Tests

there was no nuisance parameter problem. The Monte Carlo tests showed that, as with the orthogonality test, West's inequality was significantly rejected for any specification of agents' information.[19]

The LeRoy–Parke results depend on two restrictive assumptions: that dividends follow a geometric random walk and that the discount rates are constant. Neither assumption can be justified except as a crude approximation. However, the problem with more complex and realistic models for dividends and discount factors is that under such specifications the rational expectations assumption becomes less plausible. For example, it is easy to verify that dividend growth rates have nonzero autocorrelations, contrary to the LeRoy–Parke specification, but it strains one's credulity to assume that agents know a sequence of autocorrelation coefficients for dividend growth rates and price stock taking account of these deviations from randomness.

Perhaps it is a matter of taste whether one prefers simple models that are not descriptively accurate or more complex models that, because of their complexity, put the rational expectations assumption to harder use. In any case, Cochrane (1990) derived a version of the LeRoy–Parke bound on the variance of the price–dividend ratio that is consistent with time-varying discount rates and autocorrelated dividend growth rates. To evaluate the bound, Cochrane used a linear approximation similar (but not identical) to that of Campbell and Shiller. Cochrane found that the variance bound was satisfied. This result is analogous to the LeRoy–Parke finding that the model-based point estimate of the variance of the price–dividend ratio was less than its upper bound. As discussed above, LeRoy and Parke concluded that in the absence of an attempt to control the probability of type I error, this result cannot be construed as favoring the present-value relation. The same observation applies to Cochrane's finding.

Cochrane also estimated decompositions of the variance of the price–dividend ratio into the sum of expected dividend growth rates and discount factors along the lines of Campbell and Shiller. Finally, he derived bounds on the mean and standard deviation of discount factors, as in Hansen and Jagannathan (1991). This material, being related only indirectly to variance bounds, is not discussed here.

3.6 Summary

The critics had established that, as attractive as Shiller's model-free variance-bounds test was by reason of its generality, it had the defect that its econometric properties cannot be investigated. Further, under dividend models that were (represented as) reasonable, these prop-

---

[19] See LeRoy (1990) for a nontechnical exposition and for a plot of the rate of return on stock against the dividend growth rate.

erties are very unsatisfactory. The emphasis in the second-generation variance-bounds tests was on developing tests that had acceptable econometric properties under realistic dividend models. In this they largely succeeded. With the exception of Mankiw, Romer, and Shapiro (1991) and Cochrane (1990), the second-generation tests found statistically significant excess volatility. However, it is important to recall that any model-based test is necessarily a test of a joint hypothesis that includes an assumption about the process generating dividends. Consequently, there remains the possibility that the appearance of excess price volatility may be caused by misspecification of the dividend model.

4. Possible Explanations

In this section, we review the suggested explanations for excess volatility of asset prices. First, investors may overreact to dividend news. That is, market participants may not be perfectly rational and have rational expectations. According to this view, stock prices are moved by fads [Shiller (1984), DeBondt and Thaler (1985)]; or, as Keynes described it, the stock market is a beauty contest in which one wins by "anticipating what average opinion expects the average opinion to be" (1936, p. 156). DeLong et al. (1988) showed that if "noise traders" have irrationally optimistic expectations, there is an equilibrium in which stock prices are more volatile than under the present-value model. Moreover, this equilibrium is stable in the sense that noise traders do not necessarily suffer losses.

Within the paradigm of neoclassical economics, rationality is part of the inner core, insulated from falsification. Explanations for price volatility that involve irrationality are therefore inadmissible. Before jettisoning the neoclassical paradigm to explain one (admittedly important) phenomenon, one should consider explanations that do not rely on irrationality.

If stock prices have greater volatility than expected values of discounted future dividends, the reason may be rational speculative bubbles.[20] The theory and evidence on bubbles was reviewed in West (1988b) and Flood, Hodrick, and Kaplan (1986). If the price of a security contains a deterministic bubble, then the value of the bubble

---

[20] See Gilles and LeRoy (1990) for a discussion of rational speculative bubbles within a context of general equilibrium. Flood and Hodrick (1986) observed that model-free variance-bounds tests include bubbles in the null hypothesis because of the way the observable version of ex post rational price is constructed. Therefore a finding of excess volatility in a non-model-based test cannot be attributed to bubbles. Flood, Hodrick, and Kaplan (1986) noted that the model-based tests of LeRoy and Porter did not use the observable version of ex post rational price, implying that in this case the presence of bubbles could cause rejection of the variance-bounds inequality.

Variance-Bounds Tests

will grow at the rate of interest, inducing nonstationarity of the price even when dividends are stationary. The data on price–dividend ratios, however, show no evidence of the implied nonstationarity [Diba and Grossman (1988)].

Stochastic bubbles are more interesting. They grow, then collapse. A stochastic bubble, once having burst, cannot be reborn; if there exists such a bubble, it must have been present from the asset's inception. However, there may exist many stochastic bubbles on the same price, and they may start so small that they do not noticeably affect the price of the asset for a long period before taking off and then bursting. To the observer, the price will exhibit swings. If bubble bursts are independent of dividends, stock prices are more volatile than the present-value relation predicts. If bursts are correlated negatively with dividends, still greater excess price volatility results and markets appear to overreact to dividend news.

West (1987) at first favored the bubble explanation of stock-price volatility. Upon reviewing the evidence in West (1988b), however, he reversed his conclusion. One of the reasons he gave for disqualifying rational bubbles is that they cannot be the source of excess volatility in bond prices. Bubbles, whether deterministic or stochastic, can only occur when the horizon is infinite, at least in discrete-time models. With a finite horizon, a backward induction annihilates any bubble at its inception. Shiller (1979) and Singleton (1980) reported evidence of excess volatility of bond prices. Because bonds have finite maturity, such excess volatility cannot be due to bubbles. It seems likely, West argued, that stock prices show too much variability for the same reason that bond prices do; if so, the culprit is not bubbles. Against this argument, however, is the new evidence [Campbell and Shiller (1987)] that the Shiller–Singleton finding of excess volatility in the bond market is due to incorrect trend adjustment. If so, the apparent excess volatility of stock prices, even under correct trend adjustment, favors the bubble explanation.

The present-value relation as we defined it assumes that the rate of discount is constant (see note 1). Since this assumption does not necessarily hold, it is natural to study the effect of discount rate variability on asset-price volatility. To see the connection, assume that there is a single consumption good and that the representative consumer maximizes $E(\Sigma_j \beta^j u(c_{t+j}) | I_t)$ with u of the constant relative risk aversion class $u(c) = c^{1-\rho}/(1-\rho)$. Let r_{it} denote the gross one-period rate of return (1 plus the rate of return) on asset i, and let m_t be the marginal rate of substitution of consumption at $t + 1$ for consumption at t [i.e., $m_t = \beta u'(c_{t+1})/u'(c_t)$]. Assuming interior maxima, the following relation will hold [Grossman and Shiller (1981)]:

$$1 = E(m_t \times r_{it} | I_t). \tag{41}$$

If the consumer is risk neutral ($\rho = 0$), then (41) yields the constant expected rate of return (the same for all securities) $E(r_{it}) = 1/\beta$, and hence we get the present-value relation that the data reject. In this model, excess volatility (relative to that predicted by the present-value model) implies that the consumer is not risk neutral.

Grossman and Shiller (1981), LeRoy and LaCivita (1981), and Michener (1982) worked out the implications of (41) for the volatility of asset prices. All found that risk aversion implies greater price volatility than in the present-value model. These results seem encouraging, but for the utility index $\ln(c)$ to produce violation of the variance bounds of the observed magnitude, the aggregate consumption series must be much more volatile than it actually is. The very smooth actual post WWII consumption series implies an implausibly high estimated value of the coefficient of relative risk aversion ρ [Grossman and Shiller (1981)].

The Grossman–Shiller finding of a high value of ρ is consistent with that of Mehra and Prescott (1985), who based their approach on the average covariance between the marginal rate of intertemporal substitution and stock returns. They considered the implications of (41) for the equity premium, defined as the difference between the average rate of return on a diversified portfolio of stocks and the risk-free rate. Mehra and Prescott found that the average real annual yield on the Standard and Poors 500 Index was 7 percent, while that on short-term debt was less than one percent. Let the latter of these yields stand in for the risk-free rate r^* and note that (41) implies

$$r_t^* = 1/E(m_t|I_t). \tag{42}$$

We can then rewrite (41) as

$$\frac{E(r_t|I_t) - r_t^*}{r_t^*} = -\mathrm{Cov}(m_t, r_t|I_t),$$

where r_t is the gross return on the Standard and Poors 500 Index. Mehra and Prescott constructed an artificial economy parameterized so that the growth rate of consumption is a stationary random process with the same mean, variance, and serial correlation as those observed in the U.S. economy. They concluded that an elasticity of substitution between the year t and year $t + 1$ consumption good sufficiently low to yield the 6 percent average equity premium also yields real interest rates far in excess of those observed. Because the elasticity of intertemporal substitution is the reciprocal of ρ, an equity premium equal to six times the risk-free rate implies a high value of ρ, precisely the Grossman–Shiller finding. The main conclusion of Mehra and Prescott was that, even if one were to accept a high value of ρ as plausible,

Variance-Bounds Tests

the model (41) is not saved because (42) then implies too high a value for the risk-free rate.

Mehra and Prescott calibrated an artificial economy, rather than using standard econometric techniques. It is, however, possible to modify methods used in the present-value model to test the specific model with time-varying discount rates [Campbell and Shiller (1988a)]. Singleton (1987) reviewed the literature on consumption-based asset-pricing models and reported on the results of a large number of studies besides those noted here. He concluded from these results that the U.S. aggregate data reject the model. This conclusion is robust to changes in the specification of the utility function and in the number of goods assumed to be present.

Progress on these issues will almost certainly require introducing either frictions or non-von Neumann–Morgenstern preferences. Mehra and Prescott and others suggested that the equity premium puzzle points to the importance of frictions. Bewley (1982), following up on this suggestion, showed that if consumers are liquidity constrained, or can only imperfectly insure wealth fluctuations, then individual consumption streams do not fluctuate in sympathy with each other, so the aggregate consumption series will be smoother than the individual series. A definitive test of Bewley's model would require panel data on consumption, asset holdings, and income of individuals, data that are not readily available.

The main problem for the representative-agent expected-utility model is that the smoothness of the aggregate consumption series suggests a low elasticity of intertemporal substitution. A variety of other evidence suggests that the coefficient of relative risk aversion is also low (or moderate), but within the expected-utility, time-separable framework, the intertemporal elasticity of substitution is the reciprocal of the coefficient of relative risk aversion, so we cannot have both. It is possible that uncoupling risk aversion and intertemporal substitution holds the key to the volatility and equity-premium puzzles. Relaxing time separability while preserving expected utility gives enough flexibility to distinguish behavior due to temporal substitution from that due to risk aversion. However, as explained in Epstein and Zin (1989), nonseparable utility leads an expected-utility maximizer to choose consumption plans that are dynamically inconsistent: a plan chosen as optimal in one period becomes suboptimal later. Preferences that do not vary in this fashion and also induce behavior toward risk that is divorced from intertemporal choice cannot satisfy von Neumann–Morgenstern axioms; a consumer with such preferences cannot be indifferent to the timing of the resolution of uncertainty [Kreps and Porteus (1978)], as expected utility implies.

However, Kreps–Porteus or still more general preferences could reconcile the low degree of intertemporal substitution indicated by aggregate consumption with the moderate degree of risk aversion that economists consider reasonable. Epstein and Zin (1989) showed that in such a setting, the risk premium on a particular asset depends on the correlation of the asset's return both with the market portfolio—as in the capital asset pricing model—and with consumption—as in (41). Epstein and Zin (1991) conducted an empirical study with mildly encouraging results: their stock market and aggregate consumption data rejected a von Neumann–Morgenstern specification of preferences, but could not reject a Kreps–Porteus specification. Weil (1989), however, showed that Kreps–Porteus preferences did not solve the risk-premium puzzle, but instead uncover two puzzles that Mehra and Prescott could not distinguish in the expected-utility framework. In addition to the risk-premium puzzle (the risk premium is too high, in view of the evidence on risk aversion), there is a risk-free rate puzzle (the risk-free rate is too low, in view of the evidence on intertemporal substitution). To explain these puzzles, Weil concluded, a proper specification of preferences is probably less important than market frictions.

5. Conclusion

In its 15-year history, the variance-bounds literature has evolved on a circuitous path. Shiller's first-generation model-free tests provided point estimates of price variances that suggested excess volatility, but the question of statistical significance remained open. This was necessarily the case because, in the absence of an assumed model for dividends, there was no way to investigate the econometric properties of the volatility statistics. The critics remedied this deficiency by supplying models for dividends and deriving the econometric properties of the variance-bounds statistics under the joint assumption that the present-value model is correct and that the dividends equation is well specified. They identified various sources of bias in the model-free variance-bounds tests, suggesting that the apparent excess volatility of asset prices was spurious. The second-generation tests took the obvious next step: expand the null hypothesis to include a model for dividends and then specify a variance-bounds test that performs well if the null hypothesis is true. The model-based tests that resulted—for example, those of Campbell and Shiller and LeRoy and Parke—have much in common with the first-generation model-based tests.

If in some respects the variance-bounds literature evolved toward a point near to where it started; in other respects it evolved into new territory. For example, adequate solutions to the problem of detrend-

Variance-Bounds Tests

ing price data, which had been the downfall of both the first-generation papers, were offered in the second-generation papers. Also, the second-generation papers brought to the forefront several critical distinctions that, although present in the first-generation papers, were not emphasized adequately or developed there. Besides the distinction between model-based and model-free tests, one thinks of the distinction between bounds and orthogonality tests.

The evidence is in: asset prices are more volatile than is implied by the present-value equation. Because of the vast improvement in our understanding of the econometric issues attending the variance-bounds tests, there is no longer any room for reasonable doubt about the statistical significance of the excess volatility. Several possible explanations for the excess volatility were reviewed in the preceding section. The authors of this paper are not in complete agreement about how promising these lines of research are. The first author expects that taking adequate account of frictions will go a long way toward resolving the excess volatility, particularly under parameterizations of preferences that are more general than those considered so far in the variance-bounds literature. The second author hopes this is correct, but expects that when the explanation of excess volatility is found, the critical ideas will lie farther from the neoclassical paradigm.

References

Bewley, T. F., 1982, "Thoughts on Tests of the Intertemporal Asset Pricing Model," working paper, Northwestern University, July.

Billingsley, P., 1979, *Probability and Measure* (2nd ed.), Wiley, New York.

Campbell, J. Y., and R. J. Shiller, 1987, "Cointegration and Tests of Present Value Models," *Journal of Political Economy*, 95, 1062–1088.

Campbell, J. Y., and R. J. Shiller, 1988a, "The Dividend-Price Ratio and Expectations of Future Dividends and Discount Factors," *Review of Financial Studies*, 1, 195–228.

Campbell, J. Y., and R. J. Shiller, 1988b, "Stock Prices, Earnings, and Expected Dividends," *Journal of Finance*, 43, 661–676.

Cochrane, J. C., 1990, "Explaining the Variance of Price–Dividend Ratios," working paper, University of Chicago.

DeBondt, W. F. M., and R. Thaler, 1985, "Does the Stock Market Overreact?" *Journal of Finance*, 40, 793–808.

DeLong, J. B., et al., 1988, "Noise Trader Risk in Financial Markets," Discussion Paper 1416, Harvard Institute of Economic Research, December.

Diba, B., and H. Grossman, 1988, "Explosive Rational Bubbles in Stock Prices," *American Economic Review*, 78, 520–530.

Durlauf, S. N., and R. E. Hall, 1989, "Measuring Noise in Stock Prices," working paper, Stanford University.

Epstein, L. G., and S. E. Zin, 1989, "Substitution, Risk Aversion, and Temporal Behavior of Consumption and Asset Returns: A Theoretical Framework," *Econometrica*, 57, 937–969.

Epstein, L. G., and S. E. Zin, 1991, "Substitution, Risk Aversion, and the Temporal Behavior of Consumption and Asset Returns: An Empirical Analysis," *Journal of Political Economy*, 99, 263–286.

Fama, E. F., and K. R. French, 1988, "Permanent and Temporary Components of Stock Prices," *Journal of Political Economy*, 96, 246–273.

Flavin, M. A., 1983, "Excess Volatility in the Financial Markets: A Reassessment of the Empirical Evidence," *Journal of Political Economy*, 91, 929–956.

Flood, R. P., and R. J. Hodrick, 1986, "Asset Price Volatility, Bubbles, and Process Switching," *Journal of Finance*, 41, 831–842.

Flood, R. P., R. J. Hodrick, and P. Kaplan, 1986, "An Evaluation of Recent Evidence on Stock Market Bubbles," working paper, Northwestern University.

Frankel, J. A., and J. Stock, 1987, "Regression vs. Volatility Tests of Foreign Exchange Markets," *Journal of International Money and Finance*, 6, 49–66.

Froot, K. A., 1989, "Tests of Excess Forecast Volatility in the Foreign Exchange and Stock Markets," working paper, MIT.

Gilles, C., and S. F. LeRoy, 1990, "Bubbles and Charges," working paper, University of California, Santa Barbara; forthcoming in *International Economic Review*.

Grossman, S. J., and R. J. Shiller, 1981, "The Determinants of the Variability of Stock Market Prices," *American Economic Review*, 71, 222–227

Hansen, L. P., and R. Jagannathan, 1991, "Implications of Security Market Data for Models of Dynamic Economies," *Journal of Political Economy*, 99, 225–262.

Joerding, W., 1986, "Variance Bounds Tests and Simulated Stock Prices," working paper, Washington State University.

Joerding, W., 1988, "Are Stock Prices Excessively Sensitive to Current Information?" *Journal of Economic Behaviour and Organization*, 9, 71–85.

Joerding, W., 1992, "Are Stock Prices Excessively Sensitive to Current Information?: Reply," *Journal of Economic Behaviour and Organization*, forthcoming.

Keynes, J. M., 1936, *The General Theory of Employment, Interest, and Money*, Harcourt, Brace & World, New York.

Kleidon, A. W., 1986a, "Bias in Small Sample Tests of Stock Price Rationality," *Journal of Business*, 59, 237–261.

Kleidon, A. W., 1986b, "Variance Bounds Tests and Stock Price Valuation Models," *Journal of Political Economy*, 94, 953–1001.

Kleidon, A. W., and J. L. Koski, 1992, "Are Stock Prices Excessively Sensitive to Current Information?: Comment," *Journal of Economic Behaviour and Organization*, forthcoming.

Kreps, D. M., and E. L. Porteus, 1978, "Temporal Resolution of Uncertainty and Dynamic Choice Theory," *Econometrica*, 46, 185–200.

LeRoy, S. F., 1984, "Efficiency and the Variability of Asset Prices," *American Economic Review*, 74, 183–187.

LeRoy, S. F., 1989, "Efficient Capital Markets and Martingales," *Journal of Economic Literature*, 27, 1583–1621.

LeRoy, S. F., 1990, "Capital Market Efficiency: An Update," *Economic Review, Federal Reserve Bank of San Francisco*.

Variance-Bounds Tests

LeRoy, S. F., and C. J. LaCivita, 1981, "Risk Aversion and the Dispersion of Asset Prices," *Journal of Business*, 54, 535–547.

LeRoy, S. F., and W. R. Parke, 1990, "Stock Price Volatility: Tests Based on the Geometric Random Walk," working paper, University of California, Santa Barbara.

LeRoy, S. F., and R. D. Porter, 1981, "The Present-Value Relation: Tests Based on Implied Variance Bounds," *Econometrica*, 49, 555–574.

Mankiw, N. G., D. Romer, and M. D. Shapiro, 1985, "An Unbiased Reexamination of Stock Market Volatility," *Journal of Finance*, 40, 677–687.

Mankiw, N. G., D. Romer, and M. D. Shapiro, 1991, "Stock Market Forecastability and Volatility: A Statistical Appraisal," *Review of Economic Studies*, 58, 455–477.

Marsh, T. A., and R. C. Merton, 1986, "Dividend Variability and Variance Bounds Tests for the Rationality of Stock Market Prices," *American Economic Review*, 76, 483–498.

Mehra, R., and E. C. Prescott, 1985, "The Equity Premium: A Puzzle," *Journal of Monetary Economics*, 15, 145–161.

Merton, R. C., 1987, "On the Current State of the Stock Market Rationality Hypothesis," in S. Fischer (ed.), *Macroeconomics and Finance: Essays in Honor of Franco Modigliani*, MIT, Cambridge, Mass.

Michener, R. W., 1982, "Variance Bounds in a Simple Model of Asset Pricing," *Journal of Political Economy*, 90, 166–175.

Poterba, J., and L. H. Summers, 1988, "Mean Reversion in Stock Prices: Evidence and Implications," *Journal of Financial Economics*, 22, 27–59.

Scott, L., 1985, "The Present Value Model of Stock Market Prices: Regression Tests and Monte Carlo Results," *Review of Economics and Statistics*, 57, 599–605.

Shea, G. S., 1989, "Ex-Post Rational Price Approximations and the Empirical Reliability of the Present-Value Relation," *Journal of Applied Econometrics*, 4, 139–159.

Shiller, R. J., 1979, "The Volatility of Long Term Interest Rates and Expectations Models of the Term Structure," *Journal of Political Economy*, 87, 1190–1209.

Shiller, R. J., 1981, "Do Stock Prices Move Too Much to be Justified by Subsequent Changes in Dividends?" *American Economic Review*, 71, 421–436.

Shiller, R. J., 1984, "Stock Prices and Social Dynamics," *Brookings Papers on Economic Activity*, 457–498.

Shiller, R. J., 1988, "The Probability of Gross Violations of a Present Value Variance Inequality," *Journal of Political Economy*, 96, 1089–1092.

Singleton, K. J., 1980, "Expectations Models of the Term Structure and Implied Variance Bounds," *Journal of Political Economy*, 88, 1159–1176.

Singleton, K. J., 1987, "Specification and Estimation of Intertemporal Asset Pricing Models," in B. Friedman and F. Hahn (eds.), *Handbook of Monetary Economics*, North-Holland, Amsterdam.

Weil, P., 1989, "The Equity Premium Puzzle and the Risk-Free Rate Puzzle," *Journal of Monetary Economics*, 24, 401–421.

West, K. D., 1987, "A Specification Test for Speculative Bubbles," *Quarterly Journal of Economics*, 102, 553–580.

West, K. D., 1988a, "Dividend Innovations and Stock Price Volatility," *Econometrica*, 56, 37–61.

West, K. D., 1988b, "Bubbles, Fads and Stock Price Volatility Tests: A Partial Evaluation," *Journal of Finance*, 43, 636–656.

Part II
Consumption-Based Asset-Pricing Models

[10]

Stochastic Consumption, Risk Aversion, and the Temporal Behavior of Asset Returns

Lars Peter Hansen
University of Chicago

Kenneth J. Singleton
Carnegie-Mellon University

> This paper studies the time-series behavior of asset returns and aggregate consumption. Using a representative consumer model and imposing restrictions on preferences and the joint distribution of consumption and returns, we deduce a restricted log-linear time-series representation. Preference parameters for the representative agent are estimated and the implied restrictions are tested using postwar data.

I. Introduction

In the asset pricing models of Rubinstein (1976b), Lucas (1978), Breeden (1979), and Brock (1982), among others, agents effect their consumption plans by trading shares of ownership of firms in a competitive stock market. An implication of this trading is that the serial correlation properties of stock returns are intimately related to the stochastic properties of consumption and the degree of risk aversion of investors. The purposes of this paper are to characterize explicitly

This research was supported in part by NSF grant SES-8007016. Helpful comments on earlier drafts of this paper were provided by members of the workshops at Carnegie-Mellon University, the University of Washington, and the Summer NBER Institute. We wish, in particular, to acknowledge the helpful suggestions received from Scott Richard and an anonymous referee. Ravi Jagannathan and Will Roberds assisted with the computations.

the restrictions on the joint distribution of asset returns and consumption implied by a class of general equilibrium asset pricing models and to obtain maximum likelihood estimates of the parameters describing preferences and the stochastic consumption process.

The motivation for this analysis derives from two considerations. First, in general equilibrium models of stock price behavior with risk-neutral agents (i.e., linear utility), share prices will be set so that the expected return on each asset is constant. Thus, asset returns will be serially uncorrelated and, in particular, past values of consumption will be uncorrelated with current-period asset returns. LeRoy and Porter (1981) and Shiller (1981) have recently conducted tests of the linear present-value formula for stock prices, implied by this result, and in both studies the model was rejected. As Grossman and Shiller (1981) have emphasized, these rejections suggest that agents do consider consumption risk when making portfolio decisions. Second, if agents are risk averse, then the temporal covariance structure of consumption and asset returns will be nontrivial, except under very strong restrictions on the underlying production technology (see, e.g., Rubinstein [1976b], Johnsen [1978], and Sec. II below). It is this temporal covariation that we attempt to characterize here.

The framework for this analysis is a production-exchange economy of identical agents who choose consumption and investment plans so as to maximize the expected value of a time-additive von Neumann–Morgenstern utility function. In order to derive the restrictions on the joint distribution of consumption and stock returns implied by this optimizing behavior, it is necessary to specify a distribution function and to parameterize preferences. The joint distribution of consumption and returns is assumed to be lognormal, and preferences are assumed to exhibit constant relative risk aversion (CRRA). This particular form of utility was chosen in part because of its preeminent role in many previous theoretical studies of asset pricing (e.g., Merton 1973; Rubinstein 1976a). In addition, the assumptions of CRRA utility and lognormality together lead to an empirically tractable, closed-form characterization of the restrictions implied by the model.[1]

More precisely, these assumptions lead to a restricted linear time-series representation of the logarithms of consumption and asset returns. The restrictions imply that the predictable components of the

---

[1] A similar interplay among the CRRA utility function and lognormal returns was exploited by Merton (1973, 1980), Rubinstein (1976b), Breeden (1977), and Grossman and Shiller (1981), among others, to obtain closed-form solutions to their models. Breeden and Litzenberger (1978) derive a version of the CAPM for a model with CRRA utility and lognormal returns and consumption.

logarithms of asset returns are proportional to the predictable component of the change in the logarithm of consumption, with the proportionality factor being minus the coefficient of relative risk aversion. Maximum likelihood estimates of the coefficient of relative risk aversion, the subjective discount factor, and the parameters that describe the temporal evolution of consumption are obtained using this closed-form characterization of the restrictions. The model is estimated for returns on stocks listed on the New York Stock Exchange and for returns on Treasury bills, using monthly data for the period 1959:2 through 1978:12. Then likelihood ratio tests of the joint hypothesis underlying the model are conducted.

The remainder of this paper is organized as follows. In Section II the model is described and the implied time-series representation for consumption and returns is derived. In Section III, maximum likelihood estimation is discussed and estimates of the parameters are presented. Concluding remarks are presented in Section IV.

II. The Model of Stock Market Returns

Consider a single-good economy of identical consumers, whose utility functions are of the CRRA type:

$$U(c_t) = c_t^\gamma/\gamma; \quad \gamma < 1, \qquad (1)$$

where c_t is aggregate real per capita consumption and $U(\cdot)$ is the period utility function. The representative consumer in this economy is assumed to choose a stochastic consumption plan so as to maximize the expected value of his time-additive utility function,

$$E_0\left[\sum_{t=0}^{\infty} \beta^t U(c_t)\right], \quad 0 < \beta < 1. \qquad (2)$$

In (2), β is a discount factor and $U(\cdot)$ is given by (1). The mathematical expectation $E_t(\cdot)$ is conditioned on information available to agents at time t, I_t. Current and past values of real consumption and asset returns are assumed to be included in I_t.

Consumers substitute present for future consumption by trading the ownership rights of N financial and capital assets. These assets include default-free, multiperiod bonds that agents issue for the purpose of borrowing or lending among themselves and shares of ownership of firms in the economy. If firms rent capital from consumers, as in Brock's (1982) model, then the stocks of capital leased to the firms by the representative consumer will also be included among the traded assets. Let $\mathbf{w}_t$ denote the holdings of the N assets at the date t, $\mathbf{q}_t$ denote the vector of prices of the N assets in $\mathbf{w}_t$ net of any distributions, and $\mathbf{q}_t^*$ denote the vector of values of these distributions during period t. Then a feasible consumption and investment plan $\{c_t, \mathbf{w}_t\}$ must satisfy the sequence of budget constraints,

$$c_t + \mathbf{q}_t \cdot \mathbf{w}_{t+1} \leq (\mathbf{q}_t + \mathbf{q}_t^*) \cdot \mathbf{w}_t + y_t, \tag{3}$$

where y_t is the level of (real) labor income at date t.[2]

The first-order necessary conditions for the maximization of (2) subject to (3), that involve the equilibrium prices of the n assets, are (Lucas 1978; Brock 1982)

$$U'(c_t) = \beta E_t[U'(c_{t+1})r_{it+1}]; \quad i = 1, \ldots, N, \tag{4}$$

where r_{it+1} is the return on the ith asset expressed in units of the consumption good. Substituting (1) into (4) and rearranging gives

$$E_t\left[\beta\left(\frac{c_{t+1}}{c_t}\right)^\alpha r_{it+1}\right] = 1; \quad i = 1, \ldots, N, \tag{5}$$

with $\alpha \equiv \gamma - 1$. Breeden (1979) has derived an intertemporal capital asset pricing representation in a continuous-time environment. In his representation, expected excess returns on risky assets are linked to covariances of aggregate consumption and returns. Grossman and Shiller (1981) have shown how to obtain an analogous representation for the discrete time model studied here. Their representations are useful for studying the riskiness of a cross section of asset returns. The focus of this paper is instead on the link between forecastable movements in consumption and forecastable movements in asset returns. Accordingly, we proceed to derive a relation among these forecastable components implied by (5).

For the analysis of (5) that follows, it is not necessary to examine explicitly firms' production decisions, since it is not our goal to solve for an explicit representation of equilibrium prices in terms of the underlying shocks to technology. By assuming that the joint distribution of consumption and returns is lognormal, we are implicitly imposing restrictions on the production technology, however. As in many previous theoretical and empirical studies of asset pricing (see n. 1 above), we leave unspecified the exact nature of these restrictions. A formal justification of the assumption of lognormality can be provided for some economic environments for which closed-form equilibrium pricing functions have been derived. Such a justification has not been provided at the level of generality at which our empirical analysis is conducted, however. We adopt our general representation

---

[2] The inclusion of y_t does not affect our analysis if labor is supplied inelastically. Alternatively, we can introduce a period t labor supply variable, L_t, into the specification of U and let $U(c_t, L_t) = U_1(c_t) - U_2(L_t)$, where L_t is a choice variable of the consumer. For this case, $y_t = L_t W_t$, where W_t is the real wage rate at date t.

STOCHASTIC CONSUMPTION 253

to accommodate a rich temporal covariance structure which might emerge when the investment environment faced by firms is more complicated than the environments in the models of Lucas (1978) and Brock (1982) (e.g., serially correlated production shocks, costly adjustment in altering capital stocks, and gestation lags in producing new capital).

From (5) and the accompanying assumptions, a restricted linear time-series representation of the logarithms of consumption and asset returns can be derived. Suppose that observations on the first n of the N assets traded by economic agents are to be used in the econometric analysis. Let $x_t \equiv c_t/c_{t-1}$ and $u_{it} \equiv x_t^\alpha r_{it}$, $i = 1, \ldots, n$. Then (5) can be rewritten as

$$E_{t-1}(u_{it}) = 1/\beta, \quad i = 1, \ldots, n. \tag{6}$$

Next, let $X_t \equiv \log x_t$, $R_{it} \equiv \log r_{it}$, $\mathbf{Y}_t = (X_t, R_{1t}, \ldots, R_{nt})'$, $U_{it} = \log u_{it}$ ($i = 1, \ldots, n$), and ψ_{t-1} denote the information set $\{\mathbf{Y}_{t-s}: s \geq 1\}$. Further, assume that $\{\mathbf{Y}_t\}$ is a stationary Gaussian process. This distributional assumption implies that the distribution of U_{it} conditional on ψ_{t-1} is normal with a constant variance σ_i^2 and a mean μ_{it-1} that is a linear function of past observations on $\mathbf{Y}_t$. Hence,

$$E(u_{it}|\psi_{t-1}) = \exp[\mu_{it-1} + (\sigma_i^2/2)]. \tag{7}$$

Since $\psi_{t-1} \subset I_{t-1}$, we can take expectations of both sides of (6) conditional on ψ_{t-1} to obtain

$$E(u_{it}|\psi_{t-1}) = 1/\beta. \tag{8}$$

Equating the right-hand sides of equations (7) and (8) and solving for μ_{it-1} yields $\mu_{it-1} = -\log \beta - (\sigma_i^2/2)$. Define

$$V_{it} \equiv U_{it} - \mu_{it-1} = \alpha X_t + R_{it} + \log \beta + (\sigma_i^2/2), \\ i = 1, \ldots, n. \tag{9}$$

Then, $E(V_{it}|\psi_{t-1}) = 0$ and

$$E(R_{it}|\psi_{t-1}) = -\alpha E(X_t|\psi_{t-1}) - \log \beta - (\sigma_i^2/2), \\ i = 1, \ldots, n. \tag{10}$$

Equations (9) and (10) summarize the relationships among serial correlation of consumption, the level of risk aversion, and serial correlation of asset returns implied by the first-order conditions (5). Risk neutrality, for example, corresponds to the case of $\alpha = 0$, which implies that R_{it} is equal to a constant plus the serially uncorrelated error V_{it} and hence that R_{it} is serially uncorrelated, $i = 1, \ldots, n$. Alternatively, if $\alpha = -1$, then agents have logarithmic utility functions. In this case, $R_{it} - X_t = -\log \beta - (\sigma_i^2/2) + V_{it}$. Thus, the slope coefficients in the projections of R_{it} and X_t onto a subset ψ_{t-1} of I_{t-1} must be the same, and this equality must hold for the returns on all assets. More generally, (10) implies that (ignoring constant terms) the coefficients in the projection of R_{it} onto ψ_{t-1} are equal to the coefficients in the projection of X_t onto ψ_{t-1} multiplied by $-\alpha$.

To translate these observations into statements about the predictability of asset returns, it is useful to derive an expression for the coefficient of determination (R_i^2) from the projection of R_{it} onto ψ_{t-1} implied by (10). By definition,

$$R_i^2 = \frac{\text{var}\,[E(R_{it}|\psi_{t-1})]}{\text{var}\,(R_{it}|\psi_{t-1}) + \text{var}\,[E(R_{it}|\psi_{t-1})]}, \quad (11)$$

where var is the variance operator. From (10) it follows that the variances of the predictable components of $\log r_{it}$ and $\log (c_t/c_{t-1})$ are related by the expression:

$$\text{var}\,[E(R_{it}|\psi_{t-1})] = \alpha^2\,\text{var}\,[E(X_t|\psi_{t-1})]. \quad (12)$$

Substituting (12) into (11) gives

$$R_i^2 = \frac{\alpha^2\,\text{var}\,[E(X_t|\psi_{t-1})]}{\text{var}\,(R_{it}|\psi_{t-1}) + \alpha^2\,\text{var}\,[E(X_t|\psi_{t-1})]}. \quad (13)$$

From (13) it follows that a necessary condition for asset returns to have predictable components is that agents be risk averse ($\alpha \neq 0$).

Risk aversion is not a sufficient condition for predictability, however. For the special case in which the projection $E(X_t|\psi_{t-1})$ is constant, the R_i^2's are equal to zero or, equivalently, the projections of the R_{it} onto ψ_{t-1} are constants. This implication of our model is consistent with the conclusion of Rubinstein (1976b) that asset returns will be serially uncorrelated when consumption follows a logarithmic random walk and agents have CRRA preferences. When there are nontrivial predictable components in X_t and $\alpha \neq 0$, then real asset returns will also have predictable components.

The assumption that the vector process $\{\mathbf{Y}_t\}$ is stationary and Gaussian implies that the conditional expectations in (10) have linear, time-invariant representations and that the conditional variances are constant (a fact that we have exploited above). Thus, the movements in the conditional distributions of the logarithms of consumption and asset returns are completely summarized by movements in the conditional means. This distributional specification leads to a very convenient representation of the intertemporal behavior of consumption and asset returns for the purposes of empirical analyses. Once the projection $E(X_t|\psi_{t-1})$ is parameterized as a linear function of past values of $\mathbf{Y}_t$, the free parameters of (10) can be estimated by the

STOCHASTIC CONSUMPTION 255

method of maximum likelihood and the overidentifying restrictions can be tested using the likelihood ratio statistic. Since our characterization of the overidentifying restrictions relies on an assumption about the joint distribution of consumption and returns, rejection of these restrictions may result from misspecifying that distribution rather than from the empirical failure of the time-additive CRRA preference form of the asset pricing model.

Other authors have studied this asset pricing model by relying on the same distributional assumption as that employed here. Grossman and Shiller (1981) have shown how to identify preference parameters under a joint lognormality assumption on consumption and returns. They abstain from studying the intertemporal correlations of these variables and express the estimators of their preference parameters as functions of the first and second unconditional moment of two returns. Hall (1981) has independently adopted an approach that is very similar to the one employed here to estimate α for different data sets.[3] Neither of these studies considers tests of overidentifying restrictions.

III. Maximum Likelihood Estimates of the Parameters

To proceed with estimation, we assume that

$$E(X_t|\psi_{t-1}) = \mathbf{a}(L)'\mathbf{Y}_{t-1} + \mu_x, \qquad (14)$$

where $\mathbf{a}(L)$ is an $n + 1$ dimensional vector of finite order polynomials in the lag operator L. Combining equations (14) and (9) gives

$$A_0 \mathbf{Y}_t = A_1(L)\mathbf{Y}_{t-1} + \mathbf{\mu} + \mathbf{V}_t, \qquad (15)$$

where $\mathbf{V}_t = (W_t, V_{1t}, \ldots, V_{nt})'$ and $W_t \equiv X_t - E(X_t|\psi_{t-1})$. The matrix A_0 is given by

$$A_0 = \begin{bmatrix} 1 & 0 \\ \alpha & I \end{bmatrix},$$

with $\mathbf{\alpha} = (\alpha, \alpha, \ldots, \alpha)'$ and I an $n \times n$ identity matrix; the matrix lag polynomial $A_1(L)$ is given in partitioned form by

$$A_1(L) = \begin{bmatrix} \mathbf{a}(L)' \\ 0 \end{bmatrix};$$

---

[3] There are two differences in the estimation strategy employed by Hall (1981). First, he assumes that economic agents do not know the true parameter values in the forecasting equation for asset returns. Instead, they use Bayesian updating formulas as they accumulate new information over time about these parameters. Second, he expands the vector $\mathbf{Y}_t$ to include variables other than asset returns and consumption.

and the vector of constants μ is given by

$$\mu = [\mu_x, \log \beta + (\sigma_1^2/2), \ldots, \log \beta + (\sigma_n^2/2)]'.$$

From equation (9) it follows that $\{(W_{t-s}, V_{1t-s}, \ldots, V_{nt-s}); s \geq 0\}$ spans the space ψ_t. Hence, the autoregressive representation of $\mathbf{Y}_t$ is obtained by premultiplying both sides of (15) by A_0^{-1}.[4]

Now let θ denote the vector of unknown parameters containing α, β, μ_x, the parameters of $\mathbf{a}(L)$, and the elements of the covariance matrix of $\mathbf{V}_t$, denoted by Σ. It is assumed that Σ is nonsingular. Suppose that T observations on $\mathbf{Y}$ are available for estimation of θ. Then, in view of the relation (15), the joint density function of the sample, conditioned on the initial values of the variables, is given by

$$f(\mathbf{Y}_1, \ldots, \mathbf{Y}_T; \theta) = (2\pi)^{-(n+1)T/2}|\Sigma|^{-T/2} \quad (16)$$

$$\exp\left\{-(\tfrac{1}{2}) \sum_{t=1}^{T} [A_0\mathbf{Y}_t - A_1(L)\mathbf{Y}_{t-1} - \mu]'\Sigma^{-1}[A_0\mathbf{Y}_t - A_1(L)\mathbf{Y}_{t-1} - \mu]\right\}.$$

Note that (16) is also the joint density function of $(\mathbf{V}_1, \ldots, \mathbf{V}_T)$, since the Jacobian of the transformation A_0^{-1} that transforms (15) into the autoregressive representation is unity. The logarithm of the conditional likelihood function (16) is, up to a constant term,

$$L(\mathbf{Y}_1, \ldots, \mathbf{Y}_T; \theta) =$$

$$-(T/2) \log |\Sigma| - (\tfrac{1}{2}) \sum_{t=1}^{T} [A_0\mathbf{Y}_t - A_1(L)\mathbf{Y}_{t-1} - \mu]'\Sigma^{-1} \quad (17)$$

$$\times [A_0\mathbf{Y}_t - A_1(L)\mathbf{Y}_{t-1} - \mu].$$

The maximum likelihood (ML) estimate of θ is obtained by maximizing (17). Unfortunately, unless $n = 1$, the conditional log-likelihood function cannot be concentrated, because μ is a function of the parameters in Σ.

Estimates were obtained using monthly data for the period 1959:2 through 1978:12. The monthly, seasonally adjusted real consumption series, dating back to January 1959, were obtained from the CITIBASE data tape. The observations of these series were divided by the monthly estimates of population published by the Bureau of the Census to get per capita values. The ML estimates are reported for two alternative measures of consumption: nondurables plus services (NDS) and nondurables (ND). Hall (1978) and Grossman and Shiller (1981) used the former measure, while Flavin (1981) and Hall

---

[4] The following estimation and testing procedures can be modified to accommodate vector autoregressive moving average representations of $\mathbf{Y}_t$, including representations with unit roots in the moving average polynomial that might be induced by differencing.

STOCHASTIC CONSUMPTION 257

(1981) used the latter measure of consumption. We maintained the usual practice of excluding durables from measured consumption, due to the difficulty of imputing a service flow to the stock of durables.

Several monthly asset return series were studied. Return series for two levels of aggregation across common stocks were considered: an average return on all stocks listed on the New York Stock Exchange and returns on individual members of the Dow Jones Industrials. In addition to stock returns, we considered the 1-month return on Treasury bill yields. The stock return data were obtained from the Center for Research in Security Prices (CRSP) tapes, and the Treasury bill data were obtained from Ibbotson and Sinquefield (1979). Nominal returns were converted to real returns, which appear in (5), with the implicit price deflator corresponding to the measure of consumption.

Each combination of a measure of consumption and an asset return potentially corresponds to a different underlying model of economic behavior. A sufficient condition for the restrictions in equation (10) to hold for a measure of a component of aggregate consumption is that preferences be separable in consumption. Specifically, suppose that $c_{1t} + c_{2t} = c_t$ and that the function U is given by

$$U(c_{1t}, c_{2t}) = (c_{1t}^\gamma/\gamma) + U_2(c_{2t}). \tag{18}$$

Then it is appropriate to test the model with c_1 used as the measure of consumption. A separability assumption similar to that underlying (18) is implicit in all of the previous empirical studies of consumption and asset returns that use only a component of aggregate consumption. By estimating models with nondurables and nondurables plus services, we are implicitly considering two different assumptions about the separability of preferences. Similarly, the choice among stock returns amounts to choosing among different models of the return generating process. All of the returns must satisfy a condition analogous to (5), if (5) holds for individual stocks. However, both the individual and aggregated return series will not in general be lognormally distributed.

To our knowledge, this is the first empirical study of consumption and asset returns that uses monthly data.[5] The variable c_t in (5) represents the level of consumption over the period of time between decisions of economic agents. In using monthly consumption data, we assume that the representative agent makes consumption decisions at monthly time intervals. Further, we assume that the representative agent knows the return measured from the beginning of the month

---

[5] Breeden (1980) and Hall (1981) used quarterly data and Grossman and Shiller (1981) used annual data, e.g.

until the end of the month in deciding how much to consume during that month. If the appropriate decision period is shorter than 1 month or if the timing of the consumption decision is incorrectly aligned with the available information on asset returns, then our statistical model is misspecified even if the underlying economic model is correct. Of course, if the decision period is shorter than 1 month, then measurement errors will also be present (and indeed may be much larger) in studies using quarterly or annual time averages of consumption.[6] These potential measurement and timing problems are avoided by some of the tests that we have conducted.

Consider first the results for the value-weighted return on stocks listed on the New York Stock Exchange, which are summarized in table 1. Models were estimated with two, four, and six lags in the lag polynomial $La(L)$; NLAG = 2, 4, 6. For each model, the estimated values of α and β, the coefficients of determination from the unrestricted vector autoregressions of consumption (R_c^2) and returns (R_R^2), and the χ^2 statistic, $\chi^2(df)$, for testing the overidentifying restrictions implied by (10) are presented. None of the test statistics has probability values that are larger than .90 and there is a tendency for the probability values to decline with increases in NLAG.

An explanation of this inverse relationship among the probability values and the choice of NLAG can be obtained from the unrestricted autoregressive representation. Estimates of the autoregressive coefficients for the six-lag model with consumption measured as nondurables plus services (model 6) are presented in table 2. These results suggest that values of R and X beyond the second lag are not very useful in forecasting consumption. Consequently, as df increases with NLAG, there are relatively smaller increases in the χ^2 statistics.

Although the estimators of α and β are consistent even if NLAG is misspecified, the point estimates of α are quite sensitive to the choice of NLAG. At the same time, the corresponding standard errors are relatively large. Evidently, precise estimates of α cannot be obtained with the data set and choice of information set used here. Nevertheless, all of the estimated values of α displayed in table 1 are economically plausible except for model 4, which yields an estimate in the nonconcave region of the parameter space. The estimates of α for models 3 and 6 (NLAG = 6) imply a slightly larger degree of risk aversion than is implied by a logarithmic period utility function (α =

---

[6] In constructing both monthly and quarterly consumption numbers, the Department of Commerce relies on a substantial amount of interpolation between yearly benchmark observations. Monthly aggregate consumption is estimated using monthly data on important subsets of consumption, indicators of various components of aggregate consumption, and yearly benchmarks for proportional breakdowns of consumption into various categories. For a more detailed description, see Byrnes et al. (1979).

TABLE 1
Summary of Maximum Likelihood Results for Value-Weighted Aggregate Return

Model	$\hat{\alpha}$*	$\hat{\beta}$*	CONS	NLAG	R_r^2	R_R^2	χ^2†	df
1	−.325 (.828)	.9976 (.0032)	ND	2	.187	.020	6.088 (.893)	3
2	−.831 (.746)	.9985 (.0030)	ND	4	.206	.040	8.426 (.703)	7
3	−1.25 (.647)	.9993 (.0030)	ND	6	.246	.056	10.622 (.524)	11
4	.359 (1.880)	.9965 (.0048)	NDS	2	.119	.012	4.980 (.827)	3
5	−.264 (1.835)	.9979 (.0045)	NDS	4	.128	.028	6.687 (.538)	7
6	−1.509 (1.571)	1.0007 (.0042)	NDS	6	.149	.048	10.932 (.551)	11
7‡	−2.721 (3.187)	.9957 (.0627)	ND	2	.201	.074	5.130 (.838)	3
8‡	−2.671 (9.306)	.9989 (.1442)	ND	4	.445	.111	10.030 (.813)	7

* Standard errors are indicated in parentheses.
† Probability values are indicated in parentheses.
‡ Estimated using quarterly data: 1954:4–1978:4.

TABLE 2
Maximum Likelihood Estimates for Model 6
(C = Nondurables + Services, NLAG = 6)

Restricted model ($\hat{\alpha}$ = −1.5088 [1.571]; $\hat{\beta}$ = 1.007 [.0042]):
X = .0027 − .334X − .147X_{-2} + .074X_{-3} + .086X_{-4}
 (.0009) (.071) (.084) (.077) (.074)
 − .025X_{-5} + .006X_{-6} + .002R_{-1} + .012R_{-2}
 (.081) (.069) (.007) (.008)
 + .012R_{-3} + .004R_{-4} + .003R_{-5} − .014R_{-6}
 (.007) (.007) (.008) (.008)

Unrestricted model:
R = .0065 − .502X_{-1} + .033X_{-2} − 468X_{-3} − .623X_{-4}
 (.005) (.683) (.717) (.722) (.717)
 − .827X_{-5} − .239X_{-6} + .098R_{-1} − .056R_{-2}
 (.705) (.667) (.069) (.069)
 + .114R_{-3} + .073R_{-4} + .076R_{-5} − .114R_{-6}; R^2 = .048
 (.071) (.071) (.071) (.071)
X = .0028 − .334X_{-1} − .144X_{-2} + .069X_{-3} + .078X_{-4}
 (.0005) (.069) (.072) (.073) (.072)
 − .033 + .003X_{-6} + .003R_{-1} + .011R_{-2}
 (.071) (.067) (.007) (.007)
 + .012R_{-3} + .004R_{-4} + .004R_{-5} − .015R_{-6}; R^2 = .149
 (.007) (.007) (.007) (.007)

NOTE.—Standard errors are indicated in parentheses.

−1). The estimated values of β are less than, but close to, unity as expected.

The R^2's from the unrestricted, bivariate autoregressions are also reported in table 1. The R_c^2's for models 3 and 6 (NLAG = 6) are .246 and .149, respectively. It is clear that monthly differences in the logarithms of consumption (NDS and ND) are serially correlated. From expression (13) we know that the model implies that stock returns will also be serially correlated, but R_R^2's may be much smaller than R_c^2 if var $(R_{it}|\psi_{t-1})$ is large relative to the numerator of (13). Indeed, the R_R^2's displayed in table 1 are relatively small, with the R_R^2's for the models using ND as a measure of consumption larger than those for the models using NDS.

For comparison, we have estimated two quarterly models of asset returns and consumption (models 7 and 8 in table 1). The measure of consumption for both models was calculated by summing the monthly observations of nondurables over quarters. The point estimates for the coefficient of relative risk aversion are larger than those obtained for the monthly models, but again the estimates are not very precise, especially for NLAG = 4.

The discussion to this point has focused on the behavior of an aggregate average stock return. If a set of n returns on individual stocks is jointly lognormally distributed with consumption, then the restrictions in (10) can be tested using these returns. We estimated the free parameters in θ for a model including the returns on the stocks of three Dow Jones Industrials: American Brand, Exxon, and IBM. The nominal individual returns were obtained from the CRSP tapes and were converted to real returns in the same manner that the value-weighted return was converted. The assumption that the individual return series are lognormally distributed is, of course, inconsistent with the assumption that the aggregate return series is lognormal. We are, therefore, testing different models than those considered above. The likelihood ratio tests for both measures of consumption are reported in table 3 for the period February 1959 through December 1978. The three stock models are rejected by the data at essentially any significance level for both values of NLAG. Notice also that the estimated values of $|\alpha|$ increase with NLAG and are larger than unity when NLAG = 4.

The restrictions on asset returns implied by equation (10) should also hold for returns on bonds. To gain some insight into whether stocks and bonds yield qualitatively similar results, we estimated a model for the return of 1-month Treasury bills. The results are displayed in table 4. The estimates of α and β are quantitatively similar to the estimates obtained using stock return data. Note also that, for each value of NLAG, R_R^2 is much larger for the Treasury bill data

STOCHASTIC CONSUMPTION

TABLE 3

LIKELIHOOD RATIO TESTS FOR THE MODELS OF INDIVIDUAL DOW JONES RETURNS
(1959:2–1978.12)

	$\hat{\alpha}$	$\hat{\beta}$	χ^2	df	Probability
C = Nondurables:					
NLAG = 2	−.466	.995	310.3	25	1.000
NLAG = 4	−1.738	.997	334.7	49	1.000
C = Nondurables plus services:					
NLAG = 2	−.507	.995	238.5	25	1.000
NLAG = 4	−4.106	1.003	393.7	49	1.000

than the stock return data. This finding alone is evidence neither for nor against the model, even though the corresponding values of $\hat{\alpha}$ and R_c^2 are similar across tables 1 and 4. As noted above, ceteris paribus, the smaller the var $(R_{it}|\psi_{t-1})$ is, the larger R_i^2 will be, and this variance is smaller for the bond data. The most dramatic differences between the results for the stock return models 1 through 6 in table 1 and the Treasury bill models are the χ^2 statistics. For the Treasury bill models, the marginal significance levels are essentially zero, providing strong evidence against the restrictions.

Grossman and Shiller (1980) have examined the implications of the same multiperiod models of asset returns that are considered here. They noted that the parameters α and β can be identified and estimated from unconditional means and covariances of the logarithms of returns on two assets and aggregate consumption. Using yearly observations, they found values of $|\hat{\alpha}|$ substantially greater than one with correspondingly large standard errors for a variety of sample periods, including samples confined to the postwar period. For the

TABLE 4

SUMMARY OF MAXIMUM LIKELIHOOD RESULTS FOR NOMINAL RISK-FREE RETURN

Model	$\hat{\alpha}*$	$\hat{\beta}*$	CONS	NLAG	R_c^2	R_R^2	χ^2†	df
1	−.164 (.056)	.9997 (.0002)	ND	2	.218	.130	27.34 (.9999)	3
2	−.188 (.060)	.9998 (.0002)	ND	4	.212	.152	33.48 (.9999)	7
3	−.931 (.044)	1.0015 (.0004)	NDS	2	.128	.181	30.08 (.9999)	3
4	−1.289 (.088)	1.0022 (.0006)	NDS	4	.131	.198	30.82 (.9999)	7

* Standard errors are indicated in parentheses.
† Probability values are indicated in parentheses.

TABLE 5

SUMMARY OF MAXIMUM LIKELIHOOD RESULTS FOR NOMINAL
RISK-FREE AND VALUE-WEIGHTED RETURNS

Model	$\hat{\alpha}$*	$\hat{\beta}$*	CONS	NLAG	χ^2†	df
1	−30.58 (34.06)	1.001 (.0462)	ND	0	Just identified	Just identified
2	−.205	.999	ND	4	170.25 (.9999)	24
3	−58.25 (66.57)	1.088 (.0687)	NDS	0	Just identified	Just identified
4	−.209	1.000	NDS	4	366.22 (.9999)	24

* Standard errors are indicated in parentheses.
† Probability values are indicated in parentheses.

sake of comparison, we have computed maximum likelihood estimates of α and β using monthly observations on the value-weighted New York Stock Exchange return, the 1-month Treasury bill return, and the consumption of nondurables for values of NLAG equal to 0 and 4. These results are reported in table 5. The estimation procedure employed by Grossman and Shiller corresponds to the case when NLAG = 0. Consistent with their results, we found $|\hat{\alpha}|$ to be very large with a correspondingly large standard error when NLAG = 0. Consistent with our other findings, $|\hat{\alpha}|$ is approximately one when the serial correlation in the time-series data is taken into account in estimation. This shows the extent to which the precision and magnitude of our estimates rely on the restrictions across the serial correlation parameters of the respective time series.

By simultaneously considering more than one asset, we can test CRRA-lognormal models using nominal returns without having to measure aggregate consumption and the implicit consumption deflator. These tests remain valid in the presence of multiplicative shocks to preferences. To see this, consider the following generalization of the CRRA period utility function,

$$U(c_t, \lambda_t) = \frac{c_t^\gamma \lambda_t}{\gamma},$$

where λ_t is a (possibly degenerate) random shock observed by agents at time t that can be serially correlated. For this set of preferences, condition (5) becomes

$$E_t\left[\beta\left(\frac{c_{t+1}}{c_t}\right)^\alpha\left(\frac{\lambda_{t+1}}{\lambda_t}\right)r_{it+1}\right] = 1; \quad i = 1, \ldots, N. \quad (19)$$

STOCHASTIC CONSUMPTION

Under the assumption that $\{(\log \lambda_t - \log \lambda_{t-1}, \mathbf{Y}'_t)'\}$ is a stationary Gaussian process, relation (9) becomes

$$\tilde{V}_{it} = \alpha X_{it} + R_{it} + \log \lambda_t - \log \lambda_{t-1} + \log \beta + (\tilde{\sigma}_i^2/2), i = 1, \ldots, n,$$

where $E(\tilde{V}_{it}|\tilde{\psi}_{t-1}) = 0, \tilde{\psi}_{t-1} = \psi_{t-1} U\{\lambda_{t-s}: s \geq 1\}$, and $\tilde{\sigma}_i^2$ is the conditional variance of $\tilde{V}_{it}$. Thus, the difference between the logarithms of any two real returns is

$$R_{it} - R_{jt} = (\tilde{\sigma}_j^2/2) - (\tilde{\sigma}_i^2/2) + V_{it} - V_{jt}. \tag{20}$$

Since the difference $R_{it} - R_{jt}$ equals the difference between the logarithms of the nominal returns (say, $\tilde{R}_{it} - \tilde{R}_{jt}$) and the error $V_{it} - V_{jt}$ is orthogonal to the elements of the information set $\tilde{\psi}_{t-1}$, (20) implies that $\tilde{R}_{it} - \tilde{R}_{jt}$ must be uncorrelated with the elements of $\tilde{\psi}_{t-1}$ if the model is true. Therefore, the model can be tested by determining whether the slope coefficients in regressions of the difference between the logarithms of nominal returns onto variables in $\tilde{\psi}_{t-1}$ are significantly different from zero.

These tests also avoid some of the timing problems in aligning the consumption data with the return data alluded to earlier. For instance, relation (19) is also implied by a continuous time asset pricing model in which c_t is the instantaneous real per capita consumption flow, r_{it} is the return on asset i over the time interval $(t - 1, t]$, and β is related to the continuous time rate of time preference $\dot{\rho}$ via $e^{-\rho} = \beta$. Since the tests do not require observations on consumption, we are free to interpret them as tests of either discrete or continuous-time specifications. A drawback of not using consumption data is that the preference parameters α and β can no longer be identified.

We conducted these tests as follows. Let $\tilde{R}_{1t}$ denote the logarithm of the nominal Treasury bill return, $\Delta \tilde{R}_{1t} = \tilde{R}_{1t} - \tilde{R}_{1t-1}$, and $\mathbf{R}_t = (\tilde{R}_{2t} - \tilde{R}_{1t}, \tilde{R}_{3t} - \tilde{R}_{1t}, \ldots, \tilde{R}_{nt} - \tilde{R}_{1t})'$. The system of regression equations $\tilde{\mathbf{R}}_t = \tilde{\boldsymbol{\mu}} + \tilde{A}_1(L)\Delta \tilde{R}_{1t} + \tilde{A}_2(L)\tilde{\mathbf{R}}_{t-1} + \tilde{\mathbf{U}}_t$ was estimated using equation-by-equation ordinary least squares, where $\tilde{\boldsymbol{\mu}}$ is an $n - 1$ dimensional vector of constants, $\tilde{A}_1(L)$ is an $n - 1$ dimensional vector lag polynomial of order NLAG, and $\tilde{A}_2(L)$ is an $n - 1$ dimensional matrix lag polynomial of order NLAG $- 1$. Since the nominal Treasury bill return is risk free, $\Delta \tilde{R}_{1t}$ is known to agents at date $t - 1$, and it was therefore included as a right-hand-side variable in the regression equations. Likelihood ratio statistics were calculated to test the restriction $\tilde{A}_1(L) = 0$ and $\tilde{A}_2(L) = 0$ implied by CRRA-lognormal models. Two choices of returns corresponding to two different models were used in conducting the tests. For the first model, we used two returns, the second being the nominal value-weighted stock return. For the second model, we used four returns, the last three being the nominal stock returns on three Dow Jones Industrial stocks: American Brand, Exxon, and IBM. In both cases, NLAG was set equal to 2. The likelihood ratio statistics for the aggregate and individual return models are $\chi^2(5) = 16.56$ and $\chi^2(27) = 53.19$, respectively. The associated probability values are .9946 and .9981, and thus the restrictions are rejected by the data except at extremely low significance levels.

IV. Concluding Remarks

In this paper we have derived a time-series representation of consumption and asset returns that characterizes the restrictions on the temporal covariance structure of these variables implied by a class of general-equilibrium asset pricing models with time-separable, constant relative risk-averse preferences in which consumption and returns are lognormally distributed. Maximum likelihood estimation of the free parameters of most of the monthly models yielded point estimates of the coefficient of relative risk aversion that were between zero and two. The test statistics provided little evidence against the models using the value-weighted return on stocks listed on the New York exchange. In contrast, the marginal significance levels of the test statistics for the models of individual Dow Jones and Treasury bill returns were essentially zero. We also conducted tests of CRRA-lognormal models using multiple returns that are robust to mismeasurement of consumption and the deflator and accommodate certain types of shocks to preferences. These tests provided substantial evidence against the restrictions as well. In light of results reported here, we plan on pursuing models of asset returns with more general specifications of preferences and distribution-free methods of estimation and inference (Eichenbaum, Hansen, and Singleton 1982; Hansen and Singleton 1982).

References

Breeden, Douglas T. "Changing Consumption and Investment Opportunities and the Valuation of Securities." Ph.D. dissertation, Stanford Univ., 1977.
———. "An Intertemporal Asset Pricing Model with Stochastic Consumption and Investment Opportunities." *J. Financial Econ.* 7 (September 1979): 265–96.
———. "Consumption Risk in Futures Markets." *J. Finance* 35 (May 1980): 503–20.
Breeden, Douglas T., and Litzenberger, Robert H. "Prices of State-contingent Claims Implicit in Option Prices." *J. Bus.* 51 (October 1978): 621–51.
Brock, William A. "Asset Prices in a Production Economy." In *The Economics*

of Information and Uncertainty, edited by John J. McCall. Chicago: Univ. Chicago Press (for Nat. Bur. Econ. Res.), 1982.

Byrnes, James C.; Donahoe, Gerald F.; Hook, Mary W.; and Parker, Robert P. "Monthly Estimates of Personal Income, Taxes, and Outlays." *Survey Current Bus.* 59 (November 1979): 18–38.

Eichenbaum, Martin S.; Hansen, Lars Peter; and Singleton, Kenneth J. "Specification and Estimation of Representative Agent Models with Multiple Goods." Manuscript, Carnegie-Mellon Univ., 1982.

Flavin, Marjorie A. "The Adjustment of Consumption to Changing Expectations about Future Income." *J.P.E.* 89, no. 5 (October 1981): 974–1009.

Grossman, Sanford J., and Shiller, Robert J. "Preliminary Results on the Determinants of the Variability of Stock Market Prices." Manuscript, Univ. Pennsylvania, 1980.

———. "The Determinants of the Variability of Stock Market Prices." *A.E.R. Papers and Proc.* 71 (May 1981): 222–27.

Hall, Robert E. "Stochastic Implications of the Life Cycle–Permanent Income Hypothesis: Theory and Evidence." *J.P.E.* 86, no. 6 (December 1978): 971–88.

———. "Intertemporal Substitution in Consumption." Manuscript, Stanford Univ., June 1981.

Hansen, Lars Peter, and Singleton, Kenneth J. "Generalized Instrumental Variables Estimation of Nonlinear Rational Expectations Models." *Econometrica* 50 (September 1982): 1269–86.

Ibbotson, Roger G., and Sinquefield, Rex A. *Stocks, Bonds, Bills, and Inflation: Historical Returns, 1926–1978.* 2d ed. Charlottesville, Va.: Financial Analysts Res. Found., 1979.

Johnsen, Thore H. "The Risk Structure of Securities Prices: Notes on Multi-Period Asset Pricing." Manuscript, Columbia Univ., 1978.

LeRoy, Stephen F., and Porter, Richard D. "The Present-Value Relation: Tests Based on Implied Variance Bounds." *Econometrica* 49 (May 1981): 555–74.

Lucas, Robert E., Jr. "Asset Prices in an Exchange Economy." *Econometrica* 46 (November 1978): 1429–45.

Merton, Robert C. "An Intertemporal Capital Asset Pricing Model." *Econometrica* 41 (September 1973): 867–87.

———. "On Estimating the Expected Return on the Market: An Exploratory Investigation." *J. Financial Econ.* 8 (December 1980): 323–61.

Rubinstein, Mark. "The Strong Case for the Generalized Logarithmic Utility Model as the Premier Model of Financial Markets." *J. Finance* 31 (May 1976): 551–71. (*a*)

———. "The Valuation of Uncertain Income Streams and the Pricing of Options." *Bell J. Econ.* 7 (Autumn 1976): 407–25. (*b*)

Shiller, Robert J. "Do Stock Prices Move Too Much to Be Justified by Subsequent Changes in Dividends?" *A.E.R.* 71 (June 1981): 421–36.

Stambaugh, R. "Measuring the Market Portfolio and Testing the Capital Asset Pricing Model." Manuscript, Univ. Chicago, 1978.

THE EQUITY PREMIUM
A Puzzle*

Rajnish MEHRA
Columbia University, New York, NY 10027, USA

Edward C. PRESCOTT
Federal Reserve Bank of Minneapolis
University of Minnesota, Minneapolis, MN 55455, USA.

Restrictions that a class of general equilibrium models place upon the average returns of equity and Treasury bills are found to be strongly violated by the U.S. data in the 1889–1978 period. This result is robust to model specification and measurement problems. We conclude that, most likely, an equilibrium model which is not an Arrow–Debreu economy will be the one that simultaneously rationalizes both historically observed large average equity return and the small average risk-free return.

1. Introduction

Historically the average return on equity has far exceeded the average return on short-term virtually default-free debt. Over the ninety-year period 1889–1978 the average real annual yield on the Standard and Poor 500 Index was seven percent, while the average yield on short-term debt was less than one percent. The question addressed in this paper is whether this large differential in average yields can be accounted for by models that abstract from transactions costs, liquidity constraints and other frictions absent in the Arrow–Debreu set-up. Our finding is that it cannot be, at least not for the class of economies considered. Our conclusion is that most likely some equilibrium model with a

---

*This research was initiated at the University of Chicago where Mehra was a visiting scholar at the Graduate School of Business and Prescott a Ford foundation visiting professor at the Department of Economics. Earlier versions of this paper, entitled 'A Test of the Intertemporal Asset Pricing Model', were presented at the University of Minnesota, University of Lausanne, Harvard University, NBER Conference on Intertemporal Puzzles in Macroeconomics, and the American Finance Meetings. We wish to thank the workshop participants, George Constantinides, Eugene Fama, Merton Miller, and particularly an anonymous referee, Fischer Black, Stephen LeRoy and Charles Plosser for helpful discussions and constructive criticisms. We gratefully acknowledge financial support from the Faculty Research Fund of the Graduate School of Business, Columbia University, the National Science Foundation and the Federal Reserve Bank of Minneapolis.

0304-3923/85/$3.30©1985, Elsevier Science Publishers B.V. (North-Holland)

friction will be the one that successfully accounts for the large average equity premium.

We study a class of competitive pure exchange economies for which the equilibrium *growth* rate process on consumption and equilibrium asset returns are stationary. Attention is restricted to economies for which the elasticity of substitution for the composite consumption good between the year t and year $t+1$ is consistent with findings in micro, macro and international economics. In addition, the economies are constructed to display equilibrium consumption growth rates with the same mean, variance and serial correlation as those observed for the U.S. economy in the 1889–1978 period. We find that for such economies, the average real annual yield on equity is a maximum of four-tenths of a percent higher than that on short-term debt, in sharp contrast to the six percent premium observed. Our results are robust to non-stationarities in the means and variances of the economies' growth processes.

The simple class of economies studied, we think, is well suited for the question posed. It clearly is poorly suited for other issues, in particular issues such as the volatility of asset prices.[1] We emphasize that our analysis is not an estimation exercise, which is designed to obtain better estimates of key economic parameters. Rather it is a quantitative theoretical exercise designed to address a very particular question.[2]

Intuitively, the reason why the low average real return and high average return on equity cannot simultaneously be rationalized in a perfect market framework is as follows: With real per capita consumption growing at nearly two percent per year on average, the elasticities of substitution between the year t and year $t+1$ consumption good that are sufficiently small to yield the six percent average equity premium also yield real rates of return far in excess of those observed. In the case of a growing economy, agents with high risk aversion effectively discount the future to a greater extent than agents with low risk aversion (relative to a non-growing economy). Due to growth, future consumption will probably exceed present consumption and since the marginal utility of future consumption is less than that of present consumption, real interest rates will be higher on average.

This paper is organized as follows: Section 2 summarizes the U.S. historical experience for the ninety-year period 1889–1978. Section 3 specifies the set of economies studied. Their behavior with respect to average equity and short-term debt yields, as well as a summary of the sensitivity of our results to the specifications of the economy, are reported in section 4. Section 5 concludes the paper.

---

[1] There are other interesting features of time series and procedures for testing them. The variance bound tests of LeRoy and Porter (1981) and Shiller (1980) are particularly innovative and constructive. They did indicate that consumption risk was important [see Grossman and Shiller (1981) and LeRoy and LaCavita (1981)].

[2] See Lucas (1980) for an articulation of this methodology.

Table 1

Time periods	% growth rate of per capita real consumption Mean	Standard deviation	% real return on a relatively riskless security Mean	Standard deviation	% risk premium Mean	Standard deviation	% real return on S&P 500 Mean	Standard deviation
1889–1978	1.83 (Std error = 0.38)	3.57	0.80 (Std error = 0.60)	5.67	6.18 (Std error = 1.76)	16.67	6.98 (Std error = 1.74)	16.54
1889–1898	2.30	4.90	5.80	3.23	1.78	11.57	7.58	10.02
1899–1908	2.55	5.31	2.62	2.59	5.08	16.86	7.71	17.21
1909–1918	0.44	3.07	−1.63	9.02	1.49	9.18	−0.14	12.81
1919–1928	3.00	3.97	4.30	6.61	14.64	15.94	18.94	16.18
1929–1938	−0.25	5.28	2.39	6.50	0.18	31.63	2.56	27.90
1939–1948	2.19	2.52	−5.82	4.05	8.89	14.23	3.07	14.67
1949–1958	1.48	1.00	−0.81	1.89	18.30	13.20	17.49	13.08
1959–1968	2.37	1.00	1.07	0.64	4.50	10.17	5.58	10.59
1969–1978	2.41	1.40	−0.72	2.06	0.75	11.64	0.03	13.11

2. Data

The data used in this study consists of five basic series for the period 1889–1978.[3] The first four are identical to those used by Grossman and Shiller (1981) in their study. The series are individually described below:

(i) *Series P*: Annual average Standard and Poor's Composite Stock Price Index divided by the Consumption Deflator, a plot of which appears in Grossman and Shiller (1981, p. 225, fig. 1).
(ii) *Series D*: Real annual dividends for the Standard and Poor's series.
(iii) *Series C*: Kuznets–Kendrik–USNIA per capita real consumption on non-durables and services.
(iv) *Series PC*: Consumption deflator series, obtained by dividing real consumption in 1972 dollars on non-durables and services by the nominal consumption on non-durables and services.
(v) *Series RF*: Nominal yield on relatively riskless short-term securities over the 1889–1978 period; the securities used were ninety-day government Treasury Bills in the 1931–1978 period, Treasury Certificates for the

---

[3] We thank Sanford Grossman and Robert Shiller for providing us with the data they used in their study (1981).

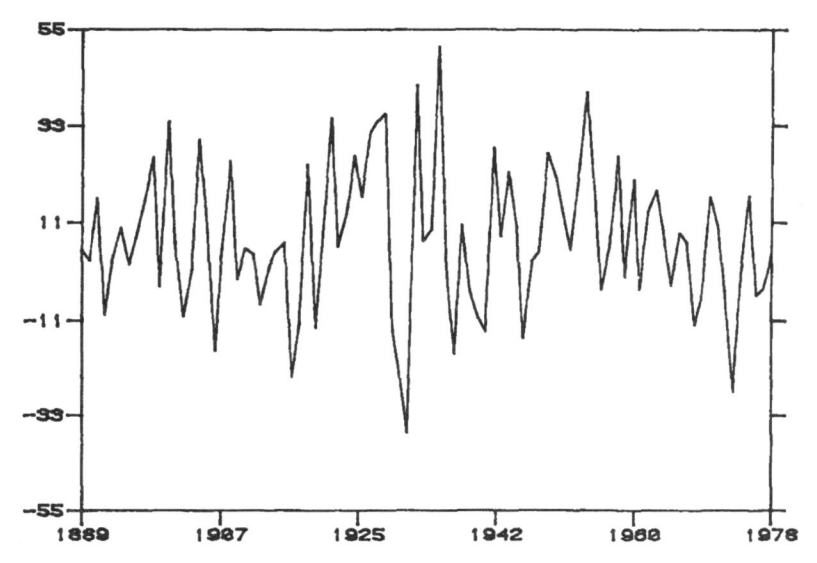

Fig. 1. Real annual return on S&P 500, 1889–1978 (percent).

1920–1930 period and sixty-day to ninety-day Prime Commercial Paper prior to 1920.[4]

These series were used to generate the series actually utilized in this paper. Summary statistics are provided in table 1.

Series P and D above were used to determine the average annual real return on the Standard and Poor's 500 Composite Index over the ninety-year period of study. The annual return for year t was computed as $(P_{t+1} + D_t - P_t)/P_t$. The returns are plotted in fig. 1. Series C was used to determine the process on the growth rate of consumption over the same period. Model parameters were restricted to be consistent with this process. A plot of the percentage growth of real consumption appears in fig. 2. To determine the real return on a relatively riskless security we used the series RF and PC. For year t this is calculated to be $RF_t - (PC_{t+1} - PC_t)/PC_t$.

This series is plotted in fig. 3. Finally, the Risk Premium (RP) is calculated as the difference between the Real Return on Standard and Poor's 500 and the Real Return on a Riskless security as defined above.

---

[4]The data was obtained from Homer (1963) and Ibbotson and Singuefield (1979).

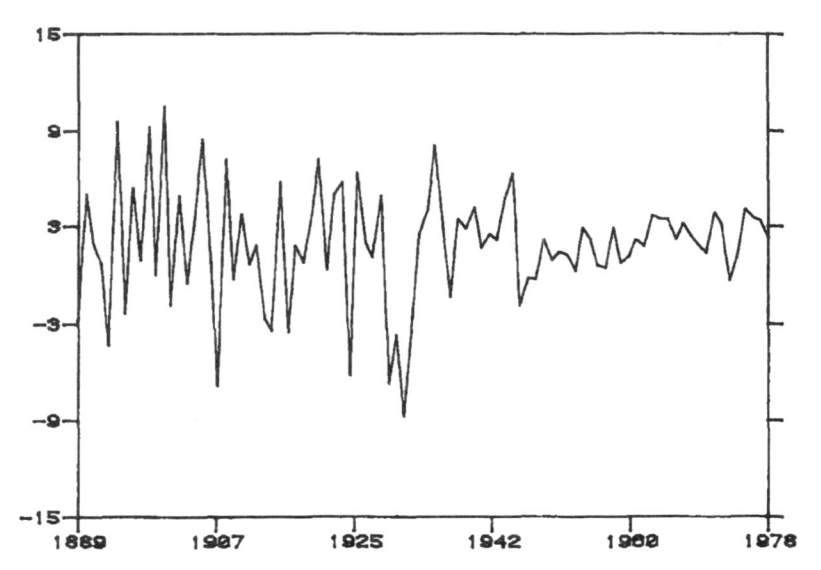

Fig. 2. Growth rate of real per capita consumption, 1889–1978 (percent).

Fig. 3. Real annual return on a relatively riskless security, 1889–1978 (percent).

3. The economy, asset prices and returns

In this paper, we employ a variation of Lucas' (1978) pure exchange model. Since per capita consumption has grown over time, we assume that the *growth rate* of the endowment follows a Markov process. This is in contrast to the assumption in Lucas' model that the endowment *level* follows a Markov process. Our assumption, which requires an extension of competitive equilibrium theory, enables us to capture the non-stationarity in the consumption series associated with the large increase in per capita consumption that occurred in the 1889–1978 period.

The economy we consider was judiciously selected so that the joint process governing the growth rates in aggregate per capita consumption and asset prices would be stationary and easily determined. The economy has a single representative 'stand-in' household. This unit orders its preferences over random consumption paths by

$$E_0\left\{\sum_{t=0}^{\infty} \beta^t U(c_t)\right\}, \qquad 0 < \beta < 1, \tag{1}$$

where c_t is per capita consumption, β is the subjective time discount factor, $E_0\{\cdot\}$ is the expectation operator conditional upon information available at time zero (which denotes the present time) and $U: R_+ \to R$ is the increasing concave utility function. To insure that the equilibrium return process is stationary, the utility function is further restricted to be of the constant relative risk aversion class,

$$U(c, \alpha) = \frac{c^{1-\alpha} - 1}{1 - \alpha}, \qquad 0 < \alpha < \infty. \tag{2}$$

The parameter α measures the curvature of the utility function. When α is equal to one, the utility function is defined to be the logarithmic function, which is the limit of the above function as α approaches one.

We assume that there is one productive unit producing the perishable consumption good and there is one equity share that is competitively traded. Since only one productive unit is considered, the return on this share of equity is also the return on the market. The firm's output is constrained to be less than or equal to y_t. It is the firm's dividend payment in the period t as well.

The growth rate in y_t is subject to a Markov chain; that is,

$$y_{t+1} = x_{t+1} y_t, \tag{3}$$

where $x_{t+1} \in \{\lambda_1, \ldots, \lambda_n\}$ is the growth rate, and

$$\Pr\{x_{t+1} = \lambda_j;\ x_t = \lambda_i\} = \phi_{ij}. \tag{4}$$

It is also assumed that the Markov chain is ergodic. The λ_i are all positive and $y_0 > 0$. The random variable y_t is observed at the beginning of the period, at which time dividend payments are made. All securities are traded ex-dividend. We also assume that the matrix A with elements $a_{ij} \equiv \beta \phi_{ij} \lambda_j^{1-\alpha}$ for $i, j = 1, \ldots, n$ is stable; that is, $\lim A^m$ as $m \to \infty$ is zero. In Mehra and Prescott (1984) it is shown that this is necessary and sufficient for expected utility to exist if the stand-in household consumes y_t every period. They also define and establish the existence of a Debreu (1954) competitive equilibrium with a price system having a dot product representation under this condition.

Next we formulate expressions for the equilibrium time t price of the equity share and the risk-free bill. We follow the convention of pricing securities ex-dividend or ex-interest payments at time t, in terms of the time t consumption good. For any security with process $\{d_s\}$ on payments, its price in period t is

$$P_t = E_t\left\{ \sum_{s=t+1}^{\infty} \beta^{s-t} U'(y_s) d_s / U'(y_t) \right\}, \tag{5}$$

as equilibrium consumption is the process $\{y_s\}$ and the equilibrium price system has a dot product representation.

The dividend payment process for the equity share in this economy is $\{y_s\}$. Consequently, using the fact that $U'(c) = c^{-\alpha}$,

$$P_t^e = P^e(x_t, y_t)$$

$$= E\left\{ \sum_{s=t+1}^{\infty} \beta^{s-t} \frac{y_t^\alpha}{y_s^\alpha} y_s \middle| x_t, y_t \right\}. \tag{6}$$

Variables x_t and y_t are sufficient relative to the entire history of shocks up to, and including, time t for predicting the subsequent evolution of the economy. They thus constitute legitimate state variables for the model. Since $y_s = y_t \cdot x_{t+1} \cdots x_s$, the price of the equity security is homogeneous of degree one in y_t, which is the current endowment of the consumption good. As the equilibrium values of the economies being studied are time invariant functions of the state (x_t, y_t), the subscript t can be dropped. This is accomplished by redefining the state to be the pair (c, i), if $y_t = c$ and $x_t = \lambda_i$. With this convention, the price of the equity share from (6) satisfies

$$p^e(c,i) = \beta \sum_{j=1}^{n} \phi_{ij}(\lambda_j c)^{-\alpha}\left[p^e(\lambda_j c, j) + c\lambda_j\right]c^{\alpha}. \tag{7}$$

Using the result that $p^e(c, i)$ is homogeneous of degree one in c, we represent this function as

$$p^e(c, i) = w_i c, \tag{8}$$

where w_i is a constant. Making this substitution in (7) and dividing by c yields

$$w_i = \beta \sum_{j=1}^{n} \phi_{ij} \lambda_j^{(1-\alpha)}(w_j + 1) \quad \text{for} \quad i = 1, \ldots, n. \tag{9}$$

This is a system of n linear equations in n unknowns. The assumption that guaranteed existence of equilibrium guarantees the existence of a unique positive solution to this system.

The period return if the current state is (c, i) and next period state $(\lambda_j c, j)$ is

$$r_{ij}^e = \frac{p^e(\lambda_j c, j) + \lambda_j c - p^e(c, i)}{p^e(c, i)}$$

$$= \frac{\lambda_j(w_j + 1)}{w_i} - 1, \tag{10}$$

using (8).

The equity's expected period return if the current state is i is

$$R_i^e = \sum_{j=1}^{n} \phi_{ij} r_{ij}^e. \tag{11}$$

Capital letters are used to denote expected return. With the subscript i, it is the expected return conditional upon the current state being (c, i). Without this subscript it is the expected return with respect to the stationary distribution. The superscript indicates the type of security.

The other security considered is the one-period real bill or riskless asset, which pays one unit of the consumption good next period with certainty.

From (6),

$$p_i^f = p^f(c, i)$$

$$= \beta \sum_{j=1}^{n} \phi_{ij} U'(\lambda_j c)/U'(c) \qquad (12)$$

$$= \beta \sum_{j=1}^{n} \phi_{ij} \lambda_j^{-\alpha}.$$

The certain return on this riskless security is

$$R_i^f = 1/p_i^f - 1, \qquad (13)$$

when the current state is (c, i).

As mentioned earlier, the statistics that are probably most robust to the modelling specification are the means over time. Let $\pi \in R^n$ be the vector of stationary probabilities on i. This exists because the chain on i has been assumed to be ergodic. The vector π is the solution to the system of equations

$$\pi = \phi^T \pi,$$

with

$$\sum_{i=1}^{n} \pi_i = 1 \quad \text{and} \quad \phi^T = \{\phi_{ji}\}.$$

The expected returns on the equity and the risk-free security are, respectively,

$$R^e = \sum_{i=1}^{n} \pi_i R_i^e \quad \text{and} \quad R^f = \sum_{i=1}^{n} \pi_i R_i^f. \qquad (14)$$

Time sample averages will converge in probability to these values given the ergodicity of the Markov chain. The risk premium for equity is $R^e - R^f$, a parameter that is used in the test.

4. The results

The parameters defining preferences are α and β while the parameters defining technology are the elements of $[\phi_{ij}]$ and $[\lambda_i]$. Our approach is to assume two states for the Markov chain and to restrict the process as follows:

$$\lambda_1 = 1 + \mu + \delta, \quad \lambda_2 = 1 + \mu - \delta,$$

$$\phi_{11} = \phi_{22} = \phi, \quad \phi_{12} = \phi_{21} = (1 - \phi).$$

The parameters μ, ϕ, and δ now define the technology. We require $\delta > 0$ and $0 < \phi < 1$. This particular parameterization was selected because it permitted us to independently vary the average growth rate of output by changing μ, the variability of consumption by altering δ, and the serial correlation of growth rates by adjusting ϕ.

The parameters were selected so that the average growth rate of per capita consumption, the standard deviation of the growth rate of per capita consumption and the first-order serial correlation of this growth rate, all with respect to the model's stationary distribution, matched the sample values for the U.S. economy between 1889–1978. The sample values for the U.S. economy were 0.018, 0.036 and -0.14, respectively. The resulting parameter's values were $\mu = 0.018$, $\delta = 0.036$ and $\phi = 0.43$. Given these values, the nature of the test is to search for parameters α and β for which the model's averaged risk-free rate and equity risk premium match those observed for the U.S. economy over this ninety-year period.

The parameter α, which measures peoples' willingness to substitute consumption between successive yearly time periods is an important one in many fields of economics. Arrow (1971) summarizes a number of studies and concludes that relative risk aversion with respect to wealth is almost constant. He further argues on theoretical grounds that α should be approximately one. Friend and Blume (1975) present evidence based upon the portfolio holdings of individuals that α is larger, with their estimates being in the range of two. Kydland and Prescott (1982), in their study of aggregate fluctuations, found that they needed a value between one and two to mimic the observed relative variabilities of consumption and investment. Altug (1983), using a closely related model and formal econometric techniques, estimates the parameter to be near zero. Kehoe (1984), studying the response of small countries balance of trade to terms of trade shocks, obtained estimates near one, the value posited by Arrow. Hildreth and Knowles (1982) in their study of the behavior of farmers also obtain estimates between one and two. Tobin and Dolde (1971), studying life cycle savings behavior with borrowing constraints, use a value of 1.5 to fit the observed life cycle savings patterns.

Any of the above cited studies can be challenged on a number of grounds but together they constitute an *a priori* justification for restricting the value of α to be a maximum of ten, as we do in this study. This is an important restriction, for with large α virtually any pair of average equity and risk-free returns can be obtained by making small changes in the process on consump-

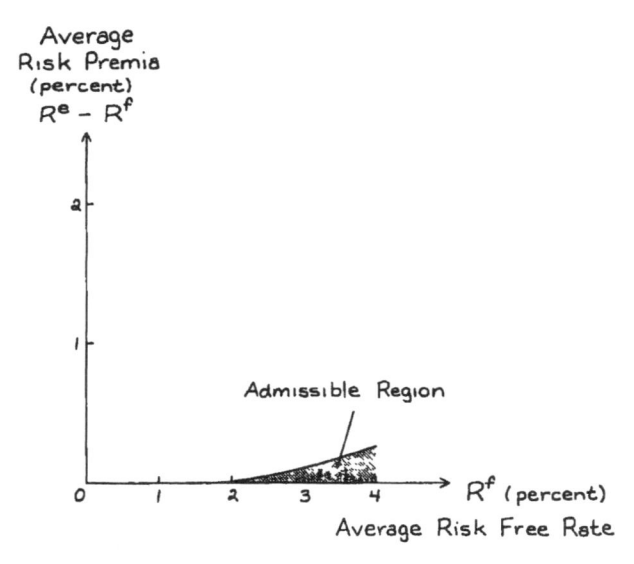

Fig. 4. Set of admissible average equity risk premia and real returns.

tion.[5] With α less than ten, we found the results were essentially the same for very different consumption processes, provided that the mean and variances of growth rates equaled the historically observed values. An advantage of our approach is that we can easily test the sensitivity of our results to such distributional assumptions.

The average real return on relatively riskless, short-term securities over the 1889–1978 period was 0.80 percent. These securities do not correspond perfectly with the real bill, but insofar as unanticipated inflation is negligible and/or uncorrelated with the growth rate x_{t+1} conditional upon information at time t, the expected real return for the nominal bill will equal R_i^f. Litterman (1980), using vector autoregressive analysis, found that the innovation in the inflation rate in the post-war period (quarterly data) has standard deviation of only one-half of one percent and that his innovation is nearly orthogonal to the subsequent path of the real GNP growth rate. Consequently, the average realized real return on a nominally denoted short-term bill should be close to that which would have prevailed for a real bill if such a security were traded. The average real return on the Standard and Poor's 500 Composite Stock

---

[5] In a private communication, Fischer Black using the Merton (1973) continuous time model with investment opportunities constructed an example with a curvature parameter (α) of 55. We thank him for the example.

Index over the ninety years considered was 6.98 percent per annum. This leads to an average equity premium of 6.18 percent (standard error 1.76 percent).

Given the estimated process on consumption, fig. 4 depicts the set of values of the average risk-free rate and equity risk premium which are both consistent with the model and result in average real risk-free rates between zero and four percent. These are values that can be obtained by varying preference parameters α between zero and ten and β between zero and one. The observed real return of 0.80 percent and equity premium of 6 percent is clearly inconsistent with the predictions of the model. The largest premium obtainable with the model is 0.35 percent, which is not close to the observed value.

4.1. Robustness of results

One set of possible problems are associated with errors in measuring the inflation rate. Such errors do not affect the computed risk premium as they bias both the real risk-free rate and the equity rate by the same amount. A potentially more serious problem is that these errors bias our estimates of the growth rate of consumption and the risk-free real rate. Therefore, only if the tests are insensitive to biases in measuring the inflation rate should the tests be taken seriously. A second measurement problem arises because of tax considerations. The theory is implicitly considering effective after-tax returns which vary over income classes. In the earlier part of the period, tax rates were low. In the latter period, the low real rate and sizable equity risk premium hold for after-tax returns for all income classes [see Fisher and Lorie (1978)].

We also examined whether aggregation affects the results for the case that the growth rates were independent between periods, which they approximately were, given that the estimated ϕ was near one-half. Varying the underlying time period from one one-hundredths of a year to two years had a negligible effect upon the admissible region. (See the appendix for an exact specification of these experiments.) Consequently, the test appears robust to the use of annual data in estimating the process on consumption.

In an attempt to reconcile the large discrepancy between theory and observation, we tested the sensitivity of our results to model misspecification. We found that the conclusions are not at all sensitive to changes in the parameter μ, which is the average growth rate of consumption, with decreases to 1.4 percent or increases to 2.2 percent not reducing the discrepancy. The sensitivity to δ, the standard deviation of the consumption growth rate, is larger. The average equity premium was roughly proportional to δ squared. As the persistence parameter ϕ increased ($\phi = 0.5$ corresponds to independence over time), the premium decreased. Reducing ϕ (introducing stronger negative serial correlation in the consumption growth rate) had only small effects. We also modified the process on consumption by introducing additional states that permitted us to increase higher moments of the stationary distribution of the growth rate without varying the first or second moments. The maximal equity premium increased by 0.04 to 0.39 only. These exercises lead us to the conclusion that the result of the test is not sensitive to the specification of the process generating consumption.

That the results were not sensitive to increased persistence in the growth rate, that is to increases in ϕ, implies low frequency movements or non-stationarities in the growth rate do *not* increase the equity premium. Indeed, by assuming stationarity, we biased the test *towards* acceptance.

4.2. Effects of firm leverage

The security priced in our model does not correspond to the common stocks traded in the U.S. economy. In our model there is only one type of capital, while in an actual economy there is virtually a continuum of capital types with widely varying risk characteristics. The stock of a typical firm traded in the stock market entitles its owner to the residual claim on output after all other claims including wages have been paid. The share of output accruing to stockholders is much more variable than that accruing to holders of other claims against the firm. Labor contracts, for instance, may incorporate an insurance feature, as labor claims on output are in part fixed, having been negotiated prior to the realization of output. Hence, a disproportionate part of the uncertainty in output is probably borne by equity owners.

The firm in our model corresponds to one producing the entire output of the economy. Clearly, the riskiness of the stock of this firm is not the same as that of the Standard and Poor's 500 Composite Stock Price Index. In an attempt to match the two securities we price and calculate the risk premium of a security whose dividend next period is actual output less a fraction of expected output. Let θ be the fraction of expected date $t+1$ output committed at date t by the firm. Eq. (7) then becomes

$$p^e(c,i) = \beta \sum_{j=1}^{n} \phi_{ij}(\lambda_j c)^{-\alpha} \left[p^e(\lambda_j c, j) + c\lambda_j - \theta \sum_{k=1}^{n} \phi_{ik} c \lambda_k \right] c^{\alpha}. \quad (15)$$

As before, it is conjectured and verified that $p^e(c,i)$ has the functional form $w_i c$. Substituting $w_i c$ for $p^e(c,i)$ in (15) yields the set of linear equations

$$w_i = \beta \sum_{j=1}^{n} \phi_{ij} \lambda_j^{-\alpha} \left[\lambda_j w_j + \lambda_j - \theta \sum_{k=1}^{n} \phi_{ik} \lambda_k \right], \quad (16)$$

for $i = 1, \ldots, n$. This system was solved for the equilibrium w_i and eqs. (10), (11), and (14) used to determine the average equity premium.

As the corporate profit share of output is about ten percent, we set $\theta = 0.9$. Thus, ninety percent of expected output is committed and all the risk is borne by equity owners who receive ten percent of output on average. This increased the equity risk premium by less than one-tenth percent. This is the case because financial arrangements have no effect upon resource allocation and, therefore, the underlying Arrow–Debreu prices. Large fixed payment commitments on the part of the firm do not reverse the test's outcome.

4.3. Introducing production

With our structure, the process on the endowment is exogenous and there is neither capital accumulation nor production. Modifying the technology to admit these opportunities cannot overturn our conclusion, because expanding the set of technologies in this way does not increase the set of joint equilibrium processes on consumption and asset prices [see Mehra (1984)]. As opposed to standard testing techniques, the failure of the model hinges not on the acceptance/rejection of a statistical hypothesis but on its inability to generate average returns even close to those observed. If we had been successful in finding an economy which passed our not very demanding test, as we expected, we planned to add capital accumulation and production to the model using a variant of Brock's (1979, 1982), Donaldson and Mehra's (1984) or Prescott and Mehra's (1980) general equilibrium stationary structures and to perform additional tests.

5. Conclusion

The equity premium puzzle may not be why was the average equity return so high but rather why was the average risk-free rate so low. This conclusion follows if one accepts the Friend and Blume (1975) finding that the curvature parameter α significantly exceeds one. For $\alpha = 2$, the model's average risk-free rate is at least 3.7 percent per year, which is considerably larger than the sample average 0.80 given the standard deviation of the sample average is only 0.60. On the other hand, if α is near zero and individuals nearly risk-neutral, then one would wonder why the average return of equity was so high. This is not the only example of some asset receiving a lower return than that implied by Arrow–Debreu general equilibrium theory. Currency, for example, is dominated by Treasury bills with positive nominal yields yet sizable amounts of currency are held.

We doubt whether heterogeneity, per se, of the agents will alter the conclusion. Within the Debreu (1954) competitive framework, Constantinides (1982) has shown heterogeneous agent economies also impose the set of restrictions tested here (as well as others). We doubt whether non-time-additivity separable preferences will resolve the puzzle, for that would require consumptions near in time to be poorer substitutes than consumptions at widely separated dates. Perhaps introducing some features that make certain types of intertemporal trades among agents infeasible will resolve the puzzle. In the absence of such markets, there can be variability in individual consumptions, yet little variability in aggregate consumption. The fact that certain types of contracts may be non-enforceable is one reason for the non-existence of markets that would otherwise arise to share risk. Similarly, entering into contracts with as yet unborn generations is not feasible.[6] Such non-Arrow–Debreu competitive equilibrium models may rationalize the large equity risk premium that has characterized the behavior of the U.S. economy over the last ninety years. To test such theories it would probably be necessary to have consumption data by income or age groups.

Appendix

The procedure for determining the admissible region depicted in fig. 4 is as follows. For a given set of parameters μ, δ and ϕ, eqs. (10)–(14) define an algorithm for computing the values of R^e, R^f and $R^e - R^f$ for any (α, β) pair belonging to the set

$$x = \{(\alpha, \beta): 0 < \alpha \leq 10, 0 < \beta < 1, \text{ and the}$$
existence condition of section 3 is satisfied$\}$.

Letting $R^f = h_1(\alpha, \beta)$ and $R^e - R^f = h_2(\alpha, \beta)$, $h: X \to R^2$, the range of h is the region depicted in fig. 4. The function h was evaluated for all points of a fine grid in X to determine the admissible region.

The experiments to determine the sensitivity of the results to the period length have model time periods $n = 2, 1, 1/2, 1/4, 1/8, 1/16, 1/64$ and $1/128$ years. The values of the other parameters are $\mu = 0.018/n$, $\delta = 0.036/\sqrt{n}$ and $\phi = 0.5$. With these numbers the mean and standard deviation of annual growth rates are 0.018 and 0.036 respectively as in the sample period. This follows because $\phi = 0.5$ implies independence of growth rates over periods. The change in the admissible region were hundredths of percent as n varied.

The experiments to test the sensitivity of the results to μ consider $\mu = 0.014$, 0.016, 0.018, 0.020 and 0.022, $\phi = 0.43$ and $\delta = 0.036$. As for the period length, the growth rate's effects upon the admissible region are hundredths of percent.

The experiments to determine the sensitivity of results to δ set $\phi = 0.43$, $\mu = 0.018$ and $\delta = 0.21, 0.26, 0.31, 0.36, 0.41, 0.46$ and 0.51. The equity premium varied approximately with the square of δ in this range.

---

[6] See Wallace (1980) for an exposition on the use of the overlapping generations model and the importance of legal constraints in explaining rate of return anomalies.

Similarly, to test the sensitivity of the results to variations in the parameter ϕ, we held δ fixed at 0.036 and μ at 0.018 and varied ϕ between 0.005 and 0.95 in steps of 0.05. As ϕ increased the average equity premium declined.

The test for the sensitivity of results to higher movements uses an economy with a four-state Markov chain with transition probability matrix

$$\begin{bmatrix} \phi/2 & \phi/2 & 1-\phi/2 & 1-\phi/2 \\ \phi/2 & \phi/2 & 1-\phi/2 & 1-\phi/2 \\ 1-\phi/2 & 1-\phi/2 & \phi/2 & \phi/2 \\ 1-\phi/2 & 1-\phi/2 & \phi/2 & \phi/2 \end{bmatrix}.$$

The values of the λ are $\lambda_1 = 1 + \mu$, $\lambda_2 = 1 + \mu + \delta$, $\lambda_3 = 1 + \mu$, and $\lambda_4 = 1 + \mu - \delta$. Values of μ, δ and ϕ are 0.018, 0.051 and 0.36, respectively. This results in the mean, standard deviation and first-order serial correlations of consumption growth rates for the artificial economy equaling their historical values. With this Markov chain, the probability of above average changes is smaller and magnitude of changes larger. This has the effect of increasing moments higher than the second without altering the first or second moments. This increases the maximum average equity premium from 0.35 percent to 0.39 percent.

References

Altug, S.J., 1983, Gestation lags and the business cycle: An empirical analysis, Carnegie-Mellon working paper, Presented at the Econometric Society meeting, Stanford University (Carnegie-Mellon University, Pittsburgh, PA).
Arrow, K.J., 1971, Essays in the theory of risk-bearing (North-Holland, Amsterdam).
Brock, W.A., 1979, An integration of stochastic growth theory and the theory of finance, Part 1: The growth model, in: J. Green and J. Scheinkman, eds., General equilibrium, growth & trade (Academic Press, New York).
Brock, W.A., 1982, Asset prices in a production economy, in: J.J. McCall, ed., The economics of information and uncertainty (University of Chicago Press, Chicago, IL).
Constantinides, G., 1982, Intertemporal asset pricing with heterogeneous consumers and no demand aggregation, Journal of Business 55, 253–267.
Debreu, G., 1954, Valuation equilibrium and Pareto optimum, Proceedings of the National Academy of Sciences 70, 588–592.
Donaldson, J.B. and R. Mehra, 1984, Comparative dynamics of an equilibrium, intertemporal asset pricing model, Review of Economic Studies 51, 491–508.
Fisher, L. and J.H. Lorie, 1977, A half century of returns on stocks and bonds (University of Chicago Press, Chicago, IL).
Friend, I. and M.E. Blume, 1975, The demand for risky assets, American Economic Review 65, 900–922.
Grossman, S.J. and R.J. Shiller, 1981, The determinants of the variability of stock market prices, American Economic Review 71, 222–227.
Hildreth, C. and G.J. Knowles, 1982, Some estimates of Farmers' utility functions, Technical bulletin 335 (Agricultural Experimental Station, University of Minnesota, Minneapolis, MN).
Homer, S., 1963, A history of interest rates (Rutgers University Press, New Brunswick, NJ).
Ibbotson, R.G. and R.A. Singuefield, 1979, Stocks, bonds, bills, and inflation: Historical returns (1926–1978) (Financial Analysts Research Foundation, Charlottesville, VA).

Kehoe, P.J., 1983, Dynamics of the current account: Theoretical and empirical analysis, Working paper (Harvard University, Cambridge, MA).

Kydland, F.E. and E.C. Prescott, 1982, Time to build and aggregate fluctuations, Econometrica 50, 1345–1370.

LeRoy, S.F. and C.J. LaCivita, 1981, Risk-aversion and the dispersion of asset prices, Journal of Business 54, 535–548.

LeRoy, S.F. and R.D. Porter, 1981, The present-value relation: Tests based upon implied variance bounds, Econometrica 49, 555–574.

Litterman, R.B. 1980, Bayesian procedure for forecasting with vector autoregressions, Working paper (MIT, Cambridge, MA).

Lucas, R.E., Jr., 1978, Asset prices in an exchange economy, Econometrica 46, 1429–1445.

Lucas, R.E., Jr., 1981, Methods and problems in business cycle theory, Journal of Money, Credit, and Banking 12, Part 2. Reprinted in: R.E. Lucas, Jr., Studies in business-cycle theory (MIT Press, Cambridge, MA).

Mehra, R., 1984, Recursive competitive equilibrium: A parametric example, Economics Letters 16, 273–278.

Mehra, R. and E.C. Prescott, 1984, Asset prices with nonstationary consumption, Working paper (Graduate School of Business, Columbia University, New York).

Merton, R.C., 1973, An intertemporal asset pricing model, Econometrica 41, 867–887.

Prescott, E.C. and R. Mehra, 1980, Recursive competitive equilibrium: The case of homogeneous households, Econometrica 48, 1365–1379.

Shiller, R.J., 1981, Do stock prices move too much to be justified by subsequent changes in dividends?, American Economic Review 71, 421–436.

Tobin, J. and W. Dolde, 1971, Wealth, liquidity and consumption, in: Consumer spending and monetary policy: The linkage (Federal Reserve Bank of Boston, Boston, MA) 99–146.

Wallace, N., 1980, The overlapping generations model of fiat money, in: J.H. Kareken and N. Wallace, eds., Models of monetary economies (Federal Reserve Bank of Minneapolis, Minneapolis, MN) 49–82.

Empirical Tests of the Consumption-Oriented CAPM

DOUGLAS T. BREEDEN, MICHAEL R. GIBBONS, and ROBERT H. LITZENBERGER*

ABSTRACT

The empirical implications of the consumption-oriented capital asset pricing model (CCAPM) are examined, and its performance is compared with a model based on the market portfolio. The CCAPM is estimated after adjusting for measurement problems associated with reported consumption data. The CCAPM is tested using betas based on both consumption and the portfolio having the maximum correlation with consumption. As predicted by the CCAPM, the market price of risk is significantly positive, and the estimate of the real interest rate is close to zero. The performances of the traditional CAPM and the CCAPM are about the same.

IN AN INTERTEMPORAL ECONOMY, Rubinstein (1976), Breeden and Litzenberger (1978), and Breeden (1979) demonstrate that equilibrium expected excess returns are proportional to their "consumption betas." This contrasts with the market-oriented capital asset pricing model (hereafter, CAPM) derived in a single-period economy by Sharpe (1964) and Lintner (1965). While tests of the CAPM by Black, Jensen, and Scholes (1972), Fama and MacBeth (1973), Gibbons (1982), and others find a positive association between average excess returns and betas using a proxy for the market portfolio, the relation is not proportional. This paper studies similar empirical issues for the consumption-oriented capital asset pricing model (hereafter, CCAPM).

Even though the relevant market portfolio includes all assets, most empirical research focuses on common stocks for which accurately measured data are available. In contrast, reported consumption data are estimates of the relevant consumption flows, and the data are subject to measurement problems not found with stock indexes. In this paper the tests of the CCAPM incorporate some adjustments for these measurement problems.

The outline of the paper is as follows. Section I provides an alternative derivation of the CCAPM. Section II examines four econometric problems associated with measured consumption: the durables problem, the problem of

---

*Duke University; Stanford University and visiting the University of Chicago (1988-1989); and University of Pennsylvania, respectively. We are grateful for the comments we have received from seminar participants at a number of universities. Special thanks go to Eugene Fama, Wayne Ferson, Bruce Lehmann, Bill Schwert, Jay Shanken, Kenneth Singleton, René Stulz, and an anonymous referee. Over the years this paper has benefited also from research assistance by Susan Cheng, Hal Heaton, Chi-Fu Huang, Charles Jacklin, and Ehud Ronn. Financial support was provided in part to all authors by the Stanford Program in Finance. Breeden (1981-1982) and Gibbons (1982-1983) acknowledge with thanks financial support provided for this research by Batterymarch Financial Management.

measured consumption as an integral of spot consumption rates, the problem that consumption data are reported infrequently, and the problem of pure sampling error in consumption measures. Time series properties of consumption measures are also discussed in Section II. Section III analyzes the empirical characteristics of estimated consumption betas for various stock and bond portfolios. The composition of the portfolio whose return has the highest correlation with the growth rate of real, per capita consumption is also discussed in Section III. This portfolio is used in some of the tests of the model. Section IV presents empirical tests of the consumption and market-oriented CAPMs based on their implications for unconditional moments. Section V concludes the paper with a review of the results obtained.

I. A Synthesis of the CCAPM Theory

The Rubinstein (1976) derivation of the CCAPM assumes that, over a discrete time interval, the joint distribution of all assets' returns with each individual's optimal consumption is normal. More generally, Breeden and Litzenberger (1978) derive the CCAPM in a discrete-time framework for the *subset of assets* whose returns are jointly lognormally distributed with aggregate consumption. Breeden's (1979) continuous-time derivation of the CCAPM applies instantaneously to all assets, based on the assumption that assets' returns and individuals' optimal consumption paths follow diffusion processes. In all these papers, utility functions are time additive.

Since the CCAPM is well known, a standard review is unnecessary. The following synthesis provides a theoretical basis that is more relevant for the subsequent empirical work. In particular, theoretical predictions are derived for easily estimated models which are based on unconditional moments of returns using discretely sampled data.

Let $\{\tilde{R}_{it}, i = 1, \cdots, M\}$ be the rates of return on risky assets from time $t - 1$ to time t. M may be less than the number of all risky assets in the economy. Let $\tilde{R}_{zt}$ be the rates of return on a portfolio whose return is uncorrelated with the growth rate in aggregate consumption. All individuals are assumed to have time-additive, monotonically increasing, and strictly concave von Neumann-Morgenstern utility functions for lifetime consumption. Identical beliefs, a fixed population with infinite lifetimes, a single consumption good, and capital markets that permit an unconstrained Pareto-optimal allocation of consumption are also assumed. From the first-order conditions for individual k's optimal consumption and portfolio plan, it follows that

$$\mathscr{E}[(\tilde{R}_{it} - \tilde{R}_{zt})[U^{k'}(\tilde{C}_t^k)/U^{k'}(C_{t-1}^k)] \mid \phi_{t-1}] = 0, \forall \ i, k, \qquad (1)$$

where ϕ_{t-1} describes the full information set at time $t - 1$. This relation holds for any sampling interval. This is well known (e.g., see Lucas (1978)).

An individual achieves an optimal portfolio by adjusting the portfolio weights and consumption plans until relation (1) holds for all assets. Breeden and Litzenberger (1978) show that, in a capital market that permits an unconstrained Pareto-optimal allocation of consumption, each individual's consumption at a given date is an increasing function of *aggregate* consumption. Furthermore, each individual's optimal marginal utility of consumption at a given date t is equal to a scalar, a_k, times a monotonically decreasing function of aggregate consumption, $g(C_t,t)$, which is identical for all individuals. The assumption that all individuals have the same subjective rate of time preference implies that the time dependence of the aggregate marginal utility function is the same for all dates, so $g(C_t,t) = f(C_t)$. Thus, in equilibrium in a Pareto-efficient capital market, the growth rate in the marginal utility of consumption would be identical for all individuals and equal to the growth rate in the "aggregate marginal utility" of consumption. That is,

$$\frac{U'(C_t)}{U'(C_{t-1})} = \frac{f(C_t)}{f(C_{t-1})} \cong 1 - [-C_{t-1}f'(C_{t-1})/f(C_{t-1})]c_t^*, \qquad (2)$$

where c_t^* is the growth rate in aggregate consumption per capita and where the approximation follows from a first-order Taylor series. The term in square brackets is aggregate relative risk aversion evaluated at C_{t-1}. If we take relative risk aversion as approximately constant and denote it as b, we can combine (1) with these approximations in (2) and find (ignoring the approximations)[1]

$$\mathscr{E}\{(\tilde{R}_{it} - \tilde{R}_{zt})(1 - b\tilde{c}_t^*)|\phi_{t-1}\} = 0. \qquad (3)$$

Since (3) is zero conditional on any information, it also holds in terms of unconditional expectations:

$$\mathscr{E}\{(\tilde{R}_{it} - \tilde{R}_{zt})(1 - b\tilde{c}_t^*)\} = 0. \qquad (4)$$

The return on an asset may be stated as a linear function of the growth rate in aggregate consumption per capita, c_t^*, plus a disturbance. This disturbance term is assumed to be uncorrelated with $\tilde{c}_t^*$ for a proper subset of assets ($i = 1, \cdots, M$). These conditions, combined with the assumption of constant unconditional consumption betas and alphas, imply

$$\tilde{R}_{it} = \alpha_{ci}^* + \beta_{ci}^*\tilde{c}_t^* + \tilde{u}_{it}^*, \quad \forall i = 1, \cdots, M,$$

$$\mathscr{E}\{\tilde{u}_{it}^*\} = 0 \quad \text{and} \quad \mathscr{E}\{\tilde{u}_{it}^*\tilde{c}_t^*\} = 0, \qquad (5)$$

where $\beta_{ci}^* \equiv \text{cov}(\tilde{R}_{it}, \tilde{c}_t^*)/\text{var}(\tilde{c}_t^*)$, $\alpha_{ci}^* \equiv \mu_i - \beta_{ci}^*\mathscr{E}\{\tilde{c}_t^*\}$, and $\mu_i \equiv \mathscr{E}\{\tilde{R}_{it}\}$. Asterisks indicate parameters in relation to true consumption growth. Later asterisks are removed to indicate parameters in relation to measured consumption growth.

For a zero consumption beta portfolio consisting of just the M assets,

$$\tilde{R}_{zt} = \gamma_0 + \tilde{\mu}_{zt}^*,$$

$$\mathscr{E}\{\tilde{u}_{zt}\} = 0,$$

$$\mathscr{E}\{\tilde{u}_{zt}\tilde{c}_t^*\} = 0. \qquad (6)$$

Substituting the right-hand side (hereafter, RHS) of (5) and (6) into relation (4) gives the CCAPM:

$$\mu_i - \gamma_0 = \gamma_1^*\beta_{ci}^*, \quad \forall i = 1, \cdots, M, \qquad (7)$$

[1] The approximation can be avoided by making an additional distributional assumption that $\text{cov}(\tilde{u}_{it}^*, \tilde{X}_t) = 0$, where $X_t \equiv f(\tilde{c}_t)/f(\tilde{c}_{t-1})$ and $\tilde{u}_{it}^*$ is defined in (5) below. All the following results go through, and $\gamma_1^* \equiv \text{cov}(\tilde{c}_{t1} - \tilde{X}_t)/E(\tilde{X}_t)$. The market price of consumption beta risk, γ_1^*, appears in equation (7) below.

where $\gamma_1^* \equiv b \, \text{var}(\tilde{c}_t^*)/[1 - b\mathscr{E}(\tilde{c}_t^*)]$. The market price of consumption beta risk, γ_1^*, increases as the variability of consumption increases. If $[1 - b\mathscr{E}(\tilde{c}_t^*)] > 0$ and $\mathscr{E}(\tilde{c}_t^*) > 0$, then γ_1^* also increases as relative risk aversion increases.

This model only gives the CCAPM for a proper subset of assets—those assets that have a conditionally linear relation with c_t^* over the measurement interval. Assets which do not satisfy (5) still are priced according to their joint distributions of payoffs with consumption, but higher order co-moments with consumption are required for pricing over discrete intervals. Since in the continuous-time model all assets' returns and consumption are locally jointly normally distributed, the CCAPM applies to all assets as long as returns can be measured over instantaneous intervals. However, since the available data are measured discretely, the CCAPM in (7) is more useful for empirical tests.

In continuous time with time-additive utility, Breeden (1979) demonstrates that Merton's (1973) intertemporal multi-beta asset pricing model is equivalent to a single-beta CCAPM. However, Cornell (1981) emphasizes that the conditional consumption beta in such a representation need not be constant. The tests presented in this paper are tests of restrictions on the unconditional co-moments of assets returns and consumption growth. As Grossman and Shiller (1982) point out, such tests do not ignore Cornell's (1981) concerns about changes in the conditional moments. An advantage of tests based on unconditional moments is that a specification of the changes in conditional moments is not required. To the extent that changes in the conditional moments could be modeled, the resulting tests may be more powerful. For examples of such tests see Gibbons and Ferson (1985), Hansen and Singleton (1983), and Litzenberger and Ronn (1986). Since the CCAPM has predictions for conditional and unconditional expectations, failure to reject the "unconditional CCAPM" is a necessary, but not sufficient, condition for acceptance of the model.

II. Econometric Problems Associated with Measured Consumption

In this section, a distinction is made between the appropriate theoretical definition of aggregate consumption per capita and the consumption reported by the Department of Commerce. Four measurement problems are examined: 1) the reporting of expenditures, rather than consumption, 2) the reporting of an integral of consumption rates, rather than the consumption rate at a point in time, 3) infrequent reporting of consumption data relative to stock returns, and 4) reporting aggregate consumption with sampling error since only a subset of the total population of consumption transactions is measured.

The CCAPM prices assets with respect to changes in aggregate consumption between two points in time. In contrast, the available data on aggregate "consumption" provide total expenditures on goods and services over a period of time. These differences between consumption in theory and its measured counterpart suggest the first two problems. First, goods and services need not be consumed in the same period that they are purchased. Second, measured aggregate consumption is closer to an integral of consumption over a period of time than to "spot" consumption (at a point in time). This second problem creates a "summation bias."

Empirical Tests of the CCAPM

While returns on stocks are available on an intraday basis, corresponding consumption data are not available. Currently, only quarterly data are provided back to 1939, and monthly reporting begins in 1959. Infrequent reporting of aggregate expenditures on consumption is the measurement problem analyzed in the third subsection. The fourth subsection demonstrates that sampling error in aggregate consumption does not bias the statistical tests.

A. Description of the Consumption Data

Exploring the empirical implications of the CCAPM for a long sample period requires aggregate consumption data from different sources. The tables in Sections III and IV focus on a time series for consumption that requires "splicing" the data at two points. Each of these three data sources is discussed in turn.

As is discussed later, powerful tests of any asset pricing model require precise estimations for the relevant betas. Precision of the estimators improves if the variability of the consumption measure increases, holding everything else constant. Since consumption was quite variable in the 1930s, we want to include this time period in our empirical work.[2] Unfortunately, aggregate consumption data are not available, except for annual sampling intervals, from 1929 to 1939. However, nominal personal income less transfer payments is available on a monthly basis from the U.S. Department of Commerce,[3] and these income numbers are used to approximate quarterly consumption for this decade.

From 1929 to 1939 a regression of annual consumption data on personal income yields

$$z_{1t} = 0.00186 + 0.56 z_{2t} + \hat{v}_t, \qquad R^2 = 0.94, \qquad (8)$$
$$\phantom{z_{1t} = }(0.39) \qquad (11.51)$$

where $z_{1t} \equiv$ annual growth of real nondurables and services consumption per capita, $z_{2t} \equiv$ annual growth of real personal income less transfer payments per capita, and t-statistics are in parentheses. The data for the above regression are deflated by the average level of the Consumer Price Index (CPI) from the U.S. Bureau of Labor Statistics. The population numbers, which are used to calculate per capita values, are from the *Statistical Abstract of the United States* and reflect the resident population of the U.S. The monthly numbers on personal income less transfer payments are used to infer the consumption numbers based on the above regression equation. From these monthly estimates of consumption, quarterly growth rates are constructed.

From 1939 through 1958 the spliced consumption data rely on quarterly expenditures on nondurable goods and services based on national income accounting. From 1939 through 1946, the data are deflated by the average level of the monthly CPI for the relevant quarter. From 1947 through 1958, real consumption data are available from the Commerce Department. Only seasonally

---

[2] There is no doubt that part of the unusual volatility of consumption during the 1930s is due to data construction, not variation in true consumption.

[3] For both the annual and monthly data, see *National Income and Product Statistics of the United States, 1929–46*. This appeared as a supplement to the *Survey of Current Business*, July 1947.

adjusted numbers for consumption are available.[4] Average total U.S. population during a quarter as reported by the Commerce Department is used to calculate the per capita numbers. Various issues of *Business Conditions Digest, Business Statistics*, and *The National Income and Product Accounts of the United States* report the relevant data.

The consumption data from 1959 to 1982 are constructed in essentially the same manner as that from 1947 to 1958.[5] However, since the government started publishing monthly numbers during this latter period, these monthly numbers are used to compute growth in real consumption per capita over a quarter. For example, growth in a first quarter is based on expenditures during March, relative to expenditures during the prior December.

In later sections, the term "spliced" consumption data refers to the data base constructed in the above manner, which combines the quarterly observations on monthly income data from 1929 to 1938, the quarterly consumption expenditures from 1939 to 1958, and the quarterly observations on the monthly consumption expenditures from 1959 to 1982.

For the whole time period (1929–1982), the consumption data are based on expenditures on nondurables plus services, following Hall (1978). This is an attempt to minimize the measurement problem associated with expenditures versus current consumption of goods and services. No attempt is made to extract the consumption flow from durable goods.[6] While monthly sampling of consumption data is available after 1958, most of the tables do not rely on this information. As the sampling interval decreases, "nondurables" become more durable. However, some of the calculations have been repeated using monthly sampling intervals, and these results are summarized in the text and footnotes.

B. Interval versus Spot Consumption (the Summation Bias)

Ignoring other measurement problems, the reported ("interval") consumption rate for a quarter is the integral of the instantaneous ("spot") consumption rates during the quarter. The CCAPM relates expected quarterly returns on assets (e.g., from January 1 to March 31) and the covariances of those returns with the change in the spot consumption rate from the beginning of the quarter to the end of the quarter. This subsection derives the relation between the desired population covariances (and betas) of assets' returns relative to spot consumption changes and the population covariances (and betas) of assets' returns relative to changes in interval consumption. The variance of interval consumption changes is shown to have only two thirds the variance of spot consumption changes, while the autocorrelation of interval consumption is 0.25 due to the integration of spot

---

[4] Since the seasonal adjustment smoothes expenditures, such an adjustment may be desirable if the transformed expenditures better resemble actual consumption. Of course, seasonal adjustment is inappropriate if it removes seasonals in true consumption.

[5] The only exception to this occurs for the population number for December 1978. This number is adjusted from the published tables because there is an obvious typographical error.

[6] Alternative treatments for this measurement problem exist in the literature. For example, Marsh (1981) postulates a latent variable model to estimate the parameters of the CCAPM. A more recent attempt is made by Dunn and Singleton (1986), using an econometric approach that relies on the specification of preferences for the representative economic agent.

rates. These latter results are reported by Working (1960) and generalized by Tiao (1972). Similar results on time aggregation have been used in studies of stock prices and corporate earnings (Lambert (1978) and Beaver, Lambert, and Morse (1980)). In an independent and contemporaneous paper, Grossman, Melino, and Shiller (1987) derive maximum-likelihood estimates of CCAPM parameters, explicitly accounting for time aggregation of consumption data. Our bias corrections are much simpler but give similar results.

Without loss of generality, consider a two-quarter period with $t = 0$ being the beginning of the first quarter and $t = T$ being the end of the first quarter. All discussion will analyze annualized consumption rates, so $T = 0.25$ for a quarter. Initially, let the change in the spot consumption rate over a quarter be the cumulative of n discrete changes, $\{\tilde{\Delta}_1^C, \tilde{\Delta}_2^C, \cdots, \tilde{\Delta}_n^C\}$ for the first quarter, and $\{\tilde{\Delta}_{n+1}^C, \tilde{\Delta}_{n+2}^C, \cdots, \tilde{\Delta}_{2n}^C\}$ for the second quarter. That is, $\hat{C}_T = C_0 + \sum_1^n \tilde{\Delta}_i^C$. Similarly, let the wealth at time T from buying one share of an asset at time 0 (and reinvesting any dividends) equal its initial price plus n random increments $\{\tilde{\Delta}_i^a\}: P_T = P_0 + \sum_i^n \tilde{\Delta}_i^a$.[7]

Changes in consumption, $\tilde{\Delta}_i^C$, are assumed to be homoscedastic and serially uncorrelated. Similar assumptions are made for the asset's return, $\tilde{\Delta}_i^a$, with variance σ_a^2. The contemporaneous covariation of an asset's return with consumption changes is σ_{ac}, and noncontemporaneous covariances are assumed to be zero. The variance of the change in the spot consumption from the beginning of a quarter to the end of the quarter is $\text{var}(\hat{C}_T - \hat{C}_0) = \text{var}(\sum_1^n \tilde{\Delta}_i^C) = \sigma_C^2 T$.

The first quarter's *reported* annualized consumption, C_{Q1}, is a summation of the consumption during the quarter, annualized by multiplying by 4 (or $1/T$):

$$C_{Q1} = (1/T) \sum_{j=1}^n C_j \Delta t = (1/T) \sum_{j=1}^n (C_0 + \sum_{i=1}^j \Delta_i^C) \Delta t. \tag{9}$$

The annualized consumption rate for the second quarter, C_{Q2}, is the same as (9), but with the first summation for j being $n + 1$ to $2n$.

Continuous movements in consumption and asset prices can be approximated by letting the number of discrete movements per quarter, n, go to infinity ($\Delta t \to 0$). Doing this, the change in reported consumption becomes[8]

$$C_{Q2} - C_{Q1} = \int_0^T (t/T) \Delta_t^C \, dt + \int_T^{2T} ((2T - t)/T) \Delta_t^C \, dt. \tag{10}$$

---

[7] The summation bias is developed for price changes and consumption changes, not rates of return and consumption growth rates. When the prior period's price and consumption are fixed, the results of this section apply. However, in tests involving unconditional moments, the prior period's price and consumption are random. Since it is difficult to derive a closed-form solution for the summation bias in terms of rates, the subsequent analysis ignores this distinction.

[8] To see this, represent C_{Q2} and C_{Q1} as in (9) and take the difference:

$$C_{Q2} - C_{Q1} = (1/T) \left\{ \sum_1^{n+1} \Delta_i^C + \sum_1^{n+2} \Delta_i^C + \cdots + \sum_1^{2n} \Delta_i^C \right\} \Delta t$$

$$- (1/T) \left\{ \sum_1^1 \Delta_i^C + \sum_1^2 \Delta_i^C + \cdots + \sum_1^n \Delta_i^C \right\} \Delta t$$

$$= n^{-1} \{\Delta_2^C + 2\Delta_3^C + \cdots + (n-1)\Delta_n^C + n\Delta_{n+1}^C + (n-1)\Delta_{n+2}^C + \cdots + \Delta_{2n}^C\}.$$

Letting n become large gives equation (10).

Given the independence of spot consumption change over time, (10) implies that the variance of reported annualized consumption changes is

$$\text{var}(\tilde{C}_{Q2} - \tilde{C}_{Q1}) = \int_0^T ((t/T)^2 \sigma_C^2) \, dt$$

$$+ \int_T^{2T} ((2T - t)/T)^2 \sigma_C^2 \, dt = (2/3)\sigma_C^2 T. \tag{11}$$

Thus, the population variance of reported (interval) consumption changes for a quarter is two thirds of the population variance for changes in the spot consumption from the beginning of a quarter to the end of the quarter. The averaging caused by the integration leads to the lower variance for reported consumption.

Next, consider the covariance of an asset's quarterly return with quarterly changes in the consumption. The covariance of the change in spot consumption from the beginning of a quarter to the end of the quarter with an asset's return over the same period is $\sigma_{ac}T$, given the i.i.d. assumption. With reported, interval consumption data, the covariance can be computed from (10):

$$\text{cov}(\tilde{C}_{Q2} - \tilde{C}_{Q1}, \tilde{P}_{2T} - \tilde{P}_T) = T^{-1} \int_T^{2T} (2T - t)\sigma_{ac} \, dt = T\sigma_{ac}/2. \tag{12}$$

Thus, from (12) the population covariance of an asset's quarterly return with reported (interval) consumption is half the population covariance of the asset's return with spot consumption changes.

Given (11) and (12), betas measured relative to reported quarterly consumption changes are ¾ times the corresponding betas with spot consumption:

$$\beta_{ac}^{sum} = \frac{(1/2)\sigma_{ac}T}{(2/3)\sigma_C^2 T} = (3/4)\beta_{ac}^{spot}. \tag{13}$$

Since the CCAPM relates quarterly returns to "spot betas," the subsequent empirical tests multiply the mean-adjusted consumption growth rates by ¾ to obtain unbiased "spot betas." The ¾ relation of interval betas to spot betas in (17) is a special case of the multiperiod differencing relation: $\beta_{ac}^{sum} = \beta_{ac}^{spot} [K - (1/2)]/[K - (1/3)]$, where K is the differencing interval. Thus, monthly data sampled quarterly (i.e., $K = 3$) should give interval betas that are $(5/2)/(8/3) = 0.9375$ times the spot betas. When quarterly consumption growth rates are calculated from monthly data, the quarterly numbers are mean adjusted and multiplied by 0.9375.

Although changes in spot consumption are uncorrelated, changes in reported, interval consumption rates have positive autocorrelation. To see this, use (10) to compute the covariance of the reported consumption change from Q1 to Q2 with the reported change from Q2 to Q3, noting that all covariance arises from the time overlap from T to $2T$:

$$\text{cov}(\tilde{C}_{Q3} - \tilde{C}_{Q2}, \tilde{C}_{Q2} - \tilde{C}_{Q1}) = \int_T^{2T} ((t-T)(2T-t)/T^2)\sigma_C^2 \, dt = (1/6)\sigma_C^2 T. \tag{14}$$

Empirical Tests of the CCAPM

The first-order autocorrelation in reported consumption is 0.25 since

$$\rho_1 = \text{cov}(\tilde{C}_{Q3} - \tilde{C}_{Q2}, \tilde{C}_{Q2} - \tilde{C}_{Q1})/\text{var}(\tilde{C}_{Q2} - \tilde{C}_{Q1}) = \frac{(1/6)\sigma_C^2 T}{(2/3)\sigma_C^2 T} = 0.25. \quad (15)$$

By similar calculations, higher order autocorrelation is zero. Table I presents the time series properties of reported *unspliced* quarterly consumption growth rates. First-order autocorrelation of quarterly real consumption growth for the entire 1939–1982 period is estimated to be 0.29, which is insignificantly different from the theoretical value of 0.25 at usual levels of significance. Higher order autocorrelations are not significantly different from zero. Thus, the model for reported consumption is not rejected by the sample autocorrelations.

Monthly growth rates of real consumption from 1959 to 1982 exhibit negative autocorrelation of −0.28, which is significantly different from zero and from the hypothesized 0.25. This may be caused by vagaries such as bad weather and strikes in major industries, which cut current consumption temporarily but are followed by catch-up purchases. Quarterly growth rates in consumption computed from the monthly series again have positive autocorrelation of 0.13, more closely in line with the value 0.0625 (or 1/16) predicted by the summation bias.[9] The longer the differencing interval, the less affected the data are by temporary fluctuations and measurement errors in consumption.

Chen, Roll, and Ross (1986) and Hansen and Singleton (1983) use monthly data on unadjusted consumption growth. Since those data's autocorrelation statistics suggest significant departures from the random-walk assumption, the statistics they present warrant re-examination. The use of larger differencing intervals should be fruitful.

C. Infrequent Reporting of Consumption: The Maximum Correlation Portfolio

Since the returns on many assets are available for a longer time and are reported more frequently than consumption, more precise evidence on the CCAPM can be provided if only returns were needed to test the theory. Fortunately, Breeden's (1979, footnote 8) derivation of the CCAPM justifies the use of betas measured relative to a portfolio that has maximum correlation with growth in aggregate consumption, in place of betas measured relative to aggregate consumption. This result is amplified below, as it is shown that securities' betas measured relative to this maximum correlation portfolio (hereafter, MCP) are equal to their consumption betas divided by the consumption beta of the MCP. If a riskless asset exists, then the consumption beta of the MCP can be changed by adjusting leverage. Our MCP excludes the riskless asset, resulting in a consumption beta of 2.9.

In the following, the first M assets have a linear relation with consumption as in equation (5). The CCAPM holds with respect to these M assets when betas are measured relative to the MCP obtained from these M assets. Second, for *any* subset N (where $N \leq M$) of these M assets, the CCAPM holds for that subset when betas are measured relative to the MCP obtained from these N assets.

---

[9] The derivation of this prediction is similar to the derivation of equation (15).

Table I
Time Series Properties of Percentage Changes in Real, Per Capita Consumption of Nondurable Goods and Services

Data are seasonally adjusted as reported by the Department of Commerce in the *Survey of Current Business*. T denotes the number of observations, while $\hat{c}$ and $\widehat{SD}(c)$ are the sample mean and standard deviation, respectively. Under the hypothesis that the observations are serially uncorrelated, the asymptotic standard errors for the sample autocorrelations are $1/\sqrt{T}$, as given by $SD^*(\hat{\rho}_k)$. Under the hypothesis that $\rho_1 = 0.25$ and $\rho_k = 0 \ \forall \ |k| > 1$, $SD(\hat{\rho}_1)$ and $SD(\hat{\rho}_k)$ report the asymptotic standard errors using the results of Bartlett (1946). The test statistic for the joint hypothesis that all autocorrelations are zero for lags 1 through 12 is given by Q_{12}, the modified Box-Pierce Q-statistic. Q_{12} is asymptotically distributed as chi-square with 12 degrees of freedom. The p-value is the probability of drawing a Q_{12} statistic larger than the current value under the null hypothesis.

Time Period	T	$\hat{c}$	$\widehat{SD}(c)$	$\hat{\rho}_1$	$\hat{\rho}_2$	$\hat{\rho}_3$	$\hat{\rho}_4$	$\hat{\rho}_6$	$SD^*(\hat{\rho}_k)$	$\widehat{SD}(\hat{\rho}_1)$	$\widehat{SD}(\hat{\rho}_k)$	Q_{12}	p-Value
\multicolumn{14}{c}{Panel A: Quarterly Consumption Data}
39Q2–82Q4	175	0.00543	0.00951	0.29	0.03	−0.00	0.07	0.02	0.08	0.07	0.08	23.93	0.02
39Q2–52Q4	55	0.00665	0.01517	0.30	0.03	−0.04	0.08	0.08	0.13	0.12	0.14	11.26	0.51
53Q1–67Q4	60	0.00463	0.00549	0.21	0.09	0.11	−0.01	−0.22	0.13	0.12	0.14	11.25	0.51
68Q1–82Q4	60	0.00511	0.00487	0.36	0.01	0.26	0.09	−0.31	0.13	0.12	0.14	25.95	0.01
\multicolumn{14}{c}{Panel B: Monthly Consumption Data}
1959–1982	287	0.00178	0.00447	−0.28	−0.02	−0.14	−0.12	−0.19	0.06	0.05	0.06	43.09	0.00
1959–1970	143	0.00199	0.00467	−0.31	−0.11	0.18	−0.08	−0.17	0.08	0.08	0.09	33.49	0.00
1971–1982	144	0.00156	0.00427	−0.24	0.07	0.09	−0.16	−0.16	0.08	0.08	0.09	20.56	0.06
\multicolumn{14}{c}{Panel C: Quarterly Sampling of Monthly Consumption Data}
59Q2–82Q4	95	0.00521	0.00568	0.13	−0.13	0.20	0.04	−0.17	0.10	0.09	0.11	13.42	0.34
59Q2–70Q4	47	0.00576	0.00506	0.13	−0.15	0.13	−0.03	−0.04	0.15	0.13	0.15	10.61	0.56
71Q1–82Q4	47	0.00468	0.00623	0.12	−0.07	0.22	−0.10	−0.26	0.14	0.13	0.15	11.40	0.50

Empirical Tests of the CCAPM

The following notation will be used throughout the paper. Let μ be the $N \times 1$ vector unconditional expected returns, let $\mathbf{1}$ be an $N \times 1$ vector of ones, and let β_c^* be the $N \times 1$ vector of unconditional consumption betas. Let $\tilde{R}_{mcp}$ be the return on the MCP that excludes the riskless asset, let $\beta_{c,nb}^*$ be the unconditional consumption beta of this "no borrowing" MCP, and let β_{mcp} be the $N \times 1$ vector of unconditional MCP betas. The $N \times N$ unconditional covariance matrix for returns is $\mathbf{V}$, which is assumed to be nonsingular.

Consider the following portfolio problem: find the minimum-variance portfolio that has a consumption beta of $\beta_{c,nb}^*$ (i.e., with no borrowing). The consumption beta of a portfolio is the product of its correlation coefficient with consumption and the portfolio's standard deviation, divided by the standard deviation of consumption. By constraining the consumption beta to be fixed and then minimizing variance, the resulting portfolio has the maximum correlation with consumption, i.e., the MCP. Mathematically, the MCP solves

$$\min_{|\mathbf{w}|}: \mathbf{w}'\mathbf{V}\mathbf{w} + 2\lambda(\beta_{c,nb}^* - \mathbf{w}'\beta_c^*), \tag{16}$$

where λ is a Lagrange multiplier. The weights (i.e., $\mathbf{w}$) in the MCP are not constrained to unity since the risky assets may be combined with a riskless asset without any effect on the correlation coefficient, the variance, or the consumption beta. Alternatively, if the weights obtained sum to the value S, those same weights multiplied by $1/S$ sum to unity and have the same correlation with consumption. Thus, the existence or nonexistence of a riskless asset does not affect the MCP analysis.

The first-order conditions imply

$$\mathbf{w}_{mcp} = \lambda \mathbf{V}^{-1} \beta_c^*. \tag{17}$$

Since $\beta_{c,nb}^* = \mathbf{w}_{mcp}'\beta_c^*$, $\lambda = \beta_{c,nb}^*/(\beta_c^{*'}\mathbf{V}^{-1}\beta_c^*)$. Pre-multiplying (17) by $\mathbf{w}_{mcp}'\mathbf{V}$ and simplifying implies $\mathbf{w}_{mcp}'\mathbf{V}\mathbf{w}_{mcp} = \lambda \beta_{c,nb}^*$. The MCP betas of risky assets are

$$\beta_{mcp} = \frac{\mathbf{V}\mathbf{w}_{mcp}}{\mathbf{w}_{mcp}'\mathbf{V}\mathbf{w}_{mcp}} = \frac{\lambda \beta_c^*}{\lambda \beta_{c,nb}^*} = \beta_c^*/\beta_{c,nb}^*, \tag{18}$$

using the facts just derived.

In words, (18) states that assets' betas measured relative to the MCP are proportional to their betas relative to true consumption. Substituting (18) into the zero-beta CCAPM and using the CCAPM to get the expected excess return on the MCP implies

$$\mu - \gamma_0 \mathbf{1} = \beta_{mcp}(\mu_{mcp} - \gamma_0), \tag{19}$$

Where $\mu_{mcp} \equiv \mathscr{E}(\tilde{R}_{mcp,t})$. Thus, the CCAPM may be restated (and tested) in terms of the MCP, and the testable implication is that the MCP is ex ante mean-variance efficient. Obviously, any zero-consumption beta portfolio also has a zero beta relative to the MCP.

The above result also suggests an intuitive interpretation of the portfolio weights for the MCP. Equation (17) implies

$$\mathbf{w}_{mcp} = \lambda \mathbf{V}^{-1} \beta_c^* = \theta \mathbf{V}^{-1} \mathbf{V}_{ac}^*, \tag{20}$$

where $\theta \equiv \lambda/\mathrm{var}(\tilde{c}) = \dfrac{\beta^*_{c,nb}}{(\beta^{*'}_c \mathbf{V}^{-1} \beta^*_c) \mathrm{var}(\tilde{c}^*)}$ and $\mathbf{V}^*_{ac}$ is the $N \times 1$ vector of covariances of returns with consumption. From (20), the MCP's weights are proportional to the coefficients in a multiple regression of consumption on the various risky assets' returns, with θ being the factor of proportionality. Actually, θ equals $\beta^*_{c,nb}$ divided by the coefficient of determination (R^2) of the multiple regression just described. To see this, note that the weights in the multiple regression, $\mathbf{w}^*_c$, are $\mathbf{w}^*_c = \mathbf{V}^{-1} \mathbf{V}^*_{ac}$, and the R^2 in the regression is

$$R^2 = (\mathbf{w}^{*'}_c \mathbf{V} \mathbf{w}^*_c)/\mathrm{var}(\tilde{c}^*) = (\beta^{*'}_c \mathbf{V}^{-1} \beta^*_c) \mathrm{var}(\tilde{c}^*), \tag{21}$$

which shows that $\theta = \beta^*_{c,nb}/R^2$. If there is a riskless asset, a unit beta MCP has weights that equal the regression's coefficients divided by the R-squared value of the regression (with any residual wealth in the riskless asset). Betas with respect to such a unit-beta MCP equal the assets' direct consumption betas (see equation (18)).

The optimization problem of (16) does not involve a constraint on means, so a MCP is not tautologically a mean-variance efficient portfolio. However, the CCAPM does imply mean-variance efficiency of that MCP in equilibrium. This implication is tested later in our paper.

D. Sampling Error In Reported Consumption

In this section the problem of pure sampling error in reported consumption is examined. These errors are assumed to be random and uncorrelated with economic variables. Continue with $\tilde{c}^*_t$ as the true growth rate of real consumption from $t - 1$ to t, and let $\tilde{c}_t$ be the reported growth rate. The measurement error, $\tilde{\epsilon}_t$, is such that

$$\tilde{c}_t = \tilde{c}^*_t + \tilde{\epsilon}_t \tag{22}$$

$$\mathscr{E}(\tilde{\epsilon}_t) = 0, \qquad \mathrm{cov}(\tilde{\epsilon}_t, \tilde{c}^*_t) = 0,$$

$$\mathrm{cov}(\tilde{\epsilon}_t, \tilde{R}_{it}) = 0, \qquad \forall\ i = 1, \cdots, N. \tag{23}$$

Substituting (22) into the CCAPM of (7) gives

$$\mu_i - \gamma_0 = \gamma^*_1 \beta^*_{ci} = \gamma^*_1 \mathrm{cov}(\tilde{R}_{it}, \tilde{c}_t - \tilde{\epsilon}_t)/\mathrm{var}(\tilde{c}^*_t) \tag{24}$$

$$= \gamma^*_1 \dfrac{\mathrm{var}(\tilde{c}_t) \mathrm{cov}(\tilde{R}_{it}, \tilde{c}_t)}{\mathrm{var}(\tilde{c}^*_t) \mathrm{var}(\tilde{c}_t)} = \gamma_1 \beta_{ci},$$

where β_{ci} is the beta asset of i with respect to reported consumption, and $\gamma_1 \equiv \gamma^*_1 [\mathrm{var}(\tilde{c}_t)/\mathrm{var}(\tilde{c}^*_t)]$. As long as the variance of the measurement error is positive, the variance of measured consumption exceeds the variance of true consumption. From (24), the slope coefficient, γ_1, in the relation between excess returns and betas with reported consumption is biased upward as an estimate of the price of risk, γ^*_1.

Sampling error in reported consumption does not cause a bias in the coefficients of a multiple regression of consumption growth on risky asset returns. However, the coefficient of determination for such a regression is downward biased. While the portfolio weights of the MCP are calculated by taking ratios of the regression coefficients divided by this R^2, the downward bias in R^2 affects all the weights in a proportional fashion. Thus, this has no effect on the subsequent tests.

Some other important measurement errors in aggregate consumption data involve interpolation (i.e., expenditures for all items are sampled every month), to which the analysis of this subsection is not applicable. This problem is similar to one faced by Fama and Schwert (1977, 1979) in their analysis of components of the CPI. Unfortunately, the interpolation problems with consumption are exacerbated by the summation bias, and it is difficult to disentangle the two effects. For example, either problem leads to serial correlation in consumption, noncontemporaneous correlation between aggregate consumption and returns, and more serious effects as sampling interval becomes shorter.[10] If the summation bias were not present, presumably an approach similar to that in Scholes and Williams (1977) would be appropriate. Perhaps the combination of interplation and summation bias explains the pattern of serial correlations in monthly data on consumption growth (see Panel B of Table I). Interpolation is yet another reason for avoiding monthly sampling intervals.

III. Empirical Characteristics of Consumption Betas and the MCP

Since existing empirical research on the CCAPM is not extensive, we summarize how consumption betas vary across different assets. Several types of assets will be studied, including government and corporate bonds and equities.

Monthly returns on individual securities are gathered from the Center for Research in Security Prices (CRSP) at the University of Chicago. Twelve portfolios of these stocks are formed by grouping firms using the first two digits of their SIC numbers. The grouping closely followed a classification used by Sharpe (1982), with the major exception being that Sharpe's "consumer goods" category is subdivided. This subdivision should increase the dispersion of consumption betas in the sample. While other groupings of stocks have been suggested (e.g., see Stambaugh (1982)), Sharpe's scheme is selected because the industry portfolios are reasonable and capture some important correlation patterns among stocks. Table II provides more details on the classification scheme. To represent the return on a "buy and hold" strategy, relative market values are used to weight the returns in a given portfolio. Every return on the CRSP tape from 1926 through 1982 is included, which should minimize problems with survivorship bias.[11]

---

[10] Interpolation should result in serial correlation in the residuals in regressions of returns on consumption growth. (Equation (26) below is such a regression.) Yet, when we examine the residuals, the autocorrelations are not striking. On the other hand, when we run a multiple regression of returns on a leading value of consumption, current consumption, and lagged consumption, we do get an interesting pattern. Generally, the coefficient on the lead value is insignificant, the coefficient on current consumption is significant and positive, and the coefficient on lagged consumption is significant and negative. Usually, the absolute value of the coefficient on the lagged value is about half the value of the coefficient on current consumption. However, as we note in the text, the significance of the coefficient on lagged consumption is predictable if only a summation bias is present.

[11] However, all firms with a SIC number of 39 (i.e., miscellaneous manufacturing industries) are excluded to avoid any possible problems with a singular covariance matrix when the CRSP value-weighted index is added to the sample.

Table II
Estimated Betas Relative to 1) Growth in Real, Per Capita Consumption[a], 2) Maximum-Correlation Portfolio for Consumption, and 3) CRSP Value-Weighted Index

NA denotes not available. The maximum correlation portfolio (MCP) is constructed from the seventeen assets given in Table III. The weights of the MCP are determined by maximizing the sample correlation between the return on the portfolio and the growth rate of real consumption; see Table III for more details.

Asset (SIC Codes)	Number of Firms 1/26	Number of Firms 6/54	Number of Firms 12/82	Spliced Consumption, Quarterly 1929–1982 ($T=215$) $\hat{\beta}_c$	$t(\hat{\beta})$	R^2	Max.-Correlation Cons. Portfolio, Monthly 1926–1982 ($T=684$) $\hat{\beta}_{MCP}$	$t(\hat{\beta})$	R^2	CRSP Value-Weighted Index Monthly 1926–1982 ($T=684$) $\hat{\beta}_{CRSP}$	$t(\hat{\beta})$	R^2
U.S. Treasury bills	—	—	—	−0.11	−1.27	0.01	0.03	3.86	0.02	0.01	2.04	0.01
Long-term govt. bonds	NA	NA	NA	−0.01	−0.02	0.00	0.07	2.53	0.01	0.07	4.93	0.03
Long-term corp. bonds	NA	NA	NA	0.24	0.91	0.00	0.07	2.52	0.01	0.08	6.62	0.06
Junk bond premium	NA	NA	NA	2.45	6.85	0.18	0.63	18.52	0.33	0.33	20.45	0.38
Petroleum (13, 29)	46	51	69	4.31	6.37	0.16	1.41	20.61	0.38	0.92	38.63	0.69
Finance & real estate (60–69)	16	43	234	5.85	6.30	0.16	1.50	18.81	0.34	1.19	75.95	0.89
Consumer durables (25, 30, 36, 37, 50, 55, 57)	69	157	180	6.86	6.80	0.18	1.79	22.03	0.42	1.29	80.79	0.91

Basic industries (10, 12, 14, 24, 26, 28, 33)	94	207	194	5.45	6.95	0.18	1.48	21.98	0.41	1.09	100.80	0.94
Food & tobacco (1, 20, 21, 54)	64	103	81	3.25	5.69	0.13	0.99	18.62	0.34	0.76	58.15	0.83
Construction (15–17, 32, 52)	5	28	53	7.36	7.06	0.19	1.57	19.16	0.35	1.20	61.22	0.85
Capital goods (34, 35, 38)	39	120	191	5.31	6.74	0.18	1.45	21.10	0.39	1.08	85.90	0.92
Transportation (40–42, 44, 45, 47)	78	85	46	5.15	4.97	0.10	1.27	13.52	0.21	1.19	49.04	0.78
Utilities (46, 48, 49)	24	102	176	3.73	6.10	0.15	1.04	19.40	0.35	0.75	46.34	0.76
Textiles & trade (22, 23, 31, 51, 53, 56, 59)	46	101	119	5.63	7.84	0.22	1.66	30.49	0.58	0.95	48.73	0.78
Services (72, 73, 75, 80, 82, 89)	3	4	57	4.21	4.18	0.08	1.65	12.97	0.20	0.80	12.82	0.19
Leisure (27, 58, 70, 78, 79)	12	31	59	7.35	6.95	0.18	1.85	23.03	0.44	1.22	49.82	0.78
CRSP value-weighted	NA	NA	NA	4.92	7.06	0.19	1.37	23.73	0.45	1.00	—	—

^{a} The spliced consumption data are scaled to adjust for the summation bias problem. Real growth in per capita consumption is multiplied by 0.75 for observations between 1939Q2 and 1959Q1, and by 0.9375 otherwise.

While methods which handle data on individual securities rather than aggregate portfolios could be dveloped, this route was not followed.[12] The dimensionality of the parameter space is enormous when analyzing a large cross-section of securities, and conventional methods for statistical inference may become unreliable. Also, a grouping procedure by industry decreases the number of statistics to be reported—probably without a disastrous loss of information.

Several types of assets should be represented, for Stambaugh (1982) finds the statistical results are not robust to the assets under study. Short-term Treasury bills, long-term government bonds, and high-grade long-term corporate bonds are included using the data in Ibbotson and Sinquefield (1982). In addition, the recent work by Chen, Roll, and Ross (1986) suggests that the difference in returns is between low-grade long-term corporate bonds (or "junk" bonds) and long-term government bonds is useful in explaining expected returns, so these returns are included as well.[13] To capture the spread between junk bonds and government bonds, a return is calculated on a portfolio which buys junk bonds by shorting long-term government bonds and then invests in short-term T-bills.[14] This portfolio's return is referred to as the "junk bond premium." Returns on junk bonds relative to government bonds primarily reflect changes in investors' perceptions concerning the probability of default. This is related to their perceptions of current and future economic conditions, which should be related to consumption growth. In fact, our statistical analysis shows a strong relation between junk bond returns and real consumption growth.

Returns are expressed in real terms and on a simple basis without continuously compounding. Returns are deflated by the Consumer Price Index, as reported by the U.S. Bureau of Labor Statistics on a monthly basis for the entire sample period. For purposes of testing the CAPM, the CRSP value-weighted index is used as the proxy for the market portfolio.[15]

Table II reports estimated betas for various assets. (The construction of the MCP is described below.) The table reveals that different measures of risk are highly correlated. In fact, the correlation between the market betas and the consumption betas (or the MCP betas) is 0.96 (or 0.94). Of course, while the risk measures are highly correlated, the rankings of the risk measures for the various assets are not exactly the same.

As discussed by Breeden (1980), industries' consumption betas should be related to price and income elasticities of demand and to supply elasticities. Goods with high income elasticities of demand should have high consumption

---

[12] Using different econometric methods, Mankiw and Shapiro (1985) have analyzed a version of the CCAPM using individual securities. However, they only rely on quarterly consumption data from 1959 to 1982.

[13] We are grateful to Roger Ibbotson, who made these data available to us. Since Ibbotson's data ended in 1978, the data are extended through 1982 using the monthly return on a mutual fund which is managed by Vanguard. This portfolio, the High Yield Bond Fund, is based on an investment strategy very similar to the one used by Ibbotson in constructing his return series.

[14] The investment in short-term T-bills is convenient but not necessary. The asset pricing models are specified assuming that the assets are held with some net capital invested.

[15] See Roll (1977) for a discussion of the potential consequences of selecting a proxy for the true market portfolio. The reader should keep in mind that one usual form of the consumption-based theory includes the market portfolio as part of the statement. Nevertheless, the theoretical results hold when any security replaces the market portfolio (Breeden (1979)).

betas, ceteris paribus. This appears to be borne out in the data, for consumer durables, construction, and recreation and leisure all have high consumption betas. While the services portfolio may have a high income elasticity, it does not have a high consumption beta. However, the number of firms in that portfolio is quite low (<5) for the first thirty years, and the R-squared is also low (0.08). Goods with lower income elasticities of demand, such as utilities, petroleum, food and agriculture, and transportation, have the lowest consumption betas of the stock portfolios.

Section II. C discusses the usefulness of a maximum correlation portfolio. Equation (20) suggests a way to calculate the weights in an MCP. Table III reports the results of running a regression of consumption growth on the returns from the twelve industry portfolios, four bond portfolios, and the CRSP value-weighted index for the period 1929–1982. Consumption growth is adjusted so that the summation bias in the estimated covariances between consumption and the returns on assets is removed. Table III gives the coefficients after they are rescaled so that they sum to one hundred percent, for the MCP in Section IV does not use the riskless asset.

The composition of the MCP in Table III helps to explain why Chen, Roll, and Ross (1986) found such an unimportant role for aggregate consumption. The MCP gives large absolute weights to long-term government bonds (−31%), the

Table III

Estimated Weights for the Maximum-Correlation Portfolio for Consumption Based on Spliced Quarterly Data from 1929–1982

All data are in real terms. (Consumption growth is scaled to adjust for the summation bias). The coefficient of determination for the above regression is 0.25, and the F-statistic for testing the joint significance of all the coefficients is 3.93 with a p-value of 0.0001. Before running real consumption growth on the returns, the data are mean adjusted. Then consumption growth is multiplied by two for observations between 1939Q2 and 1959Q1, and by 1.2 otherwise.

Asset	Weight	t-Statistic
U.S. Treasury bills	0.01	0.02
Long-term government bonds	0.54	1.05
Long-term corporate bonds	−0.31	−0.64
Junk bond premium	0.59	2.71
Petroleum	0.27	1.13
Banking, finance and real estate	−0.17	0.38
Consumer durables	0.10	0.44
Basic industries	0.33	0.90
Agriculture, food, and tobacco	−0.35	−1.45
Construction	−0.11	−0.80
Capital goods	0.03	0.11
Transportation	−0.29	−2.25
Utilities	0.18	0.72
Textiles, retail stores, and wholesalers	0.49	2.69
Services	0.08	1.39
Recreation and leisure	0.13	1.17
CRSP value-weighted index	−0.51	−0.38
	1.00	

junk bond premium (59%), and the CRSP index (−51%). Since these three variables were included as factors in Chen, Roll, and Ross (1986), aggregate consumption may be dominated as an additional factor given multicollinearity and measurement error.

The weights reported in Table III seem extreme, for the MCP involves large short positions in assets. Placing restrictions on the estimated weights would eliminate the extreme positions but could sacrifice some consistency with the underlying theory. The collinearity among the assets makes it difficult to estimate any single weight with precision, but the fitted value from the regression may be useful for our purposes. To see how the MCP tracks consumption growth, the following regression is run using spliced quarterly data from 1929 to 1982 (again, consumption has been scaled so that the reported beta is free of the summation bias):[16]

$$R_{MCP,t} = 0.00828 + 2.90 c'_t + \hat{u}_{MCP,t}, \ R^2 = 0.33, \quad (25)$$
$$(2.62) \quad\quad (10.19)$$

where t-statistics are given in parentheses. Since the MCP places no funds in the minimum-variance zero-beta portfolio, it need not have a unit beta. Even though the correlation between the MCP and consumption growth is 0.57, the theory of Section II.C still predicts that the MCP should be mean-variance efficient relative to the assets that it contains. Furthermore, the estimated risk measures when using actual consumption growth versus the MCP give similar rankings, and the sample correlation between the two sets of betas in Table II is 0.98.

Unlike the CRSP value-weighted index, the MCP has fixed weights since the entire sample period is used in the estimation reported in Table III. Constant weights are appropriate for the empirical work focuses on unconditional moments.[17] Moreover, estimating the weights by subperiods is not practical since quarterly data limit the number of available observations.

To better understand the MCP, Table IV compares it with the CRSP value-weighted index, a portfolio that has been studied extensively. According to Table III, the CRSP index has a negative weight in the MCP (−51%), yet the two portfolios are positively correlated. For the overall period, the correlation is 0.67. Furthermore, the MCP has roughly half the mean and standard deviation as the proxy for the market. Risk aversion combined with the CAPM predicts that the

---

[16] For observations between 1939Q2 and 1959Q1, $c'_t = 0.75(c_t - \bar{c})$. Otherwise, $c'_t = 0.9375\,(c_t - \bar{c})$, where $\bar{c}$ is the sample mean of c_t for the entire time period. By reducing the sizes of the consumption growth deviations, the slope coefficient is scaled up so as to be consistent (at least with regard to the summation bias).

[17] Even if second moments change conditional on predetermined information, working with a constant weight MCP is still appropriate for investigations involving unconditional moments. However, certain forms of heteroscedasticity may pose a problem for our statistical inference even in large samples. There is evidence of heteroscedasticity. We divided the overall period into four subperiods (1929Q2–1939Q1, 1939Q2–1947Q1, 1947Q2–1959Q1, and 1959Q2–1982Q4) and examined the constancy of the covariance matrix across all four periods. The covariance matrix is 18 × 18 involving the returns on seventeen assets in Table III and consumption growth. Using a likelihood-ratio test and an asymptotic approximation involving the F-distribution (Box (1949)), the F-statistic is 3.378 with degrees of freedom of 513 and 43165.7. The p-value is less than 0.001.

Table IV
Descriptive Statistics on Real Returns from Treasury Bills, the CRSP Value-Weighted Index, and the Maximum-Correlation Portfolio (MCP) for Consumption Based on Monthly Data, 1926–1982

The sample means are annualized by multiplying by 12. The sample standard deviations are annualized by multiplying by $\sqrt{12}$. (Since returns on T-bills are serially correlated, the annualized standard deviation is not the approximate standard deviation for annual holding periods.) Correlations between the CRSP return and the MCP return for the four periods are 0.67, 0.75, 0.59, and 0.41, respectively. The maximum-correlation portfolio (MCP) is constructed from the seventeen assets given in Table III. The weights of the MCP are determined by maximizing the sample correlation between the return on the portfolio and the growth rate of real consumption; see Table III for more details.

Date	Number of Observations	Mean of T-bills	t-Statistic for Mean of T-bills	Standard Deviation
1926–1982	684	0.0013	0.48	0.0204
1926–1945	240	0.0100	1.77	0.0253
1946–1965	240	−0.0082	−1.74	0.0211
1966–1982	204	0.0023	0.89	0.0106

Date	Number of Observations	Mean of CRSP Return	t-Statistic for Mean of CRSP Return	Standard Deviation
1926–1982	684	0.0767	2.88	0.2013
1926–1945	240	0.1002	1.61	0.2782
1946–1965	240	0.1039	3.70	0.1257
1966–1982	204	0.0172	0.44	0.1615

Date	Number of Observations	Mean of MCP Return	t-Statistic for Mean of MCP Return	Standard Deviation
1926–1982	684	0.0370	2.83	0.0987
1926–1945	240	0.0598	1.98	0.1351
1946–1965	240	0.0382	2.62	0.0651
1966–1982	204	0.0086	0.46	0.0786

mean of the market is positive, and the CCAPM makes the same prediction about the mean of the MCP. The point estimates for both portfolios are consistent with these predictions. However, when the standard deviation of the return is large in 1926–1945, the mean of the MCP is marginally significant while the market proxy is not.

IV. Testing the CCAPM and the CAPM

The usefulness of the risk measures in predicting expected returns is examined in this section. Two issues are studied. First, does expected return increase as the risk increases? Second, is the relation between risk and return linear? These two issues are synonymous with the question of mean-variance efficiency for a given portfolio. In addition, estimates of the expected real return on the zero-beta portfolio will be compared with the real return on a nominally riskless bill.

The empirical implications of the CCAPM in terms of aggregate consumption are examined first. Then the empirical results are extended by testing the mean-variance efficiency of the maximum-correlation portfolio. Finally, the CAPM is studied by testing the ex ante efficiency of the CRSP index.

Since the relevant econometric methodology is detailed by Gibbons (1982), only a brief development is provided here. In the case of the CCAPM, a regression similar to the market model is assumed to be a well-specified statistical model. That is, the joint distribution between the return on an asset and real growth in per capita consumption, $\tilde{c}_t$, is such that the disturbance term in the following regression has mean zero and is uncorrelated with $\tilde{c}_t$. Such an assumption justifies the following regression model:

$$\tilde{R}_{it} = \alpha_{ci} + \beta_{ci}\tilde{c}'_t + \tilde{u}_{it}, \quad \forall\ i = 1, \cdots, N, \quad t = 1, \cdots, T. \tag{26}$$

Further, it is assumed that

$$\mathscr{E}(\tilde{u}_{is}\tilde{u}_{jt}) = \begin{cases} \sigma_{ij} & \forall\ s = t, \\ 0 & \text{otherwise.} \end{cases} \tag{27}$$

Since $\tilde{c}'_t$ has already been mean-adjusted, μ_i is equal to α_{ci}.[18] Also, $\tilde{c}'_t$ has been scaled so that the summation bias is avoided.

Using the CCAPM as modified in Section II.D to account for sampling error in consumption provides

$$\mu_i = \gamma_0 + \gamma_1\beta_{ci}. \tag{28}$$

The theoretical relation in (28) imposes a parameter restriction on (26) of the form:

$$\alpha_{ci} = \gamma_0 + \gamma_1\beta_{ci}. \tag{29}$$

Pooling the time-series regressions in (26) across all N assets and then imposing the parameter restriction given in (29) provides a framework in which to estimate the expected return on the zero-beta portfolio, γ_0, and the market price of beta risk, γ_1. In addition, the parameter restriction may be tested.

There are various econometric methods for estimating the above model. Many of these are asymptotically (as T approaches infinity) equivalent to a full maximum-likelihood procedure. In the past, these alternatives have been selected because of computational considerations. However, results by Kandel (1984) and extended by Shanken (1985) make full maximum likelihood easy to implement.[19]

---

[18] Consumption growth is adjusted by its sample mean, not the unknown population mean. Our statistical inference that follows is conditional on the sample mean equal to its population counterpart. We overstate the significance of our tests as a result.

[19] Shanken (1982) shows that the full maximum-likelihood estimator may have desirable properties as the number of assets, N, used in estimating the model becomes large.

Empirical Tests of the CCAPM

Shanken establishes that the full maximum-likelihood estimators for γ_0 and γ_1 can be found by minimizing the following function:

$$L(\gamma_0, \gamma_1) = (1/(1 + (\gamma_1^2/s_c^2)))e'(\gamma)\hat{\Sigma}^{-1}e(\gamma), \qquad (30)$$

where

$e(\gamma) \equiv \bar{R} - \gamma_0 \mathbf{1}_N - \gamma_1 \hat{\beta}_c,$
$\hat{\Sigma} \equiv T^{-1} \sum_{t=1}^{T} \hat{u}_t \hat{u}_t',$
$s_c^2 \equiv T^{-1} \sum_{t=1}^{T} c_t'^2;$
$\hat{\beta}_c \equiv N \times 1$ vector with typical element $\hat{\beta}_{ci}$, where $\hat{\beta}_{ci}$ is the usual unrestricted ordinary least-squares estimator of β_{ci} in (26),
$\hat{u}_t \equiv N \times 1$ vector with typical element $\hat{u}_{it}$, where $\hat{u}_{it}$ is the residual in (26) when ordinary least squares is performed, and
$\bar{R} \equiv N \times 1$ vector with typical element $\bar{R}_i$, the sample mean of the return on asset i.

The concentrated-likelihood function is proportional to equation (30),[20] in which γ_0 and γ_1 are the only unknowns.

The first-order conditions for minimizing (30) involve a quadratic equation. The concentrated-likelihood function for the overall time period is graphed in Figure 1, a and b. Figure 1b suggests that, in the neighborhood of the maximum-likelihood estimates, γ_0 is estimated more precisely than γ_1 and there is negative correlation between the two estimates. Figure 1a has a coarser grid than Figure 1b. Figure 1a suggests that higher values for $\hat{\gamma}_1$ will not dramatically affect the maximized value of the likelihood function, but lower values for $\hat{\gamma}_1$ will have an impact.

Table V provides the point estimates of γ_0 and γ_1 along with the asymptotic standard errors. The subperiods in Table V correspond to the points where the data are spliced (see Section II. A). Unlike many studies on asset pricing models, the estimates of the expected return on the zero-beta asset are quite small. With the exception of the first subperiod, the point estimates are less than or equal to fifteen basis points (annualized), and in many cases the rate is negative (but only significant and negative in the second subperiod). This suggests that one implication of a riskless asset version of the CCAPM is consistent with the data. Table IV provides some information about the ex post real return on short-term Treasury bills during this time period, and in all cases the estimate of γ_0 is smaller than the sample mean in Table IV. Another implication of the CCAPM is that the market price of risk should be positive, for the expected return increases as the risk increases. This implication is verified for all periods, and

---

[20] Shanken derives this result by first maximizing the likelihood function with respect to β_{ci} and σ_{ij}. These estimators depend on γ_0 and γ_1. Shanken then substitutes the estimators for β_{ci} and σ_{ij} back into the original likelihood function, which then depends on only γ_0 and γ_1. This new function is the concentrated-likelihood function. After some algebra he discovers that maximizing the concentrated-likelihood function is equivalent to minimizing (30) above. Note that full maximum likelihood refers to maximizing the likelihood function with respect to σ_{ij} as well as γ_0, γ_1, and β_{ci}.

Figure 1a. Likelihood surface for the CCAPM based on quarterly consumption data. 1929-82

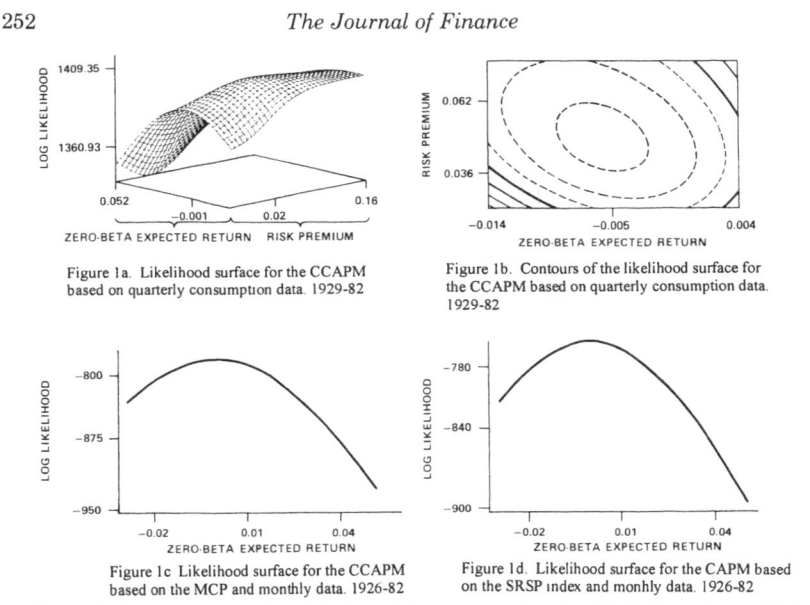

Figure 1b. Contours of the likelihood surface for the CCAPM based on quarterly consumption data. 1929-82

Figure 1c. Likelihood surface for the CCAPM based on the MCP and monthly data. 1926-82

Figure 1d. Likelihood surface for the CAPM based on the SRSP index and monhly data. 1926-82

Figure 1. The concentrated log likelihood functions, l, for the CCAPM and the CAPM. The relevant parameters of these functions are the expected annualized return on the "zero-beta" portfolio, γ_0, and the expected annualized premium for consumption-beta risk, γ_1.

the point estimate is statistically significant in most rows of Table V.[21] While the magnitude of the estimate of γ_1 seems large, Section II.D shows that γ_1 is biased upwards relative to γ_1^* by the variance of the sampling error in reported consumption. Reflecting the large standard errors in some subperiods, the variation in $\hat{\gamma}_1$ across subperiods is striking. Since γ_1 does reflect the variance in measurement error for consumption, the high value of $\hat{\gamma}_1$ in the earlier subperiods (except 1929–1939) may be the result.

The CCAPM also implies that the relation between expected returns and betas is linear, or the null hypothesis is the equality given in (29). This null hypothesis is tested against a vague alternative that the equality does not hold.

Gibbons (1982) suggests a likelihood ratio for testing hypotheses like (29). Such an approach relies on an asymptotic distribution as T becomes large. However, the methodology may have undesirable small sample properties, espe-

---

[21] In addition to the full maximum-likelihood procedure, the two step GLS estimator suggested in Gibbons (1982) was used. This estimator is not as desirable as the full maximum-likelihood approach as the number of securities approaches infinity, and it should be downward biased due to a phenomenon similar to errors-in-variables for simple regressions. Since consumption betas are measured less precisely than market betas, the difference between the full maximum likelihood and two-step GLS should be larger in this application than in past tests of the CAPM. In fact, the GLS estimate of γ_1 is usually half the value reported in Table V. On the other hand, the simulations by Amsler and Schmidt (1985) suggest that the finite sample behavior of the GLS estimator is better than the maximum-likelihood alternative. Since the sign of the estimate from either approach is the same across any row in Table V and since the significance from zero is the same across any row (except for one subperiod), the GLS results are not reported, but they are available on request to the authors.

Empirical Tests of the CCAPM

Table V
Estimating and Testing the CCAPM Using Aggregate Consumption Data

All data are annualized and in real terms ($\tilde{R}_{it}$), and consumption growth ($\tilde{c}'_t$) is adjusted to correct for the summation bias. The model is fit to seventeen assets (twelve industry portfolios, four boud portfolios, and the CRSP value-weighted index). The econometric model is

$$\tilde{R}_{it} = \alpha_{ci} + \beta_{ci}\tilde{c}'_t + \tilde{u}_{it},$$
$$\mathscr{E}(\tilde{u}_t\tilde{u}'_s) = \Sigma \text{ if } t = s, 0 \text{ otherwise.}$$
$$H_0: \alpha_{ci} = \gamma_0 + \gamma_1\beta_{ci}, \forall i = 1,\cdots,17.$$

The data are annualized by multiplying the quarterly returns by 4 and monthly returns by 12. $F(\beta_{ci} = \beta_{cj})$ is the F-statistic for testing the hypothesis that $\beta_{ci} = \beta_{cj}$ $\forall$ $i \neq j$, while $F(\beta_{ci} = 0)$ tests the hypothesis that $\beta_{ci} = 0, \forall i = 1,\cdots,17$. Both $\hat{\gamma}_0$ and $\hat{\gamma}_1$ are estimates from a full maximum-likelihood procedure, and their respective standard errors (given in parentheses below the estimates) are based on the inverse of the relevant information matrix. The likelihood ratio (LR) provides a test of the null hypothesis that expected returns are linear in consumption betas as implied by the CCAPM. The likelihood ratio is adjusted by Bartlett's (1938) correction. In all cases, p-value is the probability of seeing a higher statistic than the one reported under the null hypothesis. If the test statistics are independent across subperiods, then the last four rows can be aggregated into one summary measure. In the case of the likelihood-ratio test, the overall results yield a χ^2_{60} random variable with a realization equal to 69.06. (This yields a p-value equal to 0.198.) For the F-statistic, the overall results yield a standardized normal random variable with a realization equal to 0.68, which implies 0.25 as a p-value.

Date	Number of Observations	F-test: Betas Equal (p-Value)	F-test: Betas = Zero (p-Value)	$\hat{\gamma}_0$ (SE($\hat{\gamma}_0$))	$\hat{\gamma}_1$ (SE($\hat{\gamma}_1$))	LR Test of H_0 (p-Value)
Panel A: Spliced Quarterly Consumption Data, Adjusted for Summation Bias, 1929-1982							
1929Q2–1982Q4	215	3.874 (<0.001)	3.912 (<0.001)	−0.0061 (0.0044)	0.0478 (0.0133)	28.03 (0.021)
1929Q2–1939Q1	40	4.319 (0.001)	4.241 (0.001)	0.0484 (0.0091)	0.0329 (0.0189)	26.45 (0.034)
1939Q2–1947Q1	32	0.502 (0.908)	1.410 (0.261)	−0.2558 (0.0859)	0.5850 (0.2507)	6.84 (0.962)
1947Q2–1959Q1	48	1.006 (0.476)	1.182 (0.334)	−0.0699 (0.0469)	0.2928 (0.1865)	14.82 (0.464)
1959Q2–1982Q4	95	2.257 (0.009)	2.277 (0.008)	0.0015 (0.0028)	0.0187 (0.0062)	20.95 (0.138)
Panel B: Unspliced Quarterly Consumption Data, Adjusted for Summation Bias, 1947-1982							
1947Q2–1982Q4	144	1.342 (0.182)	1.695 (0.052)	−0.0325 (0.0256)	0.2136 (0.1430)	19.87 (0.177)
1959Q2–1982Q4	96	1.398 (0.165)	1.450 (0.137)	−0.0007 (0.0040)	0.0528 (0.0179)	16.29 (0.363)
Panel C: Unspliced Monthly Consumption Data, Adjusted for Summation Bias, 1959-1982							
1959 Feb–1982 Dec	287	1.316 (0.186)	1.581 (0.069)	−0.0008 (0.0034)	0.0804 (0.0263)	10.47 (0.789)

cially when the number of assets is large (Stambaugh (1982)). The simulation by Amsler and Schmidt (1985) indicates that Barlett's (1938) correction, which was suggested by Jobson and Korkie (1982), improves the small sample performance of the likelihood ratio even when the number of assets is large. This correction is applied in all of the following tables.[22]

For the overall period using spliced data, the linear equality between reward and risk implied by the CCAPM is rejected at the traditional levels of significance. The last column of Table V reports the statistic in Panel A. This rejection is confirmed by Shanken's (1985, 1986) lower bound statistic, which suggests that the inference is robust to the asymptotic approximation of the likelihood ratio.[23] However, as noted at the bottom of Table V, aggregation of the results for each subperiod fails to reject the CCAPM at traditional levels of significance. The subperiod of 1929–1939 is the most damaging to the model. Given the nature of the consumption data used for this time period (see Section II.A), such behavior is troubling, for the rejection of the CCAPM may be due to measurement problems. On the other hand, the F-statistics given in the third and fourth columns of Table V suggest another interpretation. These F-statistics examine the joint significance of the risk measures across the assets as well as the significance of the dispersion of the risk measures across the assets. If all the risk measures were equal, then tests of (29) would lack power, and γ_0 and γ_1 would not be identified. In the first subperiod the risk measures are estimated with the most precision, and as a result tests against the null are more powerful.

Panel A of Table V is based on spliced quarterly consumption data. That is, monthly income predicts consumption for 1929–1939, and monthly consumption forms the basis of the quarterly numbers from 1959 to 1982. The above statistics are also calculated using the unspliced quarterly data from 1947 to 1982 with quarterly sampling intervals and on unspliced monthly data from 1959 to 1982 with monthly sampling intervals. The spliced data are considered first because the time series is longer and the measurement problems are less severe for quarterly observations than for monthly.

However, the results based on the spliced data are the least favorable to the CCAPM. Panels B and C in Table V suggest that the linearity hypothesis is *never* rejected with the unspliced monthly numbers and the unspliced quarterly numbers. Shanken's upper bound test confirms this result except for the subperiod 1947Q2-1982Q4. Also, the market price of consumption beta risk is also higher (with one exception) for the unspliced results.[24]

---

[22] The Lagrange multiplier test (suggested by Stambaugh (1982)) and the CSR test (suggested by Shanken (1986)) were also computed for all time periods without dramatically different results and not reported here. Both tests are monotonic transforms of the likelihood ratio (Shanken (1985)). The choice of which statistic to report is somewhat arbitrary. Since the geometric interpretation of the likelihood ratio follows, this statistic is reported in the tables.

[23] For all results we confirmed the inferences with Shanken's (1985, 1986) tests which have upper or lower bounds based on *finite sample* distributions. If the null hypothesis is not rejected with the upper bound, then it would not be rejected using an exact distribution. Similarly, if one rejects with the lower bound, such a result holds with a finite sample distribution.

[24] These results are consistent with those of Wheatley (1986), who re-examined Hansen and Singleton's (1983) tests of the CCAPM. Using 1959–1981 data, Wheatley showed by simulation that measurement error in consumption biased their test statistics. After correcting for that bias, he was unable to reject the CCAPM.

Empirical Tests of the CCAPM

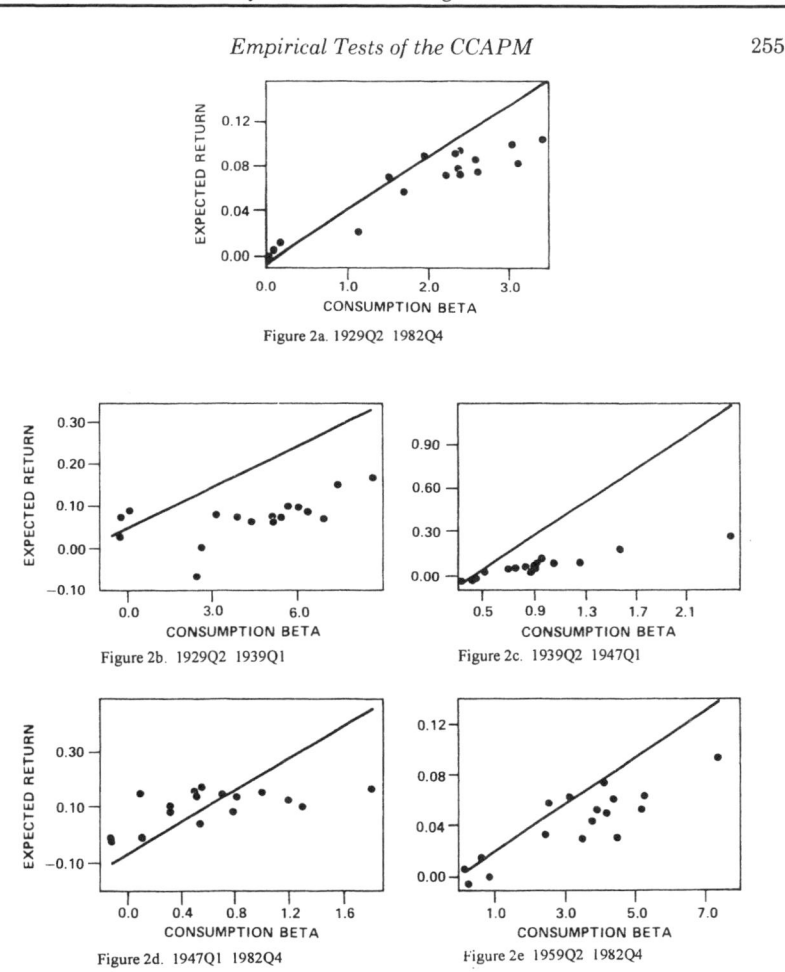

Figure 2a. 1929Q2 1982Q4

Figure 2b. 1929Q2 1939Q1

Figure 2c. 1939Q2 1947Q1

Figure 2d. 1947Q1 1982Q4

Figure 2e. 1959Q2 1982Q4

Figure 2. Scatter plots of parameter estimates with and without CCAPM restriction. All data are annualized and in real terms, and consumption growth is adjusted to correct for summation bias. Seventeen assets (twelve industry portfolios, four bond portfolios, and the CRSP value-weighted index) are used. The intercept and slope of the solid straight line in each plot are determined by the maximum-likelihood estimates for the expected return on the "zero-beta" asset and premium for consumption-beta risk, respectively (*not* the ordinary least squares fit of the points). All points should fall on this line if the CCAPM is true. The seventeen points on each plot represent unrestricted estimates of expected return, $E(\tilde{R}_{it})$, and consumption beta, β_{ci}. (Note that the scale varies across the scatter plots.)

Except for the subperiod 1929–1939 (and its effect on the results for the overall time period), Table V provides positive support for the CCAPM. To provide a more intuitive interpretation of the empirical results, Figure 2 informally examines the deviations for the null hypothesis. Figure 2 plots the unrestricted mean returns against the unrestricted estimates of betas. The straight line represents the relation estimated by maximum likelihood. Despite the rejection of the theory by formal tests, the relation between expected returns and betas is reasonably linear[25]—perhaps more than could have been anticipated given the poor quality of the consumption data. In some of the plots (e.g., Figure 2d), a straight line fit to the points would be flat. Given the measurement error in the consumption betas, this flatness is expected. To better understand the "empirical validity" of the CCAPM, the efficiency of the maximum-correlation portfolio will now be considered.

Section II. C demonstrates that the MCP is ex ante mean-variance efficient under the CCAPM. This result is derived when the covariance matrix among returns on securities and consumption growth is known, which is not the case here. Thus, all the statistical inference concerning the ex ante efficiency of the MCP is conditional on the portfolio being the desired theoretical construct. Estimation error in the portfolio weights is ignored.

Following Gibbons (1982), consider testing the efficiency of any portfolio p when the riskless asset is not observed. Assume that the following regression is well specified in the sense that the error term has a zero mean and is uncorrelated with $\tilde{R}_{pt}$:

$$\tilde{R}_{it} = \alpha_{pi} + \beta_{pi}\tilde{R}_{pt} + \tilde{u}_{it}. \tag{31}$$

If portfolio p is efficient, then the following parameter restriction holds:

$$\alpha_{pi} = \gamma(1 - \beta_{pi}), \tag{32}$$

where γ is the expected return on the portfolio which is uncorrelated with p. Similar to the econometric model of (26) and (29) above, (31) and (32) are combined and then estimated by a full maximum-likelihood procedure. Furthermore, when (32) is treated as a null hypothesis, both a likelihood ratio and an asymptotic F are calculated. In the tests that follow, the maximum-correlation portfolio or the CRSP index is used as portfolio p.[26]

Figure 1, c and d, graphs the concentrated-likelihood function relative to possible estimates of the expected return on the zero-beta portfolio in the case of the MCP and CRSP index, respectively. Table VI summarizes the statistical results for both portfolios as well. Like Table V, the third column of Table VI indicates a small expected return on the zero-beta asset. Further, the point estimate when using the MCP never exceeds that when using the CRSP index. Also, the overall period rejects the efficiency of either the maximum-correlation portfolio or the CRSP index, as indicated by the last column of the table. (This rejection would occur even without relying on asymptotic theory to approximate the sampling distribution, for the lower bound test also rejects.) Unlike Table V, the rejection of the model does not stem from just the first subperiod. These

---

[25] Like beauty, perceived linearity is in the eyes of the beholder. One reviewer of this paper thought the graphs in Figure 2 revealed remarkable nonlinearities.

[26] Panels A and B of Table VI are based on sixteen assets, not seventeen as in Table V. The regressions using the CRSP index as the dependent variable have been excluded because otherwise the covariance matrix of the residuals would be singular.

Empirical Tests of the CCAPM

Table VI
Estimating and Testing the Mean-Variance Efficiency of the Maximum-Correlation Portfolio (MCP) and the CRSP Value-Weighted Index, 1926–1982

All returns ($\tilde{R}_{it}$) are annualized and in real terms. The model is fit to sixteen assets (twelve industry and four bond portfolios). The econometric model is

$$\tilde{R}_{it} = \alpha_{pi} + \beta_{pi}\tilde{R}_{pt} + \tilde{u}_{it},$$
$$\mathscr{E}(\tilde{u}_t\tilde{u}'_s) = \Sigma \text{ if } t = s, 0 \text{ otherwise.}$$
$$H_0: \alpha_{pi} = \gamma(1 - \beta_{pi}), \forall\, i = 1, \cdots, 16.$$

$\tilde{R}_{pt}$ is either the return on MCP or a CRSP index. The maximum correlation portfolio (MCP) is constructed from the seventeen assets given in Table III. The weights of the MCP are determined by maximizing the sample correlation between the return on the portfolio and the growth rate of real consumption; see Table III for more details. The data are annualized by multiplying the monthly returns by 12. $\hat{\gamma}$ is an estimate from a full maximum-likelihood procedure, and the standard errors (given in parentheses below the estimates) are based on the inverse of the relevant information matrix. The likelihood ratio (LR) provides a test of the null hypothesis that a given portfolio is efficient. The ratio is adjusted by Bartlett's (1938) correction. The *p*-value is the probability of seeing a higher statistic than the one reported under the null hypothesis. If the tests are independent across subperiods, then the last three rows in each panel can be aggregated into one summary measure based on either the likelihood ratio or the *F*-test. These aggregate test statistics always have *p*-values less than 0.0001.

Date	Number of Observations	$\hat{\gamma}$ (SE($\hat{\gamma}$))	LR Test (*p*-Value)
Panel A: Mean-Variance Efficiency Tests on the MCP			
1926–1982	684	−0.0009 (0.0027)	26.86 (0.029)
1926–1945	240	0.0064 (0.0054)	49.21 (<0.001)
1946–1965	240	−0.0151 (0.0049)	40.96 (<0.001)
1966–1982	204	0.0016 (0.0024)	19.25 (0.203)
Panel B: Mean-Variance Efficiency Tests on CRSP Index			
1926–1982	684	0.0000 (0.0027)	26.77 (0.031)
1926–1945	240	0.0076 (0.0053)	49.98 (<0.001)
1946–1965	240	−0.0125 (0.0047)	36.62 (0.001)
1966–1982	204	0.0016 (0.0024)	19.23 (0.204)

stronger rejections are probably due to the increased number of observations, which provides more precision. The joint significance of the betas across assets, as well as the significance of the dispersion of the betas across assets, is unrelated in Table VI, but it is much higher than the comparable F-statistics reported in Table V. Unfortunately, the test of efficiency of the MCP assumes that the portfolio weights are estimated without error, which is obviously not the case. If this measurement error were taken into account, the p-value would increase (see Kandel and Stambaugh (1988)).

The likelihood ratio test in Table VI can be given a geometrical interpretation based on the position of either the MCP or the CRSP index relative to the ex post efficiency frontier (Kandel 1984)). The mean-variance frontier is a parabola. A line joining the points corresponding to any given frontier portfolio and the minimum-variance portfolio intersects the mean axis at a point corresponding to the expected return of all portfolios having a zero beta relative to the frontier portfolio. When graphed with the variance on the horizontal axis, the slope of this line is equal to half the slope of the tangent at the point corresponding to the frontier portfolio (Gonzales-Gaviria (1973), pp. 58–61).

Building on this geometric relation, Figure 3 presents a graphical interpretation of the test statistic based on the ex post frontier. The maximum-likelihood estimates of the expected return on a portfolio having a zero beta relative to a test portfolio p (either MCP or CRSP in Figure 3) is denoted as γ. A line joining the mean axis at $\hat{\gamma}$ and the ex post minimum variance portfolio intersects the ex post frontier at a point (A or B in Figure 3) corresponding to the frontier portfolio having ex post zero-beta portfolios whose mean returns are equal to $\hat{\gamma}$. Let x equal the slope of this line. This portfolio would be the test portfolio, p, if and only if the test portfolio were ex post mean-variance efficient. Now consider a line joining the point corresponding to the test portfolio, p, and $\hat{\gamma}$. Denote the slope of this line by y. The LRT is equal to $T \ln(x/y)$ and is directly testing whether the slope of the second line is significantly less than the slope of the first line. A significantly lower slope for the second line implies rejection of the null hypothesis that the test portfolio is ex ante mean-variance efficient. The results of Table IV suggest that the two lines in either Figure 3a or 3b do have statistically different slopes.

Figure 3 also provides a comparison of the inefficiency of the MCP versus the CRSP index. For example, Figure 3a provides the unconstrained ex post frontier as well as a parabola which represents the maximum-likelihood estimate of the frontier assuming that the MCP is efficient. Figure 3b provides similar information in the case of the CRSP index. The scales of Figure 3a and 3b are equal, and there is little difference between the frontier constrained so that MCP is efficient versus a case where the CRSP index is efficient. Figure 3c, which has a very fine grid, is provided to see the difference between the two constrained frontiers.

Based on Table VI and Figure 3, the relative merits of the CCAPM versus the CAPM are difficult to discern. The inefficiency of either the MCP or the CRSP index is about the same. The two models are hard to compare because they are inherently non-nested hypotheses, which makes formal inference difficult. How-

Empirical Tests of the CCAPM

Figure 3a Unconstrained ex post efficient frontier (U) versus frontier constrained so that MCP is efficient

Figure 3b Unconstrained ex post efficient frontier (U) versus frontier constrained so that CRSP Index is efficient

Figure 3c Unconstrained ex post efficient frontier (U) versus frontier constrained so that CRSP Index is efficient (CRSP CON) versus frontier constrained so that MCP is efficient (MCP CON) Note the grid is very fine

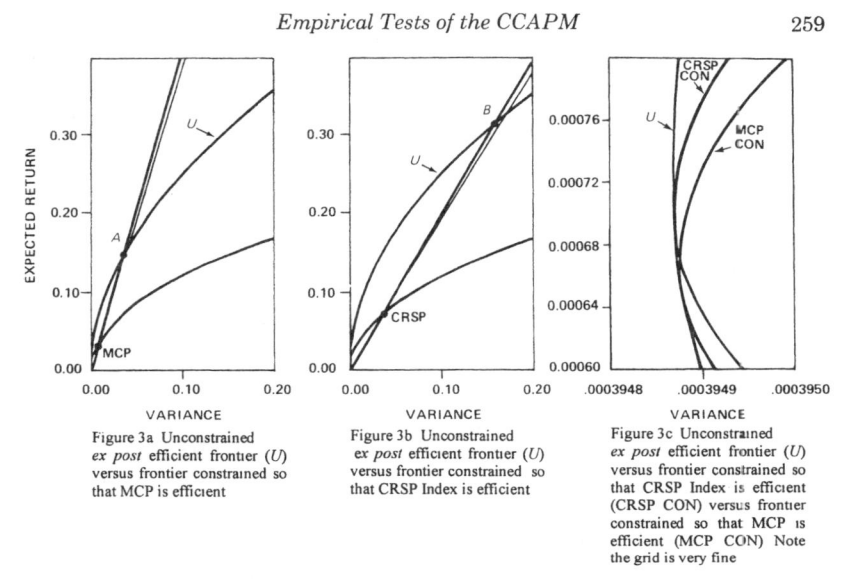

Figure 3. A geometrical interpretation of the likelihood-ratio test, LRT, of ex ante efficiency for the MCP and CRSP value-weighted index based on monthly real returns, 1926–1982. The sample means and variances are annualized by multiplying by twelve. The LRT equals $T \ln(x/y)$. x is the slope of the straight line that passes through the maximum-likelihood estimate of the expected return on the zero-beta portfolio, $\hat{\gamma}$, and the global minimum variance point of the ex post frontier. y is the slope of the straight line that passes through $\hat{\gamma}$ and the test portfolio (either the MCP or CRSP index). The ex post frontier is based on sixteen assets (twelve industry portfolios and four bond portfolios) and either the MCP or CRSP index.

ever, the apparent inefficiency of the MCP is overstated since the portfolio weights are estimated with error.

V. Conclusion

This paper tests the consumption-oriented CAPM and compares the model with the market-oriented CAPM. Two econometric problems peculiar to consumption data are analyzed. First, real consumption reported for a quarter is an integral of instantaneous consumption rates during the quarter, rather than the consumption rate on the last day of the quarter. This "summation bias" lowers the variance of measured consumption growth and creates positive autocorrelation, even when the true consumption rate has no autocorrelation. This summation bias also underestimates the covariance between measured consumption and asset returns by half the true values, with the result that measured consumption betas are ¾ of their true values. The empirical work accounts of these problems.

A second major econometric problem is the paucity of data points for consumption growth rates. Some tests use the consumption data (adjusted for the summation bias). However, alternative tests are based on the returns of the portfolio of assets (the "MCP") that is most highly correlated with the growth rate of real consumption. The CCAPM implies that expected returns should be linearly related to betas calculated with respect to the MCP. Interestingly, the MCP has a correlation of 0.67 with the CRSP value-weighted index. Apart from stocks, a major component of the MCP is the return on a "junk bond" portfolio. Thus, the correlation between average returns and the sensitivity of returns on various assets to junk bond returns, which has been discussed by Chen, Roll, and Ross (1983), may be attributed to the correlation between junk bond returns and real growth in consumption.

A number of tests of the consumption-oriented CAPM are examined. Unlike past studies on asset pricing, the estimated return on the zero-beta asset is quite small. Except for one subperiod, all the estimates are less than or equal to fifteen basis points (annualized). This suggests that some of the implications of a riskless real asset version of the CCAPM are consistent with the data. Another implication of the CCAPM is that the market price of risk should be positive; in other words, the expected return increases as the risk increases. This implication is verified for all periods, and the point estimate is statistically significant in most of the subperiods.

Based on the quarterly consumption data for the overall period, the linear equality between reward and risk implied by the CCAPM is rejected at the 0.05 level. However, a plot suggests that the relation is reasonably linear given the poor quality of the consumption data. Analysis by subperiods reveals that the time period from 1929 through 1939 seems to be the most damaging to the model. In fact, when the model is estimated by subperiods and then the results are aggregated across subperiods, no rejection occurs at the usual levels of significance. The first subperiod may be rejecting the model because the risk measures are estimated more precisely due to the large fluctuations in consumption and asset returns in the 1930s. The added precision should increase the power of tests. On the other hand, the quality of the data for this time period is particularly suspicious. While the CCAPM is by no means a perfect description of the data, we found the fit better than we anticipated.

For the overall period (1926–1982), the mean-variance efficiency is rejected for both the CRSP value-weighted index and the portfolio with maximum correlation with consumption (the MCP). This rejection occurs in a number of time periods, not just the 1929–1939 subperiod. Given that the estimated risk measures for both models are highly correlated, this similarity in the performances by the CAPM and the CCAPM is predictable. Since these tests permit the use of monthly, not quarterly, data, the rejection could be attributed to the increased power of the tests due to additional observations. On the other hand, the statistical significance of the rejection of the efficiency of the MCP is overstated since the portfolio weights are unknown and had to be estimated.

REFERENCES

Amsler, C. and P. Schmidt, 1985, A Monte Carlo investigation of the accuracy of multivariate CAPM tests, *Journal of Financial Economics* 14, 359–376.

Bartlett, M., 1938, Further aspects of the theory of multiple regression, *Proceedings of the Cambridge Philosophical Society* 34, 33–47.

———, 1946, On the theoretical specification and sampling properties of autocorrelated time series, *Supplement to the Journal of the Royal Statistical Society* 8, 27–41.

Beaver, W., R. Lambert, and D. Morse, 1980, The information content of security prices, *Journal of Accounting and Economics* 2, 3–28.

Black, F., 1972, Capital market equilibrium with restricted borrowing, *Journal of Business* 45, 444–454.

———, M. Jensen, and M. Scholes, 1972, The capital asset pricing model: Some empirical findings, in M. Jensen, ed.: *Studies in the Theory of Capital Markets* (Praeger, New York).

Box, G., 1949, A general distribution theory for a class of likelihood criteria, *Biometrika* 36, 317–346.

Breeden, D., 1979, An intertemporal asset pricing model with stochastic consumption and investment opportunities, *Journal of Financial Economics* 7, 265–296.

———, 1980, Consumption risk in futures markets, *Journal of Finance* 35, 503–520.

——— and R. Litzenberger, 1978, Prices of state-contingent claims implicit in option prices, *Journal of Business* 51, 621–651.

Chen, N., R. Roll, and S. Ross, 1986, Economic forces and the stock market, *Journal of Business* 59, 383–404.

Cornell, B., 1981, The consumption based asset pricing model: A note on potential tests and applications, *Journal of Financial Economics* 9, 103–108.

Dunn, K. and K. Singleton, 1986, Modeling the term structure of interest rates under nonseparable utility and durability of goods, *Journal of Financial Economics* 17, 27–56.

Fama, E., 1976, *Foundations of Finance* (Basic Books, New York).

——— and J. MacBeth, 1973, Risk, return and equilibrium: Empirical tests, *Journal of Political Economy* 81, 607–636.

——— and G. Schwert, 1977, Asset returns and inflation, *Journal of Financial Economics* 5, 115–146.

——— and G. Schwert, 1979, Inflation, interest, and relative prices, *Journal of Business* 52, 183–209.

Ferson, W., 1983, Expected real interest rates and consumption in efficient financial markets: Empirical tests, *Journal of Financial and Quantitative Analysis* 18, 477–498.

Gibbons, M., 1982, Multivariate tests of financial models: A new approach, *Journal of Financial Economics* 10, 3–27.

——— and W. Ferson, 1985, Testing asset pricing models with changing expectations and an unobservable market portfolio, *Journal of Financial Economics* 14, 217–236.

Gonzales-Gaviria, N., 1973, Inflation and capital asset market prices: Theory and tests, Ph.D. dissertation, Graduate School of Business, Stanford University.

Grossman, S. and R. Shiller, 1982, Consumption correlatedness and risk measurement in economies with non-traded assets and heterogeneous information, *Journal of Financial Economics* 10, 195–210.

———, A. Melino, and R. Shiller, 1987, Estimating the continuous-time consumption-based asset-pricing model, *Journal of Business and Economic Statistics* 5, 315–328.

Hall, R., 1978, Stochastic implications of the life cycle-permanent income hypothesis: Theory and evidence, *Journal of Political Economy* 86, 971–987.

Hansen, L. and K. Singleton, 1982, Generalized instrumental variables estimation of nonlinear rational expectations models, *Econometrica* 50, 1269–1286.

——— and K. Singleton, 1983, Stochastic consumption, risk aversion, and the temporary behavior of asset returns, *Journal of Political Economy* 91, 249–265.

Ibbotson, R. and R. Sinquefield, 1982, *Stocks, Bonds, Bills and Inflation: Updates* (1926–1982) (R. G. Ibbotson Associates Inc., Chicago, IL).

Jobson, J. and R. Korkie, 1982, Potential performance and tests of portfolio efficiency, *Journal of Financial Economics* 10, 433–466.

Kandel, S., 1984, The likelihood ratio test statistic of mean-variance efficiency without a riskless asset, *Journal of Financial Economics* 13, 575–592.

——— and R. Stambaugh, 1988, A mean-variance framework for tests of asset pricing models, Unpublished manuscript, Graduate School of Business, University of Chicago.

Lambert, R., 1978, The time aggregation of earnings series, Unpublished manuscript, Graduate School of Business, Stanford University.

Lintner, J., 1965, The valuation of risk assets and the selection of risky investments in stock portfolios and capital budgets, *Review of Economics and Statistics* 47, 13–37.

Litzenberger, R. and E. Ronn, 1986, A utility-based model of common stock returns, *Journal of Finance* 41, 67–92.

Lucas, R., 1978, Asset prices in an exchange economy, *Econometrica* 46, 1429–1445.

Mankiw, N. and M. Shapiro, 1985, Risk and return: Consumption beta versus market beta, Unpublished manuscript, Cowles Foundation, Yale University.

Marsh, T., 1981, Intertemporal capital asset pricing model and the term structure of interest rates, Ph.D. dissertation, University of Chicago.

Merton, R., 1973, An intertemporal capital asset pricing model, *Econometrica* 41, 867–887.

Roll, R., 1977, A critique of the asset pricing theory's test—Part 1: On past and potential testability of the theory, *Journal of Financial Economics* 4, 129–176.

Rubinstein, M., 1976, The valuation of uncertain income streams and the pricing of options, *Bell Journal of Economics and Management Science* 7, 407–425.

Scholes, M. and J. Williams, 1977, Estimating betas from nonsynchronous data, *Journal of Financial Economics* 5, 309–327.

Shanken, J., 1982, An asymptotic analysis of the traditional risk-return model, Unpublished manuscript, School of Business Administration, University of California, Berkeiey.

———, 1985, Multivariate tests of the zero-beta CAPM, *Journal of Financial Economics* 14, 327–348.

———, 1986, Testing portfolio efficiency when the zero-beta rate is unknown: A note, *Journal of Finance* 41, 269–276.

Sharpe, W., 1964, Capital asset prices: A theory of market equilibrium under conditions of risk, *Journal of Finance* 19, 425–442.

———, 1982, Factors in New York Stock Exchange security returns, 1931–1979, *Journal of Portfolio Management* 8, 5–19.

Stambaugh, R., 1982, On the exclusion of assets from tests of the two-parameter model: A sensitivity analysis, *Journal of Financial Economics* 10, 237–268.

Stulz, R., 1981, A model of international asset pricing, *Journal of Financial Economics* 9, 383–406.

Tiao, G., 1972, Asymptotic behavior of temporal aggregates of time series, *Biometrika* 59, 525–531.

Wheatley, S. 1986, Some tests of the consumption based asset pricing model, Unpublished manuscript, School of Business Administration, University of Washington, Seattle.

Working, H., 1960, A note of the correlation of first differences of averages in a random chain, *Econometrica* 28, 916–918.

[13]
Implications of Security Market Data for Models of Dynamic Economies

Lars Peter Hansen
University of Chicago, National Bureau of Economic Research, and National Opinion Research Center

Ravi Jagannathan
University of Minnesota and Federal Reserve Bank of Minneapolis

We show how to use security market data to restrict the admissible region for means and standard deviations of intertemporal marginal rates of substitution (IMRSs) of consumers. Our approach (i) is nonparametric and applies to a rich class of models of dynamic economies, (ii) characterizes the duality between the mean–standard deviation frontier for IMRSs and the familiar mean–standard deviation frontier for asset returns, and (iii) exploits the restriction that IMRSs are positive random variables. The region provides a convenient summary of the sense in which asset market data are anomalous from the vantage point of intertemporal asset pricing theory.

I. Introduction

In this paper we investigate the implications of asset market data for a rich class of models of dynamic economies. The models within this

---

Hansen gratefully acknowledges support from the National Science Foundation and Jagannathan from the Institute for Financial Studies, Carlson School of Management. Part of this research was completed while Hansen was a visiting scholar at Stanford University. Helpful comments were made by Darrell Duffie, Wayne Ferson, Larry Glosten, Allan Kleidon, Robert Litterman, Robert E. Lucas, Jr., Masao Ogaki, Scott Richard, José Scheinkman, Karl Snow, Alex Taber, Grace Tsiang, Larry Weiss, and especially John Cochrane and John Heaton. Karl Snow provided valuable assistance with the computations.

class differ with respect to the heterogeneity of consumers' preferences, the span of the payoffs on tradable securities, and the role of money in the acquisition of consumption goods. In spite of these differences, a common implication of these models is that the equilibrium price of a future payoff on any traded security can be represented as the expectation (conditioned on current information) of the product of the payoff and an appropriately interpreted intertemporal marginal rate of substitution (IMRS) of any consumer (see, e.g., LeRoy 1973; Rubinstein 1976; Lucas 1978; Breeden 1979; Harrison and Kreps 1979; Hansen and Richard 1987). This representation is a generalization of the familiar tenet from price theory that prices should equal marginal rates of substitution. To apply this principle to models of asset pricing, securities are viewed as claims to a numeraire good indexed by future states of the world.

If price data were available from a complete set of security markets, the IMRSs of all consumers could be inferred from Arrow-Debreu prices. However, economic agents may not trade in a complete set of contingent-claims markets. Furthermore, it may be practical for an econometrician to use data on only a small array of securities. Because of these limitations, asset market data alone are typically not sufficient to identify IMRSs.

One approach that has been used extensively is to identify IMRSs by restricting them to be parametric functions of data observed by an econometrician (see, e.g., Hansen and Singleton 1982; Brown and Gibbons 1985; Epstein and Zin, this issue). This approach imposes potentially stringent limits on the class of admissible asset pricing models and then tests whether the particular parameterizations are consistent with the observed asset market data.

While this parametric approach has yielded interesting insights into the empirical plausibility of particular families of models, the approach proposed in this paper goes to another extreme. We purposely enlarge the class of asset pricing models under investigation by imposing as little structure as possible on the admissible class of models. In doing so we eliminate most of the testable implications except possibly the law of one price (portfolios with the same payoffs have the same price) and the absence of arbitrage opportunities (nonnegative payoffs that are positive with positive probability have positive prices). Although we are not able to identify the IMRSs fully, we can extract information about them. When IMRSs are constant, portfolio payoffs with the same price must also have the same mean. Thus the existence of portfolios of securities with the same price but distinct expected payoffs implies that IMRSs must vary. We exploit this observation to derive greatest lower bounds on the standard deviations of IMRSs, that is, volatility bounds. These bounds are expressed most conveniently as regions of admissible mean–standard deviation pairs for the IMRSs.

The existence of volatility bounds on IMRSs was originally noted by Shiller (1982) (see also Hansen 1982). His goal was to construct a diagnostic for a particular family of asset pricing models that is insensitive to the alignment of the data. The volatility implications he deduced for IMRSs used only two asset returns and, even for the two-asset case, are weaker than those reported here.

Our reasons for examining volatility bounds are somewhat different from Shiller's. First, our nonparametric approach can serve as a useful complement to the parametric approach that is prevalent in the literature. In particular, it can assist in understanding better why particular models are rejected on the basis of statistical tests: does the parameterization admit too little variability in the IMRSs? Second, our approach provides a common set of diagnostics for a potentially large class of asset pricing models. These diagnostics can also be used to evaluate models in which IMRSs are parameterized as functions of observables as well as models for which moments can be computed from characterizations of the stochastic equilibria. Third, our approach allows us to determine which asset market data sets present the most stringent restrictions for IMRSs and consequently the most startling implications for dynamic economic models. It allows us to make these comparisons without having to focus on a parametric family of such models.

To illustrate these points, we provide an alternative characterization of the so-called *equity premium puzzle* (see, e.g., Mehra and Prescott 1985). In contrast to other characterizations, ours does not depend either on a Markov chain approximation with a small number of states or on a narrow class of asset valuation models. Figure 1 reports a restricted region for the means and standard deviations of IMRSs implied by the annual (1891–1985) time-series data on stocks and bonds used by Campbell and Shiller (1988). The shaded region gives the admissible pairs of means and standard deviations for IMRSs. As benchmarks, we also report time-series sample means and standard deviations for IMRSs implied by a representative consumer model with commonly used period utility functions of the form

$$U(c) = \frac{c^{\gamma+1} - 1}{\gamma + 1}$$

for negative values of γ. For this specification of preferences, the IMRS can be measured by forming a consumption ratio for two different points in time, raising it to the power γ and discounting. For

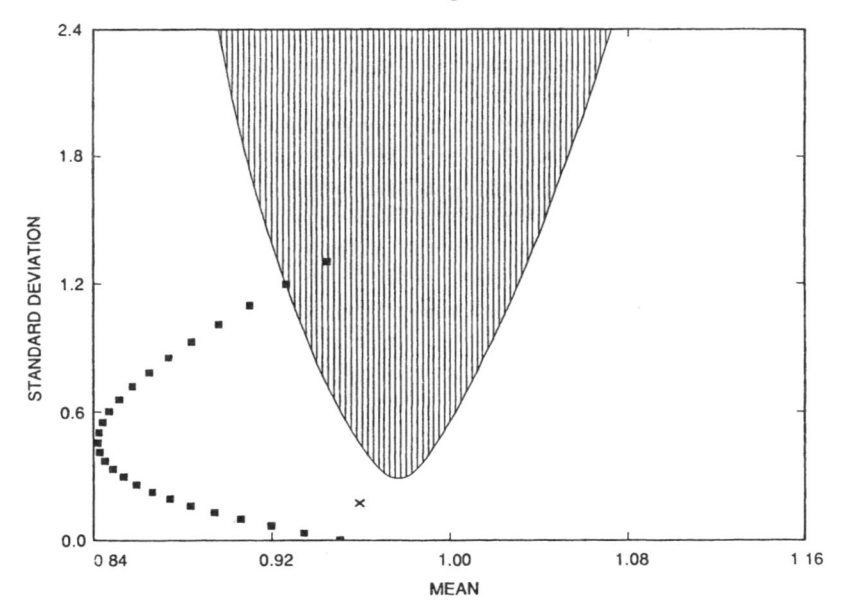

FIG. 1.—IMRS frontier computed using annual data

illustrative purposes, the annual subjective discount factor is taken to be .95. The "boxes" in the figure represent mean–standard deviation pairs for alternative values of γ ranging from zero to -30. As $|\gamma|$ increases, the volatility of the IMRS increases but the effect on the mean of the IMRS is not uniform. Initially the mean decreases but subsequently increases so that for large $|\gamma|$ the boxes are in the admissible shaded region.

Our strategy for constructing regions such as that reported in figure 1 is to construct minimum variance random variables with prespecified means that are related to asset payoffs and prices in the same manner as the IMRSs. We refer to such random variables as being on the *mean–standard deviation frontier* for IMRSs. In Section III, we construct these frontier random variables ignoring the fact that IMRSs must be positive. In this case the minimum variance random variables are simply linear combinations of the asset payoffs translated by a constant. As a by-product of this construction, we relate our analysis to two commonly used empirical paradigms in finance: mean-variance analysis and linear factor pricing. More precisely, we characterize the *duality* between the mean–standard deviation frontier for IMRSs and the familiar mean–standard deviation frontier for asset payoffs. This analysis reveals that asset payoffs on the mean–standard deviation frontier are sufficient to generate the mean–standard deviation frontier for IMRSs. Hence the dimensionality reduction techniques used in linear factor pricing models can be exploited to derive a region like that reported in figure 1.

In Section IV, we modify the analysis of Section III by incorporating the restriction that IMRSs are positive random variables. For prespecified means, we construct nonnegative random variables that behave like IMRSs and have minimum variances. These random variables are not necessarily linear functions of the payoffs but instead can be interpreted as European call and put options on portfolios of these payoffs. In contrast to the analysis in Section III, for some prespecified means there may not be any nonnegative random variables with finite second moments that behave like IMRSs. While the approach of this section yields more restrictive (and therefore more informative) volatility bounds, these sharper bounds are harder to compute.

In Section V, we illustrate the results in Sections III and IV by displaying volatility bounds computed using alternative data sets and generating mean–standard deviation pairs for alternative parametric models of IMRSs. Among other things, we use these bounds to help assess the plausibility of some parametric models of asset prices.

II. A General Model of Asset Pricing

In this section we present a general model of asset valuation. Consider an environment in which multiple consumers trade in securities markets. The preferences and information sets of these consumers may be heterogeneous. We fix both the trading period (say time 0) and the time period for the receipts of the asset payoffs (say time $\tau > 0$). Let I^j denote the information set of consumer j at time 0, and $I = \cap\, I^j$, where the intersection is taken over the consumers in the economy who trade securities. The prices of securities traded at date 0 are presumed to be in the individual information set I^j of individual j for each j and hence in I. Let P denote a set of portfolio payoffs of the numeraire good at time τ that are traded at time 0. Since the prices of the portfolio payoffs are in I, we represent these prices as a function π_I mapping P into I. Hence $\pi_I(p)$ is the price at time 0 of a portfolio that will pay p units of a numeraire good at a future date τ.

Consumers are presumed to solve optimal portfolio problems in determining their asset holdings. This imposes restrictions relating marginal rates of substitution to asset payoffs and prices. To see this, let mu_0^j and mu_τ^j denote the equilibrium marginal utilities of consumer j in terms of the numeraire consumption good at dates 0 and

τ, respectively. In equilibrium the marginal utility–scaled price must equal the expected marginal utility–scaled payoff conditioned on I^j:

$$mu_0^j \pi_I(p) = E(mu_\tau^j p | I^j) \quad \text{for all } p \text{ in } P. \tag{1}$$

As long as consumer j is not satiated at time 0, $mu_0^j > 0$ and we can divide both sides of (1) by mu_0^j, which yields $\pi_I(p) = E(pm^j|I^j)$ for all p in P, where $m^j = mu_\tau^j/mu_0^j$ is the IMRS of consumer j. Since asset prices are presumed to be observed by all consumers, it follows from the law of iterated expectations that

$$\pi_I(p) = E(pm^j|I) \quad \text{for all } p \text{ in } P. \tag{2}$$

In a world with common information sets and complete markets, marginal rates of substitution are equated across consumers ($m^j = m$ for all j). In such a world, P can be chosen to be sufficiently large that the common IMRS is uniquely determined by (2). In general, (2) does not uniquely determine m^j. As we shall see, however, (2) does restrict the unconditional moments of m^j even when markets are incomplete.[1] Since the restrictions we derive apply to all the individual IMRSs, to simplify notation we drop the j superscript on m.

We now give a more complete description of P and the associated asset pricing function π_I. We do not require that P contain all the portfolio payoffs that are traded by consumers. Omitting payoffs will, however, weaken the implications for m. As a matter of convenience, we consider the case in which there is an n-dimensional vector $\mathbf{x}$ of asset payoffs at date τ. The time 0 prices of these assets can also be represented as an n-dimensional vector, say $\mathbf{q}$, and pricing relation (2) can be expressed as

$$\mathbf{q} = E(\mathbf{x}m|I). \tag{3}$$

We are interested in the implications of (3) for the IMRS m. To investigate this relation empirically, we must have some way to replicate observations on payoffs, prices, and information over time. As in Hansen and Richard (1987), we imagine an environment in which relation (3) is replicated over time. In other words, there is a composite process $\{(m_t, \mathbf{x}_t, \mathbf{q}_t)\}$ and a sequence of information sets $\{I_t\}$ that satisfy a version of (3) for all t. Econometricians seeking to study this economy are presumed to have data on a finite record $(\mathbf{x}_t, \mathbf{q}_t)$, for $t = 1, 2, \ldots, T$, and the composite process $\{(m_t, \mathbf{x}_t, \mathbf{q}_t)\}$ is presumed to be sufficiently regular that a time-series version of a law of large numbers applies. Thus sample moments formed from the finite rec-

---

[1] In the case of incomplete markets, formula (2) abstracts from the possible existence of short sale constraints.

ords of data converge to population counterparts as the sample size T becomes large. Even though asset prices are determined τ periods prior to the realization of the asset payoffs, from the vantage point of econometricians, we model $\{q_t\}$ as a stochastic process to accommodate possible variation over time in the asset prices. In what follows we use the unconditional expectation operator E to represent the limit points of the time-series averages of the sample moments.[2]

We now impose restrictions on m, $\mathbf{x}$, and $\mathbf{q}$, which are expressed in terms of unconditional expectations.

ASSUMPTION 1. $E|m|^2 < \infty$, $E|\mathbf{x}|^2 < \infty$, $E\mathbf{x}\mathbf{x}'$ is nonsingular, and $E|\mathbf{q}| < \infty$.

The restriction that the second-moment matrix of $\mathbf{x}$ is nonsingular is made as a matter of convenience to rule out cases in which the entries of $\mathbf{x}$ are linearly dependent. Among other things, this guarantees that the law of one price holds trivially for linear combinations of $\mathbf{x}$. If the moment restrictions imposed on $\mathbf{x}$ and $\mathbf{q}$ are not satisfied for an original vector of assets, then it is often possible to scale the payoffs and prices so that these restrictions are satisfied. A special case of such scaling occurs when all the payoffs are constructed to have a unit price as in the case of measured returns to holding securities between time 0 and time τ.

Applying the law of iterated expectations to the pricing relation (3) results in the following restriction.

RESTRICTION 1. $E\mathbf{q} = E\mathbf{x}m$.

We focus on the unconditional moment restriction 1 instead of the conditional moment restriction (3) because it is typically easier to estimate unconditional moments than conditional moments. Restriction 1, however, is in general weaker than (3). Gallant, Hansen, and Tauchen (1990) show how to extend some of the analysis in this paper by exploiting characterizations of the moments of $\mathbf{x}$ conditioned on (possibly a subset of) I.

As long as consumers are not satiated in the numeraire consumption good at time τ, the IMRS should be strictly positive.

RESTRICTION 2. $m > 0$.

Restriction 2 is sufficient to imply the absence of arbitrage opportunities. That is, the restriction guarantees that nonnegative payoffs that are strictly positive with positive probability conditioned on I have positive prices. In the next two sections we explore the implica-

---

[2] Our use of the unconditional expectation operator in this context is justified formally when the time series converges appropriately to a stochastic steady state and is ergodic. In this case, unconditional expectations are computed using the stationary distribution. For processes that are asymptotically stationary but not ergodic, the limit points can often be represented as conditional expectations in which the conditioning occurs on the invariant sets for the approximating stationary stochastic process.

tions that restrictions 1 and 2 have for the mean and standard deviation of m.

So far, we have treated the case in which only a finite vector of asset payoffs and prices is investigated. In our subsequent analysis, it will be convenient to extend the pricing function and its unconditional expectation to the linear span of $\mathbf{x}$. Define $P \equiv \{\mathbf{c} \cdot \mathbf{x} : \mathbf{c} \text{ in } \mathbb{R}^n\}$.[3] In Section IV, we shall also consider derivative claims formed by taking particular nonlinear functions of payoffs in P. In light of assumption 1, each portfolio payoff in P has a finite second moment. With this in mind we define a norm on P to be $\|p\| \equiv [E(p^2)]^{1/2}$. Notice that the standard deviation of a portfolio payoff p, denoted $\sigma(p)$, is given by $\|p - Ep\|$.

Since the portfolio payoffs in $\mathbf{x}$ are linearly independent, for each p in P there is a unique $\mathbf{c}$ in $\mathbb{R}^n$ for which p is equal to $\mathbf{c} \cdot \mathbf{x}$. We extend the pricing function so that the prices of these payoffs are given by the corresponding linear combinations of $\mathbf{q}$: $\pi_I(\mathbf{c} \cdot \mathbf{x}) \equiv \mathbf{c} \cdot \mathbf{q}$. As required, π_I maps P into I. Notice that π_I is constructed so that (3) extends to the linear span $\mathbf{x}$: $\pi_I(p) = E(pm|I)$ for all p in P.

It is also of interest to define a functional π mapping portfolio prices into the *expected value* of the prices: $\pi(p) \equiv E\pi_I(p)$. Hence π maps P linearly into the real line $\mathbb{R}$. Again the law of iterated expectations implies that

$$\pi(p) = E(mp) \quad \text{for all } p \text{ in } P. \tag{4}$$

It is straightforward to show that restriction (4) is equivalent to restriction 1.

III. Implications of Restriction 1

In this section we characterize the volatility restrictions for m as implied by restriction 1. In subsection A we suppose that there is a unit payoff in P, while in subsection B we consider the more common case in which such a payoff is not included in P. Finally, in subsection C we describe how existing empirical methodologies in finance can be used to characterize these volatility restrictions.

A. Riskless Payoff

Suppose that P contains a payoff that is equal to one with probability one. In deriving implications for the volatility of m, we should first

---

[3] Notice that a larger set P of portfolio payoffs could be constructed by following more complex trading strategies in which the vector $\mathbf{c}$ is replaced by a vector of random variables in I. The theoretical analysis in Hansen and Richard (1987) is designed to accommodate this case as is the econometric analysis in Gallant et al. (1990). We focus on the linear span of $\mathbf{x}$ for pedagogical convenience and empirical tractability.

construct a random variable m^* in P that satisfies restriction 1. This amounts to finding a vector α_o in $\mathbb{R}^n$ such that

$$E\mathbf{xx}'\alpha_o = E\mathbf{q}, \tag{5}$$

where $m^* = \mathbf{x} \cdot \alpha_o$. Solving (5) for α_o gives $\alpha_o = (E\mathbf{xx}')^{-1}E\mathbf{q}$. Notice that α_o depends on the second moment of $\mathbf{x}$ and the first moment of $\mathbf{q}$. Hence m^* can be constructed from asset market data.

Consider any other random variable m satisfying restriction 1. Since P contains a unit payoff, $Em = \pi(1) = Em^*$. Consequently, all random variables m that satisfy restriction 1 have the same mean, and this mean is equal to the expected price of a unit payoff. Also, $E[\mathbf{x}(m - m^*)] = 0$ because both m and m^* satisfy restriction 1. In other words, the discrepancy between m and m^* is orthogonal to the random vector $\mathbf{x}$. Since m^* is in P, m^* is the least-squares projection of m onto P and

$$\sigma^2(m) = \sigma^2(m^*) + \sigma^2(m - m^*).$$

Therefore, we have the following relations:

$$\sigma(m) \geq \sigma(m^*), \quad Em^* = Em. \tag{6}$$

The volatility bound in (6) is as sharp as possible because m^* satisfies restriction 1 by construction.

B. No Riskless Payoff

Next we consider the more usual case in which P does not contain a unit payoff. It turns out that much of the previous analysis can be exploited in analyzing this case. Let $\mathbf{x}^a$ denote the $(n + 1)$–dimensional random vector formed by augmenting $\mathbf{x}$ with a unit payoff. Since $E\mathbf{xx}'$ is nonsingular and no linear combination of $\mathbf{x}$ is equal to one with probability one, $E\mathbf{x}^a\mathbf{x}^{a\prime}$ is also nonsingular. We build an augmented payoff space P^a containing a unit payoff by using $\mathbf{x}^a$ in place of $\mathbf{x}$.

To apply the analysis in Section IIIA, we must assign a number v to $\pi(1)$, which is the expected price of a unit payoff. Such price data may not be available, and for this reason we examine implications for an array of hypothetical expected prices. Let v be any candidate for $\pi(1)$ and π_v the corresponding extension of π from P to P^a. We then replicate the analysis in subsection A to construct a random variable m_v in P^a such that

$$E\mathbf{x}m_v = E\mathbf{q}, \quad Em_v = v. \tag{7}$$

The counterpart to volatility bound (6) is

$$\sigma(m) \geq \sigma(m_v) \tag{8}$$

for any random variable m that satisfies restriction 1 and has mean v. This volatility bound is as sharp as possible because, by construction, m_v satisfies restriction 1 and has mean v.

We replicate the construction of m_v for all real numbers v and generate an indexed collection $\{m_v : v \text{ in } \mathbb{R}\}$ of random variables, each of which satisfies restriction 1. This collection is of interest because for any m satisfying restriction 1, the ordered pair $[Em, \sigma(m)]$ is in the region

$$S \equiv \{(v, w) \text{ in } \mathbb{R}^2 : w \geq \sigma(m_v)\}. \tag{9}$$

This region summarizes the volatility implications for m implied by restriction 1. We refer to the boundary of S as being the mean–standard deviation frontier for IMRSs, and we refer to members of the set $\{m_v : v \text{ in } \mathbb{R}\}$ as being on this frontier.

It is of interest to derive an expression for $\sigma(m_v)$ that is both easy to compute and easy to interpret. The moment conditions in (7) can be rewritten as terms of the covariance of m and $\mathbf{x}$:

$$E[(\mathbf{x} - E\mathbf{x})(m_v - v)] = E\mathbf{q} - vE\mathbf{x}. \tag{10}$$

Now

$$m_v = (\mathbf{x} - E\mathbf{x})'\boldsymbol{\beta}_v + v \tag{11}$$

for some $\boldsymbol{\beta}_v$ in $\mathbb{R}^n$ because m_v is a linear combination of a unit payoff and the entries of $\mathbf{x}$ and Em_v is v. Substituting (11) into (10) and solving for $\boldsymbol{\beta}_v$ give

$$\boldsymbol{\beta}_v = \Sigma^{-1}(E\mathbf{q} - vE\mathbf{x}),$$

where Σ is the covariance matrix of $\mathbf{x}$. It follows that

$$\sigma(m_v) = [(E\mathbf{q} - vE\mathbf{x})'\Sigma^{-1}(E\mathbf{q} - vE\mathbf{x})]^{1/2}. \tag{12}$$

Notice that for a given v, $\sigma(m_v)$ depends only on the means of $\mathbf{q}$ and $\mathbf{x}$ and the covariance matrix of $\mathbf{x}$.

The standard deviation bound given in (12) has the following interpretation. Consider a risk-neutral valuation of the asset payoffs in which m is set to a constant value v for all states of the world. In this case the means of the prices should be proportional to the means of the asset payoffs with proportionality factor v. The bound in (12) is the square root of a quadratic form in the vector of deviations of the observed average prices from the average risk-neutral prices. For a fixed Σ, larger deviations from risk-neutral pricing imply larger bounds on the volatility of m. Shanken (1987) derived a related bound on the pricing error induced by using error-ridden proxies in computing the valuation of asset payoffs. When a constant v is used as a proxy for m, the bound in (12) can be viewed as a special case of Shanken's bound (see his proposition 1, pp. 93–94).

C. Relation to Empirical Models of Asset Prices

In this subsection we derive the relation between the mean–standard deviation frontier for m and the mean-variance frontier for asset returns. This latter frontier is the focal point of the static capital asset pricing model. The link we deduce between the two frontiers provides an alternative interpretation of the volatility bounds for m. We then describe how linear factor restrictions as imposed in Ross's (1976a) arbitrage pricing model (see also Chamberlain 1983; Chamberlain and Rothschild 1983; Connor 1984) can be used to characterize the mean–standard deviation frontier for m.

Define

$$R \equiv \{p \text{ in } P: \pi(p) = 1\}. \tag{13}$$

When the vector $\mathbf{q}$ is not random, R is the collection of (gross) returns on portfolios in P. More generally, R contains all the payoffs in P with *expected prices* that are equal to one.

Consider first the case in which P contains a unit payoff and $\pi(1)$ is different from zero. Then $1/\pi(1)$ is in R. A second payoff in R is $r^* \equiv m^*/\pi(m^*)$. Note that $\pi(m^*) = E[(m^*)^2]$, and hence

$$\|r^*\| = \frac{\|m^*\|}{\|m^*\|^2} = \frac{1}{\|m^*\|}. \tag{14}$$

Furthermore, Hansen and Richard (1987) established that r^* is the payoff in R that has the smallest norm (second moment). Consequently, r^* is the solution to the following optimization problem:

$$\underset{r \text{ in } R}{\text{minimize }} \sigma(r) \quad \text{subject to } Er = \mu$$

when μ is set equal to Er^*. Therefore, m^* is proportional to a particular payoff on the mean–standard deviation frontier for R.

To relate the bound for $\sigma(m)$ given in (8) to the slope of the mean–standard deviation frontier for R, note that

$$\frac{\sigma(m)}{Em} \geq \frac{\sigma(m^*)}{Em^*} = \frac{\sigma(r^*)\|m^*\|^2}{Em^*} = \frac{\sigma(r^*)}{Er^*}. \tag{15}$$

Recall that the second moment of a random variable r satisfies $E(r^2) = \sigma(r)^2 + (Er)^2$. Since P contains a unit payoff, the mean–standard deviation frontier for R is a cone with apex at $[0, 1/\pi(1)]$ and axis parallel to the horizontal axis. In order for r^* to be the minimum second-moment payoff in R, the ordered pair $[\sigma(r^*), Er^*]$ must occur at the tangency of a circle with center (0, 0) and the lower (inefficient) portion of the mean–standard deviation frontier for R. This tangency point is depicted in figure 2. Since the lower portion of the frontier is a ray from $[0, 1/\pi(1)]$ through $[\sigma(r^*), Er^*]$, the slope of this ray is the *Sharpe ratio* of the payoff r^*, $\{Er^* - [1/\pi(1)]\}/\sigma(r^*)$, and the slope of the circle with center (0, 0) that passes through $[\sigma(r^*), Er^*]$ is $-\sigma(r^*)/Er^*$. Therefore,

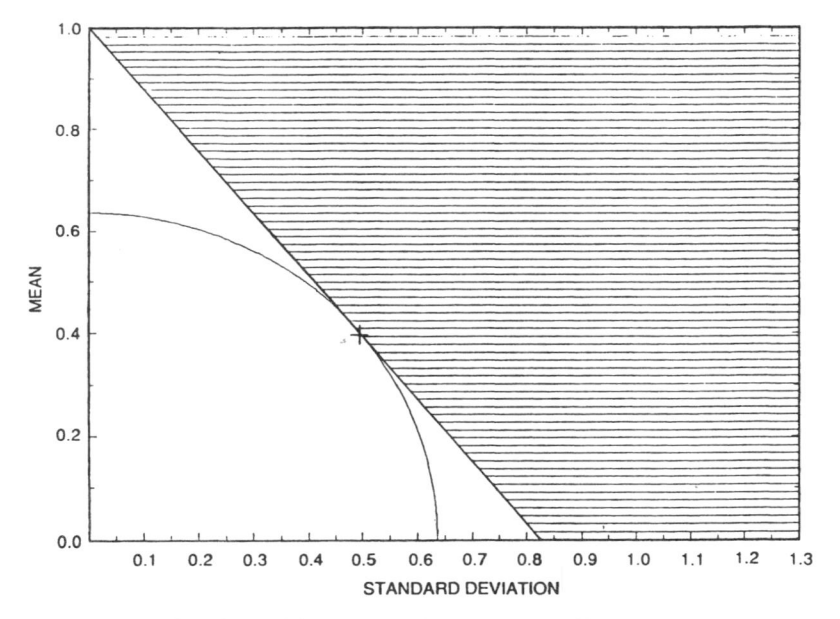

$$\frac{\sigma(r^*)}{Er^*} = \frac{[1/\pi(1)] - Er^*}{\sigma(r^*)}. \qquad (16)$$

In light of (15) and (16), the bound on the ratio $\sigma(m)/Em$ is given by the absolute value of the slope of the mean–standard deviation frontier for R.[4] These relations demonstrate the precise sense in which a

---

[4] An alternative way to derive the bound is as follows. Consider any payoff z with an expected price equal to zero. By restriction 1, $Ezm = 0$. It follows from the covariance decomposition and the Cauchy-Schwarz inequality that $\sigma(m)/Em \geq |Ez|/\sigma(z)$. The sharpest bound on $\sigma(m)/Em$ is given by the zero expected price payoff z with the largest Sharpe ratio $|Ez|/\sigma(z)$. As is well known from financial economics, the largest Sharpe ratio is equal to the absolute value of the slope of the mean–standard deviation frontier when R contains a riskless payoff.

steep slope of a mean–standard deviation frontier for asset payoffs can imply a potentially dramatic bound on the volatility of m.

Next we consider the case in which P does not contain a unit payoff and hence R does not contain an unconditionally riskless payoff. We follow the strategy used in subsection B by augmenting $\mathbf{x}$ with a unit payoff and assigning this payoff an expected price v. This results in an expansion of R to R_v, where $1/v$ is now in R_v. Let r_v^* denote the payoff in R_v with the smallest second moment. Since r_v^* is on the mean–standard deviation frontier for R_v, it is well known from static capital asset pricing theory that r_v^* is a linear combination (with coefficients that sum to one) of $1/v$ and any other distinct return on the mean–standard deviation frontier for R_v. As long as $1/v$ is not equal to the mean of the minimum variance payoff in R, we can find a payoff r_v that is on the mean–standard deviation frontier for both R and R_v. Also, for each v the variable m_v is proportional to r_v^*. Therefore, with one exception, for each random variable m_v on the mean–standard deviation frontier for IMRSs, there is a corresponding payoff r_v on the mean–standard deviation frontier for R such that m_v is a linear combination of r_v and a unit payoff. In this sense the mean–standard deviation frontier for IMRSs can be thought of as the *dual* of the mean–standard deviation frontier for R. The exceptional case occurs when $1/v$ is the mean of the minimum variance payoff in R. In this case, m_v is a linear combination of a unit payoff and a payoff that is on the mean–standard deviation frontier for the space of payoffs with expected prices equal to zero.

The impact of augmenting R with $1/v$ can be seen graphically by passing a ray from the point $(0, 1/v)$ through a tangent point on the mean–standard deviation frontier for R. One side of the mean–standard deviation frontier for the augmented set R_v is given by this tangent ray, and the other is a reflection about a horizontal ray from $(0, 1/v)$. This construction is displayed graphically in figure 3. In the special case in which $1/v$ is the mean of the minimum variance payoff in R, it is not possible to draw a tangent line to the mean–standard deviation frontier of R from the point $(0, 1/v)$. Instead the frontier for R_v is given by the two asymptotes.

Once the frontier for the augmented set R_v is obtained, the construction illustrated in figure 2 can be mimicked using R_v in place of R. Thus for any m with mean v that satisfies restriction 1,

$$\frac{\sigma(m)}{Em} \geq \frac{\sigma(m_v)}{v} = \frac{\sigma(r_v^*)}{Er_v^*} = \frac{[1/\pi(1)] - Er_v^*}{\sigma(r_v^*)}. \tag{17}$$

The relations in (17) show the connection between the volatility bound on m's with mean v and the slope of the mean–standard devia-

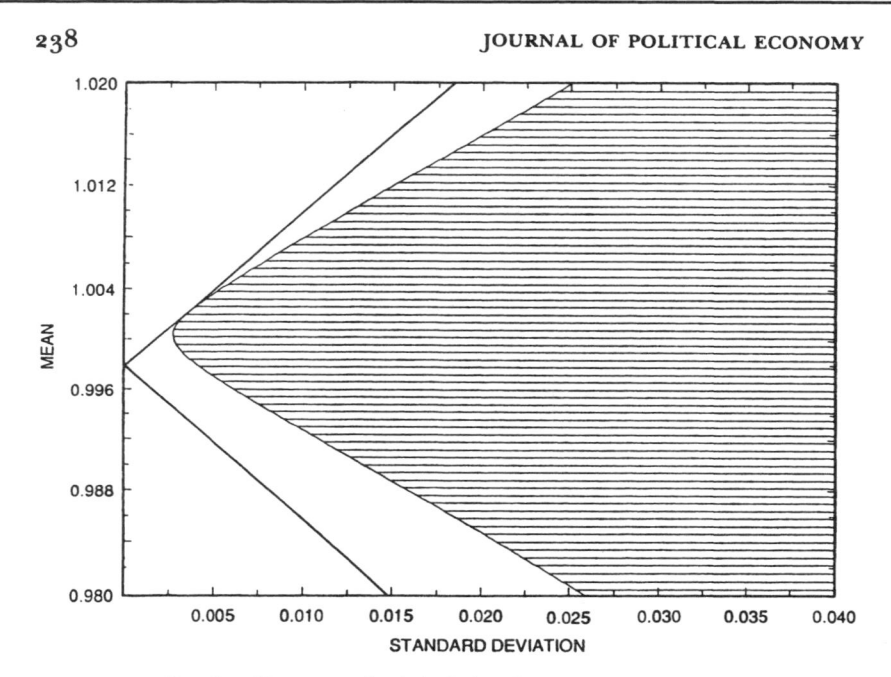

FIG. 3.—Mean–standard deviation frontiers for R and R_v

tion frontier for R_v. A steeper slope of the frontier for R_v implies a correspondingly sharper volatility bound for m.[5]

Since the mean–standard deviation frontier for R is known to have a two-fund characterization, the preceding results show that the mean–standard deviation frontier for m can be represented using two distinct frontier payoffs in R. For the general class of asset pricing models considered in this paper, there is no prediction that particular payoffs in R, say the returns on the wealth portfolios of consumers, are mean-variance efficient. Thus without additional restrictions, there is no guidance on how to reduce, a priori, a potentially large collection of portfolio payoffs into a small collection used in a time-series analysis.

One ad hoc approach that is often used to reduce the dimensionality of the collection of payoffs is factor analysis as employed in empirical arbitrage pricing models (see, e.g., Connor and Korajczyk 1988; Lehmann and Modest 1988). Suppose that P is generated by a se-

---

[5] Using conditioning information in clever ways can sharpen the volatility bounds on m by increasing the maximum Sharpe ratio of the payoffs in R_v. For example, Breen, Glosten, and Jagannathan (1989) show that information in Treasury bill returns can be used to construct a portfolio that has the same average return as the value-weighted index of New York Stock Exchange securities but is only half as variable.

quence $\{p_j\}$, where

$$p_j = \gamma_j \cdot \mathbf{f} + e_j \tag{18}$$

and $\mathbf{f}$ is a vector of common factors for all the payoffs.[6] Often, the factors $\mathbf{f}$ are in an appropriately defined span of $\{p_j\}$. Hence, it follows from the law of one price that there exists a unique vector $\pi(\mathbf{f})$ of hypothetical expected prices for the factor payoffs. One possible strategy for deducing volatility bounds on m is to use the extensive collection of payoffs $\{p_j\}$ (or possibly a subset of it) to identify the first two moments of $\mathbf{f}$ and the expected price vector $\pi(\mathbf{f})$. A region S then could be constructed from these factor moments and prices using formula (12).

In general, information is lost in going from the larger space P to the smaller space F of linear combinations of factors. Tests of factor models of asset pricing examine whether the pricing relation

$$\pi(p_j) = \gamma_j \cdot \pi(\mathbf{f}) \tag{19}$$

holds at least approximately. When (19) holds exactly, the regions S generated by P and F coincide. Therefore, if asset payoffs can be priced in terms of a small number of factors $\mathbf{f}$, there is no loss to constructing the region S from F instead of the larger space P.[7]

As argued in Hansen and Richard (1987), an unconditional factor decomposition as in (18) may not be very appealing when economic agents can use conditioning information in I to make investments. If the factor decomposition (18) is conditioned on an information set I and γ_j is a vector of random variables in I, a reduction in payoffs is more complicated but still feasible.

IV. Implications of Restriction 2

In Section III, we showed how to construct minimum variance random variables that satisfy restriction 1. These random variables may be negative with positive probability and hence may fail to satisfy

---

[6] Although our derivation of the volatility bounds for m assumed that the payoff space P is finite-dimensional, this restriction was made for pedagogical convenience. In fact the duality relation between the mean–standard deviation frontiers for m's that satisfy restriction 1 and for payoffs in R extends to environments in which P is generated by an infinite number of payoffs, say by $\{p_j\}$.

[7] In contrast to factor analytic approaches, Huberman and Kandel (1987) test whether the dimensionality of P can be reduced to a prespecified *observed* subset of security returns, namely three size-based portfolios of New York Stock Exchange securities. In this case, F can be constructed using these three returns. Huberman and Kandel find, however, that this construction of F is not adequate to span the mean–standard deviation frontier for the original P constructed using 33 size-sorted portfolios. Hence in this case the dimensionality reduction from P to F will result in weaker implications for m.

restriction 2.[8] As long as we limit ourselves to candidate IMRSs that are translations of payoffs in P, it may not be possible to ensure that frontier random variables are strictly positive or, for that matter, nonnegative.

In this section we initially replace restriction 2 by a weaker requirement that m be nonnegative. We then construct minimum variance candidates for m among the class of nonnegative random variables satisfying restriction 1. It turns out that these minimum variance random variables can be interpreted as either European call or put options on payoffs in P. Recall that when the payoff on the underlying portfolio is p and the strike price is k, a European call option entitles an investor to the payoff $\max\{p - k, 0\}$ and a put option to $\max\{k - p, 0\}$. These payoffs are clearly nonnegative, but they may be nonlinear functions of $\mathbf{x}$. The resulting volatility bounds for nonnegative random variables satisfying restriction 1 also apply when the random variables are restricted to be strictly positive (satisfy restriction 2). However, in this case the lower bounds may only be approximated rather than attained.

This section is divided into three subsections. In subsection A we suppose that there is a unit payoff in P, while in subsection B we consider the more common case in which such a payoff is not included in P. Finally, in subsection C we discuss the close connection between our analysis and work by Harrison and Kreps (1979) and Kreps (1981) on the viability of equilibrium pricing functions consistent with the absence of arbitrage opportunities.

A. Riskless Payoff

First consider the case in which there is a unit payoff in P. For each p in P, let p^+ denote $\max\{p, 0\}$. Note that for any p' in P and any nonnegative strike price k that is proportional to the unit payoff, the payoffs $p' - k$ and $k - p'$ are also in P. Therefore, the collection of all random variables p^+ for some p in P includes the payoffs on European call and put options with constant strike prices.

Suppose that we weaken restriction 2 to the requirement that m be nonnegative. By construction, all derivative claims of the form p^+ for payoffs p in P are nonnegative. It turns out that the minimum variance nonnegative random variable $\tilde{m}$ satisfying restriction 1 is given by such a derivative claim. Hence we are led to the problem of finding

---

[8] As Dybvig and Ingersoll (1982) have pointed out, naive use of m^* to compute (expected) prices of contingent claims may lead to assignment of negative (expected) prices to some positive payoffs and hence to the appearance of an arbitrage opportunity.

a vector $\boldsymbol{\alpha}_o$ in $\mathbb{R}^n$ such that

$$E[\mathbf{x}(\mathbf{x}'\boldsymbol{\alpha}_o)^+] = E\mathbf{q}, \qquad (20)$$

where $\tilde{m} = (\mathbf{x}'\boldsymbol{\alpha}_o)^+$. In what follows we shall first show that $\tilde{m}$, when it exists, has the smallest variance among all nonnegative random variables m satisfying restriction 1. We then discuss the existence and computation of a solution to (20).

To show that $\tilde{m}$ has the smallest variance, consider any other nonnegative random variable m satisfying restriction 1. Clearly, $E\mathbf{x}m = E\mathbf{x}\tilde{m}$. Exploiting the nonnegativity of m, we have that

$$E\tilde{m}m \geq \boldsymbol{\alpha}_o' E\mathbf{x}m = \boldsymbol{\alpha}_o' E\mathbf{x}\tilde{m} = E[(\tilde{m})^2]. \qquad (21)$$

It follows from the Cauchy-Schwarz inequality that $\|m\| \geq \|\tilde{m}\|$. Since P contains a unit payoff, both m and $\tilde{m}$ must have the same mean. Therefore, we have the following relations:[9]

$$\sigma(m) \geq \sigma(\tilde{m}), \quad Em = E\tilde{m}. \qquad (22)$$

Next we ask whether the volatility bound in (22) can be sharpened by requiring m to be strictly positive instead of nonnegative. If $\tilde{m}$ is strictly positive (with probability one), then clearly the answer is no. This can occur only when $\tilde{m}$ coincides with m^* computed in Section IIIA. Consider the case in which m^* is not strictly positive with probability one, and let m be any random variable satisfying restrictions 1 and 2. Then $\tilde{m}$ is zero with positive probability, and it follows from (21) that

$$0 < \|m - \tilde{m}\|^2 = \|m\|^2 - 2E\tilde{m}m + \|\tilde{m}\|^2 \leq \|m\|^2 - \|\tilde{m}\|^2.$$

Therefore, at the very least the weak inequality ($\geq$) in (22) is replaced by the strong inequality ($>$). In fact, no further improvements are possible. To see this form,

$$m_j = \left(1 - \frac{1}{j}\right)\tilde{m} + \left(\frac{1}{j}\right)m. \qquad (23)$$

Then m_j is strictly positive and $\{\sigma(m_j)\}$ converges to $\sigma(\tilde{m})$. Therefore, $\sigma(\tilde{m})$ is in fact the greatest lower bound for $\sigma(m)$ when m is restricted to satisfy restrictions 1 and 2.

---

[9] An alternative way to deduce these bounds is to exploit the fact that when payoffs on calls and puts are included in the analysis, the space of admissible payoffs is essentially complete (see Ross 1976b; Breeden and Litzenberger 1978; Arditti and John 1980; Green and Jarrow 1987). If the prices of all such payoffs were available, the counterpart to the random variable m^* in Sec. III would be strictly positive. Although this extensive collection of option price data is typically not available, we can follow Merton (1973) and use lower bounds on option prices to obtain a lower bound on the volatility of m.

Equation system (20) is nonlinear in the parameter vector α_o, and its solution cannot necessarily be represented in terms of matrix manipulations. There is a closely related optimization problem whose solution may be easier to compute. This problem entails finding a payoff in R whose truncation has the smallest second moment:

$$\min_{r \text{ in } R} \|r^+\|^2. \tag{24}$$

In Appendix A we show that (24) has a solution, although this solution may not be unique. Furthermore, a necessary and sufficient condition for $\tilde{r}$ to be a solution to (24) is

$$E(\tilde{r}^+ z) = 0 \quad \text{for all } z \text{ in } P \text{ such that } \pi(z) = 0. \tag{25}$$

We can think of (25) as being the first-order condition for optimization problem (24).

It turns out that we can construct a solution to (20) by scaling $\tilde{r}^+$ appropriately. Let

$$\tilde{m} = \frac{\tilde{r}^+}{\|\tilde{r}^+\|^2}. \tag{26}$$

This scaling is permissible because $\|\tilde{r}^+\|$ must be strictly positive as long as there exists at least one random variable m satisfying restrictions 1 and 2. To see this, suppose to the contrary that $\|\tilde{r}^+\|$ is zero. Then $-\tilde{r}$ is a nonnegative payoff with a strictly negative expected price. Such a payoff is inconsistent with restriction 2 because it implies that there exists an arbitrage opportunity.

Clearly $\tilde{m}$ as given by (26) can be represented as $(\alpha_o'x)^+$ for some α_o in $\mathbb{R}^n$. To verify that α_o solves (20), we must show that $\tilde{m}$ as given by (26) satisfies restriction 1. Let p be any payoff in P and form the payoff $z = p - \pi(p)\tilde{r}$. Note that $\pi(z) = 0$ because $\pi(\tilde{r}) = 1$. It follows from first-order condition (25) that

$$0 = E\tilde{m}z = E\tilde{m}p - \pi(p)E\tilde{m}\tilde{r}$$

$$= E\tilde{m}p - \pi(p)\frac{E[\tilde{r}^+\tilde{r}]}{\|\tilde{r}^+\|^2}$$

$$= E\tilde{m}p - \pi(p).$$

Thus $\tilde{m}$ satisfies restriction 1 as required.

This construction of $\tilde{m}$ parallels a similar construction reported in Hansen and Richard (1987) and in Section III. If one ignores restriction 2, one way to construct the random varible m^*, which has minimum variance among the class of random variables satisfying restriction 1, is to compute the minimum second-moment payoff, r^*, in R and divide it by its second moment, $\|r^*\|^2$. We have just demonstrated that a similar strategy works for constructing a random variable $\tilde{m}$ that attains the volatility bound among the class of nonnegative random variables satisfying restriction 1. Instead of computing the minimum second-moment payoff in R, we calculate the minimum *truncated* second-moment payoff, $\tilde{r}$, in R. To form $\tilde{m}$, the truncation of this payoff, $\tilde{r}^+$, is divided by the second moment of its truncation, $\|\tilde{r}^+\|^2$. Whereas $\|m^*\|$ is given by $1/\|r^*\|$, $\|\tilde{m}\|$ is given by $1/\|\tilde{r}^+\|$. Since truncating a random variable reduces its norm, as required, $\tilde{m}$ has a larger second moment than m^*. The difference in the two norms reflects the incremental contribution of restriction 2 for the volatility bound on m.

One advantage to solving optimization problem (24) instead of solving directly the nonlinear equation system (20) is that optimization problem (24) has a convex objective function $\|r^+\|^2$ and a convex constraint set R so that numerical solutions are quite feasible to obtain. Although $\tilde{r}$ is not necessarily unique, its truncation $\tilde{r}^+$ is (see App. A). A sufficient condition for $\tilde{r}$ to be unique, which is often satisfied in practice, is that no two payoffs in R have the same truncation.

B. No Riskless Payoff

Consider the more common case in which P does not contain a unit payoff. As in Section IIIB, augment **x** with a unit payoff and form an augmented payoff space P^a. Similarly, assign alternative strictly positive numbers v for $\pi(1)$ and extend π from P to P^a. Let R_v be the augmented set of payoffs with expected prices equal to one when $\pi(1)$ is assigned v. The counterpart to equation (20) is not guaranteed to have a solution, however. It turns out that there are additional limits on the admissible choices of v consistent with restriction 2.

To investigate these limits, we study the counterparts to optimization problem (24) using the augmented space of payoffs R_v in place of R. Define

$$\delta_v \equiv \inf_{r \text{ in } R_v} \|r^+\|^2. \tag{27}$$

When δ_v is positive, the bound on $\|m\|^2$ among the class of nonnegative random variables satisfying restrictions 1 and 2 with mean v is $1/\delta_v$. However, particular choices of v may result in δ_v being zero and hence $1/\delta_v$ being infinite. For instance, when there is a portfolio payoff p in P such that p is less than or equal to one with probability one and v is strictly less than $\pi(p)$, δ_v is zero. This is true because the random variable $(1 - p)/[v - \pi(p)]$ is in R_v and is less than or equal to zero with probability one. Consequently, the norm of its truncation is zero.

As noted by Merton (1973), Cox, Ross, and Rubinstein (1979), Harrison and Kreps (1979), and Kreps (1981), it is possible to obtain arbitrage bounds on the admissible (expected) prices that can be assigned to payoffs not in P. In the case of a unit payoff, the upper and lower bounds are given by

$$\overline{\pi}(1) \equiv \begin{cases} \inf\{\pi(p): p \geq 1\} & \text{if } \{p \text{ in } P: p \geq 1\} \text{ is not empty} \\ +\infty & \text{otherwise,} \end{cases}$$

$$\underline{\pi}(1) \equiv \sup\{\pi(p): p \leq 1\},$$

respectively.[10] Since the zero payoff is in P, $\underline{\pi}(1)$ is always nonnegative. The arbitrage bounds $\underline{\pi}(1)$ and $\overline{\pi}(1)$ determine the range of admissible values of Em that are compatible with m being a nonnegative random variable. Clearly, $\{v: \delta_v > 0\}$ must be a subset of the interval $[\underline{\pi}(1), \overline{\pi}(1)]$. In fact, the interiors of these sets coincide (see App. A).

When δ_v is strictly positive, there exists a minimum variance, nonnegative random variable with mean v that satisfies restriction 1 (see lemma A4 in App. A). Let this random variable be denoted $\tilde{m}_v$. The corresponding volatility bound is

$$\sigma(m) \geq \sigma(\tilde{m}_v), \tag{28}$$

and the family of random variables, $\{\tilde{m}_v: \delta_v > 0\}$, comprises the mean–standard deviation frontier for nonnegative random variables satisfying restriction 1. Thus the counterpart to the region S given in (9) is

$$S^+ \equiv \{(v, w): \delta_v > 0 \text{ and } w \geq \sigma(\tilde{m}_v)\}. \tag{29}$$

The set S^+ is convex. To see this consider two values of v for which δ_v is strictly positive, say $v(l) \leq v(u)$. Form convex combinations of the random variables $\tilde{m}_{v(l)}$ and $\tilde{m}_{v(u)}$. These convex combinations are nonnegative random variables that also satisfy restriction 1. Recall that the mean of a convex combination of random variables is equal to the convex combination of the means, and by the triangle inequality, the standard deviation of a convex combination is less than or equal to the convex combination of the standard deviations. While convex combinations of $\tilde{m}_{v(l)}$ and $\tilde{m}_{v(u)}$ are not necessarily on the mean–standard deviation frontier, the ordered pairs of their means and standard deviations must be in S^+. This is sufficient for S^+ to be convex.

---

[10] The characterization reported in Harrison and Kreps (1979) and Kreps (1981) is somewhat more complicated because they allow the counterpart to the space P to be infinite-dimensional.

Next we consider the incremental contribution of requiring m to be strictly positive as in restriction 2. It is shown in Appendix A that Em must be in the open interval $(\underline{\pi}(1), \overline{\pi}(1))$ (see lemma A6). Hence one effect of the imposition of strict positivity is that endpoints of the interval $\{v: \delta_v > 0\}$ are eliminated.

For v's in $(\underline{\pi}(1), \overline{\pi}(1))$, $\tilde{m}_v$ can be interpreted as either a call or put option on a payoff in P. More precisely, we obtain the counterpart to the result in Section IVA that $\tilde{m}_v = (\tilde{p}_v - k)^+$, where $\tilde{p}_v$ is a portfolio payoff in P and k is in $\mathbb{R}$. When k is nonnegative, $\tilde{m}_v$ is a call option on a portfolio with payoff $\tilde{p}_v$ and strike price k; when k is negative, $\tilde{m}_v$ is a put option on a portfolio with payoff $-\tilde{p}_v$ and strike price $-k$. Therefore, for any v in $(\underline{\pi}(1), \overline{\pi}(1))$, the counterpart to equation (20) has a solution.

As in subsection A, this gives us a simple check of the incremental impact of positivity on the volatility bounds given in (28). For any v in $(\underline{\pi}(1), \overline{\pi}(1))$ such that $\tilde{m}_v$ is strictly positive, the bound in (28) cannot be improved by restricting m to be strictly positive. This can occur only when $\tilde{m}_v$ coincides with m_v calculated in Section IIIB. On the other hand, for any v for which m_v is not strictly positive, the weak inequality ($\geq$) in (28) is replaced by a strong inequality ($>$).

Even though S^+ may be a proper subset of S, the region S is still of interest for a variety of reasons. First, S is easier to use in practice because a characterization of S^+ may require that a nonquadratic optimization problem be solved for each value of v. Second, the lower boundaries of S^+ and S coincide for values of v for which m_v is nonnegative. Consequently, it is advantageous to characterize S as a first step in characterizing S^+ and then check for nonnegativity of m_v. Finally, even for values of v for which m_v is negative with positive probability, the coefficients on $\mathbf{x}^a$ given in representation (11), when scaled appropriately, can be used as starting values for a numerical search routine used in computing δ_v.

In figure 4 we report plots of the regions S and S^+ for the same financial data set that was used to generate figure 1. The region S^+ is shaded, and the lower boundary of the region S is given by the dashed line below S^+. While the lower boundaries of these regions coincide for points closest to the horizontal axis, they diverge for other points. The divergence between the boundaries is greater when the volatility bounds are more restrictive. Recall that in generating the lower boundary of S, we constructed random variables m_v with mean v that satisfy restriction 1 and are linear combinations of $\mathbf{x}^a$. When these random variables have large standard deviations relative to their means, it is not surprising that they are negative with high probability. As a result, the positivity restriction 2 often has more bite when $\sigma(m_v)/v$ is larger.

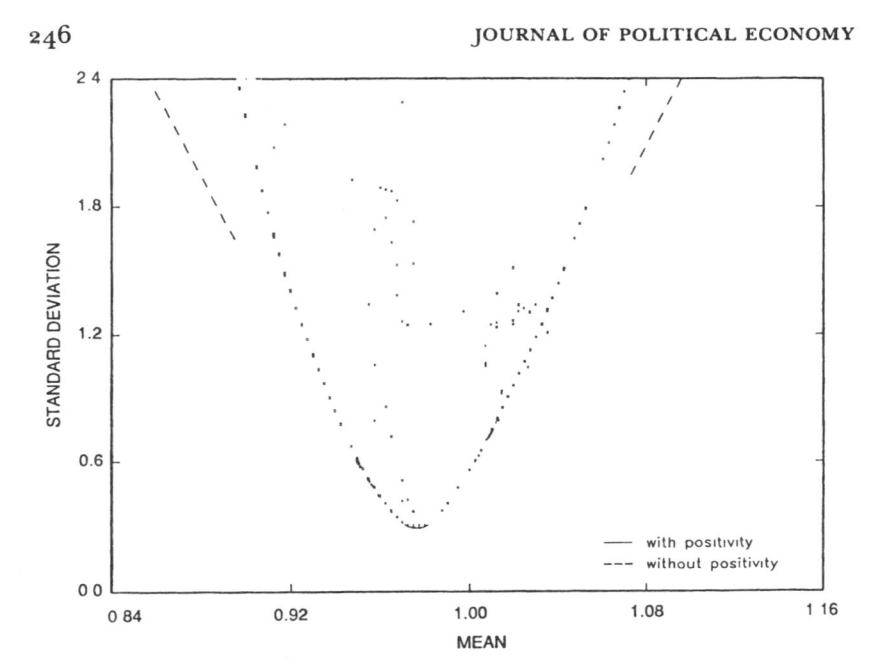

Fig. 4.—IMRS frontier with and without positivity imposed

As is true for S, the dimensionality of P can sometimes be reduced prior to the construction of S^+. Suppose that members of P have factor decompositions of the form $p = \gamma \cdot \mathbf{f} + e$, where $\mathbf{f}$ is a vector of common factors. Suppose further that the idiosyncratic components of the payoffs satisfy $E(e|\mathbf{f}) = 0$ and $\pi(e) = 0$. Hence we have exact factor pricing and each payoff p in P is a mean-preserving spread of a payoff $\gamma \cdot \mathbf{f}$ with the same price. Consequently, $\|p^+\|^2 \geq \|(\gamma \cdot \mathbf{f})^+\|^2$ because the function $[(p)^+]^2$ of p is convex. Hence when one solves (24) or (27), it suffices to restrict attention to linear combinations of the factors with expected prices equal to one. While it is evident how to use this reduction when the factors are observed, unobserved factors are problematic because it may be difficult to compute or estimate $\|(\gamma \cdot \mathbf{f})^+\|^2$ for arbitrary vectors γ. Because of the truncation of $(\gamma \cdot \mathbf{f})$, calculating $\|(\gamma \cdot \mathbf{f})^+\|$ requires knowledge of the entire probability distribution of $\mathbf{f}$, whereas typical factor analytic procedures identify only the first two moments of $\mathbf{f}$.

C. Viability of Equilibrium Pricing Functions and Arbitrage Pricing

The analysis in this section is intimately connected to general treatments of pricing derivative claims (see, e.g., Ross 1978; Harrison and

Kreps 1979; Kreps 1981). Among other things, Harrison and Kreps and Kreps consider the following question. Given a set of payoffs on primitive securities and the prices of those securities, when is it possible to extend the pricing function to a larger collection of payoffs in such a way as to preserve no arbitrage? As emphasized by Kreps, this experiment should not be construed as introducing new markets in an economy that might alter the resulting competitive equilibrium allocations. It is merely a hypothetical extension leaving intact the (expected) prices of the payoffs in P. When such an extension is possible, Harrison and Kreps and Kreps refer to the pricing function as being *viable*.

Throughout the analysis in this section, we have presumed that the family of m's that satisfy restrictions 1 and 2 is not empty. Clearly this is sufficient to eliminate arbitrage opportunities on P. Rather than assume that this family is not empty, an alternative starting point is to verify that no arbitrage opportunities exist on P and then to appeal to theorem 3 in Kreps (1981) to show that π can be extended from P to the collection L^2 of all random variables that are (Borel measurable) functions of $\mathbf{x}$ and have finite second moments.[11] The existence of an m satisfying restrictions 1 and 2 then follows from the Riesz representation theorem applied to L^2 (see also lemma 2.3 in Hansen and Richard [1987]).

V. Illustrations and Discussion

We now illustrate our analysis with alternative parametric models of m and alternative data sets on asset payoffs and prices. The model of m described in the Introduction and used to generate figure 1 assumed that consumers' preferences are separable over time and states of the world. In subsection A we investigate the impact on m of relaxing time separability. In subsection B we focus on logarithmic risk preferences but do not require these preferences to be state separable. Finally, in subsection C we describe the implications of price data on short-term Treasury bills for IMRSs and comment briefly on the implications for monetary models.

A. *Preferences That Are Not Time Separable*

Consider the following stylized version of a model with time nonseparabilities in preferences. As in the Introduction, we use a time- and

---

[11] In addition to a *no-arbitrage* restriction, Kreps also imposed a *no-free-lunch* restriction on (P, π). As demonstrated by Clark (1990), this extra restriction is not needed when P is a closed subspace of an L^2 space, as is true in our analysis.

state-separable specification of preferences for *consumption services* with a power utility function

$$E \sum_{t=0}^{\infty} \lambda^t \frac{s_t^{\gamma+1} - 1}{\gamma + 1}, \tag{30}$$

except now s_t depends on measured consumption in the current period and one previous period:

$$s_t = c_t + \theta c_{t-1}. \tag{31}$$

More general versions of this model have been investigated by Dunn and Singleton (1986), Eichenbaum, Hansen, and Singleton (1988), Gallant and Tauchen (1989), and Eichenbaum and Hansen (1990). We shall proceed as if there is a single representative consumer. As noted by Wilson (1968) and Rubinstein (1974), this assumption can be relaxed when θ is zero, the consumption allocations are consistent with the existence of complete contingent-claims markets, and all consumers have the same preferences. This aggregation result also applies more generally, say when θ is different from zero, as long as there are, in effect, complete markets in consumption services (see Eichenbaum, Hansen, and Richard 1987). When θ is positive, consumption generates positive services in the current as well as in one subsequent time period. In this case there is *intertemporal substitution* in generating consumption services from consumption goods. More precisely, there is a durable component to consumption that depreciates fully after one time period. Alternatively, when θ is negative, there is *intertemporal complementarity* in generating consumption services from consumption goods. Put somewhat differently, the term $-\theta c_{t-1}$ is a component of current-period consumption that reflects either *committed consumption* from the previous time period or *habit persistence*. Sundaresan (1989), Constantinides (1990), and Novales (1990) have argued that habit persistence may be important in explaining the relation between asset market data and economic aggregates.

For these forms of time nonseparabilities, the marginal utility of consumption is

$$mu_\tau = (s_\tau)^\gamma + \lambda \theta E[(s_{\tau+1})^\gamma | I_\tau].$$

The IMRS between time 0 and time τ is the corresponding ratio of marginal utilities scaled by λ^τ. Constructing m requires computation of the conditional expectation $E[(s_{\tau+1})^\gamma | I_\tau]$ except in the special case in which θ is zero.

To illustrate what impact positive and negative values of θ have for the volatility of m, we report calculations from Gallant et al. (1990).

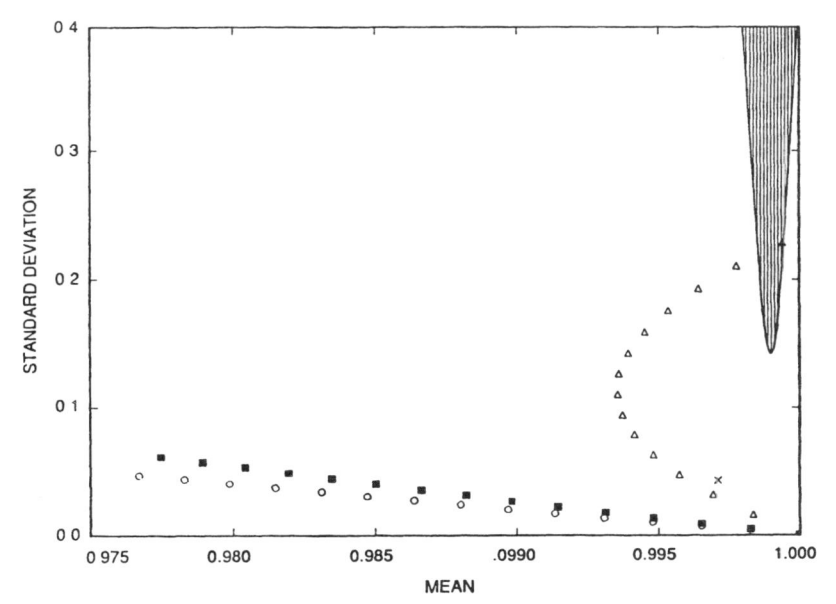

FIG. 5.—IMRS frontier computed using monthly data

For these calculations the ratio c_t/c_{t-1} is a component of a Markov process with a stochastic law of motion estimated by Gallant and Tauchen (1989) for monthly data on the consumption of nondurables and services (for more details, see Gallant and Tauchen [1989] and Gallant et al. [1990]). The estimated law of motion was then used to compute $E[(s_{\tau+1})^\gamma | I_\tau]$ required in forming a time series for m. Sample means and standard deviations were calculated for m's implied by alternative values of γ and θ.[12]

For this illustration we let $\theta = -.5$, $\theta = 0$, and $\theta = .5$. The results are reported in figure 5. The boxes are used to denote mean–standard deviation pairs for $\theta = 0$, the circles for $\theta = .5$, and the triangles for $\theta = -.5$. For each choice of θ, we let γ range from zero to -14 with decrements of minus one. In all cases the subjective discount factor λ is set to one. Smaller values of λ decrease proportionately the mean and standard deviation of m. When $\gamma = 0$, m is one for all choices of θ. In this case, $[E(m), \sigma(m)] = (1, 0)$.

Consider first the case in which $\theta = 0$. Increasing $|\gamma|$ magnifies the volatility of m but initially reduces its mean. Extrapolated much fur-

---

[12] Note that the calculations of mean–standard deviation pairs for m when $\theta = 0$ do not exploit the Markov specification estimated by Gallant and Tauchen (1989) and are consequently more robust.

ther, the curve (indexed by γ) does not turn around until $|\gamma|$ is in the vicinity of 100; afterward, increasing $|\gamma|$ enlarges the mean of m. The initial decline in the mean of m reflects the dominant role of positive growth rates in consumption. For extremely large values of $|\gamma|$, observations with negative growth rates in consumption come to dominate the sample mean, eventually resulting in a change of slope of the curve. In comparing the curves denoted by boxes in figures 1 and 5, recall that the long annual time series used to generate figure 1 contains negative growth rate observations on consumption during the depression. The absence of *bad* events in the monthly data set is responsible for the fact that the curve (indexed by γ) does not turn until the magnitude of γ is substantial.[13]

Consider next the case in which $\theta = .5$. Not surprisingly, introducing this local durability into preferences reduces the volatility of m. The quantitative effect of this smoothing does not appear to be very substantial, however. The curves for $\theta = .5$ and $\theta = 0$ are similar for the range of γ's that are plotted. Hence there is little adverse effect on the volatility of m to introducing durability by setting $\theta = .5$.

Finally, consider the case in which $\theta = -.5$. This intertemporal complementarity has the anticipated impact of increasing the volatility of m for a given value of γ. This effect is quite dramatic as indicated in figure 5. Furthermore, the value of $|\gamma|$ at which the curve turns is reduced dramatically. For $\theta = -.5$ the turning point for γ is in the vicinity of -7, and the initial decline in the mean of m is much less dramatic.

We now compare the three curves, which describe alternative mean–standard deviation pairs for parametric models of m, to a region S^+ generated using monthly data on asset payoffs and prices. The asset market data are the same as those used by Hansen and Singleton (1982) except that data revisions were incorporated and more recent data points were included. The resulting time period is 1959:3–1986:12. The first two asset payoffs are the 1-month real return on Treasury bills and the 1-month real value-weighted return on the New York Stock Exchange. Six additional time series of asset payoffs were constructed using these data by scaling the original two payoffs and prices by the one-period lagged returns and the one-period lag in the consumption ratio. For the range of hypothetical means considered, the region S described in Section III was essentially the same as the region S^+ described in Section IV.

---

[13] The *sample* volatility of m may be substantially lower than the *population* volatility if consumers anticipate that extremely bad events can occur with small probability when such events do not occur in the sample. Reitz (1988) argued that this phenomenon could explain the equity premium puzzle.

For the specification of preferences with $\theta = 0$, larger values of $|\gamma|$ initially make the mean–standard deviation pair for m further from the S^+ region because of the adverse effect on the mean of m. This is consistent with the fact that Hansen and Singleton (1982) found point estimates for γ that were close to zero but substantial evidence against the overidentifying restrictions. As emphasized by Singleton (1990), estimates of the discount factor λ are often greater than one when bond returns are included in the analysis. For a fixed γ, enlarging λ has the desired effect of increasing proportionately the mean and standard deviation of m.

From the vantage point of figure 5, the case for intertemporal complementarities in preferences is appealing. For a given value of γ, a negative value of θ increases both the mean and the standard deviation of m. However, it is quite possible for m to have a mean and standard deviation in S^+ and not satisfy restriction 1. In other words, for a given parametric specification of m, requiring $[E(m), \sigma(m)]$ to be in S^+ does not exhaust the testable implication of restriction 1. As emphasized by Gallant et al. (1990), there is substantial statistical evidence that the resulting m's violate restriction 1. In fact, empirical studies that use similar data and preference specifications, such as Dunn and Singleton (1986), Eichenbaum, Hansen, and Singleton (1988), and Eichenbaum and Hansen (1990), typically find parameter estimates that reflect intertemporal substitution ($\theta > 0$), although they find statistical evidence against the resulting parametric model of m.

B. Logarithmic Risk Preferences

In (30) and (31), suppose that θ is zero and γ is minus one. In this case, preferences are logarithmic. As noted by Rubinstein (1976), m is equal to the reciprocal of the return on the wealth portfolio of the representative consumer between time 0 and time τ (see also Brown and Gibbons 1985). Epstein and Zin (this issue) showed that this same conclusion applies to a parametric class of recursive preferences that are not state separable so long as the risk preferences remain logarithmic. Whereas in the state-separable case the return on the wealth portfolio is equal to the discounted consumption ratio, this exact relation no longer applies when state separability is relaxed. Nevertheless, the return on the wealth portfolio can still be used as a valid measure of m.

For this reason we have included a "cross" in figures 1 and 5. In figure 1, this cross denotes the sample mean–standard deviation pair for the reciprocal of the measured annual return on the Standard and Poor's 500 stock price index, and in figure 5 it represents the sample mean–standard deviation pair for the reciprocal of the measured monthly value-weighted return on the New York Stock Exchange. In both cases the means are near the points in S^+ that are closest to the horizontal axis, but the mean–standard deviation pair is outside of S^+. However, after a sampling error is taken, there is very little evidence against the null hypothesis that this model of m satisfies restriction 1.[14]

C. Treasury Bill Data and Monetary Models

We also calculated the regions S and S^+ using monthly data on 3-month holding period returns on Treasury bills. The holding period returns were constructed using bond prices on 3-, 6-, 9-, and 12-month discount bonds for 1964:7–1986:12. Nominal returns were converted to real returns using the implicit price deflator on nondurables and services. These bond price data (with the most recent periods excluded) have been used by Fama (1984), Dunn and Singleton (1986), and Stambaugh (1988), among others, to investigate time variation in risk premia and particular models of bond prices. In figure 6 we report the resulting regions S and S^+. The region S^+ is shaded, and the lower boundary of the region S is given by the dashed line below S^+. The resulting standard deviation bounds for m are quite striking. For means of m in the vicinity of one, the bound on the standard deviation is near one. Given the magnitude of these bounds, it is not surprising that restriction 2 has an important incremental contribution vis-à-vis restriction 1.

The bounds reported in figure 6 appear to us to pose quite a challenge to a large class of asset valuation models. For instance, the quarterly counterpart to the $\theta = 0$ curve in figure 5 ranges from (1, 0) to (.90, .08) as γ ranges from zero to -14. Volatility bounds of a similar magnitude were also obtained using monthly data on 1-month holding period returns for Treasury bills with maturities from 1 to 6 months. These bounds are directly comparable to the three curves plotted in figure 5. Since these latter bounds apply to IMRSs measured over a shorter time period (1 month instead of 3 months), they are even more startling. However, short-term Treasury bills are often held to maturity, and trading of these Treasury bills takes place in secondary markets except for the 3-, 6-, and 12-month bills. The bid-ask spreads for the short-term bills can be quite substantial (see Stambaugh 1988; Knez, Litterman, and Scheinkman 1989), so the

---

[14] In fig. 1, one of the two moment conditions, $E(m\mathbf{x} - \mathbf{q}) = 0$, is satisfied by construction. The other condition was tested using the method suggested in Hansen and Singleton (1982): the $\chi^2(1)$ statistic is 1.40 with probability value .24. Similarly, for fig. 5, four of the eight moment conditions are satisfied by construction. The $\chi^2(4)$ statistic for the four remaining conditions is 4.88 with a probability value of .30.

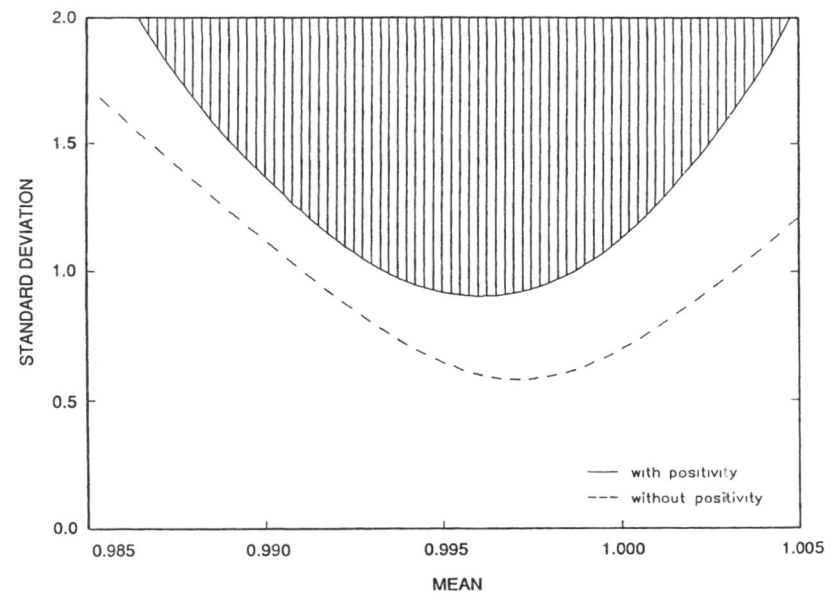

FIG. 6.—IMRS frontier computed using quarterly holding period returns

prices used in our calculations may be less reliable. These concerns should be less problematic for the results in figure 6 since they were computed using price data from the more richly traded 3-, 6-, 9-, and 12-month Treasury bills.

As emphasized by Knez et al., short-term Treasury bills often may be held to maturity as cash substitutes for particular transactions. As such, these bills may generate important liquidity services that are not measured appropriately by the implied ex post real returns. Hence the measured real returns may understate the value of the assets to the holders of the securities. Recall from Section IIIC that large standard deviation bounds for m occur when the slope of the mean–standard deviation frontier for R is steep. For the Treasury bill data, this means that the reason that the volatility bounds on m are large (as reflected in the region S) is that the expected short-term gain associated with holding longer-term bills is large relative to the increase in the standard deviation. Abstracting from the liquidity services of the short-term bills may distort the magnitude of the resulting volatility bounds on m.

Refinements of real asset pricing models to incorporate money, such as the cash-in-advance models of Svensson (1985), Lucas and Stokey (1987), and Townsend (1987), are designed to accommodate the rate of return dominance between one-period bonds and money.

However, in their current form they are not well suited to differentiate among short-term Treasury bills with different maturity dates. Although the link between measured *real* IMRSs and security market data may be confounded in these models, there is an alternative notion of the *indirect* IMRS for money that reflects the fact that the cash-in-advance constraint may not always be binding. Hence for an appropriate interpretation of m, these monetary models are compatible with restrictions 1 and 2 as long as money is not included among the vector of assets used to generate S^+.

VI. Conclusions and Extensions

In this paper we have characterized the implications of security market data for means and standard deviations of IMRSs. This exercise is important in evaluating alternative models of dynamic economies because IMRSs are the channels by which the attributes of these models impinge on asset prices. Abstracting from the restriction that IMRSs are positive, we established the connection between volatility bounds on IMRSs and mean–standard deviation frontiers for asset payoffs. Thus we showed how diagnostics commonly used in empirical finance can be translated into information about IMRSs. We also showed how to extract sharper volatility bounds by taking account of the fact that IMRSs should be positive. These sharper bounds exploit more fully the absence of arbitrage opportunities in the underlying economic environment than, say, linear factor representations of asset prices.

There are three important directions in which the ideas in this paper can be developed further. An earlier version of this paper has already provoked some work along these three lines.

i) In this paper we focused exclusively on deriving implications for IMRSs expressed in terms of population moments of asset payoffs and prices. In practice, these attributes of asset market data will not be known a priori, but can be approximated only by using time-series averages in place of population moments. This introduces sampling error into the analysis. A major drawback of the discussion in Section V is that it abstracted from the presence of approximation error introduced by using sample averages from historical time series in place of population moments. Hansen and Jagannathan (1990) show how to use large sample theory both to assess whether there is sufficient statistical evidence to reject that the bounds are degenerate (equal to zero) and to assess the magnitude of the approximation errors.

ii) The restrictions on IMRSs derived in this paper all pertain to first and second moments. More generally, it would be desirable to characterize the admissible family of distributions for IMRSs given asset market data. An additional step toward such a characterization is taken by Snow (1990), who shows how to extend the analysis in this paper to obtain bounds on other moments of the IMRSs.

iii) The diagnostics derived in this paper can be applied to any intertemporal asset pricing model for which moments of m can be computed. While the calculations in Section V were performed by first constructing hypothetical time series on m, such a construction is not necessary. All that is really essential is the ability to compute the moments of m implied by the model. As an alternative to constructing a time series on m, these moments can be deduced from the equilibrium stochastic law of motion for the model (see, e.g., Heaton 1990). Therefore, calculations like those illustrated in Section V can be performed for an extensive array of intertemporal asset pricing models including models that take account of measurement errors in consumption, seasonality, and aggregation-over-time biases.

Appendix A

Let L^2 be the Hilbert space of all random variables with finite second moments that are Borel measurable functions of $\mathbf{x}$. Let P be a closed linear subspace of L^2 and π be a continuous linear functional on P. In contrast to the analysis in the text, we allow the space P to be infinite-dimensional. Define $R \equiv \{r \in P: \pi(r) = 1\}$, $R^+ \equiv \{r^+: r \in R\}$, and $Z \equiv \{z \in P: \pi(z) = 0\}$. Throughout our analysis we assume that R is not empty. Let C denote the closure (in L^2) of R^+. In this Appendix we establish several results that support conclusions in Section IV.

Consider the following two minimum norm problems. The first problem is

$$\delta \equiv \inf_{r \in R} \|r^+\|^2. \tag{P1}$$

A closely related minimum norm problem is

$$\eta \equiv \inf_{y \in C} \|y\|^2. \tag{P2}$$

This second problem has the advantage that the inf is attained.

There are two additional problems that are closely related to (P1) and (P2). The first one is an orthogonality problem:

$$\text{Find } \hat{y} \in C \text{ such that } E(\hat{y}z) = 0 \text{ for all } z \in Z. \tag{P3}$$

As in standard minimum norm problems on Hilbert spaces, it is often the case that (P3) has the same solution as (P1) and (P2). The focal point of our analysis is the following problem:

Find $y^* \in L^2$ such that $y^* \geq 0$ and $\|y^*\|^2 = \dfrac{1}{\delta}$ and $\pi(p) = E(y^*p)$ for all $p \in P$.

$$\tag{P4}$$

We now investigate the relation among these four problems. First we establish the connection between (P1) and (P2).

LEMMA A1. There is a unique $\bar{y}$ in C such that $\|\bar{y}\|^2 = \delta = \eta$.

Proof. Let $\{r_j\}$ be a sequence in R such that $\{\|(r_j)^+\|^2\}$ converges to δ. Then for any positive integers j and k,

$$\|(r_j)^+ - (r_k)^+\|^2 = -\|(r_j)^+ + (r_k)^+\|^2 + 2\|(r_j)^+\|^2 + 2\|(r_k)^+\|^2$$

$$\leq -\|(r_j + r_k)^+\|^2 + 2\|(r_j)^+\|^2 + 2\|(r_k)^+\|^2$$

because $\|(r_j)^+ + (r_k)^+\|^2 \geq \|(r_j + r_k)^+\|^2$. Since r_j and r_k are both in R, $(r_j/2) + (r_k/2)$ is also in R. Consequently, $\|[(r_j/2) + (r_k/2)]^+\|^2 \geq \delta$, and

$$\|(r_j)^+ - (r_k)^+\|^2 \leq -4\left\|\left(\frac{r_j}{2} + \frac{r_k}{2}\right)^+\right\|^2 + 2\|(r_j)^+\|^2 + 2\|(r_k)^+\|^2 \quad \text{(A1)}$$

$$\leq -4\delta + 2\|(r_j)^+\|^2 + 2\|(r_k)^+\|^2.$$

If we take limits as $j, k \to \infty$, it follows that $\{(r_j)^+\}$ is Cauchy and hence converges to some $\bar{y}$ in C. Therefore, $\{\|(r_j)^+\|^2\}$ converges to $\|\bar{y}\|^2 = \delta$.

Since C is the closure of R^+, for any y in C there is a sequence $\{\|(r_j)^+\|^2\}$ that converges to $\|y\|^2$. Therefore, $\eta = \delta$.

Finally, let $\hat{y}$ be any member of R^+ for which $\|\hat{y}\|^2 = \delta$, and let $\{\hat{r}_j\}$ be a sequence in R such that $\{(\hat{r}_j)^+\}$ converges to $\hat{y}$. Analogous to (A1),

$$\|(r_j)^+ - (\hat{r}_j)^+\|^2 \leq -4\delta + 2\|(r_j)^+\|^2 + 2\|(\hat{r}_j)^+\|^2.$$

Since $\{\|(r_j)^+\|^2\}$ and $\{\|(\hat{r}_j)^+\|^2\}$ both converge to δ, $\{(r_j)^+\}$ and $\{(\hat{r}_j)^+\}$ have the same limit points. Therefore, $\hat{y}$ and $\bar{y}$ are equal (with probability one). Q.E.D.

Next we establish the connection between (P2) and (P3).

LEMMA A2. A solution $\bar{y}$ to (P2) is also a solution to (P3).

Proof. To prove this result we use the following inequality:

$$[(r + cz)^+]^2 \leq (r^+ + cz)^2. \quad \text{(A2)}$$

To see that it holds, first suppose that $r + cz \leq 0$. In this case the left side of (A2) is zero while the right is greater than or equal to zero. Second, suppose that $r + cz \geq 0$. Then $0 \leq r + cz \leq r^+ + cz$, which also implies (A2).

Let $\{r_j\}$ be a sequence in R such that $\{(r_j)^+\}$ converges to $\bar{y}$, and let z be any member of Z distinct from zero. Then

$$\liminf_{j \to \infty} \|(r_j + cz)^+\|^2 \leq \|\bar{y} + cz\|^2. \quad \text{(A3)}$$

The right side of (A3) is minimized by $\bar{c} = -E(\bar{y}z)/E(z^2)$. In order for $\bar{y}$ to be the solution to (P2), it must be that $\bar{c} = 0$ or, equivalently, that $E\bar{y}z = 0$. Q.E.D.

Lemma A2 has the following partial converse.

LEMMA A3. If $\hat{r}^+ \in R^+$ is the solution to (P3) and the solution to (P2) is in R^+, then $\hat{r}^+$ is the solution to (P2).

Proof. Let $\bar{r}^+$ denote the solution to problem (P2). It follows from lemma A2 that $E[(\bar{r}^+ - \hat{r}^+)(\bar{r} - \hat{r})] = 0$. Also, $(\bar{r})^+\hat{r} \leq (\bar{r})^+(\hat{r})^+$ and $\bar{r}(\hat{r})^+ \leq (\bar{r})^+(\hat{r})^+$. Hence

$$0 = E[(\bar{r}^+ - \hat{r}^+)(\bar{r} - \hat{r})] \geq E[(\bar{r}^+ - \hat{r}^+)^2] \geq 0.$$

Therefore, $\bar{r}^+ = \hat{r}^+$ (with probability one). Q.E.D.

We now use the Hahn-Banach theorem and the Riesz representation theorem to establish the existence of a solution to (P4).

LEMMA A4. If $\delta > 0$, (P4) has a solution.

Proof. The first half of this proof follows closely the proof of lemma 1 in Kreps (1981). Since $\delta > 0$,

$$\pi(p) \leq \left(\frac{1}{\delta}\right)^{1/2} \|p^+\| \quad \text{for all } p \in P. \tag{A4}$$

Among other things, inequality (A4) implies that $\pi(p) \geq 0$ whenever $p \geq 0$ because π is linear and $(-p)^+$ is zero. The right side of (A4), $(1/\delta)\|(\cdot)^+\|$, is a particular version of the sublinear function used by Kreps in applying the Hahn-Banach theorem to extend P to a larger space, say L^2. (The analogues to the spaces P and L^2 are much more general in Kreps's analysis.) Let Π denote such an extension. Then Π satisfies the counterpart to (A4):

$$\Pi(y) \leq \left(\frac{1}{\delta}\right)^{1/2} \|y^+\| \quad \text{for all } y \in L^2. \tag{A5}$$

Clearly Π is continuous and $\Pi(y) \geq 0$ whenever $y \geq 0$. It follows from the Riesz representation that there exists a $y^* \in L^2$ such that

$$\Pi(y) = E(y^*y) \quad \text{for all } y \in L^2. \tag{A6}$$

It remains to show that $y^* \geq 0$ and $\|y^*\| = (1/\delta)^{1/2}$. Consider any $r \in R$ and note that $\Pi(r^+) \geq \Pi(r) = 1$. Since (A6) is satisfied, it follows from the Cauchy-Schwarz inequality that $\|y^*\|\|r^+\| \geq E(y^*r^+) \geq 1$. Consequently,

$$\|y^*\|\delta^{1/2} = \|y^*\| \inf_{r \in R} \|r^+\|^2 \geq 1,$$

or, equivalently, $\|y^*\| \geq (1/\delta)^{1/2}$. Relations (A5) and (A6) imply

$$\|y^*\|^2 = \Pi(y^*) \leq \left(\frac{1}{\delta}\right)^{1/2} \|(y^*)^+\| \leq \left(\frac{1}{\delta}\right)^{1/2} \|y^*\|.$$

Therefore, $\|y^*\| = \|(y^*)^+\| = (1/\delta)^{1/2}$. Q.E.D.

For our next set of results we find it convenient to restrict (P, π) to satisfy the no-arbitrage condition:

For any $p \in P$ such that $p \geq 0$ and $\|p\| > 0$, $\pi(p) > 0$.

LEMMA A5. If (P, π) satisfies the no-arbitrage condition, then $\delta > 0$ and the solution to (P2) is in R^+.

Proof. Let $\{r_j\}$ be a sequence in R such that $\{(r_j)^+\}$ converges to $\tilde{y}$, where $\tilde{y}$ is a solution to (P2). Our goal is to show that there exists a convergent subsequence of $\{r_j\}$ with limit payoff $\tilde{r}$. Given this convergence, we then argue that $(\tilde{r})^+ = \tilde{y}$.

First we show that $\{\|r_j\|^2\}$ is bounded. Suppose to the contrary that $\{\|r_j\|^2\}$ is unbounded. Without loss of generality, we may assume that this sequence is increasing (otherwise we could extract a subsequence that is increasing and unbounded). Form $p_j = r_j/\|r_j\|$. Since the closed unit ball of L^2 is weakly compact, $\{p_j\}$ has a subsequence that converges weakly to a random variable $\tilde{p}$ in L^2. The weak limit of a sequence in P must be orthogonal to the orthogonal complement of P. In other words, $\tilde{p}$ must be in P. Furthermore, $\pi(\tilde{p}) = 0$ because $\pi(p_j) = 1/\|r_j\|$ and $\{\|r_j\|\}$ is increasing and unbounded. The sequence $\{(p_j)^+\}$ converges strongly to zero because $\|(p_j)^+\| = \|(r_j)^+\|/\|r_j\|$ and $\{(r_j)^+\}$ converges strongly to $\tilde{y}$. Form the orthogonal decomposition $p_j = (p_j)^+ - (-p_j)^+$.

Then $\{\|(-p_j)^+\|\}$ converges to one and a subsequence of $\{(-p_j)^+\}$ converges weakly to $-\hat{p}$. It follows from Clark (1990, lemma 2) that $-\hat{p} \geq 0$ and $\|-\hat{p}\| > 0$. This finding violates the no-arbitrage condition because $\pi(-\hat{p}) = 0$. Therefore, $\{\|r_j\|\}$ must be bounded.

Since $\{\|r_j\|\}$ is bounded, $\{r_j\}$ has a subsequence that converges weakly to a limit point $\tilde{r}$ in R. Consequently, $\{(-r_j)^+\}$ has a subsequence that converges weakly to $\tilde{y} - \tilde{r} \geq 0$. Note that

$$0 = E[(r_j)^+(-r_j)^+] = E\{(-r_j)^+[(r_j)^+ - \tilde{y}]\} + E[(-r_j)^+\tilde{y}]. \quad (A7)$$

It follows from the Cauchy-Schwarz inequality that

$$|E\{(-r_j)^+[(r_j)^+ - \tilde{y}]\}| \leq \|(-r_j)^+\|\|(r_j)^+ - \tilde{y}\| \quad (A8)$$
$$\leq \|r_j\|\|(r_j)^+ - \tilde{y}\|.$$

The right side of (A8) converges to zero because $\{\|r_j\|\}$ is bounded and $\{(r_j)^+\}$ converges strongly to $\tilde{y}$. If we take limits of the right side of (A7) along the weakly convergent subsequence, it follows that $\tilde{y} - \tilde{r}$ is orthogonal to $\tilde{y}$. Since both of these random variables are nonnegative and $\tilde{r} = \tilde{y} + (\tilde{r} - \tilde{y})$, $(\tilde{r})^+ = \tilde{y}$.

To verify that $\delta > 0$, suppose to the contrary that $\delta = 0$. In this case, $-\tilde{r} \geq 0$, implying a violation of the no-arbitrage condition because $\pi(-\tilde{r}) = -1$ and π is linear. Q.E.D.

In light of lemmas A1, A2, A3, and A5, when (P, π) satisfies the no-arbitrage condition, the solutions to (P1)–(P3) coincide and are in R^+. As is shown in Section IV, in this case a solution to (P4) is given by $y^* = \tilde{y}/\|\tilde{y}\|^2$, where $\tilde{y}$ is the solution to (P1)–(P3).

Consider now the special case in which 1 is not in P. As in Section IV, let $P^a \equiv P + \{1\}$ and define the arbitrage bounds

$$\overline{\pi}(1) \equiv \inf\{\pi(p): p \geq 1\},$$
$$\underline{\pi}(1) \equiv \sup\{\pi(p): p \leq 1\}.$$

Extend π from P to P^a by assigning v to 1, and let π_v denote the resulting extension.

LEMMA A6. Suppose that (P, π) satisfies the no-arbitrage condition. Then (P^a, π_v) satisfies the no-arbitrage condition if and only if $v \in (\underline{\pi}(1), \overline{\pi}(1))$.

Proof. Part of this result is an implication of theorem 4 in Kreps (1981). For completeness we include a simple proof.

Suppose that $v \in (\underline{\pi}(1), \overline{\pi}(1))$. Let $p + w \geq 0$ for some $p \in P$ and some $w \neq 0$. If $w > 0$, then $p/(-w) \leq 1$ and $\pi[p/(-w)] \leq \underline{\pi}(1) < v$. Hence $\pi(p) + vw > 0$. A similar argument applies to the case in which $w < 0$.

Next suppose that (P^a, π_v) satisfies the no-arbitrage condition. Then clearly $v \in [\underline{\pi}(1), \overline{\pi}(1)]$. Suppose that $v = \underline{\pi}(1)$. Consider a sequence $\{p_j\}$ such that $p_j \leq 1$, $\pi(p_j) \leq v$, and $\{\pi(p_j)\}$ converges to v. Since 1 is not in P and P is closed, the lim inf of the sequence $\{\|(1 - p_j)\|\}$ is strictly positive. Form the sequence $\{(1 - p_j)/\|(1 - p_j)\|\}$. Following the logic of the proof of lemma A5, this sequence has a weakly convergent subsequence with a nonnegative limit point of the form $1 - p_o$ for some $p_o \in P$, where $v - \pi(p_o) = 0$ and $\|1 - p_o\| > 0$. However, this violates the no-arbitrage condition. A similar argument applies when $v = \overline{\pi}(1)$. Therefore, v must be in the open interval $(\underline{\pi}(1), \overline{\pi}(1))$. Q.E.D.

Appendix B

In this Appendix we describe in more detail the series used to perform the calculations underlying each of the figures.

Figures 1 and 4.—For a description of the stock, bond, and consumption data, see table 1 of Campbell and Shiller (1988) under the heading Cowles/S&P 500, 1871–1986.

Figures 2 and 3.—Monthly observations for 1959:4–1986:12 on 1-month holding period returns on 1-, 2-, 3-, 4-, 5-, and 6-month Treasury bills were constructed using bond prices from the Fama term structure yield file of data tapes from the Center for Research in Security Prices (CRSP) at the University of Chicago. Nominal returns were converted to real returns using the implicit price deflator for consumption of nondurables and services from the Personal Consumption Expenditure data tape of the National Income and Product Accounts.

Figure 5.—Monthly observations on the 1-month return on Treasury bills and on the 1-month value-weighted return on the New York Stock Exchange were taken from the CRSP data tapes. Nominal returns were converted to real returns using the implicit price deflator for the consumption of nondurables and services. Monthly observations on eight series of asset payoffs were constructed using these two returns. The first two payoffs are the two original returns. The prices of these payoffs are one by construction. The second two payoffs were formed by multiplying the two returns by the one-period lagged value of the real Treasury bill return. The prices of these two payoffs are equal to the one-period lag of the real Treasury bill return. The third two payoffs were formed by multiplying the original two returns by the one-period lagged value of the real value-weighted return. The prices of these two payoffs are equal to the one-period lag of the real value-weighted return. Finally, the last two payoffs are the original two payoffs multiplied by the ratio of per capita real consumption in the two previous time periods. The prices of the last two payoffs are both equal to the lagged consumption ratio. The time period is 1959:3–1986:12.

The consumption series was taken from the Personal Consumption Expenditure data tape of the National Income and Product Accounts, and the total population series from the Citibase data tape.

Figure 6.—The bond prices were taken from the Fama term structure yield file of the CRSP data tapes. Four monthly time series of 3-month holding period returns were constructed from the monthly price data on 3-, 6-, 9-, and 12-month discount bonds. Nominal returns were converted to real returns using the monthly implicit deflator for consumption of nondurables and services described previously. The time period is 1964:7–1986:12.

References

Arditti, Fred D., and John, Kose. "Spanning the State Space with Options." *J. Financial and Quantitative Analysis* 15 (March 1980): 1–9.

Breeden, Douglas T. "An Intertemporal Asset Pricing Model with Stochastic Consumption and Investment Opportunities." *J. Financial Econ.* 7 (September 1979): 265–96.

Breeden, Douglas T., and Litzenberger, Robert H. "Prices of State-Contingent Claims Implicit in Option Prices." *J. Bus.* 51 (October 1978): 621–51.

Breen, William; Glosten, Lawrence R.; and Jagannathan, Ravi. "Economic Significance of Predictable Variations in Stock Index Returns." *J. Finance* 44 (December 1989): 1177–89.

Brown, David P., and Gibbons, Michael R. "A Simple Econometric Approach for Utility-based Asset Pricing Models." *J. Finance* 40 (June 1985): 359–81.

Campbell, John Y., and Shiller, Robert J. "Dividend-Price Ratios and Expectations of Future Dividends and Discount Factors." *Rev. Financial Studies* 1 (Fall 1988): 195–228.

Chamberlain, Gary. "Funds, Factors, and Diversification in Arbitrage Pricing Models." *Econometrica* 51 (September 1983): 1305–23.

Chamberlain, Gary, and Rothschild, Michael. "Arbitrage, Factor Structure, and Mean-Variance Analysis on Large Asset Markets." *Econometrica* 51 (September 1983): 1281–1304.

Clark, Stephen A. "The Valuation Problem in Arbitrage Price Theory." *J. Math. Econ.* 19 (1990), in press.

Connor, Gregory. "A Unified Beta Pricing Theory." *J. Econ. Theory* 34 (October 1984): 13–31.

Connor, Gregory, and Korajczyk, Robert A. "Risk and Return in an Equilibrium APT: Application of a New Test Methodology." *J. Financial Econ.* 21 (September 1988): 255–89.

Constantinides, George M. "Habit Formation: A Resolution of the Equity Premium Puzzle." *J.P.E.* 98 (June 1990): 519–43.

Cox, John C.; Ross, Stephen A.; and Rubinstein, Mark. "Option Pricing: A Simplified Approach." *J. Financial Econ.* 7 (September 1979): 229–63.

Dunn, Kenneth B., and Singleton, Kenneth J. "Modeling the Term Structure of Interest Rates under Non-separable Utility and Durability of Goods." *J. Financial Econ.* 17 (July 1986): 27–55.

Dybvig, Philip H., and Ingersoll, Jonathan E., Jr. "Mean-Variance Theory in Complete Markets." *J. Bus.* 55 (April 1982): 233–51.

Eichenbaum, Martin S., and Hansen, Lars Peter. "Estimating Models with Intertemporal Substitution Using Aggregate Time Series Data." *J. Bus. and Econ. Statis.* 8 (January 1990): 53–69.

Eichenbaum, Martin S.; Hansen, Lars Peter; and Richard, Scott F. "Aggregation, Durable Goods and Nonseparable Preferences in an Equilibrium Asset Pricing Model." Working Paper no. 87-9. Chicago: Nat. Opinion Res. Center, Program Quantitative Analysis, 1987.

Eichenbaum, Martin S.; Hansen, Lars Peter; and Singleton, Kenneth J. "A Time Series Analysis of Representative Agent Models of Consumption and Leisure Choice under Uncertainty." *Q.J.E.* 103 (February 1988): 51–78.

Epstein, Larry G., and Zin, Stanley E. "Substitution, Risk Aversion, and Temporal Behavior of Consumption and Asset Returns: An Empirical Analysis." *J.P.E.*, this issue.

Fama, Eugene F. "The Information in the Term Structure." *J. Financial Econ.* 13 (December 1984): 509–28.

Gallant, Ronald; Hansen, Lars Peter; and Tauchen, George. "Using Conditional Moments of Asset Payoffs to Infer the Volatility of Intertemporal Marginal Rates of Substitution." *J. Econometrics* 45 (July/August 1990): 141–79.

Gallant, Ronald, and Tauchen, George. "Seminonparametric Estimation of Conditionally Constrained Heterogeneous Processes: Asset Pricing Applications." *Econometrica* 57 (September 1989): 1091–1120.

Green, R. C., and Jarrow, R. A. "Spanning and Completeness in Markets with Contingent Claims." *J. Econ. Theory* 41 (February 1987): 202–10.

Hansen, Lars Peter. "Consumption, Asset Markets, and Macroeconomic Fluctuation: A Comment." *Carnegie-Rochester Conf. Ser. Public Policy* 17 (Autumn 1982): 239–50.

Hansen, Lars Peter, and Jagannathan, Ravi. "Detection of Specification Errors in Stochastic Discount Factor Models." Manuscript. Chicago: Univ. Chicago, 1990.

Hansen, Lars Peter, and Richard, Scott F. "The Role of Conditioning Information in Deducing Testable Restrictions Implied by Dynamic Asset Pricing Models." *Econometrica* 55 (May 1987): 587–613.

Hansen, Lars Peter, and Singleton, Kenneth J. "Generalized Instrumental Variables Estimation of Nonlinear Rational Expectations Models." *Econometrica* 50 (September 1982): 1269–86.

Harrison, J. Michael, and Kreps, David M. "Martingales and Arbitrage in Multiperiod Securities Markets." *J. Econ. Theory* 20 (June 1979): 381–408.

Heaton, John. "An Empirical Investigation of Asset Pricing with Temporally Dependent Preference Specifications." Manuscript. Cambridge: Massachusetts Inst. Tech., 1990.

Huberman, Gur, and Kandel, Shmuel. "Mean-Variance Spanning." *J. Finance* 42 (September 1987): 873–88.

Knez, Peter; Litterman, Robert; and Scheinkman, José A. "Explorations into Factors Explaining Money Market Returns." Manuscript. New York: Goldman, Sachs, Financial Strategies Group, 1989.

Kreps, David M. "Arbitrage and Equilibrium in Economies with Infinitely Many Commodities." *J. Math. Econ.* 8 (March 1981): 15–35.

Lehmann, Bruce N., and Modest, David M. "The Empirical Foundations of the Arbitrage Pricing Theory." *J. Financial Econ.* 21 (September 1988): 213–54.

LeRoy, Stephen F. "Risk Aversion and the Martingale Property of Stock Prices." *Internat. Econ. Rev.* 14 (June 1973): 436–46.

Lucas, Robert E., Jr. "Asset Prices in an Exchange Economy." *Econometrica* 46 (November 1978): 1429–45.

Lucas, Robert E., Jr., and Stokey, Nancy L. "Money and Interest in a Cash-in-Advance Economy." *Econometrica* 55 (May 1987): 491–514.

Mehra, Rajnish, and Prescott, Edward C. "The Equity Premium: A Puzzle." *J. Monetary Econ.* 15 (March 1985): 145–61.

Merton, Robert C. "Theory of Rational Option Pricing." *Bell J. Econ. and Management Sci.* 4 (Spring 1973): 141–83.

Novales, Alfonso. "Solving Nonlinear Rational Expectations Models: A Stochastic Equilibrium Model of Interest Rates." *Econometrica* 58 (January 1990): 93–111.

Reitz, Thomas A. "The Equity Risk Premium: A Solution." *J. Monetary Econ.* 22 (July 1988): 117–31.

Ross, Stephen A. "The Arbitrage Theory of Capital Asset Pricing." *J. Econ. Theory* 13 (December 1976): 341–60. (*a*)

———. "Options and Efficiency." *Q.J.E.* 90 (February 1976): 75–89. (*b*)

———. "A Simple Approach to the Valuation of Risky Streams." *J. Bus.* 51 (July 1978): 453–75.

Rubinstein, Mark. "An Aggregation Theorem for Securities Markets." *J. Financial Econ.* 1 (September 1974): 225–44.

———. "The Valuation of Uncertain Income Streams and the Pricing of Options." *Bell J. Econ.* 7 (Autumn 1976): 407–25.

Shanken, Jay. "Multivariate Proxies and Asset Pricing Relations: Living with the Roll Critique." *J. Financial Econ.* 18 (March 1987): 91–110.

Shiller, Robert J. "Consumption, Asset Markets and Macroeconomic Fluctuations." *Carnegie-Rochester Conf. Ser. Public Policy* 17 (Autumn 1982): 203–38.

Singleton, Kenneth J. "Specification and Estimation of Intertemporal Asset Pricing Models." In *Handbook of Monetary Economics*, edited by Benjamin Friedman and Frank H. Hahn. Amsterdam: North-Holland, 1990.

Snow, Karl N. "Diagnosing Asset Pricing Models Using the Distribution of Asset Returns." Manuscript. Chicago: Univ. Chicago, 1990.

Stambaugh, Robert F. "The Information in Forward Rates: Implications for Models of the Term Structure." *J. Financial Econ.* 21 (May 1988): 41–70.

Sundaresan, Mahadevan. "Intertemporally Dependent Preferences and the Volatility of Consumption and Wealth." *Rev. Financial Studies* 2 (1989): 73–89.

Svensson, Lars E. O. "Money and Asset Prices in a Cash-in-Advance Economy." *J.P.E.* 93 (October 1985): 919–44.

Townsend, Robert M. "Asset-Return Anomalies in a Monetary Economy." *J. Econ. Theory* 41 (April 1987): 219–47.

Wilson, Robert B. "The Theory of Syndicates." *Econometrica* 36 (January 1968): 119–32.

[14]

Evaluating the Effects of Incomplete Markets on Risk Sharing and Asset Pricing

John Heaton and Deborah J. Lucas
Northwestern University and National Bureau of Economic Research

We examine an economy in which agents cannot write contracts contingent on future labor income. The agents face aggregate uncertainty in the form of dividend and systematic labor income risk, and also idiosyncratic labor income risk, which is calibrated using the PSID. The agents trade in financial securities to buffer their idiosyncratic income shocks, but the extent of trade is limited by borrowing constraints, short-sales constraints, and transactions costs. By simultaneously considering aggregate and idiosyncratic shocks, we decompose the effect of transactions costs on the equity premium into two components. The direct effect occurs because individuals equate the net-of-cost margins. A second, indirect effect occurs because transactions costs result in individual consumption that more closely tracks individual income. In the simulations we find that the direct effect dominates and that the model can produce a sizable equity premium only if transactions costs are large or the assumed quantity of tradable assets is limited.

We thank George Constantinides, Darrell Duffie, Mark Gertler, Narayana Kocherlakota, Thomas Lemieux, José Scheinkman, Danny Quah, Jean-Luc Vila, and two anonymous referees for helpful comments and suggestions. We also thank participants at the Canadian Macroeconomics Study Group, the NBER Asset Pricing meeting, the Northwestern Summer Workshop in Economics, and seminar participants at Berkeley, Cornell, Duke, the Federal Reserve, McGill, the Minneapolis Federal Reserve Bank, Massachusetts Institute of Technology, New York University, Princeton, Queen's, Rutgers, Stanford, University of California, Los Angeles, Western Ontario, and Wharton. Part of this research was conducted at the Institute for Empirical Macroeconomics in the summer of 1991. We would also like to thank Makoto Saito for research assistance and the International Financial Services Research Center at MIT, the National Science Foundation, and the Sloan Foundation (Heaton) for financial assistance.

I. Introduction

Incomplete markets in the form of an inability to borrow against risky future income have been proposed as an explanation for the poor predictive power of the standard consumption-based asset pricing model.[1] With complete markets, individuals fully insure against idiosyncratic income shocks, and individual consumption (under some preference assumptions) is proportional to aggregate consumption. With limited insurance markets, however, individual consumption variability may exceed that of the aggregate, and the implied asset prices may differ significantly from those predicted by a representative consumer model. In this paper we study an economy in which agents cannot write contracts contingent on future labor income realizations. They face aggregate uncertainty in the form of dividend and systematic labor income risk and also idiosyncratic labor income risk. Idiosyncratic income shocks can be buffered by trading in financial securities, but the extent of trade is limited by borrowing constraints, short-sales constraints, and transactions costs.

The motivation for considering the interaction between trading frictions and asset prices in this environment is best understood by reviewing the findings of a number of recent papers. Telmer (1993) and Lucas (1994) examine a similar model with transitory idiosyncratic shocks and without trading costs. Surprisingly, they find that even though agents cannot insure against idiosyncratic shocks, predicted asset prices are similar to those with complete markets.[2] This occurs because when idiosyncratic shocks are transitory, consumption can be effectively smoothed by accumulating financial assets after good shocks and selling assets after bad shocks. Aiyagari and Gertler (1991) consider a related model with no aggregate uncertainty and with transactions costs in which agents trade to offset transitory idiosyncratic shocks. In this case, differential transactions costs affect the relative returns on stocks and bonds and reduce the total volume of trade. Finally, Constantinides and Duffie (1994) study the case of permanent idiosyncratic shocks.[3] Although agents have unrestricted access to financial markets, no trades occur, resulting in increased equilibrium consumption volatility. When the conditional variance of idiosyncratic shocks is assumed to increase during economic down-

---

[1] For discussions of problems with the standard model, see, e.g., Hansen and Singleton (1983), Mehra and Prescott (1985), and Hansen and Jagannathan (1991).
[2] Using a more volatile aggregate income process, Marcet and Singleton (1991) also calibrate this model. They find that the equity premium rises in the presence of frequently binding short-sales constraints.
[3] Related two-period models can be found in Mankiw (1986), Scheinkman (1989), and Weil (1992).

INCOMPLETE MARKETS   445

turns, the risk-free rate falls and the equity premium rises relative to the complete markets case.

These results suggest that the quantitative asset price predictions from this class of models will depend critically on several factors: (i) the extent of trading frictions in securities markets, (ii) the size and persistence of idiosyncratic shocks, and (iii) the correlation structure of idiosyncratic and aggregate shocks. To address points ii and iii, we develop an empirical model of individual income that captures both the size of the idiosyncratic shocks and the persistence of these shocks over time, on the basis of evidence from the Panel Study of Income Dynamics (PSID). The time-series properties of aggregate income and dividends are estimated using the National Income and Product Accounts (NIPA). This process is then used to calibrate the theoretical model.

Our theoretical model differs substantively from those discussed above by considering the effects of transactions costs in an environment with both aggregate and idiosyncratic shocks. Transactions costs play an important role because agents choose to trade frequently in order to buffer shocks to their individual income. As a result, transactions costs can have two effects on asset prices.

First, (gross) rates of return on securities may be altered because lenders require higher rates and borrowers require lower rates to compensate for transactions costs. This direct effect of transactions costs was emphasized by Amihud and Mendelson (1986), Aiyagari and Gertler (1991), and Vayanos and Vila (1995).[4]

A second, indirect, effect of transactions costs is that they limit the ability of agents to use asset markets to self-insure against transitory shocks. Consequently, individual consumption is more volatile than aggregate consumption. This affects each individual's attitude toward aggregate uncertainty. Since preferences are assumed to be *proper* in the sense discussed by Pratt and Zeckhauser (1987) and Weil (1992), an increase in individual consumption volatility increases the amount agents are willing to pay to avoid the aggregate uncertainty reflected in dividends. In equilibrium, the implied equity premium could rise in response to increases in transactions costs for this reason alone. This paper appears to be the first to evaluate the importance of this mechanism.

---

[4] In contrast, Constantinides (1986) argued that transactions costs should have only a small effect on asset returns. In his model, in which there is no idiosyncratic risk and agents trade only to rebalance their portfolios, agents avoid most transactions costs by reducing the frequency of trades. As a result, asset returns are not much affected by the presence of transactions costs. However, because of the idiosyncratic shocks that individuals face in our model, it is more costly for them to change their asset trading patterns in response to trading costs.

We calibrate the model under a variety of assumptions about the size and incidence of trading costs and the supply of tradable securities. When trading costs differ across markets, we find that agents readily substitute toward transacting in the lower-cost market. For example, if transactions costs are introduced only in the stock market, then agents trade primarily in the bond market and by this means effectively smooth transitory income shocks. In this case transactions costs have little effect on required rates of return.

When transactions costs are also introduced in the bond market in the form of a wedge between the borrowing and lending rate, then the equilibrium lending rate falls. With a binding borrowing constraint or a large wedge between the borrowing and lending rate, a transactions cost in the stock market can produce an equity premium of about half of the observed value. In this case approximately 20 percent of the predicted equity premium can be attributed to the indirect effect, so that the direct effect of transactions costs dominates. These results, however, are quite sensitive to the structure of transactions costs and the supply of tradable assets. For example, with a sufficiently large outside supply of bonds, the impact of transactions costs is reduced dramatically since the new bonds allow the agents to avoid the borrowing constraint and borrowing costs.

The remainder of the paper is organized as follows: In Section II we describe the model economy. In Section III we present the state variables, the calibrated model of income and dividends, and the parameterizations for trading costs, borrowing constraints, and short-sales constraints. Simulation results are reported in Section IV, and Section V presents concluding remarks.

II. The Model

A. *The Environment*

The economy contains two (classes of) agents who are distinguished by their labor income realizations.[5] At each time t, agent i receives stochastic labor income Y_t^i. By assumption, agents are not allowed to write contracts contingent on future labor income. We shall refer to the share of individual i's labor income in aggregate labor income as *idiosyncratic income* because the innovations to the share of income received by the first agent are perfectly negatively correlated with the share of income received by the second agent. Notice that these

---

[5] Scheinkman and Weiss (1986) consider one of the first two-agent equilibrium models in which idiosyncratic employment shocks are uninsurable.

INCOMPLETE MARKETS

shocks would be completely diversified away in a complete market setting.[6]

Agents also receive income from investments in stocks and bonds. At time t, a share of stock, with price p_t^s, provides a claim to a flow of stochastic dividends from time $t+1$ forward, $\{d_j\}_{j=t+1}^{\infty}$. The bond, with price p_t^b, provides a risk-free claim to one unit of consumption at time $t+1$. The agents trade these two securities to smooth their consumption over time. Trading is costly, with transactions cost function $\kappa(\cdot)$ in the stock market and $\omega(\cdot)$ in the bond market. Trade is also limited by short-sales and borrowing constraints.

At time t, each agent's preferences over consumption are given by

$$U_t^i \equiv E\left\{\sum_{\tau=0}^{\infty} \beta^{\tau} \frac{(c_{t+\tau}^i)^{1-\gamma}-1}{1-\gamma} \middle| \mathcal{F}(t)\right\}, \quad \gamma > 0, \tag{1}$$

where $\mathcal{F}(t)$ is the time t information set that is common across agents. This information is generated by the state, $\mathbf{Z}_t$, which is specified below. In principle, γ could be allowed to differ across agents. Since we want to interpret the two groups as similar except for realizations of idiosyncratic shocks, however, it seems appropriate to equate γ across the two groups.[7]

At each date t, agent i maximizes (1) via the choice of consumption c_t^i, stock share holdings s_{t+1}^i, and bond holdings b_{t+1}^i subject to the flow wealth constraint

$$c_t^i + p_t^s s_{t+1}^i + p_t^b b_{t+1}^i + \kappa(s_{t+1}^i, s_t^i; \mathbf{Z}_t) + \omega(b_{t+1}^i, b_t^i; \mathbf{Z}_t)$$
$$\leq s_t^i(p_t^s + d_t) + b_t^i + Y_t^i \tag{2}$$

and short-sales or borrowing constraints

$$s_t^i \geq K_t^s, \quad t = 0, 1, 2, \ldots, \tag{3}$$

$$b_t^i \geq K_t^b, \quad t = 0, 1, 2, \ldots. \tag{4}$$

The components of initial wealth d_0, s_0^i, b_0^i, and Y_0^i and market prices are taken as given.

---

[6] The structure of our model differs from that of models with an infinite number of agents who each receive independent shocks that sum to a constant by the law of large numbers, such as in Bewley (1986), Clarida (1990), Aiyagari and Gertler (1991), Huggett (1993), and Aiyagari (1994). For a comparison of these models and the model considered here, see Heaton and Lucas (1995).

[7] Dumas (1989) considers the implications of different risk aversion parameters in a complete markets setting.

B. Trading Frictions

The extent to which individuals will use asset markets to buffer idiosyncratic income shocks depends on the size and incidence of trading costs and the severity of borrowing and short-sales constraints. Since the assumed form of these frictions qualitatively affects predicted asset prices and since there is little agreement about the exact form of these costs, we consider several alternative cost structures.

Transactions Costs in the Stock Market

In most of the simulations we assume that both buyers and sellers face a quadratic transactions cost function:

$$\kappa(s_{t+1}^i, s_t^i; \mathbf{Z}_t) = k_t[(s_{t+1}^i - s_t^i)p_t^s]^2. \tag{5}$$

The parameter k_t is used to control the assumed magnitude of the cost. Because the realized cost is endogenous, the range of attainable costs is bounded; an increase in the cost parameter eventually leads to an offsetting reduction in trade. Dividing the cost function by $(|s_{t+1}^i - s_t^i|)p_t^s$ gives the trading cost as a percentage of the value of shares traded: $k_t|s_{t+1}^i - s_t^i|p_t^s$. The average of this percentage cost is reported in the simulation results.

We use a quadratic cost function primarily for computational simplicity. However, it also captures the idea that as more assets are sold, agents must sell increasingly illiquid assets. The fact that many individuals hold no stock at all suggests that there may be significant fixed costs to entering this market as well.[8] To partially address this issue, we also estimate the income process conditioning on data from families that own nonnegligible amounts of stock.

Another possible objection to the quadratic form is that small changes in stock holdings are likely to be as costly (proportionally) as large changes. A quadratic cost function makes small transactions relatively inexpensive, which encourages small transactions. However, in the discretized state space model below, agents trade every period to smooth large shocks so that the strategy of using very small transactions to avoid transactions costs is not relevant. As a robustness check, we also report results for the case in which costs are proportional to the size of the trade.

Bond Market Transactions Costs

The inside bonds in this model represent private borrowing and lending. While it seems sensible to treat transactions costs symmetrically

---

[8] The effects of transactions costs of this form have been considered by Saito (1995).

INCOMPLETE MARKETS

for sales and purchases of stock or outside bonds, this is less true for consumption loans. Typically consumers pay a substantial spread over the lending rate to borrow. Although part of the observed spread is a default premium that does not apply to the risk-free bonds of the model, a portion of the spread can be attributed to costs of financial intermediation or monitoring that must be incurred even if the debt is ex post risk-free.

To capture the asymmetry between effective borrowing and lending rates, in some of the simulations the bond transactions cost function is assumed to have the form

$$\omega(b^i_{t+1}, b^i_t; \mathbf{Z}_t) = \Omega_t \min(0, b^i_{t+1} p^b_t)^2. \tag{6}$$

The parameter Ω_t controls the magnitude of the spread between the borrowing and lending rates. By convention, borrowing at time t is represented by a negative value for b^i_{t+1}, so only the agent who borrows pays the transactions cost. As for stocks, the cost is reported as a percentage of the per capita amount transacted: $\Omega_t |b^i_{t+1}| p^b_t/2$.

We also consider the implications of a symmetric quadratic cost function in the bond market of the form

$$\omega(b^i_{t+1}, b^i_t; \mathbf{Z}_t) = \Omega_t (b^i_{t+1} p^b_t)^2. \tag{7}$$

As demonstrated below, the choice of (6) versus (7) will have a significant effect on the predicted equity premium.

Notice that with the one-period bonds of the model, the transactions cost is paid every period even when a loan is rolled over for several periods. The reason is that we interpret the transactions cost not as a trading cost, but rather as a wedge between the borrowing and lending rates due to monitoring and other costs incurred each period. For comparison, in some of the simulations we assume that the transactions cost in the bond market depends only on the change in the amount of bonds outstanding. We also consider the case in which the incurred cost is proportional to the amount borrowed.

To match the observed aggregate income and dividend process, the economy is assumed to be stationary in the aggregate income growth rate and in the dividend's share of income. As a result, the price of the stock grows over time. The borrowing constraint, K^b_t, is assumed to be linear in aggregate income, Y^a_t, which accommodates growth in the face value of bonds issued as the economy grows. The parameters k_t and Ω_t are also chosen to induce stationarity. This requires setting $k_t = k/Y^a_t$ and $\Omega_t = \Omega/Y^a_t$, where k and Ω are constants.

Finally, we refer to the case in which $\kappa(\cdot) \equiv 0$ and $\omega(\cdot) \equiv 0$ as the *frictionless model*. The frictionless model is similar to the models in Telmer (1993) and Lucas (1994) except that the income process is calibrated using income data from the PSID.

Borrowing and Short-Sales Constraints

Consumption smoothing may also be curtailed by institutional limits on the amount of borrowing. This type of credit rationing is represented by (4). Alternatively, these limits can be viewed as representing the effects of very high transactions costs.

How to calibrate the upper bound on borrowing is not obvious. For instance, the value of household collateral in the form of housing and other assets is a plausible limit on debt but is not easily measured. To bracket a plausible range, we consider an upper bound of 10 percent of average per capita income and a lower bound of 0 percent (no borrowing). In the simulations the agents rarely hit the assumed 10 percent upper bound, so a less restrictive limit would have little effect on the analysis. We also consider the effect of adding an outside supply of bonds.

No short sales are permitted in the stock market: $K_t^s = 0$ in (3). This is motivated in part by the observation that it is costly for individuals to take short positions, for instance, because of margin requirements. Allowing short sales in principle increases the effective quantity of tradable assets, but in the simulations, this constraint rarely binds.

C. Equilibrium

At time t aggregate output consists of the aggregate dividend, d_t, and the sum of individuals' labor income, $Y_t^1 + Y_t^2$. Market clearing requires

$$b_t^1 + b_t^2 = 0, \quad t = 0, 1, 2, \ldots, \tag{8}$$

$$s_t^1 + s_t^2 = 1, \quad t = 0, 1, 2, \ldots, \tag{9}$$

and

$$\sum_{i=1,2} [c_t^i + \kappa(s_{t+1}^i, s_t^i; \mathbf{Z}_t) + \omega(b_{t+1}^i, b_t^i; \mathbf{Z}_t)] = d_t + Y_t^1 + Y_t^2, \quad t = 0, 1, 2, \ldots. \tag{10}$$

Notice that (8) implies that bonds are in zero net supply. In Section IVF, we relax this assumption.

Let

$$\kappa_1(s_{t+1}^i, s_t^i; \mathbf{Z}_t) \equiv \frac{\partial \kappa(s_{t+1}^i, s_t^i; \mathbf{Z}_t)}{\partial s_{t+1}^i},$$

$$\kappa_2(s_{t+1}^i, s_t^i; \mathbf{Z}_t) \equiv \frac{\partial \kappa(s_{t+1}^i, s_t^i; \mathbf{Z}_t)}{\partial s_t^i},$$

INCOMPLETE MARKETS

$$\omega_1(b^i_{t+1}, b^i_t; \mathbf{Z}_t) \equiv \frac{\partial \omega(b^i_{t+1}, b^i_t; \mathbf{Z}_t)}{\partial b^i_{t+1}},$$

$$\omega_2(b^i_{t+1}, b^i_t; \mathbf{Z}_t) \equiv \frac{\partial \omega(b^i_{t+1}, b^i_t; \mathbf{Z}_t)}{\partial b^i_t}.$$

When the short-sales and borrowing constraints are not binding, the first-order necessary conditions from the agent's optimization problem imply that, for all i and t,

$$[p^s_t + \kappa_1(s^i_{t+1}, s^i_t; \mathbf{Z}_t)]u'(c^i_t) \\ = \beta E\{u'(c^i_{t+1})[p^s_{t+1} + d_{t+1} - \kappa_2(s^i_{t+2}, s^i_{t+1}; \mathbf{Z}_{t+1})]|\mathcal{F}(t)\} \quad (11)$$

and

$$[p^b_t + \omega_1(b^i_{t+1}, b^i_t; \mathbf{Z}_t)]u'(c^i_t) \\ = \beta E\{u'(c^i_{t+1})[1 - \omega_2(b^i_{t+2}, b^i_{t+1}; \mathbf{Z}_{t+1})]|\mathcal{F}(t)\}. \quad (12)$$

If an agent is constrained by the short-sales constraint (3), then (11) is replaced by

$$s^i_t = K^s_t. \quad (11')$$

Similarly, if the agent is constrained by the borrowing constraint (4), then (12) is replaced by

$$b^i_t = K^b_t. \quad (12')$$

At time t the unknowns are p^s_t; p^b_t; c^i_t, $i = 1, 2$; s^i_t, $i = 1, 2$; and b^i_t, $i = 1, 2$. The equations defining an equilibrium are the budget constraints (2), $i = 1, 2$; the market-clearing conditions (8), (9), and (10); and the asset pricing equations (11) or (11'), $i = 1, 2$, and (12) or (12'), $i = 1, 2$. By Walras's law, one of the market-clearing conditions or a budget constraint is redundant. We restrict our attention to stationary equilibria in which the consumption growth rate, portfolio rules, and equilibrium prices are functions of the time t state, $\mathbf{Z}_t$.

III. State Variables

The exogenous state is described by a Markov chain that gives the dynamics of aggregate income, dividend income, and individual income over time. The state, $\mathbf{Z}_t$, includes these exogenous variables and the endogenous distribution of asset holdings.

A. The Exogenous State Variables

The exogenous state of the economy is divided into two sets of variables: the aggregate state of the economy and the idiosyncratic income state. To specify the aggregate state, let Y_t^l be aggregate labor income and D_t^a be aggregate dividend income at time t. Total aggregate income is given by $Y_t^a \equiv Y_t^l + D_t^a$. Further, let $\gamma_t^a \equiv Y_t^a/Y_{t-1}^a$ and $\delta_t \equiv D_t^a/Y_t^a$ be the (gross) growth rate in aggregate income and dividend's share in aggregate income, respectively. The *aggregate* state of the economy at time t is given by $[\gamma_t^a\ \delta_t]'$.

Individual i's labor income as a fraction of aggregate labor income is given by $\eta_t^i \equiv Y_t^i/Y_t^l$. In our two-person economy, the law of motion for η_t^1 implies a law of motion for η_t^2 since $\eta_t^1 + \eta_t^2 = 1$. The entire exogenous state of the economy is given by $[\gamma_t^a\ \delta_t\ \eta_t^1]'$.

Calibrating the Markov Chain

In calibrating the Markov chain, we require that it capture the moments of aggregate labor income, dividend income, and individual income that are most likely to affect the predictions of the theoretical model. The dynamics are first summarized by an autoregression and then discretized using the method of Tauchen and Hussey (1991).

The moments of the aggregate economy are summarized by the coefficients of a bivariate autoregression for $\mathbf{X}_t^a \equiv [\log(\gamma_t^a)\ \log(\delta_t)]'$:

$$\mathbf{X}_t^a = \boldsymbol{\mu}^a + \Lambda^a \mathbf{X}_{t-1}^a + \Theta^a \boldsymbol{\epsilon}_t^a, \quad t = 1, 2, 3, \ldots, \tag{13}$$

where $\boldsymbol{\epsilon}_t^a$ is a vector of white-noise disturbances with covariance matrix $\mathbf{I}$, the matrix Θ^a is assumed to be lower triangular, and $\boldsymbol{\mu}^a$ is a vector of constants. The parameters of the vector autoregression (VAR) are estimated using annual aggregate labor income and dividend data from NIPA. We use aggregate labor income data rather than aggregate consumption data because they more closely match our individual-level labor income data. As shown in Section IVA, the implications of this income process under complete markets are similar to those using consumption data, so this choice is unlikely to substantially affect the results. Appendix A describes the data in more detail and reports the results of estimating the parameters of (13).

We also require that the Markov chain approximate the persistence and volatility of individual income. The individual income process is estimated from a sample of 860 households in the PSID that have annual income from 1969 to 1984. For each household the income process is summarized by a first-order autoregression of the form[9]

$$\log(\eta_t^1) = \bar{\eta} + \rho \log(\eta_{t-1}^1) + \epsilon_t^1. \tag{14}$$

---

[9] Appendix A describes the data in more detail and examines the results of estimating the model with several subsamples from the PSID and several alternative specifications of the income process.

For the entire sample, the average value of ρ across households is 0.529 and the average standard deviation of ϵ_t^i is 0.251.

As the base case for investigation, we choose an eight-state Markov chain for $[\gamma_t^a \; \delta_t \; \eta_t^1]'$ that closely approximates the estimated VAR[10] given by (13) and (14). The parameters of this Markov chain are given in table 1. The Markov chain appears to do a reasonable job of fitting the moments that we have chosen (App. B reports the approximation error, and table B1 summarizes the coefficients).

In the base case model, idiosyncratic labor income shocks are assumed to be independent of the aggregate growth rate and dividend realization. As discussed in Appendix A, we find little statistical evidence of a significant effect of the aggregate state on the conditional mean of individual income share. However, Mankiw (1986) and Constantinides and Duffie (1994) show that if the distribution of labor income widens in a downturn, then individuals will demand a large equity premium to hold stocks. To examine this potential effect, we also consider the cyclical distribution case (CDC) model.

The CDC model is calibrated by adding the following equation to the system, which accommodates heteroskedasticity in the individual shocks:

$$\text{Std}\left\{\epsilon_t^i \middle| \log\left(\frac{Y_t^a}{Y_{t-1}^a}\right)\right\} = \alpha_0 + \alpha_1 \log\left(\frac{Y_t^a}{Y_{t-1}^a}\right). \tag{15}$$

The parameters of (15) are again estimated using the PSID and aggregate income data. For the subsample of households that own stock, the point estimate of α_1 is -1.064, which is consistent with the conjecture that the conditional volatility of individual income is lower in high-growth states. However, this value of α_1 and the Markov chain for aggregate income of table 1 imply a standard deviation of 0.28 for ϵ_t^i in the low–income growth state and 0.22 in the high–income growth state. This difference across aggregate growth states is too small to produce a significant departure from the base case model and hence has no detectable effect on predicted asset prices.

Since data limitations may be the reason for the small estimated change in conditional variance over the cycle,[11] we calibrate the CDC model under the assumption that the standard deviation of shocks to individual income is twice as large in the low–aggregate growth state as in the high–aggregate growth state. The parameters of the resulting eight-state Markov chain, estimated as above, are reported in

---

[10] As discussed below, we adjusted the mean of the process for the share of dividends in income to be 15 percent.
[11] In particular, the short time series of individual data provide few observations of cyclical variations.

TABLE 1

A. Markov Chain Model for Exogenous State Variables: Base Case

State Number	States γ^a	δ	η^1
1	.9904	.1402	.3772
2	1.0470	.1437	.3772
3	.9904	.1561	.3772
4	1.0470	.1599	.3772
5	.9904	.1402	.6228
6	1.0470	.1437	.6228
7	.9904	.1561	.6228
8	1.0470	.1599	.6228

B. Transition Probability Matrix $[p_{ij}]$*

.3932	.2245	.0793	.0453	.1365	.0779	.0275	.0157
.3044	.3470	.0425	.0484	.1057	.1205	.0147	.0168
.0484	.0425	.3470	.3044	.0168	.0147	.1205	.1057
.0453	.0793	.2245	.3932	.0157	.0275	.0779	.1365
.1365	.0779	.0275	.0157	.3932	.2245	.0793	.0453
.1057	.1205	.0147	.0168	.3044	.3470	.0425	.0484
.0168	.0147	.1205	.1057	.0484	.0425	.3470	.3044
.0157	.0275	.0779	.1365	.0453	.0793	.2245	.3932

* p_{ij} = Pr{state j at time $t + 1$|state i at time t}.

table 2. Appendix B reports on the ability of the model to approximate several moments of the data (see table B2 for a summary). Of note is the fact that the Markov chain implies that the aggregate state does not affect the conditional mean of income shares and that the implied value of α_1 in (15) is -4.5.

So far we have assumed that the only tradable assets in positive net supply are claims to the dividend stream. Since dividends average only 3.9 percent of total income in the NIPA, this clearly understates the share of income from tradable assets. Assets such as government securities, corporate bonds, and so forth may also be sold or used as collateral for loans. To roughly approximate these additional sources of wealth, in most of the simulations we increase tradable asset holdings by grossing up the assumed fraction of dividend income so that nonlabor income is 15 percent of total income.[12] By comparison, capital's share of income in the NIPA averages about 30 percent. In

---

[12] An alternative would be to use a debt measure such as net interest payments from the corporate sector to the household sector (as measured by "net interest" from the NIPA). This is complicated, however, by the fact that these payments are not stationary, increasing from 1.4 percent of total income in 1946 and rising to 12.1 percent of total income in 1990.

INCOMPLETE MARKETS 455

TABLE 2

A. Markov Chain Model for Exogenous State Variables:
Cyclical Distribution Case

STATE NUMBER	STATES		
	γ^a	δ	η^1
1	.9904	.1403	.3279
2	1.0470	.1437	.4405
3	.9904	.1562	.3279
4	1.0470	.1600	.4405
5	.9904	.1403	.6721
6	1.0470	.1437	.5595
7	.9904	.1562	.6721
8	1.0470	.1600	.5595

B. Transition Probability Matrix $[p_{ij}]$

.4365	.2343	.0881	.0473	.1515	.0098	.0306	.0020
.2416	.3461	.0337	.0483	.1698	.1201	.0237	.0168
.0555	.0458	.3977	.3281	.0193	.0019	.1381	.0137
.0360	.0792	.1783	.3924	.0253	.0275	.1253	.1362
.1515	.0098	.0306	.0020	.4365	.2343	.0881	.0473
.1698	.1201	.0237	.0168	.2416	.3461	.0337	.0483
.0193	.0019	.1381	.0137	.0555	.0458	.3977	.3281
.0253	.0275	.1253	.1362	.0360	.0792	.1783	.3924

Section IVF we also examine several cases in which bonds are assumed to be in positive net supply and the level of dividend income is reduced accordingly.

To summarize, the exogenous state variables include γ_t^a, δ_t, and η_t^1. They evolve according to the Markov process specified above. An endogenous component of the state is portfolio composition. In our two-person economy, this is summarized by agent 1's holdings of stocks and bonds, since agent 2's holdings can be derived using the market-clearing conditions (8) and (9). We define the state vector of the economy by $\mathbf{Z}_t = \{\gamma_t^a, \delta_t, \eta_t^1, s_t^1, b_t^1\}$. Asset prices, consumption policies, and trading policies are found as a function of $\mathbf{Z}_t$.

IV. Simulation Results

In this section we report the results of Monte Carlo simulations of the economy of Section II using the exogenous driving processes described in Section III. The equilibrium is solved numerically using a modified version of the "auctioneer algorithm" described in Lucas (1994).[13]

---

[13] The numerical solution of the model introduces approximation error discussed in App. C.

A number of summary statistics are reported. First, the volatility of individual consumption implied by the model is compared to the volatility of aggregate consumption. This statistic is of interest because risk sharing is not complete, so the extent to which individual consumption behaves like aggregate consumption indicates the degree to which risk sharing is being accomplished by trading securities. We also report the mean and standard deviation of implied asset returns in order to evaluate whether the model with incomplete markets and transactions costs can explain the equity premium and risk-free rate puzzles discussed by Mehra and Prescott (1985) and Weil (1989). Finally, we report the mean and standard deviation of trading volume in order to gauge the sensitivity of trade to the level of transactions costs.

These statistics are computed under the assumption that initially each agent holds half the shares of stock and no debt. The economy evolves for 1,000 years, driven by realizations of the exogenous income process. Asset returns, trading volume, and consumption growth between years 999 and 1,000 are recorded for each history. The reported statistics are based on averages across 1,500 of these experiments.[14]

A. Representative Agent Baselines

Before turning to the implications of the incomplete markets model, we briefly examine the implications of the complete markets case. This experiment is very similar to the one undertaken by Mehra and Prescott (1985), except that they equate dividends to aggregate consumption whereas we treat consumption as the sum of measured dividends and labor income.

Table 3 contrasts the results of the complete markets case with the sample moments in the data. Columns 2 and 3 report the model predictions for $\gamma = 1.5$ and $\beta = 0.95$, where the representative agent is assumed to consume aggregate income. The representative agent results for the base and cyclical distribution cases are slightly different because the two laws of motion approximate the moments of aggregate data differently. Notice that although the consumption process differs from those used in previous studies, the implied asset prices are similar.[15] For example, the predicted average return on risk-free

[14] The long time horizon was chosen to eliminate the effect of initial conditions.
[15] The sample moments of stock returns were calculated using annual returns for the 1947–90 value-weighted return from the Center for Research in Security Prices. The moments of the bond returns were calculated using annual Treasury bill returns for 1947–90. Both are converted to real returns using the Consumer Price Index (CPI).

TABLE 3
Moments Implied by the Complete Markets Case

		Aggregate		Individual	
Moment	Data (1)	Base Case (2)	Cyclical Distribution Case (3)	Base Case (4)	Cyclical Distribution Case (5)
Consumption growth:					
Average	.020	.018	.015	.018	.015
Standard deviation	.030	.028	.028	.217	.259
Bond return:					
Average	.008	.080	.077	.055	.041
Standard deviation	.026	.009	.012	.175	.213
Stock return:					
Average	.089	.082	.078	.137	.152
Standard deviation	.173	.029	.028	.375	.441

bonds is high and close to the stock return in each case, whereas the observed average bond return is quite low.[16] Further, the standard deviation of the stock return is low relative to historical levels.

To assess whether uninsurable labor income shocks have the *potential* to explain the poor performance of the representative agent model, consider columns 4 and 5 in table 3. These statistics were calculated in a representative agent model using one of the agent's labor income dynamics as though they were the aggregate labor income dynamics. These results demonstrate that when consumption is equated to an individual's labor income process, the predicted equity premium becomes much larger. For example, in the base case model, the premium is predicted to be 8.2 percent. Consistent with the data, the predicted standard deviation of the stock return is also quite large (37.5 percent). Notice that in the CDC model, the equity premium is even higher because the idiosyncratic shocks are concentrated in periods of low aggregate growth. The predicted level of both the bond and stock returns is too high, but this can be corrected by assuming a larger value of β.[17]

In general, these results indicate that the model with uninsurable labor income has the potential to explain several of the observed moments of asset returns. Of interest, however, is how much these results change once trading is allowed in the stock and bond markets.

---

[16] This finding is sometimes referred to as the *risk-free rate puzzle* rather than the equity premium puzzle to draw attention to the fact that the standard model can match the mean return on stocks but has difficulty simultaneously producing a realistic average risk-free rate. For a discussion, see Weil (1989).

[17] We keep β at a relatively low value in order to improve the performance of the solution algorithm.

B. Frictionless Trading

Table 4 summarizes the case in which individuals can trade costlessly in both the stock and bond markets, subject to the conditions that stocks cannot be sold short and individuals can borrow only up to 10 percent of per capita income. Notice that for either income process, individual consumption is only slightly more volatile than aggregate consumption and is much less volatile than consumption under autarky (see table 3).

The result that idiosyncratic shocks are offset by asset trades when trading is costless strengthens the findings of Telmer (1993) and Lucas (1994), who perform a similar experiment under the assumption that labor income shocks are independent over time. In our model the labor income shares are correlated over time, and hence the idiosyncratic shocks are relatively persistent. This persistence, via a wealth effect, should make it more difficult to self-insure through the asset markets. A striking example of this is given by Constantinides and Duffie (1994), who construct cases in which labor income shocks follow a random walk and no smoothing occurs. However, the results of table 4 indicate that with the persistence estimated using the PSID, agents can still effectively smooth their idiosyncratic income shocks by trading.

These findings suggest that increasing the predicted consumption

TABLE 4

MOMENTS IMPLIED BY THE FRICTIONLESS MODEL

Moment	Base Case	Cyclical Distribution Case
Consumption growth:		
Average	.018	.016
Standard deviation	.044	.045
Bond return:		
Average	.077	.073
Standard deviation	.012	.017
Stock return:		
Average	.079	.073
Standard deviation	.032	.030
Bond trades (percentage of consumption):		
Average	.045	.042
Standard deviation	.060	.052
Stock trades (percentage of consumption):		
Average	.131	.146
Standard deviation	.066	.082

INCOMPLETE MARKETS 459

volatility in this model requires introducing some type of asset market frictions. This observation motivates the experiments with transactions costs and borrowing constraints that follow.

C. Transactions Costs in Both Markets and Asymmetric Bond Market Costs

In simulations not reported here, we find that imposing a friction in one market alone has a negligible impact on asset prices. As in Constantinides (1986), portfolio balance is a second-order consideration relative to intertemporal consumption smoothing. In the absence of binding borrowing or short-sales constraints, this causes the agents to trade almost exclusively in the frictionless market. Hence equilibrium returns are largely unaffected by the presence of trading costs in only one market. This motivates our focus on specifications in which either it is costly to trade in both markets or there are severe borrowing constraints.

We first examine the case in which transactions costs are given by (5) and (6). Bonds are in zero net supply, outstanding debt cannot exceed 10 percent of per capita income, and only the borrower pays the transactions costs. To examine the effect of increasing transactions costs in a balanced way across markets, Ω is varied from zero to 2.0 and k is set to $\Omega/2$. The results are summarized in figures 1–4 for the base case model and 5–8 for the CDC model.[18] In each of the figures the x-axis gives the value of Ω.

Figures 1 and 5 show the average stock return, bond return, equity premium, and net premium (described below) as a function of Ω for each of the models. Because of the cost of borrowing, there is a wedge between the lending rate and the borrowing rate. In the figures we plot the average return to lending, since the equity premium is typically measured using the average return on Treasury bills, and individuals cannot borrow at this rate. Notice that as Ω increases, the average bond return falls and the equity premium widens. As shown in figures 2 and 6, the average per capita transactions costs paid, as a percentage of the value of trade, increase in Ω. For example, in the base case model the average cost paid in the stock market rises from 0 percent to 5 percent of the value of trade.

Figures 3 and 7 report the average volume of trading in the two markets as a percentage of per capita income. As the transactions

---

[18] In constructing these results, we used a grid of values for Ω between zero and 2.0. To smooth across these grid points, we fit a third-order polynomial through the points. This curve is reported in the figures. This smoothing removes some very small variability in the figures that can be attributed to approximation error.

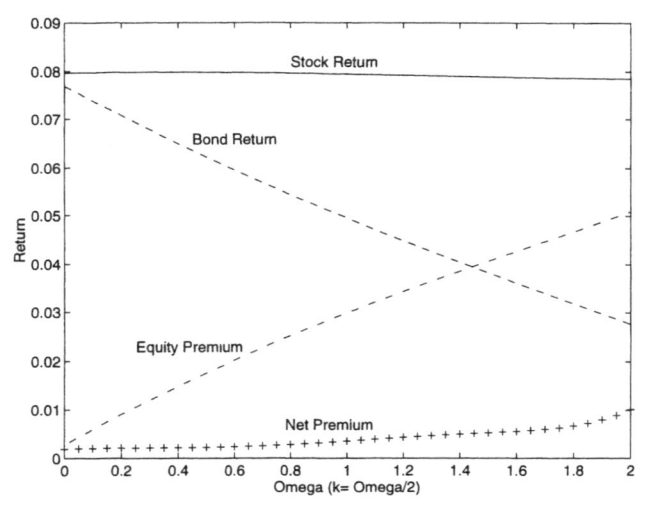

Fig. 1.—Base case, returns

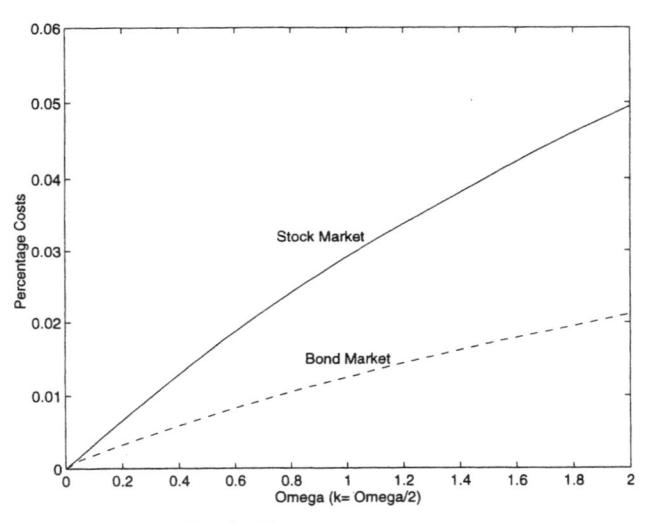

Fig. 2.—Base case, average costs

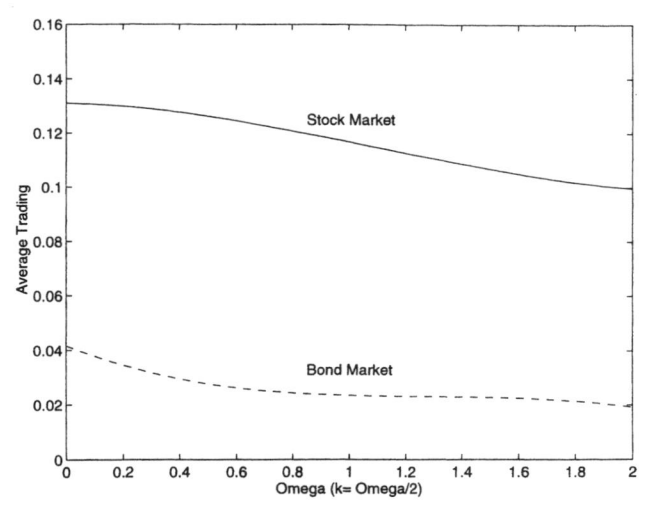

FIG. 3.—Base case, average trading

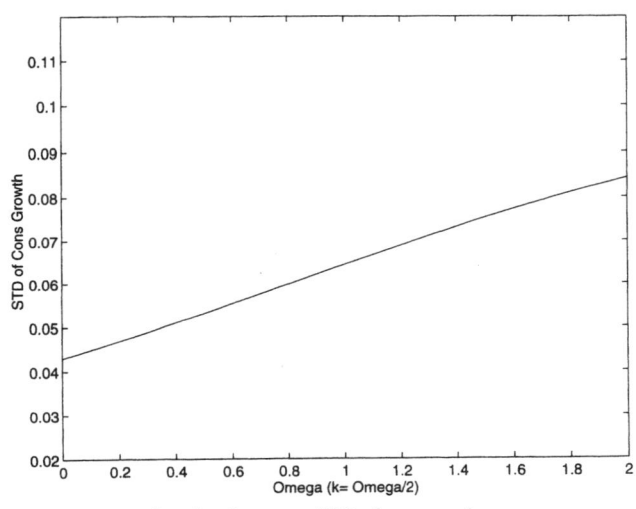

FIG. 4.—Base case, STD of consumption

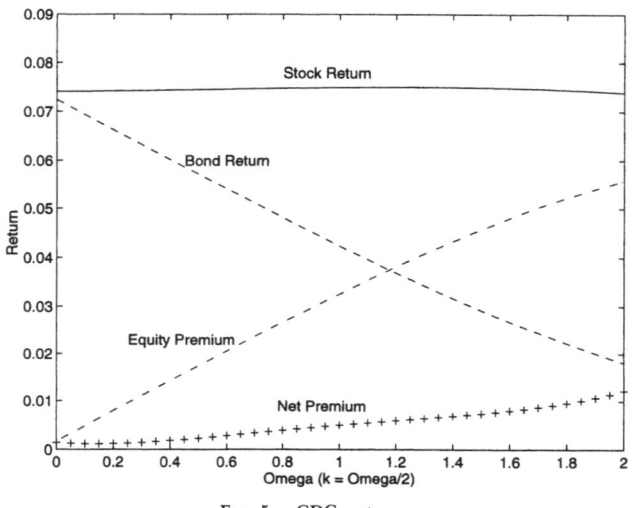

FIG. 5.—CDC, returns

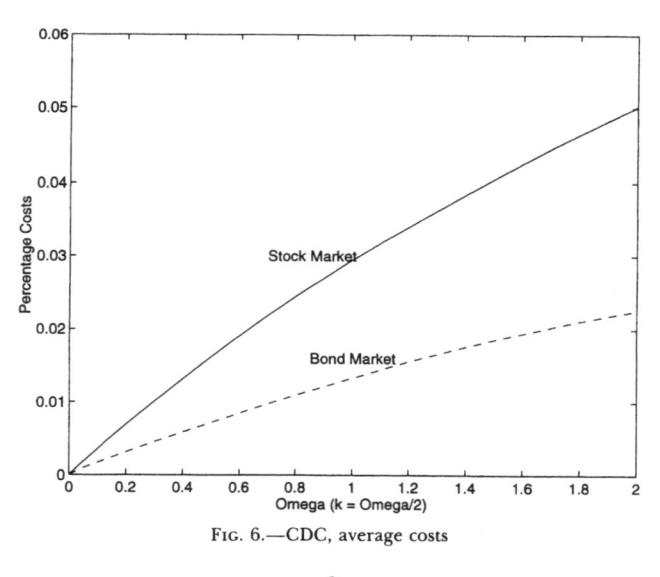

FIG. 6.—CDC, average costs

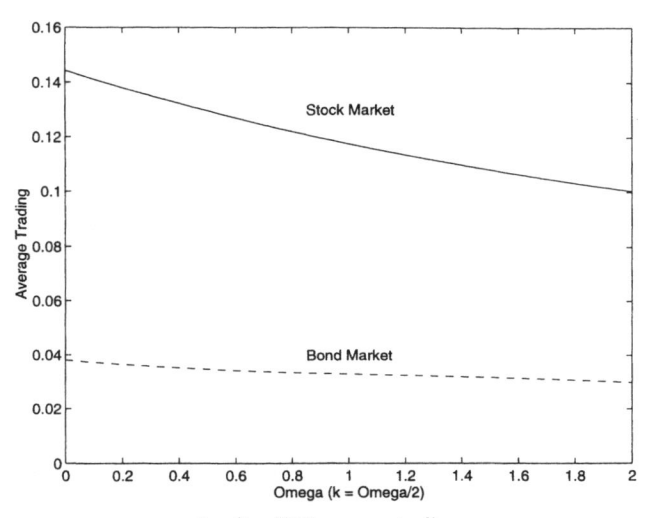

FIG. 7.—CDC, average trading

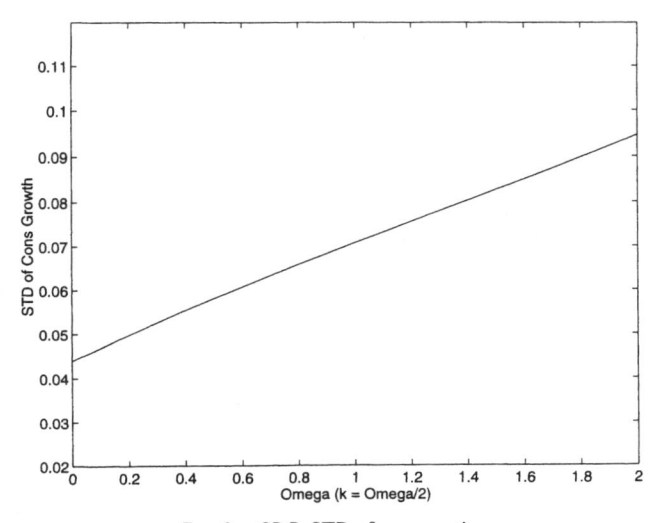

FIG. 8.—CDC, STD of consumption

costs increase, the level of trading drops off. This is reflected in a higher level of consumption volatility, measured by the standard deviation of the consumption growth of the first agent, shown in figures 4 and 8. The increased consumption volatility suggests that the model can potentially explain the volatility bounds of Hansen and Jagannathan (1991) better than a representative agent model.

Notice that an increase in transactions costs results in a decrease in the average bond return but no noticeable change in the average stock return. This result is best understood by considering the direct and indirect effects of transactions costs.

The Direct Effect of Transactions Costs

With moderate transactions costs in each market and fairly unrestrictive borrowing and short-sales constraints, agents trade almost every period to buffer the idiosyncratic income shocks. The recipient of the adverse idiosyncratic shock borrows and sells stock to partially offset the shock, whereas the recipient of the favorable shock buys both bonds and stocks to create a buffer against future adverse shocks. For shocks of the size found in the PSID, the borrower is willing to pay a relatively high borrowing rate including transactions costs. For example, consider figure 1 at $\Omega = 2$. The average return on the bond is 2.8 percent and the average transactions cost in the bond market (for a single agent over time) is 2.1 percent. Since the cost is paid only when borrowing and the cost function is quadratic, the borrower pays a marginal transactions cost of 8.4 percent (4 × 2.1 percent), which implies a net marginal borrowing rate of 10.5 percent. The lender, however, receives only a 2.8 percent rate of return on average. The lender is satisfied with this relatively low average rate of return because the high consumption variability creates a precautionary demand for assets.

In contrast, the average stock return is near 8 percent for all levels of Ω in each model. For stocks, the direct effect of transactions costs on the equilibrium stock return is difficult to predict since the buyer demands a lower price to compensate for transactions costs whereas the seller demands a higher price. Notice that the net return from buying stock is lower than 8 percent because of transactions costs, although in equilibrium the observed return is not greatly affected. The lower net return also reflects a precautionary demand for assets due to higher equilibrium consumption volatility.

In sum, for this cost specification, there is a direct effect of transactions costs that depresses the observed lending rate and hence increases the observed equity premium. Both stocks and bonds have a

INCOMPLETE MARKETS

lower net return due to a precautionary demand induced by higher consumption variability.

A similar effect of differential stock and bond market transactions costs on relative observed returns is found by Aiyagari and Gertler (1991) and Vayanos and Vila (1995). However, in those models there is no aggregate risk, so that differences in rates of return are generated solely by the differences in transactions costs across asset markets.

The Indirect Effect of Transactions Costs

To the extent that the increase in consumption volatility increases the *covariance* between individual consumption and the net returns on stock, the *net-of-transactions-costs equity premium* that agents demand widens. There are several ways to construct this net premium. First consider the payoff to an individual investor from investing in an incremental unit of the stock at time t. This investment affects the transactions costs paid at time t and at time $t + 1$. The sign and absolute magnitude of these effects depend on the portfolio positions at time t and at time $t + 1$. From (5), the marginal transactions cost that agent i pays at time t in purchasing a unit of the stock is $2k_t(s^i_{t+1} - s^i_t)(p^s_t)^2$. The marginal effect of this purchase on transactions costs at time $t + 1$ is $-2k_{t+1}(s^i_{t+2} - s^i_{t+1})(p^s_{t+1})^2$. As a result, the net (of transactions costs) one-period rate of return from an incremental unit of investment in the stock is given by

$$r^{s,\text{net}}_{t,t+1} \equiv \frac{p^s_{t+1} + d_{t+1} + 2k_t(s^i_{t+2} - s^i_{t+1})(p^s_{t+1})^2}{p^s_t + 2k(s^i_{t+1} - s^i_t)(p^s_t)^2} - 1. \qquad (16)$$

Notice that this rate of return satisfies

$$E\left\{ \beta \frac{u'(c^i_{t+1})}{u'(c^i_t)} (1 + r^{s,\text{net}}_{t,t+1}) \,\middle|\, \mathcal{F}(t) \right\} = 1, \qquad (17)$$

which is just the Euler equation for individual i (see [11]).

Similarly, the net-of-transactions-costs rate of return from an incremental unit of investment in the bond is given by

$$r^{b,\text{net}}_{t,t+1} = \begin{cases} \dfrac{1}{p^b_t} & \text{if } b^i_{t+1} \geq 0 \\[2ex] \dfrac{1}{p^b_t + 2\Omega_t b^i_{t+1}(p^b_t)^2} & \text{if } b^i_{t+1} < 0. \end{cases} \qquad (18)$$

The indirect effect of transactions costs occurs to the extent that the covariance between individual consumption and the net return on the stock changes. To assess this effect, we plot the "net premium" in figures 1 and 5. This measure of the equity premium is given by $E\{r_{t,t+1}^{s,\text{net}} - r_{t,t+1}^{b,\text{net}}\}$, which we calculate for individual 1. The net premium reflects the changing conditional covariance between $r_{t,t+1}^{s,\text{net}}$ and consumption growth as transactions costs increase. In the base case model at $\Omega = 2$, the equity premium is 5 percent but the net premium is only 1 percent. Also in the CDC model when $\Omega = 2$, the equity premium is 5.5 percent and the net premium is 1.2 percent. It appears that the indirect effect does affect the observed equity premium, although it accounts for only 20 percent of the premium in this case.[19]

Another way to measure the net-of-transactions-costs premium is to look at the difference between average stock and bond returns, subtracting out the average realized transactions costs in each market. Notice that a borrower must pay transactions costs on the entire amount borrowed each period. In contrast, the stock market transactions costs are a small percentage of the total return on stock holdings, since only a portion of the stock portfolio typically turns over each period. For instance, the average transactions costs in the stock market as a percentage of the value of the typical stock portfolio rise to a maximum of 0.4 percent when $\Omega = 2$ and $k = 1$ in the base case model. As a result, the average net return on stocks is about 7.6 percent and the average return to a lender is 2.8 percent. This implies a large net-of-transactions-costs equity premium of 4.8 percent.

Which of these two measures is a more appropriate measure of the net-of-transactions-costs premium depends on the question being asked. The first measure directly reflects the relation between risk and return in the model since it is based on marginal rates of substitution, and it is the measure we focus on in the analysis. The second measure answers the empirical question of how much investors earn on stocks relative to bonds net of transactions costs. This latter measure can be calculated using only information on returns and average transactions costs, whereas the former would also require detailed information on consumption.

Size of the Costs and Robustness to Cost Specifications

We have seen that transactions costs can be chosen so that the predicted stock and bond returns are similar to their observed values, but

---

[19] It is possible that the impact of the indirect effect would be greater with higher assumed risk aversion. An investigation of this issue presents significant computational difficulties, which we plan to address in future research.

INCOMPLETE MARKETS

the question remains whether observed transactions costs are large enough to provide a plausible explanation. A full accounting of costs should include brokerage fees, bid-ask spreads, and "price impact."[20] Aiyagari and Gertler (1991) discuss the magnitude of brokerage fees and argue (on the basis of Sharpe [1985]) that transactions costs as high as 5 percent are reasonable. Further, the average level of brokerage fees has fallen significantly over time, so a 5 percent transactions cost in the stock market may not be large by historical standards. Using the data base from the Institute for the Study of Security Markets, He and Modest (1995) estimate that the bid-ask spread implies transactions costs of 1.5 percent for an equally weighted portfolio of stocks and 0.75 percent for a value-weighted portfolio of stocks. This increases the cost of trading over and above brokerage fees. Finally, the price impact of a trade is a potentially important source of transactions costs, but it is difficult to obtain a simple measure of the size of this cost.

If marginal stock market transactions costs of 6 percent are taken as a reasonable estimate, the model still predicts a substantial equity premium. For example, when $\Omega = 1$ in the base case model, the average costs are 1.4 percent in the bond market and 2.9 percent in the stock market, and the equity premium is 3 percent. The net premium as measured using (17) and (18) is 0.5 percent, which implies that the majority of the impact of the transactions costs comes from the direct effect. Notice that to obtain an equity premium as large as 5 percent requires a marginal stock market transactions cost of 10 percent, so that without strict borrowing constraints, very large costs are needed to produce a premium close to its observed average level.

As discussed in Section III, to determine whether the qualitative results are sensitive to the quadratic form of the cost function, we consider an alternative that is close to being proportional to the value of trade. In the stock market the alternative cost function is given by

$$\kappa(s_{t+1}^i, s_t^i; Z_t) = \begin{cases} k(|s_{t+1} - s_t| - \epsilon)p_t^s + \dfrac{kp_t^s \epsilon}{2} & \text{if } |s_{t+1} - s_t| > \epsilon \\ \dfrac{k(s_{t+1} - s_t)^2 p_t^s}{2\epsilon} & \text{if } |s_{t+1} - s_t| \le \epsilon. \end{cases} \quad (19)$$

The cost function (19) is quadratic for small transactions and becomes linear afterward.[21] For ϵ small, the cost function is essentially propor-

---

[20] Price impact refers to the fact that large trades tend to move the price at which the trade occurs.
[21] The differentiability induced by smoothing the function at zero is convenient for computational purposes.

tional to the value of trade and k gives the marginal cost of a trade. The cost function in the bond market is also pseudoproportional and one-sided as in (6).

Figure 9 summarizes average predicted returns for this cost specification. Notice that the results are qualitatively similar to those for the quadratic cost function reported in figure 1. For marginal transactions costs of 2.5 percent in the stock market ($\Omega = 0.05$ and $k = 0.025$), the model with proportional costs predicts a 3 percent equity premium and an average stock return of 8 percent.

D. Transactions Costs in Both Markets and Symmetric Bond Market Costs

To investigate the effect of the incidence of transactions costs on predicted returns, we assume that borrowing costs are split equally by the lender and the borrower; the bond cost function (6) is replaced with (7). In this case, any difference between the average bond and stock return should reflect differences in risk between the two securities.

The results for expected rates of return are reported in figure 10 for the base case model and figure 11 for the CDC model. Limiting Ω to the range [0, 1] and setting k equal to Ω produces average transactions costs that are similar to those in figures 1 and 5, so that the results of the two cost structures are directly comparable. In sharp contrast to the case of one-sided borrowing costs, the equity premium rises only slightly as a result of the increase in consumption volatility.

In this case the drop in the lending rate is not due to the incidence of transactions costs but instead to an increased precautionary demand for savings. The effect of the precautionary demand can be seen most clearly by considering the net-of-transactions-costs bond return. For example, in the CDC model with $\Omega = 1$, the average bond return is 6.6 percent and the average bond market transactions cost is 2.3 percent. This is not reflected in the observed bond return because of the form of the cost function.

Again to assess the indirect effect of transactions costs on the equity premium, it is useful to construct the net premium as measured by $E\{r_{t,t+1}^{s,\text{net}} - r_{t,t+1}^{b,\text{net}}\}$, where $r_{t,t+1}^{s,\text{net}}$ is given by (16) and

$$r_{t,t+1}^{b,\text{net}} = E\left\{\frac{1}{p_t^b + 2\Omega_t b_{t+1}'(p_t^b)^2}\right\}. \tag{20}$$

Notice that consistent with the idea that in this case the net premium reflects only risk differentials, the measured equity premium and the net premium track each other closely.

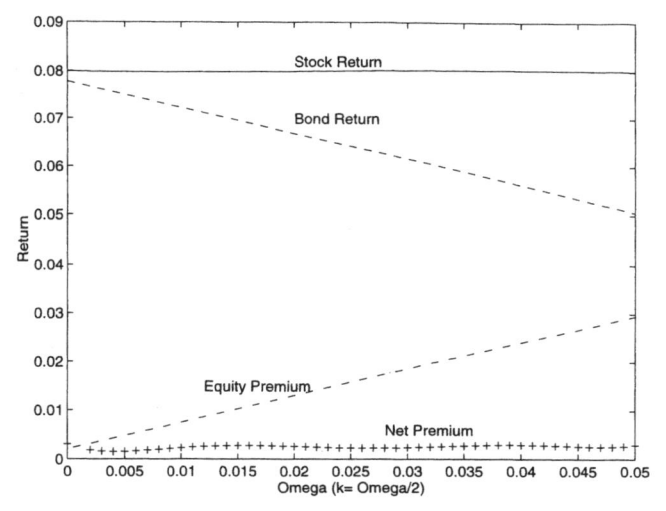

FIG. 9.—Linear costs, base case, returns

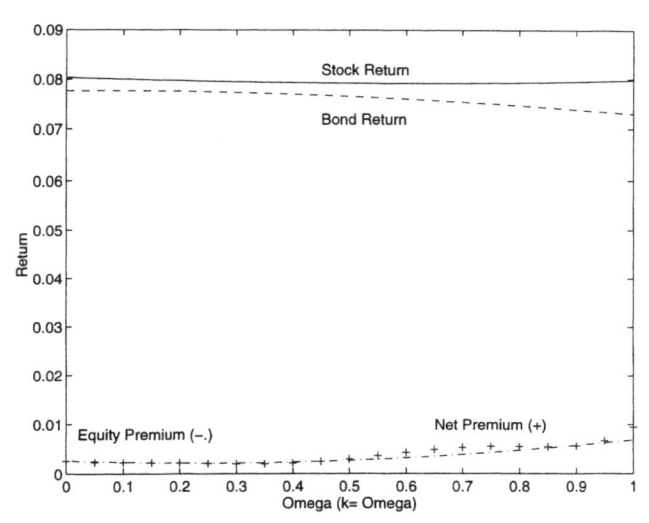

FIG. 10.—Base case, symmetric costs, returns

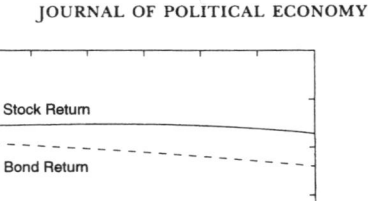

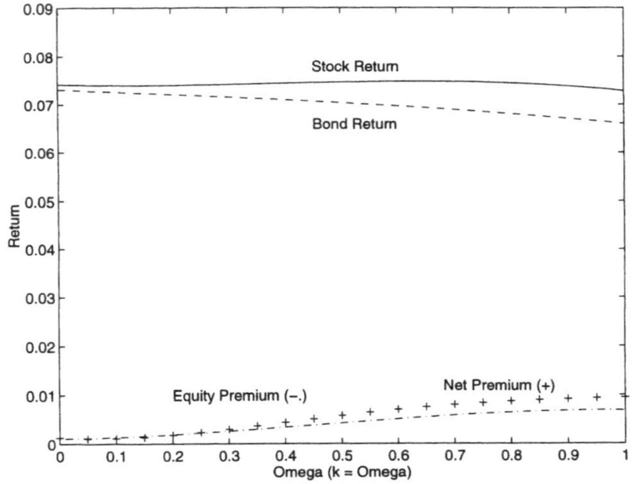

FIG. 11.—CDC, symmetric costs, returns

E. Transactions Costs in Stocks and No Borrowing

In the preceding analysis the borrowing constraint is set so that it rarely binds. There is some evidence, however, that agents may face much more severe borrowing constraints. To examine this possibility we consider a market structure that permits costly trading in stocks but precludes borrowing altogether. In this case, the (shadow) price of bonds is calculated using the marginal rate of substitution of the agent with a good idiosyncratic shock.[22] Hence, the results of this section can be interpreted as approximating a situation in which there is extreme credit rationing or there is a very large wedge between the borrowing and lending rates.

Figures 12 and 13 report average returns and costs for the base case model without borrowing,[23] in which the only cost parameter that varies is k. For any level of the costs, the effect of transactions costs on the predicted bond return and the equity premium is larger than in the case in which costly borrowing is allowed. For example,

---

[22] A second way to close the bond market would be to impose a very high cost on the borrower in the bond market, which from a pricing perspective is equivalent to the case in which borrowing is not allowed. See Heaton and Lucas (1992) for a discussion of this issue.

[23] We do not report the net premium in this case since the marginal borrowing rate is effectively infinite.

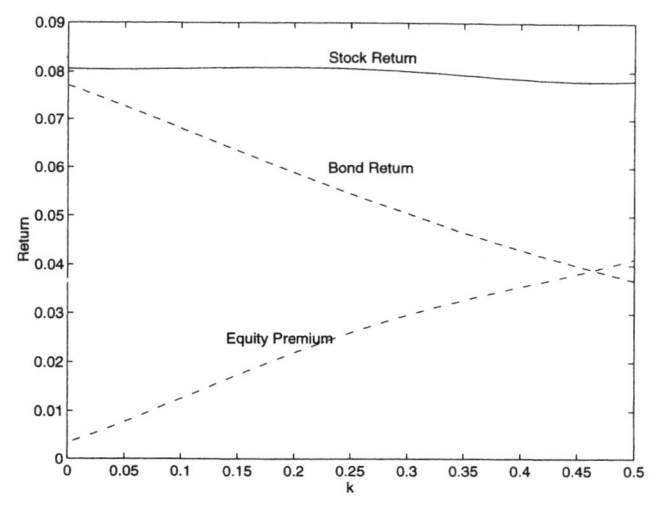

FIG. 12.—Base case, no bonds, returns

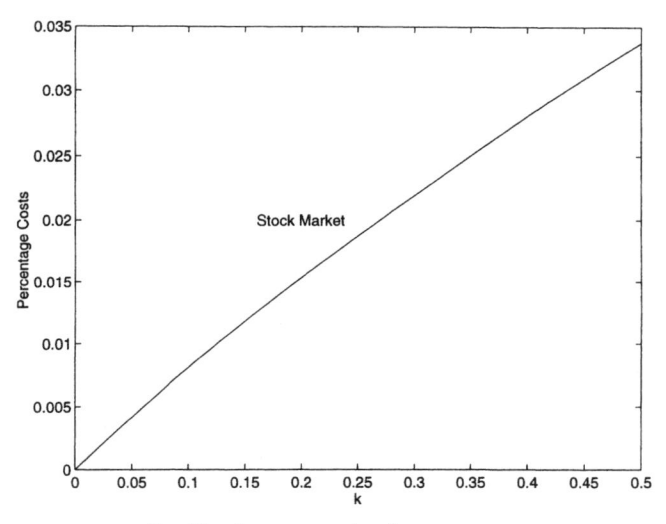

FIG. 13.—Base case, no bonds, average costs

when marginal costs in the stock market are 6 percent, the predicted equity premium is 4 percent compared to 3 percent in the case with costly borrowing (figs. 1 and 2). This occurs because agents cannot substitute toward the bond market to avoid transactions costs. The results for the CDC model are similar and are not reported.

F. The Effect of an Outside Supply of Bonds

In the preceding analysis, debt was assumed to be in zero net supply, and the assumed level of stock holdings was grossed up to reflect the income and liquidity provided by assets such as government and corporate debt. This approach has the shortcoming that it does not allow examination of the effect of differential transactions costs between stocks and outside bonds. However, we can reinterpret some of our previous analysis to capture the presence of an outside supply of government bonds.[24]

Suppose that the outside supply of bonds at time t is $2B_t$ and, as in Aiyagari and Gertler (1991) and Huggett (1993), that this debt is financed by per capita lump-sum taxation in the amount of $\tau_t = B_t - p^b_{t+1} B_{t+1}$. Assume further that there are no costs of transacting in the bond market. The budget constraint of agent i becomes

$$c^i_t + p^s_t s^i_{t+1} + p^b_t b^i_{t+1} + \kappa(s^i_{t+1}, s^i_t; \mathbf{Z}_t) + \tau_t \leq s^i_t(p^s_t + d_t) + b^i_t + Y^i_t, \quad (21)$$

and the condition for equilibrium in the bond market is $b^1_t + b^2_t = 2B_t$. Let $a^i_t \equiv b^i_t - B_t$. Then (21) can be written as

$$c^i_t + p^s_t s^i_{t+1} + p^b_t a^i_{t+1} + \kappa(s^i_{t+1}, s^i_t; \mathbf{Z}_t) \leq s^i_t(p^s_t + d_t) + a^i_t + Y^i_t, \quad (22)$$

and in equilibrium $a^1_t + a^2_t = 0$. Further, if we assume that no borrowing is allowed ($b^i_t \geq 0$), then each individual faces the restriction $a^i_t \geq -B_t$. Notice that this setup is equivalent to the case in which there is no outside supply of bonds and some borrowing is allowed, with a^i_t in place of b^i_t.

In Section IVC we showed that when the borrowing constraint is not too restrictive (e.g., when it is set at 10 percent of per capita income) and trading in at least one market is costless, agents substitute almost completely toward trading in the market without transactions costs. Consumption is effectively smoothed, and as a result, transactions costs have little impact on the equity premium. In light of the discussion in the previous paragraph, this result also will obtain when there is a sufficiently large outside supply of bonds (e.g., equal to 10 percent of per capita income) and trading these outside bonds is

---

[24] This interpretation was suggested to us by an anonymous referee.

INCOMPLETE MARKETS 473

costless. Aiyagari and Gertler (1991) report similar results from a model without aggregate shocks but with an infinite number of agents.

To examine the sensitivity of our results to the supply and composition of assets in the presence of transactions costs, in the simulations the sum of bond holdings across the two agents is set to a positive value. This treats bonds as another "tree" technology, but with a dividend equal to the equilibrium interest rate.[25] The quadratic cost function for bonds is also modified so that the buyer and seller face symmetric transactions costs based only on changes in the value of their bond holdings, so that the cost depends on $|p_t^b b_{t+1} - b_t|$ rather than $b_{t+1} p_t^b$.

To calibrate the average quantity of bonds held by households, we use median household liquid assets as measured in the PSID, which represent about 3.3 months of income for our sample.[26] To bracket this value, we set maximum per capita debt holdings to either 20 percent or 39 percent of per capita income and assume no inside debt.[27] Recall that the level of dividends in the preceding analysis was grossed up relative to measured dividends to proxy for other tradable assets. Here the assumed level of dividends is scaled back slightly to keep the total capital stock approximately constant. This is referred to below as the case of "high stock holdings." We also experiment with reducing dividends to their measured level in the data, which significantly lowers the total available liquidity in the economy.

The results can be summarized as follows: For average stock trading costs of 2.6 percent, average bond trading costs of 0.5 percent, maximum per capita debt holdings of 20 percent, and high stock holdings, the implied equity premium is 1.9 percent and the standard deviation of consumption growth is 6 percent. For the same specification except with no trading costs in the bond market, the equity premium and consumption volatility are almost identical. The reason is that the level of debt is insufficient to entirely buffer the idiosyncratic consumption shocks; debt holding is frequently at a corner, and most

---

[25] This approach abstracts from the taxes the government would levy to support this debt policy, which if included would be only a small fraction of income. As modeled, these bonds more closely resemble corporate debt.

[26] An alternative would be to use the per capita amount of government debt held by the public, which represents a much larger fraction of income. This would probably overestimate available household liquidity, however, since a substantial fraction of government debt is held by foreigners, pension funds, corporations, and other institutions.

[27] The assumption of no inside debt was made primarily for computational tractability. Whether it would represent a significant additional source of liquidity would depend on the associated transactions costs assumed and the severity of the borrowing constraint.

smoothing takes place via the stock market. When the quantity of debt is increased to 38 percent of per capita income with no change in other parameters, the equity premium falls to 1.2 percent and the standard deviation of consumption to 5.5 percent, reflecting the increased opportunities to smooth costlessly in the bond market. As in our comparison of the results of figures 10 and 12, the premium is quite sensitive to the amount of debt available and is reduced when an outside supply of bonds is introduced.

Finally, we consider the low-liquidity case in which dividends average only 5 percent of income (consistent with NIPA data) and maximum bond holdings are 20 percent of per capita income. For an average transactions cost of 2.6 percent in the stock market and no transactions costs in the bond market, the predicted equity premium increases to 2.7 percent, and the standard deviation of consumption is 8 percent. Notice that this result is similar to the base case (fig. 1), where the level of dividends was grossed up from its observed value to more closely reflect the quantity of tradable assets. As a result, even with an outside supply of bonds the model predicts a substantial equity premium for a reasonable level of transactions costs in the stock market. However, consistent with the base case, very large transactions costs in the stock market are needed to obtain an equity premium close to its historical value.

V. Concluding Remarks

In this paper we have examined asset prices and consumption patterns in a model in which agents face both aggregate and idiosyncratic income shocks, and insurance markets are incomplete. Agents reduce consumption variability by trading in a stock and bond market to offset idiosyncratic shocks, but frictions in both markets limit the extent of trade. The theoretical model is calibrated under a variety of assumptions about the form and size of frictions using an empirical model of labor and dividend income estimated using data from the PSID and the NIPA. Transactions costs in the stock and bond markets generate an equity premium and lower the risk-free rate. By simultaneously considering aggregate and idiosyncratic shocks, we can decompose this effect of transactions costs on the equity premium into two components, a direct and an indirect effect.

The direct effect occurs because individuals equate net-of-cost margins, so an asset with a lower associated transactions cost will have a lower market rate of return. The size of the direct effect varies widely with the structure of transactions costs. When the cost structures in the stock and bond markets are assumed to be similar, the direct effect is negligible. However, when we assume that the transactions

INCOMPLETE MARKETS 475

cost in the bond market takes the form of a wedge between the borrowing and lending rates or that there is a binding borrowing constraint, the direct effect can account for close to half of the observed premium. This is related to the findings of Aiyagari and Gertler (1991), which led them to conjecture that realistic transactions costs might account for about 50 percent of the observed equity premium.

A second, indirect effect occurs because transactions costs result in individual consumption that more closely tracks individual income than aggregate consumption. The higher variability of individual consumption increases the covariance between consumption and the dividend process and, hence, increases the systematic risk of the stock. While the size of the indirect effect increases with the assumed level of transactions costs, it is relatively insensitive to the incidence of transactions costs. In the base case analysis, the indirect effect accounts for about 20 percent of the premium.

Although the model can generate an equity premium and a risk-free rate that match the historical means, to do this the necessary level of transactions costs is very large. Also the results of our model are quite sensitive to the assumed quantity of tradable securities. For example, the addition of a large supply of government securities (outside bonds) dampens the effect of transactions costs substantially. All this suggests the sensitivity of asset price predictions to assumed market structure.

A further difficulty with the model is that it does not explain the observed second-moment differentials. It shares the feature of many consumption-based models that an increase in the equity premium is not accompanied by a substantial change in the relative volatility of bond and stock returns. Typically in models that fit the equity premium, the resulting volatility of the bond return is too high. In contrast, when this model is parameterized to match the equity premium, there is little increase in observed return volatilities, resulting in too little stock return volatility. It remains an open question whether there is a realistic assumption about transactions costs that can simultaneously explain the low volatility of short bond rates and the high volatility of stock returns.

Appendix A

Data Description and Estimation Results from the PSID

In this Appendix we describe the estimation (summarized in Sec. IIIA) in more detail and describe a more general model of labor income that is also investigated.

A. Specification of the Model of Individual Labor Income

Individual i's labor income as a fraction of aggregate labor income is given by η_t^i, where $\{\eta_t^i\}_{t=1,\infty}$ is assumed to be a stationary process for each i. To generalize (14), we estimate the process for η_t^i using the specification

$$\log(\eta_t^i) = \overline{\eta}^i + \zeta^i \log\left(\frac{Y_{t-1}^a}{Y_{t-2}^a}\right) + \rho^i \log(\eta_{t-1}^i) + \epsilon_t^i, \tag{A1}$$

where the $\{\epsilon_t^i\}_{t=1,\infty}^{i=1,n}$ are individual shocks that have mean zero, are independent over time and across individuals, and are independent of the lagged aggregate shock ϵ_{t-1}^a for all t; $E\{(\epsilon_t^i)^2\}^{1/2} \equiv \sigma^i$; and $\{\overline{\eta}^i\}^{i=1,n}$, $\{\zeta^i\}^{i=1,n}$, and $\{\rho^i\}^{i=1,n}$ are parameters.

The parameters $\{\overline{\eta}^i\}^{i=1,n}$ capture permanent differences in relative labor income. The parameter $\{\zeta^i\}$ reflects the degree to which individual i's relative income can be predicted by lagged aggregate shocks. Differences in ζ^i across individuals allow the aggregate shock to differentially affect the conditional mean of each individual's income. The parameter ρ^i captures the persistence in shocks to individual i's labor income.

To capture the differences in the distribution of labor income shocks over the business cycle, we also consider a linear model of the standard deviation of individual income shares as a function of the growth rate in aggregate income, which is given by (15) above.

B. Data

The PSID provides a panel of annual observations of individual and family income and other variables. In using the PSID, we take family income per family member as a measure of Y^i for our model. Because of the changing character of many of the families, we take a subsample from the PSID consisting of those families in which the head of the household was male and neither the head nor his spouse changed over the sample. This selection is used to avoid having to keep track of new families, family split-offs, and other dramatic changes in the family. The complete PSID oversamples poorer members of the U.S. population because of an inclusion of a sample of poor individuals from the Survey of Economic Opportunity. To make the sample closer to a random sample of the U.S. population, we excluded those families that were originally part of the Survey of Economic Opportunity.

Total labor income of the head of the household and his wife along with total transfers to the family are used for total family income. The transfers included unemployment compensation, workers' compensation, pension income, welfare, child support, and so on. If a family had zero income from all sources in any year, it was excluded from the sample.[28] This sample includes 860 families with income data spanning the years 1969–84. Because

---

[28] Once we conditioned on families with neither a change in the head of the household nor a change in his spouse, there were only 15 out of 875 families that had zero income in any of the sample years. As a result, the exclusion of families with zero income has little effect on the results.

families differ in size, family income per family member is created by dividing each family's income by the total number of family members in each year.[29] This measure of nominal family labor income is weighted by the CPI to obtain a measure of real labor income per family member.

The model of individual income dynamics is estimated using this full sample and also using a subsample restricted to households owning stock. The sample is split because a large segment of the population does not hold financial assets. For analysis of an asset pricing model, it is clearly more appropriate to consider the income dynamics of those individuals who participate in securities markets. Following Mankiw and Zeldes (1991), we split the sample on the basis of individual holdings of securities using questions about stock holdings in the 1984 PSID. If a family reported some holdings of stocks in 1984, it is included in the group called *stockholders*.

For both the complete sample and the stockholder sample, η_t^i is constructed as $Y_t^i/(\Sigma_{i=1}^n Y_t^i)$, where n is 860 for the complete sample and 327 for the stockholder sample.

Along with observations of individual labor income from the PSID, measures of annual aggregate labor income and dividends are taken from the NIPA for the years 1947–92, obtained from Citibase. For labor income we use "total compensation of employees." The aggregate series are weighted by the total U.S. population and the CPI in each year to obtain real per capita labor income and dividends.

C. Empirical Results

Aggregate Dynamics

A natural alternative to using the NIPA to measure aggregate labor income would be to use total income from the PSID. However, because of the limited time dimension in the PSID, the aggregate dynamics could not be estimated with much precision. Although the NIPA measure of labor income does not exactly correspond to the measure of labor income obtained from the PSID, the two measures of aggregate income are similar. For example, figure A1 gives a plot of the NIPA (from Citibase) measure of aggregate labor income growth, $\log(Y_t^a/Y_{t-1}^a)$, along with the corresponding measure constructed from the PSID from 1970 to 1984. The correlation between the two series is .88. As a result, measuring aggregate labor income from the NIPA series should do a reasonable job of capturing the aggregate labor income that corresponds to the sample from the PSID. Table A1 reports ordinary least squares estimates of the law of motion for aggregate labor income and aggregate dividends (see eq. [12]).

Estimation of (A1)

For each household in both samples from the PSID, we estimate (A1) using ordinary least squares. A summary of these findings is given in table A2,

[29] We also conducted the analysis in which family income was weighted only by the number of adults in the family. The results were similar and hence are not reported.

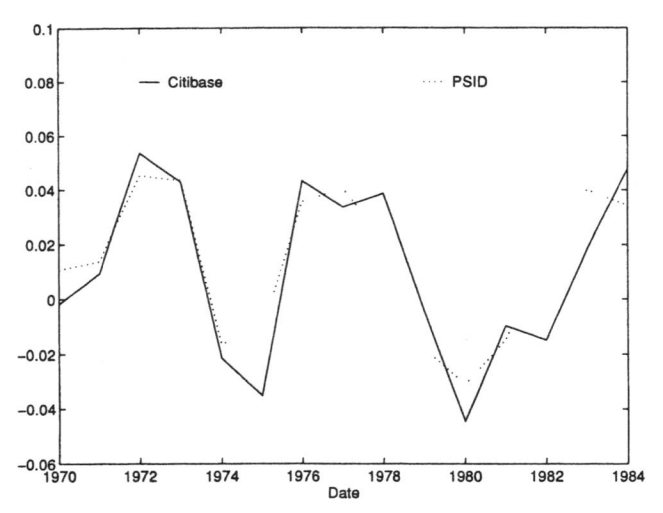

FIG. A1.—Labor income growth rates

TABLE A1
Aggregate Dynamics

$$\gamma_t^a \equiv Y_t^a/Y_{t-1}^a, \delta_t \equiv D_t^a/Y_t^a, \text{ and } \mathbf{X}_t^a \equiv [\log(\gamma_t^a)\log(\delta_t)]'$$

$$\mathbf{X}_t^a = \Lambda^a X_{t-1}^a + \Theta^a \epsilon_t^a + \mu^a$$

Ordinary Least Squares Estimates

$$\Lambda^a = \begin{bmatrix} .1487 & .0557 \\ (.1645) & (.0378) \\ \\ -.5016 & .9168 \\ (.2532) & (.0696) \end{bmatrix} \quad \Theta^a = \begin{bmatrix} .0278 & 0 \\ .0121 & .0536 \end{bmatrix}$$

$$\mu^a = \begin{bmatrix} .1961 \\ (.1228) \\ -.2607 \\ (.2260) \end{bmatrix}$$

NOTE.—Standard errors are in parentheses.

TABLE A2

INDIVIDUAL INCOME DYNAMICS WITH AGGREGATES:
CROSS-SECTIONAL MEANS AND STANDARD ERRORS
OF COEFFICIENT ESTIMATES

$$\eta_t^i \equiv Y_t^i/Y_t,$$

$$\log(\eta_t^i) = \overline{\eta}^i + \rho^i \log(\eta_{t-1}^i) + \zeta^i \log(Y_t^a/Y_{t-1}^a) + \epsilon_t^i,$$

$$\sigma^i = [E(\epsilon_t^i)^2]^{1/2}$$

	CROSS-SECTIONAL MEAN (Standard Deviation)	
COEFFICIENT	All Individuals	Stockholders
$\overline{\eta}^i$	−3.564 (2.534)	−4.222 (1.919)
ζ^i	.081 (2.542)	−1.482 (.314)
ρ^i	.527 (.349)	.299 (.155)
σ^i	.241 (.127)	.365 (3.638)

which reports sample averages of the parameter estimates along with the cross-sectional standard deviations of the estimates. The estimates reported in table A2 are consistent with results reported in MaCurdy (1982) and Abowd and Card (1989). In particular, the estimated parameter values imply that individual income growth is negatively correlated over time. Abowd and Card also argue that aggregate variation in income has little effect on the autocorrelation structure of individual income. The average parameters reported in table A2 along with the point estimates of the aggregate dynamics reported in table A1 imply that the aggregate shocks account for only 1 percent and 2 percent of the variance in individual income shares for the entire sample and the stockholder sample, respectively. For this reason, we also estimate the law of motion for the η^i's with $\zeta^i \equiv 0$ for each i. A summary of the results of this estimation is reported in table A3. Notice that for each sample the average estimated value of ρ^i and σ^i is very close to the average value reported in table A2. We conclude that aggregate shocks are not very important in explaining the conditional mean and unconditional variance of the average individual's income share.

In comparing the results for the entire sample with the results for the stockholder subsample, notice that the average autocorrelation coefficient for the stockholders is smaller than for the entire sample. Also, the variance of the idiosyncratic shock, σ^i, for the stockholders is slightly larger. However, the results generally indicate that there is significant autocorrelation in individual labor income shocks, the variance of these shocks is quite large, and there is little role for the aggregates in explaining the unconditional variance of the shocks. The base case used in experimenting with the asset pricing model is based on an approximation to the dynamics implied by the point estimates of tables A1 and A3 for the complete sample. As described below, this requires matching the first-order autocorrelation and variances in individual income shares, along with the first-order dynamics in aggregate income and dividends.

TABLE A3

Individual Income Dynamics without Aggregates:
Cross-Sectional Means and Standard Deviations
of Coefficient Estimates

$$\eta^i_t \equiv Y^i_t/Y^I_t,$$

$$\log(\eta^i_t) = \overline{\eta}^i + \rho^i \log(\eta^i_{t-1}) + \epsilon^i_t,$$

$$\sigma^i = [E(\epsilon^i_t)^2]^{1/2}$$

	Cross-Sectional Mean (Standard Deviation)	
Coefficient	All Individuals	Stockholders
$\overline{\eta}^i$	−3.354 (2.413)	−4.241 (.299)
ρ^i	.529 (.332)	.299 (.304)
σ^i	.251 (.131)	.383 (.159)

TABLE A4

Cross-Sectional Standard Deviation and Aggregates

$$\widehat{\text{Std}}[\log(\epsilon^i_t)] = \alpha_0 + \alpha_1 \log(Y^a_t/Y^a_{t-1}) + \xi_t$$

	Estimated Parameter (Standard Error)	
Coefficient	All Individuals	Stockholders
α_0	.272 (.001)	.360 (.014)
α_1	.052 (.039)	−1.064 (.419)

Note.—$\widehat{\text{Std}}[\log(\epsilon^i_t)]$ is the estimated cross-sectional standard deviation of shocks to individual income.

TABLE A5

Cross-Sectional Standard Deviation and Aggregates

$$\widehat{\text{Std}}[\log(\eta^i_t)] = \alpha_0 + \alpha_1 \log(Y^a_t/Y^a_{t-1}) + \xi_t$$

	Estimated Parameter (Standard Error)	
Coefficient	All Individuals	Stockholders
α_0	.716 (.002)	.682 (.007)
α_1	−.226 (.058)	.894 (.226)

Note.—$\widehat{\text{Std}}[\log(\eta^i_t)]$ is the estimated cross-sectional standard deviation of individual income shares.

Conditional Heteroskedasticity

To capture conditional variation in the standard deviation of individual shocks, we estimate (15) on the basis of the results of tables A2 and A3 and the residuals implied by the estimates of table A3. For each time period, the cross-sectional standard deviation of the estimated residuals, given by $(1/N) \sum_{i=1}^{N} (\hat{\epsilon}^i)^2$, is used as an estimate of the variable on the left side of (15).

The resulting parameter estimates from ordinary least squares are reported in table A4. Notice that for the entire sample of individuals the aggregate state has a positive effect on the cross-sectional standard deviation, which contradicts the presumption that individual income variance increases in economic downturns. The reason may be that our selection criterion eliminates those households with substantial changes in family composition, which may bias our estimate of the conditional heteroskedasticity of income shocks. Also the short time-series dimension of the PSID may limit our ability to recover a countercyclical pattern in the variance of shocks to individual income. In contrast, the results from the stockholder subsample provide evidence in favor of a model in which the aggregate state has a small negative effect on the cross-sectional standard deviation of labor income. The estimates are sensitive to specification, however. In table A5 we report regressions of the cross-sectional standard deviation of $\log(\eta_t^i)$ on aggregate income. Notice that in this case, when the complete sample from the PSID is used, there is evidence of a widening in the distribution of labor income during aggregate downturns. These results indicate that with the short sample period of the PSID, it is difficult to precisely quantify the cyclical movement in the distribution of labor income.

Appendix B

Markov Chain Approximation

We construct Markov chain approximations to the laws of motion (12), (14), and (15) using the method of Tauchen and Hussey (1991), with two states for each of the exogenous states. Here we report measures of how well the Markov chain approximates important aspects of the dynamics of income as summarized by the estimates reported in tables A1, A3, and A4.

Base Case

Table B1 reports the inputs for the base case model, which are the parameter vectors Λ, Θ, and μ of the trivariate VAR for aggregate income growth, the dividend share, and individual 1's (demeaned) income share. They are reported in table B1 as "Input Values" and are given by the point estimates of tables A1 and A3 for the complete sample. As discussed by Tauchen and Hussey (1991), one way to assess the Markov chain approximation error is to examine the parameter matrices Λ, Θ, and μ implied by the Markov chain. They are reported in table B1 as "Implied Values." Notice that the persistence and variability of individual 1's labor income are reasonably well

TABLE B1

PARAMETERS IMPLIED BY MARKOV CHAIN APPROXIMATION: BASE CASE

$\gamma_t^a \equiv Y_t^a/Y_{t-1}^a$, $\delta_t \equiv D_t^a/Y_t^a$, and $\mathbf{X}_t \equiv [\log(\gamma_t^a) \log(\delta_t)[\log(\eta_t^i) - E\log(\eta_t^i)]]'$,

$$\mathbf{X}_t = \Lambda \mathbf{X}_{t-1} + \Theta \epsilon_t + \mu$$

Input Values

$$\Lambda = \begin{bmatrix} .1487 & .0557 & 0 \\ -.5016 & .9168 & 0 \\ 0 & 0 & .5290 \end{bmatrix}, \quad \Theta_t = \begin{bmatrix} .0278 & 0 & 0 \\ .0121 & .0536 & 0 \\ 0 & 0 & .2508 \end{bmatrix}$$

$$\mu = \begin{bmatrix} .1961 \\ -.2607 \\ 0 \end{bmatrix}$$

Implied Values

$$\Lambda = \begin{bmatrix} .1459 & .0539 & 0 \\ -.3330 & .7331 & 0 \\ 0 & 0 & .4846 \end{bmatrix}, \quad \Theta_t = \begin{bmatrix} .0272 & 0 & 0 \\ .0118 & .0379 & 0 \\ 0 & 0 & .2194 \end{bmatrix}$$

$$\mu = \begin{bmatrix} .1902 \\ -.8594 \\ 0 \end{bmatrix}$$

approximated. However, the aggregate dynamics are not as well approximated.

Figure B1 plots the dividend's share of aggregate income from 1948 to 1992. The most substantial movement in dividends occurs in the early 1970s. The Markov chain does not capture the slow movement down and then up in the dividend share over this period. However, the overall variation in dividends is not large. Since we focus on the effects of individual income shocks, the fact that we do not completely fit the dynamics of aggregate dividends appears to be unimportant. Essentially the dividend stream in our model is given by a constant fraction of the aggregate endowment. We miss some of the slow movement in the dividend share, but this is not substantial.

The Markov chain reported in table 1 is a transformation of the Markov chain approximation to the VAR of table B1. First, as described in Section III, the dividend share is grossed up so that, on average, it accounts for 15 percent of aggregate income. This is done by multiplying the share variable by a scale factor. Second, the individual income share is scaled so that it averages 50 percent. As a result, each individual's labor income share is, on average, 50 percent. This is also accomplished by multiplying the share value by a scale factor.

Cyclical Distribution Case

We deviate from the estimated parameters of the conditional heteroskedasticity model in order to obtain a case that is significantly different from the

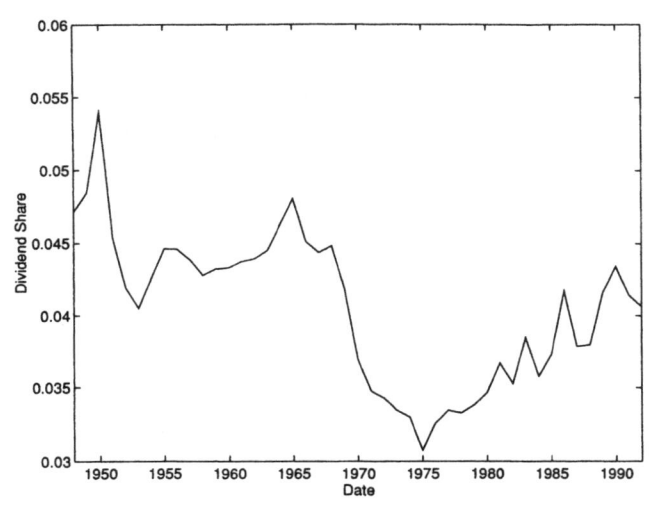

FIG. B1.—Dividend income share

TABLE B2

PARAMETERS IMPLIED BY MARKOV CHAIN APPROXIMATION:
CYCLICAL DISTRIBUTION CASE

$\gamma_t^a \equiv Y_t^a/Y_{t-1}^a$, $\delta_t \equiv D_t^a/Y_t^a$, and $\mathbf{X}_t \equiv [\log(\gamma_t^a)\log(\delta_t)[\log(\eta_t^i) - E\log(\eta_t^i)]]'$,

$$\mathbf{X}_t = \Lambda \mathbf{X}_{t-1} + \Theta_t \epsilon_t + \mu$$

Input Values

$$\Lambda = \begin{bmatrix} .1487 & .0557 & 0 \\ -.5016 & .9168 & 0 \\ 0 & 0 & .529 \end{bmatrix}, \quad \Theta_t = \begin{bmatrix} .0278 & 0 & 0 \\ .0121 & .0536 & 0 \\ 0 & 0 & \sigma_t \end{bmatrix}$$

$$\mu = \begin{bmatrix} .1961 \\ -.2607 \\ 0 \end{bmatrix}$$

Implied Values

$$\Lambda = \begin{bmatrix} .2194 & .0517 & 0 \\ -.3010 & .7322 & 0 \\ 0 & 0 & .4307 \end{bmatrix}, \quad \Theta_t = \begin{bmatrix} .0267 & 0 & 0 \\ .0116 & .0379 & 0 \\ 0 & 0 & \sigma_t \end{bmatrix}$$

$$\mu = \begin{bmatrix} .1797 \\ -.8640 \\ 0 \end{bmatrix}, \sigma_t = .2898 - 4.4504 \log(\gamma_t^a)$$

base case and provides a countercyclical distribution of income. As in the base case, there are eight states in total, with two states for aggregate income. The parameters of tables A1 and A3 are used to calibrate the conditional mean of the variables. The volatility of the individual shock is chosen to be twice as large in the low–aggregate growth state as in the high–aggregate growth state, with an average volatility chosen to match the average volatility of the shocks reported in table A3. The model maintains the feature of the base case that the conditional mean of individual income shares (as measured by a linear regression) is not affected by the aggregate state.

Table B2 reports the inputs for the Markov chain. The matrix Θ_t is time dependent because of the conditional heteroskedasticity of the shocks. The parameters of the approximating Markov chain are given in table 2 (transformed as in the base case). To check the approximations, table B2 also reports the VAR parameters implied by the approximating Markov chain. Notice that the model fits the persistence in individual income reasonably well, and the aggregate dynamics are fit about as well as in the base case. As a way of evaluating the heteroskedasticity implied by the model, table B2 also reports the implied parameters of the heteroskedasticity regression. Notice that the coefficient on aggregate income is four times larger than the estimated value reported in table A4 for the stockholder subsample.

Appendix C

Numerical Accuracy

Model equilibria are computed using the "auctioneer" algorithm of Lucas (1994). This algorithm searches for an allocation of bond and stock holdings for the two agents that clears markets in every state of the world and implies agreement between the two agents on the prices of the two securities.[30] For a given allocation, the Euler equations of the unconstrained agents define functional equations in the prices of the two securities as functions of the exogenous and endogenous state variables. These equations are approximated over a grid of 30 equally spaced points for stock and bond holdings. The resulting discrete state space is 30 × 30 × 8 since the exogenous variables take on eight different values. A fixed point to the equations is found using the fact that the Euler equations form a contraction mapping. Iterations continue until there is less than a 0.5 percent difference in the (uniformly weighted) average solution across the grid space.

Owing to the discrete approximation, it is not possible to set the prices quoted by the two agents exactly equal to one another. We report the average of the difference between the prices quoted by the two agents as a percentage of the price quoted by the first agent in tables C1 and C2. The approximation error is reported for the case of no outside bonds and asymmetric and symmetric cost functions in the bond market. For settings of the cost parameters that produce plausible marginal transactions costs, the algorithm produces

---

[30] For a complete description of the algorithm, see Lucas (1994).

INCOMPLETE MARKETS

TABLE C1

APPROXIMATION ERRORS: ASYMMETRIC COSTS (Average Percentage Difference in Prices between the Two Agents)

	BASE CASE		CYCLICAL DISTRIBUTION CASE	
VALUE OF Ω	Stock Price	Bond Price	Stock Price	Bond Price
.0	.19	.10	.23	.09
.2	.19	.36	.28	.37
.4	.14	.24	.29	.22
.6	.15	.23	.33	.22
.8	.17	.23	.39	.27
1.0	.17	.21	.53	.39
1.2	.18	.18	.59	.49
1.4	.20	.21	.65	.50
1.6	.22	.23	.71	.49
1.8	.25	.27	.86	.62
2.0	.23	.27	.90	.63

TABLE C2

APPROXIMATION ERRORS: SYMMETRIC COSTS (Average Percentage Difference in Prices between the Two Agents)

	BASE CASE		CYCLICAL DISTRIBUTION CASE	
VALUE OF Ω	Stock Price	Bond Price	Stock Price	Bond Price
.0	.26	.14	.21	.08
.2	.30	.07	.33	.17
.4	.17	.33	.34	.86
.6	.13	.28	.37	.42
.8	.12	.34	.39	.43
1.0	.12	.40	.45	.51

reasonably accurate solutions. For example, when Ω is restricted to be less than one in the asymmetric cost cases, the average errors are less than 0.5 percent.

References

Abowd, John M., and Card, David. "On the Covariance Structure of Earnings and Hours Changes." *Econometrica* 57 (March 1989): 411–45.
Aiyagari, S. Rao. "Uninsured Idiosyncratic Risk and Aggregate Saving." *Q.J.E.* 109 (August 1994): 659–84.
Aiyagari, S. Rao, and Gertler, Mark. "Asset Returns with Transactions Costs and Uninsured Individual Risk." *J. Monetary Econ.* 27 (June 1991): 311–31.
Amihud, Yakov, and Mendelson, Haim. "Asset Pricing and the Bid-Ask Spread." *J. Financial Econ.* 17 (December 1986): 223–49.
Bewley, Truman F. "Stationary Monetary Equilibrium with a Continuum of Independently Fluctuating Consumers." In *Contributions to Mathematical*

Economics in Honor of Gerard Debreu, edited by Werner Hildenbrand and Andreu Mas-Colell. Amsterdam: North-Holland, 1986.

Clarida, Richard H. "International Lending and Borrowing in a Stochastic Stationary Equilibrium." *Internat. Econ. Rev.* 31 (August 1990): 543–58.

Constantinides, George M. "Capital Market Equilibrium with Transaction Costs." *J.P.E.* 94 (August 1986): 842–62.

Constantinides, George M., and Duffie, Darrell. "Asset Pricing with Heterogeneous Consumers." Manuscript. Chicago: Univ. Chicago, Grad. School Bus., 1994.

Dumas, Bernard. "Two-Person Dynamic Equilibrium in the Capital Market." *Rev. Financial Studies* 2, no. 2 (1989): 157–88.

Hansen, Lars Peter, and Jagannathan, Ravi. "Implications of Security Market Data for Models of Dynamic Economies." *J.P.E.* 99 (April 1991): 225–62.

Hansen, Lars Peter, and Singleton, Kenneth J. "Stochastic Consumption, Risk Aversion, and the Temporal Behavior of Asset Returns." *J.P.E.* 91 (April 1983): 249–65.

He, Hua, and Modest, David M. "Market Frictions and Consumption-Based Asset Pricing." *J.P.E.* 103 (February 1995): 94–117.

Heaton, John, and Lucas, Deborah J. "The Effects of Incomplete Insurance Markets and Trading Costs in a Consumption-Based Asset Pricing Model." *J. Econ. Dynamics and Control* 16 (July/October 1992): 601–20.

———. "The Importance of Investor Heterogeneity and Financial Market Imperfections for the Behavior of Asset Prices." *Carnegie-Rochester Conf. Ser. Public Policy* 42 (June 1995): 1–32.

Huggett, Mark. "The Risk-Free Rate in Heterogeneous Agent Incomplete-Insurance Economies." *J. Econ. Dynamics and Control* 17 (September/November 1993): 953–69.

Lucas, Deborah J. "Asset Pricing with Undiversifiable Income Risk and Short Sales Constraints: Deepening the Equity Premium Puzzle." *J. Monetary Econ.* 34 (December 1994): 325–41.

MaCurdy, Thomas E. "The Use of Time Series Processes to Model the Error Structure of Earnings in a Longitudinal Data Analysis." *J. Econometrics* 18 (January 1982): 83–114.

Mankiw, N. Gregory. "The Equity Premium and the Concentration of Aggregate Shocks." *J. Financial Econ.* 17 (September 1986): 211–19.

Mankiw, N. Gregory, and Zeldes, Stephen P. "The Consumption of Stockholders and Nonstockholders." *J. Financial Econ.* 29 (March 1991): 97–112.

Marcet, Albert, and Singleton, Kenneth J. "Equilibrium Asset Prices and Savings of Heterogeneous Agents in the Presence of Incomplete Markets and Portfolio Constraints." Manuscript. Stanford, Calif.: Stanford Univ., Grad. School Bus., 1991.

Mehra, Rajnish, and Prescott, Edward C. "The Equity Premium: A Puzzle." *J. Monetary Econ.* 15 (March 1985): 145–61.

Pratt, John W., and Zeckhauser, Richard J. "Proper Risk Aversion." *Econometrica* 55 (January 1987): 143–54.

Saito, Makoto. "Limited Market Participation and Asset Pricing." Manuscript. Vancouver: Univ. British Columbia, Dept. Econ., 1995.

Scheinkman, José A. "Discussion: Market Incompleteness and the Equilibrium Valuation of Assets." In *Theory of Valuation: Frontiers of Modern Financial Theory,* vol. 1, edited by Sudipto Bhattacharya and George M. Constantinides. Totowa, N.J.: Rowman and Littlefield, 1989.

Scheinkman, José A., and Weiss, Laurence. "Borrowing Constraints and Aggregate Economic Activity." *Econometrica* 54 (January 1986): 23–45.

Sharpe, William F. *Investments.* 3d ed. Englewood Cliffs, N.J.: Prentice-Hall, 1985.

Tauchen, George, and Hussey, Robert. "Quadrature-Based Methods for Obtaining Approximate Solutions to Nonlinear Asset Pricing Models." *Econometrica* 59 (March 1991): 371–96.

Telmer, Chris I. "Asset-Pricing Puzzles and Incomplete Markets." *J. Finance* 48 (December 1993): 1803–32.

Vayanos, Dimitri, and Vila, Jean-Luc. "Equilibrium Interest Rate and Liquidity Premium under Proportional Transactions Costs." Manuscript. Stanford, Calif.: Stanford Univ., Grad. School Bus., 1995.

Weil, Philippe. "The Equity Premium Puzzle and the Risk-Free Rate Puzzle." *J. Monetary Econ.* 24 (November 1989): 401–21.

———. "Equilibrium Asset Prices with Undiversifiable Labor Income Risk." *J. Econ. Dynamics and Control* 16 (July/October 1992): 769–90.

[15]

By Force of Habit: A Consumption-Based Explanation of Aggregate Stock Market Behavior

John Y. Campbell
Harvard University and National Bureau of Economic Research

John H. Cochrane
University of Chicago, Federal Reserve Bank of Chicago, and National Bureau of Economic Research

> We present a consumption-based model that explains a wide variety of dynamic asset pricing phenomena, including the procyclical variation of stock prices, the long-horizon predictability of excess stock returns, and the countercyclical variation of stock market volatility. The model captures much of the history of stock prices from consumption data. It explains the short- and long-run equity premium puzzles despite a low and constant risk-free rate. The results are essentially the same whether we model stocks as a claim to the consumption stream or as a claim to volatile dividends poorly correlated with consumption. The model is driven by an independently and identically distributed consumption growth process and adds a slow-moving external habit to the standard power utility function. These features generate slow countercyclical variation in risk premia. The model posits a fundamentally novel description of risk premia: Investors fear stocks primarily because they do poorly in recessions unrelated to the risks of long-run average consumption growth.

Campbell thanks the National Science Foundation for research support. Cochrane thanks the National Science Foundation and the Graduate School of Business of the University of Chicago for research support. We thank Andrew Abel, George Constantinides, John Heaton, Robert Lucas, Rajnish Mehra, and especially Lars Hansen for helpful comments.

I. Introduction

A number of empirical observations suggest tantalizing links between asset markets and macroeconomics. Most important, equity risk premia seem to be higher at business cycle troughs than they are at peaks. Excess returns on common stocks over Treasury bills are forecastable, and many of the variables that predict excess returns are correlated with or predict business cycles (Ferson and Merrick 1987; Fama and French 1989). The literature on volatility tests mirrors this conclusion: price/dividend ratios move procyclically, but this movement cannot be explained by variation in expected dividends or interest rates, indicating large countercyclical variation in expected excess returns (Campbell and Shiller 1988a, 1988b; Shiller 1989; Cochrane 1991, 1992). Estimates of conditional variances of returns also change through time (see Bollerslev, Chou, and Kroner [1992] for a survey), but they do not move one for one with estimates of conditional mean returns. Hence the slope of the conditional mean-variance frontier, a measure of the price of risk, changes through time with a business cycle pattern (Harvey 1989; Chou, Engle, and Kane 1992).

As yet, there is no accepted economic explanation for these observations. In the language of finance, we lack a successful theory and measurement procedure for the fundamental sources of risk that drive expected returns. In the language of macroeconomics, standard business cycle models utterly fail to reproduce the level, variation, and cyclical comovement of equity premia.

We show that many of the puzzles in this area can be understood with a simple modification of the standard representative-agent consumption-based asset pricing model. The central ingredient is a slow-moving habit, or time-varying subsistence level, added to the basic power utility function. As consumption declines toward the habit in a business cycle trough, the curvature of the utility function rises, so risky asset prices fall and expected returns rise.

We model consumption growth as an independently and identically distributed (i.i.d.) lognormal process, with the same mean and standard deviation as postwar consumption growth. Our model can accommodate more complex consumption processes, including processes with predictability, conditional heteroskedasticity, and nonnormality. But these features are not salient characteristics of consumption data. More important, we want to emphasize that the model generates interesting asset price behavior internally, not from exogenous variation in the probability distribution of consumption growth. In this respect, our approach is the opposite of that of Kandel and Stambaugh (1990, 1991), who use fairly standard preferences but derive some of these phenomena from movement over time in the conditional moments of consumption growth.

We choose our model's functional form and parameters so that the risk-free interest rate is constant. We do this for several reasons. First, there appears to be only limited variation in the real risk-free rate in historical U.S. data, and the variation that does exist is not closely related to the business cycle or to movements in stock prices. Second, we want to show how the model can explain stock market behavior entirely by variation in risk premia without any movement in the risk-free rate. Third, many habit persistence models with exogenous consumption give rise to wild variation in risk-free rates. When production is added, consumers smooth away the consumption fluctuations. (See Jermann [1998] for a quantitative example.) A constant risk-free rate is consistent with a linear production technology and therefore suggests that our results will be robust to the addition of an explicit production sector.

We generate artificial data from the model, and then we check whether the artificial data display the patterns found in the empirical literature. The model replicates the level of the risk-free rate, the mean excess stock return (the equity premium), and the standard deviation of excess stock returns. Most important, the model fits the dynamic behavior of stock prices. It matches the level and volatility of price/dividend ratios and the long-horizon forecastability of stock returns, and it produces persistent variation in return volatility. It replicates the finding of the volatility test literature that the volatility of stock price/dividend ratios or returns cannot be accounted for by changing expectations of future dividend growth rates. The model also accounts for much of the observed low correlation between stock returns and consumption growth. Despite a lognormal forcing process, the model predicts nonnormal, negatively skewed stock prices and returns, with occasional crashes that are larger than the booms. We feed the model actual consumption data, and we find that the price/dividend ratios and returns predicted by our model provide a surprisingly good account of fluctuations in stock prices and returns over the last century. All these interesting and seemingly unrelated phenomena are in fact reflections of the same phenomenon, which is at the core of the model: a slowly time-varying, countercyclical risk premium.

A. *Habit Formation*

Habit formation has a long history in the study of consumption. Deaton and Muellbauer (1980) survey early work in the area, and Deaton (1992) gives a more recent overview. Ryder and Heal (1973),

Sundaresan (1989), and Constantinides (1990) are major theoretical papers on the subject. Habit formation captures a fundamental feature of psychology: repetition of a stimulus diminishes the perception of the stimulus and responses to it. Habit formation can explain why consumers' reported sense of well-being often seems more related to recent changes in consumption than to the absolute level of consumption. In macroeconomics, habit persistence can explain why recessions are so feared even though their effects on output are small relative to a few years' growth.

Our habit specification has three distinctive features. First, we specify that habit formation is *external*, as in Abel's (1990) "catching up with the Joneses" formulation or Duesenberry's (1949) "relative income" model. An individual's habit level depends on the history of aggregate consumption rather than on the individual's own past consumption. This specification simplifies our analysis. It eliminates terms in marginal utility by which extra consumption today raises habits tomorrow, while retaining fully rational expectations.

Second, we specify that habit moves slowly in response to consumption, in contrast to empirical specifications in which each period's habit is proportional to the last period's consumption (e.g., Ferson and Constantinides 1991). This feature produces slow mean reversion in the price/dividend ratio, long-horizon return forecastability, and persistent movements in volatility.

Third, we specify that habit adapts nonlinearly to the history of consumption. The nonlinearity keeps habit always below consumption and keeps marginal utility always finite and positive even in an endowment economy. In many models, including those of Sundaresan (1989), Ferson and Constantinides (1991), Heaton (1995), and Chapman (1998), consumption can fall below habit with undesirable consequences. Abel (1990, 1999) keeps marginal utility positive by changing utility from $u(C - X)$ to $u(C/X)$, but this specification eliminates changing risk aversion. Most important, the nonlinear habit specification is essential for us to capture time variation in the Sharpe ratio (mean to standard deviation of returns) and a constant risk-free rate.

II. The Model

A. *Preferences and Technology*

Identical agents maximize the utility function

$$E \sum_{t=0}^{\infty} \delta^t \frac{(C_t - X_t)^{1-\gamma} - 1}{1 - \gamma}. \tag{1}$$

Here X_t is the level of habit, and δ is the subjective time discount factor.

It is convenient to capture the relation between consumption and habit by the *surplus consumption ratio* $S_t \equiv (C_t - X_t)/C_t$. The surplus consumption ratio increases with consumption: $S_t = 0$ corresponds to an extremely bad state in which consumption is equal to habit; S_t approaches one as consumption rises relative to habit. The local curvature of the utility function, which we write as η_t, is related to the surplus consumption ratio by

$$\eta_t \equiv -\frac{C_t u_{cc}(C_t, X_t)}{u_c(C_t, X_t)} = \frac{\gamma}{S_t}.$$

Thus low consumption relative to habit, or a low surplus consumption ratio, implies a high local curvature of the utility function.

To complete the description of preferences, we must specify how the habit X_t responds to consumption. Through most of our analysis we use an *external habit* specification in which habit is determined by the history of aggregate consumption rather than the history of individual consumption. Define

$$S_t^a \equiv \frac{C_t^a - X_t}{C_t^a}, \tag{2}$$

where C^a denotes average consumption by all individuals in the economy. We specify how each individual's habit X_t responds to the history of aggregate consumption C^a by specifying a process for S_t^a. The log surplus consumption ratio $s_t^a \equiv \ln S_t^a$ evolves as a heteroskedastic AR(1) process,

$$s_{t+1}^a = (1 - \phi)\bar{s} + \phi s_t^a + \lambda(s_t^a)(c_{t+1}^a - c_t^a - g), \tag{3}$$

where ϕ, g, and $\bar{s}$ are parameters. (Throughout, we use lowercase letters to indicate logs.) We call $\lambda(s_t^a)$ the *sensitivity function*, and we specify it further below. Substituting (2) into (3), we see that (3) does in fact describe how habit X_t adjusts to the history of consumption $\{C_{t-j}^a\}$. Though this adjustment is nonlinear, to a first approximation near the steady state $\bar{s}$, equation (3) implies that habit x_t itself adjusts slowly and geometrically to consumption c_t^a with coefficient ϕ. In equilibrium, identical individuals choose the same level of consumption, so $C_t = C_t^a$ and $S_t = S_t^a$. Therefore, we drop the a superscripts in what follows where they are not essential for clarity.

Having described tastes, we now turn to technology. We model consumption growth as an i.i.d. lognormal process

$$\Delta c_{t+1} = g + v_{t+1}, \quad v_{t+1} \sim \text{i.i.d. } \mathcal{N}(0, \sigma^2). \tag{4}$$

It is convenient, though not essential, to use the same value g for the mean consumption growth rate and the parameter g in the habit accumulation equation (3).

We can regard equation (4) as the specification of the endowment process and close our model as an endowment economy. In principle, this interpretation does not imply a loss of generality: If the statistical model of the "endowment" is the same as the equilibrium consumption process from a production economy, then the joint asset price–consumption process is the same whether the economy is truly an endowment or a production economy. In practice, the asset pricing predictions of many habit persistence economies are strongly affected by the specification of technology. In many endowment economies with habits and random walk consumption, risk-free rates vary a great deal, as the varying surplus consumption ratio gives rise to strong motives for intertemporal substitution. When production is added to these economies, consumers make strong use of production opportunities to smooth marginal utility over time. The interest rate variation is quieted down, but the equilibrium consumption process moves far from a random walk (Jermann 1998). However, we shall pick the functional forms and parameters of our model to generate a constant real risk-free rate. Therefore, we can also close the model with a linear technology:

$$K_{t+1} = R^f(K_t + E_t - C_t),$$

$$\Delta e_{t+1} = g + v_{t+1}, \quad v_{t+1} \sim \text{i.i.d. } \mathcal{N}(0, \sigma^2),$$

where K_t and E_t denote the capital stock and an exogenous endowment or additive technology shock, respectively. This specification results in exactly the same process for consumption and asset prices as the endowment specification does. This fact suggests that the model's consumption and asset pricing implications will not be much affected if the model is closed with any standard concave specification of technology that gives easy opportunities for intertemporal transformation and thus a roughly constant risk-free interest rate.

B. Marginal Utility

Since habit is external, marginal utility is

$$u_c(C_t, X_t) = (C_t - X_t)^{-\gamma} = S_t^{-\gamma} C_t^{-\gamma}.$$

The intertemporal marginal rate of substitution is then

$$M_{t+1} \equiv \delta \frac{u_c(C_{t+1}, X_{t+1})}{u_c(C_t, X_t)} = \delta \left(\frac{S_{t+1}}{S_t} \frac{C_{t+1}}{C_t} \right)^{-\gamma}.$$

AGGREGATE STOCK MARKET BEHAVIOR

It is related to the state variable s_t and the log consumption innovation v_{t+1} by

$$M_{t+1} = \delta G^{-\gamma} e^{-\gamma(s_{t+1}-s_t+v_{t+1})} = \delta G^{-\gamma} e^{-\gamma(\phi-1)(s_t-\bar{s})+[1+\lambda(s_t)]v_{t+1}}. \quad (5)$$

We can now calculate moments of the marginal rate of substitution and find asset prices.

Slope of the Mean–Standard Deviation Frontier

The slope of the conditional mean–standard deviation frontier can be found from the conditional moments of the marginal rate of substitution. Following Shiller (1982), Hansen and Jagannathan (1991) show that the first-order condition $0 = E_t(M_{t+1} R^e_{t+1})$ for an excess return R^e implies that the Sharpe ratio of any asset return must obey

$$\frac{E_t(R^e_{t+1})}{\sigma_t(R^e_{t+1})} = -\rho_t(M_{t+1}, R^e_{t+1}) \frac{\sigma_t(M_{t+1})}{E_t(M_{t+1})} \leq \frac{\sigma_t(M_{t+1})}{E_t(M_{t+1})}, \quad (6)$$

where ρ_t denotes a conditional correlation. In our model, M is conditionally lognormal,[1] so we can find the largest possible Sharpe ratio by

$$\max_{\{\text{all assets}\}} \frac{E_t(R^e_{t+1})}{\sigma_t(R^e_{t+1})} = \{e^{\gamma^2 \sigma^2 [1+\lambda(s_t)]^2} - 1\}^{1/2} \approx \gamma\sigma[1 + \lambda(s_t)]. \quad (7)$$

This formula helps us to specify the model. To produce a time-varying Sharpe ratio, $\lambda(s)$ must vary with s. To produce risk prices that are higher in bad times, when s is low, $\lambda(s)$ and hence the volatility of s must increase as s declines.

Risk-Free Interest Rate

The real risk-free interest rate is the reciprocal of the conditionally expected stochastic discount factor

$$R^f_t = \frac{1}{E_t(M_{t+1})}.$$

---

[1] For lognormal M with mean μ and standard deviation σ,
$$\frac{\sigma(M)}{E(M)} = \frac{\sqrt{E(M^2) - E(M)^2}}{E(M)} = \frac{\sqrt{e^{2\mu+2\sigma^2} - e^{2\mu+\sigma^2}}}{e^{\mu+(\sigma^2/2)}} = \sqrt{e^{\sigma^2} - 1}.$$

From equation (5) and the lognormality of consumption growth, the log risk-free rate is

$$r^f_t = -\ln(\delta) + \gamma g - \gamma(1 - \phi)(s_t - \bar{s}) - \frac{\gamma^2 \sigma^2}{2}[1 + \lambda(s_t)]^2. \quad (8)$$

The $s_t - \bar{s}$ term reflects intertemporal substitution. If the surplus consumption ratio is low, the marginal utility of consumption is high. If there were no shocks to consumption, marginal utility would fall as the surplus consumption ratio reverts to $\bar{s}$. The consumer would then like to borrow, which would drive up the equilibrium risk-free interest rate. We can interpret the last term in equation (8) as a precautionary savings term. As uncertainty increases, consumers are more willing to save, and this willingness drives down the equilibrium risk-free interest rate.

In the data, we notice relatively little variation in risk-free rates. This means that the serial correlation parameter ϕ must be near one, or $\lambda(s_t)$ must decline with s_t so that uncertainty is high when s is low and the precautionary saving term offsets the intertemporal substitution term. The decline of $\lambda(s_t)$ with s_t is the same condition we need to get counterclyclical variation in the price of risk. We now use this insight to pick the functional form of $\lambda(s_t)$.

C. Choosing the Sensitivity Function $\lambda(s_t)$

We have not yet specified the functional form of $\lambda(s_t)$. We choose $\lambda(s_t)$ to satisfy three conditions: (1) The risk-free interest rate is constant; (2) habit is predetermined at the steady state $s_t = \bar{s}$; and (3) habit is predetermined near the steady state or, equivalently, habit moves nonnegatively with consumption everywhere.

We have already discussed the motivation for a constant risk-free interest rate. We further restrict habit behavior to keep the specification close to traditional and sensible notions of habit. We normally think that it takes time for others' consumption to affect one's habits. In our model, habit cannot be completely predetermined, or a sufficiently low realization of consumption growth would leave consumption below habit, in which case a power utility function is undefined. Hence, we require that habit be predetermined, but only at and near the steady state. Finally, the notion of habit would be strained if we allowed habit to move in the opposite direction from consumption.

These three considerations lead us to a restriction that must hold between the steady-state surplus consumption ratio $\bar{S}$ and the other parameters of the model, namely,

AGGREGATE STOCK MARKET BEHAVIOR

$$\bar{S} = \sigma\sqrt{\frac{\gamma}{1-\phi}}, \tag{9}$$

and they lead us to a specification of the sensitivity function

$$\lambda(s_t) = \begin{cases} \frac{1}{\bar{S}}\sqrt{1 - 2(s_t - \bar{s})} - 1, & s_t \leq s_{max} \\ 0 & s_t \geq s_{max}, \end{cases} \tag{10}$$

where s_{max} is the value of s_t at which the upper expression in (10) runs into zero:

$$s_{max} \equiv \bar{s} + \frac{1}{2}(1 - \bar{S}^2). \tag{11}$$

In the continuous-time limit, the s_t process never attains the region $s > s_{max}$.

This specification achieves the three objectives set out above. First, simply plugging the definition of $\bar{S}$, (9), and the definition of $\lambda(s_t)$, (10), into the formula for the risk-free rate, (8), we see that the risk-free rate is a constant:

$$r_t^f = -\ln(\delta) + \gamma g - \left(\frac{\gamma}{\bar{S}}\right)^2 \frac{\sigma^2}{2} = -\ln(\delta) + \gamma g - \frac{\gamma}{2}(1 - \phi). \tag{12}$$

Second, differentiating the transition equation (3), we obtain

$$\frac{dx_{t+1}}{dc_{t+1}} = 1 - \frac{\lambda(s_t)}{e^{-s_{t+1}} - 1} \approx 1 - \frac{\lambda(s_t)}{e^{-s_t} - 1}. \tag{13}$$

The latter approximation holds near the steady state. To obtain $dx/dc = 0$ at $s_t = \bar{s}$, we require

$$\lambda(\bar{s}) = \frac{1}{\bar{S}} - 1. \tag{14}$$

When equation (10) is evaluated at $\bar{s}$, it satisfies this condition. Third, to ensure that habit is predetermined in a neighborhood of the steady state, we add the requirement

$$\left.\frac{d}{ds}\left(\frac{dx}{dc}\right)\right|_{s=\bar{s}} = 0.$$

This condition also implies that habit moves *nonnegatively* with consumption everywhere since dx/dc is a U-shaped function of s. Taking the derivative $(d/ds)(dx/dc)$ of the expression in (13) and setting it to zero at $s = \bar{s}$, we obtain

$$\lambda'(\bar{s}) = -\frac{1}{\bar{S}}.$$

Equation (10) satisfies this condition. Since the functional form of $\lambda(s_t)$ was already determined by the first two conditions, this condition determines the constraint (9) on the parameters of the model.

Panel a of figure 1 plots the sensitivity function $\lambda(s_t)$ against the surplus consumption ratio, given the parameter values described below. The sensitivity function $\lambda(s_t)$ is a shifted square root function of $-s_t$, so λ increases to infinity as s_t declines to minus infinity, or as $S_t = e^{s_t}$ declines to zero in the figure. As we discussed above, a negative relationship between $\lambda(s_t)$ and s_t is needed to produce a constant risk-free interest rate and a countercyclical price of risk. Where $\lambda(s_t)$ hits zero, we see the upper bound of the surplus consumption ratio, S_{max}.

Panel b of figure 1 plots the derivative of log habit with respect to log consumption, as given by equation (13). The figure verifies that habit does not move contemporaneously with consumption—$dx/dc = 0$—at and near the steady state, marked by a vertical line, and that habit responds positively to consumption—$dx/dc \geq 0$—everywhere. As the surplus consumption ratio declines to zero or increases to its upper bound (the vertical dashed line), log habit starts to move one for one with log consumption in order to keep habit below consumption or the surplus consumption ratio below its upper bound.

Time Variation in the Riskless Interest Rate

Different functional forms for $\lambda(s_t)$ can of course generate riskless interest rates that vary with the state variable. For example, a natural generalization is to choose $\lambda(s_t)$ so that the interest rate is a linear function of the state s_t, rather than a constant:

$$r_t^f = r_0^f - B(s_t - \bar{s}). \tag{15}$$

The only difference this modification makes to the previous analysis is that the relation between parameters in equation (9) generalizes to

$$\bar{S} = \sigma\sqrt{\frac{\gamma}{1 - \phi - (B/\gamma)}}. \tag{16}$$

This generalization of the model produces a rich term structure of interest rates. Since the risk-free rate is a linear function of the state

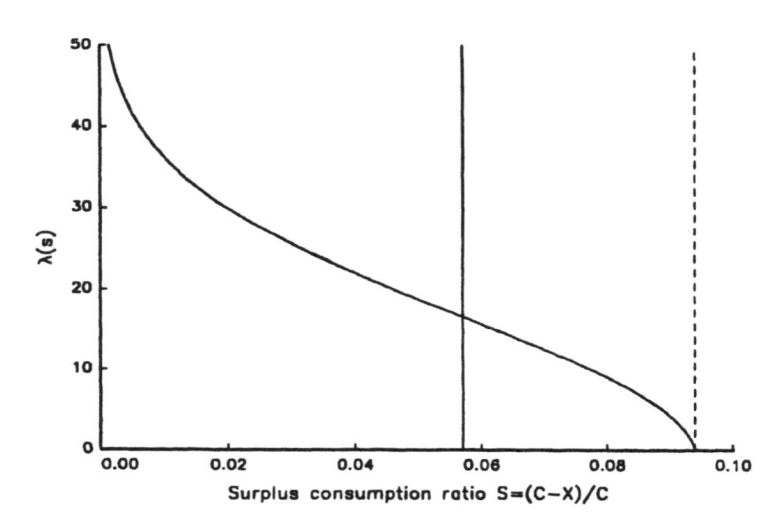

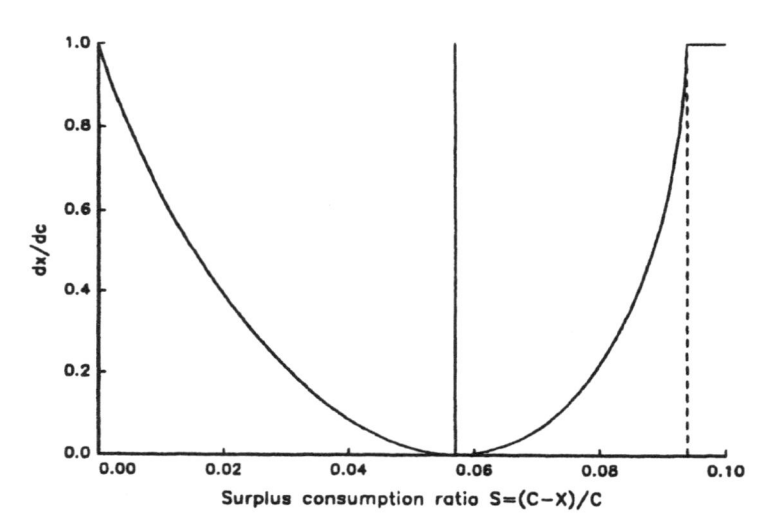

FIG. 1.—*a*, Sensitivity function $\lambda(s_t)$. *b*, Implied sensitivity of habit x to contemporaneous consumption. The vertical solid line in this and subsequent figures shows the steady-state surplus consumption ratio $\bar{S}$. The dashed vertical line shows the maximum surplus consumption ratio S_{max}.

variable s_t and since from equation (10) the conditional standard deviation of s_t is very close to a square root function of s_t, this generalized model is similar to square root models of the term structure such as the model of Cox, Ingersoll, and Ross (1985). The generalized model also implies that yield spreads are functions of the state variable s_t, so they forecast stock and bond returns about as well as the dividend/price ratio. However, adding interest rate variation in this way has very little effect on the stock market results on which we focus below. The working paper version of this article (Campbell and Cochrane 1995) includes explicit calculations of the term structure and its forecasts of stock and bond returns.

D. Prices of Long-Lived Assets

Pricing a Consumption Claim

We start by modeling stocks as a claim to the consumption stream. This is the simplest specification; it is common in the equity premium literature, allowing an easy comparison of results; and it is the natural definition of the "market" or "wealth portfolio" studied in finance theory. From the basic pricing relation and the definition of returns,

$$1 = E_t[M_{t+1} R_{t+1}], \quad R_{t+1} \equiv \frac{P_{t+1} + D_{t+1}}{P_t},$$

the price/dividend or, equivalently, the price/consumption ratio for a consumption claim satisfies

$$\frac{P_t}{C_t}(s_t) = E_t\left[M_{t+1} \frac{C_{t+1}}{C_t}\left[1 + \frac{P_{t+1}}{C_{t+1}}(s_{t+1})\right]\right]. \quad (17)$$

The surplus consumption ratio s_t is the only state variable for the economy, so the price/consumption ratio is a function only of s_t. We substitute for M_{t+1} from (5) and consumption growth from (4) and then solve this functional equation numerically on a grid for the state variable s_t, using numerical integration over the normally distributed shock v_{t+1} to evaluate the conditional expectation. Given the price/consumption ratio as a function of state, we calculate expected returns, the conditional standard deviation of returns, and other interesting quantities.

Imperfectly Correlated Dividends and Consumption

The growth rates of stock market dividends and consumption are only weakly correlated in U.S. data. This fact suggests that it may be important to model dividends and consumption separately rather than to treat them as a single process. We separate dividends and consumption in a particularly simple way in order to avoid adding state variables to our model. Surprisingly, we find that prices and returns of dividend claims behave very much like those of consumption claims, despite the low correlation between consumption and dividend growth rates.

We specify an i.i.d. process for dividend growth, imperfectly correlated with consumption growth. Letting D denote the level of dividends and d the log of dividends, we specify

$$\Delta d_{t+1} = g + w_{t+1}; \quad w_{t+1} \sim \text{i.i.d. } \mathcal{N}(0, \sigma_w^2), \text{corr}(w_t, v_t) = \rho. \quad (18)$$

The price/dividend ratio of a claim to the dividend stream then satisfies

$$\frac{P_t}{D_t}(s_t) = E_t\left[M_{t+1}\frac{D_{t+1}}{D_t}\left[1 + \frac{P_{t+1}}{D_{t+1}}(s_{t+1})\right]\right]. \quad (19)$$

We calculate this price/dividend ratio as a function of state in much the same manner as the price/consumption ratio of the consumption claim. Our appendix (Campbell and Cochrane [1998a], available from the authors) gives details of the calculation.

The correlation between consumption growth and dividend growth in the model (18) is the same at all horizons, and dividends wander arbitrarily far from consumption as time passes. It would be better to make dividends and consumption cointegrated. We have explored a model in which the log dividend/consumption ratio is i.i.d. and the correlation of one-period dividend and consumption growth rates is low as in the data. This model behaves so similarly to the basic consumption claim model that graphs of the solutions are indistinguishable. A cointegrated model with a persistent log dividend/consumption ratio would be more realistic, but this modification would require an additional state variable. Any such model is likely to make the consumption and dividend claims even more alike than in our specification (18) since it increases the correlation between dividends and consumption at long horizons.

E. *Choosing Parameters*

We compare the model to two data sets: (1) postwar (1947–95) value-weighted New York Stock Exchange stock index returns from the Center for Research in Security Prices (CRSP), 3-month Treasury bill rate, and per capita nondurables and services consumption and (2) a century-long annual data set of Standard & Poors 500

TABLE 1

PARAMETER CHOICES

Parameter	Variable	Value
Assumed:		
Mean consumption growth (%)*	g	1.89
Standard deviation of consumption growth (%)*	σ	1.50
Log risk-free rate (%)*	r^f	.94
Persistence coefficient*	ϕ	.87
Utility curvature	γ	2.00
Standard deviation of dividend growth (%)*	σ_w	11.2
Correlation between Δd and Δc	ρ	.2
Implied:		
Subjective discount factor*	δ	.89
Steady-state surplus consumption ratio	$\bar{S}$	.057
Maximum surplus consumption ratio	S_{max}	.094

* Annualized values, e.g., $12g$, $\sqrt{12}\sigma$, $12r^f$, ϕ^{12}, and δ^{12}, since the model is simulated at a monthly frequency.

stock and commercial paper returns (1871–1993) and per capita consumption (1889–1992) from Campbell (1999).

We choose the free parameters of the model to match certain moments of the postwar data. Table 1 summarizes our parameter choices. We take the mean and standard deviation of log consumption growth, g and σ, to match the consumption data. We choose the serial correlation parameter ϕ to match the serial correlation of log price/dividend ratios. We choose the subjective discount factor δ to match the risk-free rate with the average real return on Treasury bills. Since the ratio of unconditional mean to unconditional standard deviation of excess returns is the heart of the equity premium puzzle, we search for a value of γ so that the returns on the consumption claim match this ratio in the data.

We take the standard deviation of dividend growth, σ_w, from the CRSP data as well. Assigning a value ρ for the correlation between dividend growth and consumption growth is a little trickier. If dividend growth were uncorrelated with consumption growth, a claim to dividend growth would have no risk premium. However, correlations are difficult to measure because they are sensitive to small changes in timing or time aggregation. Campbell (1999) reports correlations in postwar U.S. data varying from .05 to almost .25 as the measurement interval increases from 1 to 16 quarters and correlations in long-run annual U.S. data varying from almost .2 to just over .1 as the measurement interval increases from 1 to 8 years. In the very long run, one expects the correlation to approach 1.0 since dividends and consumption should share the same long-run trends. Furthermore, these point estimates are subject to large sampling

AGGREGATE STOCK MARKET BEHAVIOR

error. The usual standard error formula $1/\sqrt{T}$ for a correlation coefficient is .1 in a century and .15 in postwar data, so we cannot convincingly reject zero or accurately measure economically interesting correlations of .2 or .3. Given these results, we do not try to match a particular correlation but choose a baseline correlation of .2 to show that the model works well even with quite a low correlation between consumption and dividends. The results are insensitive to the precise value of this correlation.

III. Solution and Evaluation

In this section we solve the model numerically and characterize its behavior. Then we simulate data by drawing shocks from a random number generator, and we show how the simulated data replicate many interesting statistics found in actual data. Finally, we feed the model historical consumption shocks to see what it tells us about historical movements in stock prices.

A. Asset Prices and the Surplus Consumption Ratio

Stationary Distribution of the Surplus
Consumption Ratio

Figure 2 presents the stationary distribution of the surplus consumption ratio. The figure plots the distribution of the continuous-time

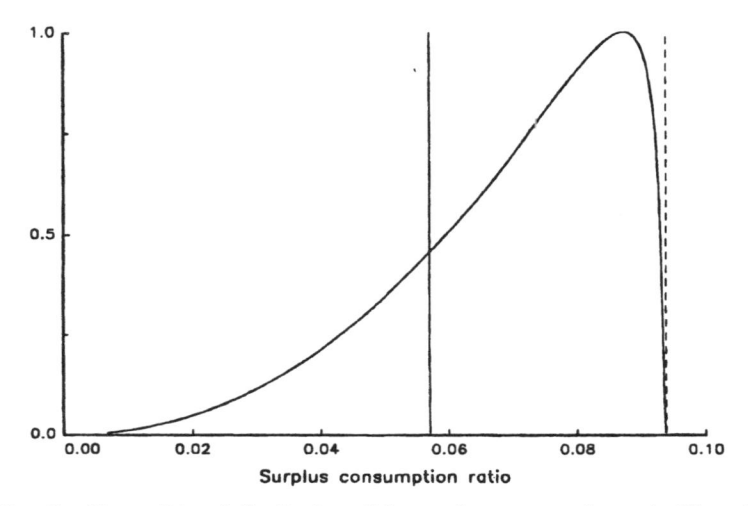

FIG. 2.—Unconditional distribution of the surplus consumption ratio. The solid vertical line indicates the steady-state surplus consumption ratio $\bar{S}$, and the dashed vertical line indicates the upper bound of the surplus consumption ratio S_{max}.

version of the process, calculated in the appendix (Campbell and Cochrane 1998a). This distribution is an excellent approximation to histograms of the discrete-time process for simulation time intervals of a year or less. With this stationary distribution and the slow mean reversion of the state variable in mind, one can get a good idea of the behavior of other quantities plotted against the state variable S_t.

The plot verifies that the unconditional distribution is well behaved: it does not pile up at the boundaries or wash out. The distribution of the surplus consumption ratio is negatively skewed. The surplus consumption ratio spends most of its time above the steady-state value $\bar{S}$, but there is an important fat tail of low surplus consumption ratios. We shall refer to a low surplus consumption ratio as a "recession" and a high surplus consumption ratio as a "boom." Thus the model predicts occasional deep recessions not matched by large booms.

Price/Dividend Ratios and the Surplus Consumption Ratio

Figure 3 presents the price/dividend ratios of the consumption claim and the dividend claim as functions of the surplus consumption ratio. These are the central quantities for our simulations; all other variables are calculated from the price/dividend ratio.

The price/dividend ratios increase with the surplus consumption

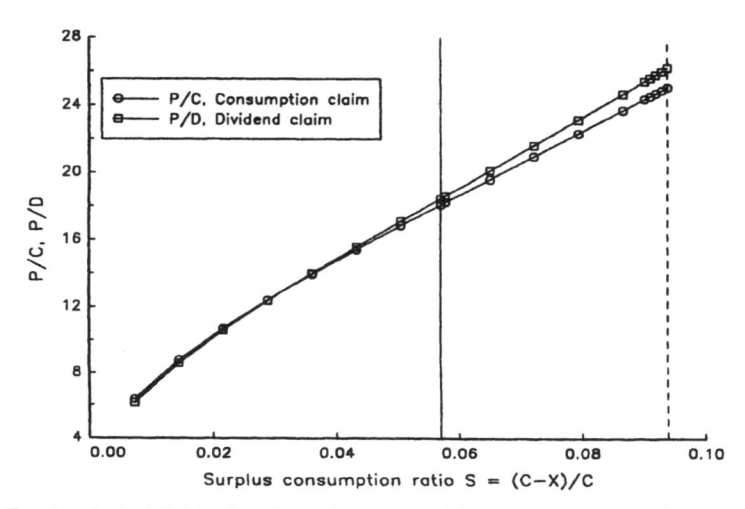

FIG. 3.—Price/dividend ratios as functions of the surplus consumption ratio

ratio. When consumption is low relative to habit in a recession, the curvature of the utility function is high, and prices are depressed relative to dividends. Since the price/dividend ratios are nearly linear functions of the surplus consumption ratio and the distribution of the surplus consumption ratio is negatively skewed, the distribution of price/dividend ratios inherits this negative skewness despite i.i.d. lognormal consumption growth.

The price/dividend ratio of the dividend claim is almost exactly the same as the price/dividend ratio of the consumption claim despite the very low (.2) correlation of dividend growth with consumption growth. Dividend growth is much more volatile than consumption growth, so the regression coefficient $\beta = \rho \sigma_{\Delta d}/\sigma_{\Delta c}$ of dividend growth on consumption growth is roughly one. The systematic or priced components of the two assets are similar, and therefore so are their prices.

Conditional Moments of Returns

Figure 4 presents the expected consumption claim and dividend claim returns and the risk-free interest rate as functions of the surplus consumption ratio. As consumption declines toward habit, expected returns rise dramatically over the constant risk-free rate.

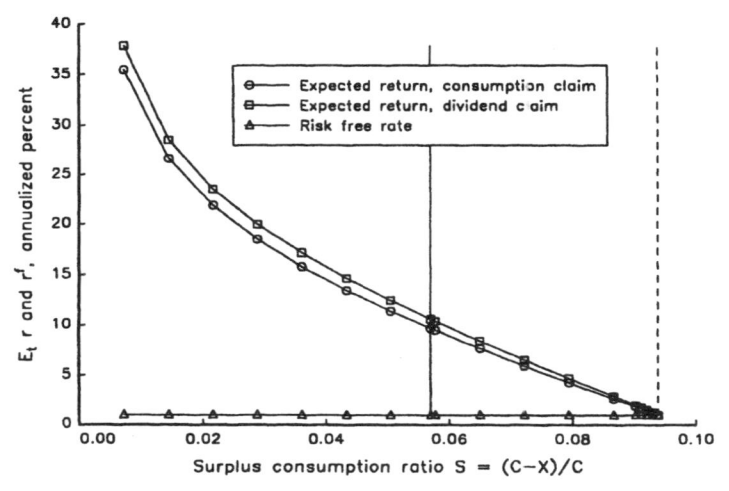

FIG. 4.—Expected returns and risk-free rate as functions of the surplus consumption ratio.

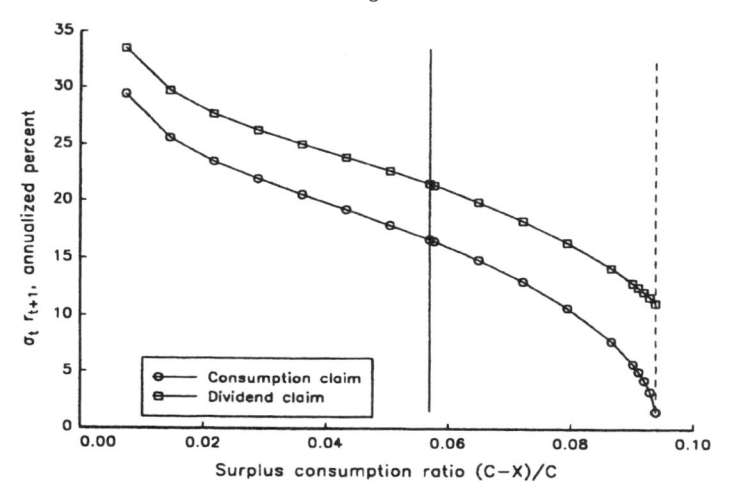

FIG. 5.—Conditional standard deviations of returns as functions of the surplus consumption ratio.

Figure 5 presents the conditional standard deviations of returns as functions of the surplus consumption ratio. As consumption declines toward habit, the conditional variance of returns increases. Thus the model produces several effects that have been emphasized in the autoregressive conditional heteroscedasticity (ARCH) literature: highly autocorrelated conditional variance in stock returns, a "leverage effect" that price declines increase volatility, and countercyclical variation in volatility.

In figure 4, the expected return of the dividend claim is almost exactly the same as that of the consumption claim. In figure 5 the dividend claim has a noticeably higher standard deviation than the consumption claim but the same dependence on the surplus consumption ratio. The return on the dividend claim is

$$R_{t+1} = \frac{P_{t+1} + D_{t+1}}{P_t} = \frac{(P_{t+1}/D_{t+1}) + 1}{P_t/D_t} \times \frac{D_{t+1}}{D_t}. \quad (20)$$

The expected returns are nearly identical because the price/dividend ratio and price/consumption ratio are nearly identical functions of state, and dividend and consumption growth are not predictable. The conditional standard deviation of the dividend claim inherits the same dependence on state through the nearly identical P/D term but adds the extra, constant, standard deviation of dividend growth.

AGGREGATE STOCK MARKET BEHAVIOR

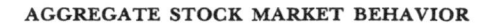

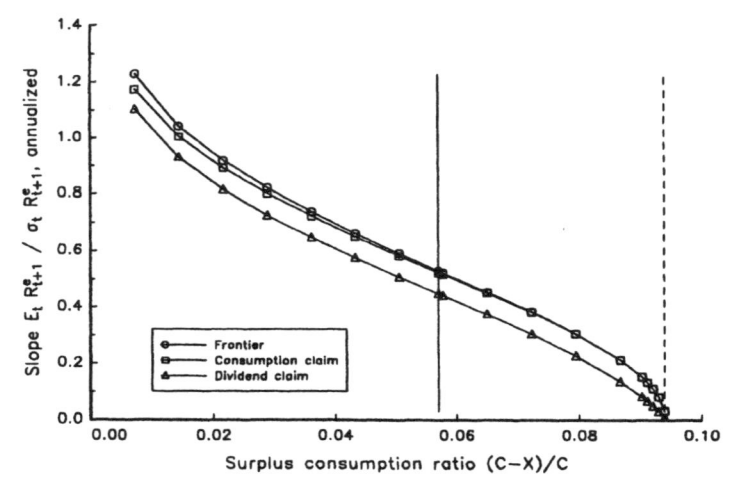

Fig. 6.—Sharpe ratios as functions of the surplus consumption ratio

Conditional Sharpe Ratios

Comparing figures 4 and 5, we see that conditional means and conditional standard deviations are different functions of the surplus consumption ratio, so the Sharpe ratio of conditional mean to conditional standard deviation of excess returns varies over time. To get a precise measure, figure 6 presents the Sharpe ratio as a function of the surplus consumption ratio. The top line is the maximum possible Sharpe ratio, calculated from the Hansen-Jagannathan bound, equation (7).

The consumption claim nearly attains the Sharpe ratio bound, implying that it is nearly conditionally mean-variance efficient. The consumption claim model has only one shock. Hence the only reason the consumption claim (or any claim whose return depends on the single shock) is not exactly conditionally mean-variance efficient is that it is nonlinearly related to the shock. For the consumption claim, the effects of such nonlinearity are slight.

The dividend claim has a slightly higher mean return and a substantially higher standard deviation since there is a second dividend growth shock as well as the consumption (discount rate) shock. Hence, the dividend claim has a somewhat lower Sharpe ratio and is less conditionally efficient. However, since the dividend payoff is correlated only .2 with the consumption claim payoff, it is surprising how close the Sharpe ratios are. In equation (20), most of the variation in the dividend claim return is due to changing risk premia and hence changing price/dividend ratios common to both assets, not to the volatility of the payoff itself.

The Sharpe ratios of both securities increase substantially when the surplus consumption ratio declines. In our model, recessions are times of low consumption relative to habit, low prices, somewhat higher standard deviations of returns, very much higher expected returns, and correspondingly high Sharpe ratios.

The top line of figure 6 is also interesting as a characterization of the discount factor. The conditional mean of the discount factor is constant, so this line plots the conditional standard deviation of the discount factor. That conditional standard deviation moves with the state variable s_t and so inherits its positive serial correlation. Thus our economic model generates a time-series model for the *second* moment of the stochastic discount factor, like an ARCH model rather than an autoregressive moving average model. The Hansen-Jagannathan analysis shows that this form is necessary in order to generate a time-varying risk premium.

B. Statistics from Simulated Data

We simulate 500,000 months of artificial data to calculate population values for a variety of statistics. In order to facilitate a comparison with historical data, we simulate the model at a monthly frequency and then construct time-averaged artificial annual data. As in the actual data, we average the level of consumption in each year. We form annual returns by taking the product of intervening monthly returns. The annual price/dividend ratio is its value at the end of the year.

We report corresponding historical statistics with some trepidation. On the one hand, it is useful to get some quantitative idea of the target. On the other hand, the historical statistics are the subject of an enormous empirical literature, and the point estimates of simplified statistics from one particular sample do little justice to the econometric and data-handling sophistication of that literature. Also, estimates should be accompanied by standard errors, but useful measures of sampling uncertainty require a far more sophisticated analysis than space allows here. This is particularly true since our model suggests that peso problems will be important; stock returns in the model are nonnormally distributed and strongly influenced by the small possibility of a severe crash or depression.

Means and Standard Deviations

Table 2 presents means and standard deviations in simulated data, with the corresponding statistics from our two historical data sets.

AGGREGATE STOCK MARKET BEHAVIOR 225

TABLE 2

MEANS AND STANDARD DEVIATIONS OF SIMULATED AND HISTORICAL DATA

Statistic	Consumption Claim	Dividend Claim	Postwar Sample	Long Sample
$E(\Delta c)$	1.89*		1.89	1.72
$\sigma(\Delta c)$	1.22*		1.22	3.32
$E(r^f)$	.094*		.094	2.92
$E(r - r^f)/\sigma(r - r^f)$	.43*	.33	.43	.22
$E(R - R^f)/\sigma(R - R^f)$	.50		.50	
$E(r - r^f)$	6.64	6.52	6.69	3.90
$\sigma(r - r^f)$	15.2	20.0	15.7	18.0
$\exp[E(p - d)]$	18.3	18.7	24.7	21.1
$\sigma(p - d)$	.27	.29	.26	.27

NOTE.—The model is simulated at a monthly frequency; statistics are calculated from artificial time-averaged data at an annual frequency. All returns are annual percentages.
* Statistics that model parameters were chosen to replicate.

The first four moments match the postwar statistics exactly because we chose parameters to fit those moments. In particular, we picked the parameter $\gamma = 2.00$ to exactly match the Sharpe ratio for log returns of 0.43 in postwar data. The model also matches the Sharpe ratio for simple returns of 0.50. A γ value of about four matches the dividend claim Sharpe ratio to the postwar value without much effect on other statistics.

We chose to match the postwar time series because they are a significantly harder target. The long historical time series feature a much larger standard deviation of consumption growth, a lower Sharpe ratio, and a higher risk-free rate. A γ of about 0.7 matches the 0.22 Sharpe ratio for log returns in the long-term data, with little effect on the other statistics.

It is noteworthy that the model can match the mean and standard deviation of excess stock returns, with a constant low interest rate and a discount factor $\delta = 0.89$ less than one, by *any* choice of parameters. These moments are the equity premium and risk-free rate puzzles, which we discuss below.

The remaining moments were not used to pick parameters, so we can use them to check the model's predictions. The choice of γ matches the *ratio* of mean return to standard deviation, but it says nothing about the *level* of mean and standard deviation of returns. The ratio 0.43 could be generated by a mean of 0.43 percent and a standard deviation of 1 percent. In fact, the mean and standard deviation of excess returns are almost exactly equal to the corresponding values in the postwar data, using either the consumption claim or the dividend claim.

The mean price/dividend ratio is a bit below that found in postwar data, but this statistic is poorly measured because the price/dividend ratio is highly serially correlated. The standard deviation of the price/dividend ratio is almost exactly the same as that found in the data. In this sense, the model accounts for the volatility of stock prices, a point we discuss in more detail below.

Autocorrelations and Cross-Correlations

Table 3 presents autocorrelations and table 4 presents cross-correlations from our simulated data, along with sample values from the historical data.

We picked the parameter ϕ to generate the .87 first-order annual autocorrelation of the price/dividend ratio seen in the table. Higher autocorrelations decay slowly, as in the data. The dividend claim has exactly the same autocorrelation pattern. This is not a surprise given that the dividend claim and consumption claim price/dividend ratios are almost identical.

Returns display a series of small negative autocorrelations that generate univariate mean reversion (Fama and French 1988b; Po-

TABLE 3

AUTOCORRELATIONS OF SIMULATED AND HISTORICAL DATA

	\multicolumn{5}{c}{Lag (Years)}
Variable and Source	1	2	3	5	7
$p - d$:					
Consumption claim	.87	.76	.66	.51	.39
Dividend claim	.87	.76	.66	.51	.39
Postwar sample	.87	.77	.70	.41	.04
Long sample	.78	.57	.50	.32	.29
$r - r^f$:					
Consumption claim	−.06	−.05	−.04	−.02	−.02
Dividend claim	−.05	−.04	−.03	−.02	−.01
Postwar sample	−.11	−.28	.15	.02	.10
Long sample	.05	−.21	.08	−.14	.11
$\sum_{i=1}^{j} \rho(r_t^e, r_{t-i}^e)$:*					
Consumption claim	−.06	−.11	−.15	−.20	−.26
Dividend claim	−.05	−.09	−.12	−.14	−.18
Postwar sample	−.11	−.39	−.24	.18	.13
Long sample	.05	−.16	−.09	−.28	−.15
$|r|$:					
Consumption claim	.09	.09	.09	.07	.05
Dividend claim	.05	.05	.05	.04	.03
Postwar sample	.08	−.26	−.10	−.08	.05
Long sample	.13	.09	.07	.14	.15

NOTE.—The model values are based on time-aggregated annual values with a monthly simulation interval. All data are annual.
* Partial sum of return autocorrelations out to lag j.

TABLE 4
CROSS-CORRELATIONS OF SIMULATED AND HISTORICAL DATA

VARIABLE AND SOURCE	LAG (Years) 1	2	3	5	7
$p_t - d_t, r^e_{t+j}$:					
Consumption claim	−.35	−.30	−.26	−.20	−.15
Dividend claim	−.28	−.24	−.20	−.16	−.12
Postwar sample	−.42	−.25	−.13	−.35	−.17
Long sample	−.20	−.21	−.10	−.19	−.08
$r^e_t, |r^e_{t+j}|$:					
Consumption claim	−.09	−.07	−.06	−.03	−.03
Dividend claim	−.06	−.04	−.04	−.03	−.02
Postwar sample	−.32	−.14	.10	−.04	−.08
Long sample	−.15	.03	.12	.02	−.01
$p_t - d_t, |r^e_{t+j}|$:					
Consumption claim	−.49	−.42	−.37	−.28	−.21
Dividend claim	−.36	−.31	−.27	−.21	−.16
Postwar sample	−.16	.09	.11	−.05	.02
Long sample	−.12	.02	−.06	−.10	−.05

terba and Summers 1988). The negative autocorrelations of returns also generate observations that price changes tend to be reversed.

Since individual long-term autocorrelations are small and poorly measured, the empirical literature focuses on a number of clever statistics designed to better measure univariate mean reversion. One such statistic is the partial sum of autocorrelation coefficients, shown in the table. The model replicates the pattern and the rough (poorly measured) magnitude found in the data. The prewar data show a stronger mean-reverting pattern, which is a well-known feature of this statistic.

The autocorrelation of absolute returns reveals long-horizon conditional heteroskedasticity in the model. The ARCH literature (for a summary see Bollerslev et al. [1992]) finds higher values for these autocorrelations in high-frequency data but values similar to our first-order autocorrelation at annual frequencies. The ARCH literature has not noted the negative 2- and 3-year autocorrelations of absolute returns in the postwar data, but these findings may be artifacts of a simplistic technique or sampling error. The dividend claim has a lower autocorrelation of absolute returns since its return is a noisier indicator of changes in the surplus consumption ratio.

The cross-correlation between the price/dividend ratio and subsequent excess returns, shown in table 4, verifies that the price/consumption ratio forecasts long-horizon returns with the right sign: high prices forecast low returns. Since high prices forecast low returns for many years in the future, the forecastability of returns increases with the horizon, as we show next. The correlations are slightly smaller for the dividend claim since its return is slightly noisier.

The cross-correlations between the price/dividend ratio or returns and subsequent absolute returns show that a low price/consumption ratio or a big price decline signals high volatility for several years ahead. This is the "leverage effect" that Black (1976), Schwert (1989), Nelson (1991), and many others have found in the data. As with the univariate autocorrelation of absolute returns, the data seem to indicate a somewhat shorter-lasting change in conditional variance than is predicted by the model, at least as viewed by this simple statistic. Again, the dividend claim behaves much like the consumption claim, despite the very low .2 correlation of dividend growth with consumption growth.

Long-Horizon Regressions

Table 5 presents long-horizon regressions of log excess stock returns on the log price/dividend ratio in simulated and historical data. We use excess returns to emphasize that risk premia rather than risk-free rates vary over time. We see the classic pattern documented by Campbell and Shiller (1988*b*) and Fama and French (1988*a*). The coefficients are negative: high prices imply low expected returns. The coefficients increase linearly with horizon at first and then less quickly; the R^2's start low but then rise to impressive values. The model's predictions for the consumption claim match closely the postwar data. The coefficients for the dividend claim are about the same, but the R^2's do not rise as fast since the dividend claim return contains the extra dividend growth noise.

TABLE 5

LONG-HORIZON RETURN REGRESSIONS

Horizon (Years)	Consumption Claim 10 × Coefficient	R^2	Dividend Claim 10 × Coefficient	R^2	Postwar Sample 10 × Coefficient	R^2	Long Sample 10 × Coefficient	R^2
1	−2.0	.13	−1.9	.08	−2.6	.18	−1.3	.04
2	−3.7	.23	−3.6	.14	−4.3	.27	−2.8	.08
3	−5.1	.32	−5.0	.19	−5.4	.37	−3.5	.09
5	−7.5	.46	−7.3	.26	−9.0	.55	−6.0	.18
7	−9.4	.55	−9.2	.30	−12.1	.65	−7.5	.23

Volatility Tests

Papers in the volatility test literature have found that stock prices move far more than can be explained by varying expectations of dividend growth and interest rates. In our model, expected dividend growth and the riskless interest rate are constant over time, so they explain *none* of the variation in stock prices. Therefore, the model implies an extreme version of the volatility test results.

To demonstrate this point, we replicate a volatility test in Cochrane (1992), which is closely related to the tests in Campbell and Shiller (1988a). A log-linearization of the accounting identity $1 = R_{t+1}^{-1} R_{t+1}$, with $R_{t+1} = (P_{t+1} + D_{t+1})/P_t$, implies that, in the absence of rational asset price bubbles,

$$\text{var}(p_t - d_t) \approx \sum_{j=1}^{\infty} \rho^j \text{cov}(p_t - d_t, \Delta d_{t+j}) \qquad (21)$$
$$- \sum_{j=1}^{\infty} \rho^j \text{cov}(p_t - d_t, r_{t+j}),$$

where $\rho \equiv (P/D)/[1 + (P/D)]$ and P/D is the point of linearization. The price/dividend ratio can vary only if it sufficiently forecasts dividend growth or returns or both.

Table 6 presents estimates of (21), using 15 years of covariances to estimate the sums in artificial data and in the two data samples. The point estimates in the data find that more than 100 percent of the price/dividend ratio variance is attributed to expected return variation. A high price/dividend ratio signals a *decline* in subsequent real dividends, so it must signal a large decline in expected returns. The forecast dividend decline is not statistically different from zero,

TABLE 6

VARIANCE DECOMPOSITIONS

Source	Returns (%)	Dividends (%)
Consumption claim	100	1
Dividend claim	99	3
Postwar sample	137	−31
Long sample	101	−10

NOTE.—Table entries are the percentage of var$(p - d)$ accounted for by dividend growth and returns,

$$100 \times \sum_{j=1}^{15} \frac{\rho^j \text{cov}(p_t - d_t, x_{t+j})}{\text{var}(p_t - d_t)},$$

$x = -r$ and Δd, respectively.

however. All of the price/dividend ratio variance is accounted for, providing evidence against the view that stock market volatility is driven by rational bubbles.

In the model, all variation in the price/dividend ratio is due to changing expected returns by construction. To within the accuracy of the log-linear approximation, the variance decomposition on artificial data reflects this fact.

The Correlation of Consumption Growth with Stock Returns

Equilibrium consumption-based models typically imply that consumption growth and stock returns are highly, if not perfectly, correlated. For example, with log utility, the return on the wealth portfolio equals consumption growth, ex post, data point for data point. This implication is the basis for many theoretical models in finance that substitute portfolio returns for consumption growth. However, this implication is seldom checked or used to test asset pricing models, for the obvious reason that it is dramatically false. As Cochrane and Hansen (1992) emphasize, the actual low correlation between stock returns and consumption growth lies at the heart of many empirical failures of the consumption-based model.

In our model, consumption growth and consumption claim returns are conditionally perfectly correlated since consumption growth is the only source of uncertainty. But the relation between consumption growth and returns varies over time with the surplus consumption ratio. Hence the unconditional correlation between consumption growth and returns is not perfect. Panel *a* of figure 7 shows this effect of conditioning information by plotting artificial data on monthly consumption growth versus returns. For a given surplus consumption ratio, such pairs lie on a line, but the slope of the line changes as the surplus consumption ratio changes. Therefore, the consumption-return pairs fill a region bounded by two straight lines, each of which corresponds to one limit of the surplus consumption ratio.

Panel *b* of figure 7 plots the correlation of consumption growth with returns in simulated annual data. As the figure shows, time aggregation further degrades the perfect conditional correlation between consumption growth and returns.

Table 7 presents several measures of the correlation between consumption growth and stock returns. In the data, there is very little contemporaneous correlation between consumption growth and returns. However, returns are negatively correlated with previous consumption growth and positively correlated with subsequent con-

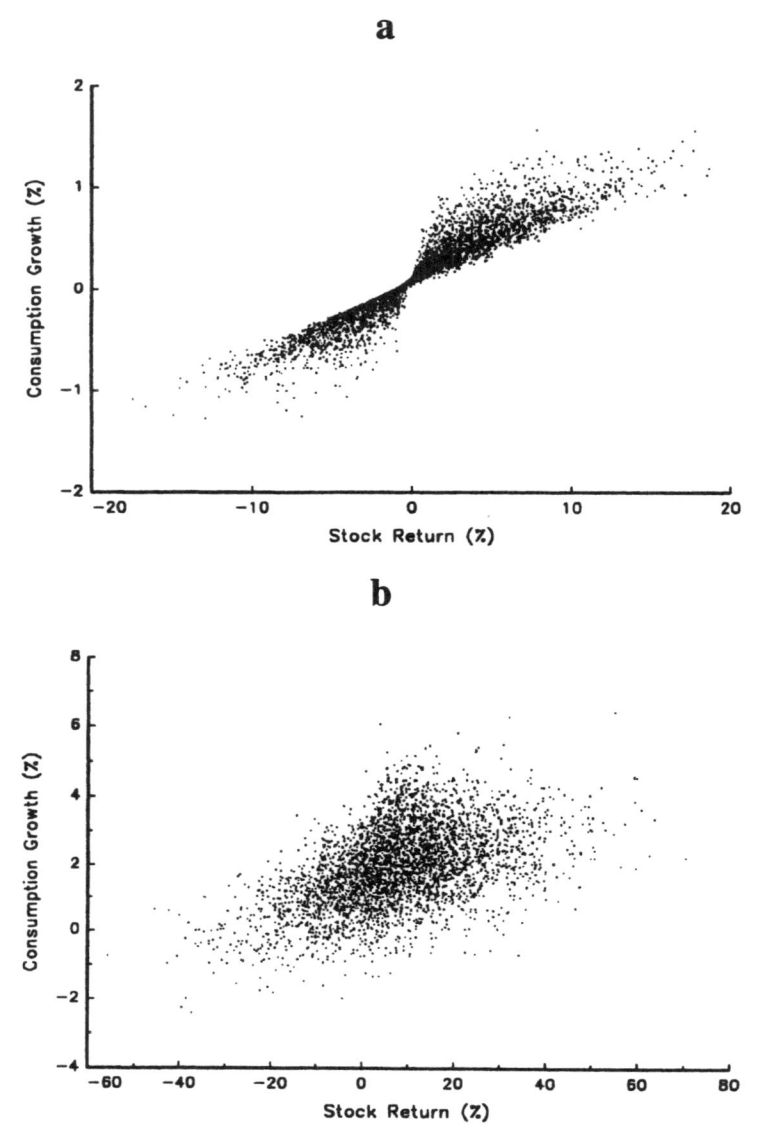

FIG. 7.—*a*, Simulated monthly consumption growth vs. monthly consumption claim returns. *b*, Growth in simulated annual consumption vs. annual consumption claim returns.

TABLE 7
CORRELATION BETWEEN CONSUMPTION GROWTH AND STOCK RETURNS

	MONTHLY MODEL		ANNUAL MODEL		DATA SAMPLE		
CORRELATION	Consumption Claim	Dividend Claim	Consumption Claim	Dividend Claim	Quarterly Postwar	Annual Postwar	Annual Long
$r_t^s, \Delta c_{t-2}$	.00	.00	−.16	−.13	−.05	−.16	−.05
$r_t^s, \Delta c_{t-1}$	.00	.00	−.19	−.15	−.10	−.34	−.08
$r_t^s, \Delta c_t$	.93	.79	.47	.40	.12	−.05	.09
$r_t^s, \Delta c_{t+1}$	.00	.00	.50	.42	.19	.37	.49
$r_t^s, \Delta c_{t-2}$	.00	.00	.00	.00	.15	−.26	.05

sumption growth. The highest correlations occur between returns and the next year's consumption growth: .37 in postwar data and .49 in long-term data. Fama (1990) interprets similar correlations of returns with output as evidence that returns move on news of future cash flows.

In our model, the unconditional correlation between monthly consumption claim returns and monthly consumption growth is .93. This value corresponds to panel *a* of figure 7. It is less than the 1.0 of the standard time-separable model but much greater than the correlations we see in the data. In addition, there is no correlation between returns and consumption growth at any lead or lag.

When we time-aggregate the artificial data to annual frequencies, the contemporaneous correlation drops to .47. Furthermore, time aggregation produces a strong positive correlation between returns and subsequent consumption growth and a negative correlation between returns and previous consumption growth, the same sign pattern that we see in the data.

Modeling stocks as a dividend claim further reduces the correlations. Since the correlation of dividend growth with consumption growth is only .2, one might expect still lower correlations. But again most return variation is driven by price variation, so the extra dividend volatility has a relatively small effect.

Thus the varying conditioning information in our model together with an explicit accounting for time aggregation goes a long way toward resolving the puzzling low correlation of consumption growth with returns and toward explaining the correlation between returns and subsequent macroeconomic variables. Adding state variables and accounting for lags and errors in data collection are likely to further help to account for the correlations of consumption with asset returns.

The Correlation of the Discount Factor with Consumption Growth and Stock Returns

The static capital asset pricing model (CAPM) often does a better job of accounting for risk premia than the consumption-based asset pricing model with power utility (Mankiw and Shapiro 1986). It turns out that this is true in our artificial data as well, even though the data are generated by a consumption-based model. Campbell and Cochrane (1998*b*) present detailed calculations. We show the basic point here by calculating the correlation between the true stochastic discount factor and consumption growth or stock returns. Discount factor proxies that are better correlated with the true discount factor produce smaller pricing errors for a given set of assets.

TABLE 8

CORRELATION OF THE STOCHASTIC DISCOUNT FACTOR WITH
CONSUMPTION GROWTH, CONSUMPTION CLAIM RETURN,
AND DIVIDEND CLAIM RETURN

	CORRELATION OF STOCHASTIC DISCOUNT FACTOR WITH:		
	Consumption Growth	Consumption Claim Return	Dividend Claim Return
Monthly	.90	.99	.83
Annual	.45	.99	.80

NOTE.—The stochastic discount factor is

$$M_{t+1} = \delta \left(\frac{C_{t+1}}{C_t} \frac{S_{t+1}}{S_t} \right)^{-\gamma}.$$

Table 8 presents the correlations. In monthly artificial data, the consumption claim return is far better correlated with the true discount factor than consumption growth is. Therefore, the static CAPM using the "wealth portfolio" return is a better approximate model. Although the discount factor is *conditionally* perfectly correlated with consumption growth, the *unconditional* correlation is low because the surplus consumption ratio varies. The stock return moves when the surplus consumption ratio changes and hence reveals more of the discount factor movement.

We might expect the relative performance of the consumption-based model to deteriorate further at longer horizons. At longer horizons, there is more movement of the surplus consumption ratio independent of consumption growth, and this movement will be revealed by stock return variation since stock prices decline when the surplus consumption ratio declines. Time aggregation further obscures the consumption signal. Table 8 confirms this intuition: at an annual frequency the correlation of the discount factor with time-averaged consumption growth has declined to .45, whereas the correlation with the consumption claim return is still .99.

At a monthly horizon, the dividend claim return is a poorer proxy than even consumption growth because dividend growth contains noise not correlated with the discount factor. When we go to a longer horizon and introduce time aggregation in consumption, however, even the dividend claim return is a far better proxy for the discount factor than consumption growth. These annual results are the relevant ones because actual monthly consumption data, unlike the simulated monthly consumption data, are time-averaged and measured with error.

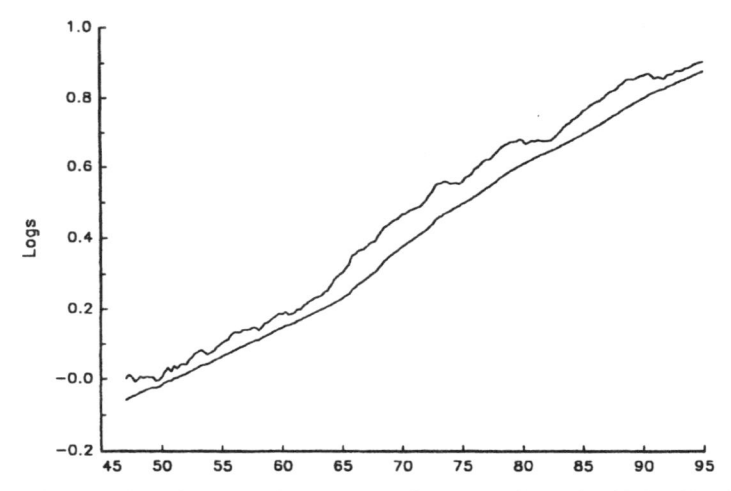

FIG. 8.—Nondurable and services consumption per capita and habit level implied by the model, under the assumption that the surplus consumption ratio starts at the steady state.

C. Interpreting Historical Consumption and Stock Price Data

Instead of simulating artificial consumption data, we now feed our model actual data on nondurables and services consumption per capita. Figure 8 presents the postwar history of consumption and the habit level implied by our model, assuming that habit starts at the steady state at the beginning of the sample.

The figure shows how habit responds smoothly to changes in consumption, trending up in the high-growth 1960s and growing more slowly in the 1970s. Cyclical dips in consumption bring consumption closer to habit. Our model will predict low price/dividend ratios and high expected returns for those periods.

Figure 9 presents the model's prediction for the price/dividend ratio of a consumption claim, together with the actual price/dividend ratio on the S & P 500 index. The prewar prediction is based on a calibration of the model to the long data set; it uses the lower mean and higher standard deviation of consumption growth of those data and a lower value of $\gamma = 0.7$ to generate the lower Sharpe ratio in that data set. We emphasize that the "model" line on the graph is produced using only consumption data and no asset market data. A similar graph using the dividend claim rather than the consumption claim is almost identical since the predicted price/divi-

FIG. 9.—Historical price/dividend ratio and model predictions based on the history of consumption.

dend ratio of the dividend claim is almost exactly the same function of state as the price/dividend ratio of the consumption claim.

To our eyes, the model provides a tantalizing account of cyclical and longer-term fluctuations in stock prices. When consumption declines for several years in a row, coming nearer to our constructed habit, stock prices fall. Model and actual price/dividend ratios fall in the sharp recessions of the late nineteenth and early twentieth centuries, and the model also captures the long-term rise and then decline from 1890 to 1915. The model accounts for the boom of the 1920s. The decline in consumption in the Great Depression was so extreme that the model predicts an even larger fall in stock prices than actually occurred. Then the model tracks the recovery during World War II, the consumption and stock boom of the 1960s (though with a lag), the secular and cyclical declines of the 1970s, and the consumption and stock market boom of the 1980's.

It is a little embarrassing that the worst performance occurs in the last few years. Growth in consumption of nondurables and services was surprisingly slow in the early 1990s, bringing consumption near our implied habit level (fig. 8), so our model predicts a fall in price/dividend ratios rather than the increase we see in the data. Possible excuses include a shift in corporate financial policy toward the repurchase of equity rather than dividend payments; an increase in the consumption of stock market investors that is not properly captured in the aggregate consumption data, due perhaps to rising income inequality in the period or the demographic effects of the baby boom generation entering peak saving years; and measurement problems such as compositional shifts of consumption away from nondurables and services.

D. Model Intuition

The Equity Premium and Risk-Free Rate Puzzles

Our model is consistent with the equity premium and a low and constant risk-free rate. It is worth seeing how the model resolves these long-standing puzzles.

With power utility $M_{t+1} = \beta(C_{t+1}/C_t)^{-\eta}$, a constant risk-free rate, and i.i.d. lognormal consumption growth with mean g and standard deviation σ, the Hansen-Jagannathan or Sharpe ratio inequality (6) specializes to

$$\frac{E(R^e)}{\sigma(R^e)} \leq \sqrt{e^{\eta^2\sigma^2} - 1} \approx \eta\sigma, \qquad (22)$$

and the log interest rate is

$$r_t^f = -\ln(\beta) + \eta g - \eta^2 \frac{\sigma^2}{2}. \qquad (23)$$

To explain a (gross return) Sharpe ratio of 0.50 with $\sigma = 1.22$ percent, the power utility model needs a risk aversion coefficient $\eta \geq 41$ by equation (22). This is Mehra and Prescott's (1985) "equity premium puzzle." One can object to $\eta \geq 41$ as an implausibly large value of risk aversion, and we discuss this interpretation below.

More important, a high value of η makes the term ηg in the risk-free rate equation very large. Thus $\eta = 41$ and $g = 1.89$ percent means that we need $\beta = 1.90$ to get a 1 percent risk-free rate. Imposing $\beta \leq 1$, one predicts a risk-free rate of more than 90 percent per year! Weil (1989) emphasizes this "risk-free rate puzzle," and Cochrane and Hansen (1992) discuss the level and variability of risk-free interest rates in high–risk aversion models.

Despite its intuitive implausibility, one might argue that setting $\beta = 1.90$ resolves the risk-free rate puzzle. However, with $\beta = 1.90$ and $\eta = 41$, equation (23) implies that the risk-free interest rate should be quite sensitive to the mean consumption growth rate, which is not the case. Real interest rates do not vary across time or countries by 40 times the variation in predicted or average consumption growth. (Equivalently, one must assume wild cross-country variation in patience β to save the model. Campbell [1999] reports such estimates.)

Our model also features high curvature. Though the power $\gamma = 2$ is low, the surplus consumption ratio is also low. So local curvature $\eta = -Cu_{cc}/u_c = \gamma/S$ is high: 35 at the steady state and higher still in states with low surplus consumption ratios. However, our model does not predict a sensitive relation between consumption growth and interest rates. Equation (12),

$$r_t^f = -\ln(\delta) + \gamma g - \left(\frac{\gamma}{S}\right)^2 \frac{\sigma^2}{2},$$

shows that the power parameter $\gamma = 2$, much lower than utility curvature γ/S, controls the relationship between average consumption growth and the risk-free interest rate.

Thus we avoid the risk-free rate puzzle: an intuitively plausible $\delta = 0.89 < 1$ is consistent with the low observed real interest rate; and the model predicts a much less sensitive relationship across countries or over time between mean consumption growth and interest rates. Furthermore, the time-varying risk-free rate version of our model (Campbell and Cochrane 1995) produces a risk-free rate that varies over time as a function of the surplus consumption ratio, whereas consumption growth is i.i.d. Therefore, this model predicts no time-series relationship at all between interest rates and expected consumption growth rates, consistent with the great difficulty the empirical literature has found in documenting any such relation in the data.

In order to remove the tension between equity premia as in (22) and risk-free rates as in (23), our model uses non-*time*-separable preferences to distinguish intertemporal substitution and risk aversion. Weil (1989), Epstein and Zin (1991), Kandel and Stambaugh (1991), and Campbell (1996) use non-*state*-separable preferences to the same effect but do not generate time-varying risk aversion.

Our solution to the risk-free rate puzzle has one other important advantage. Abel (1999) highlights the danger of accounting for an equity premium by a term premium. If a model assigned a high premium to the interest rate exposure of stock cash flows and long-term bond cash flows alike, it would account for the equity premium of stocks over short-term bonds, but it would counterfactually predict high expected returns for long-term bonds as well. Since interest rates are constant in our model, long-term bonds earn exactly the same returns as short-term bonds, and the entire equity premium is a risk premium, not a term premium.

The Long-Run Equity Premium

The equity premium puzzle is a feature of long as well as short horizons. Consumption is roughly a random walk at any horizon, so the standard deviation of consumption growth grows roughly with the square root of the horizon. The negative autocorrelation of stock returns means that k-year return variances are somewhat less than k times 1-year return variances, so the market Sharpe ratio grows, if anything, *faster* than the square root of the horizon (MaCurdy and Shoven 1992; Siegel 1994; Campbell 1996).

In our model, the k-period stochastic discount factor is

$$M_{t,t+k} = \delta^k \left(\frac{S_{t+k}}{S_t} \frac{C_{t+k}}{C_t} \right)^{-\gamma}. \tag{24}$$

Equation (6) implies that the standard deviation of this discount factor must increase roughly with the square root of the horizon to be consistent with the long-run equity premium, and even faster to generate the negative autocorrelation of stock returns.

One can think of our model as a member of a large class that adds a new state variable S_{t+1} to the discount factor. However, most extra state variables—such as recessions, labor, and instruments for time-varying expected returns ("shifts in the investment opportunity set")—are stationary. Hence, the standard deviation of their growth rates eventually stops growing with horizon. At a long enough horizon, the standard deviation of the discount factor is dominated by the standard deviation of the consumption growth term, and we return to the equity premium puzzle at a long enough run. One could of course (and many models that explain the short-run equity premium do so) posit positive serial correlation in consumption growth, so that long-run consumption growth is much more volatile than annual consumption growth. But we do not see this in the data.

Our model has a pure random walk in consumption, yet it produces negative autocorrelation in returns and therefore high Sharpe ratios at all horizons. How does it accomplish this feat with a stationary state variable S_t? The answer is that while S_t is stationary, $S_t^{-\gamma}$ is not. The variable S_t has a fat tail approaching zero (see fig. 2), so the conditional variance of $S_{t+k}^{-\gamma}$ grows without bound. We can demonstrate this behavior using the formula for the distribution of S given in the appendix (Campbell and Cochrane 1998a): As $s \to -\infty$, the leading terms in the distribution are

$$f(s) \approx e^{-\gamma |s| - 2\gamma \bar{S} \sqrt{2} |s|}.$$

We can integrate polynomials multiplied by this expression, so s is a covariance-stationary process with a well-defined unconditional mean, variance, and all higher moments. The surplus consumption ratio $S = e^s$ is also well behaved. However, while $S^{-\gamma}$ has a finite unconditional mean, since $e^{-\gamma s}f(s)$ is integrable, $S^{-\gamma}$ does not have a finite unconditional variance since $e^{-2\gamma s}f(s) \approx e^{\gamma|s|}$ explodes as $s \to -\infty$.

While the distinction between stationary S and nonstationary $S^{-\gamma}$ seems initially minor, it is in fact central. *Any* model that wishes to explain the equity premium at long and short runs by means of an additional, stationary state variable must find some similar transformation so that the equity premium remains high at long horizons.

A Recession State Variable

Equation (24) emphasizes that our model makes a fundamental change in the way we understand risk premia. Consumers do not fear stocks because of the resulting risk to wealth or to consumption per se; they fear stocks primarily because stocks are likely to do poorly *in recessions*, times of low surplus consumption ratios. While $(C_{t+1}/C_t)^{-\gamma}$ and $(S_{t+1}/S_t)^{-\gamma}$ enter symmetrically in the formula, the volatility of $(C_{t+1}/C_t)^{-\gamma}$ is so low that it accounts for essentially no risk premia. The volatility of $(S_{t+1}/S_t)^{-\gamma}$ is much larger and accounts for nearly all risk premia. Variation across assets in expected returns is driven by variation across assets in covariances with *recessions* far more than by variation across assets in covariances with consumption growth.

At short horizons, S_{t+1} and C_{t+1} move together, so the distinction between a recession state variable and consumption risk is minor; one can regard S as an amplification mechanism for consumption risks in marginal utility. At long horizons, however, S_{t+k} becomes less and less conditionally correlated with C_{t+k}; S_{t+k} depends on C_{t+k} relative to its recent past, but the overall level of consumption may be high or low. Therefore, in contrast to Rietz's (1988) model of a small probability of a very large negative consumption shock, investors fear stocks because they do badly in occasional serious recessions *unrelated* to the risks of long-run average consumption growth.

Nonstochastic Analysis

It is common in growth theory to abstract from uncertainty and compare data from actual economies to the predictions of nonstochastic models. Many stochastic business cycle models study small deviations from nonstochastic steady states, which are thought to describe means well. Our model offers an interesting laboratory to study the accuracy of this sort of approximation.

Equation (8) for the risk-free rate,

$$r_t^f = -\ln(\delta) + \gamma g - \gamma(1 - \phi)(s_t - \bar{s}) - \frac{\gamma^2 \sigma^2}{2}[1 + \lambda(s_t)]^2,$$

highlights one important danger of nonstochastic analysis. In our model, the second to last intertemporal substitution term exactly offsets the last precautionary savings term in order to produce a constant interest rate. In the absence of the precautionary savings effect of a changing $\lambda(s_t)$, interest rates would vary a great deal with the state variable s_t in our model. A researcher who analyzed data from our economy with a nonstochastic version of the model would be puzzled by the stability of the risk-free interest rate. Precautionary savings are *not* a second-order effect.

A nonstochastic analysis also has trouble with the fact that price/dividend ratios are finite. At our parameter values, the consumption growth rate (1.89 percent) is about double the interest rate (0.94 percent). Thus a risk-neutral or certainty version of our economy predicts an infinite price of the consumption and dividend streams. Only the risk-corrected prices are finite.

IV. Some Microeconomic Implications

In this section, we address several of the most important objections to the model: that it seems not to allow for any heterogeneity across consumers, that it assumes implausibly high risk aversion, and that it relies on an external-habit rather than the more common internal-habit specification.

At heart, all three objections have to do with the potential application of the model to microeconomic data. This is not our chief concern in this paper. Our goal, ambitious enough, is to find representative-agent preferences that explain the joint behavior of aggregate consumption and stock returns. These representative-agent preferences could take the same form as the underlying preferences of individual agents, but they could also result from aggregation of heterogeneous consumers with quite different preferences. As one example, Constantinides and Duffie (1996), building on Mankiw (1986), show how to disaggregate *any* representative-agent marginal utility process, including ours, to individual agents with power utility and low risk aversion in incomplete markets by allowing the cross-sectional variance of idiosyncratic income to vary with the posited marginal utility process. Nonetheless, we find external habit formation appealing as a description of individual preferences, and the representative-agent model is clearly more compelling if its preferences *can* result from aggregation of individuals with similar preferences. Therefore, we now briefly consider whether the external habit model makes sense for microeconomic data.

A. Heterogeneity

Our identical-agent model, with parameter values such that habit is only about 5 percent below consumption, seems initially to be inconsistent with cross-sectional variation in wealth and consumption. If everyone has the same habit level, then poor people with consumption more than 5 percent below average would have consumption below habit, which makes no sense in our power specification.

In fact, however, our model can at least aggregate under complete markets with heterogeneous agents and heterogeneous groups. While these aggregation results are not as general as one might like—a standard problem in representative-agent models—it is still reassuring that many of the simple aggregation arguments for power utility apply, so the model is not automatically inconsistent with the wide cross-sectional variation of individual consumption and wealth. As usual in such results, the trick is to maintain identical growth in marginal utility while allowing some heterogeneity across individuals in levels of consumption, utility, or marginal utility.

We can allow many different reference groups with different levels of wealth by letting each agent's habit be determined by the average consumption of his reference group rather than by average consumption in the economy as a whole. Then poor people with low consumption levels have the same surplus consumption ratio as rich people with high consumption, since their reference groups also have low consumption. Each agent still has an identical power utility function of the difference between his consumption and his habit, and each group's consumption *growth* still moves in lockstep. With identical surplus consumption ratios and consumption growth rates, marginal utility growth is unchanged despite the heterogeneity in group consumption levels. In the appendix (Campbell and Cochrane 1998a), we show algebraically that the representative-agent preferences are the same as those of the individuals in this economy.

We can also allow some individual heterogeneity. Suppose that each agent i receives an endowment C_t^i, which is determined from the aggregate endowment C_t^a by

$$C_t^i = \left(\frac{\xi_i}{\xi}\right)^{1/\gamma} (C_t^a - X_t) + X_t.$$

The weights ξ_i vary across individuals, and X_t is determined from the history of aggregate consumption via (3) as usual. By construction of the example, marginal utility $(C_t^i - X_t)^{-\gamma}$ is proportional across agents, so marginal utility growth is the same for all individuals despite the heterogeneity in consumption levels. Therefore, all individuals agree on asset prices and have no incentive to trade away from their endowments. To complete the example, we show in the appendix (Campbell and Cochrane 1998a) that C_t^a is in fact the average of C_t^i in each period. The combination of group and individual heterogeneity is straightforward, if algebraically unpleasant.

B. Risk Aversion

Do we achieve a model consistent with the historical equity premium by assuming implausibly high values of risk aversion? We have emphasized an interpretation of the "equity premium puzzle" in terms of aggregate observations: high risk aversion is undesirable in power utility models because it leads to counterfactual predictions for interest rates and consumption growth, and our model resolves these problems. But many people object to high risk aversion per se, even if it is consistent with all data on asset prices and economic aggregates. This objection is therefore also fundamentally a concern about microeconomic evidence.

Most intuition about risk aversion comes from surveys of individual attitudes toward bets on wealth (including introspection, which is a survey with a sample size of one). But survey evidence for low risk aversion can be hard to interpret. To avoid the implication that we are all risk neutral to small zero-beta bets, surveys focus on very large bets on wealth, outside ordinary experience, that consumers may reasonably have trouble digesting. Building on this observation, Kandel and Stambaugh (1991) subject some common thought experiments to a careful sensitivity analysis and show that high risk aversion is not as implausible as one might have believed. As is often the case, existing empirical microeconomic evidence does not give precise measurements for input into macroeconomic models.

The agents in our model *do* display high risk aversion. However, we argue that high risk aversion is inescapable (or at least has not yet been escaped) in the class of identical-agent models that are consistent with the equity premium facts at short and long runs.

Risk aversion measures attitudes toward pure wealth bets and is therefore conventionally captured by the second partial derivative of the value function with respect to individual wealth, with any other state variables held constant. In the appendix (Campbell and Cochrane 1998a), we define the value function for an individual in our economy. The value function depends on individual wealth W and on aggregate variables that describe asset prices or investment opportunities and the level of the external habit. We write it as $V(W_t, W_t^a, S_t^a)$. Risk aversion is defined as the elasticity of value with respect to individual wealth:

$$rra_t \equiv -\frac{WV_{WW}}{V_W} = -\frac{\partial \ln V_W(\cdot)}{\partial \ln W}.$$

Risk aversion, defined in this way, plays no direct role in describing our model at the aggregate level. Individual wealth, aggregate wealth, and the surplus consumption ratio always move together, so all partial derivatives of the value function are involved in generating asset prices. Risk aversion is potentially interesting only in a reconciliation with microeconomic data.

In the appendix (Campbell and Cochrane 1998a) we calculate risk aversion for our model. Risk aversion is about 80 at the steady state (twice the curvature of about 40), rises to values in the hundreds for low surplus consumption ratios, and is still as high as 60 at the maximum surplus consumption ratio. Thus risk aversion is countercyclical, like utility curvature, and it is actually higher than utility curvature everywhere. This result can be understood as follows. The envelope condition $u_c = V_W$ implies that risk aversion can be written as utility curvature times the elasticity of consumption to individual wealth, with aggregates held constant:

$$rra_t = -\frac{\partial \ln V_W(\cdot)}{\partial \ln W_t} = \frac{\partial \ln u_c(\cdot)}{\partial \ln C_t} \times \frac{\partial \ln C_t}{\partial \ln W_t} = \eta_t \frac{\partial \ln C_t}{\partial \ln W_t}. \quad (25)$$

If date t consumption moves proportionally to an individual wealth shock, risk aversion is the same as utility curvature. In our model, consumption rises *more* than proportionally to an increase in idiosyncratic wealth, so risk aversion rra_t is larger than curvature η_t. An increase in individual wealth allows the individual to permanently increase his individual consumption over habit. This increase reduces the consumer's precautionary savings, implying that consumption increases more than proportionally at first. The consumer finances the extra initial consumption by increasing consumption less than proportionally to the initial wealth shock in subsequent states with high curvature and hence high contingent claim value.

Constantinides (1990) and Boldrin, Christiano, and Fisher (1996) present models with low risk aversion that are consistent with the equity premium and low consumption volatility at short horizons. In these models, consumers adjust consumption slowly after an idiosyncratic wealth shock. The term $(\partial \ln C_t)/(\partial \ln W_t)$ in (25) is low,

C. Internal Habit Formation

So far we have specified an *external* habit: habits are set by everyone else's consumption. There are two reasons to calculate marginal utility and asset prices under the assumption that habits are internal rather than external. First, one wants to check whether the social marginal utility of consumption is always positive, despite the externality. Second, it is interesting to know whether the external-habit specification is essential to the results or whether it is just a convenient simplification.

It is possible that external rather than internal habits make little difference to aggregate consumption and asset pricing implications. With internal habits, consumption today raises future habits, lowering the overall marginal utility of consumption today. But asset prices are determined by *ratios* of marginal utilities. If internal habits simply lower marginal utilities at all dates by the same proportion, then a switch from external to internal habits has no effect on allocations and asset prices. For example, we show in the appendix (Campbell and Cochrane 1998a) that this occurs with power utility $(C - X)^{1-\gamma}$, a constant interest rate, and linear habit accumulation $X_t = \theta \sum_{j=1}^{\infty} \phi^j C_{t-j}$. Hansen and Sargent (1998) provide a similar example.

Our model adopts a nonlinear habit accumulation equation to generate an exact random walk in consumption along with a constant risk-free rate. (In the linear habit example above, consumption is close to but not exactly a random walk.) The nonlinearity in the habit accumulation process is thus the *only* reason there is any difference between the internal-habit and external-habit specification of our model. Still, it is interesting to know how big this difference is.

When habit is internal, marginal utility at time t in our model has extra terms reflecting the effect of time t consumption on time $t + j$ habits:

$$MU_t = \frac{\partial U_t}{\partial C_t} = (C_t - X_t)^{-\gamma} - E_t \left[\sum_{j=0}^{\infty} \delta^j (C_{t+j} - X_{t+j})^{-\gamma} \frac{\partial X_{t+j}}{\partial C_t} \right]. \quad (26)$$

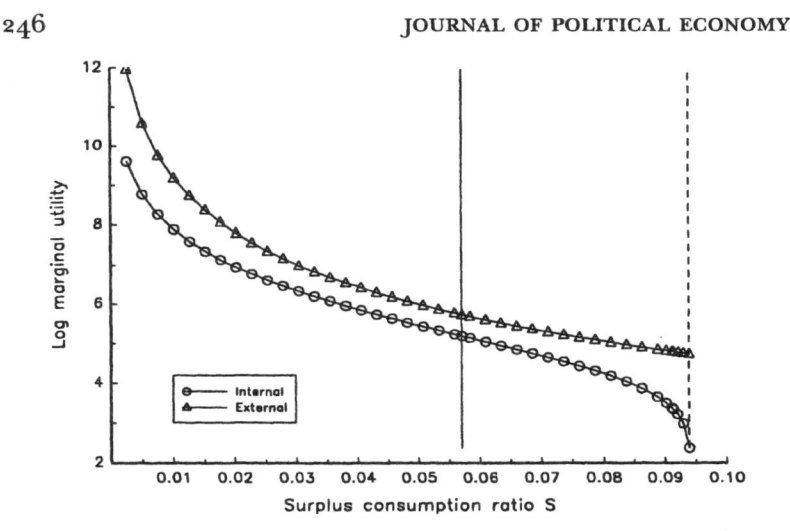

FIG. 10.—Marginal utility with internal vs. external habits. In each case, the marginal utility of consumption is given by $MU_t = C_t^{-\gamma} f(S_t)$; the figure plots $f(S_t)$.

In the appendix (Campbell and Cochrane 1998a) we show how to calculate marginal utility for the internal-habit version of our model, closed as an endowment economy with random walk consumption.

Figure 10 plots marginal utility as a function of state S in the internal- and external-habit cases of our model. Several features are worth noting. First, internal-habit marginal utility is always positive. This fact verifies that more consumption is always socially desirable despite the externality. Second, internal-habit and external-habit marginal utility are nearly proportional near the steady state $\bar{S}$, as in the linear example. This feature makes sense since the nonlinear habit accumulation process is approximately linear near the steady state. Third, internal-habit marginal utility falls away from external-habit marginal utility as the surplus consumption ratio varies far from the steady state. As we move farther from the steady state, changes in consumption have larger impacts on future habits, even immediately (as we saw above, dx/dc rises). The more an increase in consumption raises habits, of course, the less it raises utility.

The fact that this version of internal-habit marginal utility is nearly proportional to external-habit marginal utility is encouraging for the robustness of our model to the habit specification. We repeated all the analysis above and found that many features of the asset pricing predictions are maintained. The average excess return and unconditional Sharpe ratio are not much affected, and price/dividend ratios and expected returns vary with the state variable S about as before. However, the small deviations from proportionality of marginal utility turn out to have some significant effects on other predictions. Most important, this internal-habit version of our model generates risk-free rates that are higher and vary with the state variable S, so the *excess* return is a less sensitive function of the state variable and is less predictable.

However, in this comparison we close the internal-habit model as an endowment economy with a random walk in consumption rather than with a constant risk-free rate, we use the same parameter values that were selected to match moments with the external-habit specification, and we use the same habit formation process that was reverse-engineered to deliver a constant risk-free rate with random walk consumption and external habit. The asset pricing results might be closer if one reverse-engineered a different habit accumulation equation to deliver constant risk-free rates and random walk consumption with internal habits; if one closed the existing model using a constant interest rate, tolerating a possibly small predictability of consumption growth; or if one picked parameters to match moments using the internal-habit specification.

The predictions of internal- versus external-habit models for *individual* behavior can be quite different. One may be forced to the external-habit view when one wishes to integrate the lessons of aggregate and microeconomic data. If an individual with an internal or "addictive" habit and the ability to save receives an idiosyncratic windfall, he will increase consumption slowly and predictably. If an individual with an external habit receives an idiosyncratic windfall, his consumption will rise immediately. (An aggregate windfall can have different effects because it can cause asset prices to move, which is why the distinction between internal and external habits may make little difference to aggregate consumption behavior.) The vast literature on the permanent income hypothesis finds that individual consumption changes are quite difficult to predict. If anything, people spend windfalls even more quickly than predicted by the simple permanent income hypothesis, not too slowly. In analyzing risk aversion, we found that the external-habit model produces just such "overreaction" to individual wealth shocks, $\partial c_t/\partial w_t \geq 1$, because precautionary saving falls when individual wealth increases.

V. Conclusion

We have documented a broad variety of empirical successes for our consumption-based model with external habit formation. We calibrate the model to fit the unconditional equity premium and risk-free interest rate. The model then generates long-horizon predictability of excess stock and bond returns from the dividend/price ratio and mean reversion in returns; it generates high stock price and return volatility despite smooth and unpredictable dividend streams; and it generates persistent movements in return volatility. All these phenomena are linked to economic fluctuations: When consumption falls, expected returns, return volatility, and the price of risk rise, and price/dividend ratios decline. The model predicts many puzzles that face the standard power utility consumption-based model, including the equity premium and risk-free rate puzzles and the low unconditional correlation of consumption growth with stock returns. The model is consistent with an even sharper long-run equity premium puzzle that results from mean reversion in stock prices, together with low long-run consumption volatility. When we feed actual consumption data to the model, the model captures the main secular and business cycle swings in stock prices over the last century. The results are almost completely unchanged whether one uses a consumption claim or claims to volatile dividends that are very poorly correlated with consumption. The model predicts all this time variation despite a constant real interest rate and constant conditional moments for consumption and dividend growth.

In order to match these features of the data, our model posits a fundamentally novel view of risk premia in asset markets. Individuals fear stocks primarily because they do badly in recessions (times of low surplus consumption ratios), *not* because stock returns are correlated with declines in wealth or consumption.

The parameter values in our calibrated model imply that habits are only about 5 percent lower than consumption on average. This degree of habit formation may seem rather extreme. However, in this calibration we have used the sample period and the variable definitions that give the smoothest consumption and highest equity premium, we have ignored sampling variation and survivorship bias in mean returns, and we have not used standard devices to boost the equity premium such as occasional extremely bad states in the consumption distribution or frictions that concentrate stock ownership on a subset of the population. A less ambitious calibration exercise can produce similar dynamic results with a considerably higher average surplus consumption ratio.

The model gives some hope that finance can productively search for fundamental risk factors that explain at least the time-series behavior of aggregate stock returns rather than just relate some asset returns to other asset returns, leaving fundamental issues such as the equity premium as free parameters. The model also suggests that habit formation, or some other device to generate time-varying countercyclical risk premia along with relatively constant risk-free rates, is an important element for producing macroeconomic models with realistic production sectors that capture asset price movements as well as quantity dynamics.

References

Abel, Andrew B. "Asset Prices under Habit Formation and Catching Up with the Joneses." *A.E.R. Papers and Proc.* 80 (May 1990): 38–42.

———. "Risk Premia and Term Premia in General Equilibrium." *J. Monetary Econ.* 43 (February 1999): 3–33.

Black, Fischer. "Studies of Stock Price Volatility Changes." In *Proceedings of the Business and Economic Statistics Section, American Statistical Association*. Washington: American Statis. Assoc., 1976.

Boldrin, Michele; Christiano, Lawrence J.; and Fisher, Jonas D. M. "Asset Pricing Lessons for Modeling Business Cycles." Manuscript. Evanston, Ill.: Northwestern Univ., 1996.

Bollerslev, Tim; Chou, Ray Y.; and Kroner, Kenneth F. "ARCH Modeling in Finance: A Review of the Theory and Empirical Evidence." *J. Econometrics* 52 (April/May 1992): 5–59.

Campbell, John Y. "Understanding Risk and Return." *J.P.E.* 104 (April 1996): 298–345.

———. "Asset Prices, Consumption, and the Business Cycle." In *Handbook of Macroeconomics*, edited by John B. Taylor and Michael Woodford. Amsterdam: North-Holland, 1999.

Campbell, John Y., and Cochrane, John H. "By Force of Habit: A Consumption-Based Explanation of Aggregate Stock Market Behavior." Working Paper no. 4995. Cambridge, Mass.: NBER, January 1995.

———. "Appendix to 'By Force of Habit: A Consumption-Based Explanation of Aggregate Stock Market Behavior.'" Manuscript. Cambridge, Mass.: Harvard Univ.; Chicago: Univ. Chicago, 1998. (*a*)

———. "Explaining the Poor Performance of Consumption-Based Asset Pricing Models." Manuscript. Cambridge, Mass.: Harvard Univ.; Chicago: Univ. Chicago, 1998. (*b*)

Campbell, John Y., and Shiller, Robert J. "The Dividend-Price Ratio and Expectations of Future Dividends and Discount Factors." *Rev. Financial Studies* 1, no. 3 (1988): 195–227. (*a*)

———. "Stock Prices, Earnings, and Expected Dividends." *J. Finance* 43 (July 1988): 661–76. (*b*)

Chapman, David A. "Habit Formation and Aggregate Consumption." *Econometrica* 66 (September 1998): 1223–30.

Chou, Ray Y.; Engle, Robert F.; and Kane, Alex. "Measuring Risk Aversion from Excess Returns on a Stock Index." *J. Econometrics* 52 (April–May 1992): 201–24.

Cochrane, John H. "Volatility Tests and Efficient Markets: A Review Essay." *J. Monetary Econ.* 27 (June 1991): 463–85.

———. "Explaining the Variance of Price-Dividend Ratios." *Rev. Financial Studies* 5, no. 2 (1992): 243–80.

Cochrane, John H., and Hansen, Lars Peter. "Asset Pricing Lessons for Macroeconomics." In *NBER Macroeconomics Annual*, vol. 7, edited by Olivier J. Blanchard and Stanley Fischer. Cambridge, Mass.: MIT Press, 1992.

Constantinides, George M. "Habit Formation: A Resolution of the Equity Premium Puzzle." *J.P.E.* 98 (June 1990): 519–43.

Constantinides, George M., and Duffie, Darrell. "Asset Pricing with Heterogeneous Consumers." *J.P.E.* 104 (April 1996): 219–40.

Cox, John C.; Ingersoll, Jonathan E., Jr.; and Ross, Stephen A. "A Theory of the Term Structure of Interest Rates." *Econometrica* 53 (March 1985): 385–407.

Deaton, Angus. *Understanding Consumption.* New York: Oxford Univ. Press, 1992.

Deaton, Angus, and Muellbauer, John. *Economics and Consumer Behavior.* New York: Cambridge Univ. Press, 1980.

Duesenberry, James S. *Income, Saving, and the Theory of Consumer Behavior.* Cambridge, Mass.: Harvard Univ. Press, 1949.

Epstein, Larry G., and Zin, Stanley E. "Substitution, Risk Aversion, and the Temporal Behavior of Consumption and Asset Returns: An Empirical Analysis." *J.P.E.* 99 (April 1991): 263–86.

Fama, Eugene F. "Stock Returns, Expected Returns, and Real Activity." *J. Finance* 45 (September 1990): 1089–1108.

Fama, Eugene F., and French, Kenneth R. "Dividend Yields and Expected Stock Returns." *J. Financial Econ.* 22 (October 1988): 3–25. (*a*)

———. "Permanent and Temporary Components of Stock Prices." *J.P.E.* 96 (April 1988): 246–73. (*b*)

———. "Business Conditions and Expected Returns on Stocks and Bonds." *J. Financial Econ.* 25 (November 1989): 23–49.

Ferson, Wayne E., and Constantinides, George M. "Habit Persistence and Durability in Aggregate Consumption: Empirical Tests." *J. Financial Econ.* 29 (October 1991): 199–240.

Ferson, Wayne E., and Merrick, John J., Jr. "Non-stationarity and Stage-of-the-Business-Cycle Effects in Consumption-Based Asset Pricing Relations." *J. Financial Econ.* 18 (March 1987): 127–46.

Hansen, Lars Peter, and Jagannathan, Ravi. "Implications of Security Market Data for Models of Dynamic Economies." *J.P.E.* 99 (April 1991): 225–62.

Hansen, Lars Peter, and Sargent, Thomas J. "Recursive Models of Dynamic Linear Economies." Manuscript. Chicago: Univ. Chicago, 1998.

Harvey, Campbell R. "Time-Varying Conditional Covariances in Tests of Asset Pricing Models." *J. Financial Econ.* 24 (October 1989): 289–317.

Heaton, John C. "An Empirical Investigation of Asset Pricing with Temporally Dependent Preference Specifications." *Econometrica* 63 (May 1995): 681–717.

Jermann, Urban J. "Asset Pricing in Production Economies." *J. Monetary Econ.* 41 (February 1998): 257–75.

Kandel, Shmuel, and Stambaugh, Robert F. "Expectations and Volatility of Consumption and Asset Returns." *Rev. Financial Studies* 3, no. 2 (1990): 207–32.

———. "Asset Returns and Intertemporal Preferences." *J. Monetary Econ.* 27 (February 1991): 39–71.

MaCurdy, Thomas E., and Shoven, John. "Accumulating Pension Wealth with Stocks and Bonds." Working paper. Stanford, Calif.: Stanford Univ., 1992.

Mankiw, N. Gregory. "The Equity Premium and the Concentration of Aggregate Shocks." *J. Financial Econ.* 17 (September 1986): 211–19.

Mankiw, N. Gregory, and Shapiro, Matthew D. "Risk and Return: Consumption Beta versus Market Beta." *Rev. Econ. and Statis.* 68 (August 1986): 452–59.

Mehra, Rajnish, and Prescott, Edward C. "The Equity Premium: A Puzzle." *J. Monetary Econ.* 15 (March 1985): 145–61.

Nelson, Daniel B. "Conditional Heteroskedasticity in Asset Returns: A New Approach." *Econometrica* 59 (March 1991): 347–70.

Poterba, James M., and Summers, Lawrence H. "Mean Reversion in Stock Prices: Evidence and Implications." *J. Financial Econ.* 22 (October 1988): 27–59.

Rietz, Thomas A. "The Equity Risk Premium: A Solution." *J. Monetary Econ.* 22 (July 1988): 117–31.

Ryder, Harl E., Jr., and Heal, Geoffrey M. "Optimum Growth with Intertemporally Dependent Preferences." *Rev. Econ. Studies* 40 (January 1973): 1–33.

Schwert, G. William. "Why Does Stock Market Volatility Change over Time?" *J. Finance* 44 (December 1989): 1115–53.

Shiller, Robert J. "Consumption, Asset Markets and Macroeconomic Fluctuations." *Carnegie-Rochester Conf. Ser. Public Policy* 17 (Autumn 1982): 203–38.

———. *Market Volatility*. Cambridge, Mass.: MIT Press, 1989.

Siegel, Jeremy J. *Stocks for the Long Run: A Guide to Selecting Markets for Long-Term Growth*. Burr Ridge, Ill.: Irwin, 1994.

Sundaresan, Suresh M. "Intertemporally Dependent Preferences and the Volatility of Consumption and Wealth." *Rev. Financial Studies* 2, no. 2 (1989): 73–89.

Weil, Philippe. "The Equity Premium Puzzle and the Risk-Free Rate Puzzle." *J. Monetary Econ.* 24 (November 1989): 401–21.

Part III
Term Structure Models and Credit

[16]

MEASURING THE TERM STRUCTURE OF INTEREST RATES

J. HUSTON MC CULLOCH[*]

INTRODUCTION

This paper develops a technique of fitting a smooth curve, called the "discount function," to observations on prices of securities with varying maturities and coupon rates. The yield curve, instantaneous forward interest rates, mean forward interest rates, and consistent values for securities are derived from this discount function. Formulas for estimating the variances of these derived statistics are given. All formulas are worked out for a broad family of discount functions amenable to linear regression. A preferred form for the generalized discount function is described which focuses resolution in the vicinity of concentrations of observations. It is used to compare regression yield curves with those obtained by Durand[1] and those shown in the *Treasury Bulletin*.[2]

THE DISCOUNT FUNCTION

The most fundamental curve describing the term structure of interest rates, the one from which all others must be derived, is the *discount function* $\delta(m)$. It describes the present value of $1.00 repayable in m years. It is natural to suppose that the discount function is continuously differentiable. We may expect it to be monotonically decreasing.

Except for a few short-term securities called "bills," none of the zero coupon "bonds" whose prices are determined directly from the discount function exists. However, given the maturity m_0 and the coupon rate c of a security, its value p can be computed as the sum of the values of the payments that comprise it:

$$p = 100\delta(m_0) + c \int_0^{m_0} \delta(m) dm. \quad (1)$$

For simplicity, we have assumed that the coupons arrive in a continuous stream instead of semiannually. This enables us to use "and interest" prices as quoted.[3]

In order to fit a curve to the discount function by linear regression, we must postulate k continuously differentiable functions $f_j(m)$, and then express it as a constant term plus a linear combination of these functions:

$$\delta(m) = a_0 + \sum_{j=1}^{k} a_j f_j(m).$$

Since the present value of present money is unity, we must have $\delta(0) = 1$. The only way to force the curve through this point is to set $a_0 = 1$ and

$$f_j(0) = 0. \quad (2)$$

Therefore the discount function takes the form

$$\delta(m) = 1 + \sum_{j=1}^{k} a_j f_j(m). \quad (3)$$

---

[*] Department of Economics, University of Chicago. I am indebted to Professors Reuben A. Kessel, Merton H. Miller, Lester G. Telser, and Henri Theil, and Mr. Michael Mussa for helpful suggestions. The comments and suggestions of the referees were especially fruitful, and resulted in a thorough revision of the paper.

[1] David Durand, *Basic Yields of Corporate Bonds, 1900–1942* (New York: National Bureau of Economic Research, 1942).

[2] *Treasury Bulletin* (Washington, D.C.: Government Printing Office, March 1966), p. 78.

[3] A prorated share of the next coupon is added to the quoted "and interest" price to arrive at the "flat" price at which the security actually changes hands. Today, only bonds in default are quoted

The form of the functions $f_j(m)$ and the value of k are very important to the quality of our fit of the discount function. However, opinions may differ on their specification and there is no indisputably best method. Therefore we will develop all formulas at the present level of generality. Possible forms of these functions and rules for selecting k will be considered at the end of this paper.

Combining (1) and (3) we obtain

$$p = 100 \left[1 + \sum_{j=1}^{k} a_j f_j(m_0)\right]$$
$$+ c \int_0^{m_0} \left[1 + \sum_{j=1}^{k} a_j f_j(m)\right] dm$$
$$= 100 \left[1 + \sum_{j=1}^{k} a_j f_j(m_0)\right]$$
$$+ c \left[m_0 + \sum_{j=1}^{k} a_j \int_0^{m_0} f_j(m) dm\right]$$
$$= 100 + cm_0$$
$$+ \sum_{j=1}^{k} a_j \left[100 f_j(m_0) + c \int_0^{m_0} f_j(m) dm\right]. \quad (4)$$

Setting

$$y = p - 100 - cm_0 \quad (5a)$$

and

$$x_j = 100 f_j(m_0) + c \int_0^{m_0} f_j(m) dm, \quad (5b)$$

equation (4) becomes

$$y = \sum_{j=1}^{k} a_j x_j. \quad (5c)$$

Because c, m_0, and the postulated functions $f_j(m)$ are given, the right-hand side of (5c) is a linear combination, in unknown constants a_j, of known constants x_j.

---

flat. However, a few prewar bonds have to be converted to the "and interest" basis before equation (1) can be used.

We chose to start with the discount function, expressing it as a linear combination in unknowns as in (3), because we knew that the linearity of the integration operator in (1) would then also make p a linear combination in these unknowns, permitting estimation of the a_j by linear regression. Previous workers, notably Cohen, Kramer, and Waugh,[4] have instead started with the yield curve $\eta(m)$, a nonlinear transform of $\delta(m)$: $\eta(m) = -(1/m) \ln \delta(m)$. If one were to begin with this yield curve instead, so that

$$\eta(m) = a_0 + \sum_{j=1}^{k} a_j f_j(m),$$

when the value of the coupons was added to that of the principal using (1), he would obtain

$$p = 100 \exp\left\{-m_0\left[a_0 + \sum_{j=1}^{k} a_j f_j(m_0)\right]\right\}$$
$$+ c \int_0^{m_0} \exp\left\{-m\left[a_0 + \sum_{j=1}^{k} a_j f_j(m)\right]\right\} dm.$$

This expression is not linear in the a_j and therefore the a cannot be estimated by linear regression without the use of crude approximations. Consequently, we will use the approach of equation (3) and will not develop the yield curve until later.

ESTIMATION OF THE UNKNOWN PARAMETERS a_j

At any moment in time there will not be simultaneous actual sale prices for every security. This is especially true of slow-moving corporate issues. However, there often are enough securities with simultaneously standing bid and asked offers to make inferences about the term

---

[4] Kalman J. Cohen, Robert L. Kramer, and W. Howard Waugh, "Regression Yield Curves for U.S. Government Securities," *Management Science* 13, no. 14 (December 1966): B168–75.

structure. If we have such observations p_i^b and p_i^a on n securities, define mean prices $\bar{p}_i$ as $\bar{p}_i = (p_i^b + p_i^a)/2$. Let c_i and m_i be the coupon rate and term to final maturity of the ith security.

Instead of (1) holding exactly for the bid-asked mean price, we will find that

$$\bar{p}_i = 100\delta(m) + c_i \int_0^{m_i} \delta(m) dm + \epsilon_i, \quad (6)$$

where ϵ_i is an error term with positive variance. These errors can be caused by transactions costs, tax exemption, the capital gains tax treatment of deep discount bonds, callability, convertibility, ineligibility for commercial bank purchase, ability to be surrendered at par in payment of estate taxes (true of so-called flower bonds), risk of default, imperfect arbitrage, and the rigidity which will be introduced by postulating any specific form for $f_j(m)$. Thanks to transactions costs alone, the absolute value of ϵ_i could be as high as $v_i = (p_i^a - p_i^b)/2 + b$, where b is the brokerage fee of 0.5 parts per 100 for the broker-traded corporate issues and zero for dealer-quoted U.S. Governments. The difference between the maximum price to a buyer and the minimum price to the seller is $2v_i$. Because of the other sources of error, the error term will often be larger than v_i. Nevertheless, it will have a variance that is related to v_i. Since the other sources of error are more difficult to quantify, it is convenient to assume that the standard error of ϵ_i is simply proportional to v_i: S.E. $(\epsilon_i) = \sigma v_i$. The value of σ, which is to be measured, gives us an indicator of how well arbitrage is working and of the size of the factors other than coupon and maturity which enter into the value of the securities. If it is as low as 1.0, the bid-asked mean price of most of the securities observed will be within the transactions costs tolerance v_i of a value consistent with the observations on the other securities used. We probably cannot expect the fit to be any better than 1.0.[5]

Adapting (5) to the error term assumption of (6), the regression equation is:

$$y_i = \sum_{j=1}^{k} a_j x_{ij} + \epsilon_i, \quad i = 1, 2, \ldots n, \quad (7a)$$

and

$$\text{var}(\epsilon_i) = \sigma^2 v_i^2, \quad (7b)$$

where

$$y_i = \bar{p}_i - 100 - c_i m_i, \quad (7c)$$

$$x_{ij} = 100 f_j(m_i) + c_i \int_0^{m_i} f_j(m) dm, \quad (7d)$$

and

$$v_i = (p_i^a - p_i^b)/2 + b. \quad (7e)$$

We run a weighted least-squares regression on (7) to obtain estimates $\hat{a}_1$, $\hat{a}_2$, $\ldots$, $\hat{a}_k$ and $\hat{\sigma}$ of the parameters a_1, a_2, $\ldots$, a_k and σ. The discount function is then estimated by

$$\hat{\delta}(m) = 1 + \sum_{j=1}^{k} \hat{a}_j f_j(m). \quad (8)$$

We are not justified in extrapolating $\hat{\delta}(m)$ or any of its derived functions beyond the longest maturity of the securities observed. Notice that we are able to fit the discount function with a smooth curve, even though we do not have direct observations on it. We could never have done this by hand, or even by ordinary curve-fitting techniques.

I have actually fit (7) to observations on railroad bonds for fifteen selected dates from 1920 to 1938 and on U.S. Government securities for the close of every month from December 1946 to

[5] In the context of this application of linear regression, R^2 is a bad indicator of goodness of fit. It has no obvious intuitive interpretation and almost always is over .999. On the other hand, $\hat{\sigma}$ is meaningful and sensitive.

March 1966. The discount functions for two of these dates are shown in figures 1 and 2. They are displayed plus and minus their estimated standard errors of measurement. The estimated curve itself is not shown in order to avoid clutter. It lies halfway between the upper and lower edges of the band shown. Notice that the error, relative to the value of the curve, increases with time to maturity because the market is less concerned with the distant future than with the near future, and therefore does not define the curve for the distant future with as great a precision. The calculation of these errors is discussed in a later section.

Mean values of $\hat{\sigma}$ for selected subperiods are given in table 1. These figures would seem to indicate that prior to the Treasury–Federal Reserve Accord of March 4, 1951, and again after the beginning of "Operation Twist" in mid-1961, some sort of "disarbitrageur" was active in the market for U.S. Government securities. The fall in $\hat{\sigma}$ from 13.9 at the close of February 1951 to 5.6 at the close of March 1951 was especially dramatic. Since most of the same securities were still present in the market, this fall could not have been entirely due to a change in the special features of the securities.

Having estimated the parameters a_j we can estimate the true values p_i of the n securities:

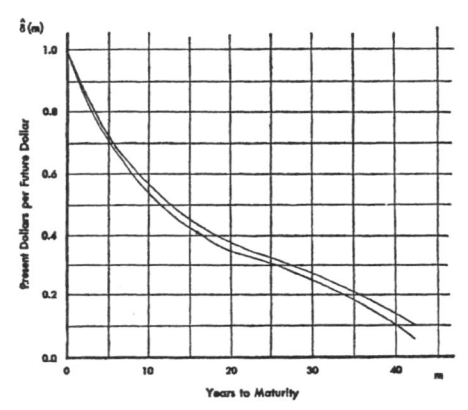

FIG. 1.—Discount function for the close of February 1922 based on bid-asked mean prices of high-grade (Moody's Aa and Aaa) railroad bonds. Convertibles and securities with any chance of being called before maturity were excluded. The band shows the best estimate plus and minus its standard error. In this regression, $n = 26$, $k = 5$, and $\hat{\sigma} = 2.67$.

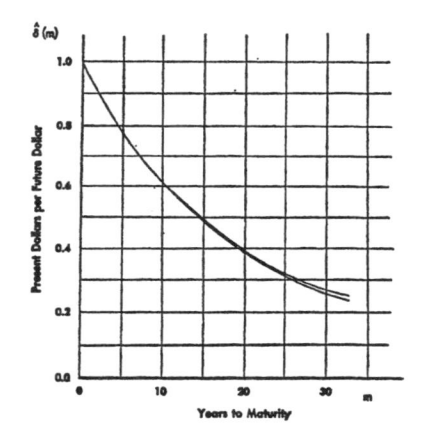

FIG. 2.—Discount function for the close of February 1966 based on bid-asked mean prices for taxable U.S. Government bills, notes, and bonds. Redemption of callable issues is assumed to be at earliest call date if price is above par and at maturity date when price is below par. In this regression, $n = 78$, $k = 9$, and $\hat{\sigma} = 7.81$. In spite of the higher value of $\hat{\sigma}$, this curve is better defined than that of fig. 1 because bid-asked spreads were smaller and because of the absence of brokerage fees.

TABLE 1

MEAN VALUES OF ERROR COEFFICIENT

Period	Type of Security	Mean Value of $\hat{\sigma}$
1920–1938	High-grade railroad bonds	2.6
1/1/47–3/1/51	Taxable U.S. government securities	15.9
4/1/51–1/1/62	Taxable U.S. government securities	4.6
2/1/62–4/1/66	Taxable U.S. government securities	9.0

MEASURING THE TERM STRUCTURE OF INTEREST RATES

$$\hat{p}_i = 100 + c_i m_i \qquad (9)$$
$$+ \sum_{j=1}^{k} \hat{a}_j \left[100 f_j(m_i) + c_i \int_0^{m_i} f_j(m) dm \right].$$

This formula can even be used to estimate the value of securities that did not enter into the regression. It is of use to dealers, banks, insurance companies, and large borrowers who need to compare the values of securities differing in coupon rate and maturity. Sophisticated users may even want to adjust (1) for taxes and for the value of special provisions. Examination of the weighted residuals, $(\bar{p}_i - \hat{p}_i)/v_i$, shows that the ineligibility for commercial bank purchase of many bonds prior to the Accord (and to a lesser degree until 1954) tended to cause negative residuals and that the special tax status of deep discount bonds tended to cause positive residuals. These properties account in large measure for the disappointingly high values of $\hat{\sigma}$ for postwar Treasury securities. Compensating for such factors should give better fits and reduce the unaccounted-for error.

FORWARD INTEREST RATES

The discount function $\delta(m)$ is an exponential decay curve whose rate of decay need not be constant. Its rate of decay is the *instantaneous forward interest rate* $\rho(m)$:

$$\rho(m) = -\frac{\delta'(m)}{\delta(m)}. \qquad (10)$$

Equivalently,

$$\delta(m) = \exp\left[-\int_0^m \rho(x) dx\right], \qquad (11)$$

and

$$\rho(m) = \lim_{h \to 0}\left[\frac{\delta(m)/\delta(m+h) - 1}{h}\right]. \qquad (12)$$

By differentiating (3) we have

$$\delta'(m) = \sum_{j=1}^{k} a_j f_j'(m). \qquad (13)$$

Consequently we can estimate (10) with

$$\hat{\rho}(m) = \frac{-\Sigma \hat{a}_j f_j'(m)}{1 + \Sigma \hat{a}_j f_j(m)}. \qquad (14)$$

Forward curves corresponding to the discount functions shown in figures 1 and 2 are depicted in figures 3 and 4, plus

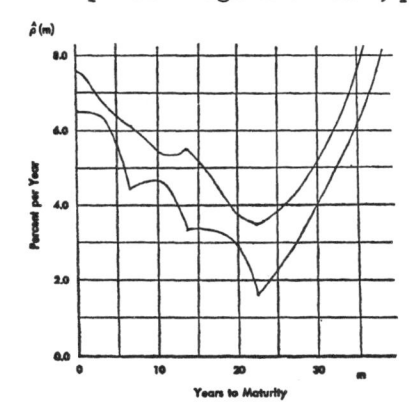

Fig. 3.—Instantaneous forward interest rate curve corresponding to the discount curve of fig. 1 for the close of February 1922.

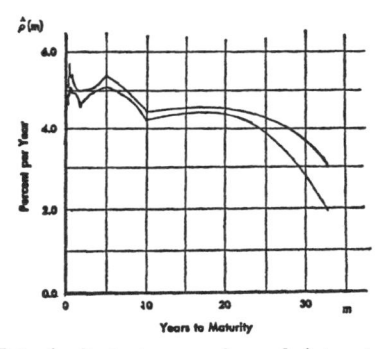

Fig. 4.—Instantaneous forward interest rate curve corresponding to the discount curve shown in fig. 2 for the close of February 1966. The high resolution at the short end is made possible by the concentration of bill observations.

and minus their standard errors of measurement. The calculation of these errors will be discussed in a later section. The "knuckles" in the bands are to be expected, unless we are willing to specify that $\delta(m)$ must be twice continuously differentiable instead of just once. As m goes to infinity, $\rho(m)$ does not necessarily approach an asymptote. Rather, its standard error of measurement will be found to grow without limit, so that its value simply fades away. This effect is more apparent in figure 4 than in figure 3, which goes off scale.

The instantaneous forward rate curve is a very important theoretical construct. However, its value for a single maturity m is of little practical concern, because it is prohibitively expensive in terms of transactions costs to make a forward contract between two points in the distant future if these points are only a small distance apart, as are m and $m + h$ in definition (12) of the instantaneous forward rate.

Only the average of $\rho(m)$ over a considerable interval in the future is of practical concern. Given any two values of m, say m_1 and m_2, the *mean forward interest rate* $r(m_1, m_2)$ is the average of $\rho(m)$ over the interval $[m_1, m_2]$:

$$r(m_1, m_2) = \frac{1}{m_2 - m_1} \int_{m_1}^{m_2} \rho(m) dm. \quad (15)$$

Equivalently,

$$r(m_1, m_2) = \frac{1}{m_2 - m_1} \ln \frac{\delta(m_1)}{\delta(m_2)}. \quad (16)$$

Computationally, (16) is more useful and can be estimated by

$$\hat{r}(m_1, m_2) = \frac{1}{m_2 - m_1} \ln \frac{\hat{\delta}(m_1)}{\hat{\delta}(m_2)}. \quad (17)$$

Notice that we have derived and estimated forward interest rates without use of the yield curve.

THE YIELD CURVE

The instantaneous forward interest rate curve $\rho(m)$ gives the rate of decay of the discount function $\delta(m)$ at each point m. The *yield curve* $\eta(m)$ is the average of that rate of decay over the interval from 0 to m. Thus,

$$\eta(m) = \frac{1}{m} \int_0^m \rho(x) dx. \quad (18)$$

Equivalent formulations are

$$\delta(m) = \exp[-m\eta(m)], \quad (19)$$

$$\eta(m) = -\frac{1}{m} \ln \delta(m), \quad (20)$$

and

$$\eta(m) = r(0, m).$$

Equation (18) states that η stands in the relation of an average curve to the marginal curve ρ. Although η and ρ are not cost curves, they still bear the same mathematical interrelationships as do average and marginal cost curves:

i) $m\eta'(m) + \eta(m) = \rho(m)$. (21)

ii) $\eta(0) = \rho(0)$.

iii) If η is $\begin{Bmatrix}\text{rising}\\ \text{falling}\end{Bmatrix}$ at m,

then ρ is $\begin{Bmatrix}\text{above}\\ \text{below}\end{Bmatrix} \eta$ at m.

iv) If $\eta'(m) = 0$, then $\eta(m) = \rho(m)$.

The yield curve, as defined in equation (20), can be estimated by

$$\hat{\eta}(m) = -\frac{1}{m} \ln \hat{\delta}(m). \quad (22)$$

Yield curves corresponding to figures 1 and 2 are given in figures 5A and 6A. The figures show the estimators plus and minus their standard errors of measurement. Notice how the standard error is large for both large and small m on the railroad curve for 1922 and is smaller for intermediate m. The computation of these errors will be discussed in a later section.

Other investigators have measured the term structure of interest rates by fitting

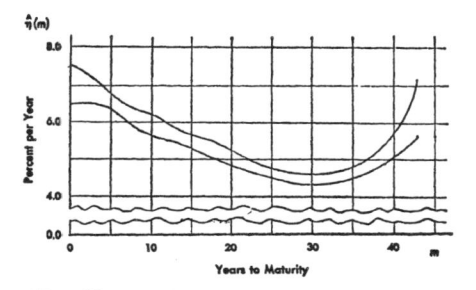

FIG. 5A.—Yield curve corresponding to the discount curve of fig. 1, for the close of February 1922.

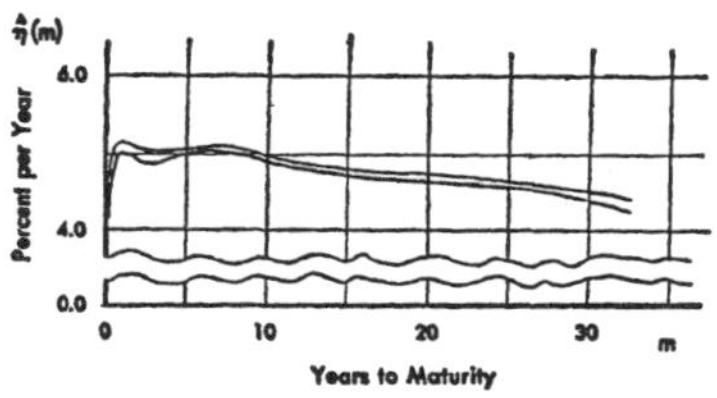

FIG. 6A.—Yield curve corresponding to the discount curve shown in fig. 2 for the close of February 1966.

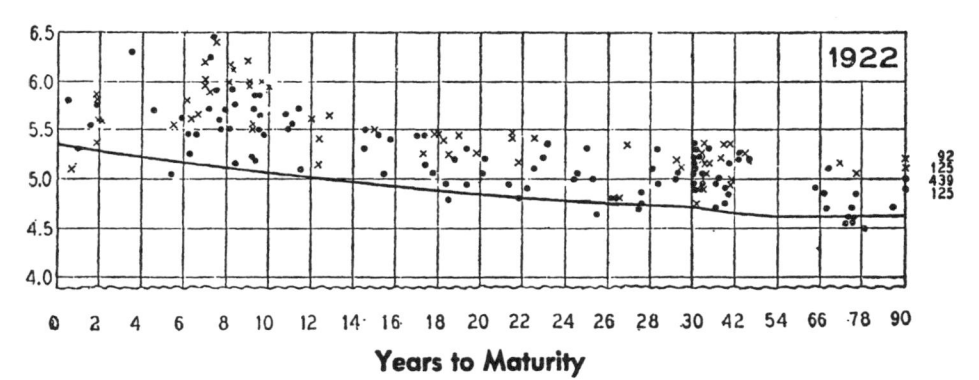

FIG. 5B.—A Durand yield curve for comparison with fig. 5A based on high-grade corporate bond transaction prices from the first quarter of 1922.

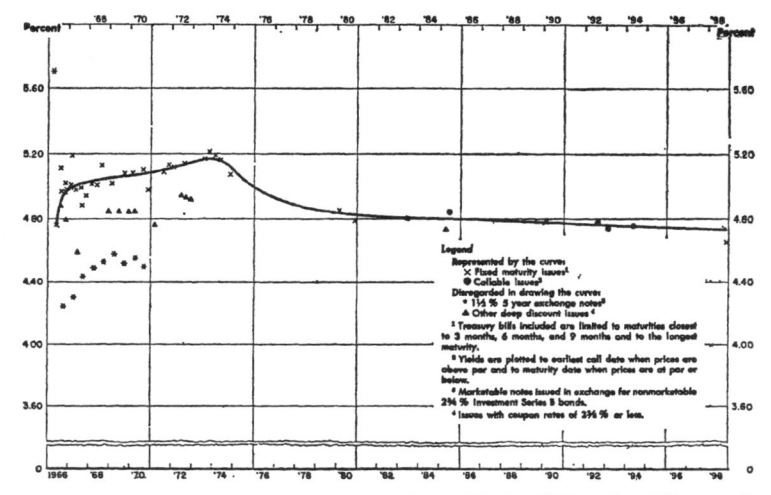

FIG. 6B.—A *Treasury Bulletin* yield curve for comparison with fig. 6A based on bid quotations for the close of February 1966. The smooth curve is fitted by eye. Market yields on coupon issues due in less than three months are excluded. Source: *Treasury Bulletin*, March 1966, p. 78.

a smooth curve to the average yields to maturity of the securities observed. Durand and the *Treasury Bulletin*[6] hand fitted the points with a French curve, while Cohen, Kramer, and Waugh[7] fit them with a linear regression. A Durand curve and a *Treasury Bulletin* curve are shown in figures 5B and 6B for comparison with figures 5A and 6A.

Both the hand and the regression approaches to directly fitting the yield curve are open to two serious objections. First, unless the yield curve $\eta(m)$ is flat, there is no reason to expect the average yield of a bond with a positive coupon rate to lie on it. The pure yield curve is defined for hypothetical bonds with zero coupon rates, so that the yield of an ordinary bond with maturity m_0 is a complicated average of $\eta(m)$ over the whole interval $[0, m_0]$, with only one of many weights at m_0, corresponding to the principal. For instance, if a bond has more than fifteen or twenty years to go before maturity, less than half of its value is due to the principal. The rest is embodied in the coupons. This averaging process washes out any shape the yield curve might have at the long end. The upward slope at the right end of the curve in figure 5A may be a pertinent example of the shape the yield curve may still have at the long end. Unfortunately, however, its measurement error becomes so large that this upward slope may or may not be statistically significant.

Second, any minor error incurred while directly fitting the yield curve will be magnified, especially for large m, if one tries to use formula (21) to calculate forward rates from the yield curve. Durand himself has insisted that his curves should not be used to derive forward rates.[8] For example, Durand's 1922 curve, shown in figure 5B, could not be used to infer the interestingly low forward rates shown in figure 3 which spanned the interval from 1937 to 1950 in the future (maturities fifteen to twenty-eight years). These remarkable rates foreshadowed the low rates for that period which were again to prevail during the later part of the Depression. When the data for 1922 were first fit, the author used a polynomial form for $\delta(m)$, and similar low forward rates resulted. In an effort to get rid of them, he devised the piecewise quadratic formulation which will be discussed in a later section, screened the data more carefully for bonds which were callable, convertible, or were not entirely risk free, and added more observations. In spite of these efforts the low forward rates persisted.

The tendency for the standard error of measurement of $\eta(m)$ to be large for small m means that Durand's "basic" yield curves are open to another objection: They are biased so that they tend to have an upward slope. In order to obtain rates for absolutely risk-free loans, he drew his yield curves to pass under the bulk of the plotted points. This would be a valid procedure if the width of the observed band of points were caused only by differences in the premiums for the risk of default. However, the observed bid-asked mean price of a virtually default-free security often differs from its predicted value by several times the transactions cost involved. In terms of yield, this difference is more important for short maturities than for long, causing Durand's plotted points to diverge for short maturities. By fitting his

[6] See nn. 1 and 2.
[7] See n. 4.

[8] David Durand, "A Quarterly Series of Corporate Basic Yields, 1952-57, and Some Attendant Reservations," *Journal of Finance* 13 (1958): 348-56.

curves under the bulk of the points instead of through them, his curves tend to have an upward-sloping shape too often. Figures 5A and 5B illustrate one case when Durand appears to have obtained an insufficiently downward-sloping yield curve. The fact that Durand's curves are biased to slope upward could provide an alternative to the "liquidity preference" explanation of why the conglomerate yield curve derived by averaging his annual curves is upward sloping.[9]

ESTIMATION OF THE ERROR OF MEASUREMENT

The observed prices of securities are seldom even close to their true values as given by (1). In fact, the observed price itself is indeterminate, since the maximum price to a buyer (the asked price plus the brokerage fee, if any) is substantially higher than the minimum price to a seller (the bid price minus any brokerage fee). Consequently, the term structure cannot be measured exactly, and any estimator of a value derived from it is subject to random measurement errors. It is important to estimate the variance of these errors and to accompany any display of statistics derived from our regression with estimates of their standard errors. This has been done graphically in the diagrams accompanying this paper by displaying a band whose upper and lower edges are the actual estimate plus and minus its standard error.

The weighted least-squares regression on (7) produces a $k \times k$ matrix C which is the estimator of the covariance matrix of the estimators $\hat{a}_i$ of the parameters a_i.

[9] See Reuben A. Kessel, *The Cyclical Behavior of the Term Structure of Interest Rates* (New York: National Bureau of Economic Research, 1965), p. 18.

If z is a k-vector of known values and $\hat{a}$ is the vector $(\hat{a}_1, \ldots, \hat{a}_k)$, the estimator of the variance of the linear combination $z^T \hat{a}$, for example, will be the quadratic form $z^T C z$.

The estimator of the variance of $\hat{\delta}(m)$, defined by (8), is therefore

$$\widehat{\mathrm{var}}\,[\hat{\delta}(m)] = z^T C z, \quad (23\mathrm{a})$$

where

$$z_j = f_j(m). \quad (23\mathrm{b})$$

Similarly, the estimator of the variance of $\hat{p}_i$, as defined in (9), is

$$\widehat{\mathrm{var}}\,(\hat{p}_i) = z^T C z, \quad (24\mathrm{a})$$

where

$$z_j = 100 f_j(m_i) + c_i \int_0^{m_i} f_j(m)\,dm. \quad (24\mathrm{b})$$

Just as (9) can help banks, dealers, and financial intermediaries estimate the proper bid-asked mean price to offer, (24) can help estimate the proper bid-asked spread with a little experience. For instance, they might make their offers differ from $\hat{p}_i$ by 1–3 S.E., depending upon how cautious they feel and the size of the offer.

The variance of the quotient of two random variables x and y with expected values Ex and Ey can be approximated by use of the formula

$$\frac{\mathrm{var}\,(x/y)}{(Ex/Ey)^2} \approx \frac{\mathrm{var}\,(x)}{(Ex)^2} + \frac{\mathrm{var}\,(y)}{(Ey)^2} - 2\frac{\mathrm{cov}\,(x, y)}{ExEy}, \quad (25)$$

provided that $\mathrm{var}\,(x) \ll (Ex)^2$, $\mathrm{var}\,(y) \ll (Ey)^2$, and that the distribution of y is positive.[10] Using (25), it can be shown

[10] Formula (25) in the case of independently distributed variables and formula (27) are commonly used in experimental physics and chemistry. Both the variance of a quotient and that of a logarithm are related to the variance of a product, which is

that the variance of $\hat{\rho}(m)$, as given in (14), is approximated by

$$\widehat{\text{var}}\,[\hat{\rho}(m)] \approx \hat{\rho}(m)^2 z^T C z, \quad (26a)$$

where

$$z_j = [f_j'(m)]/[\hat{\delta}'(m)] \quad (26b)$$
$$- [f_j(m)]/[\hat{\delta}(m)].$$

The variance of the natural logarithm of a random variable x can be approximated by

$$\text{var}\,(\ln x) \approx \frac{\text{var}\,(x)}{(Ex)^2}, \quad (27)$$

provided that var $(x) \ll (Ex)^2$, and that the distribution of x is positive. Using (27), it can be shown that the variance of the mean forward rate $\hat{r}(m_1, m_2)$, as defined in (17), is approximated by

$$\widehat{\text{var}}\,[\hat{r}(m_1, m_2)] \approx z^T C z, \quad (28a)$$

where

$$z_j = \frac{1}{m_2 - m_1}\left[\frac{f_j(m_1)}{\hat{\delta}(m_1)} - \frac{f_j(m_2)}{\hat{\delta}(m_2)}\right]. \quad (28b)$$

Again using (27), we see at once that the variance of the yield curve $\hat{\eta}(m)$, as given in (22), is approximated by

$$\widehat{\text{var}}\,[\hat{\eta}(m)] \approx \frac{\widehat{\text{var}}\,[\hat{\delta}(m)]}{[m\hat{\delta}(m)]^2}. \quad (29)$$

THE FORM OF THE FUNCTIONS $f_j(m)$

The choice of the functions $f_j(m)$ is central to the quality of our fit of the term structure. However, the selection of a form will always be a matter of judgment. Only a few hard and fast rules hold. Two of these are that the $f_j(m)$

---

rigorously derived by Leo A. Goodman, "On the Exact Variance of Products," *Journal of the American Statistical Association* (December 1960), pp. 708–13. An approximation for products analogous to (25) can be derived from his exact formula when the variances are relatively small.

must be continuously differentiable and that $f_j(0)$ must be 0.

The maturities of the securities we observe will not be uniformly distributed over the interval from 0 to m_n, the longest maturity observed, except by accident. Where concentrations of observations occur, the shape of the discount function is relatively well defined. Where observations are sparse, we are not justified in distinguishing as much shape. Therefore it will be desirable to make $f_j(m)$ depend on the distribution of the m_i in such a way as to provide greater resolution wherever maturities are clustered. In the case of U.S. Treasury securities, following this rule will place the greatest resolution at the short end, where there are many bills outstanding. This is as it should be, since participants in the market are more concerned with small differences in time in the near future than in the far future. This greater concern means that the discount function $\delta(m)$ they define by the values they place on the outstanding securities will have the most detailed shape at the short end.

A relatively naive approach is simply to set

$$f_j(m) = m^j, \quad j = 1, 2, \ldots, k. \quad (30)$$

This assumption makes $\delta(m)$ a kth-degree polynomial with unity for its constant term. A polynomial is straightforward, but it has no theoretical motivation. Its formulation does not depend on the distribution of the m_i, nor does it have a greater capability for providing resolution for values of m where the m_i are more likely to occur. As a result of its uniform resolving power, when it is used to fit a discount function which has a finely defined shape in the first 1 or 2 percent of its length and is relatively smooth thereafter, it will either ignore the short end and conform only to the remaining 98 or 99 percent, or else, if there are so many bill observations that they take over the regression, it will conform only to the short end and ignore the long end. It would take an extremely high-order polynomial to fit both the long and short ends of such a curve. Even so, this high-order polynomial would probably take on extreme values between observations at the long end, and would not be monotonically decreasing, as the discount function must be. On the other hand, a functional form which inherently permits greater resolution in the vicinity of data concentrations would be consistent with such a curve throughout its length, would require the estimation of only a few unknown parameters, and would be monotonic. The one portion would not have to be sacrificed to suit the other.

A better functional form for $\delta(m)$ than a polynomial is a continuously differentiable, piecewise quadratic function. To define such a curve we must divide the interval $(0, m_n)$ into $k - 1$ subintervals (d_j, d_{j+1}). We will have d_1 equal to 0 and d_k equal to m_n. Our $\delta(m)$ will follow a different quadratic function of m over each of the subintervals. In order for $\delta(m)$ to be continuously differentiable, the quadratics defined over adjacent subintervals (d_{j-1}, d_j) and (d_j, d_{j+1}) must have a common slope, as well as a common value, at d_j. The greater the number of subintervals covering any part of the interval $(0, m_n)$, the greater will be the resolving power of the discount function in that part of the interval. Therefore, by defining the subintervals to contain approximately equal numbers of the terminal maturities m_i, we will get greater potential resolution where the data observations are most numerous. Each of the quadratic segments will have an approximately equal number of observations to conform to. We can define the subintervals in this way by setting $d_j = m_l + \theta(m_{l+1} - m_l)$, where $l =$ greatest integer in $[(j-1)n]/(k-1)$, and $\theta = [(j-1)n]/(k-1) - l$. (We have assumed the securities to have been arranged in order by increasing terminal maturity, so that $m_j \leq m_{j+1}$.)

The set of functions of the form

$$\delta(m) = 1 + \sum_{j=1}^{k} a_j f_j(m)$$

will comprise the entire family of continuously differentiable functions which satisfy $\delta(0) = 1$ and which are piecewise quadratic over the subintervals defined above if we define the $f_j(m)$ as shown in figure 7. The first one, $f_1(m)$, starts with value zero and with a positive slope at $m = 0$, flattens until it has a zero slope at $m = d_2$, and remains constant thereafter, as in figure 7A. Intermediate ones, $f_j(m)$, where $j = 2, 3, \ldots, k - 1$, are zero up until d_{j-1}. There the slope begins to increase from zero up to some positive value at d_j. Then the slope falls from its d_j value to zero at d_{j+1}. Its value is constant thereafter. Figure 7B shows the particular case of $f_2(m)$. The last one, $f_k(m)$, is defined the same as the intermediate ones, except that it is undefined after $m = m_n$, as shown in figure 7C. Algebraically, the $f_j(m)$ are defined as follows:

$$f_1(m) = \begin{cases} m - \dfrac{1}{2d_2} m^2, & 0 \leq m \leq d_2 \\ \tfrac{1}{2} d_2, & d_2 < m \leq m_n \end{cases} \quad (31a)$$

$$f_j(m) = \begin{cases} 0, & 0 < m < d_{j-1} \\ \dfrac{(m - d_{j-1})^2}{2(d_j - d_{j-1})}, & d_{j-1} < m \leq d_j \\ \frac{1}{2}(d_j - d_{j-1}) + (m - d_j) - \dfrac{(m - d_j)^2}{2(d_{j+1} - d_j)}, & d_j < m \leq d_{j+1} \\ \frac{1}{2}(d_{j+1} - d_{j-1}), & d_{j+1} < m \leq m_n \end{cases} \quad j = 2, \ldots, k-1 \quad (31b)$$

$$f_k(m) = \begin{cases} 0, & 0 \leq m \leq d_{k-1} \\ \dfrac{(m - d_{k-1})^2}{2(m_n - d_{k-1})}, & d_{k-1} < m \leq m_n. \end{cases} \quad (31c)$$

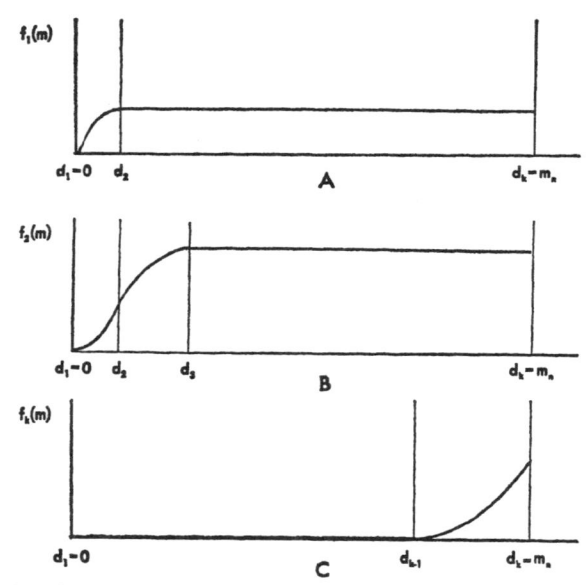

FIG. 7.—The preferred form of the $f_i(m)$. These $f_i(m)$ make (m) piecewise quadratic and continuously differentiable.

Since the vertical scales of the $f_i(m)$ are immaterial, we have arbitrarily chosen them so that

$$\delta'(d_j) = a_j f'_j(d_j)$$
$$= a_j.$$
(32)

Integration of $f_i(m)$ in order to evaluate (7d) and (24b) and differentiation in order to evaluate (14) and (26b) are matters of elementary calculus, and will be omitted here.[11]

The specification of the $f_i(m)$ given in (31) was used for the regression fits of the

[11] Other specifications of the $f_i(m)$ will generate exactly the same family of piecewise quadratic functions. The one chosen was selected only because it can have the property (32) if the scales are chosen appropriately. The other specifications will give the same $\delta(m)$ if used with the same data.

discount function shown in figures 1 and 2. Because a piecewise quadratic function has a discontinuous second derivative, the instantaneous forward interest rate curves derived from these discount functions have discontinuous first derivatives, which explains the angular shape of the bands shown in figures 3 and 4. However, as mentioned earlier, the instantaneous forward rate $\rho(m)$ is interesting mainly as a theoretical construct. Its level at one isolated value of m has little practical significance. Consequently we are not worried by the outlying values of $\hat{\rho}(m)$ and its standard error that are sometimes implied by our specification of the $f_i(m)$. In fact, our specification is sufficient to imply that mean forward rates $r(m_1, m_2)$, which are of practical concern, are continuously differentiable with respect to both m_1 and m_2. The yield curve, a special case of $r(m_1, m_2)$ with $m_1 = 0$, is therefore also continuously differentiable, as may be seen in figures 5A and 6A.

THE VALUE OF k

The number k of parameters to be estimated is another area where judgment must be used. If k is too low, we will not be able to fit the discount function closely when it takes on difficult shapes. If it is too high, the discount function may conform too closely to outliers instead of being smooth. If k is as high as n, there will be no way to estimate σ^2. In the spirit of least squares, we might try all values of k inside a range we regard as reasonable, and select that value which minimizes the unbiased estimator $\hat{\sigma}^2$ of σ^2:

$$\hat{\sigma}^2 = \frac{1}{n-k} \sum_{i=1}^{n} \left(\frac{\bar{p}_i - \hat{p}_i}{v_i} \right)^2.$$

As k increases, the residuals generally decrease, but then so do the degrees of freedom. The result is that $\hat{\sigma}^2$ declines sharply as k increases from 2 to 3 or 4, but thereafter fluctuates irregularly with a small amplitude, and often with more than one local minimum. Sometimes it shows no sign of permanently rising, even after k becomes so large that the discount function adheres to outliers.

A second approach is simply to make k a fixed function of n. We would like this function to have the following properties: First, in order to have resolution increase as the number of observations increases, our function $k(n)$ should increase with n. Second, in order to make the number of observations in the domain of each quadratic segment increase with the total number of observations, the ratio $n/k(n)$ should also increase with n. An elementary function with these properties is $k(n) =$ nearest integer to $n^{1/2}$. In practice, this formula gives approximately the same results as the first approach, without the expensive search.[12]

[12] Since the final revision of this paper, a precedent for the continuously differentiable, piecewise quadratic functional form has come to my attention (see Wayne A. Fuller, "Grafted Polynomials as Approximating Functions," *Australian Journal of Agricultural Economics* 13, no. 1 [June 1969]: 35–46).

Term Structure Modeling Using Exponential Splines

OLDRICH A. VASICEK and H. GIFFORD FONG*

I. Introduction

TERM STRUCTURE OF interest rates provides a characterization of interest rates as a function of maturity. It facilitates the analysis of rates and yields such as discussed in Dobson, Sutch, and Vanderford [1976], and provides the basis for investigation of portfolio returns as for example in Fisher and Weil [1971]. Term structure can be used in pricing of fixed-income securities (cf., for instance, Houglet [1980]), and for valuation of futures contracts and contigent claims, as in Brennan and Schwartz [1977]. It finds applications in analysis of the effect of taxation on bond yields (cf. McCulloch [1975a] and Schaefer [1981]), estimation of liquidity premia (cf. McCulloch [1975b]), and assessment of the accuracy of market-implicit forecasts (Fama [1976]). Because of its numerous uses, estimation of the term structure has received considerable attention from researchers and practitioners alike.

A number of theoretical equilibrium models has been proposed in the recent past to describe the term structure of interest rates, such as Brennan and Schwartz [1979], Cox, Ingersoll, and Ross [1981], Langetieg [1980], and Vasicek [1977]. These models postulate alternative assumptions about the nature of the stochastic process driving interest rates, and deduct a characterization of the term structure implied by these assumptions in an efficiently operating market. The resulting spot rate curves have a specific functional form dependent only on a few parameters.

Unfortunately, the spot rate curves derived by these models (at least in the instances when it was possible to obtain explicit formulas) do not conform well to the observed data on bond yields and prices. Typically, actual yield curves exhibit more varied shapes than those justified by the equilibrium models. It is undoubtedly a question of time until a sufficiently rich theoretical model is proposed that provides a good fit to the data. For the time being, however, empirical fitting of the term structure is very much an unrelated task to investigations of equilibrium bond markets.

The objective in empirical estimation of the term structure is to fit a spot rate curve (or any other equivalent description of the term structure, such as the discount function) that (1) fits the data sufficiently well, and (2) is a sufficiently smooth function. The second requirement, being less quantifiable than the first,

---

* Gifford Fong Associates

is less often stated. It is nevertheless at least as important as the first, particularly since it is possible to achieve an arbitrary good (or even perfect) fit if the empirical model is given enough degrees of freedom, with the consequence that the resulting term structure makes little sense. For a discussion of this point, see Langetieg and Smoot [1981].

A simple approach to estimation of the term structure is to postulate that bond payments occur only on a discrete set of specified dates, and assume no relationship among the discount factors corresponding to these dates (such as that they lie on a smooth curve). The discount factors can then be estimated as the coefficients in a regression with the bond payments on the given set of dates as the independent variables, and the bond price as the dependent variable. This approach has been taken by Carleton and Cooper [1976]. They include both U.S. Treasury and Federal Home Loan Bank securities in the estimation, with an adjustment for the default risk in the FHLB bonds. The resulting discount function is discrete rather than continuous, and the forward rates are found not to be smooth.

McCulloch [1971] introduced the methodology of fitting the discount function by polynomial splines. This produces estimates of the discount function as a continuous function of time. For cubic or higher order splines, the forward rates are a smooth function. Since the model is linear in the discount function, ordinary least-squares regression techniques can be used.

In addressing the effect of taxation, McCulloch [1975a] estimates the after-tax term structure of interest rates and the marginal income tax rate. Estimates of the tax rate were achieved by minimizing the standard error of the regression. This estimated tax rate is used to convert the after-tax term structure into a before-tax term structure. This procedure makes the estimated forward rates very sensitive to any estimation errors in the tax rate. Moreover, because the tax effect is estimated by best fitting to minimize large errors, the inclusion of special securities such as flower bonds tends to prejudice the results.

Langetieg and Smoot [1981] discuss extensions of McCulloch's spline methodology. These include fitting cubic splines to the spot rates rather than the discount function, and varying the location of the spline knots. Non-linear estimation procedures are required in these models.

This paper presents a different approach, which can be termed an *exponential spline fitting*. The methodology described here has been applied to historical price data on U.S. Treasury securities with satisfactory results. The technique produces forward rates that are a smooth continuous function of time. The model has desirable asymptotic properties for long maturities, and exhibits both a sufficient flexibility to fit a wide variety of shapes of the term structure, and a sufficient robustness to produce stable forward rate curves. An adjustment for the effect of taxes and for call features on U.S. Treasury bonds is included in the model.

In the next section, we provide a brief description of the basic concepts of the term structure, such as spot and forward rates, market-implicit forecasts, and the discount function. This provides some background for understanding some of the prior work, and of the model to be proposed in the last section.

II. Concepts and Terms

The spot interest rate of a given maturity is defined as the yield on a pure discount bond of that maturity. The spot rates are the discount rates determining the present value of a unit payment at a given time in the future. Spot rates considered as a function of maturity are referred to as the *term structure of interest rates*.

Spot rates are not directly observable, since there are few pure discount bonds beyond maturities of one year. They have to be estimated from the yields on actual securities by means of a *term structure model*. Each actual coupon bond can be considered a package of discount bonds, namely one for each of the coupon payments and one for the principal payment. The price of such component discount bonds is equal to the amount of the payment discounted by the spot rate of the maturity corresponding to this payment. The price of the coupon bond is then the sum of the prices of these component discount bonds. The *yield to maturity* on a coupon bond is the internal rate of return on the bond payments, or the discount rate that would equate the present value of the payments to the bond price. It is seen that the yield is thus a mixture of spot rates of various maturities. In calculation of yield, each bond payment is discounted by the same rate, rather than by the spot rate corresponding to the maturity of that payment. Decomposing the actual yields on coupon bonds into the spot rates is the principal task of a term structure model.

Spot rates describe the term structure by specifying the current interest rate of any given maturity. The implications of the current spot rates for future rates can be described in terms of the *forward rates*. The forward rates are one-period future reinvestment rates, implied by the current term structure of spot rates.

Mathematically, if $R_1, R_2, R_3 \cdots$ are the current spot rates, the forward rate F_t for period t is given by the equation

$$1 + F_t = \frac{(1 + R_t)^t}{(1 + R_{t-1})}, \qquad t = 1, 2, 3, \cdots. \tag{1}$$

This equation means that the forward rate for a given period in the future is the marginal rate of return from committing an investment in a discount bond for one more period. By definition, the forward rate for the first period is equal to the one period spot rate, $F_1 = R_1$.

The relationship of spot and forward rates described by Equation (1) can be stated in the following equivalent form:

$$(1 + R_t)^t = (1 + F_1)(1 + F_2) \cdots (1 + F_t). \tag{2}$$

This equation shows that spot rates are obtained by compounding the forward rates over the term of the spot rate. Thus, the forward rate F_t can be interpreted as the interest rate over the period from $t - 1$ to t that is implicit in the current structure of spot rates.

Just as the forward rates are determined by the spot rates using Equation (1), the spot rates can be obtained from the forward rates by Equation (2). Thus, either the spot rates or the forward rates can be taken as alternative forms of describing the term structure. The choice depends on which of these two equivalent characterizations is more convenient for the given purpose. Spot rates describe interest rates over periods from the current date to a given future date. Forward rates describe interest rates over one-period intervals in the future.

There is a third way of characterizing the term structure, namely by means of the *discount function*. The discount function specifies the present value of a unit payment in the future. It is thus the price of a pure discount riskless bond of a given maturity. The discount function D_t is related to the spot rates by the equation

$$D_t = \frac{1}{(1 + R_t)^t} \tag{3}$$

and to the forward rates by the equation

$$D_t = \frac{1}{(1 + F_1)(1 + F_2) \cdots (1 + F_t)}. \tag{4}$$

The discount function D_t considered in continuous time t is a smooth curve decreasing from the starting value $D_0 = 1$ for $t = 0$ (since the value of one dollar now is one dollar) to zero for longer and longer maturities. It typically has an exponential shape.

While the discount function is usually more difficult to interpret as a description of the structure of interest rates than either the spot rates or the forward rates, it is useful in the *estimation* of the term structure from bond prices. The reason is that bond prices can be expressed in a very simple way in terms of the discount function, namely the sum of the payments multiplied by their present value. In terms of the spot or forward rates, bond prices are a more complicated (nonlinear) function of the values of the rates to be estimated.

The concept of forward rates is closely related to that of the *market-implicit forecasts*. The market-implicit forecast $M_{t,s}$ of a rate of maturity s as of a given future date t is the rate that would equate the total return from an investment at the spot rate R_t for t periods reinvested at the rate $M_{t,s}$ for additional s periods, with the straight investment for $t + s$ periods at the current spot rate R_{t+s}. Mathematically, this can be written as follows:

$$(1 + R_t)^t (1 + M_{t,s})^s = (1 + R_{t+s})^{t+s}. \tag{5}$$

The market-implicit forecasts can be viewed as a forecast of future spot rates by the aggregate of market participants. Suppose that the current one-year rate is 12%, and that there is a general agreement among investors that the one-year rate a year from now will be 13%. Then the current two-year spot rate will be 12.50%, since

$$(1 + .1250)^2 = (1 + .12)(1 + .13).$$

The two-year rate would be set in such a way that the two-year security has the same return as rolling over a one-year security for two years. There may not be such a general agreement as to the future rate, and in any case the forecast would not be directly observable. Knowing the current one-year and two-year spot rates, however, enables us to determine the future rate for the second year that would make the two-year bond equivalent in terms of total return to a roll-over of one-year bond. This rate is the market-implicit forecast.

The market-implicit forecasts have a number of interesting properties. The first thing to note is that when a *futures contract* is available for a given future period, the rate on the futures contract is equal to the market-implicit forecast (up to a difference attributable to transaction costs). If this were not true, a riskless arbitrage can be set between a portfolio consisting of the futures contract and a security maturing at the execution date on one hand, and a security maturing at the maturity date of the contract on the other hand. Such riskless arbitrage opportunities should not exist in efficient financial markets.

Another feature of market-implicit forecasts is that the *holding period return* calculated using these forecasts is the same for any default-free security, regardless of its maturity. It is equal to the spot rate corresponding to the length of the holding period. Indeed, the total return over a holding period of length h on an issue with maturity s ($s > h$) is equal to

$$\frac{(1 + R_s)^s}{(1 + M_{h,s-h})^{s-h}}.$$

Recalling the definition of the market-implicit forecast in Equation (5), the total return over the holding period is readily calculated as

$$\frac{(1 + R_s)^s}{(1 + M_{h,s-h})^{s-h}} = \frac{(1 + R_s)^s(1 + R_h)^h}{(1 + R_s)^s} = (1 + R_h)^h.$$

Thus, the holding period return is independent of the maturity of the security, and is given by the spot rate for the holding period.

This is a characterization of the market-implicit forecasts that can actually serve as their definition. No other set of forecasts would have the property that the holding period returns over a given period are the same for securities of all maturities (including coupon bonds). In a sense, the market-implicit forecast is the most "neutral" forecast. It is the equilibrium expectation such that no maturities or payment schedules are ex-ante preferred to others.

The definition of the market-implicit forecasts as given by Equation (5) is perhaps more intuitive if stated in terms of the forward rates. It is given by the following equation:

$$(1 + M_{t,s})^s = (1 + F_{t+1})(1 + F_{t+2}) \cdots (1 + F_{t+s}). \tag{6}$$

Specifically, the market-implicit forecast of one-period rate is equal to the forward rate for that period,

$$M_{t,1} = F_t.$$

It is seen from Equation (6) that the market-implicit forecast is obtained by compounding the forward rates over the period starting at the date of the forecasting horizon and extending for an interval corresponding to the term of the forecasted rate. In other words, the market-implicit forecast corresponds to the scenario of *no change in the forward rates*. The current spot rates then change by rolling along the forward rate series.

One last thing to mention about the market-implicit forecasts is that since it is a forecast of the future spot rates, we can also infer from it the corresponding forecast of yields, discount functions, and all other characterizations of the *future term structure*. The current and future term structures have the forward rates as the one common denominator, which makes the forward rates the basic building blocks of the structure of interest rates.

III. The Model

In specification of the model proposed for estimation of the term structure, we will use the following notation:

t time to payment (measured in half years)

$D(t)$ the discount function, that is, the present value of a unit payment due in time t

$R(t)$ spot rate of maturity t, expressed as the continuously compounded semi-annual rate. The spot rates are related to the discount function by the equation

$$D(t) = e^{-tR(t)}$$

$F(t)$ continuously compounded instantaneous forward rate at time t. The forward rates are related to the spot rate by the equation

$$R(t) = \frac{-d}{dt} \log D(t).$$

n number of bonds used in estimation of the term structure
T_k time to maturity of the k-th bond, measured in half years
C_k the semi-annual coupon rate of the k-th bond, expressed as a fraction of the par value
P_k price of the k-th bond, expressed as a fraction of the par value.

The basic model can be written in the following form:

$$P_k + A_k = D(T_k) + \sum_{j=1}^{L_k} C_k D(T_k - j + 1) - Q_k - W_k + \epsilon_k \qquad (7)$$
$$k = 1, 2, \cdots, n$$

where

$$A_k = C_k(L_k - T_k)$$

is the accrued interest portion of the market value of the k-th bond,

$$L_k = [T_k] + 1$$

is the number of coupon payments to be received, Q_k is the price discount attributed to the effect of taxes, W_k is the price discount due to call features, and ϵ_k is a residual error with $E\epsilon_k = 0$.

The model specified by Equation (7) is expressed in terms of the discount function, rather than the spot or forward rates. The reason for this specification is that the price of a given bond is linear in the discount function, while it is nonlinear in either the spot or forward rates. Once the discount function is estimated, the spot and forward rates can easily be calculated.

An integral part of the model specification is a characterization of the structure of the residuals. We will postulate that the model be *homoscedastic in yields*, rather than in prices. This means that the variance of the residual error on yields is the same for all bonds. The reason for this requirement is that a given price increment, say $1 per $100 face value, has a very different effect on a short bond than on a long bond. Obviously, an error term in price on a three-month Treasury bill cannot have the same magnitude as that in price of a twenty-year bond. It is, however, reasonable to assume that the magnitude of the error term would be the same for yields.

With this assumption, the residual variance in Equation (7) is given as

$$E\epsilon_k^2 = \sigma^2 \omega_k, \quad k = 1, 2, \cdots, n \tag{8}$$

where

$$\omega_k = \left(\frac{dP}{dY}\right)_k^2 \tag{9}$$

is the squared derivative of price with respect to yield for the k-th bond, taken at the current value of yield. The derivative dP/dY can easily be evaluated from time to maturity, the coupon rate, and the present yield. In addition, we will assume that the residuals for different bonds are uncorrelated,

$$E\epsilon_k\epsilon_\ell = 0, \quad \text{for } k \neq \ell.$$

In specification of the effect of taxes, we will assume that the term Q_k is proportional to the current yield C_k/P_k on the bond,

$$Q_k = q \frac{C_k}{P_k}\left(\frac{dP}{dY}\right)_k, \quad k = 1, 2, \cdots, n. \tag{10}$$

For the call effect, the simplest specification is to introduce a dummy variable I_k, equal to 1 for callable bonds and to 0 for noncallable bonds, and put

$$W_k = wI_k, \quad k = 1, 2, \cdots, n. \tag{11}$$

Although more complicated specifications (such as those based on option pricing) are possible, the form (11) seems to work well with Treasury bonds, which invariably have the same structure of calls five years prior to maturity at par.

We will now turn to the specification of the discount function $D(t)$. Earlier approaches (cf. McCulloch [1971], [1975b]) fit the discount function by means of polynomial splines of the second or third order. While splines constitute a very flexible family of curves, there are several drawbacks to their use in fitting discount functions. The discount function is principally of an exponential shape,

$$D(t) \sim e^{-\gamma t}, \quad 0 \leq t < \infty.$$

Splines, being piecewise polynomials, are inherently ill suited to fit an exponential type curve. Polynomials have a different curvature from exponentials, and although a polynomial spline can be forced to be arbitrarily close to an exponential curve by choosing a sufficiently large number of knot points, the local fit is not good. A practical manifestation of this phenomenon is that a polynomial spline tends to "weave" around the exponential, resulting in highly unstable forward rates (which are the derivatives of the logarithm of the discount function). Another problem with polynomial splines is their undesirable asymptotic properties. Polynomial splines cannot be forced to tail off in an exponential form with increasing maturities.

It would be convenient if we can work with the logarithm $\log D(t)$ of the discount function, which is essentially a straight line and can be fitted very well with splines. Unfortunately, the model given by Equation (7) would then be nonlinear in the transformed function, which necessitates the use of complicated nonlinear estimation techniques (cf. Langetieg and Smoot [1981]).

A way out of this dilemma is provided by the following approach, which is used in our model. Instead of using a transform of the function $D(t)$, we can apply a transform to the *argument* of the function. Let α be some constant and put

$$t = -\frac{1}{\alpha} \log(1 - x), \qquad 0 \leq x < 1. \tag{12}$$

Then $G(x)$ defined by

$$D(t) = D\left(-\frac{1}{\alpha} \log(1 - x)\right) \equiv G(x) \tag{13}$$

is a new function with the following properties: (a) $G(x)$ is a decreasing function defined on the finite interval $0 \leq x \leq 1$ with $G(0) = 1$, $G(1) = 0$; (b) to the extent that $D(t)$ is approximately exponential,

$$D(t) \sim e^{-\gamma t}, \qquad 0 \leq t < \infty$$

the function $G(x)$ is approximately a power function,

$$G(x) \sim (1 - x)^{\gamma/\alpha} \qquad 0 \leq x \leq 1;$$

(c) the model specified by Equation (7) is linear in G. Thus, we have replaced the function $D(t)$ to be estimated by the approximately power function $G(x)$ which can be very well fitted by polynomial splines, while preserving the linearity of the model. Moreover, desired asymptotic properties can easily be enforced.

If $G(x)$ is polynomial with $G'(1) \neq 0$, then the parameter α constitutes the *limiting value of the forward rates*,

$$\lim_{t \to \infty} F(t) = \alpha.$$

Indeed, in that case

$$G(x) = -G'(1)(1 - x) + o(1 - x)$$

and consequently

$$D(t) = -G'(1) e^{-\alpha t} + o(e^{-\alpha t})$$

as $t \to \infty$. Using polynomial splines to fit the function $G(x)$ will thus assure the desired convergence of the forward rates. The limiting value α can be fitted to the data together with the other estimation parameters.

Let $g_i(x), 0 \leq x \leq 1, i = 1, 2, \cdots, m$ be a base of a polynomial spline space. Any spline in this space can be expressed as a linear combination of the base. If $G(x)$

is fitted by a function from this space,

$$G(x) = \sum_{i=1}^{m} \beta_i g_i(x), \qquad 0 \leq x \leq 1, \tag{14}$$

the model of Equation (7) can be written as

$$P_k + A_k = \sum_{i=1}^{m} \beta_i (g_i(X_{k1}) + \sum_{j=1}^{L_k} C_k g_i(X_{kj})) - q \frac{C_k}{P_k} \left(\frac{dP}{dY}\right)_k - wI_k + \epsilon_k, \tag{15}$$

$$E\epsilon_k = 0, \qquad E\epsilon_k^2 = \sigma^2 \omega_k, \qquad E\epsilon_k \epsilon_\ell = 0 \text{ for } k \neq \ell$$

where

$$X_{kj} = 1 - e^{-\alpha(T_k - j + 1)}, \qquad j = 1, 2, \cdots, L_k.$$

The model described by Equation (15) is used in the estimation of the term structure. It is linear in the parameters $\beta_1, \beta_2, \cdots, \beta_m, q, w$, with residual covariance matrix proportional to

$$\Omega = \begin{vmatrix} \omega_1 & & & & \\ & \omega_2 & & & \\ & & \cdot & & \\ & & & \cdot & \\ & & & & \omega_n \end{vmatrix}$$

If we write

$$U_k = P_k + A_k$$

$$Z_{ki} = g_i(X_{k1}) + \sum_{j=1}^{L_k} C_k g_i(X_{kj}), \qquad i = 1, 2, \cdots, m$$

$$Z_{k,m+1} = -\frac{C_k}{P_k}\left(\frac{dP}{dY}\right)_k$$

$$Z_{k,m+2} = -I_k$$

for $k = 1, 2, \cdots, n$, then the least-squares estimate of $\beta = (\beta_1, \beta_2, \cdots, \beta_m, q, w)'$ conditional on the value of α can be directly calculated by the generalized least-squares regression equation

$$\hat{\beta} = (Z'\Omega^{-1}Z)^{-1}Z'\Omega^{-1}U$$

where $U = (U_k)$, $Z = (Z_{ki})$. The sum of squares

$$S(\alpha) = U'\Omega^{-1}U - \hat{\beta}'Z'\Omega^{-1}U$$

is then a function of α only. We can then find the value of α that minimizes $S(\alpha)$ by use of numerical procedures, such as the three-point Newton minimization method.

Once the least-squares values of the regression coefficients $\beta_1, \beta_2, \cdots, \beta_m, q, w$ and the parameter α are determined, the fitted discount function is given by

$$\hat{D}(t) = \sum_{i=1}^{m} \hat{\beta}_i g_i(1 - e^{-\hat{\alpha}t}), \qquad t \geq 0. \tag{16}$$

As for the spline space, we choose cubic splines as the lowest odd order with continuous derivatives. The boundary conditions are $G(0) = 1$, $G(1) = 0$. The base $(g_i(x))$ should be chosen to be reasonably close to orthogonal, in order that the regression matrix

$$Z'\Omega^{-1}Z$$

can be inverted with sufficient precision.

Although the model is fitted in its transformed version given by Equation (15), it may be illustrative to rewrite it in the original parameter t. In any interval between consecutive knot points, $G(x)$ is a cubic polynomial, and therefore $D(t)$ takes the form

$$D(t) = a_0 + a_1 e^{-\alpha t} + a_2 e^{-2\alpha t} + a_3 e^{-3\alpha t}$$

on each interval between knots. The function $D(t)$ and its first and second derivatives are continuous at the knot points. This family of curves, used to fit the discount function, can be described as the *third order exponential splines*.

Since least-squares methods are highly sensitive to wrong data, we use a screening procedure to identify and exclude outliers. Observations with residuals larger than four standard deviations are excluded and the model is fitted again. This procedure is repeated until no more outliers are present.

REFERENCES

Brennan, Michael J., and Eduardo S. Schwartz. "A Continuous Time Approach to the Pricing of Bonds," *Journal of Banking and Finance*, 1979, pp. 133–155.

Brennan, Michael J., and Eduardo S. Schwartz. "Saving Bonds, Retractable Bonds, and Callable Bonds," *Journal of Financial Economics*, 1977, pp. 67–88.

Carleton, Willard R., and Ian Cooper. "Estimation and Uses of the Term Structure of Interest Rates," *Journal of Finance*, September 1976, pp. 1067–1083.

Cox, John C., Jonathan E. Ingersoll, and Stephen A. Ross. "A Theory of the Term Structure of Interest Rates," Working Paper #19, Graduate School of Business, Stanford University, 1981.

Dobson, Steven W., Richard C. Sutch, and David E. Vanderford. "An Evaluation of Alternative Empirical Models of the Term Structure of Interest Rates," *Journal of Finance*, 1976, pp. 1035–1065.

Fama, Eugene F. "Forward Rates as Predictors of Future Spot Rates," *Journal of Financial Economics*, 1976, pp. 361–377.

Fisher, L., and R. Weil. "Coping with Risk of Interest Rate Fluctuations: Returns to Bondholders from Naive and Optimal Strategies," *Journal of Business*, October 1971, pp. 408–432.

Houglet, Michel X. "Estimating the Term Structure of Interest Rates for Non-Homogeneous Bonds," dissertation, Graduate School of Business, University of California, Berkeley, 1980.

Langetieg, Terence C. "A Multivariate Model of the Term Structure," *Journal of Finance*, 1980, pp. 71–97.

Langetieg, Terence C., and Stephen J. Smoot. "An Appraisal of Alternative Spline Methodologies for Estimating the Term Structure of Interest Rates," Working Paper, University of Southern California, December 1981.

McCulloch, J. Huston. "Measuring the Term Structure of Interest Rates," *Journal of Business*, January 1971, pp. 19–31.

McCulloch, J. Huston. "The Tax Adjusted Yield Curve," *Journal of Finance*, 1975a, pp. 811–830.

McCulloch, J. Huston. "An Estimate of the Liquidity Premium," *Journal of Political Economy*, 1975b, pp. 95–118.

Schaefer, Stephen M. "Tax Induced Clientele Effects in the Market for British Government Securities," *Journal of Financial Economics*, 1981.

Vasicek, Oldrich A. "An Equilibrium Characterization of the Term Structure," *Journal of Financial Economics*, 1977, pp. 177–188.

Journal of Econometrics 31 (1986) 151–178. North-Holland

LOGIT VERSUS DISCRIMINANT ANALYSIS

A Specification Test and Application to Corporate Bankruptcies*

Andrew W. LO

University of Pennsylvania, Philadelphia, PA 19104, USA

Received April 1984, final version received November 1985

Two of the most widely used statistical techniques for analyzing discrete economic phenomena are discriminant analysis (DA) and logit analysis. For purposes of parameter estimation, logit has been shown to be more robust than DA. However, under certain distributional assumptions both procedures yield consistent estimates and the DA estimator is asymptotically efficient. This suggests a natural Hausman specification test of these distributional assumptions by comparing the two estimators. In this paper, such a test is proposed and an empirical example involving corporate bankruptcies is provided. The finite-sample properties of the test statistic are also explored through some sampling experiments.

1. Introduction

Two of the most widely used statistical procedures in empirical studies of discrete economic phenomena are discriminant analysis (DA) and logit analysis. Although distinct, these two methods are closely related as McFadden (1976) has shown. In particular, if y is a discrete variable and X is a vector of 'explanatory' continuous variables, logit and DA are alternate means of characterizing the joint distribution of (y, X). DA focuses on the distribution of the X variates conditional on y and, in practice, it is almost always assumed that the distribution of $X|y$ is normal with a common covariance matrix across the y's.[1] In contrast to DA, logit analysis involves the distribution of y conditional on the X's which is assumed to be logistic.

In addition to the essential distinction between causal and conjoint models which McFadden (1976) pointed out, logit and DA are distinguished by

---

*This is a revised version of Chapter 4 of my dissertation. I am grateful to Craig MacKinlay, Whitney Newey, and two anonymous referees for many helpful comments. I would like to thank Chris Cavanagh, Jerry Hausman, Daniel McFadden, and Jim Powell for comments on an earlier version of this paper. I also thank Data Resources Inc. for use of their computing facilities and Stephanie Hogue, Gillian Speeth, and Madhavi Vinjamuri for preparing the manuscript. The National Science Foundation and the Alfred P. Sloan Foundation provided much appreciated financial support. Any errors are of course my own.

[1] Discrimination analysis may, in principle, be performed for any distribution. Since the most common distribution employed is the multivariate normal, this will be the only case considered here.

0304-4076/86/$3.50©1986, Elsevier Science Publishers B.V. (North-Holland)

another characteristic: logit is more robust than DA. More specifically, it is easily demonstrated that logit analysis is applicable for a wider class of distributions of (y, X) than is normal DA. However, Efron's (1975) study indicates that if the normality of $X|y$ does obtain, then DA is considerably more efficient than logit.[2] Indeed, since the normal DA procedure is in fact maximum likelihood estimation when $X|y$ is normally distributed, it is asymptotically efficient in this case. Nevertheless if normality does not obtain, then the normal DA estimator is inconsistent in general whereas the logit estimator maintains its consistency under a wide class of alternative distributions of (y, X). In a related study, Amemiya and Powell (1983) show that, for purposes of classification, DA does quite well even if the X's are binary (in which case $X|y$ is clearly not normally distributed). However, they conclude that this may be more likely to hold for discrete X's than for continuous X's which are conditionally non-normal. Therefore, an important consideration in choosing between logit and DA estimation is whether or not the assumption of conditional normality obtains.

In this paper, a simple Hausman-type specification test for such departures from normality based on the logit and DA estimators is proposed. Under the null hypothesis that the explanatory variables are conditionally normal, the logit and DA estimators should be numerically close. Under the alternative joint hypothesis that $X|y$ is not normal and $y|X$ is logistic, the two estimators should differ since logit is consistent and DA is not. Due to the asymptotic efficiency of the DA estimator, the usual Hausman (1978) specification test is applicable.[3]

Since it seems that the normality of $X|y$ is at issue, why not apply the more standard tests of normality such as the Shapiro-Wilks, skewness, or kurtosis tests? The primary reason is that a rejection of normality on the basis of such tests does not leave the econometrician with a clear alternate method of analysis. Although discriminant analysis may in principle be performed for distributions of $X|y$ other than the normal, this has little practical value due to the intractibility of alternative multivariate distributions. It will be shown in section 2, however, that logit analysis is appropriate for *any* distribution of $X|y$ which is a member of the exponential family. Therefore if normality is rejected, logit analysis is the natural alternative. Since a rejection of normality will usually entail the estimation of the logit model anyway, it may be more convenient to estimate the logit model first and use those estimates to test for normality rather than perform an additional test. Furthermore, the proposed Hausman test does not involve additional manipulation of the data as do the normality tests mentioned above, but requires only the parameter estimates from logit and DA.

---

[2] In this context, efficiency is in terms of ARE.

[3] In this case, efficiency is defined as minimum-variance in the class of $CUAN$ estimators.

In section 2 the relation between logit and DA is reviewed and the parameters of interest are defined. The test statistic is derived in section 3. In section 4 an empirical example involving corporate bankruptcies is presented for which the proposed test is performed. For illustrative purposes, two simple sets of sampling experiments were performed and the results are reported in sections 5 and 6. The first set considers the finite-sample properties of the test under the null hypothesis that $X|y$ is normally distributed, and the second set studies the finite-sample power of the test under the alternative hypothesis that the distribution of $X|y$ is gamma. We conclude in section 7.

2. The relation between logit and discriminant analysis

For simplicity, only the bivariate case is considered here although the results extend readily to the general multivariate case. Let y denote a discrete dichotomous random variable which takes the values 0 or 1 and let X be a $(k \times 1)$-vector of related continuous random variables. Denote by $F(y, X)$ the joint distribution function of (y, X).

Although the joint distribution F contains all available information concerning the relation between y and X, it is often more convenient to focus on the conditional distribution of $X|y$. This is the case, for example, when dealing with the standard statistical problem of classification: Given an observation x of attributes X which is generated by one of two probability models indexed by y, decide in an optimal fashion which population x belongs to. The standard DA procedure assumes that the conditional distribution of $X|y$ is multivariate normal with mean μ_y and common covariance Σ. More formally, let $F_D(X|y)$ denote the conditional distribution function of $X|y$ and let $f_D(X|y)$ be the corresponding density function. Then normal DA requires that

$$f_D(X|y) = (2\pi)^{-k/2}|\Sigma|^{-1/2}\exp\left[-\tfrac{1}{2}(X-\mu_y)'\Sigma^{-1}(X-\mu_y)\right]. \qquad (1)$$

Under these conditions, the solution of the general classification problem takes the particularly simple form based on the well-known linear discriminant function.

The DA procedure may be related to logit analysis through a simple application of Bayes' formula. Let $f_X(X)$ denote the marginal density function of X and let

$$\pi_y = \int f_X(X) P(y|X) \, dX. \qquad (2)$$

π_y is simply the marginal distribution of y or, in DA terminology, π_y is the *a priori* probability of being a member of population y. Letting $F_L(y|X)$

denote the conditional distribution function of $y|X$ and applying Bayes' formula yields

$$F_L(y|X) = \frac{f_D(X|y)\pi_y}{f_X(X)}. \tag{3}$$

Since

$$f_X(X) = \sum_y \pi_y f_D(X|y),$$

eq. (3) may be rewritten as

$$F_L(y=1|X) = \frac{f_D(X|y=1)\pi_1}{\sum_y \pi_y f_D(X|y)} = \left(1 + \frac{\pi_0}{\pi_1}\frac{f_D(X|y=0)}{f_D(X|y=1)}\right)^{-1}. \tag{4}$$

Substituting the conditional densities of (1) into eq. (4) and simplifying then yields

$$F_L(y=1|X) = (1 + \exp[-(\alpha + \beta'X)])^{-1}, \tag{5a}$$

$$\alpha = \tfrac{1}{2}(\mu_0 - \mu_1)'\Sigma^{-1}(\mu_0 + \mu_1) - \ln(\pi_0/\pi_1), \tag{5b}$$

$$\beta = \Sigma^{-1}(\mu_1 - \mu_0). \tag{5c}$$

Eq. (5) demonstrates that the required assumptions for normal DA insure that the conditional distribution of $y|X$ is logistic. Because the converse is not true, logit analysis is a more robust procedure. In fact, as Efron (1975) observes, logit analysis is appropriate under general exponential family assumptions on $F_D(X|y)$. Specifically, let

$$f_D(X|y) = g(\theta_y, \eta)h(X, \eta)\exp[\theta_y' X], \tag{6}$$

where η is an arbitrary nuisance parameter. Note that (1) is a special case of (6). The conditional density of $y|X$ under (6) is then given by

$$F_L(y=1|X) = (1 + \exp[-(\alpha + \beta'X)])^{-1}, \tag{7a}$$

$$\alpha = \ln[g(\theta_1, \eta)/g(\theta_0, \eta)] - \ln(\pi_0/\pi_1), \tag{7b}$$

$$\beta = \theta_1 - \theta_0. \tag{7c}$$

Since logit analysis is appropriate for a wider class of distributions than normal DA, a natural test of the normality assumption against other distributions of the exponential family is a comparison of the logit and DA estimator of (α, β) using (5b) and (5c). In particular, the null and alternative hypotheses may be stated explicitly as

H_0: $f_D(X|y)$ is multivariate normal with parameters μ_y, Σ.

H_1: $f_D(X|y)$ is of the exponential family with parameters θ_y, η.

Such a specification test is developed explicitly in the next section.

3. A specification test

Suppose we have T independently and identically distributed observations $[y_1, X(y_1)], \ldots, [y_T, X(y_T)]$, where $X(y_i)$ denotes the vector of attributes associated with the response variable y_i. Define

$$T_1 = \sum_{i=1}^{T} y_i \quad \text{and} \quad T_0 = T - T_1,$$

and the index sets

$$I_0 = \{i | y_i = 0\} \quad \text{and} \quad I_1 = \{i | y_i = 1\}.$$

Then under H_0 the joint log-likelihood function of y and X is given by the sum of the conditional and marginal log-likelihood functions:

$$L(y, X) = K_0 - (T/2)\ln|\Sigma|$$
$$-\tfrac{1}{2} \sum_{i \in I_0} [X(y_i) - \mu_0]'\Sigma^{-1}[X(y_i) - \mu_0]$$
$$-\tfrac{1}{2} \sum_{i \in I_1} [X(y_i) - \mu_1]'\Sigma^{-1}[X(y_i) - \mu_1]$$
$$+ T_0 \ln \pi_0 + T_1 \ln(1 - \pi_0). \tag{8}$$

In this case, the DA estimators coincide with the full information maximum likelihood estimators and are given by

$$\hat{\mu}_0 = (1/T_0) \sum_{i \in I_0} X(y_i), \qquad \hat{\mu}_1 = (1/T_1) \sum_{i \in I_1} X(y_i), \qquad (9a)$$

$$\hat{\Sigma} = (1/T) \left[\sum_{i \in I_0} [X(y_i) - \hat{\mu}_0][X(y_i) - \hat{\mu}_0]' \right.$$

$$\left. + \sum_{i \in I_1} [X(y_i) - \hat{\mu}_1][X(y_i) - \hat{\mu}_1]' \right], \qquad (9b)$$

$$\hat{\pi}_0 = T_0/T, \qquad \hat{\pi}_1 = T_1/T. \qquad (9c)$$

By the principle of invariance, the maximum likelihood estimators of (α, β) under H_0 may be obtained from substituting the estimators (9) into eq. (5b) and (5c). Denoting such estimators by $\hat{\alpha}_{DA}$ and $\hat{\beta}_{DA}$, where the subscript indicates that they were obtained from DA estimators, we have

$$\hat{\alpha}_{DA} = (\hat{\mu}_0 - \hat{\mu}_1)' \hat{\Sigma}^{-1} (\hat{\mu}_0 + \hat{\mu}_1) + \ln \hat{\pi}_1/\hat{\pi}_0, \qquad (10a)$$

$$\hat{\beta}_{DA} = \hat{\Sigma}^{-1} (\hat{\mu}_1 - \hat{\mu}_0). \qquad (10b)$$

Although $\hat{\alpha}_{DA}$ and $\hat{\beta}_{DA}$ are consistent and asymptotically efficient under H_0, they are generally inconsistent under the alternative hypothesis H_1. A Hausman-type specification test of H_0 may then be constructed by taking the difference of the DA estimator and an alternative estimator which is consistent under both hypotheses. One such estimator may be obtained by maximizing the conditional logistic log-likelihood function of y conditioned on X:

$$L(y|X) = \sum_{i=1}^{T} \left[(\alpha + X_i'\beta)(y_i - 1) - \ln\left[1 + e^{-(\alpha + X_i'\beta)}\right] \right]. \qquad (11)$$

Denote these estimators $\hat{\alpha}_L$ and $\hat{\beta}_L$. For analytic convenience, we choose to ignore the intercept term α in the remaining analysis and base our test only on β. Let $\hat{V}_{DA}$ and $\hat{V}_L$ be the estimated asymptotic covariance matrices of the β estimators from DA and logit, respectively. Following Hausman (1978), we let $\hat{q} = \hat{\beta}_L - \hat{\beta}_{DA}$, and form the χ^2 statistic:

$$J = T\hat{q}' [\hat{V}_L - \hat{V}_{DA}]^{-1} \hat{q} \overset{A}{\sim} \chi_k^2. \qquad (12)$$

A consistent estimator of the asymptotic convariance matrix for $\hat{\beta}_L$ may be obtained from the estimated Hessian of the likelihood function in the usual way. A consistent estimator of V_{DA} may be constructed by first calculating the asymptotic covariance matrix of $\hat{\beta}_{DA}$. This is done in appendix 1 and is given by the formula

$$V_{DA} = \delta \Sigma^{-1} + 2(\mu' \otimes I)(\Sigma^{-1} \otimes \Sigma^{-1})R'Q$$
$$\times (\Sigma \otimes \Sigma)Q'R(\Sigma^{-1} \otimes \Sigma^{-1})(\mu \otimes I), \qquad (13)$$

where $\mu \equiv \mu_1 - \mu_0$, $\delta \equiv 1/\pi_0\pi_1$, and R and Q are standard selection matrices as defined in Richard (1975) (see appendix 1). A consistent estimator $\hat{V}_{DA}$ is then obtained by substituting consistent estimators $(\hat{\mu}, \hat{\Sigma}, \hat{\pi}_0, \hat{\pi}_1)$, of $(\mu, \Sigma, \pi_0, \pi_1)$ into eq. (13). Note that once the DA estimators of μ and Σ are computed, the computation of $\hat{V}_{DA}$ using (13) involves only simple matrix multiplications requiring few additional steps in standard econometric software packages such as RATS or TSP. In addition, since the estimate $\hat{V}_L$ is standard output in most maximum likelihood and logit computer programs, the proposed specification test (12) is in practice quite easy to construct. Of course, the power of this test against H_1 will differ across different members of the exponential family. To illustrate the practical value of this test, the next section considers an empirical example involving an analysis of corporate bankruptcies.

4. Empirical analysis of corporate bankruptcies

In this section, we apply the proposed specification test for logit versus DA estimation of corporate bankruptcies. Previous empirical studies of business failures using DA are too numerous to cite and the reader is referred to Scott (1981) for an excellent review and critique of the empirical literature. Logit analysis however has only recently been applied to the study of default and, to this author's knowledge, Martin (1977), Ohlson (1980) and Zavgren (1980) have been the only studies.

Because there is no single canonical source of data for bankrupt firms, the procedure for compiling the sample of failed firms requires some explanation. An initial sample of 184 firms was extracted from *Standard and Poor's COMPUSTAT* Industrial Research File using the bankruptcy deletion code '02' as the extraction criterion.[4] From this sample, firms in the financial

---

[4] The Industrial Research File is composed of data for firms deleted from the Industrial Annual file because of
    Code 01 – Acquisition or merger
    Code 02 – Bankruptcy
    Code 03 – Liquidation
    Code 04 – Other (no longer files with S.E.C., etc.).

industries (SIC code 6000 to 6999) were excluded leaving 168 firms. This subset was then reduced to 77 firms by including only those firms for which some data was available between the years 1975 and 1983 inclusively. The firms in this subset were then checked for the type of bankruptcy proceedings filed using the *Wall Street Journal Index* (WSJI) and the *Directory of Obsolete Securities* (DOS) and only those firms which explicitly filed under Chapter X or XI were included, yielding the final sample of 38 bankrupt firms. The actual year of bankruptcy filing was tabulated for each firm at this point using the WSJI and DOS. The final set of financial ratios for each firm was then constructed by using data at least one year and at most three years prior to the actual year of bankruptcy.[5]

Most DA studies of bankruptcy use samples composed of pairs of bankrupt and non-bankrupt firms matched by industry and year of failure. This procedure clearly introduces much sample-selection bias (for example, the maximum likelihood estimate of the unconditional probability of bankruptcy $\hat{\pi}_1$ will always be $\frac{1}{2}$ for a matched sample) and is discussed at greater length in Martin (1977) and Zavgren (1980). This method of constructing the sample is followed so as to render the results more readily comparable to the existing literature. A matching sample of 38 non-bankrupt firms was extracted from the *COMPUSTAT* Industrial Annual File where firms were matched by year of observation, industry and, when possible, total sales. The final sample thus consists of 38 bankrupt firms and 38 solvent firms yielding 76 data points in all. Tables 1 and 2 list these firms, the year in which the data was drawn, and the year of bankruptcy. Appendix 2 provides summary statistics.

In addition to a constant term, six explanatory variables were included in the logit estimation and are defined in table 3. The first five variables were chosen because of the frequency with which they appear in other empirical studies of bankruptcy.[6] The sixth variable included was suggested by the theoretical model developed in Lo (1984) in which the key determinant of bankruptcy was whether or not the value of cash Y plus the value of intangible assets or goodwill G exceeded the current debt obligations c. The *BANK* variable is essentially the ratio of cash plus intangibles to current debt liabilities $(G + Y)/c$.[7]

---

[5] Most empirical studies of bankruptcy use data one year prior to default. The *COMPUSTAT* database did not, however, always have data for firms one year prior to failure. In such cases, the firm was included in our sample if data were available within three years of default and rejected otherwise.

[6] Linear combinations of these variables are also often included in other empirical studies. For example, Ohlson (1980) includes the ratio of working capital to total assets in his estimation, but this is simply $CATA - CLTA$. We exclude them to preserve degrees of freedom and to avoid problems of multicollinearity.

[7] Note that the correspondence of this ratio to the ratio of accounting data used is almost certainly inexact. Therefore the results of our estimation should not be interpreted as conclusively supporting or rejecting any structural model of default.

Table 1

Sample of bankrupt firms.

Firm	Industry	Year of data	Year of bankruptcy
Frigitemp	1700	1976	1978
Tobin Packing	2010	1980	1981
CS Group	2300	1980	1982
Garland	2300	1979	1980
Lynnwear	2300	1979	1981
Nelly Don	2300	1977	1978
Poloron Products	2450	1979	1981
Brody (B.) Seating	2510	1979	1981
Saxon Industries	2600	1980	1982
Supronics	2844	1975	1977
Acme-Hamilton Mfg.	3069	1977	1978
Frier Industries	3140	1975	1978
Maule Industries	3270	1975	1976
Universal Containers	3410	1976	1978
Randal Data Systems	3573	1979	1980
Advent	3651	1979	1981
GRT	3652	1977	1979
Allied Technology	3662	1979	1980
Gladding	3662	1976	1977
Hy-Gain Electronics	3662	1977	1978
Multronics	3662	1979	1980
DAIG	3693	1980	1981
Medcor	3693	1980	1981
Allied Artists Industries	3716	1977	1979
Reinell Industries	3730	1976	1979
Gruen Industries	3870	1975	1977
Mego Industries	3940	1980	1982
Miner Industries	3940	1975	1977
Auto-Train	4013	1979	1980
Cooper-Jarrett	4210	1981	1982
Nelson Resource	4210	1978	1981
Pacific Far East Line	4400	1977	1978
Shulman Transport Enterprise	4700	1977	1978
Fireco Sales	5099	1980	1982
Gilman Services	5120	1980	1982
Ormont Drug & Chemical	5120	1975	1977
Filigree Foods	5140	1975	1976
Research Fuels	5199	1978	1979

Maximum likelihood logit estimation was performed on the 76 data points with the likelihood function given in eq. (10) using the MLOGIT computer package developed by Bronwyn H. Hall.[8] The corresponding DA estimates and test statistic were computed using FORTRAN software written by the

[8]All software was implemented on a Digital VAX 11-780 in single-precision due to MLOGIT software constraints.

Table 2

Matching sample of solvent firms.

Firm	Industry
Amelco	1700
Sunstar Foods	2010
Wolf (Howard B)	2300
Movie Star	2300
Beeline	2300
Madison Industries	2300
De Rose Industries	2450
Jensen Industries	2510
Southwest Forest Industries	2600
Roffler Industries	2844
Mark IV Industries	3069
Lama (Tony)	3140
Florida Rock Industries	3270
Plan Industries	3410
Access	3570
Esquire Radio & Electron	3651
Electrosound Group	3652
Alarm Products Intl.	3662
Communications Industries	3662
AEL Industries	3662
Watkins-Johnson	3662
Staodynamics	3693
Healthdyne	3693
Executive Industries	3716
Uniflite	3730
Talley Industries	3870
Ohio Art	3940
Empire of Carolina	3940
Falls City Industries	4210
Eazor Express	4210
Arnold Industries	4210
Overseas Shipholding Group	4400
Dereco	4700
Ronco Teleproducts	5099
Napco Industries	5120
Krelitz Industries	5120
Distribuco	5140
Nolex	5199

author.[9] Due to MLOGIT constraints, the dependent variable indicating the status of solvency was defined to be 1 if bankrupt and 2 if solvent, and solvency was the normalized alternative. The logit and DA estimation results are reported in table 4 with asymptotic standard errors enclosed in parentheses. Table 4 indicates that the same three variables are significant at the 5% level or better for both logit and DA: *OLTA*, *NITA*, and *BANK*. In contrast

---

[9] Available from author upon request.

Table 3

Variable names and definitions.

Variable	Definition
SIZE	Log(total assets/GNP price deflator)[a]
CLTA	Current debt liabilities divided by total assets
OLTA	Other debt liabilities divided by total assets
CATA	Current assets divided by total assets
NITA	Net income divided by total assets
BANK	Bankruptcy index suggested by Lo (1984)

[a] The GNP price deflator was taken from the *1984 Economic Report of the President* (table B-3) and normalized to 1.00 in 1972.

Table 4

Logit and DA estimation results.[a]

Variable	Logit estimate (Std. error)	DA estimate (Std. error)
CONSTANT	1.2140 (2.5162)	—
SIZE	−0.0441 (0.2977)	−0.0418 (0.2811)
CLTA	0.1074 (2.4177)	−2.3803 (2.0231)
OLTA	−3.3258 (1.8103)	−2.6131 (1.5159)
CATA	−1.2321 (1.9827)	−0.6597 (1.7085)
NITA	10.7971 (4.8249)	5.1616 (2.2561)
BANK	4.1642 (2.2755)	1.0686 (0.6245)

[a] Convergence criterion on each parameter = 0.001000
Convergence criterion on sums of squares = 0.000100
Value of log-likelihood at convergence = −30.662786

to Ohlson's (1980) estimated *SIZE* coefficient, which is significant at the 1% level, the *SIZE* coefficient estimated here is insignificant. This may be due to the fact that larger firms are often acquired or reorganized when faced with financial distress and do not file for bankruptcy whereas smaller firms are not included in the *COMPUSTAT* database, leaving the sample of bankrupt and solvent firms with little systematic variation in $SIZE$.[10] That the ratio of current assets to total assets $CATA$ is insignificant does not contradict the model presented in Lo (1984) since in that framework the probability of default is determined by the ratio of cash *plus intangibles* to current liabilities and is unaffected by the cash to intangibles plus cash ratio. The fact that $CLTA$ is also insignificant may seem inconsistent with our theoretical framework since current liabilities directly affects the default trigger. However, because the data were collected one to three years *prior* to bankruptcy, this result may be reasonable. This is also supported by the fact that the ratio of other debt liabilities to total assets $OLTA$ is significant since 'other liabilities' includes debt obligations maturing in the actual year of bankruptcy.

Since the solvent alternative was normalized, positive coefficients imply that higher values for the associated variable correspond to larger probabilities of solvency. In terms of the significant variables, we infer that:

(i) a larger non-current debt to total assets ratio increases the probability of default;
(ii) a larger net income to total assets ratio decreases the probability of default;
(iii) a larger cash plus intangibles to current debt ratio decreases the probability of default.

Implication (i) has been discussed above. Implications (ii) and (iii) seem to support Lo's (1984) multi-period model of bankruptcy.

Of course, all interpretations of these results in terms of the theoretical model should be seen as suggestive at best since no structural stochastic specification of bankruptcy has been made. In some cases, it may be possible to derive a logistic specification from economic behavior but this has not been considered in this study.[11] In addition, the accounting variables used may only loosely correspond to their theoretical quantities if at all. There is also the timing problem mentioned above.

Given the estimated parameters from the DA and logit analysis estimation, the Hausman test statistic is easily computed to be 6.2105 which is χ^2 with 7 degrees of freedom and has a corresponding p-value of 0.515. Since H_0 cannot be rejected at any level better than 49% it seems that the data support the normality of $X|y$ against other distributions of the exponential class. This suggests that DA is the preferred method of estimation since it is asymptotically more efficient. In fact, Efron's (1975) calculations relate the asymptotic relative efficiency (ARE) of the two procedures to the square root of the Mahalanobis distance $\Delta \equiv [(\mu_1 - \mu_0)\Sigma^{-1}(\mu_1 - \mu_0)]^{1/2}$ implying that a value of

---

[10] Since the $COMPUSTAT$ database only contains data for firms which are listed on the major exchanges, smaller privately held firms are excluded.

[11] See Palepu (1983) for an example.

4 for Δ yields an ARE of 0.343. The estimated value of Δ for the data set used here is 1.5248 which, according to eq. (1.12) in Efron (1975), corresponds roughly to an ARE of 0.968. Although the Hausman test supports the appropriateness of DA, the estimated loss in efficiency of logit under H_0 is less than 4%.

5. Finite sample properties of the specification test

In this section, we present the results of several sampling experiments which explore the finite-sample properties of the specification test statistic J proposed in section 3 under the null hypothesis that $X|y$ is normally distributed. Since these simulations involve only a single X (in addition to the constant term) and consider only one of several interesting alternative hypotheses, these simulation results are meant only to be suggestive.

Under the null hypothesis H_0, $X|y$ is normal with mean u_y and variance σ^2, where y takes on the values 0 or 1. The unconditional probabilities π_0 and π_1 are assumed to be $\frac{1}{2}$ throughout the simulations. Given numerical values for μ_0, μ_1, and σ^2, a random sample of observations $(y_1, X_1, \ldots, y_T, X_T)$ of size T is generated in the following manner: y_1 is first generated as an outcome of a Bernoulli random variate with $p = \frac{1}{2}$, then X_1 is generated as an outcome of a normal variate with mean μ_y and variance σ^2, and so on for observations 2 to T. The logit and DA estimators $\hat{\beta}_L$ and $\hat{\beta}_{DA}$, their estimated asymptotic variances $\hat{V}_L$ and $\hat{V}_{DA}$, and the test statistic J are then computed for the sample. The following parameter values were assumed and held constant throughout all experiments:

$$\mu_0 = 0.20, \qquad \sigma^2 = 0.01.$$

Experiments were then performed for sample sizes of 50, 100, 200, and 300, and for various values of the parameter μ_1. More specifically, because Efron (1975) has shown that the asymptotic relative efficiency of logit versus DA is related to the square root of the Mahalanobis distance, successive values for μ_1 were chosen to vary the Mahalanobis distance from 5.0 to 9.0 in unit increments. Asymptotic relative efficiency is particularly relevant for the specification test of section 3 because as the ARE approaches unity, the matrix difference $[\hat{V}_L - \hat{V}_{DA}]$ is more likely to be non-positive definite in finite samples even though the logit estimator is asymptotically less efficient than DA. Although performing the Hausman test in such situations is problematic, Newey (1983) has developed a general method of constraining the difference of the two covariance matrices to be positive definite. However, the use of Newey's approach involves deriving both the consistent and efficient estimators as generalized method of moments estimators and, unfortunately in the case of logit and DA, this would sacrifice much of the computational simplicity of the proposed test. In these experiments, samples which yield negative variance differences were simply discarded and replications continued until 1000 samples with positive variance differences were obtained. This procedure obviously introduces serious biases into our simulations when the logit estimator is 'close' to the DA estimator in efficiency and when the sample size is small. Indeed, the fraction of the 1000 replications with a non-positive variance difference often exceeded 20% for experiments with Mahalanobis distances less than 5.0 and are not reported here due to their unreliability.[12] However, for larger values of the Mahalanobis distance and for sample sizes above 100, the simulation results are more reliable. As an indication of the seriousness of the non-positive variance difference problem in each experiment, the number of samples with negative differences as a percentage of the total number of replications performed in reported. Tables 5a–5d summarize the results of the simulations. Each row in tables 5 corresponds to a separate independent experiment. The first column indicates the sample size T and, in parentheses, the percentage of the replications which were discarded (and replaced) because of a non-positive variance difference in the J-statistic. The second and third columns indicate the theoretical values for the Mahalanobis distance and β, respectively. The fourth and fifth columns report the mean and standard errors of the DA and logit estimates of β, respectively. Column six gives the theoretical value for the asymptotic variance of the DA estimator and column seven reports the mean and standard error of the estimated asymptotic variance. Since the J-statistic is χ^2 with one degree of freedom under the null, it may be transformed into a $N(0,1)$ random variable τ. The mean of the τ-statistic across the 1000 replications is reported in column eight with asymptotic t-statistics for the hypothesis that the true mean is zero given in parentheses. In column nine, the standard error of τ over the replications is given with asymptotic t-statistics for the hypothesis that the true standard deviation is unity given in parentheses. The last three columns report estimated 1, 5, and 10 percent tail probabilities respectively. Asymptotic t-statistics for the hypotheses that the true tail probabilities are 1, 5, and 10 percent respectively are reported in parentheses.

Tables 5a–5d are largely self-explanatory. It is clear from the tables that the difference of the variances in the J-statistic is non-positive for a significant fraction the replications when the Mahalanobis distance is small. For example, table 5a reports that for sample sizes of 50 and a Mahalanobis distance of 5.0, 1148 replications were required, with close to 13 percent of the replications having negative variance differences. However, tables 5c and 5d indicate that with sample sizes of 200 or more the number of replications with negative variance differences declines considerably, and declines monotonically as the Mahalanobis distance increases.

---

[12] The complete set of simulation results are available from the author upon request.

Table 5a

Performance of specification test for sample size of 50 observations.

T ($\bar{\tau}$-bad draws)	Δ^2	β	$\bar{\beta}_{DA}$ (Std. error)	$\bar{\beta}_L$ (Std. error)	V_{DA}	$\bar{V}_{DA}$ (Std error)	$\bar{\tau}\times10$ (t-stat.)	$\widehat{SE}(\tau)$ (t-stat.)[a]	0.01 tail (t-stat.)[b]	0.05 tail (t-stat.)	0 10 tail (t-stat.)
50 (12 9)	5.0	22 36	24 62 (5.93)	28.42 (14.66)	1400.0	1717 (730 2)	0.3496 (1.11)	0.9265 (−3.29)	0.015 (1.59)	0.025 (−3.63)	0.031 (−7.27)
50 (10 8)	6.0	24.49	26.67 (6.07)	33.88 (48.34)	1600.0	1928 (786.9)	0 4004 (1.27)	0.7810 (−9.80)	0.015 (1 59)	0.029 (−3.05)	0 039 (−6.43)
50 (10 2)	7.0	26.46	28.51 (6 22)	34.57 (21.83)	1800.0	2131 (845.6)	−0.3281 (−1.04)	1.443 (19.80)	0.0170 (2 22)	0.030 (−2 90)	0.046 (−5.69)
50 (10 4)	8.0	28.28	30.18 (6.48)	38.98 (33.50)	2000 0	2331 (928.4)	−0.6544 (−2.07)	1.262 (11.71)	0.018 (2.54)	0.030 (−2 90)	0.039 (−6.43)
50 (13.7)	9.0	30.00	31.49 (6.59)	45.30 (121.4)	2200.0	2491 (982.5)	−0.9058 (−2.86)	1 068 (3.06)	0.021 (3 50)	0.034 (−2.32)	0.043 (−6.01)

[a]Asymptotic t-statistic for the hypothesis that the true standard deviation is unity, calculated as $[(\widehat{SE}-1)(2N)]^{1/2}$, where $N=1000$ is the number of replications.
[b]Asymptotic t-statistic for the hypothesis that the true proportion is 0.01, calculated as $(\hat{p}-p)/[p(1-p)/N]^{1/2}$, where $\hat{p}$ is the estimated proportion, p is the true proportion 0.01, and $N=1000$ is the number of replications The t-statistics for the 0.05 and 0 10 tails are calculated similarly.

Table 5b
Performance of specification test for sample size of 100 observations.

T (%-bad draws)	Δ^2	β	$\hat{\beta}_{DA}$ (Std. error)	$\hat{\beta}_L$ (Std. error)	V_{DA}	$\hat{V}_{DA}$ (Std. error)	$\tau \times 10$ (t-stat)	$SE(\tau)$ (t-stat.)[a]	0.01 tail (t-stat.)[b]	0.05 tail (t-stat.)	0.10 tail (t-stat.)
100 (13.2)	5.0	22.36	23.65 (4.10)	24.76 (5.58)	1400.0	1572 (463.6)	0.0860 (0.272)	1.006 (0.262)	0.028 (5.72)	0.045 (−0.73)	0.072 (−2.95)
100 (9.0)	6.0	24.49	25.85 (4.33)	27.39 (6.69)	1600.0	1793 (526.4)	−0.5453 (−1.72)	1.435 (19.47)	0.029 (6.04)	0.049 (−0.15)	0.066 (−3.58)
100 (6.4)	7.0	26.46	27.84 (4.62)	30.05 (8.85)	1800.0	2011 (597.7)	−0.9019 (−2.85)	1.961 (42.96)	0.032 (6.99)	0.044 (−0.871)	0.065 (−3.69)
100 (4.7)	8.0	28.68	29.68 (4.86)	32.65 (10.09)	2000.0	2228 (664.0)	−0.3854 (−1.22)	1.094 (4.21)	0.028 (5.72)	0.051 (0.145)	0.065 (−3.69)
100 (3.9)	9.0	30.00	31.47 (5.07)	36.00 (19.47)	2200.0	2450 (729.2)	−0.2202 (−0.696)	1.029 (1.32)	0.027 (5.40)	0.043 (−1.02)	0.066 (−3.58)

[a] Asymptotic t-statistic for the hypothesis that the true standard deviation is unity, calculated as $[(\overline{SE}-1)(2N)]^{1/2}$, where $N=1000$ is the number of replications.

[b] Asymptotic t-statistic for the hypothesis that the true proportion is 0.01, calculated as $(\hat{p}-p)/[p(1-p)/N]^{1/2}$, where $\hat{p}$ is the estimated proportion. p is the true proportion 0.01, and $N=1000$ is the number of replications. The t-statistics for the 0.05 and 0.10 tails are calculated similarly.

Table 5c
Performance of specification test for sample size of 200 observations.

T ($\bar{\tau}$-bad draws)	Δ^2	β	$\hat{\beta}_{DA}$ (Std. error)	$\hat{\beta}_L$ (Std. error)	V_{DA}	$\hat{V}_{DA}$ (Std. error)	$\tau \times 10$ (t-stat.)	$SE(\tau)$ (t-stat.)[a]	0.01 tail (t-stat.)[b]	0.05 tail (t-stat.)	0.10 tail (t-stat.)
200 (6.2)	5.0	22.36	22.96 (2.75)	23.42 (3.43)	1400.0	1479 (295.8)	−0.8180 (−2.59)	1.409 (18.30)	0.030 (6.36)	0.055 (0.726)	0.092 (−0.843)
200 (4.2)	6.0	24.49	25.11 (2.94)	25.74 (3.96)	1600.0	1687 (339.9)	−0.447 (−1.31)	1.238 (10.65)	0.032 (6.99)	0.056 (0.871)	0.085 (−1.58)
200 (2.5)	7.0	26.46	27.09 (3.11)	27.91 (4.51)	1800.0	1896 (382.9)	0.6564 (−2.08)	1.297 (13.30)	0.032 (6.99)	0.053 (0.435)	0.084 (−1.69)
200 (1.7)	8.0	28.28	28.95 (3.28)	30.04 (5.17)	2000.0	2107 (427.3)	−0.4017 (1.27)	1.286 (12.80)	0.031 (6.67)	0.055 (0.726)	0.082 (−1.90)
200 (0.9)	9.0	30.00	30.70 (3.44)	32.15 (5.96)	2200.0	2317 (470.6)	−0.5474 (−1.73)	1.556 (24.87)	0.032 (6.99)	0.055 (0.726)	0.081 (−2.00)

[a]Asymptotic t-statistic for the hypothesis that the true standard deviation is unity, calculated as $[(\overline{SE} - 1)(2N)]^{1/2}$, where $N = 1000$ is the number of replications.
[b]Asymptotic t-statistic for the hypothesis that the true proportion is 0.01, calculated as $(\hat{p} - p)/[p(1-p)/N]^{1/2}$, where $\hat{p}$ is the estimated proportion, p is the true proportion 0.01, and $N = 1000$ is the number of replications. The t-statistics for the 0.05 and 0.10 tails are calculated similarly.

Table 5d

Performance of specification test for sample size of 300 observations.

T (%-bad draws)	Δ^2	β	$\hat{\beta}_{DA}$ (Std. error)	$\hat{\beta}_L$ (Std. error)	V_{DA}	$\hat{V}_{DA}$ (Std. error)	$\tau \times 10$ (t-stat.)	$SE(\tau)$ (t-stat.)[a]	0.01 tail (t-stat.)[b]	0.05 tail (t-stat.)	0.10 tail (t-stat.)
300 (3.2)	5.0	22.36	22.73 (2.21)	23.02 (2.75)	1400.0	1450 (233.1)	−0.8941 (−2.83)	1.196 (8.75)	0.029 (6.04)	0.057 (1.02)	0.089 (−1.16)
300 (1.1)	6.0	24.49	24.89 (2.35)	25.28 (3.15)	1600.0	1657 (267.1)	−0.8254 (−2.61)	1.193 (8.65)	0.033 (7.33)	0.053 (0.435)	0.083 (−1.79)
300 (0.4)	7.0	26.46	26.87 (2.50)	27.40 (3.61)	1800.0	1862 (302.3)	−0.7345 (−2.32)	1.264 (11.82)	0.029 (6.04)	0.052 (0.290)	0.089 (−1.16)
300 (0.1)	8.0	28.28	28.71 (2.64)	29.42 (4.13)	2000.0	2069 (337.8)	−0.3095 (−0.979)	1.081 (3.62)	0.025 (4.77)	0.051 (0.145)	0.085 (−1.58)
300 (0.0)	9.0	30.00	30.46 (2.77)	31.42 (4.74)	2200.0	2277 (372.6)	−0.3154 (−0.998 × 10⁻⁴)	1.014 (0.646)	0.022 (3.81)	0.049 (0.145)	0.076 (−2.53)

[a]Asymptotic t-statistic for the hypothesis that the true standard deviation is unity, calculated as $[(\widehat{SE} - 1)(2N)]^{1/2}$, where $N = 1000$ is the number of replications.
[b]Asymptotic t-statistic for the hypothesis that the true proportion is 0.01, calculated as $(\hat{p} - p)/[p(1-p)/N]^{1/2}$, where $\hat{p}$ is the estimated proportion, p is the true proportion 0.01, and $N = 1000$ is the number of replications. The t-statistics for the 0.05 and 0.10 tails are calculated similarly.

The majority of the experiments yielded means of the τ-statistic which were insignificantly different from zero. However, for almost all the experiments the hypothesis that the true standard deviation is unity can be rejected. Nevertheless, the estimated size of a 5 percent test for sample sizes of 100 or more are not statistically different from 0.05. Although the 5 percent test seems well-behaved, tests at the 1 and 10 percent level have estimated sizes which do differ significantly from 0.01 and 0.10, respectively, the 1 percent test rejecting too frequently and the 10 percent test rejecting less often than it should.

6. Power of the specification test

In this section we investigate the power of the proposed specification test against the alternative hypothesis that the distribution of $X|y$ is gamma with parameters (η, λ_y) where the density function is given by

$$f(X|y) = \left(X^{\eta-1}/\lambda_y^{\eta}\Gamma(\eta)\right)\exp[-X/\lambda_y]. \tag{14}$$

Since the gamma distribution approaches the normal distribution as η approaches infinity, we may examine the power of the test as the alternative hypothesis becomes 'closer' to the null by simply increasing η.

Note that, given the density function in (14), the parameter in (7c) is now given by

$$\beta = (\lambda_1 - \lambda_0)/\lambda_0\lambda_1. \tag{15}$$

In these simulations, it is also assumed throughout that the unconditional probabilities π_0 and π_1 of the dichotomous population indicator y are both $\frac{1}{2}$. As before, the construction of the random samples involve first an outcome of a Bernoulli trial with $p = \frac{1}{2}$ for y, and then a draw from a gamma distribution with parameters (η, λ_y). The parameters λ_0 and λ_1 are held at 2.0 and 4.0, respectively, for all experiments. Experiments were conducted for sample sizes of 50, 100, 200, and 300, and for values of η from 1.0 to 16.0 in unit increments. Tables 6a–6d summarize the results of these simulations.

The first column of tables 6 indicates the sample size T and, in parentheses, the percentage of negative variance-difference draws. The second and third columns display the theoretical values of η and β, respectively. Columns four and five report the means and standard errors of the DA and logit estimates of β, respectively. In column six the mean and standard error of the τ-statistic estimates are reported and the last three columns display respectively the estimated power and associated standard errors of 10, 5, and 1 percent tests.

Table 6a

Power of specification test for sample size of 50 observations[a]

T (%-bad draws)	η	β	$\bar{\beta}_{DA}$ (Std. error)	$\bar{\beta}_L$ (Std. error)	$\bar{r}$ (Std. error)	Power – 10% (Std. error)	Power – 5% (Std. error)	Power – 1% (Std. error)
50 (11.7)	2.0	0.250	0.222 (0.08)	0.284 (0.12)	0.670 (0.72)	0.054 (0.007)	0.026 (0.005)	0.011 (0.003)
50 (11.4)	4.0	0.250	0.220 (0.07)	0.278 (0.10)	0.749 (0.78)	0.071 (0.008)	0.036 (0.006)	0.012 (0.003)
50 (8.3)	6.0	0.250	0.221 (0.07)	0.285 (0.10)	0.760 (0.85)	0.098 (0.009)	0.042 (0.006)	0.012 (0.003)
50 (7.1)	8.0	0.250	0.221 (0.06)	0.296 (0.16)	0.764 (0.82)	0.089 (0.009)	0.039 (0.006)	0.008 (0.003)
50 (3.6)	10.0	0.250	0.223 (0.07)	0.308 (0.15)	0.695 (0.83)	0.074 (0.008)	0.035 (0.006)	0.007 (0.003)
50 (4.2)	12.0	0.250	0.221 (0.06)	0.319 (0.29)	0.635 (0.66)	0.039 (0.006)	0.018 (0.004)	0.006 (0.002)
50 (4.7)	14.0	0.250	0.221 (0.06)	0.325 (0.20)	0.531 (0.81)	0.042 (0.006)	0.017 (0.004)	0.010 (0.003)
50 (5.3)	16.0	0.250	0.218 (0.06)	0.329 (0.23)	0.434 (1.25)	0.023 (0.005)	0.011 (0.003)	0.007 (0.003)

[a] Standard errors for power estimates p are calculated as $[p(1-p)/1000]^{1/2}$.

Table 6b

Power of specification test for sample size of 100 observations[a]

T (%-bad draws)	η	β	$\bar{\beta}_{DA}$ (Std. error)	$\bar{\beta}_L$ (Std. error)	$\bar{r}$ (Std. error)	Power – 10% (Std. error)	Power – 5% (Std. error)	Power – 1% (Std. error)
100 (6.2)	2.0	0.250	0.211 (0.06)	0.266 (0.08)	1.063 (0.81)	0.160 (0.012)	0.093 (0.009)	0.032 (0.006)
100 (4.9)	4.0	0.250	0.212 (0.05)	0.265 (0.06)	1.237 (0.85)	0.254 (0.014)	0.130 (0.011)	0.038 (0.006)
100 (4.7)	6.0	0.250	0.210 (0.04)	0.262 (0.06)	1.263 (0.91)	0.296 (0.014)	0.162 (0.012)	0.044 (0.007)
100 (3.0)	8.0	0.250	0.209 (0.04)	0.264 (0.06)	1.196 (1.08)	0.290 (0.014)	0.153 (0.011)	0.035 (0.006)
100 (2.3)	10.0	0.250	0.212 (0.04)	0.272 (0.07)	1.148 (0.89)	0.251 (0.014)	0.133 (0.011)	0.034 (0.006)
100 (1.6)	12.0	0.250	0.212 (0.04)	0.277 (0.07)	1.113 (0.85)	0.231 (0.013)	0.111 (0.010)	0.027 (0.005)
100 (0.8)	14.0	0.250	0.211 (0.04)	0.278 (0.08)	1.003 (0.85)	0.168 (0.012)	0.079 (0.009)	0.016 (0.004)
100 (0.6)	16.0	0.250	0.211 (0.04)	0.282 (0.09)	0.878 (1.69)	0.140 (0.011)	0.066 (0.008)	0.013 (0.004)

[a] Standard errors for power estimates p are calculated as $[p(1-p)/1000]^{1/2}$.

Table 6c

Power of specification test for sample size of 200 observations.[a]

T (%-bad draws)	η	β	$\bar{\beta}_{DA}$ (Std. error)	$\bar{\beta}_L$ (Std. error)	$\bar{r}$ (Std. error)	Power – 10% (Std. error)	Power – 5% (Std. error)	Power – 1% (Std. error)
200 (2.0)	2.0	0.250	0.206 (0.04)	0.256 (0.05)	1.599 (0.75)	0.422 (0.016)	0.250 (0.014)	0.087 (0.009)
200 (2.1)	4.0	0.250	0.207 (0.03)	0.259 (0.04)	1.953 (0.90)	0.651 (0.015)	0.453 (0.016)	0.172 (0.012)
200 (1.3)	6.0	0.250	0.204 (0.03)	0.256 (0.04)	1.991 (0.85)	0.667 (0.015)	0.498 (0.016)	0.211 (0.013)
200 (0.6)	8.0	0.250	0.207 (0.03)	0.260 (0.04)	1.935 (0.82)	0.648 (0.015)	0.471 (0.016)	0.193 (0.013)
200 (0.5)	10.0	0.250	0.205 (0.03)	0.257 (0.04)	1.785 (0.84)	0.587 (0.016)	0.426 (0.016)	0.151 (0.011)
200 (0.3)	12.0	0.250	0.205 (0.03)	0.258 (0.04)	1.663 (0.88)	0.524 (0.016)	0.370 (0.015)	0.127 (0.011)
200 (0.1)	14.0	0.250	0.204 (0.03)	0.261 (0.04)	1.622 (0.83)	0.507 (0.016)	0.327 (0.015)	0.101 (0.010)
200 (0.0)	16.0	0.250	0.205 (0.03)	0.262 (0.05)	1.457 (0.82)	0.430 (0.016)	0.260 (0.014)	0.066 (0.008)

[a] Standard errors for power estimates p are calculated as $[p(1-p)/1000]^{1/2}$.

Table 6d

Power of specification test for sample size of 300 observations.[a]

T (ξ-bad draws)	η	β	$\hat{\beta}_{DA}$ (Std. error)	$\hat{\beta}_L$ (Std. error)	$\bar{r}$ (Std. error)	Power − 10% (Std. error)	Power − 5% (Std. error)	Power − 1% (Std. error)
300 (0.6)	2.0	0.250	0.206 (0.03)	0.257 (0.04)	2.009 (0.75)	0.689 (0.015)	0.481 (0.016)	0.173 (0.012)
300 (0.5)	4.0	0.250	0.204 (0.03)	0.253 (0.03)	2.407 (0.85)	0.839 (0.012)	0.709 (0.014)	0.392 (0.015)
300 (0.5)	6.0	0.250	0.204 (0.03)	0.254 (0.03)	2.470 (0.95)	0.849 (0.011)	0.740 (0.014)	0.434 (0.016)
300 (0.2)	8.0	0.250	0.203 (0.02)	0.254 (0.03)	2.469 (0.92)	0.862 (0.011)	0.731 (0.014)	0.418 (0.016)
300 (0.3)	10.0	0.250	0.204 (0.02)	0.255 (0.03)	2.248 (0.80)	0.794 (0.013)	0.649 (0.015)	0.335 (0.015)
300 (0.1)	12.0	0.250	0.205 (0.02)	0.258 (0.03)	2.148 (0.79)	0.747 (0.014)	0.598 (0.016)	0.298 (0.015)
300 (0.0)	14.0	0.250	0.204 (0.02)	0.258 (0.03)	1.983 (0.81)	0.695 (0.015)	0.558 (0.016)	0.212 (0.013)
300 (0.2)	16.0	0.250	0.204 (0.02)	0.258 (0.04)	1.827 (0.82)	0.603 (0.016)	0.443 (0.016)	0.189 (0.012)

[a] Standard errors for power estimates p are calculated as $[p(1-p)/1000]^{1/2}$.

As in the null simulations for sample sizes of 50, the results for the alternative simulations in table 6a may be unreliable due to the large proportion of negative variance-difference draws. However, for sample sizes of 100, the largest fraction of bad draws is only 6.2 percent and only 0.6 percent for 300-observation samples. By and large, the tests seem to perform well for sample sizes of 100 or larger. For example, table 6d shows that the power of the 5 percent test at $\eta = 2.0$ is 0.48 and reaches a peak of 0.74 for $\eta = 6.0$. As η increases beyond 6.0, the power declines as expected since the alternative distribution is moving closer to normality. This pattern is characteristic of all the simulations; the test power increases monotonically as η increases to 6.0 and generally declines monotonically with increases in η thereafter.

The results of section 5 and 6 seem to indicate that for sample sizes larger than 100, the proposed specification test performs well and has power against a gamma alternative in the univariate case. Of course, simulations for multivariate null and alternative distributions are required before any general conclusions concerning the test's performance may be drawn. However, it does seem that for sample sizes less than 100, the negative variance-difference problem is significant and the proposed test may not be viable in such cases.

7. Conclusion

In this paper, we have presented a specification test for the conditional normality of the attributes X and hence a test for the appropriateness of applying normal discriminant analysis under the maintained hypothesis of logistic conditional response probabilities. The specification test was performed for the analysis of corporate failures and it was concluded that the null hypothesis that DA and logit are equivalent may not be rejected.

In view of the distinction between causal and conjoint probability models which McFadden (1976) points out, the above test may be particularly useful when the estimated parameters have 'structural' interpretations or if they are to be used to forecast impacts of policy changes. For example DA is often used to forecast future bankruptcies conditional on various macroeconomic scenarios. The result of the specification test for the data set used in this paper seems to support the use of DA. The standard hypothesis tests of the structural parameters may then be performed since the data do not reject normality in favor of some other member of the exponential family.

For purposes of classification, non-normality may be less problematic as Amemiya and Powell's (1983) study suggests. Their calculations indicate that the use of normal DA when the X's are in fact binary does not appreciably increase the rate of misclassification. In this case, the proposed specification test may not have much power. However, the simulations in section 6 indicate that the test does have power against a gamma alternative. Other alternative simulations should be performed on a case-by-case basis.

Appendix 1: The asymptotic distribution of $\hat{\beta}_{DA}$

Let $\mu \equiv \mu_1 - \mu_0$ and $\hat{\mu} \equiv \hat{\mu}_1 - \hat{\mu}_0$ where $\hat{\mu}_0$ and $\hat{\mu}_1$ are given by (9a). We seek the asymptotic distribution of $\sqrt{T}(\hat{\mu} - \mu)$ first. Recall that

$$\hat{\mu}_0 = (1/T_0) \sum_{i \in I_0} X_i \Rightarrow \sqrt{T_0}(\hat{\mu}_0 - \mu_0) \stackrel{A}{\sim} N(0, \Sigma), \qquad (A.1a)$$

$$\hat{\mu}_1 = (1/T_1) \sum_{i \in I_1} X_i \Rightarrow \sqrt{T_1}(\hat{\mu}_1 - \mu_1) \stackrel{A}{\sim} N(0, \Sigma), \qquad (A.1b)$$

and $\hat{\mu}_0$ and $\hat{\mu}_1$ are independent. By definition, we have

$$\hat{\mu} - \mu = (\hat{\mu}_1 - \mu_1) - (\hat{\mu}_0 - \mu_0), \qquad (A.2)$$

and thus

$$\sqrt{T}(\hat{\mu} - \mu) = \left(\sqrt{T}/\sqrt{T_1}\right)\sqrt{T_1}(\hat{\mu}_1 - \mu_1)$$

$$- \left(\sqrt{T}/\sqrt{T_0}\right)\sqrt{T_0}(\hat{\mu}_0 - \mu_0), \qquad (A.3)$$

where $T = T_0 + T_1$. Since $\sqrt{T}/\sqrt{T_i}$ converges in probability to $1/\sqrt{\pi_i}$, $i = 0, 1$, we conclude that

$$\sqrt{T}(\hat{\mu} - \mu) \stackrel{A}{\sim} N(0, (1/\pi_0 + 1/\pi_1)\Sigma), \qquad (A.4a)$$

or

$$\sqrt{T}(\hat{\mu} - \mu) \stackrel{A}{\sim} N(0, \delta\Sigma), \qquad (A.4b)$$

where $\delta = 1/\pi_0\pi_1$. Following Richard's (1975) notation, let σ denote the $\frac{1}{2}k(k+1)$-vector of distinct elements of Σ, i.e., $\sigma = (\sigma_{11}, \ldots, \sigma_{k1}, \sigma_{22}, \ldots, \sigma_{k2}, \ldots, \sigma_{kk})'$ and let R be the $\frac{1}{2}k(k+1) \times k^2$ selection matrix such that $R'\sigma = \text{vec}(\Sigma)$. Define Q' as the Moore–Penrose inverse of R, so that $RQ' = I$. By Richard (1975) we have

$$\sqrt{T}\left(\begin{bmatrix}\hat{\mu}\\ \hat{\sigma}\end{bmatrix} - \begin{bmatrix}\mu\\ \sigma\end{bmatrix}\right) \stackrel{A}{\sim} N\left(0, \begin{bmatrix}\delta\Sigma & 0\\ 0 & 2Q(\Sigma \otimes \Sigma)Q'\end{bmatrix}\right). \qquad (A.5)$$

Pre-multiplying $(\hat{\mu}'\ \hat{\sigma}')'$ by $\text{diag}(I_k, R')$, where I_k is the kth-order identity matrix, then yields

$$\sqrt{T}\left(\begin{bmatrix}\hat{\mu}\\ \text{vec}(\hat{\Sigma})\end{bmatrix} - \begin{bmatrix}\mu\\ \text{vec}(\Sigma)\end{bmatrix}\right) \stackrel{A}{\sim} N(0, V_\gamma), \qquad (A.6a)$$

where

$$V_\gamma = \begin{bmatrix} \delta\Sigma & 0 \\ 0 & 2R'Q(\Sigma \otimes \Sigma)Q'R \end{bmatrix}. \quad (A.6b)$$

Let $\gamma \equiv (\mu', \text{vec}(\Sigma)')'$ and define $\hat{\gamma}$ similarly. Define the function f as

$$f(\gamma) \equiv \Sigma^{-1}\mu = \beta, \quad (A.7)$$

and let $J_f \equiv \partial f/\partial\gamma$ denote the Jacobian matrix of f. Applying the delta method to f then yields

$$\sqrt{T}(\hat{\beta}_{DA} - \beta) \overset{A}{\sim} N(0, J_f V_\gamma J_f'). \quad (A.8)$$

To evaluate J_f explicitly, observe that

$$J_f = \begin{bmatrix} \dfrac{\partial f}{\partial \mu} & \dfrac{\partial f}{\partial \text{vec}(\Sigma)} \end{bmatrix}. \quad (A.9)$$

But $\partial f/\partial\mu = \Sigma^{-1}$ and

$$\frac{\partial f}{\partial \text{vec}(\Sigma)} = \frac{\partial \Sigma^{-1}\mu}{\partial \text{vec}(\Sigma)} = \frac{\partial \text{vec}(\Sigma^{-1}\mu)}{\partial \text{vec}(\Sigma)}, \quad (A.10a)$$

$$= \frac{\partial(\mu' \otimes I_k)\text{vec}(\Sigma^{-1})}{\partial \text{vec}(\Sigma)}, \quad (A.10b)$$

$$= (\mu' \otimes I_k)\frac{\partial \text{vec}(\Sigma^{-1})}{\partial \text{vec}(\Sigma)}, \quad (A.10c)$$

$$= -(\mu' \otimes I_k)(\Sigma^{-1} \otimes \Sigma^{-1}). \quad (A.10d)$$

Substituting (A.10d) into (A.9) results in the relation

$$J_f = [\Sigma^{-1} - (\mu' \otimes I)(\Sigma^{-1} \otimes \Sigma^{-1})]. \quad (A.11)$$

Using (A.11) and simplifying yields the desired result

$$\sqrt{T}(\hat{\beta}_{DA} - \beta) \overset{A}{\sim} N(0, V_{DA}),$$

$$V_{DA} = J_f V_\gamma J_f' = \delta\Sigma^{-1} + 2(\mu' \otimes I_k)(\Sigma^{-1} \otimes \Sigma^{-1})$$

$$\times R'Q(\Sigma \otimes \Sigma)Q'R(\Sigma^{-1} \otimes \Sigma^{-1})(\mu \otimes I_k). \quad (A.12)$$

Appendix 2: Summary statistics of data set

	Means Matrix			Standard Deviation Matrix		
Variable	Bankrupt	Solvent	Combined	Bankrupt	Solvent	Combined
CONSTANT	1.0000	1.0000	1.0000	0.0000	0.0000	0.0000
SIZE	2.4569	2.6604	2.5586	1.1995	1.3219	1.2747
CLTA	0.5706	0.3394	0.4550	0.2114	0.2070	0.2406
OLTA	0.2974	0.2117	0.2545	0.1646	0.2581	0.2221
CATA	0.6533	0.6613	0.6573	0.2054	0.2091	0.2087
NITA	−0.1548	0.0313	−0.0618	0.2108	0.0883	0.1877
BANK	0.1086	0.6734	0.3910	0.1284	0.7954	0.6401

Correlation Matrix (Combined Sample)

	SIZE	CLTA	OLTA	CATA	NITA	BANK
SIZE	1.0000					
CLTA	−0.2855	1.0000				
OLTA	0.1556	−0.0975	1.0000			
CATA	−0.4561	0.1007	−0.3601	1.0000		
NITA	0.2155	−0.6498	−0.0618	0.0524	1.0000	
BANK	−0.0525	−0.5780	−0.1099	−0.0401	0.3370	1.0000

Correlation Matrix (Bankrupt Sample)

	SIZE	CLTA	OLTA	CATA	NITA	BANK
SIZE	1.0000					
CLTA	−0.3515	1.0000				
OLTA	0.2682	−0.4729	1.0000			
CATA	−0.3397	0.0081	−0.4535	1.0000		
NITA	0.3041	−0.5370	0.0486	0.0233	1.0000	
BANK	−0.1073	−0.3976	0.3679	−0.0556	0.2139	1.0000

Correlation Matrix (Solvent Sample)

	SIZE	CLTA	OLTA	CATA	NITA	BANK
SIZE	1.0000					
CLTA	−0.2200	1.0000				
OLTA	0.1253	−0.0700	1.0000			
CATA	0.5664	0.2438	−0.3201	1.0000		
NITA	0.0511	−0.6827	0.0586	0.1246	1.0000	
BANK	−0.1193	−0.6051	−0.0701	−0.0684	0.3072	1.0000

References

Amemiya, T., 1981, Qualitative response models: A survey, Journal of Economic Literature 19, 1483–1536.

Amemiya, T. and J. Powell, 1983, A comparison of the logit model and normal discriminant analysis when the independent variables are binary, in: Samuel Karlin, Takeshi Amemiya and Leo A. Goodman, eds., Studies in econometrics, time series, and multivariate statistics (Academic Press, New York).

Directory of Obsolete Securities (Financial Information Inc., Jersey City, NJ).

Efron, B., 1975, The efficiency of logistic regression compared to normal discriminant analysis, Journal of the American Statistical Association 70, 892–898.

Hausman, J., 1978, Specification tests in econometrics, Econometrica 46, 1251–1271.

Lo, A., 1984, Essays in financial and quantitative economics, Unpublished Ph.D. dissertation (Harvard University, Cambridge, MA).

McFadden, D., 1976, A comment on discriminant analysis 'versus' logit analysis, Annals of Economic and Social Measurement 5, 511–523.

Martin, D., 1977, Early warning of bank failure: A logit regression approach, Journal of Banking and Finance 1, 249–276.

Newey, W.K., 1983, Specification testing and estimation using a generalized method of moments, Unpublished Ph.D. dissertation (M.I.T., Cambridge, MA).

Ohlson, J.A., 1980, Financial ratios and the probabilistic prediction of bankruptcy, Journal of Accounting Research 18, 109–131.

Palepu, K., 1983, The determinants of acquisition likelihood, Working paper 9-783-056, Feb. (Harvard Business School, Cambridge, MA).

Richard, J.F., 1975, A note on the information matrix of the multivariate normal distribution, Journal of Econometrics 3, 57–60.

Scott, J.H., Jr., 1981, The probability of bankruptcy: A comparison of empirical predictions and theoretical models, Journal of Banking and Finance 5, 317–344.

Wall Street Journal Index (Dow Jones and Company, New York).

Zavgren, Christine, 1980, A probabilistic model of financial distress, Unpublished Ph.D. dissertation (University of Nebraska, Lincoln, NE).

── # The Empirical Implications of the Cox, Ingersoll, Ross Theory of the Term Structure of Interest Rates

STEPHEN J. BROWN and PHILIP H. DYBVIG*

ABSTRACT

The one-factor version of the Cox, Ingersoll, and Ross model of the term structure is estimated using monthly quotes on U.S. Treasury issues trading from 1952 through 1983. Using data from a single yield curve, it is possible to estimate implied short and long term zero coupon rates and the implied variance of changes in short rates. Analysis of residuals points to a probable neglected tax effect.

THE TERM STRUCTURE of interest rates is important to economists because the relationship among the yields on default free securities that differ in their term to maturity reflects the information available to the market about the future course of events. The Expectations Hypothesis, the Liquidity Preference Hypothesis (Hicks [10]) and the Market Segmentation Hypothesis (Culbertson [6]) are theories of the term structure that predict little more than that the implied forward rate is either equal to or not equal to the expectation of future spot rates. Cox, Ingersoll and Ross [5] (CIR) model the term structure of interest rates in a competitive equilibrium context. Their model has elements in common with the earlier hypotheses of the term structure. However, the CIR model has a rich class of empirical implications, not only for the pricing of default free securities, but also for the pricing of bond options, callable bonds and other types of financial claims.

This paper examines the extent to which the model (in its simplest one-factor form) is descriptive of the prices of U.S. Treasury Bills, Bonds and Notes traded from 1952 to 1983. Section I describes the model, and the parsimonious representation of bond prices which it implies. Section II describes the data and Section III outlines the preliminary results obtained by fitting the model to observed data. The final section outlines directions for future research.

I. The Model

The simplest form of the CIR model is based on a single factor model of interest rates. For completeness, we include a simple intuitive derivation of the model.[1] The dynamics of the interest rate process are given by

---

* Both authors are from Yale University. We wish to acknowledge the helpful comments of Jon Ingersoll, Terry Marsh, Steve Ross and workshop participants at Princeton University. All errors are our own.

[1] This exposition of the CIR model is similar to one found in an early precursor to CIR, Ingersoll [12].

$$dr = \kappa(\theta - r)dt + \sigma\sqrt{r}\,dz \tag{1}$$

where $\kappa(\theta - r)$ is the instantaneous rate if drift (a mean reversion if $0 < \kappa$) and dz is a standard Gauss–Wiener process. The variance of the interest rate process is proportional to the level of interest rates.[2] Given the instantaneous interest rate r at period t, let $P[r, t, T]$ represent the price of a riskless pure discount bond maturing at period T. From Ito's Lemma, the instantaneous rate of return on the bond is

$$dP/P = (P_r dr + 1/2 P_{rr}(dr)^2 + P_t dt)/P \tag{2}$$

$$= [\kappa(\theta - r)P_r/P + P_t/P + 1/2\sigma^2 r P_{rr}/P]dt + \sigma\sqrt{r}P_r/P\,dz, \tag{3}$$

where we substitute from (1) for dr.

In perfect markets, the instantaneous expected rate of return for any asset can be written as the instantaneous risk free return, r, plus a risk premium. In the present single factor model, the returns on all assets are locally perfectly correlated to the extent they are all correlated with the short interest rate, the only source of noise in this economy. If we write the instantaneous interest rate on the zero coupon bond as

$$dP/P = \mu(r, t, T)dt + v(r, t, T)dz, \tag{4}$$

the absence of arbitrage in this economy implies that

$$\mu(r, t, T) = r + \lambda^*(r, t)v(r, t, T) \tag{5}$$

Assuming the risk premium factor, λ^*, to be of the form $\lambda\sqrt{r}/\sigma$, substituting the expression for the expected return (5) into (3) yields

$$rP + \lambda r P_r = P_r \kappa(\theta - r) + P_t + 1/2 P_{rr}\sigma^2 r \tag{6}$$

This is the fundamental equation for the price of any asset which has a value that depends solely on the instantaneous rate, r, and the time to maturity, $T - t$ (CIR, eq. 22).

With the boundary condition that

$$P(r, T, T) = 1.0 \tag{7}$$

the solution of (6) is of the form

$$P[r, t, T] = A[t, T] \cdot e^{-B[t, T]r} \tag{8}$$

where for

$$\tau = T - t, \tag{9}$$

---

[2] This model for the interest rates is discussed in CIR [4]. Marsh and Rosenfeld [14] discuss at length the empirical evidence in favor of the model, and suggest alternative models for changes in short term interest rates. CIR [5] also consider models expressed in terms of real interest rates, where inflation uncertainty is a second factor, and Brennan and Schwartz [2] consider an alternative two factor model. The argument for considering a single factor model first is its simplicity and empirical tractability. Whether a multiple factor model will represent a significant improvement is an open empirical question, particularly since we cannot identify all the parameters of interest in the single factor model given prices of the set of bonds trading at a given point in time.

Implications of the CIR Theory

$$A[t, T] \equiv \left\{ \frac{\phi_1 \exp(\phi_2 \tau)}{\phi_2[\exp(\phi_1 \tau) - 1] + \phi_1} \right\} \quad (10)$$

$$B[t, T] \equiv \frac{\exp(\phi_1 \tau) - 1}{\phi_2[\exp(\phi_1 \tau) - 1] + \phi_1} \quad (11)$$

where

$$\phi_1 \equiv \{(\kappa + \lambda)^2 + 2\sigma^2\}^{1/2} \quad (12)$$

$$\phi_2 \equiv (\kappa + \lambda + \phi_1)/2 \quad (13)$$

$$\phi_3 \equiv 2\kappa\theta/\sigma^2 \quad (14)$$

Equations (8) through (14) define the basic CIR model we study in this paper. We estimate the parameters ϕ_1, ϕ_2, ϕ_3 and r on the basis of data on the prices of U.S. Treasury issues trading at a point in time, and therefore obtain a time series of estimates of ϕ_1, ϕ_2, ϕ_3 and r.

To gain some intuition for the model, observe that discount bond prices are a function of the instantaneous interest rate, r (the only state variable), the time to maturity, τ, and the parameters ϕ_1, ϕ_2, and ϕ_3 which are in turn related to the risk premium parameter λ and interest rate process parameters σ, κ and θ.

If we look at the price of the discount bond as a function of time to maturity, $\tau = T - t$ (leaving t, r, ϕ_1, ϕ_2, and ϕ_3 fixed), we are essentially looking at a single yield curve for such bonds trading as of period t. Specifically, the yield to maturity τ, ytm, is given by

$$\text{ytm} = -\log(P)/\tau \quad (15)$$

For small time to maturity τ, $P \approx \exp[-r\tau]$ and ytm $\approx r$. For τ large, P is of the order $\exp[-(\phi_1 - \phi_2)\phi_3\tau]$ and r_L, the discount rate on such long discount bonds is asymptotically given by[3]

$$r_L = (\phi_1 - \phi_2)\phi_3 \quad (16)$$

Of the U.S. Treasury issues trading on a given date, only Treasury Bills are pure discount issues priced by (8). Other Treasury issues are coupon bonds. Such bonds can be priced by the model if we ignore tax effects and regard each as a portfolio of discount issues, one for each coupon payment and one for the terminal payment on the bond. Consider a coupon bond that entitles the holder to the vector of remaining payments, c, to be received on the vector of dates, d. The value of such a bond at period t is given by·

$$V^*(t, c, d) = \sum_{d_i > t} c_i P(r, t, d_i) \quad (17)$$

In terms of this notation, a pure discount bond such as a Treasury Bill can be represented as a bond with a single payment to be received at the time the Bill matures.

---

[3] This is the result (expressed in terms of the definitions (10) through (14) above) given by CIR as their Equation (26). See Dybvig, Ingersoll and Ross [7] for an analysis of the properties of long interest rates.

To estimate the parameters of the model, we make the further assumption that the bond price quoted in period t, $V(t, c, d)$, deviates from the model price $V^*(t, c, d)$ by a zero-mean error, $\epsilon_{t,T}$:

$$V(t, c, d) = V^*(t, c, d) + \epsilon_{t,T} \tag{18}$$

The error in (18) is assumed to be independent and identically distributed as Normal in the cross section of bonds that cover the maturities traded at that point of time.[4]

Given the model for the observations (18), it is possible to obtain maximum likelihood estimates of the parameters ϕ_1, ϕ_2, ϕ_3 and the instantaneous interest rate, r, using nonlinear least squares procedures applied to data on the prices of bonds of different maturities trading at a given point in time.[5] From these estimates, it is possible to identify the parameter σ using

$$\sigma^2 = 2[\phi_1\phi_2 - \phi_2^2] \tag{19}$$

and the implied long rate r_L, using Equation (16), although it is not possible to separately identify the parameters θ, κ and λ.[6]

The advantage of the CIR model is that it provides for a parsimonious representation of the yield curve within the context of a relatively flexible functional form. If correct, it would simplify the comparison of bonds of different maturities. However, it represents a highly nonlinear function of the parameters to be estimated, and data from the cross section of securities representing the yield curve will suffice to identify only certain of the parameters of interest. Furthermore, little is known of the statistical properties of the data by which the model is to be estimated.

II. The Data

Data on U.S. Treasury security prices was taken from the CRSP Bond files for the period from December, 1952 through December, 1983. For each month we used data for every Treasury Bill, Note and Bond trading as of that quote date, excluding from the sample callable bonds, bonds not federally taxed, flower bonds

---

[4] This stochastic specification is motivated by the necessity to use the mean of bid and ask price quotations instead of prices that represent actual trade data. These price quotations are themselves subject to measurement errors. The assumption that the errors are i.i.d. is relatively strong. We would expect that since bonds of differing maturities trade with different frequencies the variance arising from quotation errors need not be constant across maturities. If we were to assume instead that the errors arise from possible misspecification of the pricing relation, the errors would be associated with the discount bond component $P(r, t, T)$, which would imply that the variance of the errors should increase with maturity and be correlated across bonds of different maturities. In addition, it might be reasonable to suppose that such pricing errors would be proportional rather than additive. For these reasons, it is crucial to examine carefully the specification of the error process in light of the observed data.

[5] Marsh [13] (especially pp. 427–431) proposes (but does not actually implement) an entirely different approach to obtaining maximum likelihood estimates of the model using time series of ratios of discount bond prices.

[6] This observation is exactly analogous to the observation that one can identify the variance but not the mean of the process generating stock returns from observing the prices of options trading on the stock, if the Black-Scholes option pricing formula (Black and Scholes [1]) is correct.

and bonds which had limited negotiability because of prohibitions and restrictions on commercial bank ownership. These prices were either the trading prices or the mean of bid and ask price quotations where trading prices were otherwise not available, plus the accumulated interest as of that date. For each price, the corresponding time to maturity as well as (in the case of coupon bonds) the coupon payments, the number of payments remaining and the time to next payment were computed using information available on the CRSP tape. These data were ordered by quote date and then by time to maturity, providing for 373 cross sections of default-free coupon bond prices, each of which was used to estimate the parameters of the CIR model.

III. Results

Table I reports the time series means and standard deviations of maximum likelihood estimates of three identifiable parameters of the model, for the overall period and for subperiods chosen to conform with those used by CIR [4].[7] Panel A reports estimates of the underlying process variance, σ^2. Panel B reports the model estimates of the instantaneous rate of interest. This is compared to the average of the short rates of interest obtained as the average yield of Treasury Bills with up to 14 days to maturity for each bond quote date. Panel C gives the mean estimates of the implied return on a long term discount bond based on Equation (16). This is compared to the average yield to maturity of the 14 bonds with the longest term to maturity as of each quote date. These numbers are not strictly comparable, since the average yield to maturity was computed on the basis of coupon rather than discount bonds, and therefore we should expect the observed difference between the returns to be positive (negative) when the yield curve is upward (downward) sloping.

The results reported in Panel A indicate that estimates of the implied process variance differ from the variance parameters of the process estimated by CIR [4] on the basis of a time series of week by week interest on Treasury Bills with 13 weeks to maturity. However, the numbers are of a similar order of magnitude, and the implied process variance appears more stable through time than the time series estimates. The differences may simply reflect the relative lack of precision of the time series based estimates.

In Figure 1 we report the implied standard deviation of changes in short rates given as $s = \hat{\sigma}\sqrt{r}$ (breaks in the graph represent months for which the estimate of the implied variance, $\hat{\sigma}^2$, was zero or was (slightly) negative). Month by month changes in the implied standard deviation of the short rate are quite dramatic. These are consistent with the sampling error in these estimates. On the same figure we report the annualized standard deviation of month by month changes in observed short rates estimated for each year of our sample period.[8] The fact

---

[7] We should note that the asymtotic standard errors of the individual estimates are large since the likelihood functions proved relatively flat in the region of the MLE of the parameters ϕ_1, ϕ_2, and ϕ_3.

[8] Inspection of Figure 1 would seem to suggest a regime shift in the interest rate process bracketed by the period from late 1979 through late 1982. Such a regime shift has been documented by others (Huizinga and Mishkin [11]) who associate the shift with changes in Federal Reserve operating procedures in October 1979 and October 1982.

Table I
Comparison of Estimates Based on Crossection of Bonds Trading at a Point in Time with Estimates Based on Time Series of Bond Returns[a]

A. Implied process variance compared to time series estimates

	$\bar{\sigma}^2$	Standard Deviation of σ^2	Time Series Estimate of σ^{2}[b]
1952/12–1983/12	0.2121	0.3401	NA
1967/01–1976/12	0.2041	0.2140	0.316
1967/01–1968/12	0.3178	0.3358	0.158
1969/01–1970/12	0.2651	0.2608	0.162
1971/01–1972/12	0.1212	0.0760	0.400
1973/01–1974/12	0.2266	0.2518	1.712
1975/01–1976/12	0.1758	0.1578	0.524
1977/01–1978/12	0.2102	0.4717	NA
1979/01–1980/12	0.4740	0.6572	NA
1981/01–1982/12	0.5396	0.4510	NA
1983/01–1983/12	0.2054	0.1097	NA
1952/12–1966/12	0.1196	0.2323	NA
1967/01–1970/12	0.2770	0.2742	0.172
1970/01–1976/12	0.1641	0.1612	0.412
1977/01–1983/12	0.3790	0.5118	NA

B. Implied instantaneous interest rate compared to short term Treasury Bill rate[c]

	$\bar{r}$	$\overline{r - r_f}$	$\hat{\sigma}(\hat{r} - r_f)$	t value
1952/12–1983/12	0.0534	0.0045	0.0089	9.77
1967/01–1976/12	0.0579	0.0043	0.0057	8.26
1967/01–1968/12	0.0497	0.0048	0.0049	4.80
1969/01–1970/12	0.0666	0.0057	0.0052	5.37
1971/01–1972/12	0.0425	0.0030	0.0033	4.45
1973/01–1974/12	0.0764	0.0040	0.0087	2.25
1975/01–1976/12	0.0543	0.0039	0.0049	3.90
1977/01–1978/12	0.0678	0.0058	0.0050	5.68
1979/01–1980/12	0.1163	0.0182	0.0154	5.79
1981/01–1982/12	0.1252	0.0110	0.0200	2.69
1983/01–1983/12	0.0805	−0.0039	0.0042	−3.22
1952/12–1966/12	0.0272	0.0023	0.0042	7.12
1967/01–1970/12	0.0582	0.0052	0.0050	7.21
1970/01–1976/12	0.0585	0.0036	0.0057	5.79
1977/01–1983/12	0.0999	0.0094	0.0155	5.56

[a] Time series means of CIR parameters were estimated on the basis of US Treasury Issue prices from 1952/12 to 1983/12. σ^2 represents the time series mean of estimates of the interest rate variance parameter, σ^2 (excluding from the computation those months for which the implied variance was estimated to be zero or negative).

[b] These numbers were taken from Cox, Ingersoll and Ross [4], Table 1, expressed on an annualized basis.

[c] $\bar{r}$ represents the time series mean of estimates of the implied instantaneous interest rate, and $\overline{r - r_f}$ is the mean difference between these estimates and the short term rate given as the mean yield on US Treasury Bills with at most 14 days to maturity as of each quote date.

that there is a correspondence between the two sets of numbers is remarkable given that the solid lines represent standard deviations implied by the shape of the yield curve alone at a single point in time, whereas the stars give estimates

Implications of the CIR Theory

Table I—*continued*

C. Implied long term interest rate compared to yield on long term Treasury Bond issues[d]

	$\bar{r}_L$	$\bar{r}_L - r_{fL}$	$\hat{\sigma}(\hat{r}_L - r_{fL})$	t value
1952/12–1983/12	0.0666	0.0059	0.0439	2.60
1967/01–1976/12	0.0755	0.0107	0.0741	1.58
1967/01–1968/12	0.1068	0.0558	0.1538	1.78
1969/01–1970/12	0.0659	−0.0030	0.0120	−1.22
1971/01–1972/12	0.0661	0.0073	0.0084	4.26
1973/01–1974/12	0.0576	−0.0136	0.0255	−2.61
1975/01–1976/12	0.0791	0.0057	0.0054	5.17
1977/01–1978/12	0.0771	0.0006	0.0062	0.47
1979/01–1980/12	0.1003	−0.0015	0.0037	−1.99
1981/01–1982/12	0.1266	−0.0013	0.0039	−1.63
1983/01–1983/12	0.1137	0.0037	0.0020	6.41
1952/12–1966/12	0.0390	0.0056	0.0146	4.99
1967/01–1970/12	0.0864	0.0264	0.1119	1.63
1970/01–1976/12	0.0685	−0.0001	0.0171	−0.05
1977/01–1983/12	0.1031	−0.0001	0.0047	−0.20

[d] $\bar{r}_L$ represents the time series mean of estimates of the implied long term rate, and $\bar{r} - r_{fL}$ is the mean difference between these estimates and the long term rate given as the mean yield to maturity of the 14 Treasury Bonds quoted as of the same quote date. Note that these numbers are not strictly comparable, as $\bar{r}_L$ applies to *discount bonds* and r_{fL} is estimated from long term *coupon bonds*. With a rising yield curve, we would expect the difference to be positive.

based on the time series of short rates.[9] If we consider the annual average of the implied standard deviations, $\bar{s}_t$, as a predictor of the time series estimate of the standard deviation of changes in the short rate, $\hat{s}_t$, we cannot reject the hypothesis that it is an unbiased predictor:

$$\hat{s}_t = -.00707 + .87635\, \bar{s}_t \qquad R^2 = .61 \tag{20}$$
$$(.01484) \ (.12971)$$

(Standard errors in parentheses)

with an intercept not significantly different from zero and a slope coefficient not significantly different from unity. At this level, it would appear that the model is quite well specified. Furthermore, the model may predict the variance of interest rate changes as well as or better than historical time series based estimates.

As indicated in Panel B of Table I, the model appears to systematically overestimate the implied short rates of return. The degree of overestimation is significant in all but the subperiod of the data from January to December 1983. It is interesting that this apparent misspecification matches a similar misspecification reported by Fama and MacBeth [8] in their study of equilibrium in the equity securities markets. However, it would be premature to conclude that this represents compelling evidence that the CIR model is misspecified; it may merely

---

[9] Stephen Ross has pointed out to us that, of course, if one were solely interested in using the CIR model to estimate the standard deviation parameter σ implied by the term structure, a simpler and probably more efficient alternative to full maximum likelihood estimation would be to linearize Equations (9) through (14) around σ.

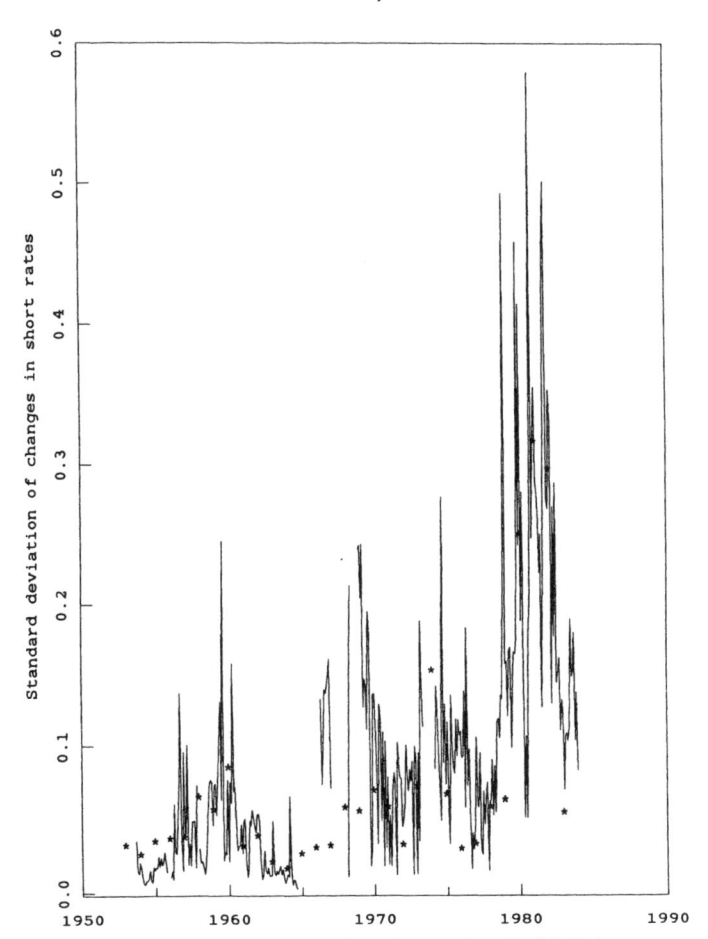

Figure 1. Implied and time series estimates of standard deviation.

indicate problems with the stochastic specification and procedures used to estimate the implied short rate.[10]

Turning to the long rates reported in Panel C of Table I, we find that while the implied long rate is greater than the yield to maturity on long term bonds for

---

[10] Computing the average term premia given as the difference between the average of forward rates implied by the monthly estimates of the model parameters for a range of terms to maturity, and the average observed short rates of return, yielded results consistent with those reported by Fama [9]. These results (available from the authors) are noteworthy in view of the fact that the CIR model is a parsimonious representation estimated on the basis of a wider class of Treasury issues than those considered by Fama.

the entire sample, the result is mixed for subperiods of the data. This result is not surprising. For the overall period, the yield curve is positively sloped ($\tilde{r} < \tilde{r}_L$) whereas in the subperiods where the implied long rate is significantly less than the yield to maturity, the yield curve is negatively sloped ($\tilde{r} > \tilde{r}_L$). In these circumstances, we would expect precisely the relationship we observe between the implied long rate and yields to maturity on long coupon bonds.

To examine in some greater depth the issue of how well specified the model is, Figure 2 reports boxplots of the standardized residuals from fitting the model using the method of maximum likelihood. These residuals are classified according to

1. whether the bond in question is a Treasury Bill or other Treasury issue,
2. whether the bond is priced to trade at a premium or at a discount (note that the latter category accounts for all Treasury Bills), and
3. whether the bond is long term or short term to maturity, where "short term" is defined as the period for which capital gains are taxed as ordinary income (six months prior to 1977 and one year subsequent to that date, except in the case of Treasury Bills the discount for which is treated as ordinary income in every year of our sample, and were thus considered "short term" issues).

To interpret these boxplots, note that 95 percent of the data falls within the extent defined by the "whiskers", while the box gives the interquartile range. The median is the dashed line within the box and the "notches" on either side of the box give the 95 percent confidence interval for the median. We see from these plots that the dispersion of the errors associated with Treasury Bills is much less pronounced than for the other Treasury issues. We see also some evidence that the distribution of errors differs across bonds trading at a premium and bonds trading at a discount in that the residuals in each case appear skewed but skewed in opposite directions. Furthermore, there is some evidence that the model appears to overpredict prices of bonds trading at a premium, to the extent that the median residual for such bonds is significantly negative.

This difference between bonds trading at a premium and bonds trading at a discount is highlighted by comparing the residuals prior to 1979 (Figure 3) with those subsequent to 1979 (Figure 4). While the general appearance of these figures is similar to Figure 2, the model appears to be seriously misspecified not only with respect to the distinction between Treasury Bills and other Treasury issues, but also between premium and discount bonds. This can be seen most clearly in the post-1979 results of Figure 4. Not only do premium and discount issues differ according to the direction of their relative skewness, but discount bond prices are very significantly *underestimated* by the model, and premium prices are as significantly *overestimated*.

Treasury Bills are traded in a more active secondary market than are other Treasury issues, and one would expect quoted prices to be much closer to trade prices for the Bills. It is not surprising that the model seems to fit better to the quoted prices for Treasury Bills than to the prices quoted for short term discount bonds which are otherwise indistinguishable from the Bills. This argument does not explain the apparent difference between discount and premium bonds. Since the two categories of bonds differ according to their United States income tax

Standardized Residuals: 521231 to 831230

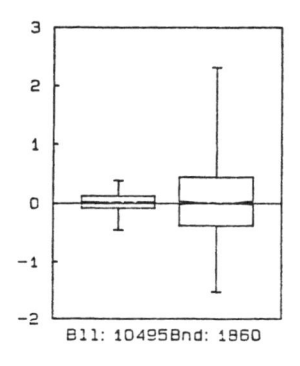

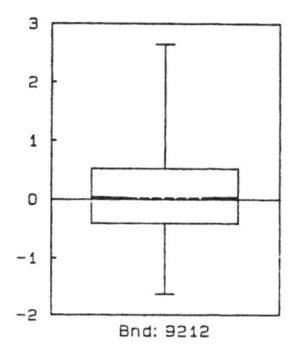

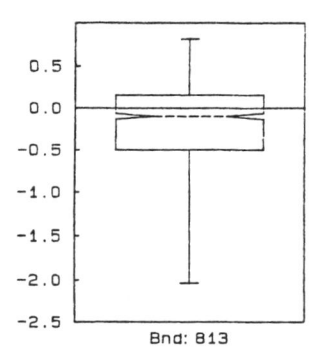

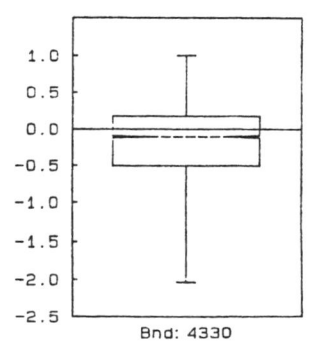

Figure 2. These boxplots (described in text) give the median, interquartile range and 95 percent confidence intervals for the residuals from the fitted model (actual price less estimated model price) standardized by the standard deviation of the residuals estimated for each month in the sample. The residuals are classified in each month by whether the bonds in question are trading at a premium or discount, and whether the remaining time to maturity would qualify any holding period gain for long term capital gains tax treatment. The numbers under each panel give the number of residuals from T Bills (Bll) and from T Bonds and Notes (Bnd) falling within each classification.

treatment, it is reasonable to explore possible tax explanations of the specification errors associated with these bonds.

The CIR model was derived without consideration of taxes, and to this point we have ignored the tax implications associated with purchase or sale of default

Standardized Residuals Prior to 1979

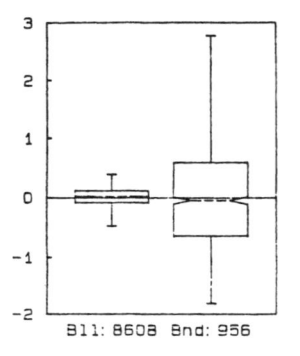

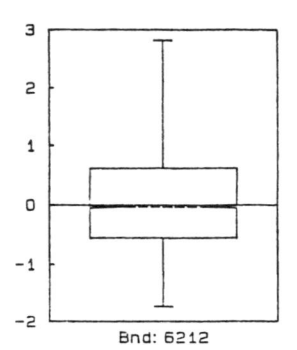

Short Term Discount Issues

Long Term Discount Issues

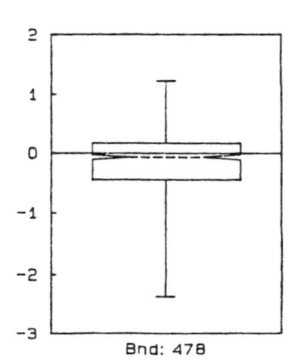

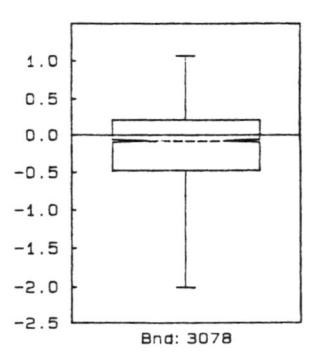

Short Term Premium Issues

Long Term Premium Issues

Figure 3. Boxplots of standardized residuals for the period 1952/12 through 1978/12.

free bonds. To see how the model might be adjusted to consider tax effects, note that the United States tax code treats premia and discounts on long term bonds held to maturity in an asymmetric fashion: the bondholder can amortize the premium over the life of the bond to reduce taxable income, whereas the discount is taxed at favorable capital gains rates when the bond is sold or redeemed. This is not the complete story, of course. There is an option effect introduced when we consider that changes in interest rates can transform a premium bond into a

Standardized Residuals After 1979

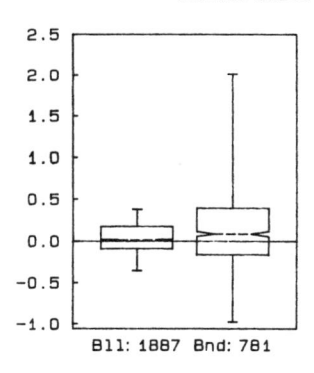

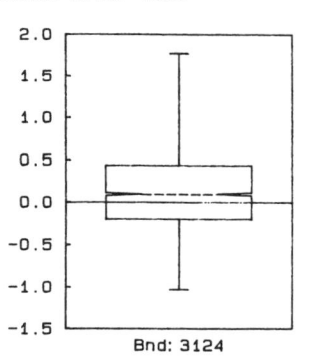

[Boxplots: Short Term Discount Issues (Bll: 1887, Bnd: 781); Long Term Discount Issues (Bnd: 3124); Short Term Premium Issues (Bnd: 308); Long Term Premium Issues (Bnd: 1281)]

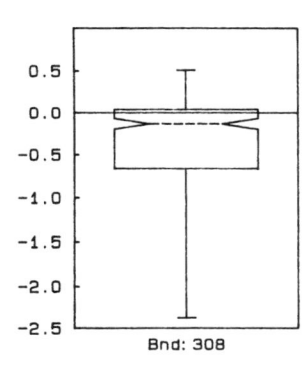

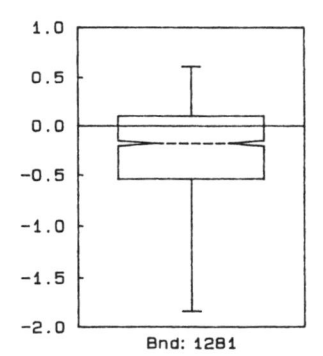

Figure 4. Boxplots of standardized residuals for the period 1979/1 through 1983/12.

discount bond.[11] However, other things equal, investors should prefer bonds trading at a discount to bonds trading at a premium.[12]

The model misspecification apparent in the comparison of premium and discount bonds presented in Figure 4 is consistent with such a tax hypothesis.

---

[11] See Constantinedes and Ingersoll [3] for an analysis of the option effect and numerical results that indicate that the effect is significant.

[12] We should note that recent changes in US tax law may serve to minimize the importance of these tax related effects.

This argument is made more compelling when we consider that the degree of model misspecification increased subsequent to 1979, a year in which there were major changes in the tax law relating the the taxable status of capital gains relative to other income. The simplest version of the tax hypothesis would predict that the degree of misspecification associated with the long term bonds should differ from that of the short term bonds. However, the degree of misspecification appears to be as severe for both classes of bonds. Clearly, more work needs to be done at both the theoretical and empirical levels to account for this phenomenon.

IV. Conclusions

The CIR model, at least in its simplest form, is readily estimable on the basis of prices of United States Treasury issues covering the maturity spectrum that are quoted at a point in time. Using data from the yield curve alone, it is possible to estimate *both* the instantaneous default free interest rate and the variance of changes in that rate. It is possible to compare such estimates implied by the prices of a cross section of bonds trading at a point of time with estimates obtained from studying the time series of short term interest rates.

While the variance of the default free return implied by the prices of different bonds trading at a point in time seems to correspond quite well to the time series variance of short interest rates, the model systematically overestimates short interest rates. Furthermore, studying the residuals from the model we find further evidence that the model is misspecified in the context of these data. The model appears to fit Treasury Bills better than it does other Treasury issues. This violates the assumption that errors in pricing are identically distributed across Treasury issues. In addition the model significantly overprices premium issues and underprices discount issues, partially consistent with a neglected tax hypothesis. Further work needs to be done to revise the specification of the model to account for these issues.

REFERENCES

1. F. Black and M. Scholes. "The pricing of options and corporate liabilities." *Journal of Political Economy* 81 (May/June, 1973), 637–659.
2. M. Brennan and E. Schwartz. "A Continuous Time Approach to the Pricing of Bonds." *Journal of Banking and Finance* 3 (July, 1979), 133–155.
3. G. Constantinedes and J. Ingersoll. "Optimal Bond Trading with Personal Taxes." *Journal of Financial Economics* 13 (September, 1984), 229–335.
4. J. Cox, J. Ingersoll and S. Ross. "Forward Rates and Expected Spot Rates: The Effect of Uncertainty." Unpublished working paper: University of Chicago, 1977.
5. ———. "A Theory of the Term Structure of Interest Rates." *Econometrica* 53 (March, 1985), 385–407.
6. J. Culbertson. "The Term Structure of Interest Rates." *Quarterly Journal of Economics* 71 (November, 1957), 485–517.
7. P. Dybvig, J. Ingersoll and S. Ross. "Do Interest Rates Converge?" Unpublished working paper: Yale University, 1985.
8. E. Fama. "Term Premiums in Bond Returns." *Journal of Financial Economics* 13 (December, 1984), 529–546.
9. ——— and J. MacBeth. "Risk, Return and Equilibrium: Empirical Tests." *Journal of Political Economy* 81 (May/June, 1973), 607–636.

10. J. Hicks. *Value and Capital.* London: Oxford University Press, 1946.
11. J. Huizinga and F. Mishkin. "Monetary Policy Regime Shifts and the Unusual Behavior of Real Interest Rates." Unpublished working paper: National Bureau of Economic Research, Cambridge, 1985.
12. J. Ingersoll. "Interest Rate Dynamics, the Term Structure, and the Valuation of Contingent Claims." Unpublished working paper: U. of Chicago, 1976.
13. T. Marsh. "Equilibrium Term Structure Models: Test Methodology." *Journal of Finance* 35 (May, 1980), 421–434.
14. T. Marsh and E. Rosenfeld. "Stochastic Processes for Interest Rates and Equilibrium Bond Prices." *Journal of Finance* 38 (May, 1983) 635–645.

DISCUSSION

WAYNE E. FERSON[*]: Brown and Dybvig [2] examine the single state variable term structure model of Cox, Ingersoll and Ross (CIR) using price data for U.S. Treasury securities of various maturities. Although the work they present is admittedly preliminary, it seems to have been thoughtfully executed and the approach is appealing for several reasons. First, given the possibility of substitution across maturities it seems reasonable that discount bonds with similar maturities should have similar yields; that is, the term structure should be "smooth." Alternative approaches to imposing smoothness in term structure estimation—such as the use of splines—do so in a relatively ad hoc way. The present approach imposes smoothness across the term structure by using a functional form for bond prices that derives from an economic model and which depends on a "small" number of parameters. A second attractive feature is that such an approach has the potential to produce estimates of parameters that may be of broader interest. As CIR and Brown and Dybvig point out, the term structure in this model embodies the information currently available to the market about the future course of events. Expected market risk premiums for example, are reflected in the term structure. Recent empirical work of Campbell [3], Keim and Stambaugh [5] and others suggests that ex ante yield curve information may have some predictive power for future rates of return on other securities. It seems to me that a confluence of these two strains might prove profitable in future research, potentially providing both guidance on functional form for predictive empirical models as well as interesting tests of pricing theory based on conditional moments.

Brown and Dybvig do not exploit an opportunity to address empirically the classical question of the term structure, namely the relation of the yield curve to expectations about future interest rates. The CIR model as estimated here does not seem to allow separate identification of term premiums; that is, the difference between forward rates and expected future spot rates. It does seem to be possible to obtain, at each date, estimates of the parameters needed to forecast the spot rate, except for the speed of adjustment coefficient. Given an estimate of these parameters (say, from a previous period time series) then ex ante forecasts of future interest rates could be formed using term structure information as of the current date. An important caveat here is that nondeterministic variation in such

[*] University of Chicago.

forecasts should by assumption depend only on the level of the spot rate. Evidence of stochastic variation in the ϕ (or σ) parameters indicates a misspecification of the model.

Another appealing feature of the approach in this paper (although readers with different tastes may disagree) is that the model can be cast explicitly in terms of nominal bonds—the bonds we observe in U.S. data. Although the CIR general equilibrium theory suggests that a single state variable model should be cast in "real" terms, it is possible to use pure arbitrage arguments and nominal stochastic process assumptions to obtain pricing results. Given a nominal, single state variable model, it is natural to wonder how this factor is related to inflation. Unfortunately, Brown and Dybvig provide no information on this issue.

To interpret the empirical results, it is useful to recall previous studies that have examined multifactor term structure models (e.g., Brennan and Schwartz [1]) and the related issues of duration and immunization using U.S. Treasury security data (e.g., Gultekin and Rogalski [4], Nelson and Schaefer [6], and others). These studies leave one with the impression that a small number of empirical "factors" is adequate to characterize the behavior of the cross-section of default-free bond price changes. Brown and Dybvig fit four parameters to the cross-section of bond prices at each date. Thus, even though the CIR model is a single state variable model, four series of empirical "factors" are generated.

Given that Brown and Dybvig essentially employ a four factor model, a certain amount of success in characterizing the term structure (e.g., the correspondence of fitted with observed slopes and average term premiums) should not be surprising. They find it remarkable that the implicit volatilities seem to predict time series estimates of short rate volatility fairly well, given that the former are derived from a cross-section of bond prices at a point in time. Persuing the analogy in footnote 6 and recalling the empirical properties of implicit volatilities derived from the option pricing model (by observing only the prices of three securities at a point in time), one would probably be discouraged if the volatilities did not predict fairly well.

There is, of course, further evidence of misspecification and much of the paper is devoted to exploring this. To avoid nitpicking, I will limit my discussion to two final comments here. One observation is that the fitted short rate typically overstates the yields of bills with less than two weeks to maturity. Possibly, the prices of the shortest bills are "too high" due to "liquidity" effects or other imperfections beyond the scope of the model. One might even argue that the shortest bills should be excluded from the sample for this reason. Is the fit of the model in the other maturity ranges adversely affected by forcing it to fit the shortest bills? A second issue is the observed systematic pricing errors. Brown and Dybvig suggest these are consistent with a "neglected tax hypothesis." I am skeptical of this explanation. I would rather think that prices of premium and discount bonds are such that a marginal investor, if one exists, does not prefer one over the other. If dealers are marginal investors in this market, then any price effects of differential taxation should be within dealer transaction cost bounds—probably narrower than bid-ask spreads. I am reminded of Schaefer's [7] earlier work in which a linear programming approach indicated that most (British) Treasury bonds seemed efficient for a zero tax bracket investor. This could not be the case if differential taxation of discount and premium issues (plausibly more severe in the U.K. than in the U.S.) produces important price effects.

REFERENCES

1. M. J. Brennan and E. Schwartz. "Conditional Predictions of Bond Prices and Returns." *Journal of Finance* 35 (May 1980), 405–416.
2. S. J. Brown and P. H. Dybvig. "The Empirical Implications of the Cox, Ingersoll, Ross Theory of the Term Structure of Interest Rates." *Journal of Finance* 41 (July 1986), 616–29.
3. J. Y. Campbell, "Stock Returns and the Term Structure," unpublished working paper, Princeton University, October, 1985.
4. N. B. Gultekin and R. J. Rogalski, "Alternative Duration Specifications and the Measurement of Basis Risk: Empirical Tests," *Journal of Business* 57 (1984), 241–264.
5. D. B. Keim and R. Stambaugh, "Predicting Returns in the Bond and Stock Markets," unpublished working paper, Universities of Pennsylvania and Chicago, December, 1985.
6. J. Nelson and S. M. Schaefer, "The Dynamics of the Term Structure and Alternative Portfolio Immunization Strategies," in *Innovations in Bond Portfolio Management: Duration Analysis and Immunization*, G. O. Bierwag, G. Kaufman and A. Toevs (eds), JAI press, 1982.
7. S. M. Schaefer, "Measuring a Tax Specific Term Structure of Interest Rates in the Market for British Government Securities," *Economic Journal* 91 (June 1981), 415–438.

[20]

The Information in Long-Maturity Forward Rates

By EUGENE F. FAMA AND ROBERT R. BLISS*

Current 1-year forward rates on 1- to 5-year U.S. Treasury bonds are information about the current term structure of 1-year expected returns on the bonds, and forward rates track variation through time in 1-year expected returns. More interesting, 1-year forward rates forecast changes in the 1-year interest rate 2- to 4-years ahead, and forecast power increases with the forecast horizon. We attribute this forecast power to a mean-reverting tendency in the 1-year interest rate.

Much of the empirical work on the term structure of interest rates is concerned with two questions. (a) Do current forward rates forecast future interest rates? (b) Do current forward rates have information about the structure of current expected returns on bonds with different maturities? Much of the empirical work on these questions uses U.S. Treasury bills and so is restricted to maturities less than a year. This paper studies the information in forward rates about future interest rates and current expected returns for annual U.S. Treasury maturities to 5 years.

Our results on expected bond returns are novel. Past tests typically fail to produce reliable inferences about the structure of expected returns for maturities beyond a year. (See, for example, Reuben Kessel, 1965; J. Huston McCulloch, 1975; and Fama, 1984b.) Using the regression approach in Fama (1984a, 1986), we are able to infer that 1-year expected returns for maturities to 5 years, measured net of the interest rate on a 1-year bond, vary through time. These expected premiums swing from positive to negative, however. On average, the term structure of 1-year expected returns on 1- to 5-year Treasury bonds is flat.

*Graduate School of Business, University of Chicago, Chicago IL 60637. The helpful comments of John Cochrane, Bradford Cornell, Wayne Ferson, Kenneth French, Merton Miller, Richard Roll, and two referees are gratefully acknowledged. This research is supported by the National Science Foundation (Fama) and the Center for Research in Security Prices (Bliss).

Differences in expected returns are usually interpreted as rewards for risk. In this view, our evidence that the ordering of expected returns across maturities changes through time implies changes in the ordering of risks. This behavior of expected returns is inconsistent with simple term structure models, like the liquidity preference hypothesis of John Hicks (1946) in which expected returns always increase with maturity. The evidence poses an interesting challenge to models like those of Robert Merton (1973), John Long (1974), Douglas Breeden (1979), and John Cox et al. (1985), that allow time-varying expected returns.

Our results on the forecast power of forward rates are also novel. Previous tests find little evidence that forward rates can forecast future interest rates. For example, Michael Hamburger and E. N. Platt (1975) and Robert Shiller et al. (1983) conclude that forward rates have no forecast power. Fama (1984a) finds some power to forecast 1-month interest rates 1 month ahead. We confirm that forward rate forecasts of near-term changes in interest rates are poor. When the forecast horizon is extended, however, forecast power increases. The 1-year forward rate calculated from the prices of 4- and 5-year bonds explains 48 percent of the variance of the change in the 1-year interest rate 4 years ahead. We argue that this forecast power is largely due to a slow mean-reverting tendency in interest rates which is more apparent over longer horizons.

The hypothesis that interest rates are mean reverting is prominent in old and new models of the term structure, for example, F. A. Lutz (1940) and Cox et al. Unlike other recent work (for example, Charles Nelson and Charles Plosser, 1982, and Fama and Michael Gibbons, 1984), our results offer supporting evidence.

I. Regression Tests: Theory

Treasury bonds with maturities longer than a year are not issued on a regular basis, and only irregularly spaced maturities are available. To estimate a term structure for regularly spaced maturities, some method of interpolation must be used. We use such a method (see the Appendix) to construct end-of-month prices for 1- to 5-year discount bonds. From the prices, we calculate forward rates, returns, and interest rates for annual maturities to 5 years.

The tests of the information in forward rates about current expected returns and future interest rates are simple regressions of future returns and changes in interest rates on current forward rates. As in most term structure work, however, even simple tests require a tedious notation.

A. Definitions of Variables

The return on an x-year discount bond bought at time t and sold at $t + x - y$, when it has y years to maturity, is defined as

(1) $h(x, y : t + x - y)$

$= \ln p(y : t + x - y) - \ln p(x : t)$,

where ln indicates a natural log, and $p(x:t)$ is the price of the bond at t. Symbols before a colon are the maturities that define a variable. The symbol after the colon is the time the variable is observed. Since most of the empirical variables are annual, time is measured in annual increments. For example, $h(5,4:t+1)$ is the 1-year return from t to $t+1$ on a 5-year bond.

The yield $r(x:t)$ on a discount bond with $1 face value and x years to maturity at t is defined as

(2) $r(x:t) = -\ln p(x:t)$.

The yield $r(1:t)$ on a 1-year bond is called the 1-year spot rate. It has a prominent role in the tests.

The time t 1-year forward rate for the year from $t + x - 1$ to $t + x$ is

(3) $f(x, x-1:t)$

$= \ln p(x-1:t) - \ln p(x:t)$

$= r(x:t) - r(x-1:t)$.

For example, $f(5,4:t)$ is the forward rate for the year from $t+4$ to $t+5$.

The time t price of an x-year discount bond that pays $1 at maturity is the present value of the $1 payoff discounted at the time t expected values (E_t) of the future 1-year returns on the bond,

(4) $p(x:t) = \exp[-E_t h(x, x-1:t+1)$

$- E_t h(x-1, x-2:t+2)$

$- \ldots - E_t r(1:t+x-1)]$.

Equation (4) is a tautology, implied by the definition of returns. It acquires testable content when we add the hypothesis that the expected returns in (4) are rational forecasts used by the market to set $p(x:t)$. Equation (4) then says that the price contains rational forecasts of equilibrium expected returns. This hypothesis about the price is the basis of the tests.

B. Forward Rates and Future Spot Rates

For example, the forward rate $f(x, x-1:t)$ can be viewed as the rate set at t on a contract to purchase a 1-year bond at $t+x-1$. Motivated by this view, the literature has long been concerned with the hypothesis that the forward rate rationally forecasts the 1-year spot rate, $r(1:t+x-1)$, to be observed at $t+x-1$. To focus on the forecast of the spot rate in $f(x, x-1:t)$, we sum the first $x-1$ expected returns in (4) and write the price as

(5) $p(x:t) = \exp[-E_t h(x,1:t+x-1)$

$- E_t r(1:t+x-1)]$.

Substituting (5) into (3) and subtracting the 1-year spot rate $r(1:t)$ gives

(6) $f(x, x-1:t) - r(1:t)$
$= [E_t r(1:t+x-1) - r(1:t)]$
$+ [E_t h(x, 1:t+x-1) - r(x-1:t)].$

We call $f(x, x-1:t) - r(1:t)$ the forward-spot spread. Our tests of the information in the forward rate $f(x, x-1:t)$ about the future spot rate $r(1:t+x-1)$ then center on the slope in the forecasting regression,

(7) $r(1:t+x-1) - r(1:t)$
$= a_1 + b_1[f(x, x-1:t) - r(1:t)]$
$+ u_1(t+x-1).$

Evidence that b_1 is greater than 0.0 implies that the forward-spot spread observed at time t has power to forecast the change in the 1-year spot rate $x-1$ years ahead.

Equation (6) holds for realized returns as well as expected values,

(8) $f(x, x-1:t) - r(1:t)$
$= [r(1:t+x-1) - r(1:t)]$
$+ [h(x, 1:t+x-1) - r(x-1:t)].$

It follows that the regression (7) is complementary to the regression

(9) $h(x, 1:t+x-1) - r(x-1:t)$
$= -a_1 + (1-b_1)[f(x, x-1:t) - r(1:t)]$
$- u_1(t+x-1).$

As indicated, the intercepts in (7) and (9) sum to 0.0, the residuals sum to 0.0 every period, and, most interesting, the slopes sum to 1.0. Thus, the slope in (7) estimates the split of variation in the forward-spot spread between the two terms of (6): (i) the forecasted change in the 1-year spot rate from t to $t+x-1$; and (ii) the premium of the $(x-1)$-year expected return on an x-year bond over the time t yield on an $(x-1)$-year bond.

Since b_1 is a constant, the estimated split of variation in the forward-spot spread does not change through time. This means that (7) can tell us that the forward-spot spread has power to forecast the change in the spot rate, but the regression fitted values only track all variation in the forecasts when the expected change in the spot rate and the expected premium in (6) always vary in fixed proportion. This is a limitation of (7) and of similar regressions outlined below.

Expression (6) for the forward-spot spread combines expected returns in (4) to focus on information in the time t price, $p(x:t)$, about the return, $h(x, 1:t+x-1)$, and the spot rate, $r(1:t+x-1)$, to be observed at the beginning of the last year in the life of the x-year bond. We turn now to a different grouping of the expected returns in (4) which focuses on the information in $p(x:t)$ about the 1-year return, $h(x, x-1:t+1)$, and the yield, $r(x-1:t+1)$, to be observed in 1 year.

C. Forward Rates and 1-Year Expected Returns

If we sum the last $x-1$ expected returns in (4), the price of an x-year bond is

(10) $p(x:t) = \exp[-E_t h(x, x-1:t+1)$
$- E_t r(x-1:t+1)].$

Substituting (10) into (3) and subtracting the spot rate $r(1:t)$ gives

(11) $f(x, x-1:t) - r(1:t)$
$= [E_t h(x, x-1:t+1) - r(1:t)]$
$+ [E_t r(x-1:t+1) - r(x-1:t)].$

Thus, when (10) is used for the price of an x-year bond, the forward-spot spread contains $E_t r(x, x-1:t+1) - r(1:t)$, the time t expected premium of the 1-year return on an x-year bond over the 1-year spot rate. But the forward-spot spread also contains the

TABLE 1—TERM PREMIUM REGRESSIONS: ESTIMATES OF (12): 1964–85
$h(x, x-1:t+1) - r(1:t) = a_2 + b_2[f(x, x-1:t) - r(1:t)] + u_2(t+1)$

Dependent	a_2	$s(a)$	b_2	$s(b)$	R^2	Residual Autos (Yearly Lag) 1	2	3	4	5
$h(2,1:t+1) - r(1:t)$	-.21	.41	.91	.28	.14	-.01	-.12	-.07	-.17	-.01
$h(3,2:t+1) - r(1:t)$	-.51	.68	1.13	.37	.11	-.18	-.12	.03	-.17	-.05
$h(4,3:t+1) - r(1:t)$	-.91	.92	1.42	.45	.11	-.23	-.10	.02	-.14	-.08
$h(5,4:t+1) - r(1:t)$	-1.06	1.31	.93	.53	.05	-.17	-.11	.03	-.17	-.10

Note: $r(1:t)$ is the 1-year spot rate observed at t; $h(x, x-1:t+1)$ is the 1-year return (t to $t+1$) on an x-year bond, and $h(x, x-1:t+1) - r(1:t)$ is the term premium in the 1-year return. $f(x, x-1:t)$ is a 1-year forward rate observed at t, and $f(x, x-1:t) - r(1:t)$ is the forward-spot spread. The regression estimates the expected value of the term premium to be observed at $t+1$, conditional on the forward-spot spread observed at t. The standard errors, $s(a)$ and $s(b)$, of the regression coefficients are adjusted for possible heteroscedasticity and for the autocorrelation induced by the overlap of monthly observations on annual returns. (See Halbert White, 1980, and Lars Peter Hansen, 1982.) The regression R^2 is adjusted for degrees of freedom. The sample size is 252, corresponding to the period January 1964 to December 1984 for the forward-spot spreads and January 1965 to December 1985 for the term premiums. If the true autocorrelations are 0.0, the standard error of the estimated residual autocorrelations is about 0.065. The data are derived from the U.S. Government Bond File of the Center for Research in Security Prices (CRSP) of the University of Chicago. See the Appendix.

expected change from t to $t+1$ in the yield on $(x-1)$-year bonds. If yields are random walks, the expected yield change in (10) is 0.0, and $f(x, x-1:t) - r(1:t)$ is $E_t h(x, x-1:t+1) - r(1:t)$. The common finding that forward rates have little power to forecast interest rates suggests a world where yields are close to random walks.

We call $h(x, x-1:t+1) - r(1:t)$ the term premium in the 1-year return on an x-year bond. Our tests of the information in time t forward rates about time t 1-year expected returns then center on the slope in the regression,

$$(12) \quad h(x, x-1:t+1) - r(1:t)$$
$$= a_2 + b_2[f(x, x-1:t) - r(1:t)]$$
$$+ u_2(t+1).$$

Evidence that b_2 is positive implies that the expected value of the term premium varies through time. Moreover, like (6), (11) holds for the realized returns, $h(x, x-1:t+1)$ and $r(x-1:t+1)$, as well as for their expected values. It follows that $1 - b_2$ in (12) is the slope in the complementary regression,

$$(13) \quad r(x-1:t+1) - r(x-1:t)$$
$$= -a_2 + (1-b_2)[f(x, x-1:t) - r(1:t)]$$
$$- u_2(t+1).$$

Evidence that b_2 differs from 1.0 means that the forward-spot spread forecasts the change in the $(x-1)$-year yield 1 year ahead.

In short, the slope in (12) splits variation in the forward-spot spread between the 1-year expected term premium and expected yield change in (11), just as the slope in (7) splits variation in the forward-spot spread between the two multiyear forecasts in (6).

II. Expected Term Premiums

A. *Time-Varying Expected Term Premiums*

Estimates of the term-premium regression (12) are in Table 1. Three slopes are more than 3.0 standard errors from 0.0, and the fourth is 1.75 standard errors from 0.0. We infer that expected term premiums in 1-year returns for maturities to 5 years vary through time, and so are typically nonzero. The results are in contrast to previous work of Kessel, McCulloch, Fama (1984b), and others that finds no convincing evidence of incremental expected returns for maturities beyond a year. The tests extend to longer maturities the conclusion of Richard Startz (1982), Shiller et al., and Fama (1984a, 1986), that expected bill returns contain time-varying maturity premiums. The evidence also confirms Shiller's (1979) claim that if bond prices are rational, the high variability of

TABLE 2—AUTOCORRELATIONS, MEANS, AND STANDARD DEVIATIONS: 1964-85

| Variable | $\bar{x}$ | $s(x)$ | \multicolumn{10}{c|}{Autocorrelations (Monthly Lag)} |
			1	2	3	4	5	6	12	24	36	48	60
Spot and Forward Rates: $r(1:t)$ and $f(x, x-1:t)$													
$r(1:t)$	7.49	2.90	.97	.93	.89	.86	.84	.81	.71	.48	.31	.15	.09
$f(2,1:t)$	7.59	2.81	.97	.94	.92	.89	.87	.86	.75	.59	.43	.24	.12
$f(3,2:t)$	7.74	2.69	.96	.94	.92	.90	.89	.87	.75	.59	.41	.23	.11
$f(4,3:t)$	7.80	2.79	.96	.94	.92	.91	.89	.87	.75	.59	.44	.24	.11
$f(5,4:t)$	7.74	2.65	.93	.90	.89	.88	.87	.85	.75	.60	.41	.26	.12
Term Premiums: $h(x, x-1:t+1) - r(1:t)$													
$h(2,1:t+1) - r(1:t)$	−.11	2.18	.90	.80	.72	.65	.59	.53	.08	−.10	−.16	−.24	−.05
$h(3,2:t+1) - r(1:t)$	−.23	3.75	.89	.79	.70	.63	.56	.49	−.01	−.09	−.11	−.22	−.09
$h(4,3:t+1) - r(1:t)$	−.46	5.13	.89	.80	.71	.64	.57	.48	−.04	−.08	−.09	−.21	−.10
$h(5,4:t+1) - r(1:t)$	−.83	6.36	.89	.80	.73	.66	.58	.49	−.04	−.07	−.08	−.22	−.13
Forward-spot spreads: $f(x, x-1:t) - r(1:t)$													
$f(2,1:t) - r(1:t)$	.10	.91	.82	.78	.71	.64	.59	.53	.34	−.12	−.42	−.22	.03
$f(3,2:t) - r(1:t)$	.26	1.13	.78	.66	.57	.51	.48	.45	.29	−.14	−.39	−.23	.01
$f(4,3:t) - r(1:t)$	.31	1.22	.74	.66	.62	.53	.51	.48	.33	−.10	−.39	−.25	−.03
$f(5,4:t) - r(1:t)$	.25	1.54	.77	.62	.53	.52	.53	.47	.32	−.14	−.34	−.23	−.09
Residuals from Spot Rate Autoregression (14)													
$e(t)$	.00	.69	.15	−.08	−.12	−.11	.08	−.08	−.07	.06	−.07	−.02	−.04

Note: $r(1:t)$ is the 1-year spot rate observed at t; $h(x, x-1:t+1)$ is the 1-year return (t to $t+1$) on an x-year bond; $f(x, x-1:t)$ is a 1-year forward rate observed at t; $e(t)$ is the residual from the first-order autoregression (14) fit to the monthly time series of $r(1:t)$. The sample size is 252, corresponding to the period January 1964 to December 1984 for the forward-spot spreads and the period January 1965 to December 1985 for the term premiums in 1-year returns. If the true autocorrelations are 0.0, the standard error of the estimated autocorrelations is about 0.065. The data are derived from the CRSP U.S. Government Bond File. See the Appendix.

yields on longer-term bonds implies time-varying expected returns.

The slopes b_2 in the term-premium regressions range from 0.91 to 1.42. All are within one standard error of 1.0. We can infer that the slopes (equal to $1 - b_2$ and with the same standard errors as b_2) in the complementary yield-change regression (13) are within one standard error of 0.0. The results suggest that when forward-spot spreads are viewed as in (11), variation in current spreads is mostly variation in the term premiums in current 1-year expected returns, and forward-spot spreads do not predict yield changes 1 year ahead. The evidence extends to longer maturities Fama's (1984a, 1986) conclusion for bills that forward rates are close to current expected returns.

B. *The Behavior of Expected Term Premiums*

Kessel, McCulloch, Fama (1984b), and others show that inferences from average returns about average expected returns on longer-maturity bonds are imprecise because of the high variability of returns. An alternative approach is to use the term-premium regressions as a license to infer average expected returns from average forward rates. The average forward rates in Table 2 show no strong tendency to increase or decrease across longer maturities. The average value of the 4- to 5-year forward rate, $f(5,4:t)$, only exceeds the average value of the 1-year spot rate, $r(1:t)$, by 0.25 percent per year.

However, the structure of forward rates varies through time, and the picture provided by average forward rates is misleading. Figure 1 plots the 5-year forward-spot spread. General patterns of variation are similar for other maturities. If forward-spot spreads are expected term premiums in 1-year returns, Figure 1 shows the general path of the variation through time of expected term premiums. The forward-spot spread is characterized by alternating runs of positive and negative values. At least after 1970, there seems to be a relation between the sign of

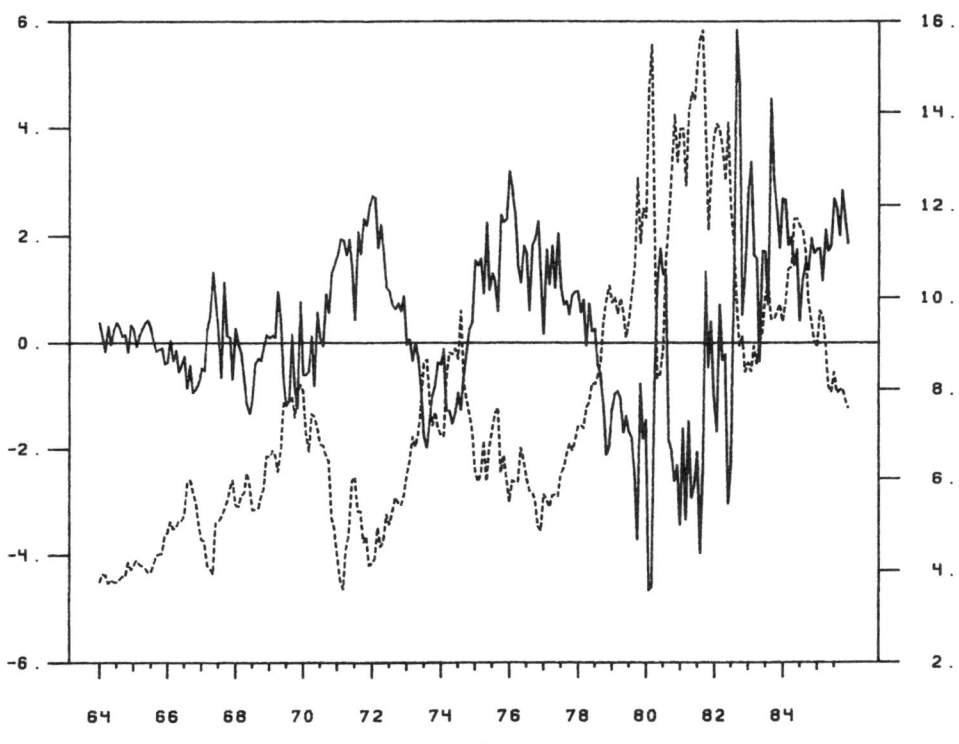

FIGURE 1. SPOT RATE AND 5-YEAR FORWARD-SPOT SPREAD

Note: 5-year forward-spot spread, $f(5,4:t) - r(1:t)$, (solid line) and the 1-year spot rate, $r(1:t)$, (dashed line). The left vertical axis is the scale (percent per year) for the spread, and the right is the scale for the spot rate. The horizontal axis is the date t.

the forward-spot spread and the business cycle. Positive forward-spot spreads in Figure 1 tend to be associated with periods of strong business activity, for example, 1970–72, 1975–78, 1983–85. Negative spreads tend to occur during the recessions of 1973–74 and 1979–82.

In short, the average forward rates in Table 2 suggest that the term structure of expected 1-year returns on 1- to 5-year Treasury bonds is on average flat. But Figure 1 shows that a flat term structure of forward rates is not typical. The path of the forward-spot spread in Figure 1 suggests that expected term premiums are typically nonzero and vary between positive and negative values. Such changes in the ordering of expected returns, and their apparent relation to the business cycle, pose an interesting challenge to term structure models that can accommodate time-varying expected returns.

C. *The Effects of Measurement Error*

A caveat about the term-premium regressions is in order. The spot rate $r(1:t)$ is obtained from a 1-year bill, but the implied prices of discount bonds used to estimate forward rates and returns for longer maturities involve interpolation that can produce measurement error. Errors in the long-maturity price $p(x:t)$ in $f(x, x-1:t)$ tend to bias the slope in the term-premium regression (12) toward 1.0, since $p(x:t)$ is the

TABLE 3—REGRESSION FORECASTS OF THE CHANGE IN THE SPOT RATE

Dependent	a	$s(a)$	b_1	$s(b_1)$	b_2	$s(b_2)$	R^2	1	2	3	4	5
$r(1:t+x-1)-r(1:t) = a + b_1[f(x,x-1;t)-r(1:t)] + u(t+x-1)$												
$r(1:t+1)-r(1:t)$	.21	.41	.09	.28			.00	−.11	−.12	−.07	−.17	−.01
$(1:t+2)-r(1:t)$	.40	.73	.69	.26			.08	.39	−.21	−.36	−.33	−.08
$r(1:t+3)-r(1:t)$	.57	.75	1.30	.10			.24	.52	.04	−.40	−46	−.25
$r(1:t+4)-r(1:t)$	1.12	.61	1.61	.34			.48	.38	.09	−.09	−.23	−.27
$r(1:t+x-1)-r(1:t) = a + b_2[[\hat{r}(1:t+x-1)-r(1:t)] + u(t+x-1)$												
$r(1:t+1)-r(1:t)$	.03	.44			.87	.40	.16	−.01	−.05	−.02	−.12	.03
$r(1:t+2)-r(1:t)$	.16	.70			.88	.20	.26	43	−.08	−.15	−.10	.04
$r(1:t+3)-r(1:t)$	.22	.84			.90	.13	.33	.57	.16	−.15	−.14	−.04
$r(1:t+4)-r(1:t)$	.37	.80			.91	.21	.36	.58	.29	.03	−.18	−.10
$r(1:t+x-1)-r(1:t) = a + b_1[f(x,x-1:t)-r(1:t)] + b_2[\hat{r}(1:t+x-1)-r(1:t)] + u(t+x-1)$												
$r(1:t+1)-r(1:t)$	.04	.43	−.14	.24	.90	.39	.16	−.03	−.04	−.00	−.09	.05
$r(1:t+2)-r(1:t)$	.15	.74	.19	.30	.82	.17	.26	.44	−.07	−.17	−.15	.01
$r(1:t+3)-r(1:t)$	.22	.84	.76	.43	.69	.22	.40	.56	.19	−.15	−.27	−.17
$r(1:t+4)-r(1:t)$	.79	.79	1.21	45	.38	.20	.51	47	.19	.03	−.14	−.23

Residual Autos (Yearly Lag) columns: 1, 2, 3, 4, 5

Note: $r(1:t)$ is the 1-year spot rate observed at t; $f(x, x-1:t)$ is a 1-year forward rate observed at t; $\hat{r}(1:t+x-1)$ is the time t forecast of $r(1:t+x-1)$ from the first-order autoregression (14) fit to the time-series of $r(1:t)$. The standard errors of the regression coefficients are adjusted for possible heteroscedasticity and for the autocorrelation induced by the overlap of monthly observations on annual or multiyear changes in the spot rate. See White and Hansen. The sample size in the regressions for the 1-year change in the spot rate, $r(1:t+1)-r(1:t)$ is 252, corresponding to the period January 1964 to December 1984 for the forward-spot spread and the period January 1965 to December 1985 for the 1-year changes in the 1-year spot rate. An additional year (12 months) of data is lost each time the forecast horizon $t + x - 1$ is extended an additional year. Thus, the regressions for the 4-year change, $r(1:t+4)-r(1:t)$, have 216 observations. Under the hypothesis that the true autocorrelations are 0.0, the standard error of the estimated residual autocorrelations is about 0.065. The data are derived from the CRSP U.S. Government Bond File. See the Appendix.

purchase price for the return $h(x, x-1:t+1)$. Errors in the short-maturity price $p(x-1:t)$ in $f(x, x-1:t)$ tend to bias the slope in (12) toward 0.0 since $-\ln p(x-1:t)$ is the yield $r(x-1:t)$ in the complementary yield change regression (13). The net effect of measurement error on the slopes in the term-premium regressions is thus difficult to predict.

It is easier to predict the effect of measurement error on the spot-rate forecasting regression (7). The future spot rate $r(1:t+x-1)$ is calculated from the price of a 1-year bill at $t + x - 1$. Measurement errors in the time t prices in $f(x, x-1:t)$ tend to bias the slope in (7) toward 0.0 and so attenuate our ability to identify forecast power in the forward rate. The forecast power we find in the tests that follow is thus in spite of any bias due to measurement error in forward rates.

III. Forecasts of 1-Year Spot Rates

A. *Estimates of the Spot-Rate Forecasting Regression*

Since slopes close to 1.0 in the estimates of the term-premium regression (12) suggest that forward rates do not forecast yields 1 year ahead, intuition suggests that they will not forecast longer-term changes in rates. Intuition is not confirmed by the slope estimates for regression (7) in Table 3. The slopes are more than 2.6 standard errors from 0.0 for all forecasts beyond a year. The forward-spot spread, $f(x, x-1:t)-r(1:t)$, forecasts the change in the 1-year spot rate, $r(1:t+x-1)-r(1:t)$, 2 to 4 years ahead. Moreover, forecast power improves with the forecast horizon: $f(3,2:t)-r(1:t)$ explains 8 percent of the variance of the change in the spot rate 2 years ahead; $f(4,3:t)-$

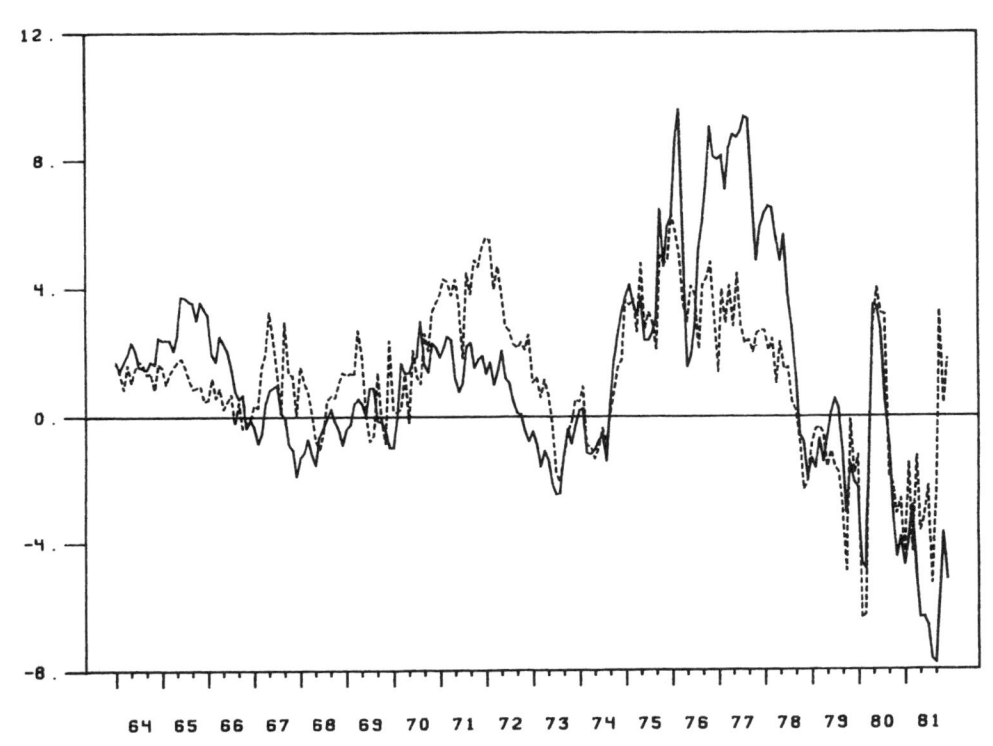

FIGURE 2. 4-YEAR CHANGE IN SPOT RATE AND FORECAST FROM REGRESSION (7)

Note: 4-year change in the spot rate, $r(1:t+4)-r(1:t)$, (solid line) and the forecasted change (dashed line) from the regression (7) of $r(1:t+4)-r(1:t)$ on the forward-spot spread $f(5,4:t)-r(1:t)$. The vertical axis is percent per year and the horizontal axis is t, the date of the forecast.

$r(1:t)$ explains 24 percent of the variance of the 3-year change; $f(5,4:t)-r(1:t)$ explains 48 percent of the variance of the 4-year change.

Figure 2 plots 4-year changes in the 1-year spot rate and the fitted values from the regression of $r(1:t+4)-r(1:t)$ on $f(5,4:t)-r(1:t)$. The figure suggests that the high R^2 (0.48) of the regression reflects consistent long-term forecast power during the sample period. We argue that long-term forecast power is largely due to slow mean reversion of the spot rate.

B. *Forecast Power and Mean Reversion*

The autocorrelations of $r(1:t)$ in Table 2 are close to 1.0 at short lags, but they decay across longer lags. The pattern suggests that month-to-month levels of the 1-year spot rate are highly autocorrelated, but the spot rate has a slow mean-reverting tendency.

The Appendix shows that if the 1-year spot rate is mean-reverting (stationary), then for long forecast horizons the expected change in the spot rate due to mean reversion explains half the variance of the actual change. If the spot rate is slowly mean reverting, the expected change explains more of the variance of the change for longer forecast horizons. Recall that the proportions of the variance of the change in the 1-year spot rate explained by forward-spot spreads in the estimates of (7) increase with the forecast horizon and reach 0.48 for 4-year changes. Thus, the forecast power of forward-spot spreads conforms to what is predicted by slow mean reversion.

It is also easy to show that if the spot rate is highly autocorrelated but slowly mean reverting, the correlation of the nonoverlapping changes, $r(1:t+T) - r(1:t)$ and $r(1:t) - r(1:t-T)$, is close to 0.0 for small values of T but approaches -0.5 for large values of T. The correlations between nonoverlapping 1-, 2-, 3-, and 4-year changes in the 1-year spot rate are -0.13, -0.29, -0.51, and -0.52.

B. Forecast Power and Mean Reversion: Direct Evidence

Although slow mean reversion of the spot rate is sufficient to explain the long-term forecast power of forward rates, the pattern of decay of the autocorrelations of the spot rate suggests that a first-order autoregression (AR1) is a reasonable model for the 1964–85 period. Fitting an AR1 to monthly observations on the 1-year spot rate yields

$$(14) \quad r(1:t) = .257 + .968 r(1:t-1) + e(t),$$
$$\quad\quad\quad\quad (.121) \quad (.015)$$

where the numbers in parentheses are standard errors. The autocorrelations of the residuals (Table 2) are generally close to 0.0. The first-order residual autocorrelation, 0.15, suggests that the model can be improved by adding a (1-month) moving-average term. Since we are interested in forecasts of spot-rate changes over long horizons, and since our goal is to document a mean-reverting tendency in a simple way, we stick with the AR1.

We use (14), fit to monthly levels of the 1-year spot rate, to forecast changes 1 to 4 years ahead. Regressions of actual on forecasted changes test for forecast power due to slow mean reversion. If the spot rate has a slow mean-reverting tendency which is well-approximated by the estimated AR1, the regressions of changes on the changes forecast by the AR1 will have forecast power that increases with the forecast horizon. Note, however, that the estimated AR1 slope, 0.968, is barely 2.0 standard errors from 1.0. If the spot rate is actually a random walk, the regressions of changes on the changes forecast by the estimated AR1 will have no power.

Regression intercepts (Table 3) that are all within 1.0 standard error of 0.0 and slopes within 1.0 standard error of 1.0 suggest that the AR1 forecasts of spot-rate changes are unbiased. As predicted by slow mean reversion, the forecasts explain more of the variance of spot-rate changes for longer forecast horizons. Explained variance rises from 16 percent for 1-year changes to 36 percent for 4-year changes. However, variance explained never reaches 50 percent, the limit predicted by mean reversion. An AR1 may not be the best model for the spot rate. But the AR1 serves its purpose. Its power to forecast spot-rate changes is consistent with a tendency toward mean reversion of the spot rate.

C. Forward Rate vs. AR1 Forecasts

Is the forecast power of forward rates solely due to mean reversion of the spot rate? Multiple regressions (Table 3) of spot-rate changes on forward-spot spreads and the AR1 forecasts are a test. In the regressions for 1- and 2-year changes, the slopes for the forward-spot spread are close to 0.0, while the slopes for the AR1 forecasts are close to 1.0 and more than 2.0 standard errors from 0.0. In the multiple regressions for 3- and 4-year changes, both the forward-spot spreads and the AR1 forecasts have marginal explanatory power (slopes almost or more than 2.0 standard errors from 0.0).

The fact that the AR1 dominates forward-spot spreads in the 1- and 2-year regressions does not necessarily mean that the market misses some of the mean reversion of the spot rate. We know from (6) that $f(x, x-1:t) - r(1:t)$ contains both the spot-rate forecast, $E_t r(1:t+x-1) - r(1:t)$, and the premium, $E_t h(x, 1:t+x-1) - r(x-1:t)$. If the expected change in the spot rate is a varying proportion of the forward-spot spread, the fixed-coefficient regression of $r(1:t+x-1) - r(1:t)$ on $f(x, x-1:t) - r(1:t)$ will not track all the information in the forward rate about the future spot rate.

There is evidence that variation in forward rates is not due entirely to forecasts of a mean-reverting spot rate. Although forward-spot spreads move opposite the spot rate, the correlations of the spot rate with 2- to 5-year forward-spot spreads (-0.25, -0.37, -0.30, and -0.42) are far from -1.0. If forward rates were driven only by mean reversion of the spot rate, our longest maturity forward rate, $f(5,4:t)$, would be much less variable than the spot rate. Table 2 shows that the standard deviation of $f(5,4:t)$ is within 10 percent of that of $r(1:t)$.

On the other hand, the fact that forward-spot spreads have marginal explanatory power relative to the AR1 in 3- and 4-year regression forecasts may reflect deficiencies of the AR1. One possibility is that reversion to a constant mean is not a complete story for the spot rate. Such mean reversion predicts that forward-spot spreads are positive (the spot rate is expected to increase) when the spot rate is low and negative when the spot rate is high. The plots of the spot rate and the 5-year forward-spot spread in Figure 1 confirm this prediction for most of the sample period. However, some of the lowest values of the spot rate occur in 1964–68. Mean reversion predicts positive expected changes in spot rates for this period, but forward-spot spreads are not systematically positive. Likewise, when the spot rate drops in 1982, it remains above its sample mean. Simple mean reversion predicts that forward-spot spreads should not become positive, but they do.

Scott Ulman and John Wood (1983) suggest that the mean of the spot rate rises with the end of the gold standard in 1971 because a fiduciary currency means higher average inflation rates. A rise in the mean of the spot rate is consistent with the behavior of the spot rate and the forward-spot spread in Figure 1. John Campbell and Shiller (1984) suggest a time-series model for the spot rate that includes mean reversion, but toward a mean that can change through time.

The important point for our purposes is that the Table 3 regressions are sufficient to conclude that much of the forecast power of forward rates is due to a mean-reverting tendency of the spot rate. But we cannot infer on the basis of our simple tests that the forecast power of forward rates is better or worse than expected on the basis of the mean reversion of the spot rate.

Finally, the autocorrelations of forward rates (Table 2) are close to those of the spot rate. The same slow decay is observed in autocorrelations of 1- to 5-year yields (not shown). A slow mean-reverting tendency is apparently a general property of interest rates—an appealing conclusion with strong roots in term structure theory. However, because the autocorrelations of interest rates are close to 1.0 for short lags, recent empirical studies (for example, Nelson-Plosser and Fama-Gibbons) tend to conclude that interest rates are approximately random walks. Our evidence that long-term changes in rates are predictable suggests that the slow decay of the autocorrelations should instead be emphasized.

V. Conclusions

A. *Forward Rates and the Term Structure of Expected Returns*

The estimates of the term-premium regression (12) allow us to infer that 1-year expected returns for U.S. Treasury maturities to 5 years, measured net of the interest rate on a 1-year bond, vary through time. Moreover, at least during the 1964–85 period, this variation of expected term premiums seems to be related to the business cycle. Expected term premiums are mostly positive during good times but mostly negative during recessions.

Differences in expected returns are usually interpreted as rewards for risk. Our evidence, like that for shorter maturities in Fama (1986), suggests that the ordering of risks and rewards changes with the business cycle. This behavior of expected returns is inconsistent with simple term structure models, like the liquidity preference hypothesis of Hicks and Kessel in which expected returns always increase with maturity. Perhaps it can be explained by models, like those of Merton, Long, Breeden, and Cox et al., that allow time-varying expected returns. The challenge is apparent.

B. Forward Rates and Future Spot Rates

Like earlier work, we find little evidence that forward rates can forecast near-term changes in interest rates. When the forecast horizon is extended, however, forecast power improves, and 1-year forward rates forecast changes in the 1-year spot rate 2 to 4 years ahead. We conclude that this forecast power reflects a slow mean-reverting tendency of interest rates.

Like any interest rate, the 1-year spot rate on Treasury bonds can be split into an expected real return and an expected inflation rate. The mean reversion of the spot rate likely implies that both of its components are mean reverting, a hypothesis with strong economic appeal. However, theory does not suggest that the processes that generate expected inflation and expected real returns should be the same, or that the processes should not change through time. We interpret the mean reversion of interest rates documented here as a tendency with details to be documented by future work.

APPENDIX

A. Data

The U.S. Government Bond File of the Center for Research in Security Prices (CRSP) has end-of-month data for all U.S. Treasury securities. We use the data to estimate end-of-month term structures for taxable, noncallable bonds for annual maturities to 5 years. The approach is outlined below. Our data are available to subscribers to the CRSP bond file.

Each month a term structure of 1-day continuously compounded forward rates is first calculated from available maturities. Bills are used for maturities to a year. To extend beyond a year, the pricing assumption is that the daily forward rate for the interval between successive maturities is the relevant discount rate for each day in the interval. Suppose daily forward rates for month t are calculated for maturities to T and the next bond matures at $T + k$. Coupons on the bond to be received prior to T are priced with the daily forward rates from t to each payment date. Coupons and the principal to be received after T are priced with the daily forward rates from t to T and with the (solved for) daily forward rate for T to $T + k$ that equates the price of the bond at t to the value of all payments. These calculations generate a step-function term structure in which 1-day forward rates are the same between successive maturities.

Prior to the recent large deficits, the number of noncallable, fully taxable bonds available each month is small. From 1964 onward there is at least one bond in each 1-year maturity interval to 5 years, but often there are few bonds beyond 5 years. We sum the daily forward rates to generate end-of-month term structures of yields for annual maturities to 5 years. The yields are used to calculate implied prices of 1- to 5-year discount bonds, from which we calculate the term-structure variables used in the tests.

B. Forecast Power and Mean Reversion

We use a first-order autoregression (AR1) to illustrate that predictable changes in the spot rate due to slow mean reversion are more apparent over longer horizons. Suppose $z(t)$ is an AR1 with parameter ϕ and mean μ,

$$(A1) \quad z(t) = \delta + \phi z(t-1) + \in(t),$$

$$\mu = \delta/(1-\phi).$$

The time t expected value of $z(t+T)$ is (see Nelson, 1973, p. 148)

$$(A2) \quad E_t z(t+T) = \mu + \phi^T [z(t) - \mu],$$

and the expected change in $z(t)$ from t to $t + T$ is

$$(A3) \quad E_t z(t+T) - z(t)$$
$$= [z(t) - \mu](\phi^T - 1).$$

If ϕ is close to 1.0, the expected change in $z(t)$ is small for small values of T. The expected change in $z(t)$ increases with the forecast horizon and approaches $\mu - z(t)$: $z(t)$ is expected to revert to its mean μ.

The variance of the expected change in $z(t)$ is

(A4) $\quad \sigma^2[E_t z(t+T) - z(t)]$

$$= \sigma^2(z)(\phi^T - 1)^2,$$

which grows with T and approaches $\sigma^2(z)$. How much of the variance of the T-period change is explained by the expected change? Since $\text{cov}[z(t+T), z(t)]$ approaches 0.0 for large values of T, the variance of $z(t+T) - z(t)$ approaches $2\sigma^2(z)$. Since $\sigma^2[E_t z(t+T) - z(t)]$ approaches $\sigma^2(z)$, for long forecast horizons the expected change in $z(t)$ due to the mean reversion of $z(t)$ explains half the variance of $z(t+T) - z(t)$. This result is a general property of stationary processes; it is not special to an AR1.

For an AR1 with ϕ close to 1.0, the ratio of the variance of the expected change to the variance of the change approaches 0.5 from below. For example, if $\phi = 0.95$, $\sigma^2[E_t z(t+1) - z(t)]$ is $0.025 \sigma^2[z(t+1) - z(t)]$. Thus, the expected change explains more of the variance of the change for longer horizons. This is a rather general implication of slow mean reversion.

REFERENCES

Breeden, Douglas T., "An Intertemporal Asset Pricing Model with Stochastic Consumption and Investment Opportunities," *Journal of Financial Economics*, September 1979, 7, 265–96.

Campbell, John Y. and Shiller, Robert J., "A Simple Account of the Behavior of Long-Term Interest Rates," *American Economic Review Proceedings*, May 1984, 74, 44–48.

Cox, John C., Ingersoll Jonathan E., Jr. and Ross, Stephen A., "A Theory of the Term Structure of Interest Rates," *Econometrica*, March 1985, 53, 385–407.

Fama, Eugene F., (1984a) "The Information in the Term Structure," *Journal of Financial Economics*, December 1984, 13, 509–28.

_____, (1984b) "Term Premiums in Bond Returns," *Journal of Financial Economics*, December 1984, 13, 529–46.

_____, "Term Premiums and Default Premiums in Money Markets," *Journal of Financial Economics*, September 1986, 17, 175–96.

_____ and Gibbons, Michael R., "A Comparison of Inflation Forecasts," *Journal of Monetary Economics*, May 1984, 13, 327–48.

Hamburger, Michael J. and Platt, E. N., "The Expectations Hypothesis and the Efficiency of the Treasury Bill Market," *Review of Economics and Statistics*, May 1975, 57, 190–99.

Hansen, Lars Peter, "Large Sample Properties of Generalized Method of Moments Estimators," *Econometrica*, June 1982, 50, 1029–54.

Hicks, John R., *Value and Capital*, 2nd ed., London: Oxford University Press, 1946.

Kessel, Reuben A., "The Cyclical Behavior of the Term Structure of Interest Rates," NBER, Occasional Paper No. 91, 1965.

Long, John, "Stock Prices, Inflation, and the Term Structure of Interest Rates," *Journal of Financial Economics*, July 1974, 1, 131–70.

Lutz, F. A., "The Structure of Interest Rates," *Quarterly Journal of Economics*, November 1940, 55, 36–63.

McCulloch, J. Huston, "An Estimate of the Liquidity Premium," *Journal of Political Economy*, February 1975, 83, 95–119.

Merton, Robert C., "An Intertemporal Capital Asset Pricing Model," *Econometrica*, September 1973, 41, 867–87.

Nelson, Charles R., *Applied Time Series Analysis*, San Francisco: Holden-Day, 1973.

_____ and Plosser, Charles I., "Trends and Random Walks in Macroeconomic Time Series," *Journal of Monetary Economics*, September 1982, 10, 139–62.

Shiller, Robert J., "The Volatility of Long-term Interest Rates and Expectations Models of the Term Structure," *Journal of Political Economy*, December 1979, 87, 1190–219.

_____, Campbell John Y. and Schoenholtz, Kermit L., "Forward Rates and Future

Policy: Interpreting the Term Structure of Interest Rates," *Brookings Papers on Economic Activity*, 1:1983, 173–217.

Startz, Richard, "Do Forecast Errors or Term Premia Really Make the Difference between Long and Short Rates?," *Journal of Financial Economics*, November 1982, *10*, 323–29.

White, Halbert, "A Heteroscedasticity-Consistent Covariance Matrix Estimator and a Direct Test for Heteroscedasticity," *Econometrica*, May 1980, *48*, 817–38.

Ulman, Scott and Wood, John H., "Monetary Regimes and the Term Structure of Interest Rates," manuscript, University of Minnesota, January 1983.

[21]

SIMULATED MOMENTS ESTIMATION OF MARKOV MODELS OF ASSET PRICES

By Darrell Duffie and Kenneth J. Singleton[1]

This paper provides a simulated moments estimator (SME) of the parameters of dynamic models in which the state vector follows a time-homogeneous Markov process. Conditions are provided for both weak and strong consistency as well as asymptotic normality. Various tradeoffs among the regularity conditions underlying the large sample properties of the SME are discussed in the context of an asset-pricing model.

Keywords: Monte Carlo simulation, generalized method of moments, geometric ergodicity, uniform strong law of large numbers, model estimation.

1. INTRODUCTION

This paper provides conditions for the consistency and asymptotic normality of a simulated moments estimator (SME) of the parameters of asset-pricing models with time-homogeneous Markov representations of the stochastic forcing process. SME's for economic models have been proposed by McFadden (1989) and Pakes and Pollard (1989) for i.i.d. environments, and by Lee and Ingram (1991) for a time series environment. The SME for time series models examined in this paper is as follows. The state vector Y_t that determines asset prices is assumed to follow a time-homogeneous Markov process whose transition function depends on an unknown parameter vector β_0. Asset prices, and possibly other relevant data, are observed as $f(Y_t, \beta_0)$, for some given function f of the underlying state and parameter vector. In parallel, a simulated state process $\{Y_s^\beta\}$ is generated (analytically or numerically) from the economic model and corresponding simulated observations $f(Y_s^\beta, \beta)$ are taken, for a given parameter choice β. The parameter β is chosen so as to "match moments," that is, to minimize the distance between sample moments of the data, $f(Y_t, \beta_0)$, and those of the simulated series $f(Y_t^\beta, \beta)$, in a sense to be made precise.

The proposed SME extends the generalized method-of-moments (GMM) estimator (Hansen (1982)) to a large class of asset-pricing models for which the moment restrictions of interest do not have analytic representations in terms of observable variables and the unknown parameter vector. We provide conditions on the transition function of Y_t and the observation function f under which the SME of β_0 is consistent, and characterize the normalized asymptotic distribution of the estimator. For two reasons, neither the regularity conditions underlying Hansen's (1982) analysis of GMM estimators for time-series models without

[1] We are grateful for several useful conversations with the co-editor Lars Hansen and Whitney Newey, and for the comments of Peter Bossaerts, Andrew Lo, Neil Pearson, Bruce Lehmann, and the referees. Singleton acknowledges funding from the National Science Foundation. Duffie acknowledges the support of a Battery March Fellowship.

simulation, which were also used by Lee and Ingram (1991) for their SME estimator, nor those imposed by McFadden (1989) and Pakes and Pollard (1989) for simulated moments estimation in i.i.d. environments, are applicable to the estimation problems posed in this paper. First, in simulating time series, pre-sample values of the series are typicallyrequired. In most circumstances, however, the stationary distribution of the simulated process, as a function of the parameter choice, is unknown. Hence, the initial conditions for the time series will generally not be drawn from their stationary distribution and the simulated process will generally be nonstationary. Second, functions of the current value of the simulated state depend on the unknown parameter vector both through the structure of the model (as in any GMM problem) and indirectly through the generation of data by simulation. The feedback effect of the latter dependence on the transition law of the simulated state process implies that the first-moment-continuity condition used by Hansen (1982), or the generalizations proposed by Andrews (1987), in establishing the uniform convergence of the sample to the population criterion functions are not directly applicable to the SME. Furthermore, the nonstationarity of the simulated series must be accommodated in establishing the asymptotic normality of the SME.

We address these difficulties by assuming geometric ergodicity as a condition on the state process ensuring that the simulated processes are asymptotically stationary with an ergodic distribution that is independent of starting values, and by imposing a damping condition on the feedback effect of parameter choice on the law of motion of the state process. Under these conditions, the nonstationarities associated with simulation are shown to be inconsequential for the asymptotic distribution of the SME.

The remainder of the paper is organized as follows. Section 2 uses a simple asset-pricing setting to illustrate in more detail the econometric issues that arise with estimation by simulation. The formal structure of the estimation problem and the definition of the simulated moments estimator are laid out in Section 3. Section 4 provides conditions for consistency, both weak and strong, the key ingredient being an appropriate extension of the uniform law of large numbers. Section 5 characterizes the asymptotic distribution of the SME, while Section 6 provides several extensions of the SME.

2. AN ILLUSTRATIVE ASSET-PRICING MODEL

In this section we describe a simple dynamic asset-pricing model that illustrates many of the econometric problems that arise in the use of simulation methods in estimation. The model is an extended version of the stochastic growth model studied by Brock (1980) and Michner (1984). After briefly describing the model, the use of simulation methods is given a more extensive motivation. Several econometric issues related to estimation using simulation are then introduced in the context of this model. This section is intended as an informal backdrop to the simulated moments estimator presented in Section 3 and analyzed in Sections 4 and 5.

Suppose that production of the single consumption commodity is determined by

(2.1) $\quad F(k_t, z_t) = z_t k_t^\phi, \quad 0 < \phi < 1,$

for some function F, where k_t is the level of the capital stock at date t and z_t is a technology shock. The firm rents capital from consumers at the rental rate r_t^k and pays out the profits to the owners of its shares in the form of dividends, d_t. In each period, the firm solves the following static optimum problem (maximization of profits)

(2.2) $\quad d_t = \arg\max_{k_t} \{z_t k_t^\phi - r_t^k k_t\}$

in order to choose the level k_t of capital to rent from the consumer. In equilibrium, this is equivalent to maximization of share market value (see, for example, Duffie (1988, Section 20)).

Given the price p_t of a share of the firm, the representative consumer faces the budget constraint

(2.3) $\quad c_t + k_{t+1} + p_t s_{t+1} = (d_t + p_t) s_t + (r_t^k + \mu) k_t,$

where c_t and s_t denote consumption and shares of claims to the dividend stream of the firm, respectively, and $(1 - \mu)$ denotes a constant depreciation rate on the capital stock. Subject to this constraint, the representative consumer chooses consumption and share holdings so as to maximize utility for the infinite-horizon consumption process $\{c_t\}$. Allowing for an unobserved (to the econometrician) taste shock $\{u_t\}$ and adopting a typical additively-separable utility criterion, the agent's problem is then

(2.4) $\quad \max_{\{c_t, k_t\}} E\left[\sum_{t=1}^{\infty} \delta^t \frac{(c_t - 1)^{1-\alpha}}{1-\alpha} u_t \right], \quad \alpha < 0,$

where α is the constant coefficient of relative risk aversion and $\delta \in (0,1)$ is a subjective discount factor.

The vector $X_t' = (z_t, u_t)$ is assumed to be a Markov process satisfying

(2.5) $\quad X_t = h(X_{t-1}, \varepsilon_t, \rho_0),$

where $\{\varepsilon_t\}$ is a two-dimensional i.i.d. stochastic process, h is a transition function, and ρ_0 is an unknown parameter vector. For the moment, we also assume that $\{X_t\}$ does not exhibit growth over time.

In order to estimate the unknown parameter vector $\beta_0 = (\phi, \alpha, \rho_0, \mu, \delta)'$, a point in some compact parameter set Θ, we proceed as follows. The economic system (2.1)–(2.5) is solved analytically or numerically for the equilibrium transition function H generating the augmented state process $Y_t = (X_t', k_t)'$, according to

(2.6) $\quad Y_{t+1} = H(Y_t, \varepsilon_{t+1}, \beta_0).$

For any admissible parameter vector $\beta \in \Theta$, we can also generate a simulated state process $\{Y_t^\beta\}$ according to the same transition function H, but using a shock sequence $\{\hat{\varepsilon}_t\}$ that is identically and independently distributed of $\{\varepsilon_t\}$; that is,

$$Y_{t+1}^\beta = H(Y_t^\beta, \hat{\varepsilon}_{t+1}, \beta).$$

From this, a history $\{Y_t^\beta\}_{t=1}^{\mathcal{T}}$ of $\mathcal{T}$ simulated equilibrium states can be generated.

Next, for some chosen observation function f, in each period t an observation $f_t^* \equiv f(Y_t, Y_{t-1}, \ldots, Y_{t-l+1})$ is made of a finite "l-history" of state information. Likewise, a corresponding observation f_t^β can be formed for each l-history of simulated states. The components of f_t^β may be known analytic functions (for example, $k_t^\beta \cdot k_{t-1}^\beta$) or determined numerically as functions of the l-history of simulated states (for example, equilibrium asset prices or consumption). Finally, the SME is a value of β chosen to minimize the distance between the sample mean of $\{f_t^\beta\}_{t=1}^{\mathcal{T}}$ and the sample mean of $\{f_t^*\}_{t=1}^{T}$, where T is the number of historical observations on f_t^*.

Several considerations motivate the simultaneous solution of the model and SME estimation of β. First, solving for the stochastic equilibrium of the model permits an assessment of the goodness-of-fit directly in terms of aspects of the joint distribution of asset returns, consumption, and capital.[2] Furthermore, estimation of asset-pricing models using Euler equations (Hansen and Singleton (1982)) is not always feasible, as in the version of this model with taste shocks. Third, temporal aggregation may lead to inconsistent GMM estimators of β_0 (Hall (1988), Hansen and Singleton (1989)), but temporal aggregation can often be accommodated using the SME.

For several reasons, this illustrative estimation problem is not a special case of either Hansen's (1982) GMM estimation problem or the simulated moments problems examined by McFadden (1989) and Pakes and Pollard (1989), or Lee and Ingram (1991). The most important difference between the estimation problem with simulated time series and the GMM estimation problem discussed by Hansen (1982) lies in the parameter dependency of the simulated time series $\{f_t^\beta\}$. In the stationary, ergodic environment studied by Hansen (1982), one observes $f(Y_t, \beta_0)$, where the data generation process $\{Y_t\}$ is fixed and β_0 is the parameter vector to be estimated. In contrast, $f_t^\beta = f(Y_t^\beta, \beta)$ depends on β not only directly, but indirectly through the dependence of the entire past history of the simulated process $\{Y_t^\beta\}$ on β. In Section 4, we present versions of uniform

---

[2] Several alternative numerical methods for solving discrete-time dynamic rational expectations models have recently been proposed in the literature; see Taylor and Uhlig (1990), Tauchen and Hussey (1991), and the references cited therein for useful summaries. Many of the algorithms discussed involve approximations to either the distributions of the forcing variables or the model itself. Additional approximations are involved when the underlying model is expressed in continuous time and a discrete-time approximation is being estimated. These approximations affect the large sample properties of the SME since, as sample size increases, one obtains a consistent estimator of the approximate model. At a minimum, the methods described in this paper apply to the approximate model if approximations are used to solve for equilibrium asset prices. They may apply to the original model if the approximation error can be made negligible as the sample size increases.

weak and strong laws of large numbers that accommodate this parameter dependency of the data generation process for simulated time series.

Furthermore, in contrast to the simulated moments estimators for i.i.d. environments, the simulation of time series requires initial conditions for the forcing variables Y_t. Even if the transition function of the Markov process $\{Y_t\}$ is stationary (that is, has a stationary distribution), the simulated process $\{Y_t^\beta\}$ is not generally stationary since the initial simulated state Y_1^β is typically not drawn from the ergodic distribution of the process. In this case, the simulated process $\{f_t^\beta\}$ is nonstationary.

A related initial conditions problem, common to the GMM and SM estimation of asset-pricing models, occurs with capital accumulation. Specifically, the current equilibrium capital stock can typically be expressed as a function of the previous period's stock plus investment in new capital. Measurements of investment are often more reliable than measurements of the stock of capital, which may not be based on compatible assumptions about depreciation. Accordingly, in constructing a time series on the capital stock to be used in estimation, one may wish to accommodate mismeasurement of the initial stock.[3]

In Section 4, we present a set of sufficient conditions for the Markov process $\{Y_t\}$ to be geometrically ergodic, which (among other things) implies that the large-sample properties of functions of Y_t are invariant to the choice of initial conditions used in simulating both exogenous (taste and technology shocks) and endogenous (e.g., the capital stock) state variables.

Throughout this discussion we have assumed that the Markov process described by (2.5) does not exhibit growth. In fact, there is real growth in output, and hence in certain asset prices. If the technology shock $\{z_t\}$, for instance, exhibits growth over time, then the implied trends for the components of Y_t are restricted by the structure of the model.[4] Conversely, the structure of the model restricts the class of admissible trend specifications. Furthermore, accommodating these trends typically requires that the implied form of the trends in Y_t is known, and that it is possible to build an adjustment for trends directly into the function f of the data and to simulate a trend-free version of the model.

Following Eichenbaum and Hansen (1988), the implied restrictions on deterministic trends in the decision variables can be imposed in estimation by appending the moment conditions associated with least squares estimation of the trend equations to the moment equations involving f^* and f^β. The subsequent discussion in this paper extends to this case using arguments similar to those in Eichenbaum and Hansen (1988) for GMM estimators of (2.11). If the forcing variables exhibit stochastic trends (unit roots), then our estimation

---

[3] See Dunn and Singleton (1986); Eichenbaum, Hansen, and Singleton (1988); and Eichenbaum and Hansen (1988) for examples of studies of Euler equations using GMM estimators in which this type of initial condition problem arises.

[4] See Eichenbaum and Hansen (1988) and Eichenbaum, Hansen and Singleton (1988) for a discussion of restrictions on trends implied by Euler equations. Singleton (1987) discusses the analogous restrictions on deterministic seasonal components of agents' decision variables.

strategy applies only if the entire model, including the forcing variables, can be transformed to a model expressed in terms of trend-free processes.

3. THE ESTIMATION PROBLEM

This section defines the simulated moments estimator. The basic primitives for the model are:
 (i) a measurable *transition function* $H: \mathbb{R}^N \times \mathbb{R}^p \times \Theta \to \mathbb{R}^N$, with compact parameter set $\Theta \subset \mathbb{R}^Q$, for some positive integers N, p, and Q;
 (ii) a measurable *observation function* $f: \mathbb{R}^{Nl} \times \Theta \to \mathbb{R}^M$, for positive integers l and M, with $M \geq Q$.

A given $\mathbb{R}^N$-valued *state process* $\{Y_t\}_{t=1}^\infty$ is generated by the difference equation

(3.1) $\quad Y_{t+1} = H(Y_t, \varepsilon_{t+1}, \beta_0),$

where the parameter vector β_0 is to be estimated, and $\{\varepsilon_t\}$ is an i.i.d. sequence of $\mathbb{R}^p$-valued random variables on a given probability space $(\Omega, \mathcal{F}, P)$. The function H may be determined implicitly by the numerical solution of a model for equilibrium asset prices. Let $Z_t = (Y_t, Y_{t-1}, \ldots, Y_{t-l+1})$ for some positive integer $l < \infty$. Estimation of β_0 is based on moments of the vector $f_t^* \equiv f(Z_t, \beta_0)$.

For certain special cases of (3.1) and f, the function mapping β to $E[f(Z_t, \beta)]$ is known and independent of t. In these cases, the GMM estimator,

(3.2) $\quad b_T = \arg\min_{\beta \in \Theta} \left[\frac{1}{T} \sum_{t=1}^T f_t^* - E[f(Z_t, \beta)] \right]' W_T \left[\frac{1}{T} \sum_{t=1}^T f_t^* - E[f(Z_t, \beta)] \right],$

for given "distance matrices" $\{W_T\}$, is consistent for β_0 and asymptotically normal under regularity conditions in, for example, Hansen (1982). The requirement that $\beta \mapsto E[f(Z_t, \beta)]$ is known, however, limits significantly the applicability of the GMM estimator to asset-pricing problems.

The simulated moments estimator circumvents this limitation by making the much weaker assumption that the econometrician has access to an $\mathbb{R}^p$-valued sequence $\{\hat{\varepsilon}_t\}$ of random variables that is identical in distribution to, and independent of, $\{\varepsilon_t\}$. Then, for any $\mathbb{R}^N$-valued initial point $\hat{Y}_1$ and any parameter vector $\beta \in \Theta$, the *simulated state process* $\{Y_t^\beta\}$ can be constructed inductively by letting $Y_1^\beta = \hat{Y}_1$ and

(3.3) $\quad Y_{t+1}^\beta = H(Y_t^\beta, \hat{\varepsilon}_{t+1}, \beta).$

Likewise, the simulated observation process $\{f_t^\beta\}$ is constructed by $f_t^\beta = f(Z_t^\beta, \beta)$, where $Z_t^\beta = (Y_t^\beta, \ldots, Y_{t-l+1}^\beta)$. Finally, the SME of β_0 is the parameter vector b that best matches the sample moments of the actual and simulated observation processes, $\{f_t^*\}$ and $\{f_t^b\}$.

More precisely, let $\mathcal{T}: \mathbb{N} \to \mathbb{N}$ define the simulation sample size $\mathcal{T}(T)$ that is generated for a given sample size T of actual observations, where $\mathcal{T}(T) \to \infty$ as

$T \to \infty$. For any parameter vector β, let

$$(3.4) \quad G_T(\beta) = \frac{1}{T} \sum_{t=1}^{T} f_t^* - \frac{1}{\mathcal{T}(T)} \sum_{s=1}^{\mathcal{T}(T)} f_s^\beta$$

denote the difference in sample moments. If $\{f_t^*\}$ and $\{f_s^\beta\}$ satisfy a law of large numbers, then $\lim_T G_T(\beta) = 0$ if $\beta = \beta_0$. With identification conditions, $\lim_T G_T(\beta) = 0$ if and only if $\beta = \beta_0$. We therefore introduce a sequence $W = \{W_T\}$ of $M \times M$ positive semi-definite matrices and define the *simulated moments estimator* for β_0 given $(H, \varepsilon, \mathcal{T}, \hat{Y}_1, W)$ to be the sequence $\{b_T\}$ given by

$$(3.5) \quad b_T = \arg\min_{\beta \in \Theta} G_T(\beta)' W_T G_T(\beta) \equiv \arg\min_{\beta \in \Theta} C_T(\beta).$$

The distance matrix W_T is chosen with rank at least Q, and may depend on the sample information $\{f_1^*, \ldots, f_T^*\} \cup \{f_1^\beta, \ldots, f_{\mathcal{T}(T)}^\beta : \beta \in \Theta\}$.

Comparing (3.2) and (3.5) shows that the SME extends the method-of-moments approach to estimation by replacing the population moment $E[f(Z_t, \beta)]$ with its sample counterpart, calculated with simulated data. The latter sample moment can be calculated for a large class of asset-pricing models. Extensions of the SME are provided in Section 6.

4. CONSISTENCY

The presence of simulation in the estimator pushes one to special lengths in justifying regularity conditions for the consistency of method-of-moments estimators that, without simulation, are often taken for granted. As illustrated in Section 2, there are two particular problems. First, since the simulated state process is usually not initialized with a draw from its ergodic distribution, one needs a condition that allows the use of an arbitrary initial state, knowing that the state process converges rapidly to its stationary distribution. Second, one needs to justify the usual starting assumption of some form of uniform continuity of the observation as a function of the parameter choice. With simulation, a perturbation of the parameter choice affects not only the current observation, but also affects transitions between past states, a dependence that compounds over time. We will present a natural (but restrictive) condition directly on the state transition function guaranteeing that this compounding effect is of a damping, rather than exploding, variety.

Initially we describe the concept of geometric ergodicity, a condition ensuring that the simulated state process satisfies a law of large numbers with an asymptotic distribution that is invariant to the choice of initial conditions. Then ergodicity of the simulated series is used to prove a uniform weak law of large numbers for $G_T(\beta)$ and weak consistency of the SME (that is, $b_T \to \beta_0$ in probability). Weak consistency is proved under a global modulus-of-continuity condition rather than the more usual local condition underlying proofs of strong consistency. Subsequently, we present Lipschitz and modulus of continuity conditions on the primitives (H, ε, f) that are sufficient for strong consistency (that is, $b_T \to \beta_0$ almost surely). Though weaker than the damping conditions typically used to verify near-epoch dependence (Gallant and White (1988)), these conditions nevertheless exclude an important class of geometrically ergodic processes. This fact is the primary reason for our initial focus on weak consistency. Finally, various tradeoffs in choosing among the regularity conditions leading to weak and strong consistency are discussed in the context of the illustrative model presented in Section 2.

4.1. Geometric Ergodicity

In order to define geometric ergodicity, let P_x^t denote the t-step transition probability for a time-homogeneous Markov process $\{X_t\}$; that is, P_x^t is the distribution of X_t given the initial point $X_0 = x$. The process $\{X_t\}$ is ρ-ergodic, for some $\rho \in (0, 1]$, if there is a probability measure π on the state space of the process such that, for every initial point x,

$$(4.1) \quad \rho^{-t} \|P_x^t - \pi\|_v \to 0 \quad \text{as} \quad t \to \infty,$$

where $\|\cdot\|_v$ is the total variation norm.[5] The measure π is the ergodic distribution. If $\{X_t\}$ is ρ-ergodic for $\rho < 1$, then $\{X_t\}$ is *geometrically ergodic*. In calculating asymptotic distributions, geometric ergodicity can substitute for stationarity since it means that the process converges geometrically to its stationary distribution. Moreover, geometric ergodicity implies strong (α) mixing in which the mixing coefficient $\alpha(m)$ converges geometrically with m to zero (Rosenblatt (1971), Mokkadem (1985)).

In what follows, for any ergodic process $\{X_t\}$, it is convenient for us to write "X_∞" for any random variable with the corresponding ergodic distribution. We adopt the notation $\|X\|_q = [E(\|X\|^q)]^{1/q}$ for the L^q norm of any $\mathbb{R}^N$-valued random variable X, for any $q \in (0, \infty)$. We let L^q denote the space of such X with $\|X\|_q < \infty$, and let $\|x\|$ denote the usual Euclidean norm of a vector x.

General criteria for the geometric ergodicity of a Markov chain have been obtained by Nummelin and Tuominen (1982) and by Tweedie (1982). We will review simple sufficient conditions established by Mokkadem (1985) for the special case of nonlinear AR(1) models, which includes our setting.

A key ingredient for ergodicity is positive recurrence,[6] for which a key condition is irreducibility. For a finite Markov chain, irreducibility means essentially that each state is accessible from each state, obviously a sufficient condition in this case for both recurrence and geometric ergodicity. Mokkadem (1985) uses the following convenient sufficient condition for irreducibility of a time-homogeneous Markov chain $\{X_t\}$ valued in $\mathbb{R}^N$ with t-step transition probability P_x^t.

---

[5] The total variation of a signed measure μ is $\|\mu\|_v = \sup_{h:\ |h(y)| \leq 1} \int h(y)\, d\mu(y)$.
[6] For a finite-state Markov chain, recurrence means essentially that each state occurs infinitely often from any given state. See, for example, Doob (1953) for some general definitions.

CONDITION B: *For any measurable $A \subset \mathbb{R}^N$ of nonzero Lebesgue measure and any compact $K \subset \mathbb{R}^N$, there exists some integer $t > 0$ such that*

$$(4.2) \qquad \inf_{x \in K} P_x^t(A) > 0.$$

It is obviously enough that $P_x^1(A)$ is continuous in x and supports all of $\mathbb{R}^N$ for each x, but this single-period "full support" condition is too strong an assumption in a setting with endogenous state variables. For example, the process for Y_t given by (2.6) fails this single-period full-support condition because the distribution of the capital stock k_{t+1} given X_t is degenerate, but often passes the weaker Condition B. To be more concrete, consider the special case of (2.1)–(2.6) with $u_t = 1$ for all t, $\mu = 0$ (100% depreciation), and $\alpha = 1$ (logarithmic utility). Also, suppose that the law of motion for the technology shock is given by

$$(4.3) \qquad \ln z_{t+1} = \zeta_z + \rho \ln z_t + \varepsilon_{t+1},$$

for constants ζ_z and ρ. Under these simplifying assumptions, the implied equilibrium asset-pricing function and law of motion for the capital stock are (Michner (1984)):

$$(4.4) \qquad p_t = \frac{\delta}{(1-\delta)} (1-\phi) z_t k_t^\phi,$$

$$d_t = (1-\phi) z_t k_t^\phi,$$

$$(4.5) \qquad k_{t+1} = \delta \phi z_t k_t^\phi.$$

If $\{\varepsilon_t\}$ is say i.i.d. normal, then $\{Y_t\}$ for this illustrative economy satisfies Condition B. More generally, Condition B is not a strong condition on models with endogenous state variables provided the endogenous state variables do not move in such a way that some states are inaccessible from others.

If the state process $\{X_t\}$ is valued in a proper subset S of $\mathbb{R}^N$, Condition B obviously does not apply, but analogous results hold if Condition B applies when substituting S everywhere for $\mathbb{R}^N$ (and relatively open sets for sets of nonzero Lebesgue measure).

A second key ingredient for ergodicity is aperiodicity. For example, the Markov chain that alternates deterministically from "heads" to "tails" to "heads" to "tails," and so on, is not geometrically ergodic, despite its recurrence.

With these definitions in hand, we can review Mokkadem's sufficient conditions for geometric ergodicity of what he calls "nonlinear AR(1) models," which includes our setting.

LEMMA 1 (Mokkadem): *Suppose $\{Y_t\}$, as defined by (3.1), is aperiodic and satisfies Condition B. Fix β and suppose there are constants $K > 0$, $\delta \in (0, 1)$, and*

$q > 0$ such that $H(\cdot, \varepsilon_1, \beta)\colon \mathbb{R}^N \to L^q$ is well defined and continuous with

(4.6) $\quad \|H(y, \varepsilon_1, \beta)\|_q < \delta \|y\|, \quad \|y\| > K.$

Then $\{Y_t\}$ is geometrically ergodic. Moreover, $\{\|Y_t^\beta\|_q\}$ and $\|Y_\infty^\beta\|_q$ are uniformly bounded over t.

Condition (4.6), inspired by Tweedie (1982), means roughly that $\{Y_t\}$, once outside a sufficiently large ball, heads back into the ball at a uniform rate.

4.2. A Uniform Weak Law of Large Numbers

Since geometric ergodicity of $\{Y_t^\beta\}$ implies α-mixing, it also implies that $\{Y_t^\beta\}$ satisfies a strong (and hence weak) law of large numbers. For consistency of the SME estimator, however, standard sufficient conditions require that a strong or weak law holds in a uniform sense over the parameter space Θ. For example, the family $\{\{f_t^\beta\}\colon \beta \in \Theta\}$ of processes satisfies the uniform weak law of large numbers if, for each $\delta > 0$,

(4.7) $\quad \lim_{T \to \infty} P\left[\sup_{\beta \in \Theta} \left| E(f_\infty^\beta) - \frac{1}{T} \sum_{t=1}^T f_t^\beta \right| > \delta \right] = 0.$

In our setting of simulated moments, $\{Z_t^\beta\}$ is simulated based on various choices of β, so continuity of $f(Z_t^\beta, \beta)$ in β (via both arguments) is useful in proving (4.7). We will use the following global modulus of continuity condition on $\{f_t^\beta\}$.

DEFINITION: The family $\{f_t^\beta\}$ is *Lipschitz, uniformly in probability*, if there is a sequence $\{K_t\}$ such that, for all t and all β and θ in Θ,

$$\|f_t^\beta - f_t^\theta\| \leq K_t \|\beta - \theta\|,$$

where $K^T = T^{-1} \sum_{t=1}^T K_t$ is bounded (with T) in probability.

LEMMA 2 (Uniform Weak Law of Large Numbers): *Suppose, for each $\beta \in \Theta$, that $\{Y_t^\beta\}$ is ergodic and that $E(|f_\infty^\beta|) < \infty$. Suppose, in addition, that the map $\beta \mapsto E(f_\infty^\beta)$ is continuous and the family $\{f_t^\beta\}$ is Lipschitz, uniformly in probability. Then $\{\{f_t^\beta\}\colon \beta \in \Theta\}$ satisfies the uniform weak law of large numbers.*

The proofs of this and all subsequent propositions in Section 4 are provided in the Appendix.

The ergodicity assumption on $\{Y_t^\beta\}$ in Lemma 2 can be replaced with Mokkadem's conditions for geometric ergodicity on the transition function H and disturbance ε_t, summarized in Lemma 1.

4.3. Weak Consistency

Next, we summarize several important assumptions that are used in our proofs of both consistency and asymptotic normality of the SME.

ASSUMPTION 1 (Technical Conditions): *For each $\beta \in \Theta$, $\{\|f_t^\beta\|_{2+\delta}: t = 1, 2, \ldots\}$ is bounded for some $\delta > 0$. The family $\{f_t^\beta\}$ is Lipschitz, uniformly in probability, and $\beta \mapsto E(f_\infty^\beta)$ is continuous.*

ASSUMPTION 2 (Ergodicity): *For all $\beta \in \Theta$, the process $\{Y_t^\beta\}$ is geometrically ergodic.*

The hypotheses of Lemmas 1 and 2 are sufficient for Assumptions 1 and 2 provided Mokkadem's conditions apply for some $q > 2$.

We impose the following condition on the distance matrices $\{W_T\}$ in (3.5).

ASSUMPTION 3 (Convergence of Distance Matrices): *Σ_0 is nonsingular and $W_T \to W_0 = \Sigma_0^{-1}$ almost surely, where (for any t)*

$$(4.8) \quad \Sigma_0 \equiv \sum_{j=-\infty}^{\infty} E\big([f_t^* - E(f_t^*)][f_{t-j}^* - E(f_{t-j}^*)]'\big).$$

For the second moments in this assumption to exist, and their sum to converge absolutely, the assumptions that $\{\|f_t^*\|_{2+\delta}: t = 1, 2, \ldots\}$ is bounded for some $\delta > 0$ and geometric ergodocity of $\{Y_t\}$ together suffice, as shown by Doob (1953, pp. 222–224). Also, as with Hansen's (1982) GMM estimator, the choice of W_0 in Assumption 3 leads to the most efficient SME within the class of SME's with positive definite distance matrices.

Notice that Σ_0 in Assumption 3 is a function of the moments of $\{f_t^*\}$ alone; in particular, Σ_0 depends neither on β nor on the moments of the simulated process $\{f_t^\beta\}$. Thus, Σ_0 can be estimated using, for instance, the approaches discussed by Andrews (1991).[7] Given the definition of Σ_0 and the fact that geometric ergodicity implies α-mixing, it follows that the Newey-West estimator is consistent for Σ_0 in our environment.

Alternatively, Σ_0 could be estimated using simulated data $\{f_t^\beta\}$. Since the rate of convergence of spectral estimators is slow and one has control over the size $\mathcal{T}(T)$ of the simulated sample, this alternative may be relatively advantageous. A two-step procedure for estimating Σ_0 is required, however, so in establishing consistency of a simulated estimator of Σ_0 one would need to account both for dependence of $\{f_t^\beta\}$ on an estimated value of β and the parameter dependence of simulated series. One approach to establishing consistency would be to

---

[7] Several estimators of Σ_0 have been proposed in the literature. See, for example, Hansen and Singleton (1982), Eichenbaum, Hansen, and Singleton (1988), and Newey and West (1987). In general, $E[f_t^* - Ef_t^*)(f_{t-j}^* - Ef_{t-j}^*)']$ is nonzero for all j in (4.8) and the Newey-West estimator is appropriate.

extend the discussion of consistent estimation of spectral density functions using estimated residuals without simulation, found in Newey and West (1987) and Andrews (1991), to the case of simulated residuals.

Under Assumptions 1–3, the criterion function $C_T(\beta)$ converges almost surely to the asymptotic criterion function $C: \Theta \to \mathbb{R}$ defined by $C(\beta) = G_\infty(\beta)' W_0 G_\infty(\beta)$.

ASSUMPTION 4 (Uniqueness of Minimizer): $C(\beta_0) < C(\beta)$, $\beta \in \Theta$, $\beta \neq \beta_0$.

Our first theorem establishes the consistency of the SME $\{b_T: T \geq 1\}$ given by (3.5).

THEOREM 1 (Consistency of SME): *Under Assumptions 1–4, the SME $\{b_T\}$ converges to β_0 in probability as $T \to \infty$.*

4.4. Strong Consistency

The Uniform Weak Law of Large Numbers (UWLLN) underlying the discussion in Sections 4.2 and 4.3 maintained the uniform continuity condition in Assumption 1. In this subsection we provide primitive conditions on H, ε, and f for a local modulus of continuity condition with simulation, and thereby explore in more depth the nature of the requirements in simulation environments for $\{f_t^\beta\}$ to satisfy the Uniform Strong Law of Large Numbers (USLLN):

$$\sup_{\beta \in \Theta} \left| \frac{1}{T} \sum_{t=1}^T f_t^\beta - E(f_\infty^\beta) \right| \xrightarrow{a.s.} 0 \quad \text{as} \quad T \to \infty.$$

The basic nature of the conditions are of three forms: continuity conditions, growth conditions, and a contraction (or "damping") condition on the transition function H that we call an "asymptotic unit-circle (AUC) condition."

Our proof of strong consistency of the SME proceeds in three steps.[8] First, we introduce the AUC condition, which assures that current shocks have a damping effect on future simulated observations. Under the AUC condition, it is shown that, for each β, there exists a stationary and ergodic process $\{Y_t^{\infty\beta}\}$ that satisfies (3.1) and can be substituted for $\{Y_t^\beta\}$ in proving consistency (and asymptotic normality) of the SME. Second, we show that the AUC condition and certain continuity and growth conditions imply a version of Hansen's (1982) modulus of continuity condition for simulation environments. Strong consistency of the SME then follows from results in Hansen (1982).

DEFINITION (The Asymptotic Unit-Circle Condition): The transition function H and shock process ε satisfy the *Asymptotic Unit-Circle Condition* if, for each

---

[8] The strategy of using a unit-circle condition with a Lipschitz coefficient that changes geometrically toward zero in proving strong consistency of the SME was suggested to us by Lars Hansen in his discussion of an earlier version of this paper.

$\theta \in \Theta$, there is some $\delta > 0$ and a sequence of positive random variables $\{\rho_\theta(\varepsilon_t)\}$ satisfying

$$(4.9) \quad \lim_{T \to \infty} \frac{1}{T} \sum_{t=1}^{T} \ln \rho_\theta(\varepsilon_t) = \alpha_\theta < 0 \quad \text{a.s.}$$

such that, whenever $\|\beta - \theta\| \leq \delta$, for any x and y,

$$\|H(y, \beta, \varepsilon_t) - H(x, \beta, \varepsilon_t)\| \leq \rho_\theta(\varepsilon_t)\|y - x\|.$$

In other words, for the AUC condition, $H(\cdot, \beta, \varepsilon_t)$ must have a Lipschitz coefficient $\rho_\theta(\varepsilon_t)$ with the property that $\prod_{s=0}^{t} \rho_\theta(\varepsilon_s)$ declines geometrically toward zero as $t \to \infty$. This is a weaker requirement than the unit-circle condition used by Gallant and White (1988) to verify near-epoch dependence of a process.

We say that f is Θ-locally Lipschitz if, for each $\theta \in \Theta$, there is a δ and a constant k such that, whenever $\|\beta - \theta\| \leq \delta$, the function $f(\cdot, \beta)$ has the Lipschitz constant k. Next, we define f to be S-smooth (sufficiently smooth) if f is Θ-locally Lipschitz and, for each $z \in \mathbb{R}^{Nl}$, the function $f(z, \cdot): \Theta \to \mathbb{R}^p$ has a Lipschitz constant $C_1(z)$, where C_1 satisfies a growth condition.[9] Obviously, if f is Lipschitz, then f is S-smooth, but a Lipschitz condition is unnecessarily strong and is not satisfied in many applications. (Take, for example, $f(z, \beta) = \beta z$.) We say that H is S-smooth if, for each $\theta \in \Theta$, there is a δ small enough that $\|\beta - \theta\| \leq \delta$ implies that, for all $y \in \mathbb{R}^N$ and $\varepsilon \in \mathbb{R}^p$,

$$\|H(y, \beta, \varepsilon) - H(y, \theta, \varepsilon)\| \leq C_2(y, \varepsilon)\|\beta - \theta\|,$$

where C_2 satisfies a growth condition.

The smoothness assumption on f and the AUC condition imply that the nonstationarity induced by the initial conditions problem can be ignored when studying the large sample properties of the SME. We establish this result in the following two lemmas.

LEMMA 3: *If (H, ε) satisfies the AUC condition, then for each β in Θ there exists a stationary and ergodic process $\{Y_t^{\infty\beta}: -\infty < t < \infty\}$ such that, for all t, $Y_t^{\infty\beta}$ is measurable with respect to $\{\hat{\varepsilon}_{t-s}: s \geq 0\}$ and $Y_{t+1}^{\infty\beta} = H(Y_t^{\infty\beta}, \hat{\varepsilon}_{t+1}, \beta)$.*

Next we argue that $\{Y_t^\beta\}$, simulated with an arbitrary initial condition, can be replaced by $\{Y_t^{\infty\beta}\}$ for the purpose of proving a USLLN.

LEMMA 4: *If f is S-smooth and (H, ε) satisfies the AUC condition, then*

$$(4.10) \quad \sup_{\beta \in \Theta} \left| \frac{1}{T} \sum_{t=1}^{T} f_t^\beta - \frac{1}{T} \sum_{t=1}^{T} f_t^{\infty\beta} \right| \xrightarrow{\text{a.s.}} 0 \quad \text{as} \quad T \to \infty,$$

where $f_t^{\infty\beta} = f[(Y_t^{\infty\beta}, Y_{t-1}^{\infty\beta}, \ldots, Y_{t-l+1}^{\infty\beta}), \beta]$.

---

[9] A real-valued function F on a Euclidean space satisfies a growth condition if there exist constants k and K such that for x, $|F(x)| \leq k + K\|x\|$.

The final step in proving strong consistency of the SME is showing that $\{f_t^{\infty\beta}\}$ satisfies a USLLN. Toward this end, for each θ in Θ and $\delta > 0$, let

$$\text{mod}_t(\delta, \theta) \equiv \sup\{\|f_t^{\infty\beta} - f_t^{\infty\theta}\| : \|\beta - \theta\| < \delta, \beta \in \Theta\}$$

denote the "modulus of continuity" of the process $\{f_t^{\infty\beta}\}$ at θ, defined ω by ω. Consider the following:

ASSUMPTION 5: *For each $\theta \in \Theta$, there is a $\delta > 0$ such that $E[\text{mod}_t(\delta, \theta)] < \infty$.*

With this, combined with our earlier assumptions, Hansen's (1982) Theorem 2.1 implies that $\{f_t^{\infty\beta}\}$ satisfies a USLLN and that $\{b_T\}$ is a strongly consistent estimator of β_0. We summarize with the following theorem.

THEOREM 2 (Strong Consistency): *Under Assumptions 3–5, the AUC condition, and the assumption that f is S-smooth, the SME $\{b_T\}$ converges to β_0 almost surely as $T \to \infty$.*

The assumption in Theorem 2 that $E[\text{mod}_t(\delta, \theta)] < \infty$ is not known to be implied by the AUC condition. However, by strengthening the statement of the AUC condition, Assumption 5 becomes redundant. Specifically, we introduce the following strong AUC condition:

DEFINITION (L^2 Unit-Circle Condition): *The transition function H and the shock process ε satisfy the L^2 Unit-Circle condition if, for each $\theta \in \Theta$, there is some $\delta > 0$ and a sequence of positive random variables $\{\rho_\theta(\varepsilon_t)\}$ satisfying $E[\rho_\theta(\varepsilon_t)^2] < 1$ such that, whenever $\|\beta - \theta\| \leq \delta$, for all x and y,*

$$\|H(y, \beta, \varepsilon_t) - H(x, \beta, \varepsilon_t)\| \leq \rho_\theta(\varepsilon_t)\|y - x\|.$$

By Jensen's inequality, $\ln E[\rho_\theta(\varepsilon_t)] > E[\ln \rho_\theta(\varepsilon_t)]$, so that the L^2 Unit-Circle Condition (L^2 UC condition) implies the AUC condition. Hence the lemmas preceeding Theorem 2 continue to hold under the L^2 UC condition.

This strengthening of the unit-circle condition leads to the following theorem.

THEOREM 3: *Under Assumptions 3–4, the assumption that H and f are S-smooth, and the L^2 UC condition, the SME is a strongly consistent estimator of β_0.*

4.5. Regularity Conditions and Dynamic Asset-Pricing Models

Weak consistency was established by assuming that the simulated processes are geometrically ergodic and that $\{f_t^\beta\}$ satisfies a uniform Lipschitz condition in β. In contrast, strong consistency was established assuming a unit-circle condition on the transition function H and an i.i.d. shock process $\{\varepsilon_t\}$. Thus, the AUC condition substitutes in part for the Lipschitz condition in Assumption 1 and in part for geometric ergodicity in Assumption 2. Indeed, the L^2 UC condition implies geometric ergodicity. On the other hand, there is an important class of geometrically ergodic processes that do not satisfy the L^2 UC condition, and this is a primary motivating reason for our analysis of weak consistency.

In order to see this, consider again the example in Section 2 and suppose that the law of motion of the technology shock is given by

$$(4.11) \quad z_t = \xi + \rho z_{t-1} + \sigma \nu_{t-1}^\gamma \varepsilon_t, \quad \gamma < 1, \quad \sigma > 0, \quad |\rho| < 1,$$

where $\nu_t = z_t$ if $z_t \geq \eta > 0$ and $\nu_t = \eta$ otherwise, and suppose that $E(\varepsilon_t) = 0$ for all t. This representation of a shock process is similar to several widely studied representations of conditionally heteroskedastic processes. Let $h(z, \varepsilon, \beta)$ denote the right hand side of (4.11). Then

$$\|h(z,\varepsilon,\beta) - h(z',\varepsilon,\beta)\|_2 = \left\|\rho + \sigma\varepsilon \frac{(\nu^\gamma - \nu'^\gamma)}{(z-z')}\right\|_2 \|z - z'\|.$$

The ratio $(\nu^\gamma - \nu'^\gamma)/(z - z')$ can be made arbitrarily large, as $\nu_t \to \eta$ for small η, in which case the factor of proportionality for $\|z - z'\|$ exceeds unity. Similarly, if ρ, σ, and the variance of ε are sufficiently large, then the unit-circle condition may be violated. This is the case, for example, if $\gamma = 1$ and $\|\rho + \sigma\varepsilon\|_2 > 1$. Furthermore, from the proofs of Lemmas 3 and 4, it is apparent that this process will not in general satisfy the AUC condition used to prove Theorem 2.

The process (4.11) is nevertheless geometrically ergodic. This can be verified easily by noting that $|\rho| < 1$ and $\|z^\gamma\|/\|z\|$ can be made arbitrarily small for large enough z when $\gamma < 1$. Thus, the process $\{z_t\}$ satisfies strong and weak laws of large numbers. If, in addition, $\{Y_t^\beta\}$ satisfies Condition B and our weak uniform continuity condition is satisfied, then weak consistency of the SME is implied by the UWLLN (Lemma 2).

Though the geometric ergodicity assumption accommodates more general processes than the AUC condition, our consistency proof based on the former requires the imposition of a uniform Lipschitz condition. This uniform continuity condition implicitly requires some damping of the effects of past shocks on current values of Y^β. We have not shown that processes of the form (4.11), for example, satisfy our uniform Lipschitz condition. Verifying this condition may well narrow the gap between the classes of models encompassed by the sets of regularity conditions used to prove weak and strong consistency of the SME.

5. ASYMPTOTIC NORMALITY

Under the unit-circle conditions introduced in Section 4.4, the stationary and ergodic process $\{Y_t^{\infty\beta}\}$ can be substituted for $\{Y_t^\beta\}$ in deducing the asymptotic distribution of the SME. Thus, the asymptotic normality of $\{b_T\}$ follows immediately under suitably modified versions of the regularity conditions imposed by Hansen (1982). If, instead, the regularity conditions used to prove weak consistency in Section 4.3 are adopted, then Hansen's (1982) conditions are no longer directly applicable because of the nonstationarity of $\{Y_t^\beta\}$. Therefore, our discussion of asymptotic normality focuses on the case of geometrically ergodic forcing processes that may not satisfy an AUC condition. The final characterization of the limiting distribution of the SME is, of course, the same for either set of regularity conditions.

In deriving the asymptotic distribution of $\{\sqrt{T}(b_T - \beta_0)\}$, we use an intermediate-value expansion of $G_T(\beta)$ about the point β_0. Accordingly, we will adopt the following assumption.

ASSUMPTION 6:
(i) β_0 and the estimators $\{b_T\}$ are interior to Θ.
(ii) f_t^β is continuously differentiable with respect to β for all t, ω by ω.
(iii) $D_0 \equiv E[\partial f_\infty^{\beta_0}/\partial \beta]$ exists, is finite, and has full rank.

Expanding $G_T(b_T)$ about β_0 gives

(5.1) $\quad G_T(b_T) = G_T(\beta_0) + \partial G^*(T)(b_T - \beta_0),$

where (using the intermediate value theorem) $\partial G^*(T)$ is the $M \times Q$ matrix whose ith row is the ith row of $\partial G_T(b_T^i)/\partial \beta$, with b_T^i equal to some convex combination of β_0 and b_T. Premultiplying (5.1) by $[\partial G_T(b_T)/\partial \beta]'W_T$, and applying the first order conditions for the optimization problem defining b_T,

(5.2) $\quad \left[\dfrac{\partial G_T(b_T)}{\partial \beta}\right]' W_T G_T(b_T) = 0 = \left[\dfrac{\partial G_T(b_T)}{\partial \beta}\right]' W_T G_T(\beta_0) + J_T(b_T - \beta_0),$

where

$$J_T = \left[\dfrac{\partial G_T(b_T)}{\partial \beta}\right]' W_T \partial G^*(T).$$

Equation (5.2) can be solved for $b_T - \beta_0$ if J_T is invertible for sufficiently large T. This invertibility is given by Assumption 5 (iii) provided $\partial G_T(b_T)/\partial \beta$ converges in probability to D_0. For notational ease, let $D_\beta f_t^\beta = (d/d\beta) f(Z_t^\beta, \beta)$ (the total derivative). Under the following additional assumptions, Lemma 2 and Theorem 4.1.5 of Amemiya (1985) imply that $\text{plim}_T \partial G_T(b_T)/\partial \beta = D_0$.

ASSUMPTION 7: *The family* $\{D_\beta f_t^\beta: \beta \in \Theta, t = 1, 2, \ldots\}$ *is Lipschitz, uniformly in probability. For all* $\beta \in \Theta$, $E(|D_\beta f_\infty^\beta|) < \infty$, *and* $\beta \mapsto E(D_\beta f_\infty^\beta)$ *is continuous.*

Under these assumptions, the asymptotic distribution of $\sqrt{T}(b_T - \beta_0)$ is equivalent to the asymptotic distribution of $(D_0' \Sigma_0^{-1} D_0)^{-1} \sqrt{T} G_T(\beta_0)$. The following theorem provides the limiting distribution of $\sqrt{T} G_T(\beta_0)$.

THEOREM 4: *Suppose $T/\mathscr{T}(T) \to \tau$ as $T \to \infty$. Under Assumptions 1–4, and 6–7,*

(5.3) $\quad \sqrt{T} G_T(\beta_0) \Rightarrow N[0, \Sigma_0(1+\tau)].$

PROOF: From the definition of G_T,

(5.4) $\quad \sqrt{T} G_T(\beta_0) = \left(\dfrac{1}{\sqrt{T}} \sum_{t=1}^{T} [f_t^* - E(f_\infty^*)] \right)$

$\qquad - \dfrac{\sqrt{T}}{\sqrt{\mathscr{T}(T)}} \left(\dfrac{1}{\sqrt{\mathscr{T}(T)}} \sum_{s=1}^{\mathscr{T}(T)} [f_s^{\beta_0} - E(f_\infty^{\beta_0})] \right).$

We do not have stationarity, but the proof of asymptotic normality of each term on the right-hand side of (5.4) follows Doob's (1953) proof of a central limit theorem (Theorem 7.5), which uses instead the stronger geometric ergodicity condition. In particular, we are using the assumed bounds on $\|f_t^\beta\|_{2+\delta}$ to conclude that asymptotic normality of f_t^* and $f_t^{\beta_0}$ (suitably normalized) follows from the geometric ergodicity of $\{Y_t\}$ and $\{Y_t^{\beta_0}\}$. (Note that, although Doob's Theorem 7.5 includes his condition D_0 as a hypothesis, the geometric ergodicity property is actually sufficient for its proof.) Our result then follows from the independence of the two terms in (5.4) and the convergence of $\sqrt{T}/\sqrt{\mathscr{T}(T)}$ to $\sqrt{\tau}$. Q.E.D.

An immediate implication of Theorem 4 is the following corollary.

COROLLARY 3.1: *Under the assumptions of Theorem 4, $\sqrt{T}(b_T - \beta_0)$ converges in distribution as $T \to \infty$ to a normal random vector with mean zero and covariance matrix*

(5.5) $\quad \Lambda = (1+\tau)\left(D_0' \Sigma_0^{-1} D_0\right)^{-1}.$

The form of the asymptotic covariance matrix Λ is familiar from the results of McFadden (1989), Pakes and Pollard (1989), and Lee and Ingram (1991). As τ gets small, the asymptotic covariance matrix of $\{b_T\}$ approaches $[D_0' \Sigma_0^{-1} D_0]^{-1}$, the covariance matrix obtained when an analytic expression for $E(f_\infty^\beta)$ as a function of β is known a priori. The proposed SM estimator uses a Monte Carlo generated estimate of this mean, which permits consistent estimation of β_0 for circumstances in which the functional form of $E(f_\infty^\beta)$ is not known. In general, knowledge of $E(f_\infty^\beta)$ increases the efficiency of the method of moments estimator of β_0. If, however, the simulated sample size $\mathscr{T}(T)$ is chosen to be large relative to the size T of the sample of observed variables $\{f_t^*\}$, then there is essentially no loss in efficiency from ignorance of this population mean. Thus, the proposed simulated moments estimator extends the class of Markov processes that can be studied using method-of-moment estimators beyond those considered previously, with potentially negligible loss of efficiency.

These results presume that the model is identified. The rank condition for the class of models considered here is Assumption 6 (iii). In many GMM problems, verifying that the choice of moment conditions identifies the unknown parameters under plausible assumptions about the correlations among the variables in the model is straightforward. However, inspection of the moment conditions used in simultaneously solving and estimating dynamic asset-pricing models may give little insight into whether Assumption 6 (iii) is satisfied. This may be especially relevant when the model is solved numerically for some of the elements of $\{Y_t^\beta\}$ as functions of the state and parameter vectors. Indeed, in this case, it may be difficult to gain much insight into which moment conditions will shed light on the values of specific parameters. We recommend that, in practice, the sensitivity of the estimates to various choices of moment conditions be examined.

Fortunately, some information about the validity of this assumption can be obtained in our environment using the simulated state $\{Y_t^\beta\}$. At a given value of β, the partial derivative matrix

$$(5.6) \quad D(\beta) = \frac{\partial \left[\frac{1}{\mathcal{T}} \sum_{t=1}^{\mathcal{T}} f_t^\beta \right]}{\partial \beta}$$

can be calculated numerically. For large values of the simulation size $\mathcal{T}$, $D(\beta)$ is approximately equal to $\partial E(f_t^\beta)/\partial \beta$. An orthogonalization of $D(\beta)$ can be examined at various values of β in order to gain some insight into whether the first order conditions defining the SME form a relatively ill-conditioned system of equations at certain points in the parameter space, including at the SME estimator of β_0.

6. EXTENSIONS AND CONCLUSIONS

The SME proposed in this paper can be extended along a variety of different dimensions. One obvious extension is to let f_t^* be a function of β. In order to accommodate this extension, we need one additional primitive, a measurable observation function $g: \mathbb{R}^{NL} \times \Theta \to \mathbb{R}^M$, where L is the number of periods of states entering into the observation $g[(Y_t, \ldots, Y_{t-L+1}), \beta]$ at time t. We can always assume without loss of generality that $L = l$. We replace the observation f_t^* on the actual state process used in the SME with the observation $g_t^{\beta_0} \equiv g(Z_t, \beta_0)$, and assume that $E[g_t^{\beta_0} - f_t^{\beta_0}] = 0$. This leads us to consider the difference in sample moments:

$$(6.1) \quad G_T(\beta) = \frac{1}{T} \sum_{t=1}^{T} g_t^\beta - \frac{1}{\mathcal{T}(T)} \sum_{s=1}^{\mathcal{T}(T)} f_s^\beta.$$

We once again introduce a sequence $\{W_T\}$ of positive semi-definite distance matrices, and define the criterion function $C_T(\beta) = G_T(\beta)' W_T G_T(\beta)$ as well as the extended simulated moments estimator $\{b_T\}$ of β_0, just as in (3.5).

SIMULATED MOMENTS ESTIMATION

In this case, we replace Σ_0 defined by (4.8) with the weighted covariance matrix, for some positive scalar weight τ,

(6.2) $\quad \Sigma_{f,g,\tau} = \tau \Sigma_0 + \Sigma_1,$

where

(6.3) $\quad \Sigma_1 = \sum_{j=-\infty}^{\infty} E\big([g_t^{\beta_0} - E(g_t^{\beta_0})][g_{t-j}^{\beta_0} - E(g_{t-j}^{\beta_0})]' \big).$

Assuming that the families $\{f_t^\beta\}$ and $\{g_t^\beta\}$ satisfy the technical conditions of Assumption 1,[10] and that $W_T \to W_0 = \Sigma_{f,g,\tau}^{-1}$ almost surely, the weak consistency of this extended SME follows from an argument almost identical to the proof of Theorem 1. Furthermore, replacing Assumption 6 (iii) by the assumption that $D_0 \equiv E[\partial g_t^{\beta_0}/\partial \beta - \partial f_\infty^{\beta_0}/\partial \beta]$ exists, is finite, and has full rank, Theorem 4 implies that $\sqrt{T}(b_T - \beta_0)$ converges in distribution to a normal random vector with mean zero and covariance matrix

(6.4) $\quad \Lambda_{f,g,\tau} = \big(D_0' \Sigma_{f,g,\tau}^{-1} D_0 \big)^{-1}.$

The new rank condition on D_0 is an identification condition which, among other things, rules out trivial sources of underidentification such as g_t^β and f_∞^β having the multiplicative representations $g^1(z_t, \beta^1)\psi(z_t, \beta^2)$ and $f^1(z_\infty, \beta^1)\psi(z_\infty, \beta^2)$, with β^1 and β^2 being distinct. Also, in contrast to the matrix Λ in (5.5), consistent estimation of $\Lambda_{f,g,\tau}$ must typically be accomplished in two steps, using both simulated and observed data.

Allowing the observation function g_t^β to depend on β is useful in many asset-pricing problems. For instance, one may wish to compare the sample mean of the intertemporal marginal rate of substitution of consumption in the data to the mean of the corresponding simulated series.

A second example arises when one or more of the coordinate functions defining g, say g_j, has the property that $h_j(\beta) = E[g_j(Z_\infty, \beta)]$ defines a known function h_j of β. If this calculation cannot be made for every j, one can mix the use of calculated and simulated moments by letting $f_j(z, \beta) = h_j(\beta)$ for all z, for any j for which h_j is known. This substitution of calculated moments for sample moments improves the precision of the simulated moments estimator, in that the covariance matrix $\Lambda_{f,g,\tau}$ is smaller than the covariance matrix Λ obtained when all moments are simulated. Errors in measurement of f_t^* are accommodated by letting $g_t^{\beta_0} = f(Z_t, \beta_0) + u_t$, where $\{u_t\}$ is an ergodic, mean-zero $\mathbb{R}^M$-valued measurement error. Note that the asymptotic efficiency of the SME is increased by ignoring the measurement error in simulation and comparing sample moments of the simulated $\{f(z_t^\beta, \beta)\}$ and $\{g_t^\beta\}$.

---

[10] Note that the uniform-in-probability Lipschitz condition for $\{g_t^\beta\}$ is qualitatively weaker than the same condition for $\{f_t^\beta\}$, since g_t^β depends only directly on β (that is, Y_t is not dependent on β).

Finally, one of the coordinate functions of the actual state observations, say g_j, may be of the form

$$g_j[(Y_t, Y_{t-1}, \ldots, Y_{t-l+1}), \beta]$$
$$= E[h_j(Y_{t+1}, \ldots, Y_{t+l+2}, \beta) | Y_t, Y_{t-1}, \ldots, Y_{t-l+1}],$$

for some h_j. It may be infeasible to calculate the function g_j explicitly, in which case the simulated observation $g_j(Z_t^\beta, \beta)$ is not available, except perhaps by numerical approximation. On the other hand, the observation of $f_j(Z_t^\beta, \beta) = h_j(Z_t^\beta, \beta)$ is often feasible and, by the law of iterated expectations, has the same mean as $g_j(Z_t^\beta, \beta)$. An important illustration of such a function g_j arises in the option pricing literature, where the European option price $g_j(Z_t^\beta, \beta)$ is the conditional expectation of the option's payoff at maturity discounted by an appropriate factor.

Grad. School of Business, Stanford University, Stanford, CA 94305-5015, U.S.A.

Manuscript received January, 1990; final revision received December, 1992.

APPENDIX

PROOF OF LEMMA 2:[11] Since Θ is compact it can be partitioned, for any n, into n disjoint neighborhoods $\Theta_1^n, \Theta_2^n, \ldots, \Theta_n^n$ in such a way that the distance between any two points in each Θ_i^n goes to zero as $n \to \infty$. Let $\beta_1, \beta_2, \ldots, \beta_n$ be an arbitrary sequence of vectors such that $\beta_i \in \Theta_i^n$, $i = 1, \ldots, n$. Then, for any $\varepsilon > 0$,

$$(\text{A.1}) \quad P\left[\sup_{\beta \in \Theta} \left| \frac{1}{T} \sum_{t=1}^T (f_t^\beta - E(f_\infty^\beta)) \right| > \varepsilon \right]$$

$$\leq P\left[\bigcup_{i=1}^n \left\{ \sup_{\beta \in \Theta_i^n} \left| \frac{1}{T} \sum_{t=1}^T (f_t^\beta - E(f_\infty^\beta)) \right| > \varepsilon \right\} \right]$$

$$\leq \sum_{i=1}^n P\left[\sup_{\beta \in \Theta_i^n} \left| \frac{1}{T} \sum_{t=1}^T (f_t^\beta - E(f_\infty^\beta)) \right| > \varepsilon \right]$$

$$\leq \sum_{i=1}^n P\left[\left| \frac{1}{T} \sum_{t=1}^T (f_t^{\beta_i} - E(f_\infty^{\beta_i})) \right| > \frac{\varepsilon}{2} \right]$$

$$+ \sum_{i=1}^n P\left[\frac{1}{T} \sum_{t=1}^T \sup_{\beta \in \Theta_i^n} |f_t^\beta - f_t^{\beta_i}| + \sup_{\beta \in \Theta_i^n} |E(f_\infty^\beta) - E(f_\infty^{\beta_i})| > \frac{\varepsilon}{2} \right],$$

where the last inequality follows from the triangle inequality. For fixed n, since $\{Y_t^{\beta_i}\}$ is ergodic and $E(|f_t^{\beta_i}|) < \infty$, the first term on the right-hand side of (A.1) approaches zero as $T \to \infty$ by the weak law of large numbers for ergodic processes.

---

[11] The strategy for proving this lemma, which was suggested to us by Whitney Newey, follows the proof strategies used by Jennrich (1969) and Amemiya (1985) to prove similar lemmas. A subsequent paper by Newey (1991) presents a more extensive discussion of sufficient conditions for uniform convergence in probability.

SIMULATED MOMENTS ESTIMATION

As for the second right-hand-side term in (A.1), the Lipschitz assumption on $\{f_t^\beta\}$ implies that there exist K_t such that

$$(A.2) \quad \sum_{i=1}^{n} P\left[\frac{1}{T}\sum_{t=1}^{T}\sup_{\beta\in\Theta_i^n}|f_t^\beta - f_t^{\beta_i}| + \sup_{\beta\in\Theta_i^n}|E(f_\infty^\beta) - E(f_\infty^{\beta_i})| > \frac{\varepsilon}{2}\right]$$

$$\leq \sum_{i=1}^{n} P\left[\sup_{\beta\in\Theta_i^n}|\beta - \beta_i|\frac{1}{T}\sum_{t=1}^{T}K_t + \sup_{\beta\in\Theta_i^n}|E(f_\infty^\beta) - E(f_\infty^{\beta_i})| > \frac{\varepsilon}{2}\right].$$

The assumption that $K^T = T^{-1}\sum_{t=1}^{T} K_t$ is bounded in probability implies that there is a nonstochastic bounded sequence $\{A_T\}$ such that $\text{plim}(K^T - A_T) = 0$. Thus, for T larger than some T^* and some bound B, the right-hand side of (A.2) is less than or equal to

$$(A.3) \quad \sum_{i=1}^{n} P\left[\sup_{\beta\in\Theta_i^n}|\beta - \beta_i||K^T - A_T| + \sup_{\beta\in\Theta_i^n}|\beta - \beta_i|B + \sup_{\beta\in\Theta_i^n}|E(f_\infty^\beta) - E(f_\infty^{\beta_i})| > \frac{\varepsilon}{2}\right].$$

By continuity of $\beta \mapsto E(f_\infty^\beta)$, we can choose n once and for all so that $|\beta - \beta_i|B + |E(f_\infty^\beta) - E(f_\infty^{\beta_i})| < (\varepsilon/4)$ for all β in Θ_i^n and all i. Thus, the limit of (A.3) as $T \to \infty$ is zero, and the result follows.
Q.E.D.

PROOF OF THEOREM 1: By the triangle inequality,

$$(A.4) \quad \left|\left(\frac{1}{T}\sum_{t=1}^{T} f_t^* - \frac{1}{\mathcal{T}}\sum_{s=1}^{\mathcal{T}} f_s^\beta\right) - [E(f_\infty^*) - E(f_\infty^\beta)]\right|$$

$$\leq \left|E(f_\infty^*) - \frac{1}{T}\sum_{t=1}^{T} f_t^*\right| + \left|E(f_\infty^\beta) - \frac{1}{\mathcal{T}}\sum_{s=1}^{\mathcal{T}} f_s^\beta\right|.$$

Assumption 2 implies that the first term on the right-hand side of (A.4) converges to zero in probability. By Lemma 2, the second term on the right-hand side of (A.4) converges in probability to zero uniformly in β. Now $\delta_T(\beta) \equiv |C_T(\beta) - C(\beta)|$ satisfies

$$(A.5) \quad \delta_T(\beta) = \left|G_T(\beta)'W_T G_T(\beta) - [E(f_\infty^*) - E(f_\infty^\beta)]'W_0[E(f_\infty^*) - E(f_\infty^\beta)]\right|$$

$$\leq |G_T(\beta) - [E(f_\infty^*) - E(f_\infty^\beta)]|'|W_T||G_T(\beta)|$$

$$+ |E(f_\infty^*) - E(f_\infty^\beta)|'|W_T - W_0||G_T(\beta)|$$

$$+ |E(f_\infty^*) - E(f_\infty^\beta)|'|W_0||G_T(\beta) - [E(f_\infty^*) - E(f_\infty^\beta)]|.$$

Therefore, letting $l_T = \sup_{\beta\in\Theta}|G_T(\beta) - [E(f_\infty^*) - E(f_\infty^\beta)]|$,

$$(A.6) \quad \sup_{\beta\in\Theta}\delta_T(\beta) \leq l_T|W_T|[\phi_0 + l_T] + \phi_0|W_T - W_0|[\phi_0 + l_T] + \phi_0|W_0|l_T,$$

where $\phi_0 \equiv \max\{|E(f_\infty^*) - E(f_\infty^\beta)|: \beta \in \Theta\}$ exists by the continuity condition in Assumption 1. Since each of the terms on the right-hand side of (A.6) converges in probability to zero, $\text{plim}_T[\sup_{\beta\in\Theta}\delta_T(\beta)] = 0$. This implies the convergence of $\{b_T\}$ to β_0 in probability as $T \to \infty$, as indicated, for example, in Amemiya (1985, page 107).
Q.E.D.

PROOF OF LEMMA 3: We fix β and t. For simplicity, we write "ε_t" for $\hat{\varepsilon}_t$. For each positive integer m, we define $\{Y_s^{m\beta}: t - m \leq s \leq t\}$ by the recursion $Y_{t-m}^{m\beta} = 0$ and

$$Y_{t-m+k+1}^{m\beta} = H(Y_{t-m-k}^{m\beta}, \beta, \varepsilon_{t-m+k+1}).$$

By construction, $Y_t^{m\beta}$ is measurable with respect to $\{\varepsilon_t, \varepsilon_{t-1}, \ldots, \varepsilon_{t-m+1}\}$. The AUC condition implies that

$$\text{(A.7)} \quad \|Y_t^{m\beta} - Y_t^{m+1,\beta}\| \leq \prod_{j=0}^{m} \rho_\beta(\varepsilon_{t-j}) \|H(0, \varepsilon_{t-m+1}, \beta)\|,$$

where

$$\frac{1}{m}\sum_{j=0}^{m} \ln \rho_\beta(\varepsilon_{t-j}) + \frac{1}{m} \ln\left(\max\left[1, \|H(0, \varepsilon_{t-m+1}, \beta)\|\right]\right) \xrightarrow{\text{a.s.}} \alpha_\beta < 0.$$

Hence,

$$\text{(A.8)} \quad \left[\prod_{j=0}^{m} \rho_\beta(\varepsilon_{t-j})\right]^{1/m} \|H(0, \varepsilon_{t-m+1}, \beta)\|^{1/m} \xrightarrow{\text{a.s.}} e^{\alpha_\beta} < 1.$$

This, in turn, implies that, given $\delta \in (e^{\alpha_\beta}, 1)$, there is some event Λ with $P(\Lambda) = 1$ and, for each $\omega \in \Lambda$, some integer $N(\omega, \delta)$ such that

$$\left[\prod_{j=0}^{m} \rho_\beta(\varepsilon_{t-j}(\omega))\right] \|H(0, \varepsilon_{t-m+1}(\omega), \beta)\| < \delta^m, \quad m \geq N(\omega, \delta).$$

Next, at arbitrary $\omega \in \Lambda$ and $m > n \geq N(\omega, \delta)$,

$$\|Y_t^{m\beta} - Y_t^{n\beta}\| \leq \|Y_t^{m\beta} - Y_t^{m-1,\beta}\| + \|Y_t^{m-1,\beta} - Y_t^{m-2,\beta}\| + \cdots + \|Y_t^{n+1,\beta} - Y_t^{n\beta}\|$$

$$\leq \prod_{j=0}^{m-1} \rho_\beta(\varepsilon_{t-j}) \|H(0, \varepsilon_{t-m}, \beta)\| + \cdots + \prod_{j=0}^{n} \rho_\beta(\varepsilon_{t-j}) \|H(0, \varepsilon_{t-n+1}, \beta)\|$$

$$\leq \delta^{m-1} + \delta^{m-2} + \cdots + \delta^n = \frac{\delta^{n-1}(1 - \delta^{m-n+1})}{1 - \delta} \leq \frac{\delta^{n-1}}{1 - \delta}.$$

It follows that, at each $\omega \in \Lambda$, $\{Y_t^{m\beta}(\omega)\}$ is a Cauchy sequence in m. We conclude that $\lim_{m \to \infty} Y_t^{m\beta} = Y_t^{\infty\beta}$ exists almost surely. The limit process $\{Y_t^{\infty\beta}: -\infty < t < \infty\}$, constructed for each t in this manner, satisfies the difference equation (3.1) by construction and $Y_t^{\infty\beta}$ is clearly measurable with respect to $\{\varepsilon_{t-s}: s \geq 0\}$. Since $\{\varepsilon_t\}$ is an i.i.d. sequence, the stationarity and ergodicity of $\{Y_t^{\infty\beta}\}$ follows immediately. $\quad$ Q.E.D.

PROOF OF LEMMA 4: Fix $\theta \in \Theta$ and without loss of generality set $l = 1$. For any $\beta \in \Theta$ such that $\|\beta - \theta\| < \delta_\theta$,

$$\left|\frac{1}{T}\sum_{t=1}^{T} f_t^\beta - \frac{1}{T}\sum_{t=1}^{T} f_t^{\infty\beta}\right| \leq k(\theta)\frac{1}{T}\sum_{t=1}^{T} \|Y_t^\beta - Y_t^{\infty\beta}\|$$

$$\leq k(\theta)\frac{1}{T}\sum_{t=1}^{T}\left[\prod_{j=0}^{t} \rho_\theta(\varepsilon_j)\right]\|Y_0^\beta - Y_0^{\infty\beta}\|,$$

where $k(\theta)$ is given by the S-smoothness assumption. The AUC condition implies that $(1/T)\sum_{t=1}^{T}[\prod_{j=0}^{t}\rho_\theta(\varepsilon_j)]$ converges almost surely to zero. Thus, given $\eta > 0$, there is an event Λ_θ with $P(\Lambda_\theta) = 1$ such that, for each ω in Λ_θ, there is some $T_\theta(\omega, \eta)$ with

$$\text{(A.9)} \quad \Delta_T^\beta \equiv \left|\frac{1}{T}\sum_{t=1}^{T} f_t^\beta - \frac{1}{T}\sum_{t=1}^{T} f_t^{\infty\beta}\right| \leq \eta, \quad T \geq T_\theta(\omega, \eta),$$

provided $\|\beta - \theta\| \leq \delta_\theta$.

Since Θ is compact, it has a finite subset Θ^* defining a finite subcover of "δ_θ neighborhoods," $\theta \in \Theta^*$. Letting $\Lambda^* = \bigcap_{\theta \in \Theta^*} \Lambda_\theta$ and $T^*(\omega, \eta) = \max_{\theta \in \Theta^*} T_\theta(\omega, \eta)$, it follows that $\Delta_T^\beta \leq \eta$, $T \geq T^*$, for all β in Θ, which leads to (4.7).   Q.E.D.

PROOF OF THEOREM 4: As noted above, the L^2 UC condition implies the AUC condition, so the conclusions of Lemmas 3 and 4 continue to hold. Thus, the consistency of $\{b_T\}$ for β_0 will be established by showing that, for each $\theta \in \Theta$, $E[\text{mod}_t(\delta, \theta)] < \infty$ for some $\delta > 0$. As before, we write "ε_t," for "$\tilde{\varepsilon}_t$."

Fix $\theta \in \Theta$. For purposes of the proof, we can assume without loss of generality that $l = 1$. Since f is S-smooth, there is a $\delta > 0$ such that, for $\|\beta - \theta\| \leq \delta$ and for each t,

$$\|f(Y_t^{\infty\beta}, \beta) - f(Y_t^{\infty\theta}, \theta)\| = \|f(Y_t^{\infty\beta}, \beta) - f(Y_t^{\infty\beta}, \theta) + f(Y_t^{\infty\beta}, \theta) - f(Y_t^{\infty\theta}, \theta)\|$$

$$\leq C_1(Y_t^\beta)\|\beta - \theta\| + k(\theta)\|Y_t^{\infty\beta} - Y_t^{\infty\theta}\|.$$

It follows that

(A.10) $\quad \text{mod}(\delta, \theta) \leq \delta \sup_t C_1(Y_t^\beta) + k(\theta) \sup_{\|\beta-\theta\|\leq\delta} \|Y_t^{\infty\beta} - Y_t^{\infty\theta}\|.$

Letting $\alpha_t = \|Y_t^{\infty\beta} - Y_t^{\infty\theta}\|$, the L^2 UC condition and S-smoothness of H imply that

(A.11) $\quad \alpha_t \leq \rho_\theta(\varepsilon_t)\alpha_{t-1} + C_2(Y_{t-1}^\theta, \varepsilon_t)\delta.$

By recursively substituting α_{t-k}, using (A.11), we have for any T

$$\alpha_t \leq \prod_{s=t-T}^{t} \rho_\theta(\varepsilon_s)\alpha_{t-T} + \delta \sum_{s=t-T}^{t} C_2(Y_s^{\theta\infty}, \varepsilon_s) \prod_{\tau=s+1}^{t} \rho_\theta(\varepsilon_\tau).$$

Now, $X_T \equiv \prod_{s=t-T}^{t} \rho_\theta(\varepsilon_s)$ converges to zero in L^2 since $E[\rho_\theta(\varepsilon_t)^2] < 1$ and $\{\varepsilon_t\}$ is i.i.d. Since $\|\alpha_{t-T}\|_2 \leq \|Y_{t-T}^{\beta\infty}\|_2$ is bounded, the Cauchy-Schwarz inequality implies that

$$\|X_T \alpha_{t-T}\|_1 \leq \|X_T\|_2 \|\alpha_{t-T}\|_2 \to_T 0,$$

so, in L^1,

$$\alpha_t \leq \delta \lim_{T\to\infty} \sum_{s=t-T}^{t} C_2(Y_s^{\infty\theta}, \varepsilon_s) \prod_{\tau=s+1}^{t} \rho_\theta(\varepsilon_\tau).$$

The right-hand side is independent of β, and taking expectations, using the independence of $\{\varepsilon_t\}$ and the Cauchy-Schwarz inequality, we have

$$E\left[\sup_{\|\beta-\theta\|\leq\delta} \|Y_t^{\infty\beta} - Y_t^{\infty\theta}\|\right] \leq \delta E\left[\sum_{s=-\infty}^{t} C_2(Y_s^{\infty\theta}, \varepsilon_s) \prod_{\tau=s+1}^{t} \rho_\theta(\varepsilon_s)\right]$$

$$\leq \frac{\delta K}{1 - \bar{\rho}},$$

where $\bar{\rho} = \|\rho_\theta(\varepsilon_t)\|_2 < 1$ and where K is a bound on $\|C_2(Y_s^{\infty\beta}, \varepsilon_s)\|_2$ implied by the growth condition on C_2 and the fact that $\|Y_t^{\infty\beta}\|_2$ and $\|\varepsilon_t\|_2$ are bounded.

The last term in (A.10) therefore has a finite mean. To establish that the first term on the right-hand side of (A.10) has a finite mean, first note that $C_1(Y_t^{\infty\beta}) \leq d_1 + d_2\|Y_t^{\infty\beta}\|$, for constants d_1, d_2. Furthermore,

(A.12) $\quad \sup_{\|\beta-\theta\|\leq\delta} \|Y_t^{\infty\beta}\| \leq \|Y_t^{\infty\theta}\| + \sup_{\|\beta-\theta\|\leq\delta} \|Y_t^{\infty\beta} - Y_t^{\infty\theta}\|,$

and both terms on the right-hand side of (A.12) have finite means.

Combining these results with Hansen's (1982) Theorem 2.1 gives the desired result.   Q.E.D.

REFERENCES

AMEMIYA, T. (1985): *Advanced Econometrics*. Cambridge, Massachusetts: Harvard University Press.
ANDREWS, D. (1987): "Consistency in Nonlinear Econometric Models: A Generic Uniform Law of Large Numbers," *Econometrica*, 55, 1465–1472.
—— (1991): "Heteroskedasticity and Autocorrelation Consistent Covariance Matrix Estimation," *Econometrica*, 59, 817–858.
BROCK, W. (1980): "Asset Prices in a Production Economy," in *The Economics of Uncertainty*, ed. by J. J. McCall. Chicago: University of Chicago Press.
DOOB, J. (1953): *Stochastic Processes*. New York: John Wiley and Sons.
DUFFIE, D. (1988): *Security Markets: Stochastic Models*. Boston: Academic Press.
DUNN, K., AND K. SINGLETON (1986): "Modeling the Term Structure of Interest Rates under Non-separable Utility and Durability of Goods," *Journal of Financial Economics*, 17, 27–55.
EICHENBAUM, M., AND L. HANSEN (1988): "Estimating Models with Intertemporal Substitution Using Aggregate Time Series Data," manuscript, University of Chicago.
EICHENBAUM, M., L. HANSEN, AND K. SINGLETON (1988): "A Time Series Analysis of Representative Agent Models of Consumption and Leisure Choice under Uncertainty," *Quarterly Journal of Economics*, 103, 51–78.
GALLANT, R., AND H. WHITE (1988): *A Unified Theory of Estimation and Inference for Nonlinear Dynamic Models*. Oxford: Basil Blackwell.
HALL, R. (1988): "Intertemporal Substitution in Consumption," *Journal of Political Economy*, 96, 339–357.
HANSEN, L. (1982): "Large Sample Properties of Generalized Method of Moment Estimators," *Econometrica*, 50, 1029–1056.
HANSEN, L., AND K. SINGLETON (1982): "Generalized Instrumental Variables Estimation of Nonlinear Rational Expectations Models," *Econometrica*, 50, 1269–1286.
—— (1989): "Efficient Estimation of Linear Asset Pricing Models with Moving Average Errors," manuscript, Stanford University.
JENNRICH, R. (1969): "Asymptotic Properties of Non-Linear Least Squares Estimators," *Annals of Mathematical Statistics*, 40, 633–643.
LEE, B., AND B. INGRAM (1991): "Simulation Estimation of Time Series Models," *Journal of Econometrics*, 47, 197–205.
MCFADDEN, D. (1989): "A Method of Simulated Moments for Estimation of Discrete Response Models without Numerical Integration," *Econometrica*, 57, 995–1026.
MICHNER, R. (1984): "Permanent Income in General Equilibrium," *Journal of Monetary Economics*, 14, 297–305.
MOKKADEM, A. (1985): "Le Modèle Non Linéaire AR(1) Général. Ergodicité et Ergodicité Géometrique," *Comptes Rendues Academie Scientifique Paris*, 301, Série I, 889–892.
NEWEY, W. (1991): "Uniform Convergence in Probability and Stochastic Equicontinuity," *Econometrica*, 59, 1161–1167.
NEWEY, W., AND K. WEST (1987): "A Simple, Positive Definite, Heteroskedasticity and Autocorrelation Consistent Covariance Matrix," *Econometrica*, 55, 703–708.
NUMMELIN, E., AND P. TUOMINEN (1982): "Geometric Ergodicity of Harris Recurrent Markov Chains with Applications to Renewal Theory," *Stochastic Processes and Their Applications*, 12, 187–202.
PAKES, A., AND D. POLLARD (1989): "The Asymptotics of Simulation Estimators," *Econometrica*, 57, 1027–1058.
ROSENBLATT, M. (1971): *Markov Processes. Structure and Asymptotic Behavior*. New York: Springer-Verlag.
SINGLETON, K. (1987): "Asset Prices in a Time-Series Model with Disparately Informed, Competitive Traders," Chapter 12 in *New Approaches to Monetary Economics*, ed. by W. Barnett and K. Singleton. Cambridge: Cambridge University Press, pp. 249–272.
TAUCHEN, G., AND R. HUSSEY (1991): "Quadrature Based Methods for Obtaining Approximate Solutions to Nonlinear Asset Pricing Models," *Econometrica*, 59, 371–396.
TAYLOR, J., AND H. UHLIG (1990): "Solving Nonlinear Stochastic Growth Models: A Comparison of Alternative Solution Methods," *Journal of Business and Economic Statistics*, 8, 1–18.
TWEEDIE, R. (1982): "Criteria for Rates of Convergence of Markov Chains, with Application to Queuing and Storage Theory," in *Probability, Statistics, and Analysis*, edited by J. F. C. Kingman and G. E. H. Reuter. Cambridge: Cambridge University Press.

[22]

A Test of the Cox, Ingersoll, and Ross Model of the Term Structure

Michael R. Gibbons
Krishna Ramaswamy
University of Pennsylvania

We test the theory of the term structure of indexed-bond prices due to Cox, Ingersoll, and Ross (CIR). The econometric method uses Hansen's generalized method of moments and exploits the probability distribution of the single-state variable in CIR's model, thus avoiding the use of aggregate consumption data. It enables us to estimate a continuous-time model based on discretely sampled data. The tests indicate that CIR's model for index bonds performs reasonably well when confronted with short-term Treasury-bill returns. The estimates indicate that term premiums are positive and that yield curves can take several shapes. However, the fitted model does poorly in explaining the serial correlation in real Treasury-bill returns.

We are grateful to Robert Bliss, Doug Breeden, John Cox, Phil Dybvig, Steve Heston, Andrew Lo, Craig MacKinlay, David Modest, George Pennacchi, Ken Singleton, Chester Spatt (editor), Rob Stambaugh, Suresh Sundaresan, and the anonymous referee for their comments, and to Tong-sheng Sun and Robert Whitelaw for research assistance. We thank participants of seminars at Carnegie Mellon University, Columbia University, Dartmouth College, the French Finance Association, Indiana University, INSEAD, London Business School, MIT, NBER, Northwestern University, Stanford University, the University of Chicago, and the University of Illinois. Financial support was provided by the Center for the Study of Futures Markets at Columbia University and the Geewax-Terker Program in Investments. This research was started while Gibbons was on a fellowship provided by Batterymarch Financial Management. We are responsible for any errors. This article replaces the authors' working paper, "The Term Structure of Interest Rates: Empirical Evidence." Address correspondence and reprint requests to Michael Gibbons, The Wharton School of The University of Pennsylvania, 2300 SH-DH, Philadelphia, PA 19104.

The relation between the yields on default-free loans and their maturities has long been a topic of interest to financial economists. The focus of the early work on the term structure of interest rates was on the relation between the interest rate expected to prevail at a future date and the implied forward rate embedded in the yield curve.[1] The earliest empirical studies focused on the historical shapes of the yield curves and their relation to stages of the business cycle.

The intertemporal capital asset pricing model pioneered by Merton (1973) and the rational expectations equilibrium model due to Lucas (1978) have led researchers to consider equilibrium models of the term structure of interest rates. The term structure model developed in Cox, Ingersoll, and Ross (1985a, 1985b) represents an equilibrium specification that is completely consistent with stochastic production and with changing investment opportunities. This model provides testable implications for the prices of bonds whose payoffs are denominated in real terms—closed-form expressions are provided for the endogenously derived real prices in terms of a single-state variable (the instantaneously riskless real rate). The evolution of this variable is determined endogenously, and this permits empirical testing of the pricing implications as well as the restrictions on the dynamics of the term structure.

In this article we conduct an empirical test of the Cox, Ingersoll, and Ross (1985b; henceforth CIR) model of the term structure. Our method has the following advantages. First, we formulate a test of the implications from a continuous-time model based on discretely sampled data, and this test is designed to avoid misspecification arising from this temporal aggregation. Second, while our test centers on a stochastic Euler equation similar to tests in other studies [e.g., Hansen and Singleton (1982)] that employ Hansen's (1982) generalized method of moments (GMM), we avoid the use of data on aggregate consumption. This enables us to avoid many of the measurement problems that accompany the use of these series. Third, our econometric procedure is such that no stochastic specification of the process for the aggregate price level is necessary. Under an assumption made in CIR concerning the effect of the aggregate price level on bond prices, we test a necessary implication for a broad class of pricing models that differ by the assumptions regarding the process for inflation. Fourth, our econometric method is fully consistent with the underlying theory even though the investment opportunity set is not constant over time. There is increased interest in asset pricing when conditional distributions are not constant; however, much of

---

[1] For a review of traditional hypotheses regarding the term structure, see Cox, Ingersoll, and Ross (1981). Breeden (1986) also provides a synthesis of several strands in the literature. Melino (1986) provides a review of the evidence, focusing on the expectations hypothesis.

Test of the CIR Model

the empirical work is based on theory that is not completely specified as to why some moments are fixed while others are changing [for example, Gibbons and Ferson (1985), Ferson, Kandel, and Stambaugh (1987)]. By contrast, our econometric method requires no additional assumption beyond those maintained in the theory.

Other empirical research has examined the CIR model. Most of this work has focused on the nominal prices of U.S. government securities. Using a general framework, Stambaugh (1988) relies on nominal Treasury-bill data to reject a single latent-variable model of conditional expected returns, but he finds that the data are consistent with a model with two or three latent variables. Brown and Dybvig (1986) have examined the fit of nominal Treasury-bill prices to CIR's single-state formulation,[2] and Pearson and Sun (1990) extended Brown and Dybvig's method to CIR's models with explicit processes for inflation. Heston (1991) also uses the CIR model to find the nominal price of a nominal Treasury bond; however, his statistical method is a modification of the econometric approach that we develop here.[3] Aït-Sahalia (1992) also develops an econometric approach for nominal data, but he relies on nonparametric methods.

The inability of a single-state-variable model to fit the nominal value of a nominal government bond has led to the development of models with multiple-state variables.[4] For example, Brennan and Schwartz (1982) and Nelson and Schaefer (1983) consider some two-state-variable models, where the factor risk premiums are specified exogenously. Multiple-state-variable models of the term structure are of considerable interest, especially when one attempts to price nominal bonds, but it is not clear to us that one must abandon the study of single-state-variable specifications of the term structure of real rates. In fact, it might well be the case that the term structure of rates embedded in indexed bonds is adequately described by one forcing variable while the behavior of nominal bonds of various maturities is driven by a vector of forcing variables. One obvious disadvantage of multiple-state-variable formulations for the pricing of indexed bonds is that this makes the valuation problem complicated and often intrac-

---

[2] Using an econometric method similar to that in Brown and Dybvig (1986), Brown and Schaefer (1990) examine the fit of indexed gilts in the United Kingdom to CIR's single-state model. This cross-sectional method, like the method of finding implied volatilities associated with the Black–Scholes model, has the virtue of simplicity, but it does not examine the dynamic information in the data. Furthermore, it does not identify all the parameters in the model.

[3] Heston's (1991) specification allows him to work with bond returns in excess of the return on a short-term Treasury bill; this procedure does not rely on inflation data. However, he approximates an instantaneous holding period with a discrete holding period, and he does not identify all of CIR's parameters.

[4] Constantinides (1992) models the nominal term structure in the spirit of Brown and Dybvig (1986). His theoretical development makes his "SAINTS" model amenable to the econometric framework that we suggest in later sections.

table, and this leads us to examine the CIR model as a parsimonious, and hopefully useful, description of the term structure.

The article is organized as follows. Section 1 lays out a general framework for real and nominal bond prices, and Section 2 summarizes the CIR model. In Section 3 we discuss the design of the econometric method; Section 4 describes the data. Section 5 contains the main empirical results and summarizes the successes and failures of the CIR model. Section 6 concludes.

1. A General Framework for Nominal and Real Bond Prices

In this section we describe a general framework for pricing real and nominal discount bonds of various maturities. The discussion here applies, strictly speaking, to the general treatment in CIR (1985a). The development of the empirical test relies heavily on the arguments in this section.

The framework within which CIR develop their continuous-time valuation model can briefly be described as follows: there are infinitely lived and identical individuals who maximize the discounted expected utility of consumption of a single good, which is produced stochastically from a finite number of technologies, each exhibiting constant stochastic returns to scale. The individuals' wealths are totally invested in these firms, and they each choose a consumption rule and an investment allocation rule in maximizing their expected utility. The values of the firms in the economy evolve continuously as a vector Itô process, whose drift rate and covariance matrix depend on the evolution of a vector of state variables. The evolution of this vector of state variables is itself governed by a system of stochastic differential equations; therefore, the future investment opportunities in this model are stochastic. The environment is competitive and frictionless; a riskless asset (which is in zero net supply) and the firms' shares are available for continuous trading with no transaction costs or taxes.

The CIR model uses additional assumptions that we discuss later; the above framework is sufficient to permit a simple exposition of the valuation model. From the first-order conditions for the representative individual's maximizing problem, it follows that the current (date t) real price of a claim that pays one unit of the consumption good at date $t + \tau$, written $P_t(\tau)$, is given by

$$P_t(\tau) = E_t\left[\delta^\tau \frac{U'(\tilde{c}(t+\tau))}{U'(c(t))}\right]. \quad (1)$$

In (1), $U(c(s))$ is the utility of the optimal consumption flow $c(s)$ at date s, δ is the rate of time preference, and $E_t[\cdot]$ denotes the conditional expectation where the subscript t reflects the conditioning

Test of the CIR Model

information set. Note that this is the expected marginal rate of substitution, and it corresponds to the real price of an indexed (or real) bond that is default free. Denoting by $\pi(s)$ the money price of one unit of the consumption good at date s, the real price at date t of a nominal bond that pays $1 at date $t + \tau$ is

$$N_t(\tau) = E_t\left[\delta^\tau \left\{\frac{U'(\tilde{c}(t+\tau))}{U'(c(t))}\right\} \frac{1}{\tilde{\pi}(t+\tau)}\right], \tag{2}$$

which is the expected real payoff weighted by the marginal rates substitution. Hence, the nominal price of a nominal unit discount bond, $N_t^*(\tau)$, can be written

$$N_t^*(\tau) \equiv \pi(t) N_t(\tau) = E_t\left[\delta^\tau \left\{\frac{U'(\tilde{c}(t+\tau))}{U'(c(t))}\right\} \frac{\pi(t)}{\tilde{\pi}(t+\tau)}\right]. \tag{3}$$

We can rewrite (3) as

$$N_t^*(\tau) = P_t(\tau) E_t\left(\frac{\pi(t)}{\tilde{\pi}(t+\tau)}\right)$$

$$+ \delta^\tau \text{Cov}_t\left(\frac{U'(\tilde{c}(t+\tau))}{U'(c(t))}, \frac{\pi(t)}{\tilde{\pi}(t+\tau)}\right), \tag{4}$$

where $\text{Cov}_t(\cdot,\cdot)$ is the covariance operator conditional on information at time t. Relations (1) and (4) give the prices of real and nominal discount bonds as a function of maturity. From these equations we can readily deduce conventional yield curves in real and nominal terms. It is important to note from (1) that the real yield

$$y_t(\tau) \equiv -\ln(P_t(\tau))/\tau \tag{5}$$

is observable and achievable in a τ-period strategy only if there is an indexed bond available to investors. The availability of a nominally riskless pure discount bond ensures, however, that the nominal yield

$$y_t^*(\tau) = -\ln(N_t^*(\tau))/\tau \tag{6}$$

is observable and achievable in a τ-period strategy.

Every model of the nominal term structure must specify the conditional moments in (4). One way to achieve this is to put sufficient structure on the model to specify the joint, conditional distribution of the marginal rate of substitution $U'(\tilde{c}(t+\tau))/U'(c(t))$ and the inverse of the inflation rate $\pi(t)/\tilde{\pi}(t+\tau)$. The specification of the joint distribution (between the marginal rate of substitution and the inverse of the inflation rate) calls for an explanation of the precise way in which money enters the economic environment. Indeed the first-order condition (1) may not be the appropriate condition in models that explicitly incorporate money into either preferences or transactions technology.

CIR (1985b) implicitly assume[5] in their nominal bond-pricing examples that the covariance in (4) is zero. Since our work is a test of the CIR framework for the pricing of real indexed bonds, we follow CIR in assuming that changes in the price level have no effect on the real variables in the model.[6] The resulting expression for the nominal discount bond price is therefore

$$N_t^*(\tau) = P_t(\tau) E_t \left[\frac{\pi(t)}{\tilde{\pi}(t + \tau)} \right], \qquad (7)$$

which readers will recognize can be transformed (by taking logarithms) into a version of the Fisherian hypothesis on interest rates.

We can employ relation (7) to specify the real return on a nominal bond over any holding period. To fix matters, define the gross real return on a nominal discount bond from date t to date $t + u$ as

$$_t\tilde{R}_{t+u}(\tau) \equiv \frac{\tilde{N}_{t+u}^*(\tau - u)/\tilde{\pi}(t + u)}{N_t^*(\tau)/\pi(t)} \qquad \text{for } u \leq \tau$$

$$= \frac{\tilde{P}_{t+u}(\tau - u)}{P_t(\tau)} \cdot \frac{E_{t+u}(1/\tilde{\pi}(t + \tau))}{E_t(1/\tilde{\pi}(t + \tau))}, \qquad (8)$$

where the first fraction in (8) is the gross return on an *indexed* discount bond with maturity τ held from t to $t + u$. Relation (8) is the object that is at the heart of our computations. The numerators of the two fractions on the right-hand side (RHS) of relation (8) depend on the information set at date $t + u$. By taking the date t conditional expectation of (8) and recalling the CIR approach where the stochastic process for inflation is exogenous and independent of the pricing of indexed bonds, we can see that the expected gross, real, holding-period return on a nominal discount bond is equal to the conditional expectation of the return on its (hypothetical) indexed counterpart:

$$E_t(_t\tilde{R}_{t+u}(\tau)) = E_t \left(\frac{\tilde{P}_{t+u}(\tau - u)}{P_t(\tau)} \right), \qquad \text{for } u \leq \tau. \qquad (9)$$

From relation (8) we can also compute the products of the gross, real returns to discount bonds of various maturities. These computations lead to specifications of comoments, which are closely related to autocovariances and serial cross-covariances of returns. From (8) we can write, for $0 < u \leq v < w \leq v + \tau_2$ and $u \leq \tau_1$,

---
[5] See CIR (1985b, p. 402). Benninga and Protopapadakis (1983) discuss this assumption in the context of the Fisherian hypothesis in a discrete-time framework.

[6] An alternative model is developed and tested in Pennacchi (1991), where the instantaneous real rate and expected inflation are found to be correlated.

Test of the CIR Model

$$_t\tilde{R}_{t+u}(\tau_1)_{t+v}\tilde{R}_{t+w}(\tau_2) = \frac{\tilde{P}_{t+u}(\tau_1 - u)}{P_t(\tau_1)} \cdot \frac{\tilde{P}_{t+w}(v + \tau_2 - w)}{P_{t+v}(\tau_2)}$$

$$\cdot \frac{E_{t+u}(1/\tilde{\pi}(t + \tau_1))}{E_t(1/\tilde{\pi}(t + \tau_1))} \cdot \frac{E_{t+w}(1/\tilde{\pi}(t + v + \tau_2))}{E_{t+v}(1/\tilde{\pi}(t + v + \tau_2))}.$$
(10)

If $v \geq u$, then the holding periods $[t, t + u]$ and $[t + v, t + w]$ are nonoverlapping. We now combine nonoverlapping holding periods with the CIR approach, where the stochastic process for inflation is exogenous and independent of the pricing of indexed bonds. These two assumptions allow us to write the conditional expectation of the product of the gross, real returns on the nominal discount bonds as the conditional expectation of the product of the gross, real returns on their indexed counterparts:

$$E_t(_t\tilde{R}_{t+u}(\tau_1)_{t+v}\tilde{R}_{t+w}(\tau_2))$$

$$= E_t\left(\frac{\tilde{P}_{t+u}(\tau_1 - u)}{P_t(\tau_1)} \cdot \frac{\tilde{P}_{t+w}(v + \tau_2 - w)}{P_{t+v}(\tau_2)}\right).$$
(11)

Equation (11) follows from Equation (10) because the first two factors are uncorrelated with the last two factors in (10); furthermore, the last two fractions in (10) are eliminated by iterating expectations over coarser information sets (because they involve the same random variable in the numerator and the denominator). We can compute conditional expectations of the product of three (or more) gross, real, holding-period returns by extending the above arguments and keeping the holding periods nonoverlapping.[7]

If we knew the relevant information upon which the expectations in (9) and (11) are based, then these equations provide a natural basis for econometric work.[8] However, if the state variables in the relevant information set are unobservable, then we need to pursue an alternative path to develop the econometric framework, to which we now in turn.

It is easy to see in relations (9) and (11) that, by the law of iterated expectations, the unconditional expectation of the corresponding quantities would also be equal. First, take the unconditional expectation of (9):

---

[7] If $v < u$, then the holding periods overlap. This will lead to nonzero correlation between $E_{t+u}(1/\tilde{\pi}(t + \tau_1))$ and $E_{t+u}(1/\tilde{\pi}(t + v + \tau_2))$. Without an explicit process for inflation, we cannot calculate this correlation. Thus, restricting our attention to nonoverlapping holding periods is an important ingredient in our econometric modeling.

[8] If the relevant information were known, then Equations (9) and (11) could generate a set of orthogonality conditions much like that in Hansen and Singleton (1982). Of course, this presumes that the relevant information is observed and that the proper specification for the impact of this information on bond returns and products of bond returns is available.

$$E\{E_t({}_t\tilde{R}_{t+u}(\tau))\} = E\left\{E_t\left(\frac{\tilde{P}_{t+u}(\tau - u)}{P_t(\tau)}\right)\right\}, \quad \text{for } u \leq \tau. \quad (12)$$

$$\equiv \Phi_1(u, \tau; \beta). \quad (13)$$

In the RHS of (12) the real indexed bond prices $P_t(\tau)$ and $P_{t+u}(\tau - u)$ depend on conditioning information (the state variables) at dates t and $t + u$, respectively. Knowledge of the functional form of these real indexed bond prices, together with knowledge of the probability densities of the state variables allows us to pass to relation (13), where the unconditional expectation has been taken. The resulting function $\Phi_1(u, \tau; \beta)$ is the unconditional first moment of the real holding-period return on a nominal bond, and β is a vector of parameters.

Next, take the unconditional expectation of (11):

$$E\{E_t({}_t\tilde{R}_{t+u}(\tau_1)\,{}_{t+v}\tilde{R}_{t+w}(\tau_2))\}$$

$$= E\left\{E_t\left(\frac{\tilde{P}_{t+u}(\tau_1 - u)}{P_t(\tau_1)} \cdot \frac{\tilde{P}_{t+w}(v + \tau_2 - w)}{P_{t+v}(\tau_2)}\right)\right\} \quad (14)$$

$$\equiv \Phi_2(u, v, w, \tau_1, \tau_2; \beta). \quad (15)$$

In the RHS of (14) the real indexed bond prices $P_s(\cdot)$ depend on conditioning information (the state variables) at dates $s = t, t + u, t + v$, and $t + w$. Again, knowledge of the functional form of these real indexed bond prices, together with the knowledge of the probability densities of the state variables, allows us to pass to relation (15), where the unconditional expectation has been taken. The resulting function $\Phi_2(u, v, w, \tau_1, \tau_2; \beta)$ is the unconditional second moment of the product of two nonoverlapping, real, holding-period returns on nominal bonds. It is easy to see that we can extend these calculations to compute unconditional moments of higher order.[9]

The functions $\Phi_1(\cdot)$ and $\Phi_2(\cdot)$ do not depend on the unobservable state variables because these variables were integrated out as part of the transition from conditional to unconditional expectations.[10] In models where explicit formulas are available for these functions, relations (13) and (15) provide a basis for empirical tests. We pursue this method.

Before we study the exact specification of the CIR model, it is useful to recognize that all the examples of nominal discount bill valuation in CIR (1985b) employ the same model for the real price $P_t(\cdot)$ and

---

[9] The functions $\Phi_1(\cdot)$ and $\Phi_2(\cdot)$ are computed in the Appendix by using the specific distributional results in CIR. Throughout the rest of the article we use Φ, to represent an expectation computed under the CIR framework.

[10] Note that the arguments of $\Phi_1(\cdot)$ and $\Phi_2(\cdot)$ depend only on the maturities of the nominal bonds being considered and not on calendar time.

Test of the CIR Model

treat the process for inflation as exogenous and independent of $P_r(\cdot)$. These examples all share the common testable implication for the real price of a riskless real bond. We derive a test of the central implication of the CIR model, which is about the term structure of real prices of bonds with real payoffs, in a way that is robust to misspecifications of the process for inflation. Of course, our test is a necessary implication for any model of the term structure that uses the same real price $P_r(\cdot)$ as in CIR even if the alternative model differs from CIR in the process for inflation.

2. The CIR Model of the Term Structure

In their principal model, CIR (1985b) derive the formula for the real price of an indexed bond, assuming a single-state variable $x(t)$ and logarithmic preferences. In their framework, $x(t)$ follows an autoregressive process with a conditional variance of the instantaneous change proportional to x. Further, the means, variances, and covariances of the rates of return on the production technologies are proportional to the level x.

CIR then show that the instantaneous riskless rate of interest, $r(t)$, which corresponds to the expected rate of change of the marginal utility of wealth, has a one-to-one correspondence with the state variable $x(t)$.[11] Hence, the stochastic process for $r(t)$ inherits the properties of the process for $x(t)$; its process can be written as

$$dr = \kappa(\theta - r)\,dt + \sigma\sqrt{r}\,dz, \tag{16}$$

where $\{z(t), t > 0\}$ is a standard Wiener process, κ is the speed of adjustment of r to its long-run mean, θ, and σ is a positive scalar. The stochastic differential equation for the instantaneous riskless rate implies the date-t conditional distribution of $r(s)$, $s > t$, is a transform of a noncentral χ^2 and the steady-state distribution is a gamma [see Feller (1951)]. The CIR pricing formula for the real unit discount bond is

$$P_t(\tau) = A(\tau)\exp\{-B(\tau)r(t)\}, \tag{17}$$

where $A(\cdot)$ and $B(\cdot)$ are given by

$$A(\tau) \equiv \left[\frac{2\gamma \exp\{(\kappa + \lambda + \gamma)\tau/2\}}{D(\tau)}\right]^{2\kappa\theta/\sigma^2}, \tag{18}$$

$$B(\tau) \equiv \frac{2[\exp(\gamma\tau) - 1]}{D(\tau)}, \tag{19}$$

---

[11] In what follows we use the term "state" variable for r, even though that applies strictly to x.

$$D(\tau) \equiv \{\kappa + \lambda + \gamma\}\{\exp(\gamma\tau) - 1\} + 2\gamma, \tag{20}$$

and

$$\gamma \equiv \sqrt{(\kappa + \lambda)^2 + 2\sigma^2}. \tag{21}$$

The parameter λ determines the risk premium; this follows from the fact that the instantaneous expected return on any default-free bond in the CIR model is

$$r + \frac{\lambda r}{P_t(\tau)} \frac{\partial P_t(\tau)}{\partial r} = r - \lambda B(\tau) r. \tag{22}$$

The risk premium is positive whenever $\lambda < 0$. Other comparative statics properties of the discount bond price are given in CIR (1985b, p. 393).

While the parameters θ, κ, λ, and σ have a natural role to play in the context of CIR's pricing model, we have adopted an alternative parametrization that we find more intuitive and more convenient for the numerical work that follows. We transform κ to a parameter that has a natural interpretation from discrete-time autoregressive models. We eliminate λ by focusing on a parameter describing the asymptote of the term structure. The scalar parameter σ is replaced by a parameter that measures the standard deviation of the steady-state distribution for r. The transformation will allow the reader to interpret the model for discrete-time intervals. Furthermore, these parameters are related to the yield curve, which is a more familiar object. It is noteworthy that we do estimate the parameters by using the implications of the continuous-time process for discrete sampling intervals; we do not rely on approximations of instantaneous holding periods for returns.

Here is a brief description of the transformed parameters. We define an autoregressive parameter for the interest rate process, ρ, given by $\rho \equiv \exp(-\kappa/12)$, instead of working with κ directly. The parameter ρ is the coefficient of a regression of the intercept of the yield curve for indexed bonds on the intercept of last month's yield curve. It is easier to interpret the unconditional standard deviation of the intercept of the yield curve, σ_U,

$$\sigma_U \equiv \sqrt{\sigma^2 \theta / 2\kappa}, \tag{23}$$

rather than σ. Instead of using λ directly, we focus on the effect of λ on the long-run yield (y_∞), which is independent of the level of the state variable and is the asymptote of the CIR yield curve as maturity increases. The transformation to this long-run yield is

$$y_\infty = \frac{2\kappa\theta}{\kappa + \lambda + \gamma}. \tag{24}$$

Test of the CIR Model

We find the long-run mean, θ, of the intercept of the yield curve easy to interpret, and we have not transformed this parameter. In summary, the vector of parameters that we estimate for the CIR model is β, where

$$\beta \equiv (\theta \quad \rho \quad y_\infty \quad \sigma_U). \tag{25}$$

This transformation of the parameters also has the property that given β we can invert to find the original CIR parameters.[12]

3. The Econometric Method

Recognizing the definition of a yield on an indexed bond given in Equation (5), we find that Equation (17) implies

$$y_t(\tau) = -\frac{\log(A(\tau))}{\tau} + \frac{B(\tau)}{\tau} r(t), \tag{26}$$

which is linear in the unobserved variable. This implies that the correlations between the yields of indexed bonds of different maturities are all unity. Therefore, applying the CIR model [viz., relation (17)] to nominal data on nominal bonds leads to a rejection of the model, for casual empiricism (ignoring the effects of measurement error) suggests that nominal yields are not perfectly correlated.

A sufficient history of properly measured prices of indexed bonds would, however, enable a direct test of the CIR model. Brown and Schaefer (1990) test the CIR model with data on indexed bonds in the United Kingdom.[13] Although the state variable is not observed in this case, nonlinear cross-sectional regressions employing (26) permit the estimation of some, but not all, of the underlying parameters. Their procedure "inverts" the CIR formula for r and some of the other parameters from a cross section of prices, just as if we backed out the stock price and the implied volatility by using the Black–Scholes model.[14]

While these nonlinear cross-sectional regressions are tractable, they cannot connect directly the estimated parameters with the time-series properties of the bond prices. For example, θ, ρ, and σ_U are not linked to the sample mean, autocorrelation, and standard deviation of r, which is estimated over time from each cross-sectional regression.

---

[12] Straightforward algebra will verify that the transformation is one-to-one.

[13] Also see Brown and Dybvig (1986), who apply the CIR model to nominal prices of nominally riskless bonds.

[14] The methods in Brown and Schaefer (1990) and in Brown and Dybvig (1986) are similar; however, the former article analyzes indexed bonds, whereas the latter focuses on nominal bonds. Both papers can only identify the following functions of the CIR parameters: $\kappa + \lambda$, $\kappa\theta$, and σ.

We will integrate the dynamic properties of the CIR model with its cross-sectional implications for bonds of differing maturities. The use of CIR's specification of the stochastic evolution of the state variable lends the test considerable sharpness, for we are able to exploit this information in testing the overidentifying restrictions and arriving at parameter estimates.

Our objective is to test the CIR model of indexed bond prices from data on nominal bonds. While our test will be robust to measurement error, we do not incorporate an explicit model of measurement error to rationalize why real bond prices are not perfectly dependent—unanticipated inflation will preclude perfect correlation in our view of the data.

In Section 3.1 we discuss an econometric procedure that allows us to compare the implications of the CIR model with the sample characteristics. This method has certain distinguishing features that are outlined in Section 3.2.

3.1 Comparing population and sample moments

The econometric technique corresponds to the GMM procedure developed by Hansen (1982) and employed in Hansen and Singleton (1982), Brown and Gibbons (1985), and elsewhere. Our procedure[15] differs from the standard GMM application in that (1) we avoid the use of consumption data or data on aggregate wealth (the "market") and (2) we exploit the availability of a functional form within the CIR model for the relevant densities of the unobserved state variable, r.

Before we can apply the GMM approach, we must calculate some population moments for real returns on nominal bonds. These population moments are characteristics that we expect to see in the data if the CIR model were true. Our procedure involves a comparison of the implied population moments with the corresponding sample moments as a way to estimate the CIR parameters and to judge the model's descriptive validity.

Recall from relation (12) that the expectation of the gross real return from owning a nominal discount bond of maturity τ from t to $t + u$ is given by

$$E\{E_t({}_t\tilde{R}_{t+u}(\tau))\}$$
$$= E\left\{E_t\left(\frac{\tilde{P}_{t+u}(\tau - u)}{P_t(\tau)}\right)\right\}, \quad \text{for } u \leq \tau \qquad (27)$$

---

[15] Our approach can be extended to other contexts as long as one can find a set of moment conditions that do not depend on consumption data or unobservable state variables. In our case the underlying model does not require consumption data but does depend on an unobservable state variable. However, given the stochastic process for this unobservable variable, we can integrate to find moment conditions that do not depend on the state variable.

Test of the CIR Model

$$= E\left\{E_t\left(\frac{A(\tau-u)}{A(\tau)}\exp(-B(\tau-u)\tilde{r}(t+u)+B(\tau)\tilde{r}(t))\right)\right\} \quad (28)$$

$$= \frac{A(\tau-u)}{A(\tau)}E\{\exp(-B(\tau-u)\tilde{r}(t+u)+B(\tau)\tilde{r}(t))\} \quad (29)$$

$$\equiv \Phi_1(u;\tau;\beta), \quad (30)$$

where the CIR formula has been used to pass to (28). Note that the expectation in the RHS of relation (29) is taken using the joint, unconditional distribution of the random variables $\{\tilde{r}(t), \tilde{r}(t+u)\}$. CIR's model specifies this joint density from relation (16). The conditional distribution of $\tilde{r}(t+u)$ given $r(t)$ is noncentral chi-square, and the unconditional distribution of $\tilde{r}(t)$ is a gamma. Clearly, the expectation in (29) defines the moment generating function for this bivariate distribution. The Appendix provides an explicit calculation of $\Phi_1(u, \tau; \beta)$, and it shows how the unobservable variables $\tilde{r}(t)$ and $\tilde{r}(t+u)$ have been integrated out.

Following an identical argument, we can use the CIR formula to compute the expectation of the product of two gross, real, nonoverlapping returns from nominal discount bonds. In the following expression we examine this product, where the first return (from a bond with maturity τ_1) is from t to $t+u$ and the second return (from a bond with maturity τ_2) is from $t+v$ to $t+w$:[16]

$$E\{E_t({_t\tilde{R}_{t+u}(\tau_1)}\,{_{t+v}\tilde{R}_{t+w}(\tau_2)})\}$$

$$= E\left\{E_t\left(\frac{\tilde{P}_{t+u}(\tau_1-u)}{P_t(\tau_1)}\cdot\frac{\tilde{P}_{t+w}(\tau_2-(w-v))}{P_{t+v}(\tau_2)}\right)\right\} \quad (31)$$

$$= A(\tau_1-u)A(\tau_2-(w-v))A^{-1}(\tau_1)A^{-1}(\tau_2)$$

$$\cdot E\{\exp(-B(\tau_1-u)\tilde{r}(t+u)+B(\tau_1)\tilde{r}(t)$$

$$-B(\tau_2-(w-v))\tilde{r}(t+w)+B(\tau_2)\tilde{r}(t+v))\} \quad (32)$$

$$\equiv \Phi_2(u,v,w,\tau_1,\tau_2;\beta). \quad (33)$$

The expectation in the RHS of (32) involves the calculation of the moment generating function of a joint distribution of four random variables, which is tedious but straightforward (see the Appendix for details).

Relations (30) and (33) serve as restrictions on the first moment and the second comoments of the gross real returns on nominal bonds. These moments are expressed solely as functions of the matur-

---

[16] We assume that $t < t+u \le t+\tau_1$, $t+v < t+w \le t+v+\tau_2$, and $u \le v$. If $u = v$, then the left-hand side of relation (31) represents the expected real growth from a sequential investment in 2 discount instruments.

ities of the bonds and of the vector of parameters, given the CIR model.

We are now in a position to apply Hansen's GMM. Suppose that we have data on the real gross returns on nominal discount bonds of maturity τ_i, $i = 1,2,...,n$. Define the following functions of the data and the moments and nonoverlapping comoments:

$$h_{1t}(u, \tau_i; \beta) \equiv {}_tR_{t+u}(\tau_i) - \Phi_1(u, \tau_i; \beta), \quad (34)$$

$$h_{2t}(u, v, w, \tau_i, \tau_j; \beta) \equiv {}_tR_{t+u}(\tau_i)_{t+v}R_{t+w}(\tau_j)$$
$$- \Phi_2(u, v, w, \tau_i, \tau_j; \beta). \quad (35)$$

Now stack these into a vector:

$$g_T(\beta) \equiv \begin{pmatrix} \frac{1}{T}\sum_t h_{1t}(u, \tau_1; \beta) \\ \frac{1}{T}\sum_t h_{1t}(u, \tau_2; \beta) \\ \vdots \\ \frac{1}{T}\sum_t h_{1t}(u, \tau_n; \beta) \\ \frac{1}{T}\sum_t h_{2t}(u, v, w, \tau_1, \tau_1; \beta) \\ \frac{1}{T}\sum_t h_{2t}(u, v, w, \tau_1, \tau_2; \beta) \\ \frac{1}{T}\sum_t h_{2t}(u, v, w, \tau_i, \tau_j; \beta) \\ \vdots \end{pmatrix}, \quad i,j = 1,2,...,n, \quad (36)$$

where n is the number of maturities for the available bills. Alternatively, $g_T(\beta) \equiv (1/T)\Sigma_t h_t(\beta)$, where $h_t(\beta)$ is a vector built by stacking $h_{1t}(\cdot)$ and $h_{2t}(\cdot)$ in the obvious way given Equation (36). The vector $g_T(\beta)$ has dimension $l \times 1$, and we assume that $l > 4$ so that the number of restrictions exceeds the number of parameters to be estimated. The model's implications [from relations (30) and (33)] are

$$E(g_T(\beta)) = 0, \quad (37)$$

so we choose β to make the sample counterparts to these moments close to zero. Hansen's procedure involves choosing β from a feasible[17] region B:

$$\min_{\beta \in B} Tg_T(\beta)'\Omega^{-1}g_T(\beta), \quad (38)$$

---

[17] The restrictions from the CIR model are ρ, θ, $\sigma_{rr} > 0$, and $\rho < 1$.

Test of the CIR Model

where the weighting matrix Ω is the asymptotic covariance matrix of the vector of sample moment conditions. Given the CIR model and mild regularity conditions, the minimand in (38) has, asymptotically, a χ^2 distribution with $l - 4$ degrees of freedom; this is the test employed below. Hansen (1982) provides the sufficient conditions for the consistency and asymptotic normality of $\hat{\beta}$ as well.

We now turn our attention to the appropriate way to construct Ω in Equation (38) so as to account for the serial dependence in the observations. In many applications of GMM, $h_t(\beta)$ is orthogonal to all past information, including information in the lagged values of $h_t(\beta)$. This orthogonality follows directly from the rationality assumption that agents use all past information in setting market prices; this was the appropriate assumption in the context of the models investigated by Hansen and Singleton (1982) and Brown and Gibbons (1985).[18] In our application, $h_t(\beta)$ is not the deviation of the realized return from its *conditional* expectation, but it is the deviation of the realization from its unconditional expectation, $\Phi_1(\cdot)$ or $\Phi_2(\cdot)$. The CIR model predicts that the deviations of bond returns (or of the products of these returns) from the unconditional expectations *will* be serially correlated because these deviations depend on $r(t)$. Our inability to observe $r(t)$ precludes us from constructing $h_t(\beta)$ as a deviation from an expectation conditional on $r(t)$, so we cannot remove this source of serial correlation in the data.

The maintained assumptions from the CIR model permit us to specify some elements of the weighting matrix. However, to determine other elements (for example, the variances along the diagonal), we need additional—and for our purposes unnecessary—assumptions about the process on inflation and the variance of any measurement error. Therefore, it is not possible to specify the exact form of the weighting matrix, although we expect serial correlation to be present in the observations.

To account for general forms of serial dependence and heteroskedasticity (at least asymptotically), we adopt the Newey-West procedure. The asymptotic justification for GMM requires only that the weighting matrix be a consistent estimator[19] of the asymptotic covari-

---

[18] In this earlier work, as in Hansen and Hodrick (1980), serial correlation in $h_t(\beta)$ could only be present because overlapping observations are used. In our application overlapping observations arise only when we focus on ex post real yields for bonds held till maturity when we perform sensitivity analysis in Section 5.2. For example, if we had examined on a monthly basis the returns on a three-month bill held till maturity, there would be two months of overlap in consecutive observations.

[19] Many applications of GMM require a two-step procedure to find the optimal set of estimates for β. The first step involves minimizing the objective in Equation (38), setting Ω equal to the identity matrix. The resulting set of estimates for β are then used to construct a second Ω matrix, not equal to the identity matrix. However, in our case the first step can be avoided due to the special structure of our orthogonality conditions, $h_t(\beta)$. Our orthogonality conditions can be written as a set of sample moments that do not depend on β and functions of β that do not depend on the sample.

633

ance matrix; the following weighting matrix, given in Newey and West (1987), is a consistent estimator that is always positive definite:

$$\Omega = \hat{\Omega}_0 + \sum_{j=1}^{m} \omega(j, m)[\hat{\Omega}_j + \hat{\Omega}'_j], \qquad (39)$$

$$\omega(j, m) \equiv 1 - \left[\frac{j}{m+1}\right], \qquad (40)$$

$$\hat{\Omega}_j \equiv \frac{1}{T} \sum_{t=j+1}^{T} (b_t - \bar{b})(b_{t-j} - \bar{b})'. \qquad (41)$$

Asymptotic justification for the Newey–West procedure relies on m growing at least at the rate $T^{0.25}$. The covariance matrix of the asymptotic distribution of the GMM estimator for β is consistently estimated by

$$\text{Var}(\hat{\beta}) = [D'(\hat{\beta})\Omega^{-1}D(\hat{\beta})]^{-1}, \qquad (42)$$

where

$$D(\hat{\beta}) \equiv \frac{1}{T} \sum_t \frac{\partial b_t}{\partial \beta}\bigg|_{\beta=\hat{\beta}}. \qquad (43)$$

3.2 The econometric procedure: some additional features

Part of the motivation for our method should be clear. We have developed a procedure that determines the implications of a continuous-time model for discretely sampled data. Further, we are not required to observe state variables or measures of aggregate consumption in comparing the theory with the data. However, there are additional reasons that encouraged us to follow this approach. These are outlined in this subsection.

First, our moment conditions are robust to the usual forms of measurement error. Even if the gross real returns are measured in error, the measurement error has no impact on the expectation of Equation (34) as long as the error has a mean equal to zero. Furthermore, assuming the measurement error is serially uncorrelated, the expectations of the comoments in Equation (35) also remain valid in the presence of such error.[20] Similarly, serial cross-comoments implicit in Equation (35) should not be affected by measurement error that is uncorrelated across bonds of different maturities. This robustness to measurement error increases our confidence in the point estimates for β that are provided in Section 5.

---

[20] Even if the measurement error had serial correlation induced by a moving-average process of low enough order, the computation in Equations (31) through (33) remains valid for lags that are sufficiently long to remove the dependency induced by the measurement error.

Test of the CIR Model

Second, we have avoided a moment condition that is related to the variance of the real returns. If we had specified a process for inflation, we could calculate a moment condition corresponding to the population variance of the gross return on a bond as a function of the parameters underlying the inflation process as well as β. This variance would be sensitive to measurement error unless we deliberately modeled the process for this error. Such an approach would have more potential for misspecification (it would lead to a joint test of the CIR model and the assumed inflation process), and it is not clear to us that it would offer any real advantages over the moment restrictions chosen here. Rather than test models of inflation, we wanted to follow a path that would allow us to investigate models for real bond returns.

Finally, there is nothing in the procedure that requires us to examine monthly (say) holding period returns on bonds of various maturities. In construction of the moment conditions there are very few restrictions on the holding period of the gross real returns on the nominal discount bonds. We could, for example, choose the holding period to correspond to the maturity of each bond. Selecting a holding period in this way leads to an examination of the data on the ex post real yields-to-maturity of these bills. However, in this case the estimation must take into account the fact that there is substantial overlap (for example, with monthly data and yields on 12-month bills, there are 11 months of overlap); therefore, we must choose higher values for m in the Newey-West procedure. We will examine the sensitivity of our empirical results when we use real yields instead of real returns.

4. Data Description and Summary Statistics

The empirical results reported here are based on monthly data from 1964 through 1989. Data on U.S. Treasury bills were obtained from the government bond files of the Center for Research in Security Prices at the University of Chicago. The Bureau of Labor Statistics' series on the consumer price index, corrected for the home ownership interest component, was kindly provided to us by John Huizinga. From these two sources, gross holding-period real returns $_tR_{t+u}(\tau)$ were constructed for each month t for maturities (τ) of 1, 3, 6, and 12 months.

We have avoided long-term bonds. Researchers, who believe that the observed variability in the long-term yields is prima facie evidence against the CIR model (in which the long-term yield is constant) will feel that excluding long-term maturities will decrease the power of our test. Such a reaction is based on the variability of nominal yields on nominal bonds. Quite naturally, models of nominal bond prices have gone beyond single-factor models (like CIR) to allow for more variability of long-term nominal yields. However, we are not studying

635

a model of nominal bond prices; we only seek to explain the expected behavior of real returns on nominal bonds.

Because we work with real returns on nominal bonds, the argument for inclusion of long-term bonds is less clear. For example, the variability in long-term nominal yields may be induced only by slow mean reversion in inflation expectations, not from a state variable that affects real yields on indexed bonds. Thus, the argument for increasing the power by including data on long-term nominal bonds is by no means obvious. Even ignoring power considerations, we suspect that the precision of our estimators may not increase if we include long-term bonds; as Dunn and Singleton (1986) argue, the variation in long-term bond returns is large, so it may be more difficult to estimate parameters from long-term bond data.

Despite plausible arguments about the questionable value of long-term data, we would still be inclined to incorporate longer maturities. Such an inclusion is natural in testing models of indexed yields, especially as part of a sensitivity analysis. However, there are other considerations. First, the nonlinear structure of the pricing formula for coupon bonds would place a heavy burden on the algorithm that searches for the parameter values satisfying the moment conditions. Second, the problems from overlapping observations (which become an issue when we employ data on yields till maturity) become progressively worse as we use bonds with long maturities. Third, the analysis of long-term bonds should incorporate tax considerations and implicit options. Any differential taxation of the income and capital gains components would affect the pricing of coupon bonds.[21] Any call features on these instruments would preclude a simple approach to the valuation of coupon-bearing bonds.[22] Also note that, although prices of stripped, single-payment certificates derived from coupon-bearing Treasuries are now available, we lack a sufficient history of these for our purposes.

Although CRSP reports Treasury-bill prices for several maturities at each month's end, this study restricts the maturities to 1-, 3-, 6-, and 12-month bills. The last three of these are the most heavily traded "on-the-run" bills; therefore, their month-end prices from CRSP are most likely to be current and simultaneous quotations. Treasury bills for other maturities are usually not as heavily traded, and the potential for nonsynchronous prices and measurement error is greater with these. It should be recognized that the CPI series that is employed has consumption goods' prices that are usually sampled during a

---

[21] Tax considerations might lead to differences in valuation among government securities; for a discussion, see Constantinides and Ingersoll (1984).

[22] During the late 1970s and early 1980s most of the long-term bonds were callable. This makes it difficult to construct a continuous series for returns on a constant maturity long-term bond.

Test of the CIR Model

month, and therefore there is error induced in taking the CPI as a month-end price level for the purpose of computing the real return. This error is unavoidable, whereas the measurement errors in nominal bill prices may be reduced, as is argued here, by employing the bills on-the-run.

The estimation procedures employed in this study build on the moments of the real, monthly holding-period returns. At each month's end, it is generally not possible to find Treasury bills with maturities of exactly 1, 3, 6, and 12 months; however, there are bills with maturities surrounding these. We constructed the prices of these Treasury bills by linearly interpolating between the annualized yields of the two bills that immediately surrounded the desired maturity.[23]

Figure 1A–D plot these data series, which are used for the bulk of the tests reported in Section 5. Panel A of Table 1 provides summary statistics of the real return series for the four maturities. The statistics reported in panel A indicate that the mean and standard deviation of the real monthly returns increases as the maturity increases; the correlations between bills of adjacent maturity are high relative to the others. The autocorrelations at lag 1 are highest for the one-month bills. While the autocorrelations decay quickly at higher lags for 6-month and 12-month bills, they are slow to die out for the 1-month and 3-month bills.

The consumer price index, which we use in computing the real returns, may induce autocorrelations in the series. This will occur if the reported index values are computed from prices that are sampled for different goods in sequence and for an individual good periodically. We are careful to employ comoment conditions that only use nonoverlapping returns; we discuss this in Section 5.2 where we report on some diagnostics.

Panel B provides the statistics on the real yields (to maturity) of the same bills. Because the monthly data for real yields to maturity involve overlapping intervals, the autocorrelations should at least reflect that degree of overlap. For instance, the autocorrelations for the yields on three-month bills should be affected by the overlap for at least two months; the autocorrelations beyond lag 2 should reflect the autocorrelation in the structural model underlying the nonoverlapping returns. For the one-month yield series there is no overlap, and the significant autocorrelations are, in the absence of measurement error, an indication of the structural model underlying the

---

[23] In the actual estimation we denoted time in units of years and treated a 1-month bill as if it matured in 30/365 years, a 3-month bill as if it matured in 90/365 years, a 6-month bill as if it matured in 180/365 years, and a 12-month bill as if it matured in 345/365 years. Since we rely on end-of-month prices of Treasury bills, we found the average maturity of the longest bill was approximately 345 days.

The Review of Financial Studies / v 6 n 3 1993

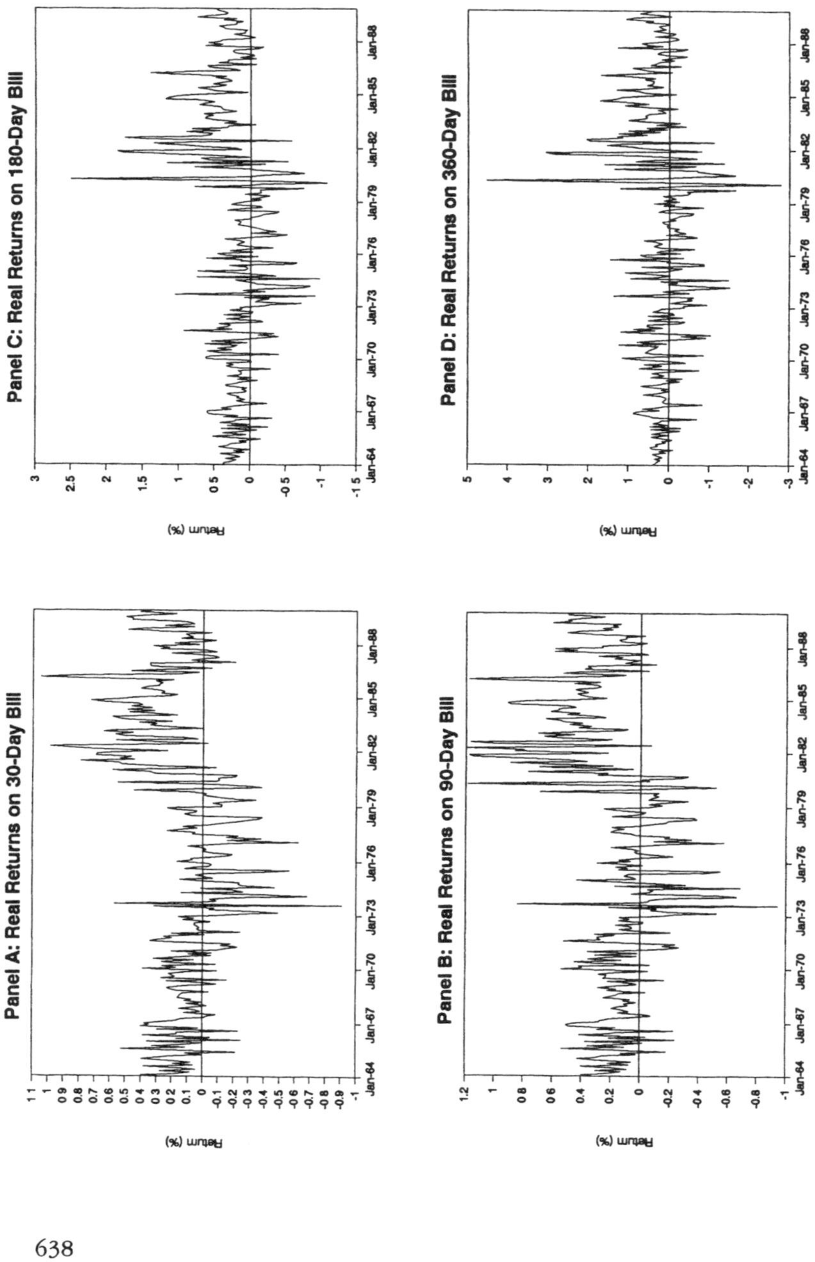

Test of the CIR Model

returns. For the other series, the autocorrelations are significant for lags well beyond their degree of overlap, and this also provides information on the process that generates the data. The contemporaneous correlation coefficients also display the effects of overlapping intervals across the observations; its greatest impact is, as expected, in the computed correlation between 6- and 12-month yields, which is 0.95.

Figure 1 and Table 1 indicate that the average ex post real return was large over this sample period by historical standards. For example, Fama (1975) reports that the average real returns on one-month Treasury bills was 7 basis points per month, from 1953 to 1971; our higher value of 13 basis points reflects recent experience. Despite the high average return, Figure 1 shows that the ex post returns were negative over some periods, especially in the late 1970s.[24]

The variation in the ex post real return series (the standard deviation of 0.273 percent per month for bills of one-month maturity) is substantially greater than that reported in Fama and Gibbons (1982, 1984) for the sample period 1953–1977. Again, the sample period in our work includes relatively volatile periods.

5. Results

In Section 5.1 we report the empirical results, using the moments given in Equations (30) and (33). To test these with the data, we limit our attention to a few discount bonds and some specific moments. As mentioned in Section 4, we rely on monthly returns on Treasury bills maturing in 1, 3, 6, and 12 months. Section 5.2 investigates the sensitivity of our findings from alternative econometric specifications. Using the CIR model, Section 5.3 focuses on a particular implication about the dynamics for real returns on bonds; here we report that the CIR framework is inconsistent with a time-series feature in the data.

5.1 Empirical tests with 14 moments

To estimate the parameters and test the implications of the CIR model, we need to summarize the historical data by using some sample moments and then compare these sample moments with values implied by the theory. For a given set of parameter values, Equation (30) provides the theoretical prediction about the first moment, while

---

[24] This is not inconsistent with the CIR model, which predicts that the ex ante real rate, over all holding periods, is nonnegative. This feature of their model is an outcome of their decision to model the process on the single-state variable as they did; there are no a priori reasons to expect a nonnegative ex ante real rate, unless the consumption good could be stored costlessly.

←

Figure 1
Real returns on Treasury bills, monthly holding periods, 1964–1989

Table 1
Summary statistics

Maturity (months)	Mean (%)	SD (%)	Correlations (months) 1	3	6	12	ρ(1)	ρ(2)	ρ(3)	ρ(4)	ρ(5)	Autocorrelations ρ(6)	ρ(7)	ρ(8)	ρ(9)	ρ(10)	ρ(11)	ρ(12)
							Panel A: Real monthly holding-period returns (% per month)											
1	0.130	0.273	1.000	0.934	0.773	0.583	0.561	0.446	0.369	0.387	0.365	0.361	0.377	0.371	0.396	0.370	0.361	0.414
3	0.193	0.322	0.934	1.000	0.925	0.774	0.529	0.379	0.344	0.368	0.418	0.362	0.379	0.345	0.367	0.318	0.350	0.359
6	0.221	0.434	0.773	0.925	1.000	0.937	0.375	0.172	0.159	0.175	0.289	0.207	0.275	0.160	0.237	0.194	0.245	0.166
12	0.213	0.699	0.583	0.774	0.937	1.000	0.274	0.001	0.010	0.041	0.189	0.086	0.224	−0.006	0.123	0.067	0.150	0.077
							Panel B: Real yields to maturity: overlapping data											
1	0.130	0.273	1.000	0.811	0.704	0.671	0.561	0.446	0.369	0.387	0.365	0.361	0.377	0.371	0.396	0.370	0.361	0.414
3	0.466	0.720	0.811	1.000	0.908	0.858	0.888	0.734	0.608	0.581	0.580	0.574	0.564	0.563	0.555	0.556	0.566	0.545
6	1.063	1.395	0.704	0.908	1.000	0.950	0.947	0.875	0.806	0.745	0.695	0.657	0.637	0.652	0.658	0.659	0.652	0.613
12	2.179	2.886	0.671	0.858	0.950	1.000	0.968	0.926	0.891	0.861	0.836	0.805	0.689	0.776	0.755	0.733	0.713	0.661

Data: Monthly observations, 1964/1–1989/12, on U.S. Treasury bills of 1-, 3-, 6-, and 12-month maturities

Test of the CIR Model

Equation (33) supplies similar restrictions regarding second moments. We must first decide which first and second moments to use.

Obviously, we would like to use sample moments that provide a good summary of the historical data and capture the important stylized facts about this time period. We would also like to use enough moments to generate overidentifying restrictions to confront the theory. However, we recognize that the GMM approach has only asymptotic justification, so we want to avoid the use of too many moments, especially if the information in one moment may be the same as in other moments.

We selected the first moment for all bond maturities, which seems like an obvious choice. Equation (30) generates four first-moment conditions since we have four maturities.

The choice of the sample comoments to be used in Equation (33) is not as straightforward. Even with a small number of bond maturities, the restrictions implied by Equation (33) permit an unmanageably large number of comoment conditions, obtained by varying the lag structure. Since the sampling characteristics of the comoments are probably superior at small lags, we focused on short lags.

We rely on the comoments to capture the dynamic characteristics of the historical data. The degree of mean reversion in the sample is an obvious summary of the temporal behavior. Thus, we examined one serial comoment for each maturity, providing four additional moment conditions. The serial comoment, even though it is not a central moment, is closely related to the first-order autocovariance.

We also wanted a measure of the correlation among bond returns of differing maturities. Because we do not specify a process for inflation, we are unable to compute a theoretical value for the contemporaneous correlation among bonds with different maturities. As a substitute for the correlation, we rely on serial cross-comoments that provide some information about the association among bond returns, as well as some information about the dynamics of bond returns. We correlated a lagged return on a 1-month bill with the returns on the other three maturities, and we correlated a lagged return on a 12-month bill with the returns on the other three maturities. These comoments correspond to using the shortest-maturity bill to predict the subsequent returns on the other bills and to using the longest bill to predict the subsequent returns on the other bills. This information represents six cross-comoments for fitting the CIR model.

To summarize, we have four first moments (one for each maturity), four serial comoments (one for each maturity), and six cross-comoments. The second column in panel B of Table 2 provides a list of the specific moments.[25]

---

[25] So far, we have chosen to interpret the expectations in Equations (30) and (33) as noncentral moments of returns. In panel B of Table 2, we have scaled these moments up by 100,000. Now

In applying GMM we sought a set of parameter estimates for β that fixed the 14 population moments as close as possible to the sample moments given in the third column of panel B of Table 2. The system is overidentified, so it is not possible to match perfectly the sample moments in the third column. We minimized the quadratic form given in Equation (38) to determine the optimal set of estimates.[26] Table 2 provides a test statistic for the overidentifying restrictions. Essentially, this test measures whether deviations from the sample moments in Table 2 are small, as would be expected if the theory is true. The deviations are measured by using the quadratic form in Equation (38), which is distributed χ^2_{10} under the null hypothesis. Since the fitted moments implied by the parameter estimates in the fifth column in panel B of Table 2 seem very close to the actual sample moments in the third column, it is not surprising to find that the CIR model cannot be rejected at traditional levels of significance. Having failed to reject the overidentifying restrictions, we now turn our attention to the parameter estimates that generated column 5 of Table 2.

The parameter estimates along with standard errors based on asymptotic theory are given in the second column in panel A of Table 2. All the estimates are more than two standard errors away from zero. The point estimate for θ is large in magnitude (154 basis points per annum) relative to a similar parameter estimated by Fama (1975). This probably reflects our use of more recent history, where the real return on Treasury bills has been high by historical standards. Indeed, Section 5.3 confirms the importance of the particular time period selected when results by subperiods are presented. For the reader's convenience, the last column in panel A of Table 2 also reports the implied parameter estimates for the parametrization using the original CIR notation.

The degree of mean reversion as measured by ρ is quick relative to the random walk model in Fama and Gibbons (1982), which implies that ρ should be close to unity. This parameter provides some guidance for the speed of adjustment of the intercept of the "real" yield curve in a CIR world. Equation (19) in CIR (1985b, p. 392) provides a formula for the conditional expectation of r. Restating their equation

---

columns 3 and 5 can be given another interpretation. Column 3 represents the historical average of the real wealth one obtained by investing $100,000 in bonds. Column 5 indicates the expected real wealth from the same investment strategy, assuming that the CIR model is correct with particular parameter values. In the case of the four first moments, the holding period of the investment is one month. In the case of the 10 second moments, the holding period is two months. Each investment strategy differs by the maturity of the bond(s) purchased during the holding period.

[26] The optimization was done using conjugate gradient methods as implemented in *Mathematica*. We also confirmed our results with the numerical minimization routine in *Gauss*, which uses an algorithm based on the Broyden-Fletcher-Goldfarb-Shanno positive-definite secant update method.

Test of the CIR Model

Table 2
Test of the CIR model of the term structure using the generalized method of moments

Panel A: Parameter estimates[1]

Yield-curve-based parameters	Estimate	Standard error	Correlations θ	ρ	y_∞	σ_U	CIR model-based parameters
θ (%)	1.54	0.26	1.000	0.436	0.893	0.295	$\theta = 1.54\%$
ρ	0.35	0.06	0.436	1.000	0.422	0.050	$\kappa = 12.43$
y_∞ (%)	3.01	0.30	0.893	0.422	1.000	0.225	$\lambda = -6.08$
σ_U (%)	1.23	0.44	0.295	0.050	0.225	1.000	$\sigma = 0.49$

Test statistic: χ^2_{10} 13.39; (p value .203)

Panel B: Sample moments[2] and fitted moments ($\times 100{,}000$)

Moment number	Definition	Sample mean	Standard error	Fitted value
1	$E[\tilde{R}_{t+1}(1)]$	100129.09	26.18	100153.71
2	$E[\tilde{R}_{t+1}(3)]$	100192.28	30.05	100214.95
3	$E[\tilde{R}_{t+1}(6)]$	100220.20	34.78	100241.13
4	$E[\tilde{R}_{t+1}(12)]$	100212.81	47.96	100247.70
5	$E[\tilde{R}_{t+1}(1)_{t+1}\tilde{R}_{t+2}(1)]$	100258.63	51.85	100307.68
6	$E[\tilde{R}_{t+1}(3)_{t+1}\tilde{R}_{t+2}(3)]$	100384.85	59.54	100430.28
7	$E[\tilde{R}_{t+1}(6)_{t+1}\tilde{R}_{t+2}(6)]$	100441.32	68.51	100482.72
8	$E[\tilde{R}_{t+1}(12)_{t+1}\tilde{R}_{t+2}(12)]$	100427.41	93.51	100495.87
9	$E[\tilde{R}_{t+1}(1)_{t+1}\tilde{R}_{t+2}(3)]$	100321.96	55.42	100369.02
10	$E[\tilde{R}_{t+1}(1)_{t+1}\tilde{R}_{t+2}(6)]$	100349.96	58.89	100395.25
11	$E[\tilde{R}_{t+1}(1)_{t+1}\tilde{R}_{t+2}(12)]$	100342.67	69.33	100401.83
12	$E[\tilde{R}_{t+1}(12)_{t+1}\tilde{R}_{t+2}(1)]$	100342.74	68.52	100401.70
13	$E[\tilde{R}_{t+1}(12)_{t+1}\tilde{R}_{t+2}(3)]$	100406.24	73.64	100463.05
14	$E[\tilde{R}_{t+1}(12)_{t+1}\tilde{R}_{t+2}(6)]$	100434.40	79.90	100489.29

Data: Real, monthly holding period returns on 1-, 3-, 6-, and 12-month U.S. Treasury bills (1964–89); Moment conditions: 4 first moments, 4 autocovariances and 6 serial cross-covariances (lagged 1 month); Newey-West lag: $m = 4$ for weighting matrix; $_t\tilde{R}_{t+u}(\tau)$: Gross real return on τ-month Treasury-bill held from t to $t + u$.

[1] θ is the mean of the steady-state distribution of r in CIR's model, and also the intercept of the "steady-state" yield curve (% per year); ρ is its autoregressive parameter at 1 month lag, where $\rho = \exp(-\kappa/12)$; y_∞ is the asymptote of the CIR yield curve (% per year); and σ_U is the unconditional value for the standard deviation of r (% per year).

[2] The standard errors reported in the fourth column are the square roots of the diagonal elements of the inverse of the Newey-West weighting matrix.

using a parametrization based on ρ gives[27]

$$E[r(s) | r(t)] = \theta + \rho^{(s-t)}[r(t) - \theta]. \quad (44)$$

Thus, employing the estimates from Table 2, if the current intercept of the yield curve is 2.77 percent (which is one standard deviation above the mean of the steady-state distribution), we expect it to be 1.98 percent in one month hence and almost equal to the long-run mean of 1.54 percent by six months. This adjustment seems quick

---

[27] In Equation (44), $s - t$ is measured in units of months, not years. Recall that we defined $\rho \equiv \exp(-\kappa/12)$; we omitted the 12 in Equation (44).

given recent behavior in the bond market when the short-term real rate was high by historical standards and remained at that high level for sustained periods. This rapid speed of adjustment is the focus of Section 5.3, so we postpone discussion of this point till that subsection.

The estimate of σ_U provides a measure of the unconditional standard deviation of r; for our sample period we estimate σ_U to be 1.23 percent. Consistent with recent experience when real rates were high, the estimates in Table 2 do imply that the conditional variation in the instantaneous rate is sensitive to the level of the rate. For example, using Equation (19) in CIR (1985b, p. 392), we find that the standard deviation of $r(s)$ conditional on $r(t) = 2.77$ percent is 1.36 percent and 1.23 percent for $s - t$ equal to 1 and 12 months, respectively. The sensitivity of the conditional variance of $r(s)$ to the current level of $r(t)$ is based on just the 14 sample moments in Table 2. None of these 14 moments are the sample variances, nor do any of the 14 moments relate directly to the predictability of the variance based on current interest rates. Section 5.2 will examine an extension to Table 2 where we build in some information about the sample correlation between the conditional variance and predictors of this variance.

In Table 2, the value for y_∞ is positive and, consistent with traditional theories of term premiums, greater than θ. Thus, the steady-state yield curve is positively sloped with a spread between the asymptote and the intercept of the yield curve of 3.01% − 1.54% = 1.47%. Theories of the term structure also make predictions about the expected returns on short versus long bonds. The last column of panel A of Table 2 reports a value for λ that is negative. Equation (22) indicates that the instantaneous expected return on any pure discount indexed bond is equal to $r - \lambda B(\tau) r$ and $B(\tau) > 0$ for all τ. This implies the instantaneous expected real return on a bond is positively related to its interest sensitivity (and its maturity). To provide an alternative perspective, panel A of Figure 2 provides a plot of the unconditional expected returns over a discrete holding period of one month for bonds of various maturities. Based on the parameter estimates in Table 2, the curve for unconditional expected returns asymptotes at 2.98 percent for maturities in excess of one year. In fact, the unconditional expected return is 2.90 percent even for bonds with six months to maturity. Fama (1984) reports that the unconditional sample average returns on Treasury bills are not monotonically increasing after six months till maturity. Our parameter estimates for the CIR model suggest that expected returns may effectively asymptote around six months till maturity.

To summarize, we have failed to reject some overidentifying restrictions implied by the CIR model, and we have parameter estimates

Test of the CIR Model

Panel A. Unconditional Expected Real Returns

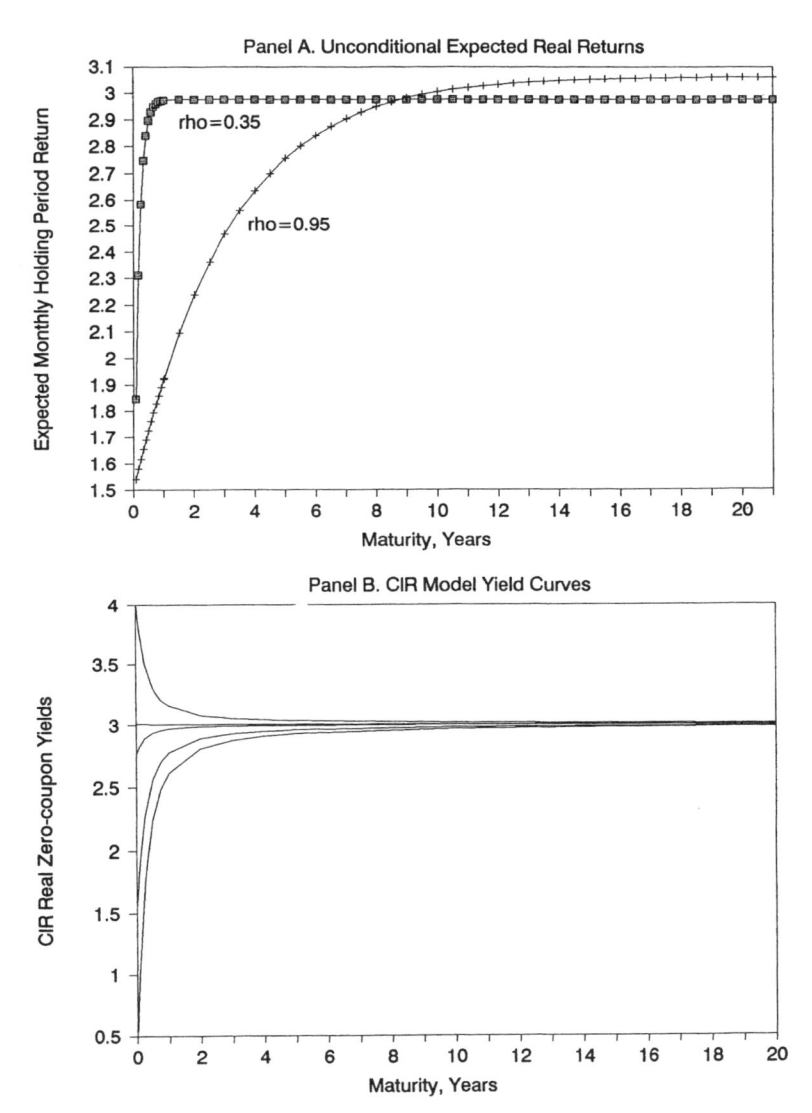

Panel B. CIR Model Yield Curves

Figure 2
Expected returns and yields to maturity for the CIR model, using GMM parameter estimates
The parameter values used in both figures are $\theta = 1.54$ percent, $\rho = 0.35$, $y_\infty = 3.01$ percent, and $\sigma_v = 1.23$ percent. Panel A shows the annualized expected monthly holding period returns for indexed bonds of various maturities for two values of the autocorrelation coefficient: $\rho = 0.35$ (the fitted estimate) and $\rho = 0.95$. Panel B shows yield curves for the CIR model for each of five values of the instantaneous rate $r = 4.00, 3.00, 2.77, 1.54,$ and 0.50 percent.

that we view as plausible.[28] Panel B of Figure 2 attempts to summarize the results in another way by providing a plot of the term structure of real yields on indexed bonds for different values of the current instantaneous rate (using the parameter estimates in the second column of Table 2). In principle the CIR model can generate a hump in the yield curve; however, our particular parameter estimates preclude a humped shape for any value of r. All yield curves in Figure 2B asymptote to y_∞, which is 3.01 percent. For cases where the r is such that it is less than the long-run yield, the term structure is uniformly increasing.

5.2 Sensitivity analysis

While our econometric framework is conceptually straightforward, the optimization requires a solution to a difficult nonlinear problem. With many of our initial runs, we had a difficult time finding the proper set of starting values in order to achieve convergence. Furthermore, we experienced some situations where the minimum was apparently found, yet our numerical calculation of the Hessian suggested that it was not positive definite. We have attempted to do a thorough search over the parameter space to confirm our estimates reported in Table 2.[29] Figure 3 provides some graphical evidence on the shape of the objective function. In these graphs we held three of the four parameter estimates fixed at the values given in Table 2; then we graphed the objective against the fourth parameter along the horizontal axis. In all cases, the graphs suggest a (locally) unimodal objective function around the optimal estimate of that fourth parameter.

We also examined the robustness of our results to various changes in the econometric specification. Table 3 summarizes the results for the sensitivity analysis. The first row of Table 3 repeats the results in Table 2 to allow for easier comparison between the initial result and alternative specifications. The sensitivity checks can be classified into four groups:

1. Lags used in the Newey–West weighting matrix (reported in row 2 of Table 3).

---

[28] One potential application of these parameter estimates is in bond portfolio management. Note, however, that any normative implications of our estimates are best drawn for the management of *indexed* bond portfolios. For portfolios of indexed bonds, one could compute measures of risk, just as discussed in CIR (1979). One could also supply our estimates to assessing the expected real returns to nominal bond portfolios (panel A of Figure 2 is relevant) but not to assessing their risk. For a discussion of contingent-claim pricing with CIR-type models, see Chen and Scott (1992).

[29] Since the value of the χ^2 statistic is small, this is also a good indication that the algorithm has successfully found a global minimum. Of course, the converse does not necessarily follow. A large value for this statistic need not imply a local minimum has been found, for a large value is also consistent with the case that the theoretical model is misspecified.

Test of the CIR Model

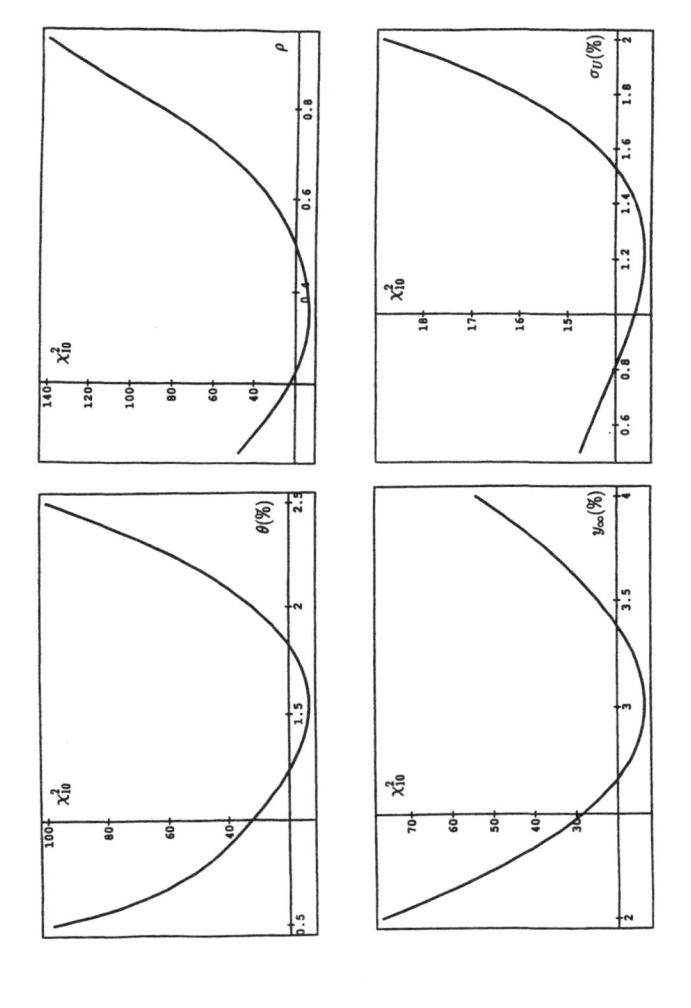

**Figure 3
Value of the objective function vs. individual parameter values**
For all four graphs, the minimum value of objective function (χ^2_{10}) occurs are $\theta = 1.54$ percent, $\rho = 0.35$, $y_\infty = 3.01$ percent, and $\sigma_{y_i} = 1.23$ percent. Each graph plots one of the four parameters, holding the other three fixed at the optimal estimate.

2. Lag structure in the second moment conditions (reported in row 3 of Table 3).
3. Additional overidentifying restrictions based on third moments (reported in row 4 of Table 3).
4. Moment conditions using yields to maturity (reported in row 5 of Table 3).

Each category will now be discussed.

Lags used in the Newey–West weighting matrix. The choice of the number of lags (i.e., m) to use in the Newey–West weighting matrix [see Equation (39)] is somewhat arbitrary. In Table 2 we set $m = 4$ to account for our inability to observe and hence condition on r, which has a first-order autoregressive structure. We varied m between 0 and 8; the second row of Table 3 illustrates the effect of increasing m to 8. Fortunately, our results in Table 2 were not significantly affected by alternative choices of m. The point estimates and the standard errors in the first two rows of Table 3 are similar.

Lag structure in the second moment conditions. Two considerations motivated the lag structure used in Table 2. First, analyzing too many lags may lead to small sample problems with GMM. Second, using short lags is probably superior to using long lags because the statistical precision is greater for the former. Nevertheless, we did experiment with alternative lag lengths, but we found little change from the results in Table 2. The parameter estimate with the greatest sensitivity to the lag structure is σ_U. The third row of Table 3 illustrates the impact of using comoments at a two-month lag rather than at the one month used in Table 2.

While we examined other lag structures as well,[30] the two-month lag is perhaps the most interesting. The advantage of the two-month lag over the one-month lag is threefold. First, we are more confident that the actual dating of the returns does not overlap (either due to nonsynchronous trading or the measurement problems with inflation). Second, extending the lag length in the comoments minimizes problems associated with autocorrelated measurement error as long as the measurement error follows a moving-average process of small order. Finally, a slightly longer lag length guarantees that the information that we viewed as known in deriving Equation (33) is in fact known by the market when it sets the prices of bonds. The results in the third row are reassuring because they suggest that our findings in the first row are robust to such measurement errors.

---

[30] For example, when we used a 12-month lag instead of the 1-month lag in Table 2 we found the standard error on σ_U increased, and we could no longer reject the hypothesis that σ_U was equal to zero, even though the point estimate did not change substantially.

Test of the CIR Model

Table 3
Sensitivity analysis of econometric estimates for the CIR model monthly data on real returns and yields, 1964–1989[1]

Case description	T-bill maturities	No. of moments	Newey-West lag	θ	ρ	y_∞	σ_U	Test statistic (p value)
Returns	1, 3, 6, 12	14	4	1.54 (0.26)	0.35 (0.06)	3.01 (0.30)	1.23 (0.45)	$\chi^2_{10} = 13.39$ (.20)
Returns	1, 3, 6, 12	14	8	1.50 (0.29)	0.36 (0.06)	3.00 (0.32)	1.16 (0.43)	$\chi^2_{10} = 13.10$ (.22)
Returns[2]	1, 3, 6, 12	14	4	1.40 (0.27)	0.36 (0.06)	3.02 (0.32)	1.81 (0.62)	$\chi^2_{10} = 9.54$ (.48)
Returns (3d moments)	1, 3, 6, 12	22	12	1.40 (0.22)	0.32 (0.05)	2.77 (0.26)	0.86 (0.20)	$\chi^2_{18} = 19.39$ (.37)
Yields	3, 6, 12	10	22	0.03 (10.68)	0.08 (0.72)	1.27 (0.43)	1.34 (26.08)	$\chi^2_6 = 14.04$ (.03)
Returns	3, 6, 12	10	4	3.05 (0.39)	0.62 (1.21)	3.05 (0.35)	1.64 (5.42)	$\chi^2_6 = 50.62$ (.00)
Returns (1964–76)	1, 3, 6, 12	14	4	0.85 (0.00)	0.48 (0.00)	2.34 (0.00)	0.00 (0.00)	$\chi^2_{10} = 7.30$ (.70)
Returns (1977–89)	1, 3, 6, 12	14	4	2.64 (0.36)	0.30 (0.06)	4.53 (0.43)	1.65 (0.92)	$\chi^2_{10} = 7.57$ (.67)

[1] All returns are for monthly holding periods; all yields are for holding periods equal to the bill's maturity.

[2] In row 3, the comoment restrictions employ a two-month lag; in all other rows, the lag is one month.

Additional overidentifying restrictions based on third moments.

Not only is the selection of the lag length somewhat arbitrary, but it is also unclear why we should limit our analysis to just the first and second moments. In Section 3, we computed the first and second moments for real returns on nominal bonds under the CIR specification; in the appendix we generalize these results to higher-order moments.

Originally we relied on just the first and second moments because of the difficulty in estimating higher-order moments precisely. We extended our investigation to include a noncentral third moment, for it is closely related to the conditional standard deviation of r, which is not constant in the CIR world. Clearly, the conditional variance is related to third moments, since it reflects the expectation of the square of the random variable times a lagged value of the random variable. We estimated the model while keeping the original 14 moments reported in Table 2 and adding eight more sample moments. These eight additional moments had the form

$$E[{}_t\tilde{R}_{t+1}(\tau_1)_{t+1}\tilde{R}_{t+2}(\tau_2)_{t+2}\tilde{R}_{t+3}(\tau_2)], \tag{45}$$

where τ_1 was set equal to 1 month or 12 months and τ_2 was set equal to each of the four maturities. That is, we correlated the lagged value

649

of the 1-month (or 12-month) bill return with the serial comoments of all four maturities.

As in Table 2, the p value of the χ^2_{18} statistic (reported in row 4 of Table 3) exceeded .05 despite the additional constraints placed on the model. Except for σ_U (and to a lesser extent y_∞), the estimates in Table 2 were not significantly affected by the additional eight moment conditions. As reported in the fourth row of Table 3, the estimate for σ_U was reduced to 0.86, but its standard error decreased as well. The reduction in the standard error provides some evidence that the third moment has information for the conditional variance of r and hence for σ_U.

Moment conditions using yields to maturity. Our final sensitivity check examines the measurement error associated with on-the-run bills versus off-the-run bills. Table 2 is based on returns where the holding period is one month. For example, a six-month bill is purchased and then sold when its maturity is five months. When this bill is sold, it is no longer on-the-run, and the measurement error in the market price of the security is somewhat greater due to decreased trading activity in the market. We can circumvent this problem by computing real returns from holding an on-the-run bill until it matures. Such a procedure for measuring real returns also provides the ex post real yield on the bond. We adopt the terminology of real yields to distinguish this analysis from the cases involving one-month holding periods. Since the Treasury does not auction one-month bills, we also exclude this maturity in this sensitivity check in an attempt to provide a clean set of prices for measuring real yields based only on on-the-run instruments.

The fifth row in Table 3 summarizes the results when real yields are used instead of one-month holding period returns. At first glance it seems that the measurement error may be important, for the point estimates of θ, ρ, and y_∞ are smaller than those in the earlier rows in this table. However, the standard errors are now much larger, so the discrepancies are less significant than they first appear. The increase in the standard errors is to be expected because the real yields generate a large amount of overlap in the observations. For example, the real yield on a 12-month bill has 11 months of overlap with the adjacent observation of the same series. This overlap requires us to extend the Newey–West lag structure of the weighting matrix; in Table 3 we report results where m is equal to 22.[31]

While the point estimates in the fifth row are bothersome, we found

---

[31] We tried alternative values for m that did not reduce the degree of discrepancy between the row for yields and the other rows in Table 3.

650

Test of the CIR Model

the χ_6^2 statistic for the overidentifying restrictions noteworthy: this is the first instance where we reject the CIR model at usual levels of significance. One could attribute all the results in the fifth row to small sample properties of the econometric procedure due to the presence of extreme amounts of serial dependence induced by overlapping observations. A second explanation may claim that the measurement error in off-the-run bill prices is substantial and biases against rejection. A third possibility may be the presence of one-month bills in the first four rows of Table 3 and the exclusion of this maturity in the fifth row.

To investigate this last possibility, we extended the results in Table 2 to a setting where returns with a one-month holding period are used but where we excluded the one-month bill. This exclusion decreases the number of moment conditions to 10 and reduces the number of overidentifying restrictions. The results of this case are reported in the sixth row of Table 3. As expected the precision of the estimates in the final row is greater than that reported in the fifth row, since the one-month holding period returns eliminate the overlap in the data. Now the point estimate for y_∞ is comparable to that in the earlier rows. However, the estimates for θ and ρ are different from the earlier rows. The inclusion of the one-month maturity provides a more reasonable estimate for θ, which may be expected since θ represents the expected return on a very-short-term maturity bond. This is consistent with the evidence in Fama (1984), where he finds that short-term bills are important in detecting term premiums. Measurement error in the off-the-run prices for bills is not an adequate explanation of the discepancy between the fifth row and the earlier rows, for the χ_6^2 continues to reject the overidentifying restrictions in the sixth row, where off-the-run prices are used in computing the real returns. (Again, we defer our discussion of ρ till the next subsection.)

The sixth row of Table 3 suggests that the CIR model fits the very short end of the term structure better than it fits the intermediate range. Row 6 of Table 3 provides an empirical discrepancy between the CIR model and the data. The next subsection discusses another deficiency of the CIR model.

5.3 A deficiency of the model

Based on the previous subsections, one might conclude that the CIR model provides a reasonable characterization of the real returns on nominal bills, at least for maturities of 12 months or less. For the most part the estimates are reasonable, the implied shape of the term structure for indexed bonds is plausible, and the results are not very sensitive to the exact specification of the econometric model.

The estimate of the autocorrelation, ρ, is the most troubling. As noted in Section 5.1 above, the yield on indexed bonds reverts rather quickly to θ given the parameter estimates in Table 2. We also found in the last row of Table 3 some evidence suggesting higher estimates for ρ for alternative econometric specifications. Figure 2A illustrates the impact on unconditional expected returns when ρ is increased from 0.35 to 0.95. As ρ increases, the graph for unconditional expected returns displays curvature even for long maturities.

Other implications from the low value of ρ show up in Table 4. Using the fitted values for the first moment and for the serial comoment in Table 2, we backed out the implied value for the autocovariance. Then we divided these theoretical autocovariances by the *sample* variances to compute a standardized measure of serial dependence based on autocorrelations; these numbers are reported in Table 4. (Since we have not specified a process for inflation, the theory does not make a prediction about variances, and we must rely on sample variances. We divided by the sample variance in order to produce numbers that are a little easier to interpret.) In all rows of Table 4, we are reporting implied autocorrelations for real returns on bonds with a one-month holding period.

The results in Table 4 are striking in that the autocorrelations are quite close to zero, die off quickly, and for most maturities the autocorrelations are negative. In contrast, panel A of Table 1 suggests that the sample autocorrelations are substantially different from zero, die off slowly, and are always positive. It is difficult to reconcile the results in Tables 1 and 4, where the model performs poorly, with Table 2 where the model seemed to fit well. Recall that in Section 5.2 we reported that the fit of the model was not affected by the use of an alternative lag structure. The very low value of the implied autocorrelation in Table 4 for even 12-month lags does not seem to be a problem in fitting the model despite the high sample autocorrelation reported in Table 1.

One possible explanation is that the parameter estimates in Table 2 are not based on moments reflecting the sample variance. As noted, the CIR model has no implications for the variance without specifying the inflation process. The predicted autocorrelations in Table 4 would be much higher if the variances used in transforming autocovariances into autocorrelations were lower. However, even if the standard deviations were lower, the signs in Tables 1 and 4 are troublesome. How can the model fit so well in Table 2 and yet the sign of the implied autocorrelation be wrong? The answer to this question may be that we used noncentral second moments. As a result, little penalty is attached to situations where the central second moments (after adjusting for the first moment) have the wrong sign relative to the sample central second moments.

652

Test of the CIR Model

Table 4
Theoretical autocorrelations using covariances implied by CIR parameter estimates and sample standard deviations

| Bill maturity (months) | Autocorrelations at lag |
	1	2	3	4	5	6	7	8	9	10	11	12
1	0.030	0.011	0.004	0.001	0.000	0.000	0.000	0.000	0.000	0.000	0.000	0.000
3	-0.074	-0.023	-0.009	-0.003	-0.001	-0.000	-0.000	-0.000	-0.000	-0.000	-0.000	-0.000
6	-0.073	-0.026	-0.009	-0.003	-0.001	-0.000	-0.000	-0.000	-0.000	-0.000	-0.000	-0.000
12	-0.031	-0.011	-0.004	-0.001	-0.000	-0.000	-0.000	-0.000	-0.000	-0.000	-0.000	-0.000

Monthly returns on 1-, 3-, 6-, and 12-month bills, using CIR formula prices. CIR parameter estimates used: $\theta = 0.0154$, $\kappa = 12.43$, $\lambda = -6.08$, and $\sigma = 0.49$. Sample standard deviations, from monthly data, 1964–1989: 0.273%, 0.322%, 0.434%, and 0.699% for 1-, 3-, 6-, and 12-month bills respectively.

Casual empiricism is also troubling for this aspect of the CIR model. During the middle 1980s expected real returns were high (by the standards of the fitted CIR model) and stayed high (i.e., mean reversion was slow). Given that we estimate θ to be about 1.54 percent, how can we have a time period like the 1980s where the "expected" real returns on short-term bonds are substantially greater than implied by this value of θ? It is hard to imagine that, when nominal interest rates were high during the middle 1980s, anyone forecast inflation to be a comparably high number, especially for short-term bond maturities. If the mean reversion is quick (or ρ is only 0.35), it would seem that the "expected" real return should have been smaller.[32]

One obvious solution is to increase the number of parameters in the CIR model and allow for a time-dependent value of θ, an extension that is discussed in Cox, Ingersoll, and Ross (1985b). Because we do not have a good model for θ, we have not pursued this line of inquiry; however, we do report parameter estimates from two subperiods in the last two rows of Table 3. Not surprisingly, θ and y_∞ are much lower in the first subperiod (1964–1976) than in the second subperiod. In line with Fama (1975), $\hat{\theta}$ is 85 basis points. We also find $\hat{\sigma}_U$ is essentially zero in the first subperiod[33] and much higher in the second. Again, Fama (1975) concludes that one cannot reject the hypothesis that the real return is constant during a time period that largely overlaps with our first subperiod. The higher value of σ_U is to be expected in a time period that includes the change in Fed monetary policy. While $\hat{\rho}$ is higher in the first subperiod than in the second, it is not close to unity, which is implied by a view that the real rate follows a random walk. In both subperiods, the overidentifying restrictions are not rejected.

6. Conclusion

We have presented a test of the model of the term structure developed in Cox, Ingersoll, and Ross (1985b). It is important to keep in mind that this model pertains to real bond prices; a multifactor model for nominal bond prices is completely consistent with a single-factor model for the yields on indexed bonds.

Our tests and estimates indicate that the model, which is at once quite complicated in its structure and rather simplistic in its dependence on a single-state variable, performs reasonably well when confronted with data on short-term Treasury bills. The parameter esti-

---

[32] Here we used "expected" to mean the forecasts of interest rates and inflation held by investors in the market, not necessarily the conditional expectation calculated by a formal theoretical model.

[33] Based on our experience with the estimation, the low standard errors reported for the first subperiod are a by-product of the low value of σ_U.

Test of the CIR Model

mates indicate that on average the term premium is positive, and they allow for both upward- and downward-sloping term structures for indexed bonds.

Much work remains to be done. We hope to exploit our estimates in a procedure that extracts a time series for the unobservable economic variable [i.e., $r(t)$]; to extend the tests to a class of models that explicitly incorporate inflation and find the nominal price of a nominally riskless bond; and to conduct empirical tests of models that find the nominal prices of derivative securities. We hope that future research will extend the theory to a general equilibrium setting wherein money has a useful economic function. Such an extension would provide an endogenous process for inflation and for interest rates.

Appendix

In this Appendix we compute the restrictions on the first and second moments of $_t\tilde{R}_{t+u}(\tau)$, denoted $\Phi_1(u, \tau; \beta)$ and $\Phi_2(u, v, w, \tau_1, \tau_2; \beta)$ as defined in Equations (30) and (33), respectively. These computations permit the application of the GMM procedure.

Consider first the computation of $\Phi_1(u, \tau; \beta)$ in (30):

$$\Phi_1(u, \tau; \beta)$$

$$\equiv \frac{A(\tau - u)}{A(\tau)} E\{\exp(-B(\tau - u)\tilde{r}(t + u) + B(\tau)\tilde{r}(t))\}. \quad (A1)$$

The expected value in the RHS of (A1) needs to be computed. Now defining $p \equiv B(\tau - u)$ and $q \equiv B(\tau)$, we can write

$$E\{\exp(-p\tilde{r}(t + u) + q\tilde{r}(t))\}$$

$$= E\{\exp(q\tilde{r}(t))E_t[\exp(-p\tilde{r}(t + u))]\}, \quad (A2)$$

$$E_t\{\exp(-p\tilde{r}(t + u))\}$$

$$= C_1 \exp[C_2 r(t)], \quad (A3)$$

where

$$C_1 \equiv \left[1 - \frac{p\sigma^2(1 - \exp\{-\kappa u\})}{2\kappa}\right]^{-2\kappa\theta/\sigma^2}, \quad (A4)$$

and

$$C_2 \equiv \left[\frac{2\kappa p}{2\kappa - p\sigma^2[1 - \exp\{-\kappa u\}]}\right]. \quad (A5)$$

Equation (A3) relies on the fact that the distribution of $\tilde{r}(t + u)$ given

$\tilde{r}(t)$ is noncentral χ^2 [see CIR (1985b)], whose properties are given in Johnson and Kotz (1970, Chapter 28). Use (A3) in (A2) to obtain

$$E\{\exp(-p\tilde{r}(t+u) + q\tilde{r}(t))\} = C_1 E\{\exp((q+C_2)\tilde{r}(t))\}. \quad (A6)$$

The distribution to be used in computing the RHS of (A6) is the unconditional or steady-state distribution of $r(t)$, which is a gamma. From CIR (1985b) and Johnson and Kotz (1970, Chapter 17)

$$E[\exp(\phi\tilde{r}(t))] = [1 - \phi/\omega]^{-\nu} \quad \text{for } \phi < \omega, \quad (A7)$$

where

$$\omega \equiv 2\kappa/\sigma^2, \quad (A8)$$

and

$$\nu \equiv 2\kappa\theta/\sigma^2. \quad (A9)$$

The function $\Phi_1(u, \tau; \beta)$ is found from substituting (A7) in (A6) and the result in (A1).

The condition on the second moment of $_t\tilde{R}_{t+u}(\tau)$ follows from relation (33). This involves the computation of the unconditional expectation involving $r(t)$, $r(t+u)$, $r(t+v)$, and $r(t+w)$ of the following form:

$$E\{\exp[a\tilde{r}(t) + b\tilde{r}(t+u) + c\tilde{r}(t+v) + d\tilde{r}(t+w)]\}, \quad (A10)$$

where a, b, c, and d are of the form $B(\tau_i)$ and hence a function of the parameters and the holding periods. This expectation can be rewritten as

$$E\{\exp[a\tilde{r}(t)] \cdot E_t\{\exp[b\tilde{r}(t+u)]$$
$$\cdot E_{t+u}\{\exp[c\tilde{r}(t+v)]$$
$$\cdot E_{t+v}\{\exp[d\tilde{r}(t+w)]\}\}\}\}. \quad (A11)$$

By repeated use of the arguments in (A2) through (A7), this expectation can be expressed as a function of a, b, c, and d (which are defined in terms of CIR parameters), the holding periods (t to $t+u$ and $t+v$ to $t+w$), and the times to maturity. This defines $\Phi_2(u, v, w, \tau_1, \tau_2; \beta)$ which is the unconditional expectation defined in the comoment (33). The restriction in the computation of this comoment is that the holding periods should not overlap.

This procedure can be extended to compute comoments of higher order; for example, in computing the third comoment the expectation would involve six distinct values for r at different times t. Section 5.2 summarizes our empirical results using an additional eight third moments.

656

Test of the CIR Model

References

Aït-Sahalia, Y., 1992, "Nonparametric Pricing of Interest Rate Derivative Securities," working paper, Department of Economics, Massachusetts Institute of Technology.

Benninga, S., and A. Protopapadakis, 1983, "Real and Nominal Interest Rates under Uncertainty: The Fisher Theorem and Term Structure," *Journal of Political Economy*, 91, 856–867.

Breeden, D. T., 1986, "Consumption, Production, Inflation and Interest Rates: A Synthesis," *Journal of Financial Economics*, 16, 3–40.

Brennan, M. J., and E. S. Schwartz, 1982, "An Equilibrium Model of Bond Pricing and a Test of Market Efficiency," *Journal of Financial and Quantitative Analysis*, 41, 301–329.

Brown, D. P., and M. R. Gibbons, 1985, "A Simple Econometric Approach for Utility-Based Asset Pricing Models," *Journal of Finance*, 40, 359–381.

Brown, R. H., and S. M. Schaefer, 1990, "The Real Term Structure of Interest Rates and the Cox, Ingersoll, and Ross Model," working paper, London Business School.

Brown, S. J., and P. H. Dybvig, 1986, "The Empirical Implications of the CIR Theory of the Term Structure of Interest Rates," *Journal of Finance*, 41, 617–630.

Chen, R-R., and L. Scott, 1992, "Pricing Interest Rate Options in a Two-Factor Cox, Ingersoll and Ross Model of the Term Structure," *Review of Financial Studies*, 5, 613–636.

Constantinides, G. M., 1992, "A Theory of the Nominal Term Structure of Interest Rates," *Review of Financial Studies*, 5, 531–552.

Constantinides, G. M., and J. E. Ingersoll Jr., 1984, "Optimal Bond Trading with Personal Taxes," *Journal of Financial Economics*, 13, 299–335.

Cox, J. C., J. E. Ingersoll Jr., and S. A. Ross, 1979, "Duration and the Measurement of Basis Risk," *The Journal of Business*, 52, 51–62.

Cox, J. C., J. E. Ingersoll Jr., and S. A. Ross, 1981, "A Re-Examination of the Traditional Hypotheses of the Term Structure of Interest Rates," *Journal of Finance*, 36, 769–799.

Cox, J. C., J. E. Ingersoll Jr., and S. A. Ross, 1985a, "An Intertemporal General Equilibrium Model of Asset Prices," *Econometrica*, 53, 363–384.

Cox, J. C., J. E. Ingersoll Jr., and S. A. Ross, 1985b, "A Theory of the Term Structure of Interest Rates," *Econometrica*, 53, 385–407.

Dunn, K., and K. R. Singleton, 1986, "Modeling the Term Structure of Interest Rates under Nonseparable Utility and Durability of Goods," *Journal of Financial Economics*, 17, 27–56.

Fama, E. F., 1975, "Short-Term Interest Rates as Predictors of Inflation," *American Economic Review*, 65, 269–282.

Fama, E. F., 1984, "Term Premiums in Bond Returns," *Journal of Financial Economics*, 13, 529–546.

Fama, E. F., and M. R. Gibbons, 1982, "Inflation, Real Returns and Capital Investment," *Journal of Monetary Economics*, 8, 297–324.

Fama, E. F., and M. R. Gibbons, 1984, "A Comparison of Inflation Forecasts," *Journal of Monetary Economics*, 13, 327–348.

Feller, W., 1951, "Two Singular Diffusion Problems," *Annals of Mathematics*, 54, 173–182.

Ferson, W., S. Kandel, and R. F. Stambaugh, 1987, "Tests of Asset Pricing with Time Varying Expected Risk Premiums and Market Betas," *Journal of Finance*, 42, 201–220.

Gibbons, M. R., and W. Ferson, 1985, "Testing Asset Pricing Models with Changing Expectations and an Unobservable Market Portfolio," *Journal of Financial Economics*, 14, 217–236.

Hansen, L. P., 1982, "Large Sample Properties of Generalized Method of Moments Estimators," *Econometrica*, 50, 1029–1054.

Hansen, L. P., and R. J. Hodrick, 1980, "Forward Exchange Rates as Optimal Predictors of Future Spot Rates: An Econometric Analysis," *Journal of Political Economy*, 88, 829.

Hansen, L. P., and K. J. Singleton, 1982, "Generalized Instrumental Variables Estimation of Nonlinear Rational Expectations Models," *Econometrica*, 50, 1269–1286; Errata, 52, 267–268.

Heston, S., 1991, "Testing Continuous-Time Models of the Term Structure of Interest Rates," working paper, School of Organization and Management, Yale University.

Johnson, N. L., and S. Kotz, 1970, *Distributions in Statistics: Continuous Univariate Distributions 2*, Houghton Mifflin, Boston, MA.

Lucas, R. E., Jr., 1978, "Asset Prices in an Exchange Economy," *Econometrica*, 46, 1426–1446.

Melino, A., 1986, "The Term Structure of Interest Rates: Evidence and Theory," Working Paper 1828, National Bureau of Economic Research.

Merton, R. C., 1973, "An Intertemporal Capital Asset Pricing Model," *Econometrica*, 41, 867–887.

Nelson, J., and S. M. Schaefer, 1983, "The Dynamics of the Term Structure and Alternative Portfolio Immunization Strategies," in George Kaufman, G. O. Bierwag, and A. Toevs (eds.), *Innovation in Bond Portfolio Management*, J.A.I. Press, Greenwich, CT.

Newey, W. K., and K. D. West, 1987, "A Simple, Positive Definite Heteroscedasticity and Autocorrelation Consistent Covariance Matrix," *Econometrica*, 55, 703–708.

Pearson, N. D., and T. S. Sun, 1990, "An Empirical Examination of the Cox, Ingersoll, Ross Model of the Term Structure of Interest Rates Using the Method of Maximum Likelihood," working paper, Graduate School of Management, University of Rochester.

Pennacchi, G. C., 1991, "Identifying the Dynamics of Real Interest Rates and Inflation: Evidence Using Survey Data," *Review of Financial Studies*, 4, 53–86.

Stambaugh, R. F., 1988, "The Information in Forward Rates: Implications for Models of the Term Structure," *Journal of Financial Economics*, 21, 41–70.

# An Econometric Model of the Term Structure of Interest-Rate Swap Yields

DARRELL DUFFIE and KENNETH J. SINGLETON*

ABSTRACT

This article develops a multi-factor econometric model of the term structure of interest-rate swap yields. The model accommodates the possibility of counterparty default, and any differences in the liquidities of the Treasury and Swap markets. By parameterizing a model of swap rates directly, we are able to compute model-based estimates of the defaultable zero-coupon bond rates implicit in the swap market without having to specify a priori the dependence of these rates on default hazard or recovery rates. The time series analysis of spreads between zero-coupon swap and treasury yields reveals that both credit and liquidity factors were important sources of variation in swap spreads over the past decade.

ALTHOUGH *PLAIN VANILLA* FIXED-for-floating interest-rate swaps comprise a major segment of the fixed-income derivative market, notably few econometric models for pricing swaps have been developed in the literature. Perhaps the primary reasons for this are: (i) swap contracts embody default risk and hence equilibrium or arbitrage-free term structure models developed for default-free government bond markets are not directly applicable to the swap market; (ii) empirical modeling of the default event underlying credit spreads on defaultable bonds and swaps has met with limited success at explaining the time-series properties of spreads; and (iii) swap spreads are likely to depend on other factors such as liquidity that are not directly related to default events. Also, until recently, data have not been widely available. In this article we develop a multifactor econometric model of the term structure of U.S. fixed-for-floating interest-rate swap yields that accommodates many of the institutional features of swap markets. Specifically, using results in Duffie and Singleton (1996), we show that the fixed payment rate of a swap, assuming that the floating rate is London Interbank Offering Rate (LIBOR), can be expressed in terms of present values of net cash flows of the swap contract

---

* Graduate School of Business, Stanford University. We are grateful for the research assistance of Qiang Dai, Stephen Gray, and Raj Tewari, and comments from Mark Fisher, Dilip Madan, Ming Huang, René Stulz, an anonymous referee, and seminar participants at the National Bureau of Economic Research, the University of Wisconsin Finance Symposium, the University of California, Berkeley, the University of Chicago, Duke University, the University of Arizona, and the University of California at San Diego, and from an anonymous referee. Data were kindly provided by Goldman Sachs and Co. Financial support was provided by the Stanford GSB Financial Research Initiative. These results appeared in preliminary form under the title "Econometric Modeling of Term Structures of Defaultable Bonds." An extended version of the valuation models from that article now appears in Duffie and Singleton (1996).

discounted by a default and liquidity-adjusted instantaneous short rate. In other words, there is an adjusted short rate process that allows us to develop a term structure model for the swap market in much the same way that models have been developed for government yield curves. Default and liquidity risks are "collapsed" into a risk-adjusted short rate for computing the present values of future risky cash flows.

Our formulation provides a model-based alternative to swap valuation models used by many investment and commercial banks. Many financial institutions use forward interest rates obtained by interpolating between fixed maturity points on the swap yield curve to derive discount rates for future net cash payments between the swap counterparties. Instead, we develop an arbitrage-free multifactor term structure model of swap yields that leads to an implied, model-based zero curve that can be used in valuing off-the-run swaps, caps, floors, and swaptions (options to enter into a swap contract at some future date). Moreover, unlike interpolation schemes, our model provides a framework for dynamically hedging swaps or derivatives, such as caps and swaptions.

The literature on swap rates has typically focused on the spreads between swap rates and the corresponding point on the U.S. Treasury (default-free) term structure. Sun, Sundaresan, and Wang (1993) examine the average swap spreads to Treasuries and the bid/ask spreads by the credit class of the swap counterparties. Brown, Harlow, and Smith (1994) regress swap spreads on various contemporaneous measures of credit risk and the hedging costs of market makers. They find that both sets of variables are correlated with swap spreads. Neither study develops a dynamic swap-pricing model.

Grinblatt (1995) argues that liquidity risk is a more plausible explanation for swap spreads than credit risk. Liquidity enters his model as a convenience yield to Treasuries associated with their relative liquidity and potential to go "on special" in markets for repurchase agreements (repo). Holders of Treasury bonds that go on special can effectively receive a special dividend by borrowing at below market rates, using Treasuries as collateral. (See also Duffie (1996) and Jordan and Jordan (1997).) Using the one-month LIBOR-Treasury bill spread as a proxy for the convenience yield of holding government securities, Grinblatt calibrates a Vasicek (1977) representation of riskless and convenience-yield processes. He finds that this model explains about 35 to 40 percent of the variation in swap spreads for maturities of two through ten years.

Instead of focusing on swap spreads to the default-free term structure, we focus on swap yields directly. Our goal is to develop a model of the swap market without needing to be precise a priori about the economic mechanisms that generate swap spreads. Subsequent to fitting the model, we can study the properties of the defaultable zero-coupon yields implied by the swap market with the goal of a better understanding of the economic factors that determine their spreads to Treasury zero-coupon yields. In Section I we show that, under the assumption that the counterparties have symmetric probabilities of default, a swap is "priced" by the present value of its cash flows discounted by a risk- and liquidity-adjusted short-rate process. Once one adopts a parameterization of this short-rate process, the parameters of the model can be estimated without having to specify a priori the functional forms for default probabilities, liquidity premiums, and so forth. Virtually any of the models examined previously for government yield curves can be used to model the default-adjusted short-rate process, including affine processes for riskless rates and Heath-Jarrow-Morton (1992) type models of forward rates.[1]

In Section II, we discuss the econometric models of defaultable swap yields. The econometric model of swaps studied in Section III presumes that the liquidity- and default-adjusted short-rate process is the sum of two independent square-root diffusions (a *2-factor* model). Assuming that the model exactly prices swaps at two points along the swap yield curve, and using the fact that the conditional distribution of discretely sampled data from a square-root diffusion has a noncentral chi-square density (e.g., Cox, Ingersoll, and Ross (1985) (CIR)), we estimate our model using the joint likelihood of swap yields at several maturities.[2] The distributions of the fitted swap yields, evaluated at the maximum likelihood (ML) estimates, have sample moments that are similar to those of the corresponding actual yields. Moreover, deviations between the actual and fitted swap yields are on average zero, with standard deviations between four and seven basis points for seven years of weekly data. (By comparison, the bid/ask spreads in the swap market averaged about four basis points during our sample period.)

In light of the close fit of the model to swap yields, we proceed to compute the implied risky zero-coupon bond yields, evaluated at the ML estimates, and compute their spreads to the corresponding U.S. Treasury zero-coupon yields. In Section IV the properties of these swap zero-coupon yield spreads are studied in the context of a multivariate vector autoregression (VAR) in an attempt to shed some light on the relative importance of liquidity and credit factors in the determination of swap spreads. Included in the VAR, along with swap zero spreads, are variables proxying for corporate credit risk and liquidity in the U.S. Treasury market. The results suggest that both liquidity and credit factors affect the temporal behavior of swap zero spreads, but that the responses of swap spreads to changes in these factors follow very different time paths. Liquidity effects are short-lived, whereas responses to credit shocks are weak initially and then increase in importance over a horizon of several months. Concluding remarks are presented in Section V.

I. Valuation of Swaps

Consider a set of M plain vanilla fixed-for-floating swaps. The mth swap has τ_m years to maturity. The floating side is reset semi-annually to the six-month LIBOR rate from six months prior. The fixed side pays a coupon c^m at the reset

---

[1] See Duffie and Singleton (1996) for a discussion of alternative formulations of risky discount rate processes for valuing defaultable bonds.

[2] This likelihood function has the same form as that studied by Chen and Scott (1993) and Pearson and Sun (1994) in their studies of the U.S. Treasury market.

dates. Let r_t^L denote the LIBOR rate set at date t for loans maturing six months in the future and $\text{PV}(t, t + \tau_m)$ denote the present value of the promised net cash flows between the parties to the swap agreement. We assume that, at the inception date of the swap,

$$0 = \text{PV}(t, t + \tau_m) = \sum_{j=1}^{2\tau_m} E_Q \left[\exp\left(-\int_t^{t+0.5j} R_s \, ds \right) (c_t^m - r_{t+0.5(j-1)}^L) | \mathcal{F}_t \right], \quad (1)$$

where E_Q denotes expectation under an equivalent martingale measure for the information sets $\{\mathcal{F}_t : t \geq 0\}$ commonly available to investors, and where R is an instantaneous discount-rate process defined below. In other words, the fixed-side coupon c_t^m of the swap is set at date t so that the present value of the net cash flows exchanged by the counterparties at the reset dates is zero at the inception of the swap, as in equation (1).

We let

$$B_t^\tau = E_Q \left[\exp\left(-\int_t^{t+\tau} R_s \, ds \right) | \mathcal{F}_t \right], \quad (2)$$

be the discount factor at time t for maturity τ associated with the short term discount-rate process R. We further assume that risky zero-coupon bonds are priced at the appropriate LIBOR rate in the interbank lending market, or

$$B_t^{0.5} = (1 + r_t^L)^{-1}. \quad (3)$$

Using equations (1) to (3), the present value of the floating rate payments is $1 - B_t^{\tau_m}$. Combining these observations, the coupon rate c_t^m on the fixed side of the swap can be expressed as

$$c_t^m = \frac{1 - B_t^{\tau_m}}{\sum_{j=1}^{2\tau_m} B_t^{0.5j}}. \quad (4)$$

In deriving the valuation model (1)–(4), we have implicitly made several important assumptions. Since the assumptions needed for equation (4) to hold are significantly weaker than those typically made (see, for example, Litzenberger (1992) for a review of the literature), we briefly examine in more depth the key underlying assumptions.

The discounting in equations (1) and (2) takes the same form as the discounting of default-free cash flows by the riskless, instantaneous interest rate in risk-neutral representations of standard term-structure models. However, models for pricing default-free cash flows are not directly applicable to swaps and LIBOR contracts, because they are defaultable instruments. Nevertheless, using results in Duffie and Singleton (1996), we can interpret R as a default-adjusted discount rate and, under this interpretation plus additional assumptions outlined below, equation (1) correctly prices a defaultable swap.

More precisely, we define a defaultable claim to be a pair $((X, T), (X', T'))$ of contingent claims. The underlying claim (X, T) is the obligation of the issuer. The secondary claim (X', T') defines the stopping time T' at which the issuer defaults, and the payment X' to be received at default. This means that the actual claim (Z, τ) generated by a defaultable claim $((X, T), (X', T'))$ is defined by

$$\tau = \min(T, T'); \qquad Z = \begin{cases} X & \text{if } T < T', \\ X' & \text{if } T \geq T', \end{cases} \qquad (5)$$

where τ is the stopping time at which Z is actually paid.

The ex-dividend price process V of any given contingent claim (Z, τ) is defined by $V_t = 0$ for $t \geq \tau$ and

$$V_t = E_Q\left[\exp\left(-\int_t^\tau r_u \, du\right) Z \bigg| \mathcal{F}_t\right], \qquad t < \tau. \qquad (6)$$

Evaluation of the pricing formula (6) is complicated in practice by the possibility of default reflected in the payoff Z in equation (5) and that the probability of default will in general be correlated with the short-rate process, r. To circumvent these difficulties, we follow Duffie and Singleton (1996), and assume that the secondary claim (X', T') is defined by a hazard rate process h and a fractional default loss process λ. The hazard rate h can be thought of as the arrival intensity of a Poisson process whose first jump occurs at default. The state dependent process λ defines the fraction of *market value* of the claim that is lost upon default.

Under mild technical regularity conditions given in Duffie and Singleton (1996), the valuation of this defaultable claim can proceed as if the promised payoff X is default-free; however, with discounting at a default-adjusted discount rate R instead of the riskless rate. Consequently, under those conditions,

$$V_t = E_Q\left[\exp\left(-\int_t^T R_s \, ds\right) X \bigg| \mathcal{F}_t\right], \qquad t < T', \qquad (7)$$

where

$$R_t = r_t + h_t \lambda_t. \qquad (8)$$

We interpret each promised cash flow of a swap contract as one of the promised payments X described above. By repeatedly applying the same logic used to derive equation (7) to the sequence of promised cash flows of a swap, we get equation (1). This approach to swap valuation avoids a more complicated structural model based on knowledge of the asset-liability structure of the swap counterparties, in the style of Merton (1974), and as applied to the swaps market by Rendleman (1992). The development of a structural model seems impractical given our data on generic market swap rates.

Beyond default, a factor that may affect swap spreads to Treasuries is the relative liquidities of the two markets (Grinblatt (1995), Bansal and Coleman (1996)). Therefore, we also include a convenience yield l_t that allows for the effect of differences in liquidity and repo specialness between the Treasury and swap markets. With this modification, the swap rate c^m at inception solves equation (1) with the adjusted discount rate,

$$R_t = r_t - l_t + h_t \lambda_t. \tag{9}$$

The processes r, l, h, and λ are assumed to be adapted to the investors' information sets $\{\mathcal{F}_t : t \geq 0\}$ and thus, may be state dependent and mutually correlated.[3]

For the case of swaps, modeling the default time as an inaccessible stopping time, such as a Poisson arrival, seems reasonable because default events, when they do occur, are rarely fully anticipated even a short time before the default. Changing expectations concerning the likelihood of default are captured by the stochastic properties of the hazard rate process h. Indeed, under the risk-neutral measure Q, the conditional probability at time t of default over the next "instant" of time of length Δt is approximately $h_t \Delta t$.

The cash payment upon default of a swap contract is based on the market value of the remaining obligations under the terms of the swap and negotiations between the counterparties. If $PV(t_d, t_d + \tau)$ denotes the value of a τ-year swap just prior to the default time t_d, then $(1 - \lambda_{t_d}) PV(t_d, t_d + \tau)$ is the present value at t_d of the cash flows ultimately generated by the negotiated settlement. The liability, if any, according to standard International Swap Dealers Association (ISDA) swap contracts, is based on midmarket quotes for similar swaps.

One implication of these observations is that the assumption of no default made by Smith, Smithson, and Wakeman (1988) and Sun, Sundaresan, and Wang (1993), among others, in justifying expressions like equation (4) is unnecessarily strong. The market yields on A-rated LIBOR issues are occasionally at large spreads to U.S. Treasuries, and these spreads fluctuate substantially over time, as we show in Section III. The pricing relation (4) implicitly allows this spread to arise from a spread between R and the riskless rate r.

There are still, however, some important implicit assumptions remaining in the valuation model (1)–(4).

A. *Exogenous Default Risk*

The valuation model (1), for a given "credit spread" process λh, implicitly presumes that revaluation of the swap is not in itself a significant determinant

---

[3] These processes are also assumed to be jointly measurable in state and time and to satisfy mild integrability conditions described in Duffie and Singleton (1996).

of default likelihood or losses on default. (This endogeneity would exist, for example, with a swap that constitutes the major part of the liabilities of one of the counterparties.) If we were to compensate by allowing λh to depend endogenously on the market value of the swap, the linear valuation model (1) would be replaced with a nonlinear model described by Duffie and Singleton (1996). This would significantly complicate the econometric model. The fact that we are using generic market quotes for swap rates, and do not depend on the valuation of a swap between two particular counterparties, presumably mitigates the impact of any endogeneity of the default spread $\lambda_t h_t$. Moreover, based on the numerical results noted below, the impact of moderate dependence of λh on the market value of the swap is minimal.

B. "Refreshed" A-Quality Counterparties

Discounting the net cash flows at all maturities with the same risky discount rate R in equation (1) presumes that the counterparties maintain the credit rating underlying generic (say, A-rated) swaps *for the life of the swap contract*. More precisely, at the inception of the swap, the counterparties are presumed to have a credit rating of A. Subsequently, there is the possibility that either counterparty could be upgraded or downgraded (or suffer a change in liquidity) during the life of the swap. This itself is not a problem for pricing in terms of equation (1), since the potential of a change in quality can be captured in the stochastic process for the discount rate R. However, we do not have data on the yields for specific swaps as they mature, but rather on new A-rated swaps with constant maturity. Thus, we are valuing a hypothetical swap priced by dealers who presume that the counterparties will maintain the quality of newly issued A-rated debt over the life of the swap. This problem is not unique to our model of swaps, but rather is inherent in any model of new-issue rates on defaultable debt from a fixed credit class. Based on the numerical calculations of Duffie and Huang (1996) and Li (1995), described below, there is a minimal impact on swap rates of reasonably anticipated variation in the credit quality of the counterparties over the life of the swap.[4]

C. Symmetric Credit-Quality

Related to the point above is the possibility that the two counterparties to a given swap may have different credit quality, either at the inception of the swap, or subsequently. Extending our model to allow for asymmetric credit qualities of the counterparties would add substantial complexity to the pricing model. For example, even if the credit-spreads $\lambda_A h_A$ of counterparty A and $\lambda_B h_B$ of counterparty B are each given exogenously, and do not depend on the market value of the swap, in keeping with the exogenous default risk assumption above, there is nevertheless the effect of an endogenous dependence of the

---

[4] For alternative models of the impact of default risk on swap rates, see Abken (1993), Cooper and Mello (1991), Hull and White (1992), Jarrow and Turnbull (1995), Li (1995), Rendleman (1992), Solnik (1990), Sorensen and Bollier (1994), and Sundaresan (1991).

credit spread on market value if $\lambda_A h_A \neq \lambda_B h_B$. This arises from the fact that the current market value of the swap determines the firm whose default risk currently "matters," based on netting provisions and the implications of one-way or two-way ("no fault") payment schemes that are usually adopted in standard swap agreements. (If the market value of the swap to counterparty A is positive at a given point in time, then it is the default risk of counterparty B that "matters" at that moment.) Therefore, the relevant credit spread would become dependent on the market value of the swap, and the nonlinear valuation model discussed in Duffie and Singleton (1996) would apply. Duffie and Huang (1996) and Li (1995) explore theoretically and numerically this aspect of swap contracts, and find that the degree of asymmetry in credit quality is a relatively minor determinant of swap rates for typical interest-rate swaps. (For example, with less than a 10-year maturity and under typical parameters, there is a correction of roughly 1 basis point or less in swap yields for a credit risk asymmetry generating a 100 basis point difference in bond yields.) Thus, there is likely to be negligible misspecification error from proceeding under the assumption that swaps are priced as if there is symmetric counterparty risk.

D. *Homogeneous LIBOR-Swap Market Credit Quality*

There is no reason in theory that LIBOR corporate bond rates, including the floating-rate swap payments r_t^L, should be determined by discounting at the same liquidity and credit adjusted short-rate R used for discounting net swap payments in equation (1). The default scenarios may be different in the two markets, recovery rates may differ, and the liquidities of the two markets are also typically different. That is, it is a nontrivial assumption that equation (3) holds.[5]

Although these remarks have focused on the default component of swap spreads, there is no presumption in equation (9) that default is the only or even the primary determinant of swap spreads. Empirically, l_t, which represents other factors that determine effective carrying costs such as liquidity, may be as (or more) important a source of variation in $R_t - r_t$.

II. Econometric Models of Defaultable Swap Yields

In this section we discuss two alternative formulations of the adjusted short-rate R that may lead to econometrically identified models of the term structure of swap rates. One approach is to focus directly on R and assume that $R_t = \rho(Y_t)$, where Y is a Markov state vector determining the default-adjusted short-rate. Under this approach, no attempt is made to distinguish between

---

[5] There are many alternative variable-rate payments streams which, in place of LIBOR floating rate payments, would justify equation (4). It is the fact that the LIBOR floating rate payments are made in arrears that forces us to make equation (3) as an assumption. Specifically, if equations (1), (2), and (4) are to hold at all maturities, and the floating rate payment at time t is known at time $t - 0.5$, then one can show by induction, using the law of iterated expectations, that the floating rate payment at time $t + 0.5$ must be $(B_t^{0.5})^{-1} - 1$.

the contributions of the riskless rate r and the premium $z_t \equiv -l_t + h_t\lambda_t$. The second approach parameterizes both r and the mean loss rate z.[6]

A. Formulations of the Adjusted Short-rate R

Consider again a generic, defaultable, contingent claim with a promised payoff of X at maturity date T. Suppose that there is a Markov (under the equivalent martingale measure Q) state-variable process Y, such that the promised contingent claim is of the form $X = g(Y_T)$, for some function g, and the default and liquidity adjusted short-rate process R is of the form $\rho(Y_t)$, for some function $\rho(\cdot)$. Then equation (7) implies that the claim to payment of $g(Y_T)$ at time T has a price at time t, assuming that the claim has not defaulted by time t, of

$$V(Y_t, t) = E_Q\left[\exp\left(\int_t^T -\rho(Y_s)\, ds\right)g(Y_T)\,\bigg|\, Y_t\right]. \tag{10}$$

We emphasize that the riskless short-rate r_t and mean loss rate $z_t = h_t\lambda_t - l_t$ do not enter directly into this pricing model (10), but rather enter implicitly through the default-adjusted short-rate $R_t = \rho(Y_t)$. For a given defaultable bond market, parameterizing R directly, as opposed to separate parameterization of r and z, provides less information concerning the mean-loss-rate process z. On the other hand, this formulation permits empirical characterizations of the adjusted short-rate process R without a need to commit to a formulation of the credit spread. One reason that this may be attractive is that there may be nondefault factors that determine spreads between bonds of various classes. Many of these nondefault factors can be accommodated in our model by appropriate reinterpretation of the adjusted short-rate process. This robustness is an attractive feature of this econometric modeling strategy when one is most concerned with characterizing the distribution of R, say for the purpose of computing a zero-coupon yield curve implied by a defaultable yield curve. On the other hand, the robustness highlights the fact that nothing can be learned directly about the default processes h and λ from this approach.

Pursuing this approach further, suppose $Y_t = (Y_{1t}, \ldots, Y_{nt})'$, for some n, solves a stochastic differential equation of the form

$$dY_t = \mu(Y_t)dt + \sigma(Y_t)dB_t, \tag{11}$$

where B is a standard Brownian motion in $\mathbb{R}^n$ under Q, and where μ and σ are well behaved functions on $\mathbb{R}^n$ into $\mathbb{R}^n$ and $\mathbb{R}^{n \times n}$, respectively. Then we know from the "Feynman–Kac formula" that, under technical conditions (Friedman

---

[6] Even if liquidity effects are absent, it is in general not possible to separately identify the components h_t and λ_t of z_t using swap data (or corporate bond data) alone. See Duffie and Singleton (1996) for a more in-depth discussion of this point and a demonstration that the hazard and recovery rates can be identified from price information on credit derivatives.

(1975), Krylov (1980)), equation (11) implies that V solves the backward Kolmogorov partial differential equation

$$\mathcal{D}^{\mu,\sigma}V(y, t) + \rho(y)V(y, t) = 0, \quad (y, t) \in \mathbb{R}^n \times [0, T], \quad (12)$$

with the boundary condition

$$V(y, T) = g(y), \quad y \in \mathbb{R}^n, \quad (13)$$

where

$$\mathcal{D}^{\mu,\sigma}V(y, t) = V_t(y, t) + V_y(y, t)\mu(y) \\ + (1/2) \text{trace}[V_{yy}(y, t)\sigma(y, t)\sigma(y, t)']. \quad (14)$$

The identification problem for this modeling strategy is identical to that of standard term-structure models for default-free yield curves. All of the models for the short-rate process r that have been successfully studied for default-free term structures are also identified, at least in principle, in the case of defaultable bonds, simply by replacing r with R. For instance, consider the special case $Y_t = (Y_{1t}, \ldots, Y_{nt})'$, where $Y_1, Y_2, \ldots, Y_n$ are independent (under risk-neutral probabilities) square-root processes. That is, equation (11) applies with $\mu_i(y) = \kappa_i(\theta_i - y_i)$, $\sigma_{ii}(y) = \bar{\sigma}_i\sqrt{y_i}$, for positive constants κ_i, θ_i, and $\bar{\sigma}_i$, and $\sigma_{ij}(y) = 0$, for $i \neq j$. By taking ρ to be affine, one can then apply the CIR solution to equations (13) and (14) with $g(y) \equiv 1$, allowing defaultable zero-coupon bond prices to be computed in closed form. This is the approach taken in Section III.

The same valuation model (12)–(13) applies, under mild technical conditions given in Duffie and Singleton (1996), when the underlying state-variable process Y is a jump-diffusion, or more general continuous-time Markov process. One merely replaces $\mathcal{D}^{\mu,\sigma}$ with the infinitesimal generator of Y. Allowing for "jumps" in credit quality may be useful if one has in mind the potential for unusually large credit-quality events.

B. Modeling the Mean Loss Rate Process

A second modeling strategy is to parameterize the behavior of the joint process $(r, z)'$ for the default-free short-rate and the risk-neutral mean-loss-rate processes, respectively.[7] Perhaps the simplest example of this strategy would be a case in which one studies the joint distribution of the returns on a defaultable bond and the associated reference Treasury bond used in pricing the defaultable bond. To be concrete, suppose that $w = (r, z)'$ follows an "affine" diffusion

$$dw_t = (\alpha_0 + \alpha_1 w_t)dt + \beta(w_t)dB_t, \quad (15)$$

---

[7] In this section we abstract from the nondefault factors discussed in Section V. If one has observable proxies for liquidity, for example, then the following discussion is extendable to accommodate a liquidity premium in R.

where

- B is a standard Brownian motion in $\mathbb{R}^2$ under an equivalent martingale measure Q;
- $\alpha_0 \in \mathbb{R}^2$; α_1 is a 2×2 matrix;
- For each i and j, there is a fixed constant γ_{0ij} and fixed vector γ_{1ij} in $\mathbb{R}^2$ such that $[\beta(w)\beta(w)']_{ij} = \gamma_{0ij} + \gamma_{1ij} \cdot w$; and,
- α_0, α_1, and β satisfy joint restrictions[8] for the existence of solutions to equation (15).

Let G_t^n and C_t^n denote the prices of default-free and defaultable zero-coupon bonds of maturity n, respectively. The identification of z as the instantaneous default premium comes from the simultaneous estimation of the implied pricing equations for G_t^n and C_t^n, and the imposition of identifying restrictions on α_1 and $\beta(\cdot)$. The latter are necessary because, as shown in Dai and Singleton (1996), there will in general be a family of observationally equivalent affine models in the absence of normalizations in α_1 and β.

As an illustrative example of how information about z can be inferred from data on G_t^n and C_t^n, consider the special case (15) in which

$$dr_t = (\alpha_{0r} + \alpha_{1r}r_t)dt + \sqrt{\gamma_{0rr} + \gamma_{1rr}r_t}\, dB_t^{(1)} \qquad (16)$$

$$dz_t = (\alpha_{0z} + \alpha_{1z}z_t)dt + \sigma_{21}\sqrt{\gamma_{0rr} + \gamma_{1rr}r_t}\, dB_t^{(1)} + \sqrt{\nu'_{1zz}w_t}\, dB_t^{(2)}. \qquad (17)$$

The default-free term structure is driven by CIR-style factor r in equation (16). Thus, the parameters α_{0r}, α_{1r}, γ_{0rr}, and γ_{1rr} are identifiable from information on one or more government bond prices G_t^n. The parameters of r and the default process z are therefore identifiable from information on default-free and defaultable bond prices (see Dai and Singleton (1996) and Duffie and Singleton (1996)). Note that the Brownian motions associated with r_t and z_t may be correlated, though σ_{21} and ν_{1zz} must satisfy certain existence conditions.

Two special cases of equation (16) and (17) are: (i) $R_t = Br_t + \varepsilon_t$, with ε_t and r_t being independent diffusions, and (ii) either h_t or λ_t is specified a priori at a fixed value with the other assumed to follow a given affine diffusion. In the first case, if the issuer is exposed to interest rate risk, then this could be captured by the assumption that $B > 1$. In the second case, the strategy of modeling h and λ simplifies to this case of modeling z. One example of the latter simplification is the model of Nielsen and Ronn (1995), who assume that λ_t is fixed at either 0.0 or 0.5, the drift of z is independent of r, and the diffusion coefficient for z is constant over time. The affine example (17) illustrates how these special cases can easily be extended to allow the mean-loss-rate process z to depend on the riskless interest rate. This seems like a potentially important extension in light of the documented cyclicality in credit spreads.

---

[8] These restrictions are basically that the state vector does not enter a region where the square root of a negative number would have to be taken, as for example with a negative drift at zero in the Cox–Ingersoll–Ross (1985) model. See Duffie and Kan (1996) for details.

More generally, credit spreads are related to business-cycle variables, as shown by Bernanke (1990), Friedman and Kuttner (1993), Jaffee (1975), and Stock and Watson (1989). Among the more prominent variables are general stock market returns, measures of output growth, panel data on consumer sentiment, and levels of capital investment. There may be value in pursuing empirical models in which the risk-neutral expected loss rate process z is linked with such macro-economic variables. The state vector in equation (15) could be expanded to introduce macro information, although the observation frequency of many macro series might limit the applications of such models.

These examples lead to closed or nearly closed-form expressions for the defaultable zero-coupon bond prices in terms of r_t and z_t. However, there is no need to restrict attention to analytic solutions for zero-coupon prices. As long as it is computationally feasible to compute the zero-coupon bond prices numerically at the same time that the objective function defining the estimator of the unknown parameters governing the process (r, z) is being optimized, our preceding comments continue to apply. Of course, the correlation structure of r_t and z_t must be such that the parameters are identified.

III. An Econometric Model of Swap Yields

In this section we implement the first estimation strategy under the assumption that the default and liquidity adjusted short-rate process R is a linear combination of independent square-root diffusion models. To be concrete, let Y_t be a Markov state vector that determines the current risk-adjusted short-rate R. Let $\mathcal{B}(Y_t, \beta_o)$ denote a vector of M_1 prices of defaultable zero-coupon bonds and $\mathcal{C}(Y_t, \beta_o)$ denote a vector of M_2 yields on newly issued swaps ($M = M_1 + M_2$) implied by the term-structure model. The parameter vector governing the probability model of the state process Y is β_o. Also, let ε_t denote an M-vector of measurement errors contaminating the observed counterparts of the zero prices and yields, $\mathbf{B}_t$ and $\mathbf{C}_t$, that are independent of the state vector Y. Then the econometric model takes the form

$$\begin{pmatrix} \mathbf{B}_t \\ \mathbf{C}_t \end{pmatrix} = \begin{pmatrix} \mathcal{B}(Y_t, \beta_o) \\ \mathcal{C}(Y_t, \beta_o) \end{pmatrix} + \varepsilon_t(\beta_o), \qquad (18)$$

where, for ease of notation, we combine the parameter vectors governing the probability laws of Y and ε into β_o.

The special case of equation (18) that we will study is the multifactor square root model, with

$$dY_t^i = \kappa_i(\theta_i - Y_t^i)dt + \sigma_i \sqrt{Y_t^i}\, dW_t^i, \qquad i = 1, 2, \qquad (19)$$

where (W^1, W^2) is a standard Brownian motion in $\mathcal{R}^2$. The positive scalars κ_i, θ_i, and σ_i have interpretations in terms of mean-reversion, steady-state mean, and volatility, respectively, that have been developed by Feller (1951) and, in the context of term-structure models, by CIR. In our empirical analysis we take

A Model of Term Structure of Interest Rate Swap Yields

$R_t \equiv Y_t^1 + Y_t^2$, in which case equation (2) and the independence of $\{Y^1\}$ and $\{Y^2\}$ under an equivalent martingale measure Q leave us with

$$B_t^\tau = \prod_{i=1}^{2} p_i(Y_t^i, \tau), \qquad (20)$$

where $p_i(y, \tau) = a_i(\tau) e^{-b_i(\tau) y}$, for

$$a_i(\tau) = \left[\frac{2\gamma_i \exp[(\gamma_i + \kappa_i + \lambda_i)\tau/2]}{(\gamma_i + \kappa_i + \lambda_i)[\exp(\gamma_i \tau) - 1] + 2\gamma_i} \right]^{\alpha(i)}, \qquad (21)$$

and

$$b_i(\tau) = \frac{2[\exp(\gamma_i \tau) - 1]}{(\gamma_i + \kappa_i + \lambda_i)[\exp(\gamma_i \tau) - 1] + 2\gamma_i}. \qquad (22)$$

$\alpha(i) = 2\kappa_i \theta_i / \sigma_i^2$, $\gamma_i = ((\kappa_i + \lambda_i)^2 + 2\sigma_i^2)^{1/2}$, and λ_i denotes a risk-premium coefficient explained by CIR. That is, each p_i is the form of a zero-coupon bond price in a univariate square-root diffusion model, and the price of a defaultable zero-coupon bond in our setting is the product of the univariate bond price formulas.

The state vector Y is unobservable. Therefore we proceed, as do Chen and Scott (1993), under the assumption that two elements of the yield vector $\{\mathcal{B}, \mathcal{C}\}$ (corresponding to the number of state variables) are measured without error. The pricing model can then be inverted to express the state variables as functions of these swap and LIBOR yields. Alternatively, we could allow all of the swap and LIBOR yields to be measured with error and then estimate the model using the simulated method of moments (Duffie and Singleton (1993)). While this strategy would permit us to obtain consistent estimates of the parameters, the state variables would be unknown. In principle, the latent state variables could be estimated using filtering methods. However, the approach taken here has several potential advantages for pricing, including having direct observations of the state variables and forcing the model to fit a subset of the swap yields exactly. These are important considerations for valuing derivative claims based on swap or LIBOR yields. The model we examine assumes that the two- and ten-year swap rates satisfy

$$c_t^\tau = \frac{1 - B_t^\tau}{\sum_{j=1}^{2\tau} B_t^{0.5j}}, \qquad (23)$$

for $\tau = 2, 10$, with no measurement errors.[9] Since equation (23) is a nonlinear function of the state variables, inferring (Y_t^1, Y_t^2) from (c_t^2, c_t^{10}) must be done numerically for each observation.

---

[9] We also examine a model in which the price of a six-month LIBOR instrument satisfies

$$-\ln B_t^{0.5} = \ln a_1(0.5) - b_1(0.5) Y_t^1 + \ln a_2(0.5) - b_2(0.5) Y_t^2,$$

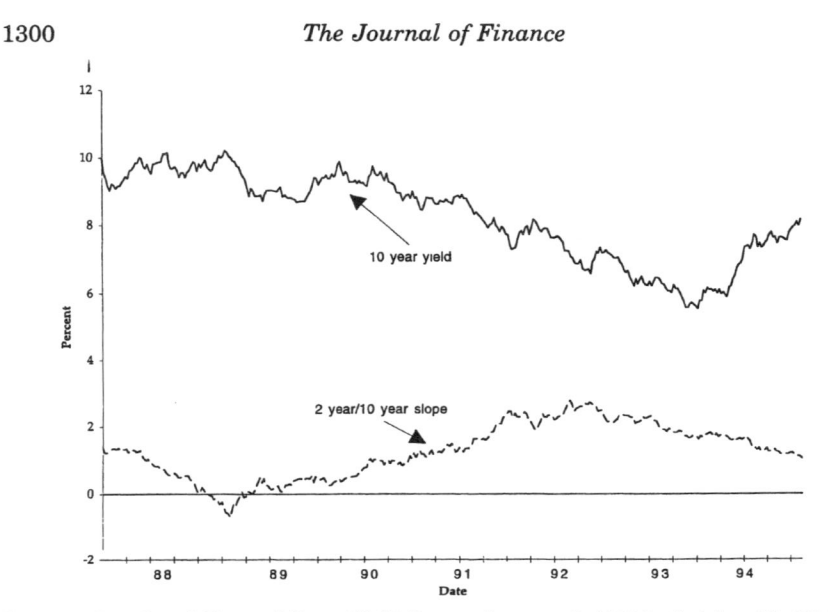

Figure 1. **Level and Slope of Swap Yield Curve, January 8, 1988 to October 28, 1994.**

We make one additional modification to the model. The square-root parameterization precludes negative state variables. As can be seen from Figure 1, the swap data exhibit substantial changes in the level of rates (as measured by the 10-year yield) and in the magnitude and sign of the slope of the swap curve (as measured by the 2-to-10-year yield spread). We are unable to find admissible parameters for equations (19) and (20) for which the implied state variables Y^1 and Y^2 are positive for the entire sample period. Therefore, we modify the discount process to be

$$R_t = Y_t^1 + Y_t^2 - \bar{y}, \tag{24}$$

where $\bar{y}$ is a positive constant.[10] With this modification, the zero-coupon bond price defined by equation (20) is replaced with

$$B_t^\tau = \prod_{i=1}^{2} p_i(Y_t^i, \tau) e^{\bar{y}\tau}. \tag{25}$$

and in which equation (23) holds for $\tau = 10$. We will comment briefly in the next section on the comparative fit of this alternative model.

[10] While the inclusion of $\bar{y}$ solves the problem of negative state variables, the adjusted discount rate R may be negative. In this respect, the modified model differs from the original square-root diffusion specification. Pearson and Sun (1994) study the same specification of a two-factor model of the riskless rate r underlying U.S. Treasury yields (their "extended model").

A Model of Term Structure of Interest Rate Swap Yields

The conditional densities of the state variables $\{Y^1, Y^2\}$ are well known to be non-central chi-square (e.g., CIR). Through a change of variables, these conditional densities can be rewritten as functions of observed swap yields multiplied by the Jacobian of the transformations (23). The Jacobian is nonlinear and time dependent.

Measurement errors are included in virtually all econometric models of the term structure because the state vector Y usually has low dimension (say, three or less) relative to the number of yields M.[11] Thus, without additional sources of uncertainty, the models would imply deterministic relations among prices or yields that are clearly violated in the data. In constructing the likelihood function, we assume that the (nonzero) measurement errors $\{\varepsilon_t\}$ for the swap yields follow univariate AR(1) processes, with innovations that are normally distributed and that may have nonzero correlation. The log-likelihood function of the swap yield data is the sum of the log-density of the noncentral chi squares of the square-root processes, adjusted for $\bar{y}$, and the log-density of the multivariate normal associated with the measurement errors.[12] We estimate the swap-pricing model using weekly data from January 4, 1988 through October 28, 1994. The swap yields are constructed as follows. Weekly data on constant-maturity U.S. Treasury bond yields are constructed by concatenating yields on the current, on-the-run Treasury bonds for maturities two, three, five, seven, and ten years. The average of the quoted bid/ask swap spreads are then added to these Treasury series to obtain the swap yield data.[13] The short-term, six-month rate is taken to be the dollar LIBOR.

Initially, several one-factor models are fitted with models indexed by the point on the swap curve used to extract the single state variable Y^1. In all of these models, the yield curves evaluated at the ML estimates tended to be too flat on average compared to the actual yield curves. Additionally, the one-factor models are unable to fit simultaneously the volatilities of changes in yields at the long and short ends of the swap curve. Finally, the deviations between actual and model-implied slopes of the swap curve exceeds 50 basis points for several extended periods. These poor results motivate our focus on a two-factor model. The state variables for the two-factor model are extracted from the two- and ten-year swap rates. The rates assumed to be measured with error are those on the three-, five-, and seven-year swaps. LIBOR is excluded from the econometric model, but is used subsequently in assessing the fit of the

---

[11] Pearson and Sun (1994) proceed instead by forming portfolios of bond yields and using as many portfolios as there are unobserved state variables. Thus, their model is presumed to fit the portfolio yields exactly. We have chosen to work directly with a cross-section of maturities of swap yields in order to assess the fit at specific points along the swap yield curve, and because a cross-section of swap yields embodies more information about the yield curve than portfolios do.

[12] See Chen and Scott (1993) and Pearson and Sun (1994) for further discussion of the likelihood function for square-root diffusions. Our likelihood function differs from that used by Chen and Scott because of the nonlinear Jacobian for swap yields. We are grateful to Qiang Dai for developing the approximation to a noncentral chi square distribution used in our numerical routines.

[13] The swap data are taken from the Telerate brokers screens and represent average bid and ask rates quoted by several large dealers.

Table I
Estimates of the 2-Factor Model Weekly Data, January 8, 1988 to October 28, 1994

The parameters κ_i, θ_i, and σ_i govern the diffusion for the ith state variable Y^i,

$$dY^i = \kappa_i(\theta_i - Y^i)dt + \sigma_i \sqrt{Y^i}\, dB^i,$$

and λ_i is the associated risk premium. The ρs are the autocorrelations of the measurement errors. Two sets of standard errors are given in parentheses below each estimate: the first is based on the outer product of the score and the second is based on the usual Hessian of the likelihood function.

Panel A: Parameters of Diffusion Y^1			
κ_1	θ_1	σ_1	λ_1
0.544	0.374	0.023	−0.036
(0.052)	(0.444)	(0.014)	(0.046)
(0.029)	(0.017)	(0.0005)	(0.012)

Panel B: Parameters of Diffusion Y^2			
κ_2	θ_2	σ_2	λ_2
0.003	0.258	0.019	−0.004
(0.016)	(0.874)	(0.014)	(0.012)
(0.0004)	(0.010)	(0.001)	(0.0006)

Panel C: Autocorrelations of Measurement Errors		
ρ_{3yr}	ρ_{5yr}	ρ_{7yr}
0.777	0.836	0.871
(0.025)	(0.017)	(0.013)
(0.007)	(0.006)	(0.005)

two-factor model.[14] Thus, the model has twelve parameters: four each for the two state variables (including the risk premia), the adjustment parameter $\bar{y}$, and the autocorrelations of the measurement errors for the three-, five-, and seven-year swap rates. The parameter estimates are displayed in Table I, with two sets of standard errors. The first is computed from the outer product of the score vector, and the second is based on the usual Hessian matrix of the log-likelihood function. As is commonly the case in small samples, the two estimates differ.

The estimate of the adjustment factor $\bar{y}$ is 0.58, with estimated standard errors of 0.58 and 0.003. Together with the estimates of the α_i, this implies an

---

[14] Initially, we keep LIBOR in the econometric model as a defaultable zero rate measured with error, but find that the model does not fit well relative to the version with LIBOR omitted altogether. Therefore, we report estimates for the version excluding LIBOR, and then examine the fit of the implied six-month swap rate to LIBOR. The previous draft of this article also presents estimates for a two-factor model in which the state variables are extracted from LIBOR and the ten-year swap rate. The fitting errors for the various swap rates are large and often over 60 basis points.

A Model of Term Structure of Interest Rate Swap Yields

estimated long-run mean $\bar{R} \equiv \theta_1 + \theta_2 - \bar{y}$ of R, of 5.2 percent.[15] The risk-premium coefficients λ_1 and λ_2 are both negative, which implies that the term premiums are, on average, positive as maturity increases. In the context of pricing defaultable bonds, excess returns over the instantaneous rate R may reflect term premiums in the underlying riskless term structure and/or an increasing term structure of average credit spreads. The term structure of credit spreads for the zero-coupon bond yields implied by our swap-pricing model are subsequently examined in more depth.

The estimates of the mean-reversion parameters κ_1 and $\kappa_1 + \lambda_1$ are much larger than the corresponding estimates for the second factor. Indeed, $\kappa_2 + \lambda_2$ is very close to zero, suggesting that there is at most weak mean reversion in the second state variable. To interpret these findings, it is informative to examine the relations between the state variables extracted from the swap yields ($\hat{Y}^1$ and $\hat{Y}^2$) and the swap yield curve.[16] The sample correlation of $\Delta \hat{Y}^1$ with changes in the slope of the swap curve (10 year–2 year) is approximately -0.99, whereas the correlation of $\Delta \hat{Y}^1$ with changes in the ten-year swap yield is -0.02. The correlations between $\Delta \hat{Y}^2$ and changes in the two-, five-, and ten-year swap yields are 0.60, 0.78, and 0.93, respectively. Thus, the first factor behaves like the negative of the slope of the swap curve, and the second factor behaves like the level of the ten-year swap yield.

Further insights come from examination of the coefficients of the zero-coupon bond yields, $-\ln B_t^\tau/\tau$ (see Fig. 2). From equation (20), these coefficients are $-\ln a_i(\tau)/\tau$ and $b_i(\tau)/\tau$ for the two state variables and $\bar{y} = 0.580$ for the scale factor in equation (25). Across the maturity τ, the coefficients $b_2(\tau)/\tau$ are nearly constant at unity and the coefficients $-\ln a_2(\tau)/\tau$ are essentially zero. It follows that the second factor represents a parallel shift in the entire zero-coupon yield curve induced by changes in Y_t^2. In light of the correlations of swap yields and state variables, this parallel-shift factor is well proxied empirically by the ten-year swap yield. In particular, the likelihood function is not maximized by selecting a short-term swap rate as the "level" risk factor. Figure 2 also shows that $b_1(\tau)/\tau$ declines with maturity τ such that positive shifts in Y_t^1 induce flattenings in the slope of the swap curve. The findings that ΔY_t^1 and slope of the swap curve were nearly perfectly negatively correlated, and that $b_1(\tau)/\tau$ is large for small values of τ, suggest that slope changes were associated with greater variation in short-maturity yields during the sample period. The rapid decline in $b_1(\tau)/\tau$ as τ increases is induced in the square-root model largely by a fast rate of mean reversion of the first factor (large κ_1).

The terms $-\ln a_1(\tau)/\tau$ and $-\ln a_2(\tau)/\tau$ contribute to an inherent upward slope in the zero-coupon yield curve, as the former is upward sloping, while the latter

---

[15] We are grateful to Mark Fisher for pointing out that the long-run mean of the R was implausible in an earlier version of this article. Upon exploring the likelihood frontier further, we found a higher value of the likelihood function with the parameter estimates reported in Table 1 and, in particular, a plausible value for $\bar{R}$. The time-series properties of the fitted swap rates are unchanged compared to the previous results.

[16] More precisely, time series on the two state variables are computed by inverting the pricing model evaluated at the ML estimates for each date of the sample.

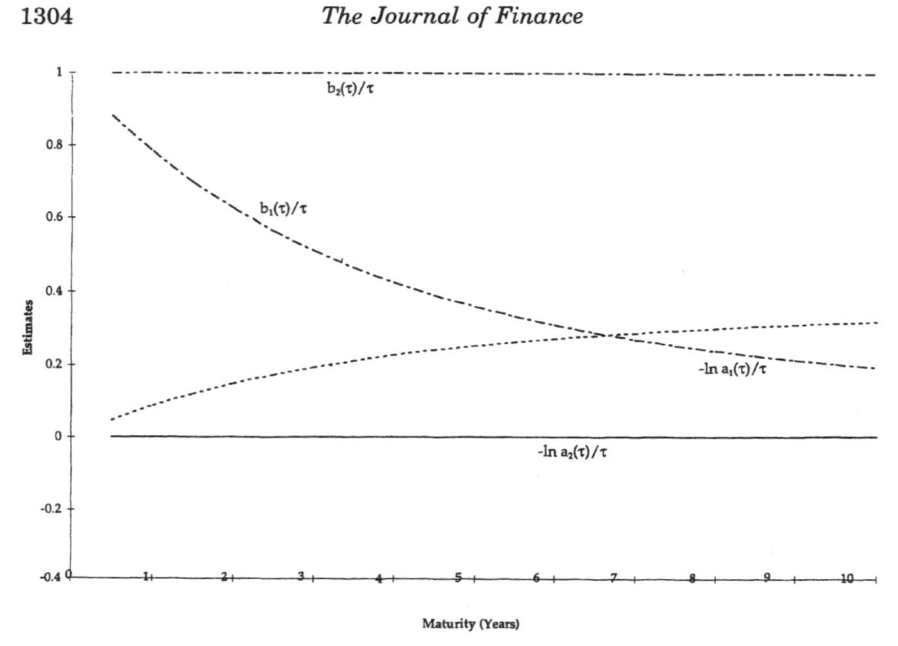

Figure 2. Coefficients of Zero Coupon Bond Yields. The price of a zero-coupon bond with maturity τ is

$$B_t^\tau = a_1(\tau)a_2(\tau)\exp\{b_1(\tau)Y^1(t) + b_2(\tau)Y^2(t)\},$$

where $Y^1(t)$ and $Y^2(t)$ are the two state variables. Therefore, the yield on a τ-year zero is

$$-\ln B_t^\tau/\tau = -\ln a_1(\tau) - \ln a_2(\tau) + (b_1(\tau)/\tau)Y^1(t) + (b_2(\tau)/\tau)Y^2(t).$$

This figure displays these coefficients of zero yields as functions of maturity τ.

is essentially zero for all maturities. Thus, Y^2 represents nearly a pure parallel shift. On the other hand, at $(Y^1 = 0, Y^2 = 0)$, the yield curve is upward sloping due to the contribution of $-\ln a_1(\tau)/\tau$. It follows that the slope of the yield curve is matched by adjusting Y^1, and that matching an inverted swap curve may require a large value of Y^1 to offset the upward slope induced by $-\ln a_1(\tau)/\tau$. In such cases, the resultant large value of $Y_t^1 b_1(\tau)/\tau$ (the mean of $\hat{Y}_t^1$ is 0.37) would clearly overstate the level of zero-coupon yields, especially for short maturities. Thus, in order to simultaneously explain the temporal behavior of the slope of the swap curve and fit the average level of the curve, an adjustment is necessary. This explains the economically significant value of the constant adjustment of $\bar{y}$.

Descriptive statistics of the swap yields implied by the models are presented in Table II. Panel A displays the sample means of the levels and first differences of the historical swap yields ("Sample"), as well as the swap yields implied by the two-factor model. The model matches the average three-, five-,

Table II
Descriptive Statistics from the 2-Factor Model

Panel A displays the sample means of the historical (Sample = c^τ) and model-implied (Model = $\hat{c}^\tau$) rates on the London Interbank Offering Rate (LIBOR) and n-year swap contracts for $n = 3, 5$, and 7. Results for the level of rates (c^τ) and changes in rates (Δc^τ) are presented. Panel B presents the analogous results for the sample standard deviations of swap yields. Panel C restricts attention to changes in swap rates (Δc^τ) and displays sample historical and model-implied skewness and kurtosis statistics. Panel D displays the standard deviations of the differences between the actual and model-implied swap rates ($c^\tau - \hat{c}^\tau$), and of ($\Delta c^\tau - \Delta \hat{c}^\tau$).

	$\tau =$	LIBOR	3 yr	5 yr	7 yr
		Panel A: Means of Swap Yields			
Sample	c^τ	6.303	7.308	7.756	8.028
	Δc^τ	−0.005	−0.004	−0.005	−0.005
Model	$\hat{c}^\tau$	6.131	7.309	7.754	8.018
	$\Delta \hat{c}^\tau$	−0.004	−0.004	−0.005	−0.005
		Panel B: Standard Deviations of Swap Yields			
Sample	c^τ	2.262	1.772	1.526	1.403
	Δc^τ	0.143	0.148	0.141	0.136
Model	$\hat{c}^\tau$	2.385	1.756	1.512	1.379
	$\Delta \hat{c}^\tau$	0.183	0.143	0.134	0.132
		Panel C: Skewness/Kurtosis for Changes in Swap Yields			
Sample	Skewness	−0.316	0.156	0.218	0.216
	Kurtosis	4.357	3.001	3.071	3.148
Model	Skewness	0.052	0.141	0.152	0.145
	Kurtosis	3.096	2.875	2.878	2.931
		Panel D: Standard Deviations of Fitting Errors (Basis Points)			
	$c^\tau - \hat{c}^\tau$	32.34	4.88	7.16	6.21
	$\Delta c^\tau - \Delta \hat{c}^\tau$	11.35	2.83	3.26	2.71

and seven-year swap yields, as well as the standard deviations of yield changes (Panel B), to the third decimal place. The skewness and kurtosis statistics of the changes in implied swap yields displayed in Panel C are slightly smaller than the corresponding statistics for their sample counterparts.

The two- and ten-year swap yields are fit exactly by construction. As such, perhaps the most challenging rate to fit is the five-year rate, because changes in the curvature of the swap curve may result in movements of the five-year swap rate that are independent of the longer and shorter ends of the swap curve. Figure 3 shows that the model fits the five-year swap yield to within 20 basis points (bp) over the seven-year sample period. The mean error is only 0.3 basis points. Moreover, the sample standard deviation of the measurement error for the level of the five-year swap rate is 7.2 bp (see Table II, Panel D), which was approximately the bid/ask spread during this sample period.

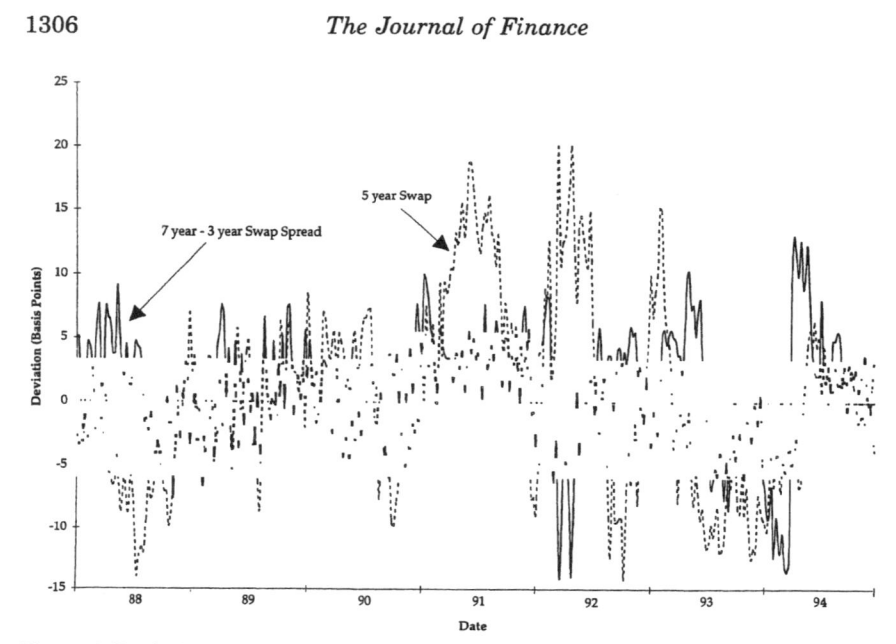

Figure 3. Deviation from Fit of Swap Yields. The dashed line is the deviation, in basis points, between the actual and fitted five-year swap rates. The solid line is the corresponding deviation for the seven-year minus three-year swap yield spread.

Figure 3 shows that there is also a close correspondence between movements in the slope of the swap yield curve (7 year–3 year) and the implied slope from the model. The maximal deviation is less than 16 basis points over the sample period, with an average fitting error of 1.1 basis points. The most challenging period to fit seems to have been the trough in swap rates during late 1993 and early 1994.

As additional evidence on goodness-of-fit, we report in Table III, Panel A, the results from regressing the changes in the actual swap yields against their fitted counterparts. If the two-factor model describes the data well, then we would expect that the estimated intercept ($\hat{\alpha}$) and slope ($\hat{\beta}$) to satisfy $\hat{\beta} \cong 1$ and $\hat{\alpha} \cong 0$, and that the R^2s would be close to one. Regressions are run with differenced yields, because of the high degree of persistence in levels and our desire to evaluate the model's explanatory power for changes. Only the results for LIBOR differ markedly from those predicted by the model. Indeed, it is striking that the model explains approximately 95 percent of the variation of the *changes* in individual swap rates over the seven year sample period. Table II expresses this finding in terms of basis points: The standard deviations of the fitting errors for the swap yields are between 4.48 bp and 7.16 bp.

Table III
Correlations of Actual and Fitted Yields

Panel A displays the results from regressing the changes in actual swap rates (Δc_t^τ) on the changes in fitted rates from the model ($\Delta \hat{c}_t^\tau$). Standard errors of the estimated intercept ($\hat{\alpha}$) and slope ($\hat{\beta}$) are given in parentheses. R^2 is the coefficient of determination and s.e.e. is the standard error of the residual. Results are reported for the London Interbank Offering Rate (LIBOR) rate and three year, five year, and seven year swap rates. Panel B displays the corresponding results for a model fit with Treasury bond yields instead of swap rates (i.e., c_t^τ is the yield on a τ-year Treasury bond).

$$\Delta c_t^\tau = \alpha + \beta \Delta \hat{c}_t^\tau + \varepsilon_t$$

Maturity	$\hat{\alpha}$	$\hat{\beta}$	R^2	s.e.e.
Panel A: Actual Against Fitted Swap Yield Regressions				
LIBOR	−0.0024	0.615	0.62	0.089
	(0.005)	(0.026)		
3 Year	0.0002	1.014	0.96	0.028
	(0.002)	(0.010)		
5 Year	0.0002	1.018	0.95	0.033
	(0.002)	(0.013)		
7 Year	−0.0000	1.014	0.96	0.027
	(0.001)	(0.011)		
Panel B: Actual Against Fitted U.S. Treasury Yield Regressions				
3 Year	0.0045	1.039	0.95	0.032
	(0.001)	(0.013)		
5 Year	0.0006	1.050	0.97	0.137
	(0.001)	(0.012)		
7 Year	0.0001	1.035	0.97	0.023
	(0.001)	(0.012)		

For comparison, we reestimate our two-factor square-root model using constant-maturity U.S. Treasury yields for the same sample period.[17] As with swaps, the state variables are extracted from the two- and ten-year maturity instruments. The results from regressing actual on fitted changes in Treasury yields are displayed in Table III, Panel B. The Treasury model also fits reasonably well during this sample period, although the $\hat{\beta}$ coefficients are significantly different from one at conventional significance levels.[18] Also, the standard error of the estimate (s.e.e.) for the five-year rate is larger for the Treasury model than for the swap model.

Although LIBOR is not used in estimation, the six-month rate implied by the two-factor model can be computed at the ML estimates and compared to LIBOR. The results in Table II show that the fitted LIBOR is about 18 basis

---

[17] The constant maturity Treasury data are not exactly what one would want in order to fit a model for the U.S. Treasury curve, since the maturities of, say, the ten-year note change slightly and the on-the-run issue is not always at par between auctions. However, for our purposes of providing a benchmark for assessing fit, the data seem adequate.

[18] Standard errors are computed using Hansen's (1982) correction for serial correlation and heteroskedasticity with five noncontemporaneous terms in the asymptotic covariance matrix.

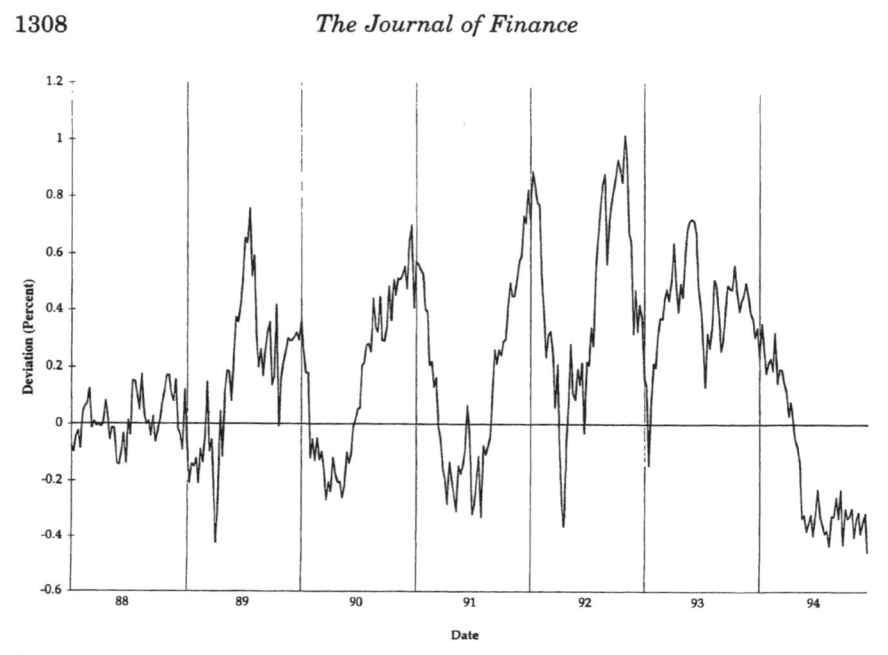

Figure 4. Deviation from Fit to LIBOR. This figure displays the deviation in percent between the actual and fitted six-month LIBOR rates.

points too small on average and is more volatile than the actual LIBOR. Moreover, the standard deviation of the fitting error is over 32 bp. The sample skewness of changes in LIBOR is negative and its kurtosis is larger than three, whereas the sample estimates of skewness for the changes in swap rates are positive and the sample estimates of their kurtoses are close to three. Not surprisingly, the model does not explain the negative skewness and excess kurtosis of LIBOR. Together, these findings suggest that the six-month LIBOR series has distinctive characteristics that are not shared by the longer end of the swap yield curve. One interpretation of these results is that LIBOR loans are of a somewhat lower average perceived quality than multi-year swaps of the same credit rating. This is consistent with the evidence presented in Sun, Sunderson, and Wang (1993). Alternatively, it may be that additional non-credit factors are needed to describe simultaneously the distributions of the long and short ends of the swap curve.

The poor fit for LIBOR is displayed in Figure 4. The deviations between actual and implied LIBOR fluctuate substantially and often exceed 50 basis points in absolute value. Moreover, there is evidently a seasonal pattern to the spreads between actual and fitted LIBOR. In every year, there are large changes in the deviations around the calendar year end. (The vertical lines in Figure 4 occur on the last Friday of each calendar year that markets were open.) For end-of-year 1989, 1990, and 1991, the actual-fitted LIBOR differences fell substantially. On the other hand, for end-of-year 1992 and 1993, there is evidence of an increase, although in the latter case the increase was quickly followed by a steady decline in the actual-fitted LIBOR deviations. These calendar effects may well be related to balance-sheet adjustments by financial institutions concerned about capital requirements or the risk profile of their securities positions at year end. Whatever the source, these results suggest that using a short-term rate as the state variable describing the entire yield curve may lead to misleading conclusions about the shapes of the distributions of long-term swap yields.

How can the poor fit for LIBOR and relatively good fit for the swap rates be reconciled with the assumption of homogeneous LIBOR-swap market credit quality? One reconciliation comes from noting that equations (4) and (18) can be used to fit the model-based discount factors B_t^τ implicit in the swap market without reference to the LIBOR market. If, as seems common in practice, swap traders also extract estimates $\hat{B}_t^\tau$ for pricing swaps from swap rates (and Eurodollar futures) without reference to LIBOR markets, then our econometric analysis provides a model-based construction of the traders' implied discount factors, $\hat{B}_t^\tau$, and pricing rules for the longer end of the swap curve. At the short end of the swap curve, additional factors are evidently necessary for our model to generate consistent pricing of Eurodollar futures and LIBOR contracts, that is, to reproduce the discount factor at the short end of the swap zero-coupon curve.

Regressing the deviation between actual and fitted LIBOR (DEVLIB) on the first and second lagged values of itself gives[19]

$$\text{DEVLIB}_t = \underset{(0.006)}{0.002} + \underset{(0.05)}{0.787}\ \text{DEVLIB}_{t-1} + \underset{(0.05)}{0.173}\ \text{DEVLIB}_{t-2},$$

with an R^2 of 0.90 and with a standard error of the residual of 11.5 basis points. The finding that a low-order autoregression explains most of the within-sample variation in the misfitting of LIBOR suggests that the two-factor model could be modified to include a third factor to accommodate the dynamics of the entire swap-LIBOR yield curve. Evidence consistent with this conjecture is presented in Dai and Singleton (1996).

IV. Analysis of Implied Swap Zero-coupon Yields

We turn next to an examination of the properties of the zero-coupon bond yields implied by our swap-pricing model. Specifically, we compute the yields $(-\log B_t^\tau/\tau) \times 100$ (i.e., continuously compounded yields) with B_t^τ given by equation (20), evaluated at the ML estimates. Our focus is on the spreads of these zero-coupon bond yields to their counterparts in the U.S. Treasury market. In light of the small residuals from fitting swap yields, the defaultable zero-coupon yields implied by our model should serve as reasonable proxies for

---

[19] While supported by a different model and different data, our characterization is largely consistent in this regard with independent work by Brown and Schaefer (1993), who study U.S. Treasury yields.

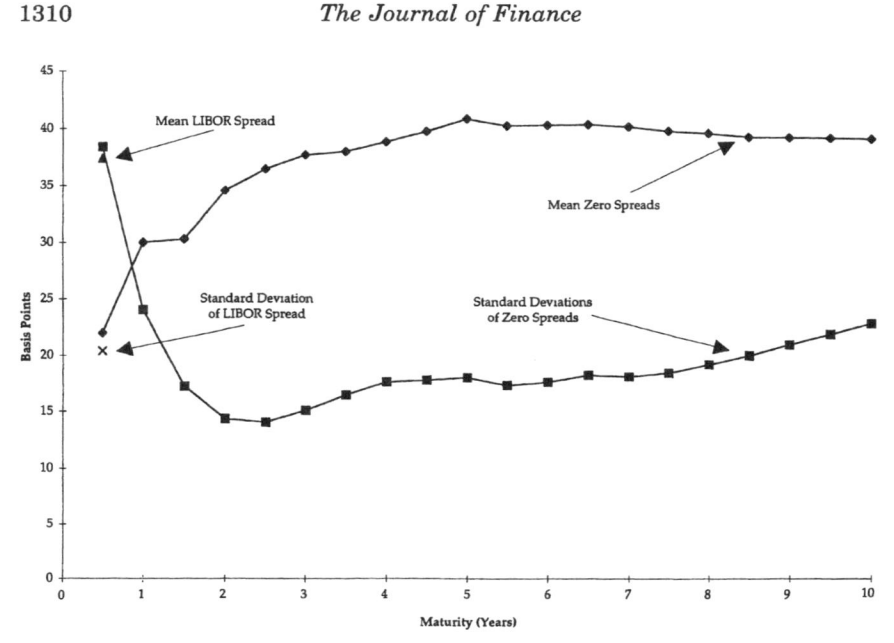

Figure 5. Means and Standard Deviations of Swap Zero Spreads, Sample Period: January 15, 1988 to October 2, 1994. The squares and diamonds are the standard deviations and means, respectively, of the spreads between zero rates implied by the swap and Treasury yield curves. The mean and standard deviation of the actual LIBOR spread to the six-month Treasury bill rate are displayed as a triangle and ×, respectively.

the defaultable zero yields implicit in the swap curve. For the U.S. Treasury zero-coupon (continuously compounded) yields, we use the discount function implied by a statistical spline model.[20] Whereas the swap curve is sparsely reported and a model-based interpolation scheme seems desirable, the substantial data on coupon U.S. Treasury yields and strip prices are used to compute Treasury zero rates.

Figure 5 displays the term structure of sample means and standard deviations of spreads between the swap and U.S. Treasury zero-coupon bond yields. The term structure of zero-coupon default spreads is, on average, upward sloping during this sample period from about 20 basis points for the six-month spread, up to about 40 basis points for the five-year spread. Beyond five years to maturity, the spreads are nearly constant at about 38 to 40 basis points. The large difference between the means of the fitted and actual spreads of LIBOR to the six-month Treasury rate is consistent with the presence of money-market effects noted earlier that are not captured by the two-factor swap model.

---

[20] The U.S. Treasury zero-coupon yields, provided to us by Goldman Sachs, are computed from the coupon yields and strip prices.

Spread volatilities tend to be increasing beyond two years to maturity, although volatilities are roughly flat between four and seven years. (The volatilities of the zero-coupon bond yields decline with maturity for both the U.S. Treasury and swap markets.) As maturity declines below two years, the spread volatilities increase. The actual six-month spread volatility in the money market is much lower than the implied spread volatility from the swap model.

To assess the relative importance of the liquidity (l_t) and credit ($\lambda_t h_t$) components of R_t in determining swap zero spreads, vector autoregressions (VAR) are estimated for zero spreads, and for proxies for credit and liquidity. For square-root diffusions, the zero yields are linear in the state variables Y^1 and Y^2. If the U.S. Treasury curve can similarly be described by a discount function that is the sum of square-root diffusions (Table III suggests that this is approximately so during our sample), then these VARs can be interpreted as linear projections of differences in linear combinations of the state variables driving the risk-adjusted and Treasury interest rate processes onto the variables in the VAR.[21] The variables included in the VARs are the six-month Treasury zero yield (TB6), the spread between the generic three-month repo rate for the ten-year Treasury note and the repo rate of the current on-the-run Treasury note (REPOSP), the spread between rates on BAA- and AAA-rated commercial paper (CPS), and the spread between the ten-year zero rates implied by the swap and Treasury markets (ZEROSP10).[22] All VARs are fit with eight lags.[23]

The inclusion of TB6 is intended to capture the effect of the level of riskless interest rates on spreads for defaultable zero-coupon bonds. CPS captures default risk. Liquidity differences may also contribute to a spread between BAA- and AAA-rated corporate paper, but we expect these differences to be small compared to the effects of the different credit rating.

REPOSP captures the specialness of the Treasury coupon note with the same maturity as the zero being studied. Of course, there is not a simple linear mapping between the specialness of a coupon bond and the spread in the underlying zero markets. However, the specialness in the coupon market is the only set of repo data available, and it should shed some light on the importance of repo effects on swap spreads. An increase in REPOSP (the repo rate on the

---

[21] Although, under the assumptions of our model, the innovations in the VARs are not normal, the model does imply linear expectations conditional on past Ys. Therefore, we expect that the VAR analysis will provide suggestive descriptive evidence about the contributions of liquidity and credit factors to swap spreads.

[22] We also fit VARs including the spread between Moody's corporate bond yield indices for BAA- and AAA-rated credits as an additional conditioning variable to proxy for the long-term corporate spreads. However, this spread is not a significant explanatory variable for zero swap spreads, and the proportion of variance of ZEROSP10 explained by the Moody's series is essentially zero.

[23] A Bayesian prior on the lag distribution, which has the own first lag of each variable equal to unity and all of the other coefficients equal to zero, is also imposed. Specifically, in the notation of the manual of the RATS statistical computer package, we use a symmetric prior with TIGHT = 0.15, and 0.5 as the weight on the other variables in the equation. The results are not substantially different with and without the prior.

Table IV
F-Tests of Exclusion Restrictions in VARs

This table displays the values of the *F*-statistics for testing the hypothesis that all of the coefficients on lagged values of the variable indicated under "Variable Excluded" are zero in the equation with the dependent variable given in the left-hand column. Marginal significance levels of the statistics are given in parentheses. Two VARs are estimated: VAR4 includes the 6-month Treasury bill rate (TB6), the spread between generic and on-the-run repo rates for ten-year Treasury bonds (REPOSP), the spread between BAA- and AAA-rated commercial paper (CPS), and the spread between the ten-year zero rates implied by the swap and Treasury markets (ZEROSP10); VAR3 includes TB6, REPOSP, and ZEROSP. The * indicates exclusion of CPS from the VAR3 model.

	Variable Excluded							
	TB6		REPOSP		CPS		ZEROSP10	
Dependent Variable	VAR4	VAR3	VAR4	VAR3	VAR4	VAR3	VAR4	VAR3
TB6	476.5	580.53	2.26	1.54	3.71	*	1.06	.573
	(0.000)	(0.000)	(0.024)	(0.143)	(0.000)		(0.395)	(0.800)
REPOSP	0.54	0.53	72.22	70.37	0.97	*	1.18	1.08
	(0.827)	(0.830)	(0.000)	(0.000)	(0.459)		(0.352)	(0.374)
CPS	0.80	*	1.62	*	79.51	*	0.65	*
	(0.602)		(0.121)		(0.000)		(0.738)	
ZEROSP10	1.42	1.68	2.93	3.06	0.65	*	96.92	101.7
	(0.187)	(0.104)	(0.004)	(0.003)	(0.739)		(0.000)	(0.000)

current bond falls relative to the generic repo rate) implies that holders of the current-issue bond receive an extra "dividend." Therefore, as argued by Grinblatt (1995), when the on-the-run Treasury notes go on special, Treasury prices tend to rise and swap spreads tend to increase.[24]

Table IV presents the *F*-tests and their marginal significance levels for the exclusion of all eight lags for each of the explanatory variables in the VAR. Two sets of results are displayed: those for a four-variable VAR (VAR4) and those for a three-variable VAR (VAR3) that excludes the commercial paper spread CPS. The test statistics for the ZEROSP10 equation, presented in the last row of Table IV, suggest that the histories of ZEROSP10 and REPOSP have significant predictive power for ZEROSP10. The *F*-statistic for TB6 is smaller, but significant at the 10 percent level in VAR3. The corporate spread CPS has little explanatory power for ZEROSP10 in VAR4, which is why we also examine VAR3. Interestingly, REPOSP is significant or borderline significant at conventional significance levels in all equations of both VARs.

---

[24] Alternatively, Evans and Parente-Bales (1991) and Brown, Harlow, and Smith (1994) argue that the costs to dealers of hedging net swap exposure is a key factor underlying changes in swap spreads. Brown, Harlow, and Smith (1994) use the repo rate as a proxy for a dealer's costs of hedging a swap exposure with Treasury bonds. A net exposure to the pay fixed side of a swap would be hedged by taking a long position in government bonds, which are financed in the repo market. Thus, they argue that a decrease in the repo rate, by lowering hedging costs, *reduces* swap spreads. The sign of this effect depends, however, on whether in the aggregate dealers are net pay- or receive-fixed counterparties in the market, so this hedging effect may change in sign over time.

Within a VAR system, the effects of, say, CPS on the future course of ZEROSP10 depends not only on the direct effects tested by the F-statistics in Table IV, but also on the indirect effects of CPS on the other variables in the VAR. CPS has significant predictive power, for example in the equation for TB6. In light of these feedback effects within the multivariate system, it is instructive to examine the impulse-response functions for the VARs. Each function traces out the effects of an innovation in one of the variables on the future course of ZEROSP10, taking into account the dynamic interactions among the variables. The shock is positive with magnitude equal to the estimated standard deviation of the variable's own residual in the VAR. The results for VAR4 are displayed in Figure 6.[25]

The impulse-response patterns are traced out over one year (52 weeks) subsequent to the impulse. In order to assess the significance of the responses, standard-error bands are computed using Monte Carlo methods. Specifically, assuming that residuals in the VARs are i.i.d. normal random variables, draws are made from the posterior distribution of the VAR coefficients and the impulse-response functions are computed based on these coefficients. These calculations are repeated 1000 times to derive upper and lower two-standard error bands for the responses over 52 weeks.[26]

The cumulative effects of these shocks are summarized in Table V, which presents the decomposition of variance. For a given VAR, each row must sum to one as the sum represents the total variation in ZEROSP10. For example, after 4 weeks, about 6.29 percent of the variation of the change in ZEROSP10 due to shocks in the four variables in VAR4 is due to TB6, about 2.74 percent is due to REPOSP, and so on.

The largest response of ZEROSP10 over the first weeks following the shocks is due to its own shock (Figure 6d). However, the effects of the own shocks die out relatively quickly. An increase in the level of interest rates (TB6) implies an increase in the ten-year zero spread (Figure 6a). This is consistent with the view that credit spreads widen during market sell-offs and narrow during rallies. Given the strong positive correlation between generic repo and TB6, this pattern is also consistent with an explanation based on dealer hedging costs under the presumption that dealers have more pay-fixed than receive-fixed swaps on their books. Interest rate "level" effects peak after about six

---

[25] The interpretation of the patterns is subject to the usual caveat that the VAR residuals are transformed to orthogonal shocks before computing the impulse responses. The variables are ordered in VAR4 as presented in Figure 6: TB6, REPOSP, CPS, and ZEROSP10. So contemporaneous correlation between the residual for TB6 and the residuals for the other three variables is attributed to variation in TB6, etc.

[26] In order to make determination of the posterior distribution tractable, we work directly with the unconstrained VAR and assume normally distributed innovations. So, in particular, the Bayesian priors imposed in calculating the results in Tables IV and V are not imposed here. As noted previously, the results are similar with and without the priors imposed. Of course, these are estimated standard-error bands based on normal innovations. Their use is therefore subject to the usual caveats that the sample may be small or the distribution for the Monte Carlo analysis may be misspecified.

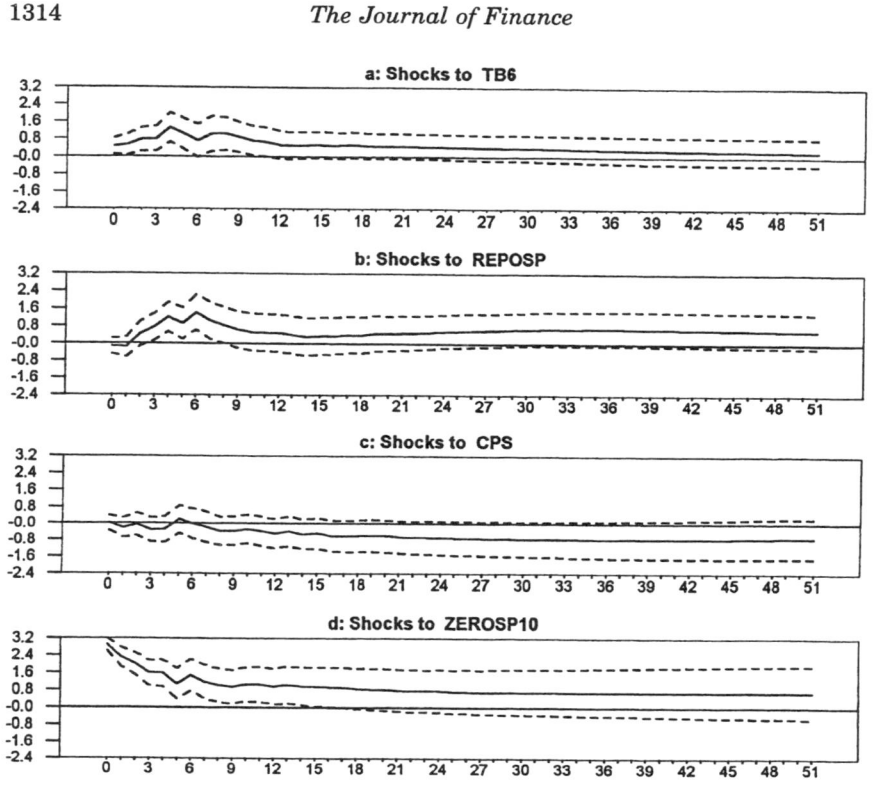

Figure 6. Impulse Responses for ZEROSP10 in VAR4. A four-variable VAR is estimated with the six-month Treasury bill rate (TB6), the spread between generic and on-the-run repo rates for ten-year Treasury bonds (REPOSP), the spread between BAA- and AAA-rated commercial paper (CPS), the spread between the ten-year zero rates implied by the swap and Treasury markets (ZEROSP10). Then each variable is perturbed by a one-standard deviation shock in its (orthogonalized) innovation. This figure displays the responses of ZEROSP10 to these shocks in basis points, over a period of 52 weeks following the impulses. The dashed lines represent plus and minus two-standard error bands around the impulse responses, estimated by Monte Carlo.

months (Table V, column *TB6*), and over a two-year horizon explain about 11 percent of the variation in ZEROSP10.

A shock to REPOSP initially has a small negative effect, but after the first couple of weeks the effect turns positive and statistically significant. This positive effect is predicted by Grinblatt's (1995) model of liquidity premiums, with REPOSP representing the convenience yield associated with treasury bonds. Over an eight-week horizon, REPOSP explains about 11.9 percent of the variation in ZEROSP10. Subsequently, variation in ZEROSP10 due to REPOSP levels off at about 20 percent.

The immediate impact of an increase in CPS on the zero-coupon spreads is essentially zero. After a few weeks, the zero-coupon spreads tend to decrease (Figure 6c); a widening of the credit spread between BAA and AAA commercial

A Model of Term Structure of Interest Rate Swap Yields

Table V

Decompositions of Variance of Ten-Year Swap Zero Spreads

Each column displays the proportion of the variance of the error from forecasting ZEROSP10 k weeks ahead, due to the impulse in the variable indicated in the column heading. k ranges from 1 to 104 weeks (2 years). Two VARs are estimated: VAR4 includes the 6-month Treasury bill rate (TB6), the spread between generic and on-the-run repo rates for ten-year Treasury bonds (REPOSP), the spread between BAA- and AAA-rated commercial paper (CPS), and the spread between the ten-year zero rates implied by the swap and Treasury markets (ZEROSP10); VAR3 includes TB6, REPOSP, and ZEROSP10. The * indicates exclusion of CPS from the VAR3 model.

Weeks Ahead	TB6 VAR4	TB6 VAR3	REPOSP VAR4	REPOSP VAR3	CPS VAR4	CPS VAR3	ZEROSP10 VAR4	ZEROSP10 VAR3
1	2.44	2.01	0.69	0.90	0.12	*	96.75	97.09
4	6.29	6.07	2.74	2.85	0.13	*	90.85	91.08
8	12.00	14.29	11.90	11.37	0.16	*	75.95	74.39
26	15.26	32.10	12.88	11.60	6.98	*	64.88	56.30
52	12.84	41.91	18.60	15.90	16.75	*	51.81	42.19
78	11.55	45.09	19.88	17.94	19.46	*	49.11	36.98
104	11.12	46.47	20.13	18.80	20.29	*	48.46	34.73

paper leads to a narrowing of the spread between the swap zero and Treasury zero yields. A potential explanation for these patterns is that relative credit spreads (BAA − AAA) tend to widen during recessions when default probabilities increase, even though commercial paper spreads to Treasury bill rates tend to narrow. Thus, a widening of CPS might be expected to eventually lead to a narrowing of ZEROSP10 if the former reflects a weakening of the economy. That commercial paper spreads to Treasuries are informative leading indicators of the U.S. business cycle has been documented by Friedman and Kuttner (1993), among others. CPS might reasonably be expected to have a similar property.

Interpreting the effects of REPOSP and CPS on ZEROSP10 as liquidity and credit shocks, respectively, the patterns suggest that liquidity shocks are relatively important over short horizons of a few months. After about 9 weeks, the response to REPOSP shocks is insignificant out to about six months. Credit shocks have little impact on ZEROSP10 over the first couple of months. Over longer horizons the effect of CPS grows in importance to the point that after two years it accounts for the second largest percentage (20.3 percent) in the variance decomposition for ZEROSP10. The estimated standard error bands for CPS include zero for all of the horizons out to one year, however, so credit effects on swap spreads may be weaker than the numbers in Table V suggest.

For comparison, we also estimate a three-variable VAR including TB6, REPOSP, and ZEROSP10, in that order. The impulse responses are displayed in Figure 7. The effect of TB6 is larger than in VAR4 (compare Figures 6a and 7a). This is reflected in the variance decompositions (Table V), where 42 percent of the variation in ZEROSP10 over a one-year horizon is attributable

a: Shocks to TB6

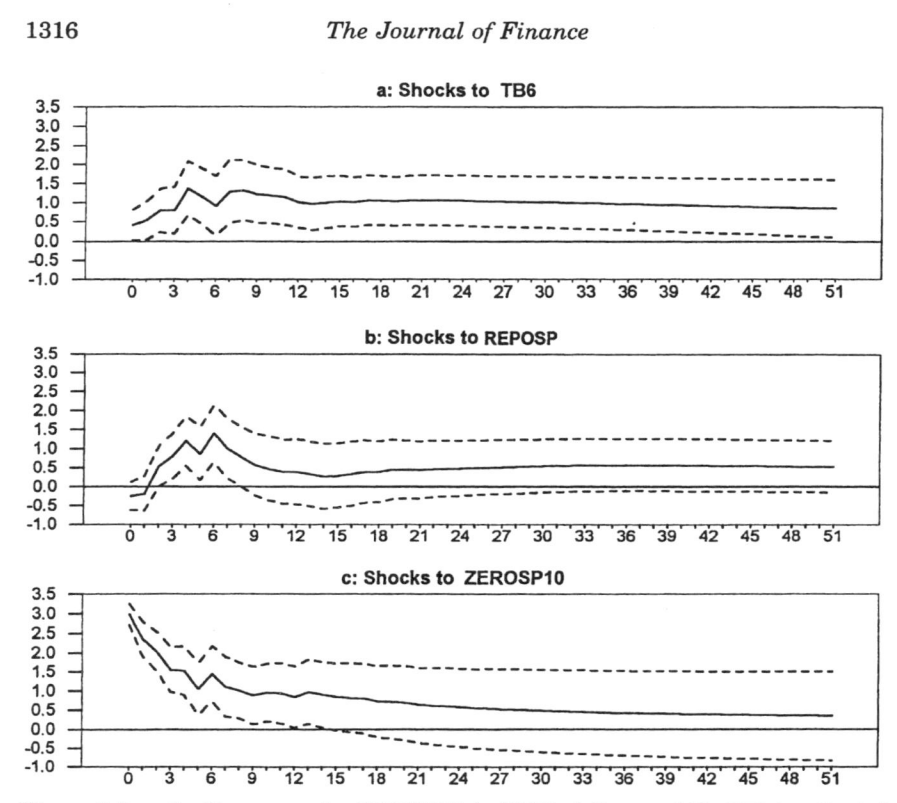

b: Shocks to REPOSP

c: Shocks to ZEROSP10

Figure 7. Impulse Responses for ZEROSP10 in VAR3. A three-variable VAR is estimated with the six-month Treasury bill rate (TB6), the spread between generic and on-the-run repo rates for ten-year Treasury bonds (REPOSP), the spread between the ten-year zero rates implied by the swap and Treasury markets (ZEROSP10). Then each variable is perturbed by a one-standard deviation shock in its (orthogonalized) innovation. This figure displays the responses of ZEROSP10 to these shocks in basis points, over a period of 52 weeks following the impulses. The dashed lines represent plus and minus two-standard error bands around the impulse responses, estimated by Monte Carlo.

to shocks in TB6 in VAR3, compared with 12.8 percent in VAR4. Evidently, omission of CPS leads to the variation in ZEROSP10 attributed to CPS in VAR4 being attributed to TB6 in VAR3.[27] An explanation for these results can be gleaned from the results for CPS in VAR4. The F-statistics in Table IV suggest that there is statistically significant feedback from CPS to TB6. Furthermore, the impulse responses for TB6 show a negative effect of a shock to CPS on TB6: A widening of the commercial paper credit spread is associated

---

[27] Notice also that the variation in ZEROSP10 due to its own shock over a two-year horizon declines in VAR3, and this reduction is attributed largely to TB6. However, in light of the wide standard error bands around the impulse response function of ZEROSP10 over horizons beyond six months (Figure 7), this change may not be of much significance.

A Model of Term Structure of Interest Rate Swap Yields

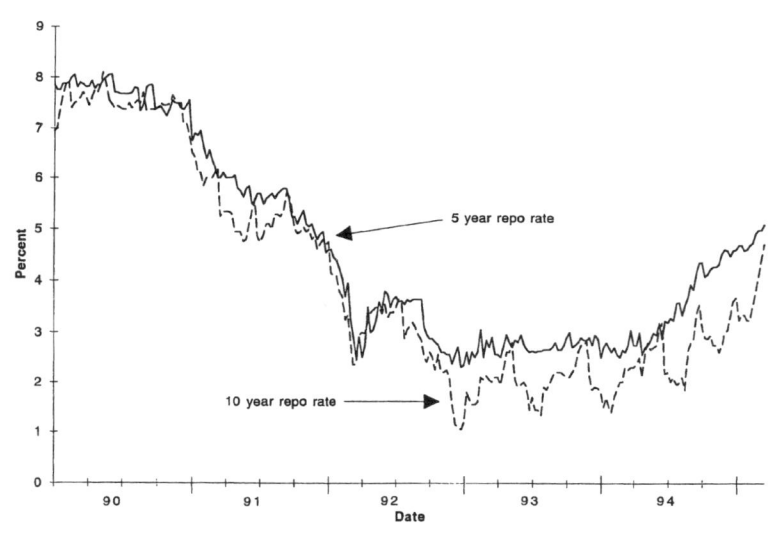

Figure 8. Repo Rates for On-the-Run Ten- and Five-Year U.S. Treasuries.

with a subsequent decline in TB6. Associating widening relative credit spreads within the commercial paper market with a weakening economy, it appears that, as the economy weakens, Treasury rates tend to decline, relative commercial paper spreads tend to widen, and the ten-year swap zero spread tends to narrow. Thus, both changes in Treasury bill rates and relative commercial paper credit spreads reflect the weakness of the economy and any induced changes in the likelihood of default by swap counterparties. When CPS is omitted from the VAR, the predictive power of changing relative commercial paper credit spreads for ZEROSP10 is largely captured by TB6.

The effects of shocks in ten-year repo specialness, as proxied by REPOSP, are striking and much larger than when the corresponding repo spread for five-year Treasury bonds is used instead. Upon further exploration, we find a strong seasonal pattern in the on-the-run ten-year repo rate (actual rate level) that is not present in the five-year rate (see Figure 8). Especially since 1992, there is a strong quarterly seasonal effect, and the ten-year repo rate (REPO10) generally is below the five-year repo rate. This suggests that the liquidity premium (convenience yield) is typically larger in the ten-year sector and more seasonally variable.

Several factors contribute to the quarterly seasonality in REPO10. First, the peaks and troughs in REPO10 correspond approximately to the calendar end-of-quarter dates. Efforts by Treasury bond dealers to cover their short positions at the ends of quarters puts pressure on bond borrowing and, hence, downward pressure on repo rates (*Wall Street Journal*, September 29, 1992, page C-20). Short covering is encouraged by the treatment of short-sales as liabilities on dealers' quarterly balance sheet reports. Additionally, the Treasury note futures contracts expire on the last date of the end of each calendar quarter. Borrowing bonds for delivery could contribute to this pressure in the repo market.[28]

However, the squeezes in the bond-borrowing market do not occur every quarter. Indeed, during 1992 and 1993, the largest dips in REPO10 were during March and September. The explanation for this pattern appears to be the reopening of ten-year note issues by the U.S. Treasury. Specifically, in May and November of 1992 and 1993 the U.S. Treasury reopened the bond from the previous auction instead of auctioning a new note. The presence of an outstanding stock of the reopened note on the reopening date implied that these issues did not go on special as much as the typical newly auctioned notes. Consequently, the repo rate for the on-the-run ten-year Treasury did not fall as much as for regular auction dates (February and August). The induced seasonal patterns in REPO10 appear to be associated with significant widenings of ten-year swap spreads after regular auctions. These Treasury "experiments" show a strong liquidity effect on swap spreads.

Within the four-variable VAR examined, nearly half of the variations of swap spreads over two-year horizons are explained by their own shocks. This suggests that an understanding of the time-series properties of swap zero spreads may require a deeper investigation of swap-specific market activity. For example, segmentation of supply or demand pressures by maturity sector due to institutional or accounting considerations may be important.

The behavior of swap spreads is also complicated by the state tax abatement on Treasuries, which would widen spreads, but in a manner whose dynamics are not clear or easily measured.

V. Conclusion

We have shown that, under the assumption of symmetric counterparty credit risk, swaps can be priced using standard term-structure models based on a risk- and liquidity-adjusted discount rate. For the purpose of our econometric analysis, we assume that the adjusted discount rate can be expressed as the sum of two independent square-root diffusions. Upon maximizing the likelihood function for about seven years of weekly data, we find that this two-factor model fits many aspects of the swap term structure well. The primary exception is the very short end of the swap curve represented by six-month LIBOR. The deviations between the actual and the model-implied LIBORs evidence strong seasonality and tend to be particularly large near the calendar year-ends. The autocorrelation structure of the deviations suggests that a third factor might allow one to fit the short-term LIBOR curve and the swap curve simultaneously. This possibility, as well as the implications of using two versus three-factor models for valuing LIBOR-based derivatives, are left for future research.

---

[28] Other factors include the heavy use of ten-year treasury notes as hedges against positions in mortgage-related securities and the more frequent auction cycle of five- versus ten-year notes.

A Model of Term Structure of Interest Rate Swap Yields

Using the estimated model, we then examine the dynamic properties of the spreads between the zero-coupon bond yields implicit in the swap curve and their Treasury counterparts. Three notable findings are as follows. First, the specialness of the on-the-run ten-year Treasury note has a positive effect on zero spreads that tends to peak in the first few weeks following the impulse, and over long horizons explains about 20 percent of the variation in ten-year zero spreads. These patterns suggest that liquidity advantages to trading in Treasury markets, and Treasury repo specials, have a substantial effect on swap spreads, consistent with Grinblatt's model, but that this liquidity-repo effect is not the only, or indeed the primary, determinant of swap-Treasury zero spreads. The same conclusion applies to the actual swap spreads as well.

This is not surprising given that the convenience yield l_t captures *relative* liquidity and repo effects in swap and treasury markets. The demands for the pay- and receive-fixed sides of swaps change over time and along the maturity spectrum. These demands may well have important effects on swap spreads. Indeed, the second conclusion that we draw from the VAR analysis is that, for the explanatory variables considered, between 35 percent and 48 percent of the variation in the ten-year zero spread over a two-year horizon is explained by its own shocks. That is, after accounting for the proxies of hedging costs, liquidity, and credit effects, a substantial fraction of the variation in swap spreads is left unexplained. Brown, Harlow, and Smith (1994) find weak evidence of an effect on the supply of corporate debt in the long end of the swap curve. Further exploration of these and other supply effects in the swap market, as well as the impact of asymmetric tax treatments, seems worthwhile.

Finally, interpreting the spreads between AAA and BAA commercial paper rates as a credit spread, we find that changes in the market's perception of credit risk have large effects on zero spreads at the ten-year maturity. The response of swap spreads to credit shocks is very different from the response to liquidity shocks. The credit effects are weak initially, then peak about six to seven months after the impulse. In terms of contribution to variation in zero spreads, the relative importance of credit shocks increases for over two years, reaching 20 percent.

However, the confidence intervals for the impulse responses for credit shocks are wide. Moreover, the channel by which CP credit spreads are correlated with swap spreads appears to be through the short-term riskless rate. The patterns are consistent with economic downturns leading to both a deterioration in credit quality (widening of relative credit spreads) and lower interest rates. Although an increasing hazard rate of default during economic downturns is one interpretation of these patterns, further analysis of the links between default hazard rates, the levels of riskless interest rates, and swap spreads seems necessary to conclude more definitively that business-cycle impacts on credit quality represent an economically significant credit component to variation in swap spreads.

These findings will prove useful in guiding a more complete model of swap spreads. Within our theoretical framework, the next step would be to parameterize the process generating the riskless rate (r_t), the liquidity convenience yield (l_t), and the credit adjustment ($\lambda_t h_t$). The VAR analysis suggests that the parameterizations adopted should be sufficiently flexible to allow very different dynamic structures for l_t and $\lambda_t h_t$, and allow for correlation between default hazard rates and the riskless interest rate. The VAR analysis may also provide potentially useful insights into the dynamic relations among our liquidity and credit proxies that may prove useful in parameterizing the correlations between the processes for r_t, l_t, and $\lambda_t h_t$. As elaborated by Duffie and Singleton (1996), one is unable to disentangle the separate influences of λ_t and h_t using only yield data, without the benefit of default incidence or loss data, or of data on the prices of instruments that respond nonlinearly with the prices of the underlying bonds at default, such as credit-spread options.

REFERENCES

Abken, Peter, 1993, Valuation of default-risky interest-rate swaps, *Advances In Futures And Options Research* 6, 93–116.

Bansal, Ravi, and William Coleman II, 1996, A monetary explanation of the equity premium, term premium, and risk-free rate puzzles, *Journal of Political Economy* 104, 1135–1171.

Bernanke, Ben S., 1990, Clearing and settlement during the crash, *Review of Financial Studies* 3, 133–151.

Brown, Kevin, W. V. Harlow, and Donald J. Smith, 1994, An empirical analysis of interest-rate swap spreads, *Journal of Fixed Income* 3, 61–78.

Brown, Roger, and Steven Schaefer, 1993, Interest rate volatility and the term structure of interest rates, *Philosophical Transactions of the Royal Society: Physics, Science, and Engineering* 347, 563–576.

Chen, Ren-Raw, and Louis Scott, 1993, Maximum likelihood estimation for a multifactor equilibrium model of the term structure of interest rates, *Journal of Fixed Income* 3, 14–31.

Cooper, Ian A., and Antonio S. Mello, 1991, The default risk of swaps, *Journal of Finance* 46, 597–620.

Cox, John C., Jonathan E. Ingersoll, Jr., and Stephen A. Ross, 1985, A theory of the term structure of interest rates, *Econometrica* 53, 385–407.

Dai, Qiang, and Kenneth J. Singleton, 1996, Specification analysis of affine term structure models, Working paper, Graduate School of Business, Stanford University.

Duffie, Darrell, 1996, Special repo rates, *Journal of Finance* 51, 493–526.

Duffie, Darrell, and Ming Huang, 1996, Swap rates and credit quality, *Journal of Finance* 51, 921–949.

Duffie, Darrell, and Rui Kan, 1996, A yield-factor model of interest rates, *Mathematical Finance* 6, 379–406.

Duffie, Darrell, and Kenneth J. Singleton, 1993, Simulated moments estimation of Markov models of asset prices, *Econometrica* 61, 929–952.

Duffie, Darrell, and Kenneth J. Singleton, 1996, Modeling term structures of defaultable bonds, Working paper, Graduate School of Business, Stanford University.

Evans, Ellen, and Gioia Parente-Bales, 1991, What drives interest rate swap spreads, in Carl R. Beidleman, Ed.: *Interest-Rate Swaps* (Business One Irwin Pubs., Homewood, Illinois).

Feller, William, 1951, Two singular diffusion problems, *Annals of Mathematics* 54, 173–182.

Friedman, Benjamin, and Kenneth Kuttner, 1993, Why does the paper-bill spread predict real economic activity? in *NBER Macro Annual*, 213–253.

Grinblatt, Mark, 1995, An analytic solution for interest-rate swap spreads, Working paper, UCLA Anderson Graduate School of Management.

Hansen, Lars, 1982, Large sample properties of generalized method of moments estimators, *Econometrica* 61, 929–335.

Heath, David, Robert Jarrow, and Andrew Morton, 1994, Bond pricing and the term structure of interest rates: A new methodology for contingent claims valuation, *Econometrica* 60, 77–106.

Hull, John, and Alan White, 1992, The impact of default risk on the prices of options and other derivative securities, *Journal of Banking and Finance* 19, 299–322.

Jaffee, Dwight, 1975, Cyclical varations in the risk structure of interest rates, *Journal of Monetary Economics* 1, 309–325.

Jarrow, Robert, and Steven Turnbull, 1995, Pricing options on financial securities subject to default risk, *Journal of Finance* 50, 53–86.

Jordan, Brad, and Susan Jordan, 1997, Special repo rates—an empirical analysis, *Journal of Finance*, Forthcoming.

Krylov, N., 1980, *Controlled Diffusion Processes* (Springer Verlag, New York).

Li, Haitao, 1995, Pricing of swaps with default risk, Working paper, Yale University, School of Organization and Management.

Litzenberger, Robert H., 1992, Swaps: Plain and fanciful, *Journal of Finance* 47, 831–850.

Merton, Robert, 1974, On the pricing of corporate debt: The risk structure of interest rates, *Journal of Finance* 29, 449–470.

Nielsen, S., and Ehud Ronn, 1995, The valuation of default risk, in *Corporate Bonds and Interest Rate Swaps* (University of Texas at Austin, Austin).

Pearson, Neil, and Tong-Sheng Sun, 1994, An empirical examination of Cox, Ingersoll, and Ross model of the term structure of interest rates using the method of maximum likelihood, *Journal of Finance* 49, 929–359.

Rendleman, Robert J., Jr., 1992, How risks are shared in interest rate-swaps, *Journal of Financial Services Research* 7, 5–34.

Smith, Clifford W., Jr., Charles W. Smithson, and Lee MacDonald Wakeman, 1988, The market for interest rate swaps, *Financial Management* 17, 34–44.

Solnik, Bruno, 1990, Swap-pricing and default risk: A note, *Journal of International Financial Management and Accounting* 2, 79–91.

Sorensen, Eric H., and Thierry F. Bollier, 1994, Pricing swap default risk, *Financial Analysts Journal* 50, 23–33.

Stock, James, and Mark Watson, 1989, New indexes of coincident and leading economic indicators, in *NBER Macro Anual*, 351–395.

Sun, Tong-Sheng, Suresh Sundaresan, and Ching Wang, 1993, Interest rate swaps—an empirical investigation, *Journal of Financial Economics* 34, 77–99.

Sundaresan, Suresh, 1991, Valuation of swaps, in *Recent Development in International Banking* (Elsevier, North Holland).

Vasicek, Oldrich, 1977, An equilibrium characterization of the term structure, *Journal of Financial Economics* 5, 177–188.

Name Index

Abel, A.B. 409, 439
Abken, P. 605
Abowd, J.M. 397
Ackley, G. 84, 198
Aït-Sahalia, Y. 561
Aiyagari, S.R. 362, 363, 365, 383, 385, 390, 391, 393
Albrecht, W.S. 108
Altug, S.J. 283
Amemiya, T. 479, 501, 550, 554, 555
Amihud, Y. 363
Amsler, C. 25, 39, 312, 314
Anderson, T.W. 47, 49
Andrews, D. 536, 545, 546
Arditti, F.D. 339
Arrow, K.J. 84, 128, 161, 283

Babiak, H. 125, 195
Baillie, R.T. 141
Ball, R. 108
Bansal, R. 604
Bartlett, M. 300, 314
Basu, S. 25
Beaver, W. 297
Benninga, S. 564
Bernanke, B.S. 610
Bewley, T.F. 250, 365
Bhattacharya, S. 105
Billingsley, P. 231
Black, F. 108, 284, 291, 429, 509, 561, 569
Blanchard, O.J. 40, 135, 186, 191, 192, 198
Blume, M.E. 283, 287
Boldrin, M. 445
Bollerslev, T. 407, 428
Bollier, T.F. 605
Bosworth, B. 67–8
Box, G.E.P. 29, 308
Brainard, W.C. 25
Brealey, R. 191, 196
Breeden, D.T. 127, 257, 258, 260, 265, 291, 292, 294, 299, 306, 324, 339, 522, 531, 560
Breen, W. 336
Brennan, M.J. 468, 506, 520, 561
Brock, W.A. 127, 257, 259, 260, 261, 287, 536
Brown, D.P. 324, 349, 561, 569, 570, 573
Brown, K. 600, 621, 631

Brown, R. 561, 569, 621
Brown, S.J. 519, 520
Byrnes, J.C. 266

Campbell, J.Y. 133, 134, 139, 143–4, 147, 151, 242–4, 246, 250, 251, 325, 357, 407, 417, 418, 419, 421, 429, 430, 434, 439, 440, 443–5, 446, 447, 519, 531
Card, D. 397
Carleton, W.R. 469
Chamberlain, G. 114, 333
Chapman, D.A. 409
Chen, R.-R. 299, 306, 307–8, 320, 586, 601, 611, 613
Chou, R.Y. 407
Christiano, L.J. 445
Clarida, R.H. 365
Clark, S.A. 345, 356
Cochrane, J.C. 244, 246, 247, 407, 417, 418, 421, 430, 431, 434, 438, 439, 440, 443–5, 446, 447
Cohen, K.J. 456, 462
Cohn, R. 25, 167
Coleman II, W. 604
Connor, G. 333, 336
Constantinides, G.M. 127, 287, 346, 362, 363, 371, 376, 377, 409, 442, 445, 517, 561, 576
Cooper, I. 469, 605
Cootner, P. 25
Copeland, B.L., Jr 77, 78, 108, 109
Cornell, B. 294
Cowles, A. 34, 36, 85, 144, 145
Cox, J. 342, 417, 468, 506–8, 510, 519, 522, 523, 531, 559, 560, 561, 562, 564, 566–8, 570, 571, 582–4, 594, 596, 601, 609
Culbertson, J. 506

Dai, Q. 609, 613, 621
Deaton, A. 193, 408
DeBondt, W.F.M. 172–3, 186, 187, 247
Debreu, G. 280, 287
DeLong, J.B. 247
Diba, B.T. 136, 199, 248
Dickey, D.A. 117, 119, 123, 142
Dobson, S.W. 468
Dolde, W. 283

Donaldson, J.B. 287
Doob, J. 542, 545, 551
Duesenberry, J.S. 409
Duffie, D. 362, 371, 376, 442, 536, 599, 600, 601, 602, 603, 604, 605, 606, 607, 608, 609, 611, 632
Dumas, B. 365
Dunn, K. 296, 346, 350, 539, 576
Durand, D. 455, 462–3
Durlauf, S.N. 134, 220, 227, 240–42
Dybvig, P.H. 338, 508, 519, 520, 561, 569

Efron, B. 479, 481, 489–90
Eichenbaum, M.S. 272, 346, 349, 539, 545
Engle, R.F. 134, 140, 142, 143, 144, 147, 407
Epstein, L.G. 250, 251, 324, 349, 439
Evans, E. 624

Fama, E.F. 6, 15, 19, 25, 28, 30, 67, 75, 83, 105, 125, 127, 134, 166, 168, 243, 244, 291, 303, 350, 407, 427, 429, 434, 468, 512, 513, 522, 523, 525, 526, 531, 579, 582, 584, 591, 594
Feller, W. 567, 610
Ferson, W.E. 294, 407, 409, 519, 561
Fischer, S. 68, 76, 168
Fisher, J.D.M. 445
Fisher, L. 285, 468
Fisher, M. 615
Flavin, M.A. 68, 104, 175, 191, 192, 217, 224, 225, 227, 228, 231, 232, 235, 238, 239, 245, 264
Flood, R.P. 202, 210, 247
Foster, G. 75, 108
Frankel, J.A. 241
French, K.R. 168, 243, 244, 407, 427, 429
Friedman, B. 607–8, 610, 627
Friedman, M. 4
Friend, I. 283, 287
Froot, K.A. 241
Fuller, W.A. 104, 117, 119, 122, 123, 124, 125, 142, 467

Gallant, R. 329, 330, 346, 347, 349, 542, 547
Gertler, M. 362, 363, 365, 383, 385, 390, 391, 393
Gibbons, M.R. 291, 294, 310, 312, 316, 324, 349, 523, 531, 561, 570, 573, 579, 582
Gilles, C. 247
Glosten, L.R. 336
Gonzales-Gaviria, N. 318
Goodman, L.A. 464
Granger, C.W.J. 8, 28, 125, 134, 140, 142, 143, 147, 201
Green, R.C. 339

Gregory, A.W. 138
Grinblatt, M. 600, 604, 626, 631
Grossman, H.I. 199, 248
Grossman, S.J. 39–40, 52, 84, 89–90, 97, 98, 100, 101–2, 109, 112, 126, 127, 136, 210, 229, 232, 248–9, 257, 260, 263, 264, 265, 269–70, 275, 276, 294, 297
Gultekin, N.B. 420

Hall, A.D. 213
Hall, B.H. 486
Hall, R.E. 44, 220, 227, 240–42, 263, 264–5, 296, 538
Hamburger, M.J. 522
Hamilton, J.D. 135, 199
Hannan, E.J. 8, 200, 204
Hansen, L.P. 103, 107, 126, 133, 134, 141, 193, 194, 197, 198, 199, 200, 201, 210, 213, 246, 272, 294, 299, 314, 324, 325, 328, 329, 330, 333, 337, 340, 345, 346, 348–9, 350, 352, 362, 382, 412, 425, 431, 438, 446, 525, 528, 535, 536, 538, 539, 545, 546, 548, 549, 557, 560, 565, 570, 572–3, 619
Harlow, W.V. 600, 621, 631
Harrison, J. 324, 338, 342, 344–5
Harvey, C.R. 407
Hasza, D.P. 117
Haugh, L. 17
Hausman, J. 479, 483
Heal, G.M. 408
Heath, D. 601
Heaton, J. 353, 365, 388, 409
Heston, S. 561
Hicks, J. 506, 522, 531
Hildreth, C. 283
Hodrick, R.J. 103, 202, 210, 247, 573
Homer, S. 277
Houglet, M.X. 468
Huang 605, 606
Huberman, G. 337
Huggett, M. 365, 390
Huizinga, J. 510, 575
Hull, J. 605
Hussey, R. 370, 399, 538

Ibbotson, R.G. 112, 265, 277, 306
Ingersoll, J.E., Jr 338, 417, 468, 506–8, 510, 517, 519, 559, 560, 561, 562, 564, 566–8, 570, 571, 576, 582–4, 594, 596, 601, 609
Ingram, B. 535, 536, 538, 551

Jaffee, D. 610
Jagannathan, R. 246, 336, 352, 362, 382, 412, 425, 446

Jarrow, R.A. 339, 605, 610
Jenkins, G.M. 29
Jennrich, R. 554
Jensen, M.C. 25, 165, 291
Jermann, U.J. 408, 411
Jobson, J. 314
Joerding, W. 112, 126, 231
John, K. 339
Johnsen, T.H. 257
Johnson, N.L. 596
Jordan, B. 600
Jordan, S. 600

Kahneman, D. 161
Kan, R. 609
Kandel, S. 310, 318, 337, 407, 439, 444, 561
Kane, A. 407
Kang, H. 118, 119
Kaplan, P. 202, 210, 247
Kehoe, P.J. 283
Keim, D.B. 519
Kessel, R.A. 463, 522, 525, 526, 531
Keynes, J.M. 67, 68, 83, 191, 247
Kleidon, A.W. 68, 75, 77, 86, 93, 94, 95, 104, 127, 133, 175, 191, 192, 206, 207, 208, 217, 223, 224, 229–35, 239, 244
Kling, A. 108
Knez, P. 350, 351
Knowles, G.J. 283
Korajczyk, R.A. 336
Korkie, R. 314
Koski, J.L. 231
Kotz, S. 596
Kramer, R.L. 456, 462
Krasker, W. 63
Kreps, D.M. 250, 324, 338, 342, 344, 355, 356
Kroner, K.F. 407
Krylov, N. 608
Kuttner, K. 610, 627
Kydland, F.E. 283

LaCivita, C.J. 126, 249, 275
Lambert, R. 297
Langetieg, T.C. 468, 469, 475
Leamer, E.E. 78, 169
Lee, B. 535, 536, 538, 551
Leftwich, R.W. 108, 123
Lehmann, B.N. 336
LeRoy, S.F. 5, 15, 16, 17, 23, 25, 31, 39, 40, 43, 63, 68, 84, 97, 104, 105, 106, 126, 127, 128, 167, 174, 176, 191, 194, 198, 217, 220, 222, 223, 228, 232, 234, 235, 237, 239, 240, 241, 243, 244–6, 247, 249, 251, 257, 275, 324
Li, H. 605, 606

Lintner, J. 72–3, 75, 105, 127, 291
Lippens, R.E. 141
Litterman, R. 350
Little, I.M.D. 108
Litzenberger, R.H. 126, 258, 291, 292, 294, 339, 602
Lo, A. 485, 489
Long, J. 522, 531
Long, J.B., Jr 167
Lookabill, L.L. 108
Lorie, J.H. 285
Lucas, D.J. 362, 365, 367, 373, 376, 388, 402
Lucas, R.E., Jr 15, 127, 208, 209, 257, 260, 261, 275, 279, 324, 351, 560
Lutz, F.A. 523

MacBeth, J. 291, 512
MaCurdy, T.E. 397, 440
Mankiw, N.G. 78, 133, 134, 145, 149, 176–82, 185, 227, 231, 237–40, 247, 306, 362, 371, 395, 434, 442
Marcet, A. 362
Markowitz, H. 183
Marsh, T.A. 102, 105, 109, 133, 168, 175, 187, 191, 192, 197, 206, 239, 296, 507, 509
Martin, D. 484, 485
McCallum, B. 197
McCulloch, J.H. 468, 469, 474, 522, 525, 526
McFadden, D. 478, 501, 535, 536, 538, 551
McKeown, J.C. 108
McMahon, P.C. 141
Meese, R.A. 117
Mehra, R. 126, 249–50, 251, 280, 287, 325, 362, 374, 438
Melino, A. 141, 297, 560
Mello, A.S. 605
Mendelson, H. 363
Merrick, J.J., Jr 407
Merton, R.C. 67, 68, 76, 78, 102, 105, 109, 127, 133, 168, 170, 175, 181, 183, 186, 187, 191, 192, 197, 206, 218, 239–40, 258, 284, 294, 339, 342, 522, 531, 560, 603
Michener, R.W. 126, 249
Michner, R. 536, 543
Miller, M.H. 31, 72, 84, 104, 109, 123, 183, 187
Miron, J.A. 149
Mishkin, F. 510
Modest, D.M. 336, 385
Modigliani, F. 25, 31, 72, 84, 104, 109, 123, 150, 160, 167, 173, 183
Mokkadem, A. 542, 543, 544, 545
Morse, D. 297
Morton, A. 601

Muellbauer, J. 408
Myers, S.C. 75, 191, 196

Nelson, C.R. 108, 117, 118, 119, 121, 523, 531
Nelson, D.B. 429
Nelson, J. 520, 561
Newbold, P. 201
Newey, W.K. 201, 213, 490, 545, 546, 554, 574
Nielsen, S. 609
Novales, A. 346
Nummelin, E. 542

Ohlson, J.A. 169, 484, 485, 488
Ouliaris, S. 134, 142, 143, 146, 147

Pagan, A.R. 213
Pakes, A. 128, 535, 536, 538, 551
Palepu, K. 489
Parente-Bales, G. 621
Parke, W.R. 237, 240, 244–6, 251
Pearson, N. 561, 601, 612, 613
Penman, S.H. 169
Pennacchi, G.C. 564
Perron, P. 143, 145, 146, 147
Pesando, J. 32, 39
Pfleiderer, P. 105
Phillips, P.C.B. 134, 142, 143, 145, 146, 147
Pierce, D.A. 8
Platt, E.N. 522
Plosser, C.I. 108, 117, 118, 119, 121, 523, 531
Pollard, D. 535, 536, 538, 551
Porter, R.D. 5, 15, 16, 17, 23, 25, 28, 31, 40, 43, 60, 63, 68, 84, 104, 105, 106, 128, 167, 174, 176, 191, 194, 198, 217, 220, 222, 223, 228, 232, 235, 238, 239, 241, 243, 244, 245, 257, 275
Porteus, E.L. 250
Poterba, J.M. 243, 244, 427–8
Powell, J. 479, 501
Pratt, J.W. 363
Prescott, E.C. 126, 249–50, 251, 280, 283, 287, 325, 362, 374, 438
Protopapadakis, A. 564

Quah, D. 135, 136
Quinn, B.G. 200, 204

Reitz, T.A. 348
Rendleman, R.J., Jr 604, 605
Richard, J.F. 484, 502
Richard, S.F. 324, 328, 330, 333, 337, 340, 345, 346
Rietz, T.A. 441
Roberts, H.V. 107

Rogalski, R.J. 520
Roll, R. 168, 299, 306, 307–8, 320
Romer, D. 133, 176–82, 185, 227, 231, 237–40, 247
Ronn, E. 126, 294, 609
Rosenblatt, M. 542
Rosenfeld, E. 507
Ross, S.A. 105, 299, 306, 307–8, 320, 333, 339, 342, 344, 417, 468, 506–8, 510, 512, 519, 559, 560, 562, 564, 566–7, 568, 570, 582–4, 594, 596, 601, 609
Rothschild, M. 333
Rozanov, Y.A. 8
Rubinstein, M. 15, 127, 257, 258, 262, 291, 292, 324, 342, 346, 349
Ryder, H.E., Jr 408

Saito, M. 366
Samuelson, P.A. 16, 28, 67, 79, 162, 163, 164, 168, 183, 186
Sargent, T.J. 133, 134, 141, 193, 194, 198, 199, 200, 446
Sawa, T. 149
Schaefer, S.M. 468, 520, 561, 569, 621
Scheinkman, J.A. 350, 362, 364
Schmidt, P. 312, 314
Schoenholz, K.L. 133
Scholes, M. 187, 291, 303, 509, 561, 569
Schwartz, E. 468, 507, 520, 561
Schwert, G.W. 303, 429
Scott, J.H., Jr 484
Scott, L. 586, 601, 611, 613
Scott, L.O. 107, 133, 153, 240
Shanken, J. 310–11, 314, 332
Shapiro, M.D. 133, 176–82, 185, 227, 231, 237–40, 247, 306, 434
Sharpe, W. 127, 183, 291, 303, 385
Shea, G.S. 227, 238–9
Shefrin, H. 187
Shiller, R.J. 4, 7, 39–40, 41, 42, 43, 51, 52–7, 58, 59, 62–3, 67, 68–70, 75, 76, 77, 78, 79, 84, 85, 86, 89–90, 97, 98, 100–102, 103, 104, 105, 106, 109, 112, 115, 123, 126–7, 128, 133, 139, 141, 143–4, 145, 147, 149, 150, 151, 152, 162, 163, 167, 171, 173, 174–5, 176–7, 182, 185, 191, 192, 198, 199, 206, 208, 210, 211, 212, 214, 217, 219, 222, 224, 227, 228, 229, 232, 234, 235, 238, 242–4, 246, 247, 248–9, 250, 251, 257, 260, 263, 264, 265, 269–70, 275, 276, 294, 297, 325, 357, 407, 412, 429, 430, 522, 525, 531
Shoven, J.B. 25, 440
Siegel, J.J. 27, 440
Sims, C.A. 14

Singleton, K.J. 10, 11, 39, 41, 43, 47, 51, 57–9, 64, 65, 117, 126, 197, 200, 201, 210, 228, 248, 250, 272, 294, 296, 299, 314, 324, 346, 348–9, 350, 362, 538, 539, 545, 560, 565, 570, 573, 576, 599, 601, 602, 603, 604, 605, 606, 607, 608, 609, 611, 621, 632
Sinquefield, R.A. 112, 265, 277, 306
Smith, C.W., Jr 604
Smith, D.J. 600, 621, 631
Smithson, C.W. 604
Smoot, S.J. 469, 475
Snow, K.N. 353
Solnik, B. 605
Sorensen, E.H. 605
Stambaugh, R.F. 303, 306, 314, 318, 350, 407, 439, 444, 519, 561
Startz, R. 525
Statman, M. 187
Steigerwald, D. 242
Stock, J.H. 55, 127, 134, 142, 241, 610
Stokey, N.L. 351
Summers, L.H. 160, 164, 170, 175, 186, 187, 192, 198, 199, 208, 243, 244, 428
Sun, T.-S. 561, 600, 601, 604, 612, 613, 620
Sundaresan, M. 346
Sundaresan, S.M. 409, 600, 604, 605, 620
Sutch, R.C. 468
Svensson, L.E.O. 351

Tauchen, G. 329, 346, 347, 370, 399, 538
Taylor, J. 538
Telmer, C.I. 362, 367, 376
Thaler, R. 172–3, 186, 187, 247
Tiao, G. 297
Tirole, J. 86, 103, 161, 186, 199
Tobin, J. 67, 70, 84, 167, 183, 283
Townsend, R.M. 351
Tuominen, P. 542
Turnbull, S. 75, 605
Tversky, A. 161
Tweedie, R. 542, 544

Uhlig, H. 538
Ulman, S. 531

Van Horne, J.C. 160, 186
Vanderfold, D.E. 468
Vasicek, O.A. 468, 600
Vayanos, D. 363, 383
Veall, M.R. 138
Vila, J.-L. 363, 383

Wakeman, L.M. 604
Wallace, N. 288
Wang, C. 600, 604, 620
Watson, M.W. 40, 135, 144, 186, 191, 192, 194, 198, 610
Watts, R.L. 108, 123
Waugh, W.H. 456, 462
Weil, P. 251, 362, 363, 374, 375, 438, 439
Weil, R. 468
Weiss, L. 25, 364
West, K.D. 78, 106–7, 133, 134, 135, 140, 144, 145, 176, 182, 193, 194, 197, 199, 200, 201, 207, 213, 217, 236–7, 245, 247, 248, 545, 546, 574
Weston, J.F. 109
Wheatley, S. 314
White, A. 605
White, H. 149, 525, 528, 542, 547
Whiteman, C.H. 135, 199
Williams, J. 303
Wilson, G.T. 14, 17
Wilson, R.B. 346
Wold, H. 29
Wood, J.H. 531
Woodward, S.E. 15
Working, H. 107, 118, 297

Zavgren, C. 484, 485
Zeckhauser, R.J. 363
Zeldes, S.P. 395
Zin, S.E. 250, 251, 324, 349, 439